uctural and Functional ymmetry of Plasma Membranes

rlington House, London,
and online

–25 April 2025

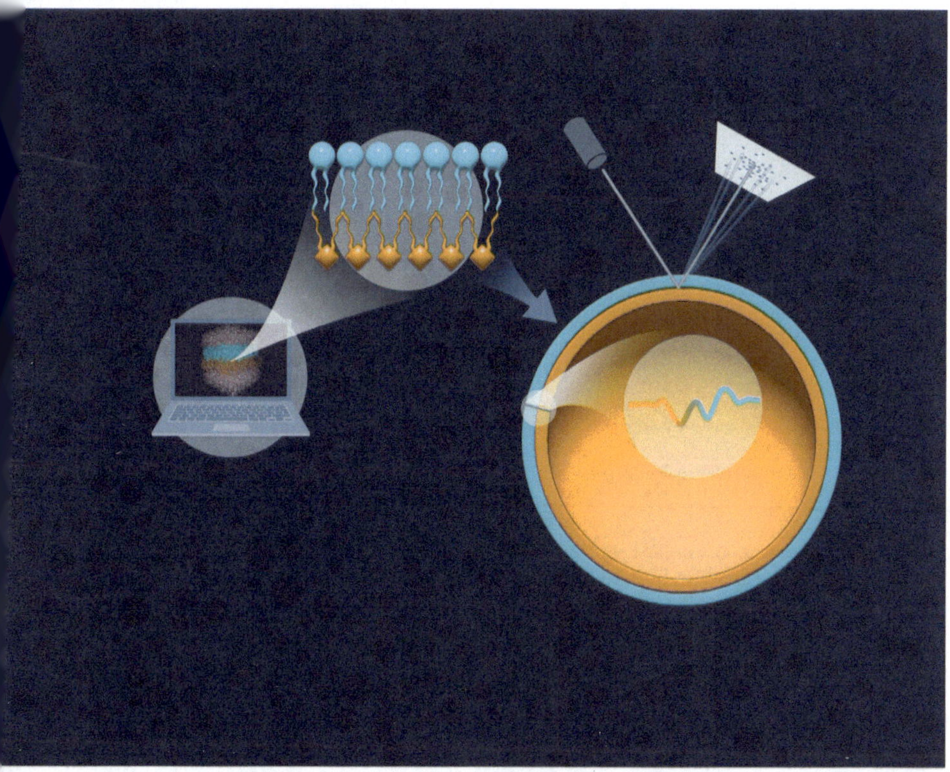

ARADAY DISCUSSIONS

olume 259, 2025

ROYAL SOCIETY
OF CHEMISTRY

The **Faraday Community for Physical Chemistry** of the Royal Society of Chemistry, previously the Faraday Society, was founded in 1903 to promote the study of sciences lying between chemistry, physics and biology.

Faraday Discussions (Print ISSN 1359-6640, Electronic ISSN 1364-5498) is published 8 times a year by the Royal Society of Chemistry, Thomas Graham House, Science Park, Milton Road, Cambridge, UK CB4 0WF.

Volume 259 ISBN 978-1-83707-037-4

2025 annual subscription price: print+electronic £1342 US $2363; electronic only £1279, US $2250. Customers in Canada will be subject to a surcharge to cover GST. Customers in the EU subscribing to the electronic version only will be charged VAT.

All orders, with cheques made payable to the Royal Society of Chemistry, should be sent to the Royal Society of Chemistry Order Department, Royal Society of Chemistry, Thomas Graham House, Science Park, Milton Road, Cambridge, CB4 0WF, UK
Tel +44 (0)1223 432398; E-mail **orders@rsc.org**

If you take an institutional subscription to any Royal Society of Chemistry journal you are entitled to free, site-wide web access to that journal. You can arrange access via Internet Protocol (IP) address at **www.rsc.org/ip**

Customers should make payments by cheque in sterling payable on a UK clearing bank or in US dollars payable on a US clearing bank.

Printed in the UK

MIX Paper from responsible sources FSC® C013604

ROYAL SOCIETY OF CHEMISTRY

Faraday Discussions

Faraday Discussions are unique interna discussion meetings that focus on rapic developing areas of chemistry and its interfaces with other scientific disciplin

Information for Authors

♾ The paper used in this publication meets the requirements of ANSI/NISO Z39.48–1992 (Permanence of Paper).

Registered charity number: 207890

ructural and Functional symmetry of Plasma embranes

ıday Discussions

v.rsc.org/faraday_d

ıeral Discussion on Structural and Functional Asymmetry of Plasma Membranes was held in
on, UK and online on the 23rd, 24th and 25th of April 2025.

NTENTS

1359-6640; ISBN 978-1-083707-037-4

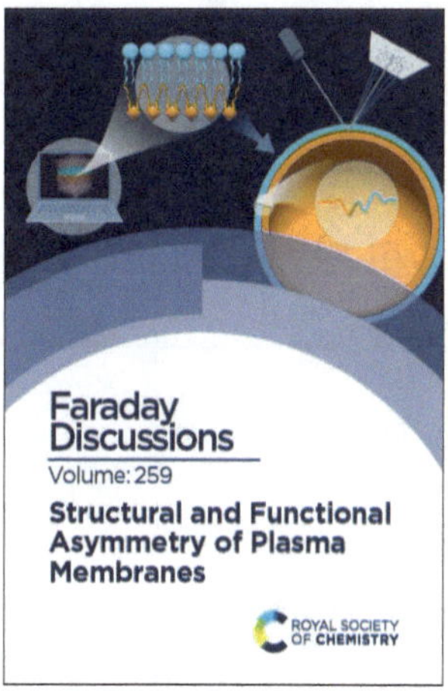

Cover
See Frederick A. Heberle and Milka Doktorova, *Faraday Discuss.*, 2025, **259**, 300–320.

Simulations of asymmetric bilayers enable analysis of the sensitivity of experimental techniques to interleaflet imbalances in liposomes *in vitro*.

Image reproduced with permission of Jerker Lokrantz from Doktorova *et al.*, *Faraday Discuss.*, 2025, **259**, 300–320.

TRODUCTORY LECTURE

PERS AND DISCUSSIONS

SPONSORS

BBA

Biochimica et Biophysica Acta

Biomembranes

POSTER SPONSORSHIP

ROYAL SOCIETY OF CHEMISTRY | FARADAY COMMUNITY FOR PHYSICAL CHEMISTR

PCCP

CONCLUDING REMARKS

ADDITIONAL INFORMATION

Faraday Discussions

ROYAL SOCIETY OF CHEMISTRY

PAPER

Spiers memorial lecture: Experimental discovery of asymmetric bilayers, and a recent asymmetry example

Gerald W. Feigenson

Received 17th March 2025, Accepted 2nd April 2025
DOI: 10.1039/d5fd00041f

This introductory *Faraday Discussions* lecture considers the wide range of subjects that involve asymmetric bilayer membranes. The *Discussions* organizers have chosen four themes that describe the current interest in asymmetric bilayers. The main part of this introductory lecture starts with the discovery from over 50 years ago that the plasma membrane of mature, human erythrocytes has different composition and properties in each leaflet of the plasma membrane. We also comment on what is newly recognized in the asymmetric membrane field. The asymmetric bilayer is a remarkable state of matter. Newly recognized, we can describe it as a "new state of matter". Like other biological matter, it evolved to be in its particular form. There is new appreciation for the properties of the asymmetric bilayer, and there has been progress in understanding these emerging properties. Asymmetric lipid bilayers have a remarkable range of characteristics that involve the physically and chemically special connection of the two different monolayer leaflets. Much of this work is an explanation of how the "van Deenen researchers", over 50 years ago, quantitatively measured the phosphatidylcholine, sphingomyelin, and phosphatidylethanolamine content of the erythrocyte plasma membrane exoplasmic leaflet. Their use of surface chemistry principles was important in proper quantitation, and this introductory lecture explains these research findings in detail. We also present a detailed discussion of one remarkable example of a newly-discovered behavior of asymmetric bilayers, which we term "induced order". This discovery has an absolute requirement for microscopy image data to reveal the superposition of induced order with the thermodynamic order of liquid-ordered + disordered phase separation.

1. Introduction

This introductory *Faraday Discussions* lecture considers the surprisingly wide range of subjects that involve asymmetric bilayer membranes. The *Discussions* organizers have chosen four themes that capture the current interest in asymmetric bilayers, and I will return to these themes towards the end. I will start the

Biochemistry and Biophysics, Cornell University, USA. E-mail: gwf3@cornell.edu

main part of this discussion with the discovery over 50 years ago that the plasma membrane of mature, human erythrocytes, referred to as red blood cells or here as "rbc", have different compositions and properties in each leaflet of their plasma membrane. I will also comment on what is newly recognized in the asymmetric membrane field.

This year, 2025, is the 100-year anniversary of Gorter and Grendel's proposal that biological membranes are lipid bilayers: "We propose to demonstrate in this paper that the chromocytes of different animals are covered by a layer of lipoids just two molecules thick".[1] Then, in about a decade after most researchers had accepted this central role for the bilayer in plasma membranes (PM) of living cells, the plasma membrane would already be found to have different phospholipid concentrations in its two bilayer leaflets, and therefore different properties in the two leaflets. We now describe this kind of bilayer as "asymmetric", and thus different from the symmetric bilayers that had been the focus of so much earlier study.

Many interesting questions involve what is increasingly understood to be the remarkable state of matter of an asymmetric bilayer. Like other biological matter, it evolved to be in its particular form as a bilayer of oriented amphiphilic molecules having polar/charged moieties in water, with nonpolar moieties packed together out of the water. What special capabilities did evolution pack into the asymmetric bilayer? We are here for these *Faraday Discussions* because there is new appreciation for the properties of the asymmetric bilayer. There has been progress in understanding these emerging properties, as we will see in the papers presented in this discussion. Asymmetric lipid bilayers have a remarkable range of recently appreciated characteristics that involve the physically and chemically special connection of the two monolayer leaflets. We are only now recognizing asymmetric bilayers as a new state of matter, with new properties that are very much in the process of being revealed.[2–4] Although I focus here on cell plasma membranes, in this discussion meeting we will explore different kinds of asymmetric bilayers that are not in cells, and a range of their applications to new questions.

When I started graduate school in 1969, I knew of two famous centers for study of lipid bilayers, one with Laurens L. M. van Deenen[5] at its center in Utrecht, and one with Thomas Thompson at its center at the UVa in Charlottesville.[6] I started reading papers from both groups. I soon encountered the name of Dennis Chapman[7,8] (1927–1999). In the 1960s his interest in spectroscopy and crystallography was turning toward surface chemistry, and a meeting with L. L. M. van Deenen led to Chapman's new focus on phospholipid phases, which he recognized as having the liquid-crystalline type of order.[9] Chapman investigated the influence of hydrocarbon chain lengths, double bonds, headgroup types, and cholesterol fraction on "bilayer fluidity". Chapman had studied phase behavior of glycerides and soaps by infrared spectroscopy. He found that outside surfaces of cells had lipids with no net charge, but that the inner surface of cells was negative. His stated goal was to make biomedical devices that were bio-compatible.[7] He was also a pioneer in the use of differential scanning calorimetry of model membranes.[8]

In living cells the most important case of an asymmetric bilayer, and indeed the only closely studied case in living cells, is the cell plasma membrane.[10–19] Yet finding out the compositions of each leaflet of a PM has proven difficult, and includes efforts over more than 50 years. Complicating these efforts, researchers wish to elucidate the PM only of living cells, because their PM changes when the cell dies.[20–22]

A potential goal for these discussions: given an increasingly complex picture of the asymmetric bilayer, it would be useful to have starting points on which we can all agree. On the other hand, we are not assembled here in order to agree! But I will attempt to sketch some starting points.

Significant progress is happening now,[4] 50 years after researchers first recognized asymmetry of rbc plasma membrane bilayers.[13] Much bilayer physical chemistry had been worked out by Chapman, by Thompson and many others, before plasma membrane asymmetry was established, including general characteristics of lipid order and motion and cholesterol behaviors.[6,8] These types of order are characteristic of the class of order termed "liquid crystal".[9] Note that this class of order is not at all necessarily a liquid or a crystal, although many researchers even today continue to mistake the name of this category as a literal description of phase order and motion. And for asymmetric bilayers there must be new categories of order, such as lipids of the cytoplasmic leaflet that are organized by an apposing Lo domain in the exoplasmic leaflet into "induced order domains", with probable but as yet unknown connections of such induced order domains to ordered proteins of, for example, an actin–myosin polymer.[23,24]

2. How asymmetry is established

I will refer now to two vexing issues with experimental asymmetry research of the plasma membranes found in nature: The first difficulty is the purity of the PM for study. For all living cells, with perhaps the single exception of the rbc, obtaining a pure preparation of the plasma membrane has not been possible, mainly because of contamination with the more numerous other membranes inside the cell, which are typically present in many-fold excess over the PM.[25]

The second problem is to chemically analyze each leaflet of this asymmetric rbc membrane. Methods used to analyze each leaflet typically include either (1) chemically reactive small molecules, or (2) phospholipase enzymes. As researcher Alain Zachowsky concisely summarized regarding measurement of rbc asymmetry, "None of the methods utilized is 'perfect'. Each one has its own drawbacks and advantages".[26] But so far, all are in at least broad agreement with the asymmetric leaflet compositions determined by the van Deenen researchers 50 years ago, as shown in Fig. 1.

Methods to quantitatively measure the rbc lipid asymmetry of PM leaflets have provided somewhat less quantitation than is sought for some proposed research that intrigues us, about how the two different leaflets interact.[4,27] I will attempt to clarify this troubling puzzle of asymmetry quantitation that is affecting our future progress in understanding this new state of matter.

Many of us have done both chemical synthesis and chemical analysis, myself included. I can report that "organic chemistry quantitation" can be difficult to achieve at ±2%, with some exceptions, for example, thin layer chromatography, TLC. When steps in addition to TLC are required, even ±5% can be a challenge. For quantitating asymmetry of a plasma membrane, we have an additional and significant problem that I will address here: the difficulty to establish that the catalyzed hydrolysis reactions of the exoplasmic leaflet lipids have gone to

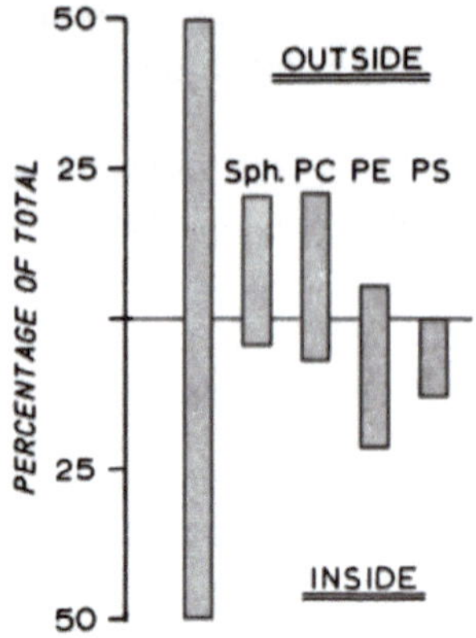

Fig. 1 From Verkleij *et al.* 1973 (ref. 13) originally BBA, 323 : 190 Fig. 13. Note that error bars, which are likely significant, are not available.

completion. This confronts us with a different kind of error that must be addressed, a systematic error.

With living cells, precise lipid analysis involves knowing that the PM is intact, and has not become leaky nor rearranged its leaflet compositions. The rbc, with its readily detected content of hemoglobin, is perhaps the best possible living cell for affirming an intact plasma membrane (Fig. 2).

Another aspect is that even the most simple plasma membrane has many chemical components, and also presents the investigator with a material that is inherently difficult to work with, being inhomogeneous in having both an aqueous part, and a water insoluble part that is organized into the two monolayer leaflets. Researchers seek a lipid analysis of each monolayer, exoplasmic and cytoplasmic.

The starting point for lipid quantitation is what is most accurately measured, the total of each PM lipid, without regard to leaflet assignment. This measurement for each lipid does not require any enzyme or reactive chemical agent, merely the solubilization of all rbc lipids in organic solvent, typically in methanol and chloroform.[17] One such 2-dimensional chromatogram of the extracted rbc lipids by the van Deenen researchers is shown in Fig. 3. Thus, for many decades, researchers have known with perhaps ±2% accuracy the total quantity of each phospholipid, as well as the total cholesterol, in the rbc plasma membrane.

Next, and more challenging by far, is to find for each lipid the fraction that is in the exoplasmic leaflet, and then by subtraction from the known total for that

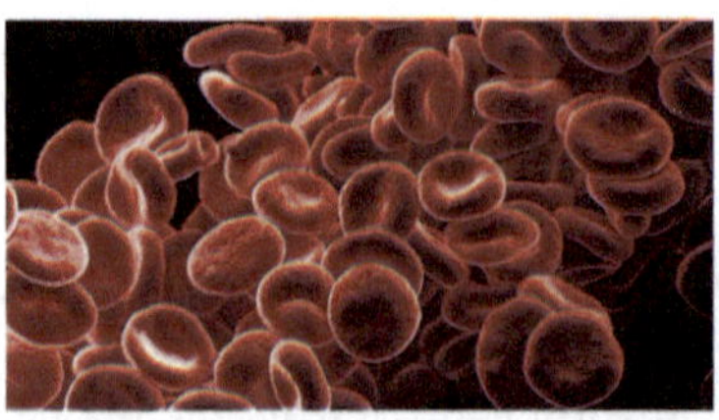

Fig. 2 A field of red blood cells.

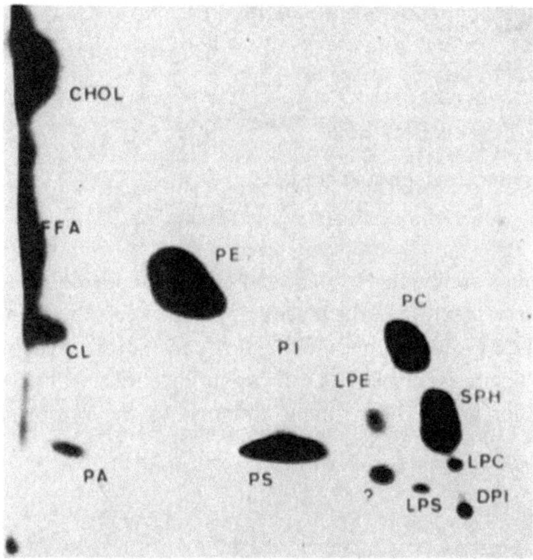

Fig. 3 2D TLC of MeOH then $CHCl_3$ extraction of red blood cell lipids showing each phospholipid. Originally Fig. 3 of Roelefsen & Zwaal 1976.[18] By phosphate assay of each scraped spot,[17] quantitation is achieved.

phospholipid, find its fraction in the cytoplasmic leaflet. Incomplete lipase-catalyzed reaction of phospholipids in the exoplasmic leaflet would lead to the unreacted fraction of those lipids being wrongly assigned to the cytoplasmic leaflet, thus potentially a significant systematic error.

In his study in 1972, Mark Bretscher found little PE (phosphatidylethanolamine) and no PS (phosphatidylserine) in the exoplasmic rbc leaflet, based on little reaction with a non-penetrating, strongly radioactive probe that he devised, FMMP shown in Fig. 4. Bretscher had synthesized FMMP originally for protein study.[11] FMMP reacts most strongly with primary amines, thus is especially sensitive for PE and PS. Gordesky and Marinetti in 1973 (ref. 12) used the now more familiar non-penetrating probe TNBS, Fig. 5, to find no PS in the exoplasmic PM leaflet, but they did find about 30% of total PE to be in the exoplasmic leaflet.

Thus, a key point for rbc quantitative lipid asymmetry analysis is the need for 100% hydrolysis of all PC and all SM (sphingomyelin) in the exoplasmic leaflet *via*

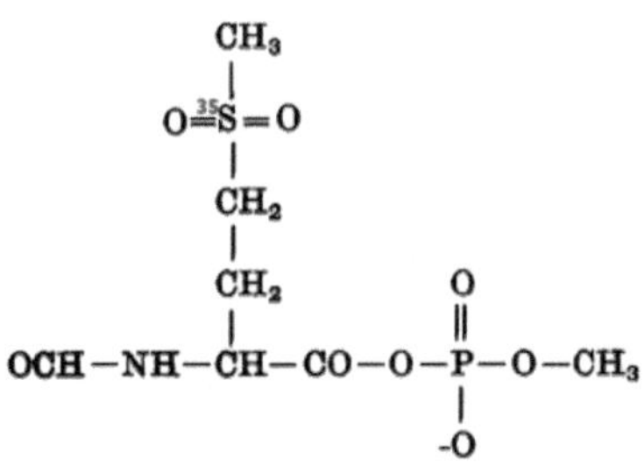

Fig. 4 FMMP = formyl methionyl sulphone methyl phosphate.

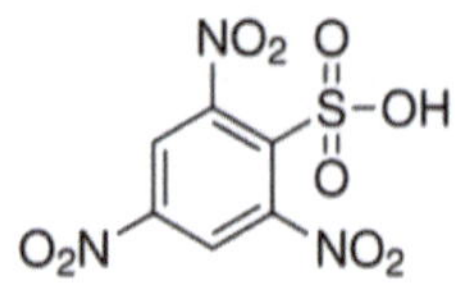

Fig. 5 TNBS = trinitrobenzene sulfonic acid.

the lipase enzymes shown in Fig. 6, added to the outside of the rbc. Fig. 7 shows where these lipase reactions take place on the rbc surface. If 100% hydrolysis is not achieved, then given the method for the quantitation of each lipid from TLC-determined totals for each, some PC (phosphatidylcholine) and SM actually in the exoplasmic leaflet would be wrongly assigned to the cytoplasmic leaflet. Yet

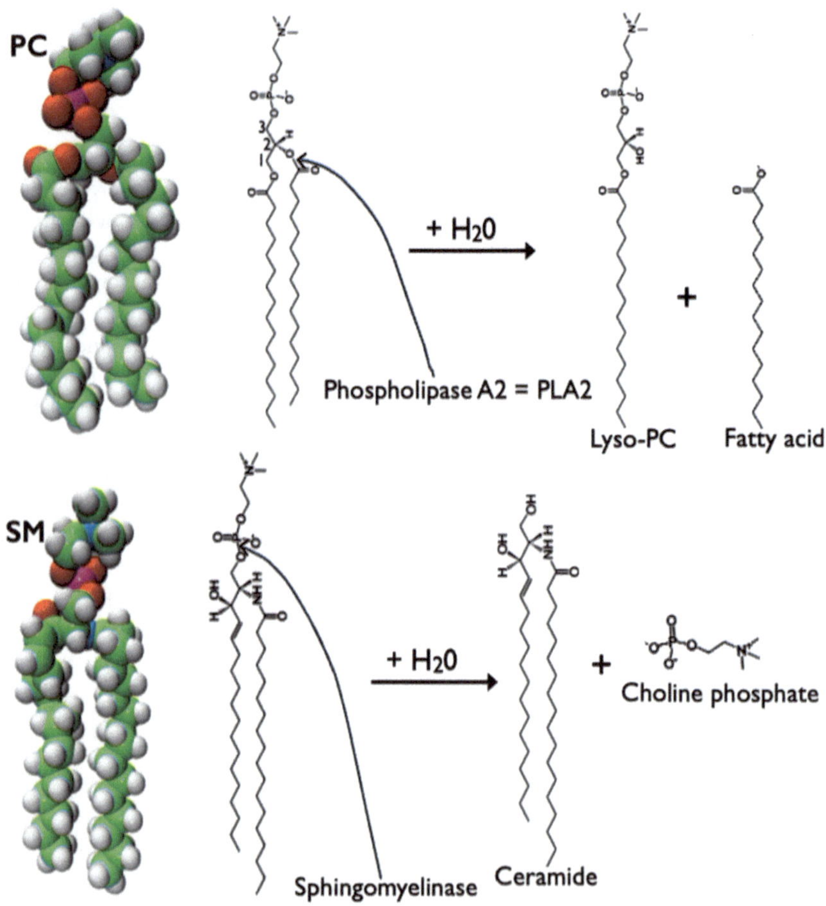

Fig. 6 Specificity of phospholipases A2 and sphingomyelinase, both with central roles in the decade of asymmetry research, 1973–1983. PLA2 and its hydrolysis products lyso-PC and fatty acid remain confined to the exoplasmic leaflet monolayer. SMase and its hydrolysis product ceramide also remain in the exoplasmic leaflet, but product choline phosphate diffuses away into water. Molecular models from Avanti Polar Lipids.

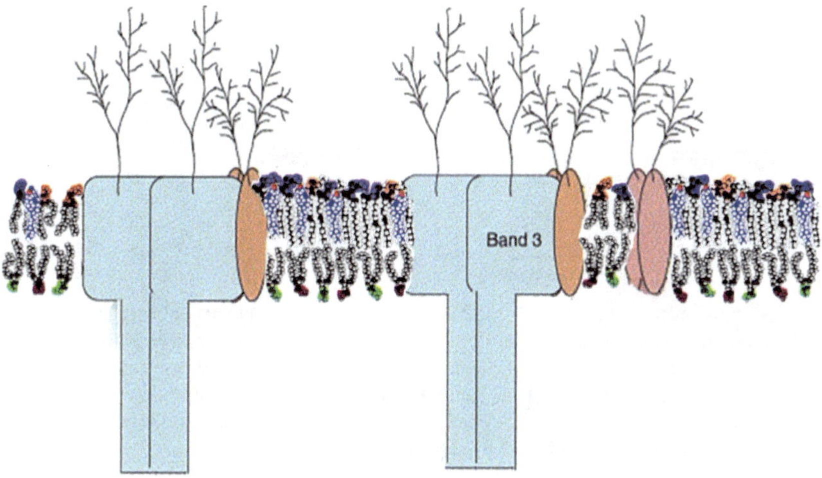

Fig. 7 Cartoon of asymmetric rbc plasma membrane, showing phospholipids to be mostly PC and SM (orange and blue) in the exoplasmic leaflet, and mostly PE and PS (red and green) in the cytoplasmic leaflet.

a significant problem for those of us now reading the many papers from the van Deenen group is that key experimental details, for example, regarding manipulations of rbc's, lipid extraction and analysis, and especially monolayer studies of lipase enzymes, are scattered among many papers in ref. 13–19, 26 and 28–30 but also in other publications from the van Deenen groups.

Putting aside these inconveniences for us readers, these researchers made the important discovery that the lipase reaction catalyzed by PLA2 does not go to completion unless the exoplasmic leaflet monolayer pressure is low enough for the PLA2 to function. This enzyme must interact with the monolayer, and monolayer pressure is a variable. Fortuitous for these measurements, a pressure reduction occurs during the next step,[31] SMase action, as measured by monolayer studies with a Wilhelmy plate and Langmuir trough.[28–30] In brief, the sequence of first PLA2 action, followed by SMase action, followed by additional PLA2 catalysis, is important, as illustrated in Fig. 8. This behavior of exoplasmic monolayer leaflet pressure increasing from PLA2 reaction product formation, followed by exoplasmic leaflet pressure decreasing because of SMase reaction product formation, is thus a reasonable, but not directly proven proposal,[33–37] even with the catalysis observations *per se* not in question. More recent PLA2 studies include molecular dynamics simulations to reveal monolayer pressure to be involved at several stages of PLA2 insertion into the bilayer and at several steps of enzymatic activity.[33,35–37]

To state again the most important difficulty for lipid quantitation, enzyme-catalyzed lipase reactions with lipids on the PM surface must go to completion. Otherwise, unreacted lipids on the surface, especially PC and PE, would be mistakenly assigned to be inner leaflet lipids. However, separate use of PLA2 and SMase is not sufficient for these monolayer reactions to go to completion. The van Deenen groups showed that pressure builds up in the exoplasmic leaflet as reaction products fatty acid and lyso-PC accumulate: PLA2-catalyzed hydrolysis

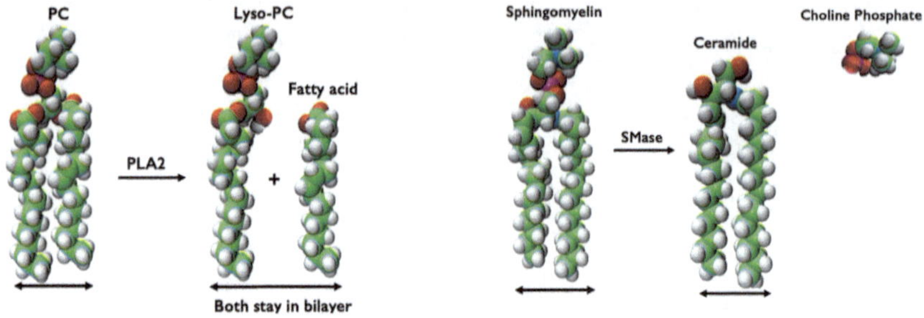

Fig. 8 Illustrations of molecular models (from Avanti Polar Lipids website) showing that PLA2-catalyzed hydrolysis reaction of PC increases monolayer pressure with both products remaining in the bilayer, whereas SMase enzymatic reaction decreases monolayer pressure, with choline phosphate, soluble in water, leaving the monolayer, and product ceramide[32] being a more condensed lipid than is SM. Double headed arrows give approximate relative indication of the lipid area.

slows, then stops, interpreted initially to be caused by lack of PLA2 penetration into the monolayer. Recent studies confirm the decreased activity of PLA2 at high monolayer pressure, but offer additional explanations[32–36] beyond simply PLA2 monolayer penetration. The monolayer pressure seems to be significantly relieved by SMase action[13–19] as illustrated in the pictures in Fig. 8. Hence, the importance of the sequence of steps to be first PLA2 catalysis of PC hydrolysis, where van Deenen researchers discovered leaflet pressure unavoidably builds up, followed by SMase catalysis, where leaflet pressure drops. These studies showing a role for monolayer pressure in chemical reactions within a cell plasma membrane might be the first such chemical implications for monolayer pressure in cell biology. An excellent discussion of monolayer pressure can be found in ref. 38.

3. Phospholipid quantitation numbers

In this way, 76% of the total PC was found to be in the exoplasmic leaflet, including by use of the most efficient PLA2 for rbc lipids, from *Naja naja*, the Indian king cobra.[13] 82% of the total PM SM is degraded by SMase from *Staph. aureus*, and is thus assigned to the exoplasmic leaflet. But the source of the PLA2 also matters, with PLA2 from bee venom digesting only 58% of total rbc PC,[4,17] leading to an incorrect conclusion that the unhydrolysed PC is entirely in the cytoplasmic leaflet. Further, by use of the lipase sequence of first, catalyzed PC hydrolysis by PLA2 from *Naja naja*, then catalyzed hydrolysis of SM by SMase which reduces monolayer pressure, followed by further PLA2-catalyzed hydrolysis of the remaining PC, 20% of all plasma membrane PE is then also hydrolyzed. Therefore another important asymmetry result is that approximately 20% of all plasma membrane PE seems to be in the exoplasmic leaflet,[13,14] in agreement with the external TNBS reaction with PE.[12] From these experiments, PLA2 catalyzed hydrolysis of substrate PE might be more efficient in the unusual new monolayer of lyso-PC, fatty acid, and ceramide.

Without attention to these leaflet pressure effects on PC and PE hydrolysis catalyzed by PLA2, not only would about 8% of the total PC not be assigned to the exoplasmic leaflet, but all of that unreactive 8% would be assigned to the

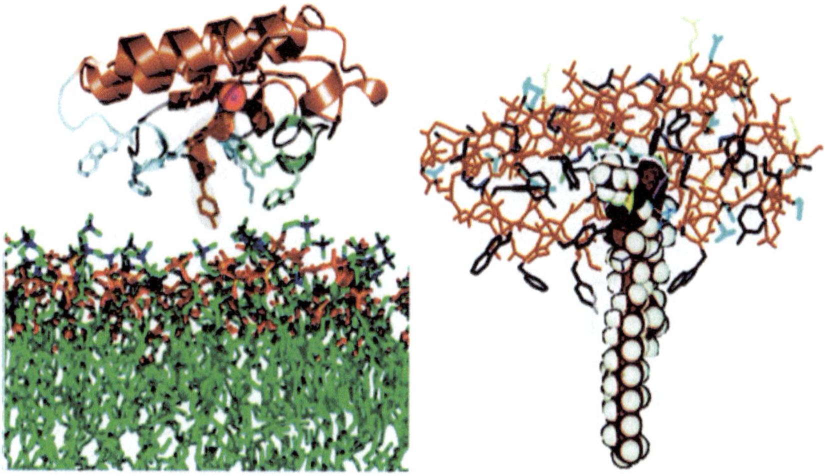

Fig. 9 PLA2 and its target lipid, modified from Fig. 4 and 5 of Dennis *et al.*.[34] (A) Water-soluble PLA2 penetrates exoplasmic monolayer to bind its PC substrate; (B) 2-chain PC with sn-2 acyl linkage in active site of PLA2 enzyme.[33–37] Surrounding bilayer not shown.

cytoplasmic leaflet. Use of PLA2 from bee venom,[17] being a less efficient PLA2 when applied to the rbc, would further miss-assign 18% of the total PC to be in the cytoplasmic leaflet, an especially large error when a researcher seeks to establish numbers of phospholipids in each leaflet.[4,39] Lack of attention to these issues of PLA2 source and monolayer pressure thus causes systematic errors, not random errors, in assignment of PC and PE to the two leaflets.

For counting all phospholipids in each leaflet, the van Deenen studies yield approximately equal numbers of phospholipids in each leaflet.[13] For counting all phospholipids in each leaflet, original data from Lorent *et al.* (2020)[4] together with added quantification of other minor lipids and GPI anchored proteins to estimate the total lipid abundance of both leaflets from Doktorova *et al.* (2025)[39] found 2× more phospholipids in the cytoplasmic leaflet.

The van Deenen researchers concluded that PLA2 fails to penetrate a monolayer to reach its substrate when monolayer pressure is too high. Specifically, their monolayer studies showed that for *N. naja* PLA2, at film pressure 35 erg per sq cm = 35 mN m^{-1}, PLA2 stops working.[28–30] This behavior of PLA2 at higher monolayer pressures has been confirmed over the ensuing decades, even while the molecular-level explanation is more complex than simply PLA2 penetration of the monolayer (Fig. 9).[33–37]

4. Asymmetric bilayers beyond the rbc plasma membrane

Did the van Deenen researchers get it right?! Reading their papers 50 years ago, I did not notice the importance of the sequence: to first add PLA2 to the rbc, then wait about one hour for catalysis to proceed, then to add SMase to catalyze SM hydrolysis, then finally to allow PLA2 hydrolysis to proceed again. This attention to pressure in a monolayer leaflet calls attention to surface chemistry. This author needed immensely useful discussions with a former van Deenen group member, Gerrit van Meer, in order to clarify these important experimental details.

In the many decades that followed studies by the van Deenen researchers, techniques have been described that yield more convenient forms for study of plasma membranes, such as latex beads or glass slides coated with adherent plasma membrane, or inducing cells to create closed vesicle "GPMVs" formed from plasma membrane.[40–43] Yet, no improvement of the method for PLA2-catalyzed hydrolysis of PC in the rbc exoplasmic leaflet seems to have been published.

What is the interest in asymmetric bilayers now, in 2025? We will be hearing and discussing the enormous range of thinking, computation, and wet-lab experiments regarding asymmetric bilayers. For eukaryotic, and especially mammalian plasma membranes, the exoplasmic leaflet has a large fraction of high melting temperature phospholipids, most notably sphingomyelin, whereas such lipids are rare in other cell membranes. Studies of multicomponent mixtures having a high melting phospholipid have shown that rich phase behavior is typical.[44] An interesting case of asymmetric bilayer behavior occurs when one leaflet is phase-separated, but not the other leaflet. One example is when liquid-ordered (Lo) and disordered (Ld) domains coexist in an exoplasmic leaflet.

5. The unexpected phenomenon of induced order

A specific result from asymmetric bilayer imaging studies, with clear implications for asymmetric cell plasma membrane behaviors, is precisely this case of Lo + Ld coexistence in the exoplasmic leaflet of a model mammalian plasma membrane. We have found that induced order domains are formed in a model cytoplasmic leaflet of the apposed monolayer. This is work done by my former colleague Dr Thais Enoki, currently a Young Investigator at the University of Sao Paulo Institute of Physics. This and other work from Dr Enoki reveal the immense value of image data to establish the properties of asymmetric bilayers, because with different dyes localized to each leaflet, individual properties of each leaflet are clearly marked, and moreover clearly show the spatial relationships of the Lo phase in one leaflet to the induced domain in the apposed leaflet.

A plasma membrane case of special interest is when an exoplasmic leaflet has coexisting Lo + Ld phases ("raft + non-raft"), with the cytoplasmic leaflet being a one-phase disordered mixture of phospholipids, predominantly PE and PS, along with cholesterol. A behavior of such asymmetric bilayers is that the Lo phase in the exoplasmic leaflet can induce an ordered phase in the otherwise disordered phase of the cytoplasmic leaflet.[23,45–50] This behavior of "induced order" across leaflets was first observed in model systems more than 15 years ago by researchers in Lukas Tamm's group, by separately imaging each leaflet of asymmetric bilayers.[45,46] This induced order seems to be a normal and characteristic property of asymmetric bilayers: an ordered Lo phase induces "Lo-like" order in DOPC/chol, whereas an ordered Lβ (gel) phase induces "Lβ-like" order in DOPC in the other leaflet.[50]

By use of multicomponent lipid mixtures to model the cytoplasmic leaflet, we plan to find out which of these lipids become the induced ordered domains, and which remain disordered. By abundance, the predominant cytoplasmic leaflet lipids are PS, PE and cholesterol. We also want to find out the behavior of the quantitatively minor but multifunctional phospholipid PIP2 in these asymmetric bilayers.[51] These questions involve, for example, whether or not PS in a cytoplasmic leaflet is concentrated into a more negatively charged and ordered domain than previously appreciated, and whether PIP2 co-localizes with such concentrated PS. Other researchers are now asking similar questions by use of data that do not require imaging.[52–55]

We have so far obtained data only with the simple system, DOPC/chol, that nonetheless is an informative model for a disordered cytoplasmic leaflet.[48–50,56] Our long term goal is to understand the more complex mixture of PS and PE, both

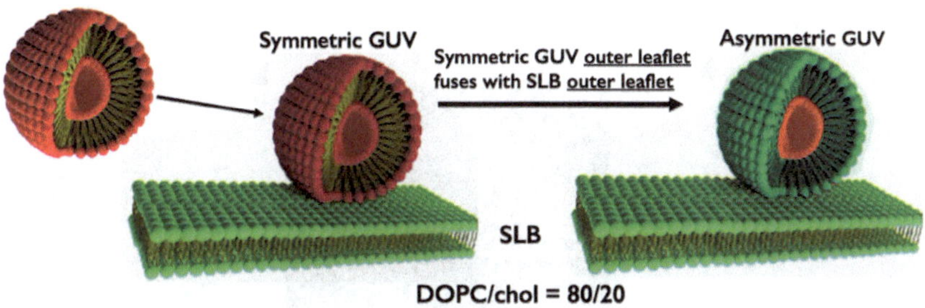

Fig. 10 aGUV preparation by use of hemifusion. The symmetric GUV hemifuses with lipids of a supported lipid bilayer. The original GUV inner leaflet is retained, a new aGUV outer leaflet forms with lipids of the SLB composition. From ref. 49 (Fig. 1).

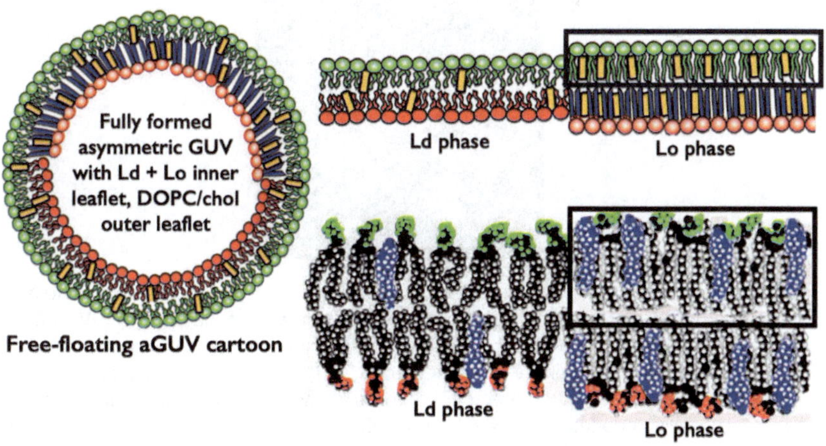

Fig. 11 Left: schematic of an aGUV. The upper right image illustrates the two compositionally different leaflets, with cholesterol as yellow rectangles. This outer leaflet was formed from an SLB of DOPC/chol = 80/20 and includes the green fluorescent dye, TopFluor-PC. Lower right image: the phospholipids and cholesterols shown as more realistic molecular models. The thermodynamically stable Lo and Ld phases are indicated, and the induced order domains are enclosed by black rectangles.[23] Modified from Fig. 2 of ref. 23.

having PUFA-containing chains, together with PIP2 and cholesterol in a disordered cytoplasmic multicomponent model mixture. DOPC is a useful model for a phospholipid with PUFA-containing chains, in part because DOPC has been found to have phase behaviors in multi-component mixtures that are rather like the phase behaviors of PUFA-chain phospholipids,[57] and in part because DOPC, unlike lipids having PUFA chains, is stable to oxygen and to intense confocal microscope illumination.[58]

To obtain good image data it is useful to prepare asymmetric bilayers that are sufficiently large, tens of microns. Generic hemifusion of a supported lipid bilayer with a GUV having the composition that becomes the outer leaflet of the aGUV, is shown in Fig. 10. Confocal fluorescence microscopy images below in Fig. 12 show induced order domains in a DOPC/chol leaflet being precisely superimposed across from the thermodynamically stable Lo domains of an aGUV leaflet of (DSPC or SM)/DOPC/chol. With this preparation, the symmetric GUV's stable Lo +

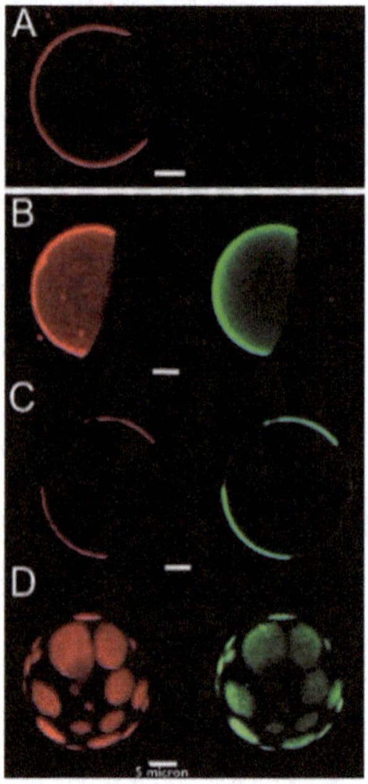

Fig. 12 Fluorescence image data. One symmetric GUV in (A), and three examples of induced order (B–D). Confocal images show green domains of the outer aGUV leaflet to be precisely superimposed on the red domains of the inner leaflet. (A) Symmetric GUV of DSPC/DOPC/chol having phase-separated Lo + Ld. Image is the GUV equator. Red DiD indicates the Ld phase, the connecting Lo phase is dark. Lo order is intrinsic thermodynamic order, not induced; (B) Z-stack of an aGUV with one large Ld and one large Lo domain. DiD in the inner leaflet labels Ld red, leaving Lo dark. TFPC in the outer leaflet labels Ld green, leaving induced order domain dark. This induced order domain is entirely DOPC/chol; (C) aGUV equator showing 3 phase domains. The inner leaflet red DiD presents Ld, leaving Lo dark. The outer leaflet green TFPC prefers Ld, leaving the induced order domain dark; (D) Z-stack of aGUV showing many phase domains. DiD in the inner leaflet labels Ld red, leaving Lo dark. TFPC in the outer leaflet labels Ld green, leaving the induced order regions dark. Scale bars indicate 5 microns. Modified from Fig. 3 and 5 of ref. 59.

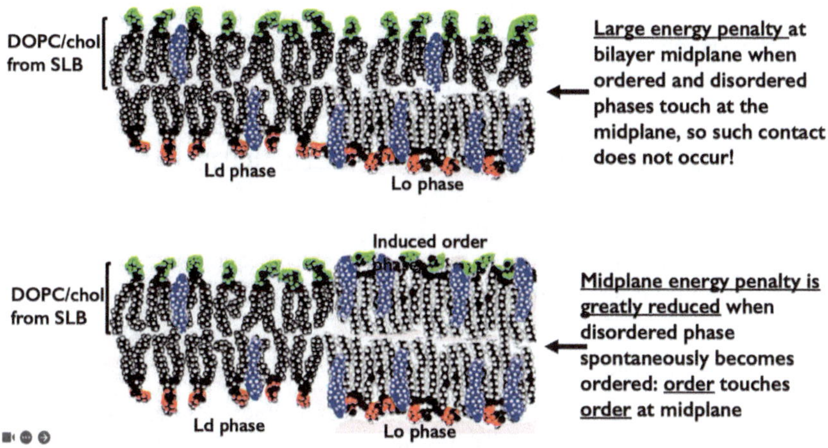

Fig. 13 Importance of the asymmetric bilayer midplane area. The upper image shows contact at the bilayer midplane of disordered with ordered phases that would incur a large energy penalty. The lower image shows induced order domains to appear spontaneously to reduce this energy penalty. This would be the same midplane energy penalty that in symmetric GUVs keeps phase-separated domains in their familiar perfect superposition.[44]

Ld coexistence becomes the phase-separated inner leaflet of a large asymmetric vesicle (Fig. 11).

Confocal imaging data of the DOPC/chol leaflet, Fig. 12, show that the brightly green-colored Ld region is superimposed over the brightly red-colored Ld regions of the phase-separated (DSPC or SM)/DOPC/chol inner leaflet of the aGUV. The minimally-colored Lo domain of (DSPC or SM)/DOPC/chol and the induced order domain of DOPC/chol are also superimposed. Here, because of the implications for cell biology, we emphasize Lo domains causing the order, which lead to "Lo-like" induced order. However, when the cholesterol fraction of the phase-separated monolayer is low, below about 0.16, the coexisting phases are Lβ + Ld, and the induced order domain has Lβ-like order.[57]

The two monolayer aGUV leaflets meet only at the bilayer midplane. This midplane is almost entirely methyl and methylene groups. This "chemically simple" midplane area seems to be the site that determines the energy of the asymmetric bilayers.[23] One way to illustrate clearly this midplane role is shown in Fig. 13.

5.1 Summary of induced order image data in asymmetric bilayers

(1) Hemifusion of a symmetric GUV with a supported lipid bilayer creates an aGUV model of a cell plasma membrane, which enables the obtaining of image information by confocal fluorescence microscopy.

(2) An Lo phase domain induces order in an otherwise highly disordered apposing DOPC/chol monolayer. The mechanism could be that induced order reduces a large energy penalty at the bilayer midplane for contact of order with disorder.

(3) Implication: membrane rafts in a PM exoplasmic leaflet could induce order in a cytoplasmic leaflet that has mostly PUFA-containing lipids plus cholesterol.

(4) If the cytoplasmic leaflet has induced order domains coexisting together with disordered domains, then PE, PS, PIP2 and membrane proteins would

partition between these two different phase environments. We do not yet know whether cytoplasmic PE or cytoplasmic PS becomes the induced order domain in actual cytoplasmic leaflets.

6. Organization of these discussions into four themes

Here I introduce the four session themes of this meeting, and add my own thoughts for each theme (shown in italics).

6.1. Plasma membrane asymmetry and lipid homeostasis

The omics era has provided much data on phospholipids, sphingolipids, sterols and proteins located in different membranes, although information on their functional relevance remains limited. This session will address challenges such as measuring trans-leaflet lipid distribution, including the spatiotemporal relation to overall lipid composition, and dissecting secondary lipid messengers from lipid 'players' with respect to bulk membrane properties.

(a) For wet lab studies, obtaining a purity of membrane without losing membrane integrity (for example, plasma membrane completely purified away from endoplasmic reticulum), is exceedingly difficult.[60]

(b) To establish spatiotemporal relationships, image data are helpful.[49]

6.2. Engineering plasma membrane mimics

Quantification of physicochemical properties of realistic plasma membrane mimics hinges on the controlled production of asymmetric bilayers in the presence and absence of integral or peripheral membrane proteins. In this theme, recently developed asymmetric membrane preparation procedures and how these might be improved will be discussed.

(a) Lipid exchange via *exchange proteins can be useful for both preparation and analysis of asymmetric bilayer models.*[61,62]

(b) New methods involve new uncertainties. Almost any recently developed asymmetric membrane preparation could be tested for a close result to the known asymmetry established for red blood cells.

6.3. Structure and dynamics of asymmetric membranes

Principles from physics and physical chemistry have now been placed within a framework that enables calculation of both membrane tension and membrane curvature. In this session we will seek to obtain physicochemical insights from lipid-only plasma membrane mimics. The session will include minimum realistic mimics of mammalian or bacterial plasma membranes, as well as the role of cholesterol in tuning and maintaining membrane asymmetry.

(a) For wet-lab studies, surface chemistry, especially monolayer methods with a Langmuir trough or similar device, can provide quantitation of monolayer pressure for individual monolayer compositions.[63]

(b) Some preparation methods involve organic solvents or detergents that can remain in the bilayer.[64]

6.4. Proteins in asymmetric membranes

This session will focus on integral proteins functionally reconstituted into asymmetric membranes. Topics will include folding and reconstitution of proteins in asymmetric lipid membranes, control of protein directionality, and stability of lipid asymmetry in the presence of integral proteins.

(a) Membrane models include the parameter of bilayer thickness, and when biological matching to proteins or else leaflet coupling is paramount, DPPC chains might not be long enough. DPPC chains are shorter than those of Hi-Tm PM lipids found in mammalian cells. Instead, DSPC but especially SM are good Hi-Tm lipid choices that closely match natural Hi-Tm lipid length.[65]

(b) When organic solvents are used for protein solubilization, the protein can be trapped for weeks in a non-physiological structure.[66]

Data availability

No primary research results, software or code have been included and no new data were generated or analysed as part of this review.

Conflicts of interest

There are no conflicts of interest to declare.

References

1 E. Gorter and F. Grendel, *J. Exp. Med.*, 1925, **41**(4), 439–443.
2 S. L. Foley, M. Varma, A. Hossein and M. Deserno, *Emerging Top. Life Sci.*, 2023, **7**(1), 95–110.
3 R. Lipowsky, R. Ghosh, V. Satarifard, A. Sreekumari, M. Zamaletdinov, B. Rózycki, M. Miettinen and A. Grafmüller, *Biomolecules*, 2023, **13**, 926–987.
4 J. H. Lorent, K. R. Levental, L. Ganesan, G. Rivera-Longsworth, E. Sezgin, M. Doktorova, E. Lyman and I. Levental, *Nat. Chem. Biol.*, 2020, **16**, 644–652.
5 L. L. M. van Deenen, *Prog. Chem. Fats Other Lipids*, 1966, **8**, 1–127.
6 M. B. Sankaram and T. E. Thompson, *Proc. Natl. Acad. Sci. USA*, 1991, **88**, 8686–8690.
7 D. Chapman, *Langmuir*, 1993, **9**, 39–45.
8 D. Chapman, *Q. Rev. Biophys.*, 1975, **8**, 185–235.
9 P.-G. de Gennes and J. Prost, *The Physics of Liquid Crystals*, Oxford University Press, 1993, vol. 83.
10 M. S. Bretscher, *Nature New Biol.*, 1972, **236**, 11–12.
11 M. S. Bretscher, *J. Mol. Biol.*, 1972, **71**, 523–528.
12 S. E. Gordesky and G. V. Marinetti, *Biochem. Biophys. Res. Commun.*, 1973, **50**, 1027–1031.
13 A. J. Verkleij, R. F. A. Zwaal, B. Roelofsen, P. Comfurius, D. Kastelijn and L. L. M. van Deenen, *Biochim. Biophys. Acta*, 1973, **323**, 178–193.
14 C. M. Colley, R. F. A. Zwaal, B. Roelofsen and L. L. M. van Deenen, *Biochim. Biophys. Acta*, 1973, **307**, 74–82.
15 R. Verger, M. C. E. Mieras and G. H. de Haas, *J. Biol. Chem.*, 1973, **248**(11), 4023–4034.

16 R. F. A. Zwaal, B. Roelofsen and C. M. Colley, *Biochim. Biophys. Acta*, 1973, **300**, 159–182.
17 R. F. A. Zwaal, B. Roelofsen, P. Comfurius and L. L. M. van Deenen, *Biochim. Biophys. Acta*, 1975, **406**, 83–96.
18 B. Roelofsen and R. F. A. Zwaal, *Methods in Membrane Biology*, ed. E. D. Korn, Springer Science, 1976, ch. 2.
19 J. A. F. Op den Kamp, *Annu. Rev. Biochem.*, 1979, **48**, 47–71.
20 J. F. R. Kerr, A. H. Wyllie and A. R. Currie, *Br. J. Cancer*, 1972, **26**, 239–257.
21 A. H. Wyllie, *J. Clin. Pathol.*, 1974, **27**, 35–4210.
22 V. A. Fadok, D. R. Voelker, P. A. Campbell, J. J. Cohen, D. L. Bratton and P. M. Henson, *J. Immunol.*, 1992, **148**(7), 2207–2216.
23 G. W. Feigenson, J. Huang and T. A. Enoki, *J. Am. Chem. Soc.*, 2023, **145**, 21717–21722.
24 I. Meenakshi, S. M. Rao, S. Mayor and R. Sowdhamini, *Front. Cell Dev. Biol.*, 2023, **11**, 1168050.
25 T. D. Pollard, W. C. Earnshaw, J. Lippincott-Schwartz and G. Johnson, *Cell Biology E-Book*, Elsevier Health Sciences, 2022.
26 A. Zachowski, *Biochem. J.*, 1993, **295**, 1–14.
27 M. Doktorova, J. L. Symons, J. Schegel, E. Sezgin, H. Wang, E. R. Lyman, K. R. Levental and I. Levental, *Biophys. J.*, 2024, **123**(3), 512a.
28 R. Demel, W. S. M. Geurts van Kessel, R. F. A. Zwaal, B. Roelofsen and L. L. M. van Deenen, *Biochim. Biophys. Acta*, 1975, **406**, 97–107.
29 R. A. Demel, *Methods Enzymol.*, 1974, **32**, 539–544.
30 R. Verger and G. H. de Haas, *Chem. Phys. Lipids*, 1973, **10**, 127–136.
31 F. Pattus, A. J. Slotboom and G. H. de Haas, *Biochemistry*, 1979, **18**(13), 2691–2697.
32 A. Alonso and F. Goni, *Annu. Rev. Biophys.*, 2018, **47**, 633–654.
33 S. A. Khan and M. A. Ilies, *Int. J. Mol. Sci.*, 2023, **24**, 1353–1388.
34 E. A. Dennis, J. Cao, Y. H. Hsu, V. Magrioti and G. Kokotos, *Chem. Rev.*, 2011, **111**, 6130–6185.
35 E. A. Dennis, *J. Biol. Chem.*, 2022, **298**, 101873.
36 S. Suladze, S. Cinar, B. Sperlich and R. Winter, *J. Am. Chem. Soc.*, 2015, **137**, 12588–12596.
37 A. S. Alekseeva, P. E. Volynsky, N. A. Krylow, V. P. Chernikov, E. L. Vodovozova and I. A. Boldyrev, *Biochim. Biophys. Acta*, 2021, **1863**, 183481.
38 B. Roelofsen and J. A. Op den Kamp, *Membrane Biogenesis*, Springer, Berlin Heidelberg, 1988, pp. 1–13.
39 M. Doktorova, J. L. Symons, X. Zhang, H.-Y. Wang, J. Schlegel, J. H. Lorent, F. A. Heberle, E. Sezgin, E. Lyman, K. R. Levental and I. Levental, *Cell*, 2025, DOI: 10.1016/j.cell.2025.02.034.
40 V. D. Vacquier, *Dev. Biol.*, 1975, **43**, 62–74.
41 M. G. Wetzel and E. D. Korn, *J. Cell Biol.*, 1969, **43**, 90–104.
42 K. R. Levental and I. Levental, *Curr. Top. Membr.*, 2015, **75**, 25–57.
43 B. G. Monasterio, N. Jimenez-Rojo, A. B. Garcia-Arribas, H. Riezman, F. Goni and A. Alonso, *Int. J. Mol. Sci.*, 2020, **21**, 2643.
44 J. Korlach, P. Schwille, W. W. Webb and G. W. Feigenson, *Proc. Natl. Acad. Sci. USA*, 1999, **96**, 8461–8466.
45 V. Kiessling, C. Wan and L. K. Tamm, *Biochim. Biophys. Acta*, 2009, **1788**, 64–71.

46 C. Wan, V. Kiessling and L. K. Tamm, *Biochemistry*, 2008, **47**(7), 2190–2198.
47 Q. Wang and E. London, *Biophys. J.*, 2018, **115**(4), 664–678.
48 E. London, *Acc. Chem. Res.*, 2019, **52**(8), 2382–2391.
49 T. A. Enoki and G. W. Feigenson, *Biophys. J.*, 2019, **117**(6), 1037–1050.
50 T. A. Enoki, J. Wu, F. A. Heberle and G. W. Feigenson, *Biochim. Biophys. Acta*, 2021, **1863**(6), 183586.
51 Y. Wen, V. M. Vogt and G. W. Feigenson, *Annu. Rev. Biochem.*, 2021, **90**, 681–707.
52 F. S. Leomil, M. Stephen, S. Pramanik, K. A. Riske and R. Dimova, *Langmuir*, 2024, **40**, 4719–4731.
53 E. London, *Membranes*, 2022, **12**, 870.
54 M. Krompers and H. Heerklotz, *Membranes*, 2023, **13**, 267.
55 G. Pabst and S. Keller, *Trends Biochem. Sci.*, 2024, **49**, 333–345.
56 G. W. Feigenson and T. A. Enoki, *Biophys. J.*, 2023, **122**(6), 925–930.
57 T. M. Konyakhina and G. W. Feigenson, *Biochim. Biophys. Acta*, 2016, **1858**, 153–161.
58 N. F. Morales-Penningston, J. Wu, E. R. Farkas, S. L. Goh, T. M. Konyakhina, J. Y. Zheng, W. W. Webb and G. W. Feigenson, *Biochim. Biophys. Acta*, 2010, **1798**, 1324–1332.
59 T. A. Enoki and G. W. Feigenson, *Biochim. Biophys. Acta*, 2022, **1864**, 183995.
60 D. F. H. Wallach and P. S. Lin, *Biochim. Biophys. Acta*, 1973, **300**, 211–254.
61 P. E. DiCorleto, J. B. Warach and D. B. Zilversmit, *J. Biol. Chem.*, 1979, **254**, 7795–7802.
62 K. W. A. Wirtz and L. L. M. van Deenen, *Trends Biochem. Sci.*, 1977, **2**, 49–51.
63 J. J. Giner-Casares, G. Brezesinski and H. Mohwald, *Curr. Opin. Colloid Interface Sci.*, 2014, **19**, 176–182.
64 D. Lombardo and M. A. Kiselev, *Pharmaceutics*, 2022, **14**, 543.
65 R. G. Cutler and M. P. Mattson, *Mech. Ageing Dev.*, 2001, **122**, 895–908.
66 W. R. Veatch and E. R. Blout, *Biochemistry*, 1974, **13**, 5257–5264.

Faraday Discussions

ROYAL SOCIETY OF CHEMISTRY

PAPER

Effect of a scramblase activator upon lipid and probe scrambling and membrane domain formation in HEK 293T cells†

Shinako Kakuda and Erwin London *

Received 28th December 2024, Accepted 5th February 2025

DOI: 10.1039/d4fd00211c

Lipid asymmetry, a difference between the lipid composition of the inner and outer leaflets (monolayers) of a membrane, is a characteristic of mammalian plasma membranes. In artificial lipid vesicles, asymmetry can either suppress or induce the formation of coexisting ordered and disordered lipid domains depending on lipid composition. In mammalian plasma membrane preparations, loss of asymmetry induces the formation of ordered domains. In this report, we studied the effect of a scramblase activator, the ionophore BrA23187 (BrA) plus Ca^{2+}, upon ordered domain (lipid raft) formation in human embryonic kidney 293T cells. Addition of BrA induced a decrease in FRET between the plasma membrane outer leaflet probe TMADPH, which partly associates with ordered domains, and ODRB, which localizes largely in liquid disordered domains. This is consistent with the formation of coexisting ordered and disordered domains in the plasma membrane. In addition, upon BrA addition, the plasma membrane outer leaflet probe Pro12A exhibited a decrease in the generalized polarization (GP) suggesting a decrease in outer leaflet membrane order, perhaps due to a decrease in outer leaflet cholesterol However, there are other explanations for these observations. To test if BrA induced scrambling of fluorescent membrane probes, which would complicate interpretation of the experiments described above, we measured the effect of BrA upon extractability of outer leaflet probes with MβCD (in most cases, MβCD was more effective for extraction than BSA). These experiments showed that at most a small fraction of probes migrate to the inner leaflet upon addition of BrA. Other experiments raise the possibility that BrA binding to membranes may directly influence ordered domain formation and properties or alter fluorescence by direct interactions with TMADPH, and thus not reflect changes in domain formation.

Introduction

Membrane lipid asymmetry refers to the difference in lipid composition between the inner and outer leaflets (lipid monolayers) of artificial or natural membranes

Dept. of Biochemistry and Cell Biology, Stony Brook University, Stony Brook, NY 11794, USA. E-mail: Erwin.London@stonybrook.edu

† Electronic supplementary information (ESI) available. See DOI: https://doi.org/10.1039/d4fd00211c

containing a lipid bilayer. A distinct lipid composition is found in mammalian plasma membranes, with sphingolipids, including sphingomyelin (SM), and phosphatidylcholine (PC) enriched in the outer (exofacial) leaflet, and the anionic phospholipids phosphatidylserine (PS) and phosphatidylinositol (PI) plus the zwitterionic lipid phosphatidylethanolamine (PE) enriched in the inner (cytofacial, cytoplasmic-facing) leaflet.[1]

The lipids found in the inner and outer leaflets have different propensities to form an ordered state. In particular, outer leaflets enriched in sphingolipids have a high propensity to form the liquid-ordered (L_o) state when the membrane has a substantial concentration of cholesterol. In contrast, the inner leaflet is enriched in lipids with a higher degree of acyl-chain unsaturation, which tend to form the liquid-disordered (L_d) state.[2] It has been found that, at least in the resting state, the lipids in the outer leaflet of the plasma membranes have a higher degree of order than those in the inner leaflet,[1] and although cholesterol distribution between leaflets remains controversial, recent evidence suggests a higher cholesterol concentration in the outer leaflet under usual conditions.[3–8] For example, in a high-resolution structure of an aquaporin, containing bound lipids,[8] it has been observed that the outer leaflet lipids in contact with a protein are both in a more ordered configuration (with more extended acyl chains) and are more cholesterol-enriched than the inner leaflet lipids in contact with a protein.

Studies on artificial and natural membranes have shown that asymmetry can greatly impact the formation of ordered lipid domains (lipid rafts).[9–11] This may be biologically important because a transient loss of lipid asymmetry can occur during signal transduction.[12] In addition, changes in asymmetry can influence membrane protein conformation.[13] In artificial lipid membranes in which only one leaflet is composed of lipids that spontaneously form ordered domains, it has been observed that, depending upon lipid composition, lipid asymmetry can either induce the formation of ordered domains in a leaflet formed of lipids that do not spontaneously form ordered domains,[14–18] or suppress ordered domain formation in a leaflet formed of lipids that would spontaneously form ordered domains in a symmetric vesicle.[9,17–19] The latter behavior has been observed in both lipid vesicles with a composition mimicking that of plasma membranes,[9] and plasma membrane vesicles from rat basophilic leukemia 2H3 (RBL-2H3) cells.[10]

In this report, we studied the effect of altering living cell plasma membrane lipid asymmetry upon membrane organization by adding a lipid scramblase activator to cells. Using plasma membrane outer leaflet probes, changes indicative of domain formation and altered membrane order were observed. Alternate explanations for the origin of these changes are discussed.

Experimental

Materials

Lipids: 1-oleoyl-2-(6-[(7-nitro-2-1,3-benzoxadiazol-4-yl)amino]hexanoyl)-*sn*-glycero-3-phosphocholine (6NBD-PC) from Avanti Polar Lipids (Alabaster, AL) was stored at 1 mM in DMSO. Brain sphingomyelin (SM), 1-palmitoyl-2-oleoyl phosphatidylcholine (POPC), dioleoylphosphatidylcholine (DOPC), and cholesterol were from Avanti Polar Lipids. All lipids were stored at −20 °C.

Probes: 1-(4-trimethylammoniumphenyl)-6-phenyl-1,3,5-hexatriene *p*-toluenesulfonate (TMADPH) was purchased from the Molecular Probes division of Invitrogen (Carlsbad, CA), and stored at 24 μM in ethanol. 3-(*N*-dodecyl-2-(methyl(6-propionylnaphthalen-2-yl) amino)acetamido) propane-1-sulfonate (Pro12A) was from Cytoskeleton, Inc (Denver, CO), and stored at 20 μM in (DMSO). 6-Dodecanoyl-2-dimethylaminonaphthalene (Laurdan) was purchased from Invitrogen (Carlsbad, CA), and stored at 32 μM in ethanol. Octadecyl rhodamine B (R18, ODRB) was from Invitrogen (Carlsbad, CA), and stored at 100 μM in ethanol or at 8.6 mM in ethanol for cell experiments. All probes were stored at −20 °C.

Reagents: Alexa Fluor 488-labeled Annexin V (Annexin V-F488) was from Invitrogen and stored at 4 °C. BrA (4-bromo-A23187, 4-bromo-calcimycin) was from Millipore Sigma (Burlington MA) and MedChemExpress (Monmouth Junction, NJ). BrA was dissolved at 2 mM in ethanol or DMSO. No difference in behavior between experiments was noted depending on solvent or manufacturer. Stored at −20 °C. Fluo-4 AM in DMSO was from Invitrogen. Methyl-β-cyclodextrin (MβCD) was from Millipore Sigma (Burlington MA). Bovine Serum Albumin (BSA), fatty acid free was from Chem Cruz (Dallas TX). Dulbecco's PBS (DPBS) (200 mg l^{-1} KCl, 200 mg l^{-1} KH_2PO_4, 8 g l^{-1} NaCl, and 2.16 g l^{-1} Na_2HPO_4), was from ThermoFisher Scientific (Waltham, MA). All other chemicals were reagent grade.

Cell lines and cell culture

Human embryonic kidney (HEK) 293T cells (ATCC CRL-11268) were maintained in DMEM (Gibco), 10% FBS, 1% pen strep (Gibco) at 37 °C, and 5% CO_2, subcultured every two days, and used between passages 4–25.

Measurement of time dependence of FRET in living cells before and after BrA treatment

Cells were detached from plates with DPBS. Two samples were prepared each with 1×10^6 cells in 1 ml DPBS. We estimate that each of these samples would have a phospholipid concentration of roughly 50–100 μM (data not shown). For the "F sample" which contains FRET donor and acceptor, the FRET acceptor ODRB was added to a concentration of 12 μM, while the "F_o sample" which contains FRET donor but not acceptor lacked ODRB. Both samples were incubated at 37 °C for 10 min and cells pelleted (200×*g*/3 min) to remove unbound ODRB. The cell pellets were resuspended in 1 ml of ABB (Annexin-binding buffer: 10 mM HEPES, 140 mM NaCl, and 2.5 mM $CaCl_2$), and then FRET donor TMADPH was added to both F and F_o samples to a final concentration of 0.07 μM, and incubated at RT for 10 min in the dark before fluorescence was measured. TMADPH intensity was read every 2 min for up to 10 min at RT (∼23 °C) and then BrA was added to both F and F_o samples to the desired final concentration (maximum 8 μM). Readings were taken every 2 min (up to 10 min). TMADPH fluorescence intensity was measured using a Horiba QuantaMaster Spectrofluorometer (Horiba Scientific, Edison, NJ) with an excitation wavelength (E_x) of 358 nm, emission wavelength (E_m) of 427 nm, and the bandpass set to 3 nm for both excitation and emission. The fluorescence intensity of the background samples (lacking TMADPH but containing an equivalent organic solvent) was negligible (less than 1%). Then the fraction of fluorescence unquenched by FRET, F/F_o, was calculated. (FRET efficiency is $1 - F/F_o$).

Preparation of large unilamellar vesicles

Large unilamellar vesicles (LUVs) composed of various combinations of SM, POPC, DOPC and cholesterol were prepared. Lipids initially dissolved in chloroform were mixed, dried by a stream of nitrogen and then subject to high vacuum for 1 h. Multilamellar vesicles (MLVs) were prepared by adding 1 ml of DPBS and incubated at 70 °C for 5 min. LUVs were prepared from MLVs with 7 freeze–thaw cycles. To avoid losses of lipid during filtration, the samples were not extruded to achieve uniform size. The final lipid concentration was 1 mM.

Measurement of the temperature dependence of FRET in lipid vesicles

FRET measurements on vesicles were carried out as previously described.[10] Briefly, the vesicles contained 50 μM lipid dispersed in 1 ml of ABB. TMADPH was the FRET donor and ODRB was the FRET acceptor. First, the F sample containing ODRB (0.4 μM) and the F_o sample lacking ODRB were incubated at 37 °C for 1 h in the dark. Next, TMADPH was added to both F and F_o samples (final concentration of 0.07 μM) and then incubated at room temperature for 10 min in the dark. TMADPH fluorescence intensity was measured as described above. The data was collected at every 4 °C interval while increasing the temperature from 16 to 64 °C. Between readings, the temperature was increased at a rate of 4 °C/6 min. The fraction of fluorescence not quenched by FRET is equal to F/F_o. (FRET efficiency is $1 - F/F_o$).

Determining exposure of PS on the cell surface by spectrofluorimetry and fluorescence microscopy

To evaluate exposure of PS on the outer leaflet of the plasma membrane, the fluorescence of conjugated Annexin V (Annexin V-F488) bound to the cell surface PS was measured. 4×10^6 cells were suspended in 4 ml of DPBS at 37 °C for 10 min and centrifuged. The cell pellets were then suspended in 4 ml of ABB and 1 ml aliquots added to four 1.5 ml tubes. BrA was added to each cell suspension at concentrations of 0, 2, 4 or 8 μM and then incubated at RT for 6 min. Next, 0.2 ml from each sample was transferred to 1 ml cold ABB and the cells were pelleted. Annexin V-F488 was diluted twenty-fold (according to the manufacturer's directions) with ABB. Then, 50 μl was added and incubated at RT for 5 min in the dark. After washing two times with ABB, cells were suspended in 1 ml of ABB. Cell-surface-bound Annexin V-F488 fluorescence intensity was measured with $E_x = 488$ nm and $E_m = 510$ nm, and the bandpass was set at 5 nm for both excitation and emission.

For imaging living cell PS exposure, cells were seeded on 5 μg ml^{-1} poly-D-lysine-coated glass bottom culture dishes (MatTek corporation). The next day, the cells were washed with ABB and stimulated with 8 μM of BrA at RT for 10 min. After washing, 200 μl of Annexin V-F488 contained in ABB solution was added and incubated at RT for 10 min in the dark. After removing the Annexin V solution, 1 ml of ABB was added. The samples were imaged with a Zeiss LSM 980 confocal microscope. Annexin V-F488 was excited at 488 nm and the emission collected between 503 and 700 nm.

Estimated internalization of 6NBD-PC by extraction of outer leaflet 6NBD-PC and dithionite treatment

The flip of 6NBD-PC added to cells across the plasma membrane was determined from the accessibility to extraction by externally added BSA. To do this, 4×10^6

cells were suspended in 4 ml of DPBS at 37 °C for 10 min and then pelleted. The cell pellets were resuspended with 4 ml of ABB containing 0.5 μM of 6NBD-PC and incubated at RT for 10 min. 1 ml aliquots were transferred to four 1.5 ml tubes. BrA was added to each cell sample giving concentrations of 0, 2, 4 and 8 μM and then incubated at RT for 6 min. Then 0.2 ml from each sample was added to 0.2 ml of pre-chilled 5 mg of fatty acid-free-BSA in ABB and incubated on ice for 10 min to extract outer leaflet 6NBD-PC. After washing two times with 1 ml ice-cold ABB and then adding 1 ml of RT ABB, the 6NBD-PC fluorescence was measured at $E_x = 464$ nm and $E_m = 531$ nm, with the bandpass set at 5 nm for both excitation and emission. The background values for samples without 6NBD-PC were low and were not subtracted from the values with 6NBD-PC.

The flip of 6NBD-PC was confirmed by the loss of exposure of 6NBD-PC to reduction by sodium dithionite. To do this, 1 ml samples of BrA-treated cells with 6NBD-PC, with or without BSA extraction, were used, as described above. The samples were then mixed with 25 μl of 1 M sodium dithionite in 1 M Tris–HCl buffer (pH 10). 6NBD-PC fluorescence intensity was recorded before dithionite addition and at 2 min intervals after dithionite addition. Dithionite abolishes NBD fluorescence by reduction of the NBD groups.

Laurdan and Pro12A GP measurements in cells and lipid vesicles

2×10^6 cells in 2 ml of DPBS were incubated at 37 °C for 10 min and then pelleted. The cell pellets were resuspended to 2 ml of ABB for two samples. LUV with various lipid compositions were prepared as described above and then diluted to 100 μM lipid with ABB. In one cell or vesicle sample, BrA was added at 8 μM and in the other (control) an equivalent amount of solvent was added. After incubation in cuvettes at RT for 6 min, fluorescence backgrounds were measured. Then the fluorescence probe was added (0.08 μM for Laurdan or 0.05 μM for Pro12A), incubated at 23 °C for 10 min, and fluorescence measured using $E_x = 395$ nm and $E_m = 440$ and 490 nm for both probes. The bandpass was set at 5 nm for both excitation and emission. GP was calculated using the formula $GP = (F_{440} - F_{490})/(F_{440} + F_{490})$. The background values from samples lacking Laurdan or Pro12A were subtracted.

BSA extraction of fluorescent probes in living cells

From 1.5×10^6 cells in 1.5 ml DPBS three 0.5 ml samples were prepared. One was the background (cells only) and to the other two was added 0.07 μM TMADPH. After incubation at RT for 10 min in the dark, the cells were then pelleted, and 1 ml of ABB was added to each cell pellet. To the cells with a fluorescent probe either 0.5 ml of 10 mg ml^{-1} BSA in ABB or 0.5 ml ABB was added, and then all samples were incubated at RT for 20 min. The cells were then pelleted, and 1 ml ABB was added. The TMADPH fluorescence intensity measurements used $E_x = 358$ nm, $E_m = 427$ nm, and the bandpass was set at 3 nm for both excitation and emission. Background fluorescence from the sample lacking a fluorescence probe was subtracted from the reported fluorescence.

For other probes the same experimental procedure was followed. The concentrations of the probes and the parameters for the fluorescence measurements were as follows: for Pro12A the concentration was 0.07 μM, $E_x = 395$ nm, $E_m = 440$ and 490 nm, bandpass = 5 nm; for 6NBD-PC the concentration was 0.4

μM, $E_x = 464$ nm, $E_m = 531$ nm, bandpass = 5 nm; for ODRB the concentration was 12 μM, $E_x = 556$ nm, $E_m = 578$ nm, bandpass = 3 nm.

For the ODRB samples, the cells were incubated at 37 °C for 10 min and then pelleted to remove unbound ODRB. Then the cells were resuspended in 1 ml of ABB and 0.1 ml of Triton X-100 (w/v) was added to abolish any ODRB self-quenching before measuring the fluorescence.

MβCD extraction of fluorescent probes in living cells with or without BrA treatment

From 2.5×10^6 cells in 2.5 ml DPBS, five 0.5 ml samples were prepared. Two were backgrounds (cells only and with BrA) and to the other three was added, for example, 0.07 μM TMADPH. After incubation at RT for 10 min in the dark, BrA was added to one sample containing a fluorescent probe, and one background sample, to a concentration of 8 μl, and incubated at RT for 6 min in the dark. Next, 55 μl of 100 mM MβCD in PBS was added to one sample with a fluorescent probe only, and one sample with a fluorescent probe and BrA. All samples were then incubated at RT for 20 min. Finally, the cells were pelleted, washed once with 1 ml ABB, and suspended in 1 ml ABB for fluorescence measurements. Background fluorescence from the samples lacking a fluorescence probe, with or without BrA, was subtracted from the reported values. The same protocol was used for the other probes, with concentrations and fluorescence measurements as above.

Fluorescence microscopy images of probe localization in living cells

All imaging was carried out at room temperature (~23 °C). Cells were seeded on glass bottom culture dishes (MatTek Corporation) a day before imaging. The next day, the cells were washed once with DPBS and then incubated with TMADPH (3 μM), ODRB (3 μM), Pro12A (1 μM), or Laurdan (20 μM) for 5 min at RT in the dark. After labeling, the cells were washed once with ABB, and imaging was performed after 20–30 min incubation at RT. The cells were imaged using a Zeiss LSM 980 confocal microscope. For TMADPH, Pro12A and Laurdan, the excitation was at 405 nm and the emission collected between 403 and 621 nm. For ODRB, the excitation was at 568 nm and the emission collected between 560 and 747 nm.

Staining cytosolic Ca^{2+} with Fluo-4 AM

Cells were plated on poly-D-lysine-coated 48-well plates. The next day, after washing the cells with DPBS three times, Fluo-4 AM in DMSO was diluted 1000-fold into 0.2 ml ABB or ABB without Ca^{2+}. The samples were incubated at RT for 10 min and then 8 μM BrA, or an equivalent amount of solvent, was added and the samples incubated at RT for another 10 min. The images were taken after one wash with ABB with or without Ca^{2+} using an Olympus IX 83 with a GFP filter cube.

Measurement of ODRB binding to cells

To 2×10^6 cells in 2 ml DPBS, ODRB was added to a concentration of 12 μM, then they were incubated at 37 °C for 10 min, divided into two equal samples, and then the cells were pelleted. To one sample, 1 ml ABB was added. To the other, 1 ml ABB without Ca^{2+} but with 0.4 mM EGTA (ethylene glycol-bis(β-aminoethyl ether)-

N,N,N',N'-tetraacetic acid) was added. Next, BrA was added to a concentration of 8 μM. The samples were incubated for 6 min at RT. Then the cells were pelleted, and the supernatant removed carefully. The cell pellets were suspended in 1 ml ABB, either with Ca^{2+} or without Ca^{2+} but with 0.4 mM EGTA. ODRB fluorescence was measured after taking 400 μl of the supernatant or 400 μl of the cell suspension, adding 500 μl of ABB, or ABB without Ca^{2+}, respectively, and 100 μl of 10% (w/v) Triton X-100. ODRB fluorescence was measured as described above.

Results and discussion

BrA-induced loss of membrane asymmetry in HEK 293T cells

To induce the loss of plasma membrane lipid asymmetry, HEK 293T cells were incubated with Ca^{2+} and an ionophore that allows Ca^{2+} entry into the cytoplasm. A high cytosolic Ca^{2+} concentration activates TMEM16F[20] and other lipid scramblases. We chose HEK 293T cells because they exhibit strong scramblase activity following the treatment with an ionophore plus Ca^{2+}.[21] To test for the loss of membrane asymmetry, we adopted widely used fluorescence-based lipid scrambling assays. The first was used to detect scrambling-induced PS exposure on the outer leaflet of the plasma membrane, because in the absence of scrambling, PS is a cytosolic leaflet lipid. To do this, the amount of cell-surface-bound fluorescently conjugated Annexin V, a PS binding protein, was measured. The second assay was used to monitor the decrease of 6NBD-PC localization in the plasma membrane outer leaflet. In all of these studies, we used the non-fluorescent ionophore BrA (4-bromo-A23187) to avoid interference from the fluorescence that would occur if ordinary A23187 was used, and ABB, a Ca^{2+}-containing Annexin-binding buffer (see Experimental).

We confirmed that addition of Ca^{2+} and BrA increased cytosolic Ca^{2+} levels using the Ca^{2+} indicator Fluo-4 AM. ESI Fig. 1† shows a dramatic increase in cytosolic Ca^{2+} levels when the cells were incubated with both the Ca^{2+} and BrA, which should activate scramblase.

To assay the scrambling of PS, the cells were incubated with different concentrations of BrA and then the binding of PS to Annexin V-F488 was measured. Fig. 1A shows that there was a BrA dose-dependent increase in outer leaflet PS. As shown by confocal microscopy images (Fig. 1B), a significant population of cells showed PS exposure on the cell surface. This observation is consistent with the A23187 activation of NK cells and mouse Ba/F3 cells after which a significant subpopulation of the cells exhibited PS exposure.[22,23]

To monitor the scrambling of outer leaflet PC, we measured the decrease in extractability of the fluorescent lipid 6NBD-PC by BSA and MβCD added to the cells.[24] BSA has a high (but not 100%) extraction efficiency. PC is normally enriched in the plasma membrane outer leaflet, and 6NBD-PC added to cells inserts in the outer leaflet. Movement to the inner leaflet prevents extraction. We found that the amount of cell-associated 6NBD-PC protected from BSA or MβCD extraction increased when BrA was added to the cells (Fig. 2A and Table 1). This indicates that a substantial fraction of 6NBD-PC translocated across the membrane in a BrA dose-dependent fashion.

To confirm the decrease in the amount of outer leaflet 6NBD-PC, the accessibility of 6NBD-PC to reduction by externally added sodium dithionite was measured.[25] Dithionite reduces the NBD nitro group and thereby abolishes NBD fluorescence. Outer leaflet NBD lipids are accessible to rapid reduction by

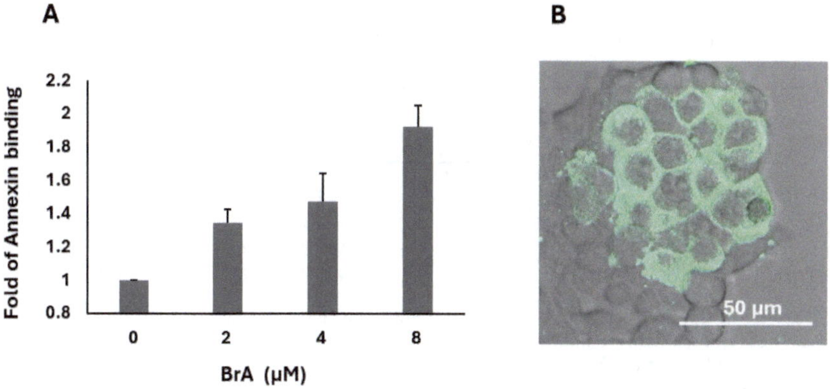

Fig. 1 BrA-induced PS exposure in the cells. (A) HEK 293T cells were incubated with BrA at the indicated concentrations and externalized PS was assayed by the binding of Annexin V-F488. The graph shows the mean values of Annexin fluorescence and standard deviations for $n = 3$. The value for the samples without BrA was normalized to 1. (B) Microscope image showing the cells exhibiting Annexin V-F488 binding after treatment with 8 μM BrA.

dithionite. Fig. 2B and C and ESI Fig. 2† show that minutes after dithionite addition, there was almost complete reduction of the NBD group in the absence of BrA, even if the samples were pretreated with BSA to remove some of the outer leaflet 6NBD-PC.

However, in the presence of BrA there was substantial protection of 6NBD-PC from dithionite. The protection was further increased after BSA extraction of some of the outer leaflet 6NBD-PC, which increased the fraction of NBD in the inner leaflet (compare Fig. 2B and C). This confirms that some of the 6NBD-PC migrated out of the outer leaflet upon BrA addition, and that the extraction step indeed removes outer leaflet 6NBD-PC. It should be noted that dithionite can cross membranes to leak into cells. Extrapolation to time zero after the addition of BrA indicated that the initial protection from dithionite is even greater than shown in Fig. 2B and C (see ESI Fig. 2†).

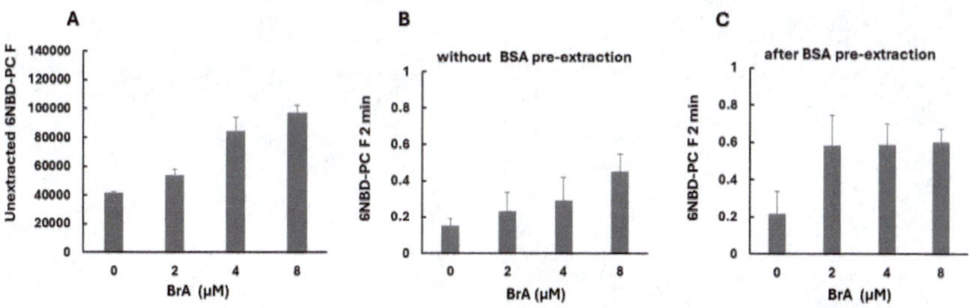

Fig. 2 Assays of BrA-induced scrambling of 6NBD-PC added to the cells. (A) Effect of BrA upon protection of 6NBD-PC from extraction by BSA. *Y*-axis shows the amount of 6NBD-PC fluorescence unextracted by BSA in arbitrary units. (B and C) Protection from dithionite after the addition of BrA without pre-extraction of 6NBD-PC (B), or after pre-extraction by BSA (C). Fraction of the initial 6NBD-PC fluorescence before dithionite addition that is unreduced 2 min after addition of 25 mM sodium dithionite is shown. Samples contained 0.5 μM 6NBD-PC added to HEK 293T cells suspended in ABB. Mean values ($n = 3$) and standard deviations are shown.

Table 1 Extraction of membrane probes by BSA and MβCD at RT.[a] Mean values for the fraction of probe molecules that were unextracted and standard deviations for $n = 3$ are shown

	BSA −BrA	MβCD −BrA	MβCD +BrA[b]
6NBD-PC[c]	0.182 ± 0.028	0.163 ± 0.022	0.327 ± 0.026
TMADPH	0.576 ± 0.125	0.300 ± 0.047	0.333 ± 0.026
ODRB	0.946 ± 0.158	0.482 ± 0.082	0.499 ± 0.037
Pro12A[d]	0.235 ± 0.081	0.092 ± 0.012	0.192 ± 0.029

[a] There was less extraction of probes when MβCD extraction was carried out on samples on ice (data not shown). [b] 8 μM BrA. [c] Extraction of 6NBD-PC and the other fluorescent probes were carried out at RT for these experiments. [d] Emission measured at 440 nm. Values were similar when the emission was measured at 490 nm.

BrA-induced decrease in FRET between plasma membrane outer leaflet probes TMADPH and ODRB

The effect of BrA treatment upon membrane domain formation was investigated using Förster resonance energy transfer (FRET). TMADPH was used as the FRET

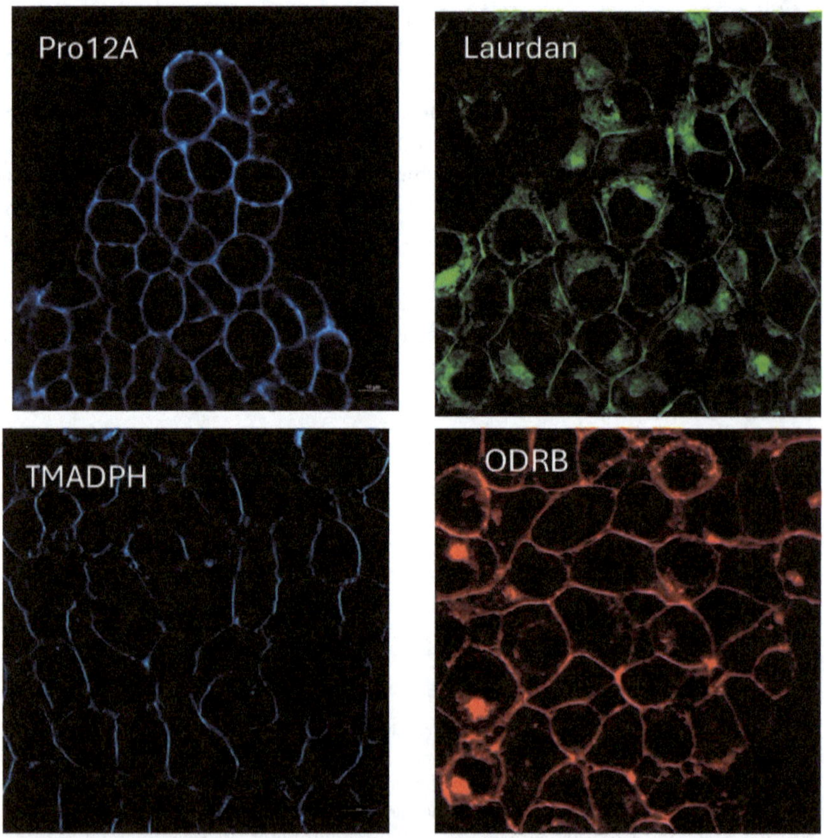

Fig. 3 Confocal microscopy images of fluorescent probe localization in HEK 293T cells at RT. Images taken 20–30 min after washing out excess probes. Laurdan image in psuedo-color.

donor and ODRB as the FRET acceptor. Both are cationic molecules that associate with the outer leaflet of plasma membranes when added to cells,[26] although ODRB, in which the positive charge density is low, can cross membranes.[27] Microscopy confirmed their localization in plasma membranes (Fig. 3). Prior studies have shown that upon formation of coexisting ordered and disordered domains in membranes there is a segregation of some TMADPH molecules into the ordered domains, while ODRB very largely remains in disordered domains. This results in a decrease in the FRET, which can be detected by the increase in the ratio of donor fluorescence (F) relative to donor fluorescence in the absence of an acceptor (F_o), *i.e.*, an increase in F/F_o.[10] Fig. 4 shows that when BrA was added to the HEK 293T cells there was a rapid decrease in the FRET (increase in F/F_o). The increase was BrA-dose-dependent (Fig. 4A). A similar, but more rapid, increase in F/F_o (and larger at 4 μM) was observed when BrA was added to the cells at 37 °C (Fig. 4B). The increase in F/F_o was blocked when BrA was added to cells incubated in ABB without Ca^{2+} but with EGTA (Fig. 4C).

The change in FRET is consistent with the formation of ordered domains upon addition of BrA. However, we were concerned that there might be other variables that could affect F/F_o. For example, if BrA in the presence of Ca^{2+} reduced the binding of ODRB to membranes, this would increase F/F_o specifically in such samples. However, ODRB binding experiments did not detect an increase in the amount or fraction of ODRB in the supernatant when cells incubated with ODRB were pelleted in the presence of BrA and Ca^{2+} relative to the amount or fraction of ODRB in the supernatant for samples incubated in the presence of BrA without Ca^{2+} but with 0.4 mM EGTA (ESI Fig. 3†). Other parameters not reflecting domain formation that could also affect FRET are described below.

Effect of BrA addition upon membrane order detected by Pro12A

The effect of BrA addition upon membrane order was also measured. The probes used were Laurdan, which resides largely, but not fully, in the plasma membrane[28] (see Fig. 3), and the anionic Laurdan derivative Pro12A, which does not cross membranes[28] and so localizes in the plasma membrane outer leaflet

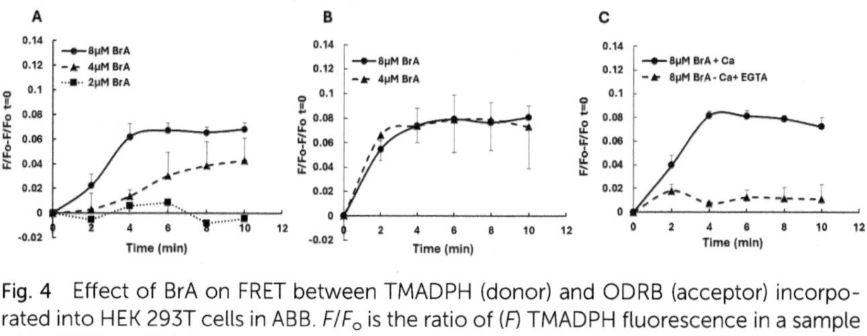

Fig. 4 Effect of BrA on FRET between TMADPH (donor) and ODRB (acceptor) incorporated into HEK 293T cells in ABB. F/F_o is the ratio of (F) TMADPH fluorescence in a sample with ODRB to (F_o) fluorescence in a sample lacking ODRB. (A) Increase in F/F_o at room temperature after addition of different BrA concentrations. Time of addition of BrA defined as $t = 0$. (B) Increase in F/F_o at 37 °C after addition of BrA. (C) Increase in F/F_o after addition of BrA with Ca^{2+}, or without Ca^{2+} but with 0.4 mM EGTA present. BrA was added immediately after $t = 0$. Mean values ($n = 3$) and standard deviations are shown. See ESI Table 1.†

when added to cells. Generalized polarization (GP), a measure of the relative fluorescence emission intensity at short and long wavelengths for these probes,[29] was used to assay membrane order. GP values are high in an ordered membrane environment.

Experiments with artificial vesicles confirmed that both Laurdan GP and Pro12A GP are very sensitive to both ordered state formation and cholesterol content. Vesicles composed of POPC, which should be in the disordered L_d state at RT, brain sphingomyelin (SM), which should be in an ordered state (gel without cholesterol, liquid-ordered (L_o) with cholesterol), and SM/POPC, which should have a combination of disordered (L_d) and ordered domains, were prepared, all with and without cholesterol. Table 2 shows that Pro12A and Laurdan GP show similar, but not identical, sensitivities to membrane composition. For both probes, GP, and thus membrane order, increased with increasing formation of an ordered state. The GP measured using Pro12A and Laurdan in SM/POPC/cholesterol vesicles, which form coexisting L_d and L_o domains,[30] was intermediate between the GP values for POPC/cholesterol vesicles and SM/cholesterol vesicles. The intermediate GP values suggest that both Laurdan and Pro12A localize within both L_o and L_d domains and do so to a similar extent. In addition, GP was higher in samples with cholesterol relative to those without cholesterol, even for POPC in which the membrane should be in the L_d state whether cholesterol was present or not. This is not unexpected, because even in the L_d state, addition of cholesterol increases membrane order.[31] The addition of BrA to these vesicles using the same BrA concentration as used in cells induced a small increase in GP in most cases. This shows that BrA has an effect on membrane order that is independent of activating a lipid scramblase, or there is a BrA-probe interaction.

Table 3 shows the level of membrane order in HEK 293T cell membranes with and without BrA treatment. In the absence of BrA, the GP of both Laurdan and Pro12A was similar to that in cholesterol-containing artificial vesicles with coexisting ordered and disordered domains (*i.e.*, SM/POPC/cholesterol). The presence of BrA slightly decreased the Pro12A GP, but slightly increased the Laurdan GP.

Table 2 Effect of vesicle lipid composition on GP values with or without 8 μM BrA. Mean values and standard deviations for n = 3 are shown

Laurdan			Pro12A		
	−BrA	+BrA		−BrA	+BrA
POPC	−0.049 ± 0.091	0.012 ± 0.046	POPC	−0.064 ± 0.010	−0.062 ± 0.017
1 : 1 SM/POPC	0.179 ± 0.033	0.269 ± 0.026	1 : 1 SM/POPC	0.094 ± 0.007	0.063 ± 0.019
SM	0.302 ± 0.016	0.487 ± 0.043	SM	0.136 ± 0.032	0.093 ± 0.014
2 : 1 POPC/chol	0.391 ± 0.053	0.435 ± 0.054	2 : 1 POPC/chol	0.358 ± 0.027	0.418 ± 0.047
1 : 1 : 1 SM/POPC/chol	0.511 ± 0.019	0.551 ± 0.001	1 : 1 : 1 SM/POPC/chol	0.530 ± 0.019	0.570 ± 0.040
2 : 1 SM/chol	0.650 ± 0.034	0.650 ± 0.022	2 : 1 SM/chol	0.682 ± 0.032	0.687 ± 0.037

Table 3 Effect of BrA on GP values in cells. Mean values and standard deviations for $n = 3$ are shown

	−BrA	+BrA[a]
Laurdan	0.489 ± 0.004	0.542 ± 0.005
Pro12A	0.482 ± 0.028	0.441 ± 0.027

[a] 8 µM BrA.

The difference in the behavior of these probes is probably a reflection of the difference in their membrane locations. The Pro12A GP should reflect membrane order only in the plasma membrane outer leaflet, while Laurdan would probe the average order in both leaflets. The behavior of Pro12A is thus consistent with a loss of cholesterol and/or a decrease in average order in the outer leaflet under conditions in which overall membrane order is maintained, but is also consistent with scrambling of Pro12A (see below).

Effects of BrA upon extractability/localization of membrane probes

Interpretation of FRET and GP data for the experiments described above could be complicated by changes in probe localization when BrA is added to the cells. Therefore, we considered the possibility that BrA activation of a scramblase would alter the fraction of fluorescent probes in the inner and outer leaflet. To investigate the leaflet localization of probes, their accessibility to extraction was assayed. For these experiments, a suitable molecule for extracting probes from the outer leaflet had to be identified. To do this, we compared extraction by BSA and by MβCD. Except for 6NBD-PC, BSA did not extract membrane probes from the cells as well as MβCD did (Table 1). Therefore, MβCD was used in subsequent experiments.

Table 1 shows that 10 mM MβCD extracted roughly 70% of the membrane-bound TMADPH and 90% of Pro12A as judged from the decrease in cell-bound fluorescence. Only about 50% of the ODRB added to the outer leaflet was extracted by MβCD. This may mean ODRB is located equally in both the inner and outer leaflets. Indeed, studies have shown that ODRB is able to cross membranes under some conditions.[27,32] Concluding that ODRB is equally localized in the outer and inner plasma membrane leaflet is valid if ODRB does not flip across the membrane quick enough for inner leaflet ODRB to be extracted by MβCD. Alternately, the low level of ODRB extraction could result from strong ODRB binding to membranes relative to its binding to MβCD. Residual TMADPH and Pro12A in the membranes after MβCD extraction could also reflect incomplete extraction of outer leaflet probes.

If there was BrA-induced scrambling that moved probes into the inner leaflet, their extraction from the outer leaflet would decrease unless the BrA-induced flip-flop of the molecules was extremely rapid. This decrease in extraction was observed for 6NBD-PC. However, there was only a very small, if any, decrease in the accessibility of TMADPH or ODRB to extraction after incubation with BrA (Table 1). This suggests that these probes are not significantly scrambled after addition of BrA, *i.e.*, not subject to flip-flop due to BrA activation of scramblase. Pro12A did show a somewhat larger decrease in accessibility to extraction, which might reflect a scramblase-induced Pro12A flip to the inner leaflet, or a direct effect of membrane-bound BrA such that it decreases the ability of MβCD to extract Pro12A from the membranes.

Direct effects of BrA upon membrane domain formation

The decrease in FRET (increase in F/F_o) upon BrA addition to the cells could reflect direct induction of ordered domain formation resulting from the binding of BrA to the plasma membrane. To test if this was possible, we compared membrane domain formation in 1 : 1 : 1 SM/DOPC/cholesterol vesicles and 2 : 1 DOPC/cholesterol vesicles with and without BrA. Fig. 5 shows the temperature dependence of FRET for these samples. The F/F_o curves for DOPC/cholesterol, which do not form ordered domains in the temperature range studied, show a baseline for the samples without ordered domain formation (Fig. 5A and B). The F/F_o curves for the SM/DOPC/cholesterol vesicles show higher F/F_o at low temperature than those for DOPC/cholesterol due to ordered domain formation.[33] At higher temperatures these ordered domains melt, abolishing domain formation such that homogeneous bilayers form, and F/F_o values become similar to those in the DOPC/cholesterol vesicles, which lack ordered domains.[34] It appears

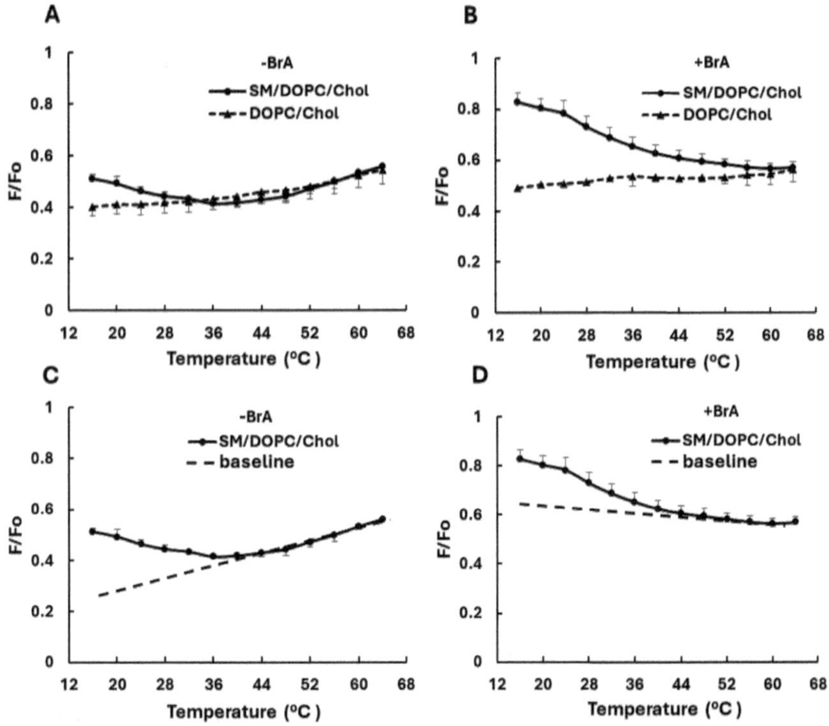

Fig. 5 Effect of BrA upon FRET between TMADPH and ODRB in lipid vesicles. F/F_o *vs.* temperature for 50 μM lipid vesicles composed of 1 : 1 : 1 SM/DOPC/cholesterol or 2 : 1 DOPC/cholesterol for samples in the absence of BrA (A and C), or in the presence of 8 μM BrA (B and D). (A) and (B) show superimposed curves for the two lipid compositions. (C) and (D) show F/F_o for 1 : 1 : 1 SM/DOPC/cholesterol vesicles with and without BrA, superimposed over baselines estimated from F/F_o *vs.* temperature trends at high temperature. Vesicles were suspended in ABB. Mean values ($n = 3$) and standard deviation are shown.

that the ordered domains are more thermally stable with BrA (Fig. 5A and B). However, interpreting the effect of BrA on domain formation depends on what baseline is chosen for FRET. If we assume that the baseline without domains is represented by the slope of the F/F_0 curves above the temperature at which domains normally form (about 45 °C[34]) (Fig. 5C and D), then there appears to be little if any effect of BrA on domain stability. There is also very little effect of Ca^{2+} on FRET with or without BrA, indicating it has little effect on domain formation (ESI Fig. 4†).

Evidence for a direct interaction between ionophore and FRET probes that influences probe fluorescence

Another variable that can influence fluorescence is direct interaction between BrA, which is anionic, and the FRET probes, which are cationic. Suggestive of such an interaction, there was a quenching of TMADPH fluorescence by 20–30% upon addition of BrA to samples without ODRB. There was no quenching of TMADPH fluorescence in the presence of ODRB, which could reflect competition between TMADPH and ODRB for binding to BrA. If this is correct, the change in TMADPH fluorescence in the presence BrA might not reflect a change in domain formation.

Discussion of the effect of loss of lipid asymmetry in HEK 293T cells

Prior studies support the conclusion that under basal conditions in the mammalian plasma membrane, the outer leaflet is in an L_o-like state, while the inner leaflet is more L_d-like.[1,35] In addition, under basal conditions the outer leaflet of the plasma membrane is enriched in sphingolipids and, although still controversial,[3–7] may be enriched in cholesterol. The extent to which laterally segregated ordered and disordered domains coexist in individual leaflets of the plasma membrane under basal conditions is unclear, and may depend on cell type and experimental conditions. As in the case of asymmetric lipid vesicles with a SM/POPC/cholesterol outer leaflet and POPC/cholesterol inner leaflet, there might be no or little lateral domain formation in cells under many conditions. As noted in the Introduction, under *in vitro* conditions, the loss of asymmetry in SM/POPC/cholesterol vesicles, which somewhat mimic the plasma membrane, induces lateral separation into L_o and L_d domains.[9] In addition, studies on plasma membrane vesicles from mammalian cells have also shown that a loss of lipid asymmetry strongly promotes the formation of ordered lipid domains.[10]

The aim of this study was to explore whether a loss of asymmetry in intact cells can induce lateral lipid domain formation in the plasma membrane of intact cells. In particular, we wished to determine if scramblase-activating conditions, *i.e.*, incubation with an ionophore in the presence of extracellular Ca^{2+}, would induce lateral segregation of lipids into coexisting L_o and L_d domains. Under these conditions, a decrease in FRET consistent with a segregation of a population of TMADPH molecules from ODRB molecules, and thus lateral domain formation, was observed. This could reflect an increase in the formation and/or size of coexisting L_o and L_d domains in the plasma membrane upon loss of membrane lipid asymmetry.[34,36]

GP values for Pro12A suggest the loss of asymmetry is also accompanied by a decrease in average outer leaflet membrane order, which could reflect a decrease in outer leaflet cholesterol, without a significant change in overall membrane

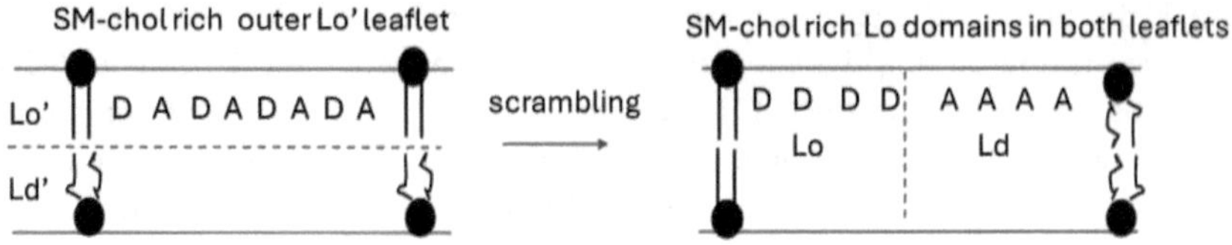

Fig. 6 Schematic illustration of possible changes in membrane organization upon addition of BrA. Only a few lipids are shown. The space between the solid lines represents a lipid bilayer. The dashed lines indicate boundaries between different physical states. D and A are FRET donor and acceptor, respectively. L_o and L_o' and L_d and L_d' are used to indicate that the L_o and L_d states do not have to be identical in properties before and after scrambling. Note: The FRET acceptor level in the inner leaflet is not shown. An alternate model is that the FRET acceptor is distributed in both leaflets before and after BrA addition.

order, as shown by Laurdan GP. It should be noted that, in addition to the random loss of asymmetry, domain formation could be impacted by changes in the outer leaflet PC/SM ratio upon activation of scrambling, as has been observed in the presence of active TMEM63B.[37] A model summarizing the hypothesis for the changes that occur under scramblase-activating conditions is schematically illustrated in Fig. 6.

Potential biological implications

A reversible loss of lipid asymmetry may be associated with the types of signal transduction that induce Ca^{2+} influx into the cytoplasm and thus might activate scramblases.[12,38,39] In processes in which ordered domain (lipid raft) formation is triggered by and/or promotes clustering and activation of raft-associating receptors and associated proteins, such as in immune cells,[40–42] activation-triggered Ca^{2+} influx could induce a loss of asymmetry and thus initiate a positive feedback loop that increases the level of raft formation and activation.

Factors that complicate interpretation of FRET and membrane order experiments

Two factors complicate interpretation of the experiments in these studies. One is the possibility that activation of scrambling results in scrambling of the fluorescent probes used. The rationale for such scrambling is that the scramblase action may arise partially from membrane thinning.[43] Thinning reduces the size of the energy barrier for flipping of membrane molecules between leaflets and could potentially induce flipping of fluorescent probes as well as lipids. If flipping of probes occurs, the changes in FRET and membrane order that fluorescent probes detect upon BrA addition could arise from BrA-induced probe scrambling. For example, scrambling of TMADPH and/or ODRB between the inner and outer leaflets of the plasma membrane would decrease FRET by increasing the average distance between donor and acceptor. In addition, probe scrambling could decrease Pro12A GP by moving Pro12A from a high-order outer leaflet to a lower-order inner leaflet. In contrast, the environment sensed by a probe expected to localize in both leaflets, such as Laurdan, might not be significantly affected by scrambling. As judged by the accessibility to extraction, there is very little change in probe exposure on the outer leaflet upon BrA addition. Nevertheless, we cannot rule out a small extent of probe scrambling upon addition of BrA, especially in the case of Pro12A.

A second factor complicating interpretation is that BrA binding to membranes might directly impact ordered domain formation and/or the physical properties in addition to altering lipid asymmetry. There was an increase in F/F_o seen when BrA was added to the artificial vesicles composed of SM/DOPC/cholesterol. BrA has a chemical structure that should not pack well in L_o domains, and so is very likely to localize in L_d domains. At the high BrA concentrations used in these studies, this could change the properties of L_d domains to decrease the miscibility of the L_d and L_o lipids and/or increase domain size to make them large enough to detect by FRET.[34] The experiments on artificial lipid vesicles described above are ambiguous as to whether BrA actually induces domain formation. In addition, F/F_o could increase due to changes in the partition of ODRB or TMADPH between L_o and L_d domains such that TMADPH molecules are more segregated from ODRB molecules when BrA is added.

However, the change in F/F_o in artificial vesicles upon BrA addition was observed both in the presence and absence of Ca^{2+} (ESI Fig. 5†), while the increase observed in the cells was Ca^{2+}-dependent (Fig. 4C), indicating that the FRET change in cells is likely to reflect the change in lipid asymmetry. Nevertheless, the change in F/F_o remains a concern as it could reflect a direct probe-BrA interaction, as noted above.

Conclusions

The studies in this report show that adding the scramblase activator BrA to HEK 293T cells induces lipid scrambling and an apparent increase in FRET-detected membrane domain formation. However, alternate explanations for the apparent FRET changes are possible. BrA has, at most, modest effects upon the leaflet localization of the non-lipid fluorescent membrane probes used with the exception of Pro12A. The bulk FRET approach may be useful for detecting plasma membrane ordered domain formation induced by biological phenomena that activate lipid scrambling in which artifacts due to use of an ionophore are not an issue.

Data availability

All data supporting the findings of this study are available within the paper and its ESI.†

Conflicts of interest

There are no conflicts to declare.

Acknowledgements

These studies were supported by NIH grant GM122493. We thank Bingying Xia for measuring the amount of lipids in cells by ^{31}P NMR.

References

1 J. H. Lorent, K. R. Levental, L. Ganesan, G. Rivera-Longsworth, E. Sezgin, M. Doktorova, E. Lyman and I. Levental, Plasma membranes are asymmetric in lipid unsaturation, packing and protein shape, *Nat. Chem. Biol.*, 2020, **16**(6), 644–652.

2 D. A. Brown and E. London, Structure and origin of ordered lipid domains in biological membranes, *J. Membr. Biol.*, 1998, **164**(2), 103–114.

3 K. C. Courtney, K. Y. Y. Fung, F. R. Maxfield, G. D. Fairn and X. Zha, Comment on 'Orthogonal lipid sensors identify transbilayer asymmetry of plasma membrane cholesterol', *eLife*, 2018, **7**, e38493.

4 M. Mondal, B. Mesmin, S. Mukherjee and F. R. Maxfield, Sterols are mainly in the cytoplasmic leaflet of the plasma membrane and the endocytic recycling compartment in CHO cells, *Mol. Biol. Cell*, 2009, **20**(2), 581–588.

5 L. M. Solanko, D. P. Sullivan, Y. Y. Sere, M. Szomek, A. Lunding, K. A. Solanko, A. Pizovic, L. D. Stanchev, T. G. Pomorski, A. K. Menon and D. Wustner, Ergosterol is mainly located in the cytoplasmic leaflet of the yeast plasma membrane, *Traffic*, 2018, **19**(3), 198–214.

6 S. L. Liu, R. Sheng, J. H. Jung, L. Wang, E. Stec, M. J. O'Connor, S. Song, R. K. Bikkavilli, R. A. Winn, D. Lee, K. Baek, K. Ueda, I. Levitan, K. P. Kim and W. Cho, Orthogonal lipid sensors identify transbilayer asymmetry of plasma membrane cholesterol, *Nat. Chem. Biol.*, 2017, **13**(3), 268–274.

7 M. Doktorova, J. L. Symons, X. Zhang, H. Y. Wang, J. Schlegel, J. H. Lorent, F. A. Heberle, E. Sezgin, E. Lyman, K. R. Levental and I. Levental, Cell membranes sustain phospholipid imbalance *via* cholesterol asymmetry, *bioRxiv*, 2024, preprintDOI: 10.1101/2023.07.30.551157.

8 P. L. Chiu, J. D. Orjuela, B. L. de Groot, C. Aponte Santamaria and T. Walz, Structure and dynamics of cholesterol-mediated aquaporin-0 arrays and implications for lipid rafts, *eLife*, 2024, **12**, RP90851.

9 J. W. St Clair, S. Kakuda and E. London, Induction of Ordered Lipid Raft Domain Formation by Loss of Lipid Asymmetry, *Biophys. J.*, 2020, **119**(3), 483–492.

10 S. Kakuda, P. Suresh, G. Li and E. London, Loss of plasma membrane lipid asymmetry can induce ordered domain (raft) formation, *J. Lipid Res.*, 2022, **63**(1), 100155.

11 E. London, Ordered Domain (Raft) Formation in Asymmetric Vesicles and Its Induction upon Loss of Lipid Asymmetry in Artificial and Natural Membranes, *Membranes*, 2022, **12**(9), 870.

12 M. Doktorova, J. L. Symons and I. Levental, Structural and functional consequences of reversible lipid asymmetry in living membranes, *Nat. Chem. Biol.*, 2020, **16**(12), 1321–1330.

13 Q. Lin and E. London, The influence of natural lipid asymmetry upon the conformation of a membrane-inserted protein (perfringolysin O), *J. Biol. Chem.*, 2014, **289**(9), 5467–5478.

14 Q. Lin and E. London, Ordered raft domains induced by outer leaflet sphingomyelin in cholesterol-rich asymmetric vesicles, *Biophys. J.*, 2015, **108**(9), 2212–2222.

15 F. A. Heberle, D. Marquardt, M. Doktorova, B. Geier, R. F. Standaert, P. Heftberger, B. Kollmitzer, J. D. Nickels, R. A. Dick, G. W. Feigenson, J. Katsaras, E. London and G. Pabst, Subnanometer Structure of an Asymmetric Model Membrane: Interleaflet Coupling Influences DomainProperties, *Langmuir*, 2016, **32**, 5195–5200.
16 T. A. Enoki and G. W. Feigenson, Improving our picture of the plasma membrane: Rafts induce ordered domains in a simplified model cytoplasmic leaflet, *Biochim. Biophys. Acta, Biomembr.*, 2022, **1864**(10), 183995.
17 C. Wan, V. Kiessling and L. K. Tamm, Coupling of cholesterol-rich lipid phases in asymmetric bilayers, *Biochemistry*, 2008, **47**(7), 2190–2198.
18 M. D. Collins and S. L. Keller, Tuning lipid mixtures to induce or suppress domain formation across leaflets of unsupported asymmetric bilayers, *Proc. Natl. Acad. Sci. U. S. A.*, 2008, **105**(1), 124–128.
19 T. A. Enoki and F. A. Heberle, Experimentally determined leaflet-leaflet phase diagram of an asymmetric lipid bilayer, *Proc. Natl. Acad. Sci. U. S. A.*, 2023, **120**(46), e2308723120.
20 J. Suzuki, T. Fujii, T. Imao, K. Ishihara, H. Kuba and S. Nagata, Calcium-dependent phospholipid scramblase activity of TMEM16 protein family members, *J. Biol. Chem.*, 2013, **288**(19), 13305–13316.
21 T. Le, Z. Jia, S. C. Le, Y. Zhang, J. Chen and H. Yang, An inner activation gate controls TMEM16F phospholipid scrambling, *Nat. Commun.*, 2019, **10**(1), 1846.
22 N. Wu, H. Song and A. Veillette, Plasma membrane lipid scrambling causing phosphatidylserine exposure negatively regulates NK cell activation, *Cell. Mol. Immunol.*, 2021, **18**(3), 686–697.
23 S. Nagata, J. Suzuki, K. Segawa and T. Fujii, Exposure of phosphatidylserine on the cell surface, *Cell Death Differ.*, 2016, **23**(6), 952–961.
24 J. Suzuki, M. Umeda, P. J. Sims and S. Nagata, Calcium-dependent phospholipid scrambling by TMEM16F, *Nature*, 2010, **468**(7325), 834–838.
25 J. C. McIntyre and R. G. Sleight, Fluorescence assay for phospholipid membrane asymmetry, *Biochemistry*, 1991, **30**(51), 11819–11827.
26 B. Chazotte, Labeling the plasma membrane with TMA-DPH, *Cold Spring Harb. Protoc.*, 2011, **2011**(5), pdb.prot5622.
27 G. B. Melikyan, B. N. Deriy, D. C. Ok and F. S. Cohen, Voltage-dependent translocation of R18 and DiI across lipid bilayers leads to fluorescence changes, *Biophys. J.*, 1996, **71**(5), 2680–2691.
28 D. I. Danylchuk, E. Sezgin, P. Chabert and A. S. Klymchenko, Redesigning Solvatochromic Probe Laurdan for Imaging Lipid Order Selectively in Cell Plasma Membranes, *Anal. Chem.*, 2020, **92**(21), 14798–14805.
29 T. Parasassi, G. De Stasio, G. Ravagnan, R. M. Rusch and E. Gratton, Quantitation of lipid phases in phospholipid vesicles by the generalized polarization of Laurdan fluorescence, *Biophys. J.*, 1991, **60**(1), 179–189.
30 R. S. Petruzielo, F. A. Heberle, P. Drazba, J. Katsaras and G. W. Feigenson, Phase behavior and domain size in sphingomyelin-containing lipid bilayers, *Biochim. Biophys. Acta, Biomembr.*, 2013, **1828**(4), 1302–1313.
31 M. Mihailescu, R. G. Vaswani, E. Jardon-Valadez, F. Castro-Roman, J. A. Freites, D. L. Worcester, A. R. Chamberlin, D. J. Tobias and S. H. White, Acyl-chain methyl distributions of liquid-ordered and -disordered membranes, *Biophys. J.*, 2011, **100**(6), 1455–1462.

32 T. I. Rokitskaya, G. A. Korshunova and Y. N. Antonenko, Effect of Alkyl Chain Length on Translocation of Rhodamine B n-Alkyl Esters across Lipid Membranes, *Biophys. J.*, 2018, **115**(3), 514–521.

33 P. Pathak and E. London, The Effect of Membrane Lipid Composition on the Formation of Lipid Ultrananodomains, *Biophys. J.*, 2015, **109**(8), 1630–1638.

34 P. Pathak and E. London, Measurementof lipid nanodomain (raft) formation and size in sphingomyelin/POPC/cholesterol vesicles shows TX-100 and transmembrane helices increase domain size by coalescing preexisting nanodomains but do not induce domain formation, *Biophys. J.*, 2011, **101**(10), 2417–2425.

35 H. J. Sharpe, T. J. Stevens and S. Munro, A comprehensive comparison of transmembrane domains reveals organelle-specific properties, *Cell*, 2010, **142**(1), 158–169.

36 T. A. Enoki, F. A. Heberle and G. W. Feigenson, FRET Detects the Size of Nanodomains for Coexisting Liquid-Disordered and Liquid-Ordered Phases, *Biophys. J.*, 2018, **114**(8), 1921–1935.

37 Y. Miyata, K. Takahashi, Y. Lee, C. S. Sultan, R. Kuribayashi, M. Takahashi, K. Hata, T. Bamba, Y. Izumi, K. Liu, T. Uemura, N. Nomura, S. Iwata, S. Nagata, T. Nishizawa and K. Segawa, Membrane structure-responsive lipid scrambling by TMEM63B to control plasma membrane lipid distribution, *Nat. Struct. Mol. Biol.*, 2025, **32**, 185–198.

38 D. Holowka, M. Wilkes, C. Stefan and B. Baird, Roles for Ca2+ mobilization and its regulation in mast cell functions: recent progress, *Biochem. Soc. Trans.*, 2016, **44**(2), 505–509.

39 Y. Baba, M. Matsumoto and T. Kurosaki, Calcium signaling in B cells: regulation of cytosolic Ca2+ increase and its sensor molecules, STIM1 and STIM2, *Mol. Immunol.*, 2014, **62**(2), 339–343.

40 P. Varshney, V. Yadav and N. Saini, Lipid rafts in immune signalling: current progress and future perspective, *Immunology*, 2016, **149**(1), 13–24.

41 M. Dykstra, A. Cherukuri, H. W. Sohn, S. J. Tzeng and S. K. Pierce, Location is everything: lipid rafts and immune cell signaling, *Annu. Rev. Immunol.*, 2003, **21**, 457–481.

42 J. Dinic, A. Riehl, J. Adler and I. Parmryd, The T cell receptor resides in ordered plasma membrane nanodomains that aggregate upon patching of the receptor, *Sci. Rep.*, 2015, **5**, 10082.

43 M. E. Falzone, Z. Feng, O. E. Alvarenga, Y. Pan, B. Lee, X. Cheng, E. Fortea, S. Scheuring and A. Accardi, TMEM16 scramblases thin the membrane to enable lipid scrambling, *Nat. Commun.*, 2022, **13**(1), 2604.

Faraday Discussions

ROYAL SOCIETY OF CHEMISTRY

PAPER

Immune cell activation produces locally scrambled foci of plasma membrane lipids

Daryna Sputay,[a] Milka Doktorova,[b] Sze Ham Chan,[a] Emma Han Guo,[a] Hong-Yin Wang,[a] Joseph H. Lorent,[c] Ilya Levental *[a] and Kandice R. Levental *[a]

Received 20th December 2024, Accepted 11th February 2025
DOI: 10.1039/d4fd00205a

Most eukaryotic cells maintain a large disparity in lipid compositions between the cytosolic and external leaflets of the plasma membrane (PM) bilayer. This lipid asymmetry is maintained by energy-consuming flippase enzymes that selectively drive phospholipids into the cytosolic leaflet, often against large concentration gradients. Scramblases, activated by intracellular Ca^{2+} or apoptotic signaling, shuttle phospholipids down their concentration gradient to release lipid asymmetry. Such scrambling is typically evidenced by exposure of phosphatidylserine (PS) to the external leaflet and is associated with many physiological processes, most notably blood clotting and cell death, but also activation of immune cells. Here, we show that both PS and phosphatidylethanolamine (PE) appear on the PM external leaflet following immune receptor-mediated activation of mast cells. We also observe similar effects in T cells. Importantly, in contrast to wholesale release of PM asymmetry induced by calcium ionophores or apoptosis, we show that scrambling in activated immune cells is focal, with small, stable regions of surface exposed PS. These scrambled foci are calcium dependent, have lower lipid packing than their surrounding outer leaflet, and are reversible. These observations of local, transient scrambling during physiological activation of healthy immune cells suggest important roles for the lateral and transbilayer organization of membrane lipids.

Introduction

Membranes are involved in nearly all aspects of cellular physiology, with their functions determined by the composition and organization of their constituent lipids and proteins. A ubiquitous feature of eukaryote plasma membranes (PM) is a highly disparate distribution of lipids between the two leaflets of the bilayer, termed lipid asymmetry.[1] The outer (exoplasmic) leaflet is generally enriched in

[a]*Department of Molecular Physiology and Biological Physics, University of Virginia, USA. E-mail: il2sy@virginia.edu; krl6c@virginia.edu*
[b]*Department of Biochemistry and Biophysics, Stockholm University, Science for Life Laboratory, Sweden*
[c]*Cellular and Molecular Pharmacology (FACM), Louvain Drug Research Institute, UCLouvain, Belgium*

choline-headgroup lipids with relatively saturated acyl chains.[1,2] The inner (cytoplasmic) leaflet of PM is composed of lipids with negatively charged headgroups (phosphatidylserine (PS) and phosphatidic acid (PA)), amine headgroup phosphatidylethanolamine (PE), and phosphoinositides (PI and PIPs). The high concentration of anionic phospholipids (PLs) in the PM inner leaflet contributes to recruitment of membrane-proximal signaling proteins, which can interact with the cytoplasmic leaflet *via* electrostatic interactions with polybasic protein domains.[3–5] These electrostatic protein–lipid interactions can regulate signaling pathways that control cell physiology.[4] Further, it was recently shown that the distinct compositions of the two bilayer leaflets produce distinct biophysical properties, with unsaturated-lipid-rich inner leaflets being less tightly packed and more fluid (*i.e.* diffusive) than the more saturated outer leaflets.[2,6] Lipid asymmetry also produces other biophysical effects: *e.g.* asymmetric membranes can be stiffer than counterparts with identical lipid compositions but symmetric distributions.[7] This effect has been attributed to leaflet stresses resulting from imbalanced total abundance of phospholipids between leaflets,[8,9] termed differential stress.

Building and maintaining lipid asymmetry is energetically expensive, requiring ATP-dependent transport of billions of phospholipids against sharp concentration gradients.[10] Specific transmembrane protein transporters termed flippases (often P4-ATPases) sustain the transbilayer gradient of phospholipids, constitutively translocating them against a concentration gradient.[11] Lipid channels called scramblases rapidly redistribute phospholipids down their concentration gradient in an ATP-independent and apparently weakly selective manner to release membrane asymmetry.[12] This process can be reported by visualization of the externalized PS on the exoplasmic leaflet using extracellular fluorescent PS-binding probes like AnnexinV (AnxV) or LactadherinC2 (LactC2).[13,14] In apoptosis, the caspase-activated scramblase Xk-related protein 8 (XKR8) is activated to shuttle PS across the whole PM, apparently fully destroying PM asymmetry.[15–17] Several family members of the transmembrane protein 16 (TMEM16) family (most notably TMEM16F) are Ca^{2+}-activated lipid channels, which also translocate PS to the outer leaflet in various cellular contexts.[18–20] While these proteins have been extensively implicated in cellular function and disease, the physiological roles of steady-state membrane asymmetry and its regulated release remain poorly understood. In this context, it has been shown that lipid scrambling affects the biophysical properties of membrane leaflets.[21] PM outer leaflets become less tightly packed, consistent with their gaining lipids with unsaturated acyl chains from the inner leaflet. Further, the overall properties (*e.g.* lipid packing, permeability, and/or stiffness) of the scrambled PM bilayer appear to be distinct from a steady-state asymmetric one.[21]

Historically, most research into lipid asymmetry focused on apoptosis and other forms of cell death, where its release exposes PS on the outer leaflet, marking dead cells for clearance by macrophages.[18,22,23] More recently, it has become clear that lipid scrambling also occurs during physiological processes in healthy cells. Regulated PM scrambling has been reported in cell–cell fusion,[24–28] release of extracellular vesicles,[29,30] blood clotting[20,23] and immune cells signaling.[31–34] These effects can have critical consequences, as scramblase deficiency leads to Scott syndrome, a blood coagulation disorder characterized by impaired membrane scrambling during platelet activation. Scrambling is also important for repairing plasma membrane damage by bacterial pore-forming agents: TMEM16F-mediated lipid scrambling increases PM membrane fluidity after damage and promotes membrane blebbing and shedding of extracellular vesicles containing damaged lipids, supporting PM repair and cell survival.[35] Finally, exposure of PE has been implicated in cell division, regulating the final steps of cytokinesis.[36]

Lipid scrambling (evidenced by PS exposure) upon activation of signaling through stimulation of cell surface immune receptors has been reported in multiple immune cell types, including T cells,[31] B cells,[32] neutrophils,[33] monocytes,[37] natural killer cells,[38] and mast cells.[34] The specific functions and mechanisms by which this lipid redistribution is implicated in functional immune signaling remain poorly understood. In T cells, scrambling through TMEM16F may reduce the electrostatic potential of the inner PM leaflet, potentially enhancing T cell receptor (TCR) phosphorylation[39] *via* electrostatic protein–membrane interactions. In support of this hypothesis, reducing expression of TMEM16F eliminates PM scrambling and decreases TCR phosphorylation, halting signal amplification.[39] In notable contrast, others have shown that TMEM16F terminates, rather than amplifies TCR signaling, to avoid T cell exhaustion.[40] Thus, while TMEM16F-mediated PM scrambling has important effects on immune cell signaling, the precise effects remain unresolved.

Here, we use fluorescence microscopy to characterize membrane changes during receptor-mediated activation of mast and T cells. We find that activated immune cells exhibit a distinct mode of PM scrambling from apoptotic scrambling, characterized by distinct foci of externalized PS and PE that persist for 1–2 h before resolving back into an asymmetric state. These foci are dependent on extracellular calcium and exhibit decreased lipid packing compared to their surrounding PM. These observations reveal that immune cell activation induces localized and transient lipid scrambling that locally change the biophysical properties of the PM outer leaflet. These effects are notably distinct from wholesale PM scrambling, prompting hypotheses about the physiological functions of lipid asymmetry and its regulated release.

Results

There have been extensive observations of PS scrambling induced by apoptosis or calcium ionophores.[16,18,20] We reproduced these findings in rat basophilic leukemia (RBL) cells, by imaging recombinant PS probes binding to the external leaflet of the PM after staurosporine treatment to induce apoptosis or ionophore to increase cytosolic [Ca^{2+}]. Staurosporine treatment for 4 h induced major changes to cell morphology, including membrane blebbing, consistent with apoptosis. As expected, these cells stained brightly with AnxV, a calcium-dependent PS-binding protein (Fig. 1A, top) and with recombinant LactC2, a calcium-independent and monomeric probe for surface-exposed PS (Fig. 1B, left). Treating RBL cells with the ionophore A23187 to increase cytosolic calcium also produced extensive PS exposure across the entire cell surface (Fig. 1A and B).

Antigen-induced activation of mast cells produces focal PM scrambling

RBL cells are a popular model for mast cell signaling because their signaling can be activated by physiologically relevant stimuli. Namely, these cells bind IgE

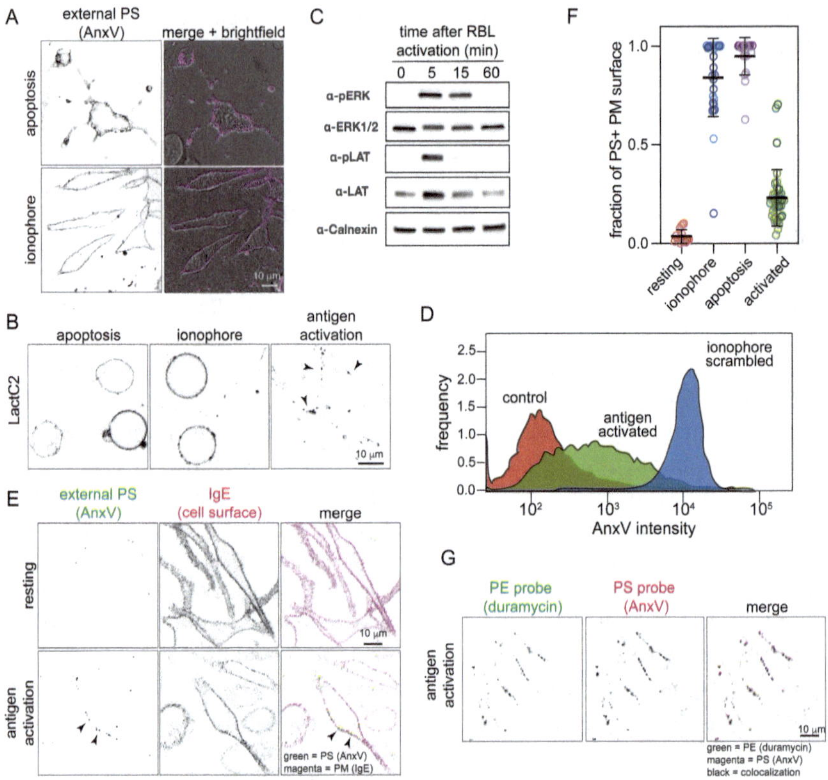

Fig. 1 Antigen-mediated activation of mast cells induces focal scrambling. (A) Top: apoptosis induction in RBL-2H3 cells by staurosporine leads to wholesale loss of PM asymmetry, reported by staining with AnnexinV-AF647 (AnxV). Bottom: similar whole-cell scrambling is induced by treatment with the calcium ionophore A23187. Left panel shows the fluorescent signal intensity inverted (see note below at end of this figure legend), and right panel is an image with the brightfield (gray scale) and fluorescence (magenta pseudocolor) merged. (B) Similar effects are observed by staining with the monomeric PS probe LactC2. Antigen activation results in focal exposure of PS as observed by LactC2 punctae (arrowheads, right). (C) RBL activation induced by treating anti-DNP IgE primed cells with the crosslinking antigen DNP-BSA. Western blotting reveals the phosphorylation of ERK and LAT expected of antigen-mediated activation. (D) Flow cytometry comparison of AnxV staining intensity between non-activated, activated (for 30 min), and ionophore scrambled (after 5 min) RBL cells. (E) RBL cells primed with anti-DNP IgE (IgE-AF555) show focal scrambling. Activation (by DNP-BSA) produced foci of AnxV binding on the PM (arrow heads), indicative of exposed PS. IgE-555 shown as a fiducial marker to visualize the PM. In the inverted merged images, PM (marked by IgE-555) is magenta, and AnxV is green. (F) Quantification of the fraction of the RBL PM surface positive for PS-probe LactC2. Dots represent individual cells with cells from independent repeats shown in different shades ($N \geq 2$). (G) Simultaneous staining for PE (duramycin-PEG-GFP, left) and PS (AnxV-647, middle) on RBL cells following antigen activation shows colocalization (black, right) of the two probes on the outer PM leaflet. Note: in all fluorescence images in this figure, the inverse of the signal is shown for better contrast on screens and paper. Darker pixels represent higher fluorescent signal. In E and G, merged pseudocolored images are shown on the right.

antibodies *via* the antibody receptor FcεRI on their PMs. Addition of specific multivalent antigens crosslinks the receptor, inducing a cascade of signals that ultimately lead to calcium mobilization, degranulation, and cytokine secretion.[41] Here, we used a classical approach to activate RBL cells by priming with IgE against the hapten dinitrophenol (DNP) and then treating with the multivalent antigen DNP-BSA. This treatment induced phosphorylation of Linker for Activation of T-cells (LAT) and extracellular signal-regulated kinase (ERK), as expected of mast cell activation (Fig. 1C). As increased cytosolic calcium is a central feature of immune cell activation,[42] we hypothesized that it may also activate lipid scrambling, similar to calcium ionophore treatment. Indeed, flow cytometry revealed that RBL activation leads to increased staining with externally applied AnxV (Fig. 1D), similar to previous reports;[34] however, AnxV staining of activated RBLs was substantially lower (more than an order of magnitude) than ionophore-induced scrambling, suggesting a distinct form of scrambling. Using confocal microscopy to visualize externalized PS (*via* AnxV) on activated RBLs, we found PS staining in bright, stable foci (Fig. 1E, bottom). Similar foci were also observed with LactC2 (Fig. 1B, right). The contrast between activation- and apoptosis/ionomycin-induced scrambling can be quantified *via* the fraction of the PM surface labeled with the PS-binding probe (Fig. 1F, LactC2 was used for this quantification because it is monomeric). Resting cells had essentially undetectable LactC2 binding (consistent with minimal PS exposure), while apoptotic and ionophore-treated cells had PS exposed over the entire PM outer leaflet. In contrast, activated RBL cells showed external PS labeling over ~20% of the PM surface (Fig. 1F). The PS-positive foci also contained externalized phosphatidylethanolamine (PE), as revealed by co-labeling of AnxV-positive foci with the PE-binding probe duramycin (Fig. 1G).

Since other immune cells use similar calcium-mediated signaling during their activation, we hypothesized that they may also locally scramble their PMs. To test this prediction, we used a well-established method of activating Jurkat T-cells by plating them onto coverslips coated with an anti-CD3 (TCR) antibody (OKT3);[43] activation was evidenced by phosphorylation of LAT and ERK (Fig. 2A), as in mast cells (Fig. 1C). As in activated RBL cells, total internal reflection fluorescence (TIRF) microscopy showed bright foci of AnxV staining on the basal (*i.e.* OKT3-contacting) surface of activated Jurkat cells, revealing local sites of PS scrambling (Fig. 2B). Interestingly, a more coverslip-distal plane imaged by epifluorescence had minimal AnxV staining (Fig. 2B, right), suggesting that scrambling only occurred on the cell surface that was transducing the activating signal.

These findings reveal that immune-receptor mediated activation of mast and T cells causes focal release of PM asymmetry evidenced by scrambling of major inner leaflet phospholipids.

Scrambled foci have reduced lipid packing

It was previously shown that outer leaflets of scrambled PMs have lower lipid packing than outer leaflets of asymmetric PMs of cells at steady-state;[2] similar effects were reported in apoptotic cells.[44,45] Thus, we hypothesized that the PS-positive foci produced during immune cell activation would also have lower lipid packing. We measured lipid packing in the exoplasmic PM leaflet by imaging fluorescence emission lifetime of the polarity-sensitive probe Di-4-ANEPPDHQ

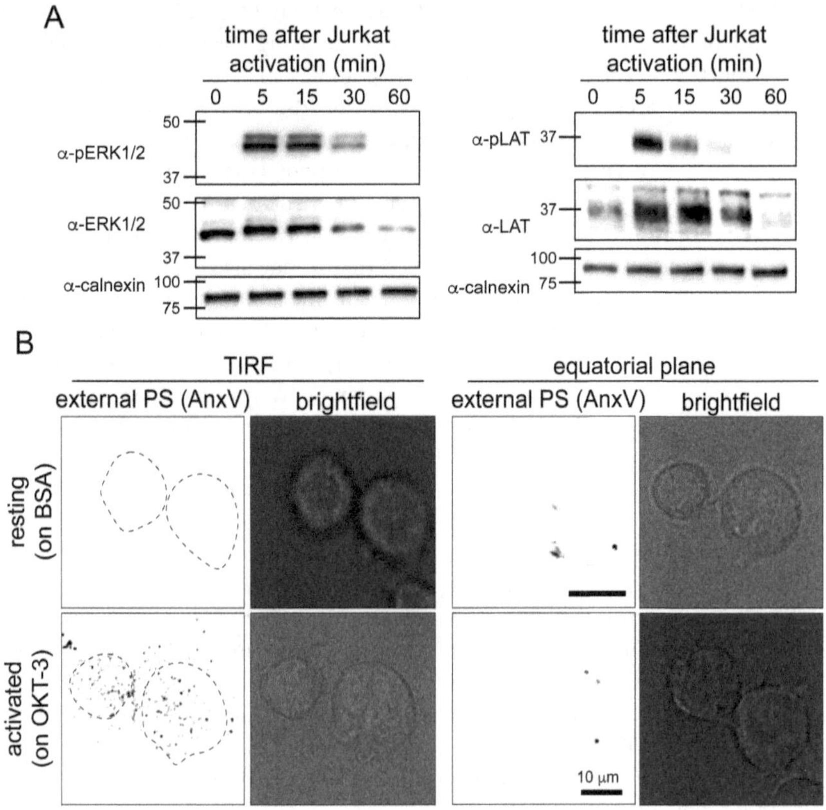

Fig. 2 Activation of T cells induces focal scrambling. (A) Jurkat activation induced by treating cells with anti-CD3. Western blotting reveals the phosphorylation of ERK and LAT expected of antigen-mediated activation. (B) Jurkat T cells exhibit focal PM scrambling when activated by an anti-CD3 (*i.e.* OKT3) coated glass surface. Foci are visible on the basal PM imaged in TIRF, whereas no PS exposure is observed on the cell PM mid-plane. All fluorescence images in this figure are inverted, with darker pixels representing more fluorescent signal.

(Di4), a widely used proxy for lipid packing.[2,21,46,47] Wholesale PM scrambling by ionomycin or apoptosis reduced outer leaflet lipid packing, evidenced by greatly reduced Di4 lifetimes (Fig. 3A and B). This effect is consistent with the introduction of unsaturated lipids from the inner leaflet. Di4 lifetime was also notably reduced in activated RBL cells, but the magnitude of the reduction was much smaller than for ionophore or apoptosis treatments (Fig. 3A). This effect was due to foci of reduced Di4 lifetime, with the remainder of the PM remaining similar to resting cells (Fig. 3B). This effect could be quantified by measuring Di4 lifetime along the cell periphery: in both resting and ionophore-treated cells, lifetime was generally homogeneous whereas the PM of activated cells alternated between a 'resting-like' bulk and 'scrambled-like' foci (Fig. 3C). This lateral heterogeneity, summarized as the range in measured lifetime values, was significantly greater in activated cells than either resting or fully scrambled (Fig. 3D). We confirmed that these foci of lower lipid packing were sites of PS scrambling by their co-staining

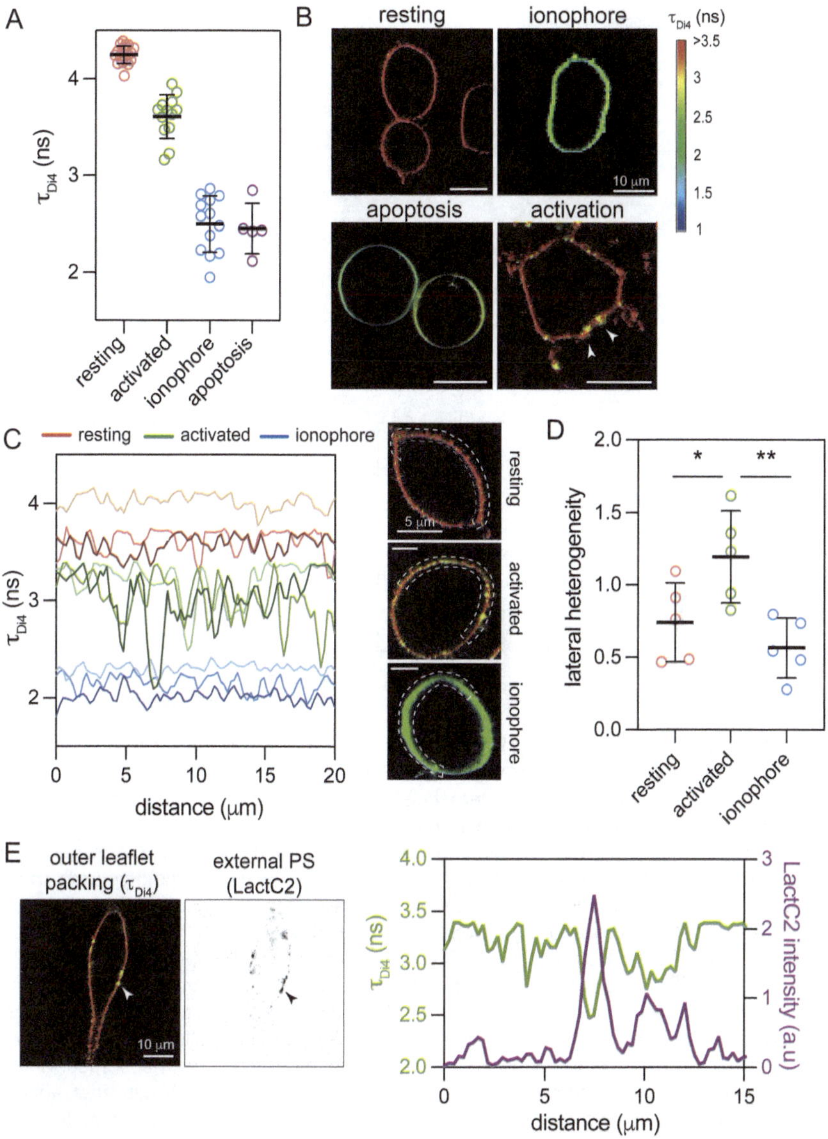

Fig. 3 Scrambled foci have reduced lipid packing. (A) Measurement of PM outer leaflet packing *via* Di4 fluorescence lifetime imaging (FLIM) in mast cells. Scrambling by apoptosis or ionophore dramatically reduces outer leaflet packing, while activation has a smaller but nevertheless prominent effect. Dots represent mean lifetime values from individual cells. (B) Representative Di4 fluorescence lifetime imaging microscopy (FLIM) images show that activated RBL cells have foci of reduced lipid packing (arrows). Ionophore and apoptotic cell PMs have a uniformly lower lifetime (*e.g.* are less packed). (C) Di4 lifetime profiles of resting (shades of red), activated (green), and ionophore-scrambled cells (blue) illustrate PM packing changes along the surface, with activated mast cells exhibiting regions of lower lipid packing within a bulk PM whose packing is similar to resting cells. Representative PM line traces are taken from the boxed region of the confocal FLIM images shown on right. (D) The lateral heterogeneity in lipid packing is greater in activated mast cells than either resting or fully scrambled. (E) Foci of lower lipid packing revealed by Di4 lifetime are colocalized with foci of externalized PS (LactC2 staining, black). Di4 lifetime and LactC2 intensity along the membrane surface in the region indicated by the arrow are shown on the right.

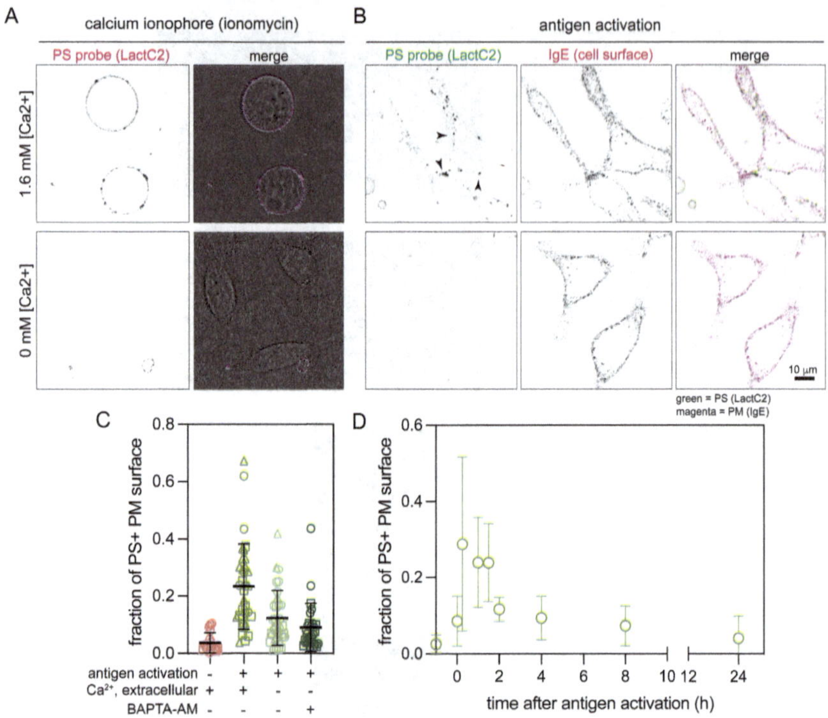

Fig. 4 Calcium dependence and kinetics of scrambling in activated mast cells. (A and B) PS externalization due to ionophore treatment (ionomycin, (A)) or antigen activation (B) in RBL-2H3 cells is dependent on external calcium, as visualized by LactC2 binding. Cells treated in calcium-free buffer had minimal PS externalization in both conditions. (A) Left column shows LactC2-mCherry binding, right shows a merge with LactC2 fluorescence and brightfield. (B) Left column shows LactC2-mClover binding, middle column shows IgE-555 as a fiducial marker to visualize the PM, right, shows merged LactC2 (green) and IgE (magenta); colocalization of external PS (LactC2) and the PM marker (IgE-555) are shown as black. (C) Quantification of PM surface fraction positive for staining by the PS-probe LactC2. Absence of extracellular calcium reduced scrambling by ~50% and further chelation of intracellular calcium (by BAPTA-AM) had only a minimal additional effect. Dots represent individual cells with different shapes representing cells from independent repeats ($N \geq 2$). (D) Temporal profile of the PS-probe positive PM surface fraction after antigen-mediated RBL activation. PS externalization peaked by ~15 min after antigen addition and remained stable for ~1.5 h, before gradually returning to completely asymmetric by ~4 h.

with the PS marker LactC2 (Fig. 3E). Thus, immune cell activation produced focal PM regions with externalized inner leaflet lipids and reduced lipid packing.

Activation-induced scrambling requires external calcium and is reversible

We hypothesized that scrambled foci produced during immune cell activation result from the transient increases in cytosolic calcium that are a hallmark of immune signaling.[48] Calcium fluxes during signaling may be sufficient to activate the non-apoptotic scramblase TMEM16F, which is gated by cytosolic calcium.[18,20] To evaluate this possibility, we treated cells without external calcium and imaged scrambling by external LactC2. As expected, the absence of external calcium completely blocked scrambling by the ionophore A23187 (Fig. 4A). Similarly, absence of external calcium significantly reduced the frequency of LactC2-positive foci in activated RBLs (Fig. 4B). Quantification of this effect revealed that lack of external calcium reduced LactC2-positive PM area by ~50%. Further chelation of cytosolic calcium by BAPTA-AM did not have a notable effect, suggesting that extracellular calcium is a major mediator of this effect (Fig. 4C).

Finally, we hypothesized that since activation-induced PM scrambling was occurring in healthy cells undergoing physiological signaling, it would be reversed after the activating stimulus was resolved. To test reversibility, we imaged PS exposure in RBL cells by staining with LactC2 at various time points after inducing activation. We observed that PS puncta peaked within 15 min after activation and plateaued at that level for ~1.5 h before reversing back to steady-state levels by ~4 h (*i.e.* almost no detectable staining, Fig. 4D).

Discussion

The release of steady-state lipid asymmetry, *i.e.* lipid scrambling, has been implicated in several physiological contexts, most notably apoptosis[15–17] and blood clotting,[20,23] but also cell–cell fusion[24–28] and immune signaling.[31–34] Here, we report that activation of immune cells produces reversible, calcium-dependent scrambling of the PM, distinguished by local foci of surface-exposed PS, PE, and possibly other inner leaflet phospholipids (Fig. 1 and 2). Observations of low lipid packing in these foci (Fig. 3) are consistent with these findings, as inner leaflet lipid classes like PS and PE are typically more unsaturated than outer leaflet

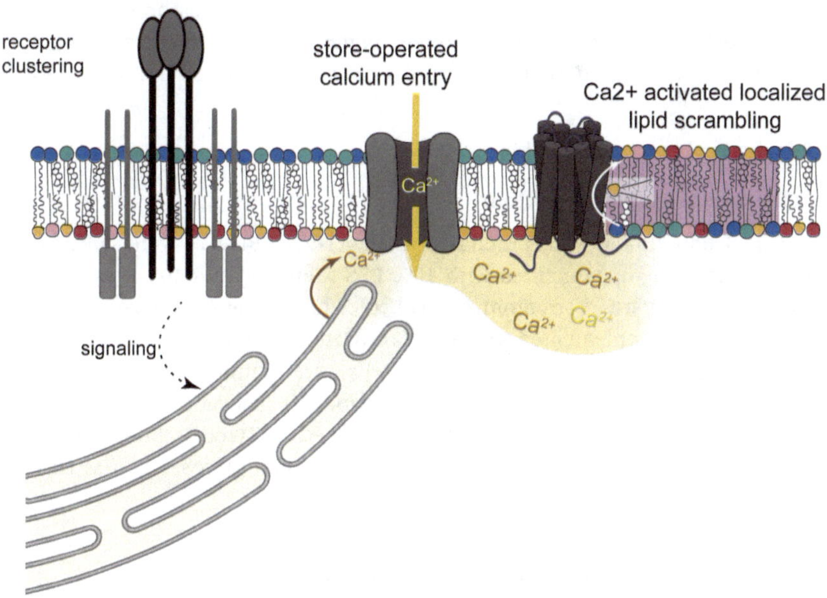

Fig. 5 Schematic of hypothesized cellular mechanism of lipid scrambling downstream of immune activation. Engagement of immune receptors activates calcium mobilization through IP3 and SOCE. The resulting increase in intracellular calcium activates PM resident scramblases to locally release lipid asymmetry.

lipids,[2,49,50] thus their scrambling would introduce unsaturated lipids into the typically tightly packed outer leaflet.[2]

The calcium-dependence of the scrambled foci (Fig. 4A and B) leads us to speculate that they are mediated by calcium-induced activation of members of the scramblase TMEM16 family[18] by calcium mobilized during immune activation. Namely, engagement of antigen receptors leads to phosphorylation and consequent activation of phospholipase C, resulting in production of inositol-1,4,5-trisphosphate (IP3). IP3 in turn activates the ER-resident IP3 Receptor (IP3R) to trigger release of Ca^{2+} from ER stores (Fig. 5). This ER Ca^{2+} release often triggers further store-operated Ca^{2+} entry (SOCE) through PM channels, an often-necessary Ca^{2+} influx pathway in immune cells in response to antigen receptor engagement.[48] As extracellular Ca^{2+} is required for formation of scrambled foci, SOCE is its likely mechanism.

Unlike lipid scrambling associated with cell death (*e.g.* apoptosis), the scrambled foci we report are reversible (Fig. 4D). However, the foci are long-lived, persisting notably longer than either receptor proximal signaling (*i.e.* phosphorylation of LAT) and even the more distal signaling of ERK phosphorylation (*e.g.* Fig. 1C). How this scrambling is resolved is unclear. Non-exclusive possibilities include flippase-mediated rectification of asymmetry and/or shedding or endocytosis of scrambled membrane regions. Another mystery is why the scrambled lipids persist as foci for >1 h, contrary to the expectation that fast lipid diffusion would quickly eliminate local heterogeneity. It is possible that fast scrambling rates locally outcompete diffusion and enzyme-mediated flipping, producing local structure. A non-exclusive alternative is that scrambled regions are somehow isolated from the rest of the PM.

Most importantly, the cellular consequences of this local scrambling are yet unknown. We speculate that the major changes in local membrane properties (lipid packing, charge, perhaps protein composition) in these foci regulate signaling at the PM. These mechanisms remain to be resolved in future work.

Materials and methods

Cell lines and culture conditions

RBL-2H3 and Jurkat were purchased from ATCC. RBL-2H3 were cultured in EMEM medium supplemented with 10% FBS (Genesee) and 100 U mL^{-1} penicillin and 100 μg mL^{-1} streptomycin (Gibco), while Jurkat cells were cultured in RPMI medium supplemented with 10% FBS (Genesee) and 100 U mL^{-1} penicillin and 100 μg mL^{-1} streptomycin (Gibco). Cells were routinely checked for mycoplasma using an ATCC mycoplasma detection kit. Activation and ionophore treatments were performed with reagents diluted in Tyrode's buffer (135 mM NaCl, 5 mM KCl, 1.8 mM $CaCl_2$, 1 mM $MgCl_2$, 5.6 mM glucose, 20 mM HEPES) (pH 7.4) or Ca^{2+}-free Tyrode's buffer.

Fluorescence lifetime imaging (FLIM) and other microscopy

Cells were incubated with 1 μg mL^{-1} Di-4-ANEPPDHQ (Di4, ThermoFisher D36802) at 4 °C for 10 min in Tyrode's buffer. Cells were briefly washed twice in the same buffer at ambient temperature before imaging at room temperature using a Leica SP8 confocal microscope with white light laser and TCSPC capabilities with Falcon software. We acquired images using 63× water-immersion objective with 485 nm excitation and emission of 550–800 nm. To prevent any contaminating signal from internalization of the dye, all images were acquired within 20 minutes of staining. The data was fit to two lifetimes with an exponential reconvolution fit and we report the calculated mean intensity-weighted lifetime. For LactC2 imaging, recombinant LactC2-mCherry was added and images were collected with 587 nm excitation and 760 nm emission. All confocal imaging was performed using Leica SP8 confocal microscope using 63× water immersion objective. Total internal reflection fluorescence (TIRF) microscopy was performed using Leica DMi8 Thunder Imager and 63× oil-immersion objective.

Apoptosis and ionophore-treatment scrambling induction

Apoptosis was induced by incubating cells in full medium with 1 μM staurosporine (STS) (S6942, Sigma-Aldrich) for 4 hours at 37 °C. Calcium ionophores A23187 (Cayman Chemical Company, 11016) and ionomycin (Thermo Scientific, J60628) were used to elevate intracellular [Ca^{2+}] and induce PM scrambling – both were freshly dissolved in DMSO and used with a final concentration of 3 μM in Tyrode's buffer for 15 minutes at 37 °C.

Immune cell activation by receptor crosslinking

To induce RBL-2H3 activation, cells were first primed with 1 μg mL^{-1} monoclonal anti-dinitrophenyl mouse IgE (D8406, Sigma-Aldrich) in full medium. Then cells were activated by treating with 1 μg per mL 2,4-dinitrophenylated bovine serum albumin (DNP-BSA) (Invitrogen, A23018) dissolved in PBS as a crosslinking antigen. Activation was performed in Tyrode's buffer or Ca^{2+}-free Tyrode's buffer (Fig. 3A and B). To label the plasma membrane, the IgE was labeled with AlexaFluor-555 using the Alexa Fluor 555 Antibody Labeling Kit (A20187, Thermo Scientific), with the degree of labeling determined to be 3.9 labels per antibody. Jurkat T cells were activated by plating onto a glass coverslip surface coated with monoclonal anti-human CD3 antibody (OKT-3, 317301, BioLegend). Coverslip glass surface was incubated with 5 μg mL^{-1} of OKT-3 in PBS for 30 minutes at 37 °C prior to imaging.[43]

Calcium-chelation/quenching

BAPTA-AM (B6769 Thermo Scientific) was used to chelate intracellular Ca^{2+}. BAPTA-AM was freshly dissolved in DMSO and used at final concentration of 10 μM in Ca^{2+}-free Tyrode's buffer. Cells were pre-treated with BAPTA-AM for 5 minutes before activation and chelator was also present during DNP-BSA antigen activation of RBL-2H3 cells.

Lipid-specific probes

PS-specific probe Annexin V conjugated with Alexa Fluor 647 (AnxV-647) was purchased from Thermo Scientific (A13204). LactadherinC2-mClover construct was expressed and purified in-house, using Co-NTA agarose purification resin (31405, Cube Biotech) and reconstituted in 20 mM HEPES and 150 mM NaCl, pH 7.4 buffer. The expression plasmid was a gift from S. Grinstein (Hospital for Sick

Children, Toronto, Canada). The PE-specific probe duramycin-PEG-GFP[51] was incubated for 30 minutes with RBL cells during DNP-BSA-mediated FcεRI activation at final concentration of 10 μg mL^{-1} in Tyrode's buffer. Purified recombinant probe was a generous gift from Ming Zhao (University of Wisconsin–Madison).

Western blot

All primary antibodies were used in 1 : 1000, all secondary – 1 : 10 000, diluted in 5% BSA in TBS-T blocking buffer. Primary antibodies: anti-ERK1/ERK2 (E.657.8) (Invitrogen, MA5-15227), anti-phospho-ERK1/ERK2 (Tyr204) (B.742.5) (Invitrogen, MA-15174), anti-LAT (phospho Y200) antibody [EP983(2)Y] (Abcam, ab68139), anti-LAT (E3U6J) (Cell Signaling, 45533S), anti-Calnexin (Abcam, ab22595). Secondary antibodies: anti-rabbit IgG, HRP linked antibody (Cell Signaling, 7074S), anti-mouse IgG, HRP-linked antibody (Cell Signaling, 7076).

Flow cytometry

RBL cells were first primed with 1 μg mL^{-1} monoclonal anti-dinitrophenyl mouse IgE (D8406, Sigma-Aldrich) in full medium at 37 °C for 1 h. Then cells were activated by treating with 1 μg mL^{-1} DNP-BSA in Tyrode's buffer for 15 min at 37 ° C. A parallel set of cells was incubated with ionomycin in Tyrode's buffer for 15 min at 37 °C. AnxV-488 was added to the cells for 5 min before performing flow cytometry on an LSR Fortessa. Viable cells were gated on forward- and side-scatter, and the AnxV intensity was plotted as a histogram of viable cells.

Data availability

All data relevant to this publication are included in the figures. Primary datasets including imaging data and raw Western blots are freely publicly available *via* UVA Dataverse (https://doi.org/10.18130/V3/RTP9A2), a public institutional repository that is part of the Libra Scholarly Repository.

Conflicts of interest

There are no conflicts to declare.

Acknowledgements

TIRF microscopy was performed at the Advanced Microscopy Facility at the University of Virginia. Funding for this work was provided by the NIH (GM134949, AI183581, F32GM134704, F31AI181453) and the Owens Foundation.

References

1 P. F. Devaux, Static and dynamic lipid asymmetry in cell membranes, *Biochemistry*, 1991, **30**(5), 1163–1173.
2 J. H. Lorent, *et al.*, Plasma membranes are asymmetric in lipid unsaturation, packing and protein shape, *Nat. Chem. Biol.*, 2020, **16**(6), 644–652.

3 Y. Zhou, *et al.*, Membrane potential modulates plasma membrane phospholipid dynamics and K-Ras signaling, *Science*, 2015, **349**(6250), 873–876.

4 T. Banerjee, *et al.*, Spatiotemporal dynamics of membrane surface charge regulates cell polarity and migration, *Nat. Cell Biol.*, 2022, **24**(10), 1499–1515.

5 H. M. Hankins, *et al.*, Role of flippases, scramblases and transfer proteins in phosphatidylserine subcellular distribution, *Traffic*, 2015, **16**(1), 35–47.

6 A. Gupta, *et al.*, Plasma membrane asymmetry of lipid organization: fluorescence lifetime microscopy and correlation spectroscopy analysis, *J. Lipid Res.*, 2020, **61**(2), 252–266.

7 Y. Elani, *et al.*, Measurements of the effect of membrane asymmetry on the mechanical properties of lipid bilayers, *Chem. Commun.*, 2015, **51**(32), 6976–6979.

8 A. Hossein and M. Deserno, Stiffening transition in asymmetric lipid bilayers: The role of highly ordered domains and the effect of temperature and size, *J. Chem. Phys.*, 2021, **154**(1), 014704.

9 S. L. Foley, *et al.*, Elastic and thermodynamic consequences of lipid membrane asymmetry, *Emerging Top. Life Sci.*, 2023, **7**(1), 95–110.

10 S. Shukla and T. Baumgart, Enzymatic trans-bilayer lipid transport: Mechanisms, efficiencies, slippage, and membrane curvature, *Biochim. Biophys. Acta, Biomembr.*, 2021, **1863**(3), 183534.

11 K. Balasubramanian and A. J. Schroit, Aminophospholipid asymmetry: A matter of life and death, *Annu. Rev. Physiol.*, 2003, **65**, 701–734.

12 K. M. Kodigepalli, *et al.*, Roles and regulation of phospholipid scramblases, *FEBS Lett.*, 2015, **589**(1), 3–14.

13 T. Yeung, *et al.*, Membrane phosphatidylserine regulates surface charge and protein localization, *Science*, 2008, **319**(5860), 210–213.

14 T. Yeung, *et al.*, Receptor activation alters inner surface potential during phagocytosis, *Science*, 2006, **313**(5785), 347–351.

15 K. Segawa, *et al.*, Caspase-mediated cleavage of phospholipid flippase for apoptotic phosphatidylserine exposure, *Science*, 2014, **344**(6188), 1164–1168.

16 J. Suzuki, *et al.*, Xk-related protein 8 and CED-8 promote phosphatidylserine exposure in apoptotic cells, *Science*, 2013, **341**(6144), 403–406.

17 U. Sivagnanam, S. K. Palanirajan and S. N. Gummadi, The role of human phospholipid scramblases in apoptosis: An overview, *Biochim. Biophys. Acta*, 2017, **1864**(12), 2261–2271.

18 T. Sakuragi and S. Nagata, Regulation of phospholipid distribution in the lipid bilayer by flippases and scramblases, *Nat. Rev. Mol. Cell Biol.*, 2023, **24**, 576–596.

19 J. Suzuki, *et al.*, Calcium-dependent phospholipid scramblase activity of TMEM16 protein family members, *J. Biol. Chem.*, 2013, **288**(19), 13305–13316.

20 J. Suzuki, *et al.*, Calcium-dependent phospholipid scrambling by TMEM16F, *Nature*, 2010, **468**(7325), 834–838.

21 M. Doktorova, *et al.*, Cell membranes sustain phospholipid imbalance via cholesterol asymmetry, *Cell*, 2025, DOI: 10.1016/j.cell.2025.02.034.

22 J. Ousingsawat, *et al.*, Contribution of TMEM16F to pyroptotic cell death, *Cell Death Dis.*, 2018, **9**(3), 300.

23 R. F. Zwaal, P. Comfurius and E. M. Bevers, Surface exposure of phosphatidylserine in pathological cells, *Cell. Mol. Life Sci.*, 2005, **62**(9), 971–988.

24 Y. Zhang, *et al.*, TMEM16F phospholipid scramblase mediates trophoblast fusion and placental development, *Sci. Adv.*, 2020, **6**(19), eaba0310.

25 J. M. Whitlock, *et al.*, Anoctamin 5/TMEM16E facilitates muscle precursor cell fusion, *J. Gen. Physiol.*, 2018, **150**(11), 1498–1509.

26 S. K. Verma, *et al.*, Cell-surface phosphatidylserine regulates osteoclast precursor fusion, *J. Biol. Chem.*, 2018, **293**(1), 254–270.

27 C. M. Rival, *et al.*, Phosphatidylserine on viable sperm and phagocytic machinery in oocytes regulate mammalian fertilization, *Nat. Commun.*, 2019, **10**(1), 4456.

28 M. Tsuchiya, *et al.*, Cell surface flip-flop of phosphatidylserine is critical for PIEZO1-mediated myotube formation, *Nat. Commun.*, 2018, **9**(1), 2049.

29 E. Cocucci and J. Meldolesi, Ectosomes and exosomes: shedding the confusion between extracellular vesicles, *Trends Cell Biol.*, 2015, **25**(6), 364–372.

30 J. M. Whitlock and H. C. Hartzell, Anoctamins/TMEM16 Proteins: Chloride Channels Flirting with Lipids and Extracellular Vesicles, *Annu. Rev. Physiol.*, 2017, **79**, 119–143.

31 K. Fischer, *et al.*, Antigen recognition induces phosphatidylserine exposure on the cell surface of human $CD8^+$ T cells, *Blood*, 2006, **108**(13), 4094–4101.

32 S. R. Dillon, *et al.*, Annexin V binds to viable B cells and colocalizes with a marker of lipid rafts upon B cell receptor activation, *J. Immunol.*, 2000, **164**(3), 1322–1332.

33 S. C. Frasch, *et al.*, Phospholipid flip-flop and phospholipid scramblase 1 (PLSCR1) co-localize to uropod rafts in formylated Met-Leu-Phe-stimulated neutrophils, *J. Biol. Chem.*, 2004, **279**(17), 17625–17633.

34 S. Martin, *et al.*, Immunologic stimulation of mast cells leads to the reversible exposure of phosphatidylserine in the absence of apoptosis, *Int. Arch. Allergy Immunol.*, 2000, **123**(3), 249–258.

35 N. Wu, *et al.*, Critical Role of Lipid Scramblase TMEM16F in Phosphatidylserine Exposure and Repair of Plasma Membrane after Pore Formation, *Cell Rep.*, 2020, **30**(4), 1129.

36 K. Emoto and M. Umeda, An essential role for a membrane lipid in cytokinesis. Regulation of contractile ring disassembly by redistribution of phosphatidylethanolamine, *J. Cell Biol.*, 2000, **149**(6), 1215–1224.

37 A. MacKenzie, *et al.*, Rapid secretion of interleukin-1β by microvesicle shedding, *Immunity*, 2001, **15**(5), 825–835.

38 N. Wu, H. Song and A. Veillette, Plasma membrane lipid scrambling causing phosphatidylserine exposure negatively regulates NK cell activation, *Cell. Mol. Immunol.*, 2021, **18**(3), 686–697.

39 A. Connolly, *et al.*, TMEM16F mediates bystander TCR-CD3 membrane dissociation at the immunological synapse and potentiates T cell activation, *Sci. Signaling*, 2021, **14**(675), eabb5146.

40 Y. Hu, *et al.*, Scramblase TMEM16F terminates T cell receptor signaling to restrict T cell exhaustion, *J. Exp. Med.*, 2016, **213**(12), 2759–2772.

41 A. M. Gilfillan and C. Tkaczyk, Integrated signalling pathways for mast-cell activation, *Nat. Rev. Immunol.*, 2006, **6**(3), 218–230.

42 M. Vig and J. P. Kinet, Calcium signaling in immune cells, *Nat. Immunol.*, 2009, **10**(1), 21–27.

43 H. Y. Wang, *et al.*, Coupling of protein condensates to ordered lipid domains determines functional membrane organization, *Sci. Adv.*, 2023, **9**(17), eadf6205.

44 R. Kreder, *et al.*, Solvatochromic Nile Red probes with FRET quencher reveal lipid order heterogeneity in living and apoptotic cells, *ACS Chem. Biol.*, 2015, **10**(6), 1435–1442.

45 Z. Darwich, *et al.*, Detection of apoptosis through the lipid order of the outer plasma membrane leaflet, *Biochim. Biophys. Acta*, 2012, **1818**(12), 3048–3054.

46 D. M. Owen, *et al.*, Quantitative imaging of membrane lipid order in cells and organisms, *Nat. Protoc.*, 2012, **7**(1), 24–35.

47 D. M. Owen, *et al.*, Fluorescence lifetime imaging provides enhanced contrast when imaging the phase-sensitive dye di-4-ANEPPDHQ in model membranes and live cells, *Biophys. J.*, 2006, **90**(11), L80–L82.

48 S. Feske, Immunodeficiency due to defects in store-operated calcium entry, *Ann. N. Y. Acad. Sci.*, 2011, **1238**, 74–90.

49 K. R. Levental, *et al.*, ω-3 polyunsaturated fatty acids direct differentiation of the membrane phenotype in mesenchymal stem cells to potentiate osteogenesis, *Sci. Adv.*, 2017, **3**(11), eaao1193.

50 K. R. Levental, *et al.*, Lipidomic and biophysical homeostasis of mammalian membranes counteracts dietary lipid perturbations to maintain cellular fitness, *Nat. Commun.*, 2020, **11**(1), 1339.

51 S. Hou, S. E. Johnson and M. Zhao, A One-Step Staining Probe for Phosphatidylethanolamine, *ChemBioChem*, 2015, **16**(13), 1955–1960.

Faraday Discussions

ROYAL SOCIETY OF CHEMISTRY

PAPER

Nonequilibrium asymmetry in the living cell membrane†

Ajay Kumar Bansal and Madan Rao *

Received 22nd December 2024, Accepted 29th January 2025
DOI: 10.1039/d4fd00207e

We will discuss how sustained nonequilibrium processes operating at the plasma membrane (PM) determine the dynamical organisation (both lateral and transverse) of lipids, their maintenance and control, under physiological conditions. These nonequilibrium processes include active contractile stresses arising from the inevitable interaction of the inner leaflet of the PM with the adjoining actomyosin cortex, and active flipping of specific lipids. Recently, we showed that the inner leaflet phosphatidylserine (PS) interacts with the actomyosin cortex and engages in a strong transbilayer coupling across the leaflets. Here we develop an active Flory–Huggins theory for the mesoscale segregation of liquid-ordered (lo)–liquid-disordered (ld) domains in an asymmetric membrane bilayer, that incorporates both active contractile stresses at the inner leaflet and transbilayer coupling across the leaflets. The interplay between chemical potential gradients, transbilayer coupling and active stresses drives a rich pattern of mesoscale lo domains – static, strongly fluctuating and moving active emulsions – even at temperatures beyond the equilibrium phase transition temperature. We study conditions under which domain registry and slippage could be observed. We end with a discussion on the role of active flippases on PS in maintaining the active mesoscale organisation.

1 Introduction

An enduring theme in modern cell biology has been the search for organizing principles that underlie the spatiotemporal distribution of proteins and lipids on the plasma membrane of living cells.[1] This acquires greater significance by the fact that such dynamic organisation has a profound impact on cell physiology and function. Indeed, numerous studies have revealed that molecular organisation at the cell surface, ranging from 10 to 10^3 nm in space and 10^{-3} to 10 s in time, have direct functional consequences for signaling reactions, molecular sorting and endocytosis.[2,3]

Simons Centre for the Study of Living Machines, National Centre for Biological Sciences, Bangalore 560 065, India. E-mail: rao.madan@gmail.com; Fax: +91 80 2363 6662; Tel: +91 80 2366 6270

† Electronic supplementary information (ESI) available. See DOI: https://doi.org/10.1039/d4fd00207e

The plasma membrane (PM) comprises a multitude of lipid species.[1] The pioneering fluid-mosaic,[4] lipid-shell[5] and lipid-raft models[6] posit that the lateral organisation of PM lipids is solely determined by in-plane intermolecular interactions between lipids and/or lipid–proteins. The raft model and its variants suggest that the cell membrane exhibits stable compositional heterogeneities at physiological temperatures, as a consequence of being below the equilibrium liquid-order (lo)–liquid-disorder (ld) phase transition. However, such exclusively equilibrium models fail to explain, why Giant Plasma Membrane Vesicles (GPMV) extracted from blebbing cells are completely mixed at physiological temperatures (37 °C), and in fact need to be brought down to below 20 °C to show lo–ld phase segregation.[7]

Further, the PM is strikingly asymmetric in its lipid composition,[8] this asymmetry being crucial to its function. Despite this, it is only recently that investigators have turned their attention to the transverse organisation of the asymmetric cell membrane.

Over the past two decades, there has been increasing support from theoretical studies and cell-based experiments for the role of both nonequilibrium forces and transbilayer interactions in defining the lateral and transverse organisation of the cell membrane.[9–19] The nonequilibrium forces are an inevitable consequence of the juxtaposition of the actomyosin cortex[11,17,20,21] with the inner leaflet of the cell membrane. This nonequilibrium force asymmetry is accompanied by compositional asymmetry – thus specific components of the inner leaflet (*e.g.*, PS) interact with both the actomyosin cortex and with saturated lipids on the outer leaflet *via* a strong transbilayer coupling.[15,16] Viewed this way, the nonequilibrium force and compositional asymmetries are intimately related.

In this paper, we first review previous theoretical studies that demonstrate how nonequilibrium contractile stresses[22] from the actomyosin cortex and strong transbilayer coupling *via* PS[15,16] drive nanoscale and mesoscale lo–ld segregation of cell membrane lipids at physiological temperatures, well above the putative equilibrium lo–ld phase transition.[10,15,16,23] We refer to this dynamic organisation as an active emulsion. We then present a new theoretical approach based on an active Flory–Huggins theory for a multicomponent bilayer with compositional asymmetry of saturated/unsaturated lipids similar to that determined in typical mammalian cells.[24] The theory depends on the maintenance of adequate levels of PS in the inner leaflet, allowing it to simultaneously connect to the actomyosin cortex and forge a strong transbilayer coupling. The maintenance of the PS asymmetry is controlled by another active process in the cell membrane involving active flippases and scramblases.[25] Thus lipid asymmetry and its control are crucial for the lateral nano and meso scale organisation of lipids.

2 Nonequilibrium force asymmetry in the living cell membrane

The PM of metazoan cells is an asymmetric multicomponent bilayer, juxtaposed at the inner leaflet with an actomyosin cortex. The inner leaflet of the PM is thus naturally subject to nonequilibrium active contractile stresses from the actomyosin cortex, provided the membrane components interact with elements of the actomyosin cortex. There is mounting evidence that this is true, for instance, the inner leaflet PS binds to myosin,[15,16,26] PIP2 lipids bind to actin[27] and many membrane proteins have the capacity to bind either directly or indirectly to actin[28] by recruiting actin-binding domains such as ezrin, ankyrin, or α-catenin.[29,30]

These considerations form the basis of the active composite model of the cell surface,[10,11,15,16,31] where membrane components that couple, directly or indirectly, to cortical actomyosin constituents are subject to fluctuating contractile stresses that locally drive contractile flows,[10,32–35] leading to their dynamic nano and meso scale clustering.[9–11,15,16,34,36]

The actomyosin cortex is a thin layer of width ≈180 nm (ref. 37) comprising formin-nucleated dynamic linear actin filaments[9,10,15,26] and ARP2/3 nucleated actin mesh, with multiple myosin species, such as non-muscle Myo-II minifilaments and Myo-I. This layer adheres to the inner leaflet of the PM, through a variety of membrane anchored proteins such as ERM proteins,[27] and lipids such as $PI(4,5)P_2$ and PS. As a result, contractile stresses generated by myosin in the cortical fluid are transmitted to the PM.[38]

Using an active hydrodynamic theory and agent based simulations,[10,33–35] we describe how gradients of myosin density drive contractile cortical flows that in turn advect both actin filaments and myosin. These contractile cortical flows result in a constitutive patterning of actin filaments in the form of static and moving asters, with myosin concentrated at the aster cores. Due to turnover of both actin and myosin, these asters continuously remodel at high ATP levels, giving rise to a nonequilibrium steady state with characteristic positional and orientational correlations. These can be computed precisely from the active hydrodynamic theory and agent based simulations[10,11,15,23,33–36] and measured in *in vitro* reconstitution experiments of a minimal cortical system consisting of actomyosin on a supported bilayer.[39]

This dynamical patterning of actomyosin has a profound effect on both the nano and meso scale organisation and dynamics of cell membrane molecules that bind to it.

2.1 Nanoscale organisation

The in-plane actomyosin cortical flows advect membrane molecules that bind to the actomyosin cortex, such as inner leaflet lipids like PS or inner leaflet proteins or even transmembrane proteins. Being contractile, these local cortical flows drive localised "nanoscale" clustering of these membrane molecules. These nanoclusters are transient on account of the dynamic remodelling of asters due to the continual turnover of actin and myosin.

This kind of clustering forms even in the absence of attractive interactions between the membrane components and is entirely nonequilibrium in origin. The presence of attractive interactions enhances the lifetime of the nanoclusters.[40]

Can membrane components in the outer leaflet, such as lipid tethered GPI-anchored proteins, that do not have a direct connection to the actomyosin cortex, also form active nanoclusters? In a detailed atomistic MD study,[16] we showed that the clustering of inner leaflet PS (that may arise from its binding to actomyosin) creates a local lo environment enriched in cholesterol and saturated lipids at physiological temperatures. This in turn drives a local nano patch of saturated lipids and cholesterol in the outer leaflet, *via* a strong transbilayer coupling. Indeed this transbilayer coupling requires adequate levels of cholesterol and is very sensitive to the chemistry of the acyl chains of PS – a small reduction in the chain length and degree of saturation can lower the transbilayer coupling and destroy the nanoclustering in the upper leaflet. These MD results were elegantly confirmed by cell-based experiments,[16] which also demonstrated the binding of PS to myosin.[26] We emphasize that this nanoclustering in cells is insensitive to temperature changes between 40–23 °C.[9,10]

Active nanoclustering appears to be a general phenomenon;[17] it is exhibited by transmembrane proteins that directly bind to actin, such as model proteins that consist of actin-binding motifs at their cytoplasmic tails (TMABD[10]) and by naturally occurring proteins such as E-cadherin and CD44 that recruit actin-binding modules to bind to cortical actin.[13,41] Further, lipid-tethered membrane proteins, such as outer leaflet GPI-anchored proteins (GPI-AP),[10,42] glycolipids[14] and inner leaflet Ras-GTPases[19] also exhibit actin-dependent nanoclusters.

Our active hydrodynamic theory and agent based simulations[10,33–35] make precise predictions for the statistical distribution and fluctuations of the density of nanoclustering in the nonequilibrium steady state. These have been experimentally verified in high resolution fluorescence experiments in living cells.[9,10,21,43]

Most importantly, the principle of actomyosin driven clustering of membrane molecules was emphatically verified in *in vitro* reconstitutions of the active composite, *viz.* actomyosin layered atop a planar bilayer membrane,[39] albeit at larger scales.

Taken together this not only identifies a novel molecular organiser of local plasma membrane composition, namely cortical actomyosin, but also a new mechanism, highlighting the role of non-equilibrium, energy consuming 'active' processes[22] in driving local compositional heterogeneities.[31] In addition it points to the key role of inner leaflet PS, which couples both to the dynamic actomyosin cortex and facilitates the strong transbilayer coupling that drives molecular clustering in the outer leaflet.

2.2 Mesoscale organisation

Can the same organising principles that drive lateral organisation of lipids at the nanoscale, also drive mesoscale segregation at physiological temperatures?

We first addressed this question using a kinetic Monte Carlo simulation[36] that described the mesoscale segregation in a two-component planar membrane driven by the usual Kawasaki dynamics augmented by active actomyosin contractile forces on one of the components. In addition, we formulated an active Flory–Huggins (AFH) theory[23] to study the segregation of a two-component saturated/unsaturated lipid planar membrane, with the saturated lipid subject to local active stresses computed from the hydrodynamics of the adjoining actomyosin cortex.

The results obtained by these two methods were consistent. Despite being a symmetric mixture of saturated and unsaturated lipids, the interplay between the chemical potential and active stresses drives a segregation instability where the segregated lo-domains grow and saturate to a finite size. This is true both above and below the equilibrium phase transition temperature T_c. Indeed if the initial condition is a fully segregated lo–ld configuration with a single interface at

$T < T_c$, the active stresses drive an instability of the interface leading to breakup into lo-domains of finite size. Owing to the continual active driving from the actomyosin cortex, the mesoscale domains can be highly dynamic, especially at high active stresses, resulting in intermittent coalescence and breakup of domains. We find that the active segregation is characterised by anomalous growth dynamics and fluctuations that are unique to the nonequilibrium steady state.[23,36] In addition, the interaction of the active cortical flows on the membrane components leads to a nonreciprocal effective interaction between proximal domains, leading to their propulsion, phoretic reorientation and rapid coalescence.[23]

In the next section, we extend these earlier ideas to the asymmetric bilayer and include the key role of the inner leaflet PS, which both couples to the dynamic actomyosin cortex and facilitates strong transbilayer coupling, thus transmitting the active stresses to the outer leaflet. As before we study this using kinetic Monte Carlo dynamics and an AFH model for a multi-component asymmetric bilayer membrane comprising inner-leaflet PS and saturated and unsaturated lipids in both leaflets.

3 Nonequilibrium forces and transbilayer coupling in the asymmetric cell membrane

We first formulated a kinetic Monte Carlo approach to describe the lo–ld segregation in a 4-component asymmetric bilayer, where the inner leaflet PS, driven by active contractile flows, drives the lo–ld segregation in both the leaflets *via* lateral interactions and a strong transbilayer coupling.[15] This allowed us to make a systematic study of the patterning of the various components in the two leaflets and analyse their co-registry. We then successfully verified the predictions of this theory using a variety of high resolution fluorescence imaging modalities in living cells,[15] which estimated the typical meso domain size to be in the range 350–500 nm.[9,15,31] Thus the combination of active stress fluctuations arising from a coupling to cortical actomyosin, lateral lipidic interactions involving cholesterol and saturated lipids such as SM, and strong transbilayer coupling involving inner leaflet PS, drive the segregation of mesoscale lo domains in both leaflets at physiological temperatures (above the equilibrium lo–ld phase transition temperature). These actively segregated mesoscale domains are aptly described as active emulsions,[15] as we will see below.

We now formulate an AFH model for a multi-component asymmetric bilayer membrane comprising inner-leaflet PS, with saturated and unsaturated lipids in both leaflets.

3.1 AFH theory for an asymmetric lipid bilayer

We consider a bilayer comprising saturated and unsaturated lipids distributed asymmetrically in the two leaflets. In addition, the inner leaflet consists of PS, a saturated lipid with long acyl chains. As discussed previously, inner leaflet PS interacts with the actomyosin cortex and so is subject to nonequilibrium active stresses. The saturated lipids and PS engage in a transbilayer coupling across the leaflets.

Denoting the saturated lipids, unsaturated lipids and PS as species $\alpha = 1, 2, 3$, respectively, and representing the outer/inner leaflets as +/−, the Helmholtz free-energy functional describing its equilibrium configurations can be written as,[44]

$$F = \int \sum_{\alpha,\beta} \Big[\rho_\alpha^+(\mathbf{x}) U_{\alpha\beta}^+\left(\mathbf{x}-\mathbf{x}'\right)\rho_\beta^+\left(\mathbf{x}'\right) \\ + \rho_\alpha^-(\mathbf{x}) U_{\alpha\beta}^-\left(\mathbf{x}-\mathbf{x}'\right)\rho_\beta^-\left(\mathbf{x}'\right) + \rho_\alpha^+(\mathbf{x}) w_{\alpha\beta}\left(\mathbf{x},\mathbf{x}'; z_+, z_-\right)\rho_\beta^-\left(\mathbf{x}'\right)\Big] \mathrm{d}\mathbf{x}\mathrm{d}\mathbf{x}' \\ + T\int \sum_\alpha \left[\rho_\alpha^+(\mathbf{x}) \ln \rho_\alpha^+(\mathbf{x}) - \rho_\alpha^+(\mathbf{x}) + \rho_\alpha^-(\mathbf{x}) \ln \rho_\alpha^-(\mathbf{x}) - \rho_\alpha^-(\mathbf{x})\right] \mathrm{d}\mathbf{x}. \quad (1)$$

where $\rho_\alpha^\pm$ are the areal densities of the α component, $U^\pm$ are the non-local, short-range 2-body attractive interactions along each leaflet, and w is a short-range 2-body attractive transbilayer coupling across the leaflets with a range smaller than the bilayer thickness. It is convenient to express this as an FH free-energy functional, written in terms of the area fractions, $\phi_\alpha^\pm = \rho_\alpha^\pm/\rho^\pm$, with $\rho^\pm = \sum_\alpha \rho_\alpha^\pm$, which we take to be incompressible,

$$F = \int \Big(\chi_{\alpha\beta}^\pm(T,\rho^\pm)\phi_\alpha^\pm\phi_\beta^\pm + \kappa_{\alpha\beta}^\pm(T,\rho^\pm)\nabla\phi_\beta^\pm \cdot \nabla\phi_\beta^\pm + \\ \nu_\alpha^\pm(T,\rho^\pm)\phi_\alpha^\pm - W_{\alpha\beta}(T,\rho^\pm,\phi^\pm)\phi_\alpha^+\phi_\beta^- + \ldots T\phi_\alpha^\pm \ln\phi_\alpha^\pm \Big)\mathrm{d}\mathbf{x}, \quad (2)$$

where the interaction parameters in eqn (2) and (1) are linearly related. We comment on the form of the transbilayer coupling W, which arises from an enhanced van der Waals interaction when the saturated lipid chains across the bilayer interdigitate. Interdigitation increases when the saturated lipid chains are stretched as happens in the lo-phase. This feedback can be incorporated by allowing W_{11}, W_{13} to depend on $\phi^\pm$.

3.2 Two-component asymmetric bilayer

To get a feeling for the underlying nonequilibrium physics of the active segregation in the asymmetric bilayer, we suppress the $\alpha = 3$ index and treat PS as a regular saturated lipid residing in the inner leaflet. We allow for all saturated lipids in the inner leaflet to interact with the actomyosin cortex. We will restore the identity of PS as an explicit component in Section 4.2.

For this effective 2-component lipid membrane, $\phi_1^\pm = 1 - \phi_2^\pm \equiv \phi^\pm$, and so the FH free-energy functional can be rewritten as

$$F = \int \Big(-\chi^\pm(T,\rho^\pm)\phi^{\pm 2} + \kappa^\pm(T,\rho^\pm)\left(\nabla\phi^\pm\right)^2 + \nu^\pm(T,\rho^\pm)\phi^\pm \\ -W(T,\rho^\pm,\phi^\pm)\phi^+\phi^- + \ldots T\phi^\pm \ln\phi^\pm + T(1-\phi^\pm)\ln(1-\phi^\pm)\Big)\mathrm{d}\mathbf{x} \quad (3)$$

This free-energy functional describes segregation at temperatures below T_c, which depends on W and $\chi^\pm$. The mean field estimate of T_c for a symmetric bilayer is given by (see Appendix)[45]

$$T_c = \frac{1}{4}\left(\chi^+ + \chi^- + \sqrt{(\chi^+ - \chi^-)^2 + W^2}\right). \quad (4)$$

To describe the dynamics, we define the individual component velocities $\mathbf{v}_1^\pm$, $\mathbf{v}_2^\pm$, and a barycentric hydrodynamic velocity at each leaflet,

$$\mathbf{v}^{\pm} = \phi^{\pm}\mathbf{v}_1^{\pm} + (1 - \phi^{\pm})\mathbf{v}_2^{\pm}, \tag{5}$$

which we take to be incompressible, $\nabla \cdot \mathbf{v}^{\pm} = 0$. The area fraction $\phi^{\pm}$ at each leaflet obeys a continuity equation,

$$\frac{\partial \phi^{\pm}}{\partial t} = -\nabla \cdot \left(\phi^{\pm}\mathbf{v}_1^{\pm}\right) = \nabla \cdot \left(\left(1 - \phi^{\pm}\right)\mathbf{v}_2^{\pm}\right). \tag{6}$$

Force balance at the two leaflets implies,[46,47]

$$b\phi^{+}\left(\mathbf{v}_1^{+} - \mathbf{v}^{-}\right) + \gamma\phi^{+}\left(\mathbf{v}_1^{+} - \mathbf{v}^{+}\right) = -\frac{1}{2}\nabla \cdot \Pi_1^{+} + \phi^{+}\nabla p^{+} \tag{7}$$

$$b\phi^{-}\left(\mathbf{v}_1^{-} - \mathbf{v}^{+}\right) + \gamma\phi^{-}\left(\mathbf{v}_1^{-} - \mathbf{v}^{-}\right) = -\frac{1}{2}\nabla \cdot \Pi_1^{-} + \nabla \cdot \Sigma_1^{-} + \phi^{-}\nabla p^{-} \tag{8}$$

$$b\left(1 - \phi^{\pm}\right)\left(\mathbf{v}_2^{\pm} - \mathbf{v}^{\mp}\right) + \gamma\left(1 - \phi^{\pm}\right)\left(\mathbf{v}_2^{\pm} - \mathbf{v}^{\pm}\right) = -\frac{1}{2}\nabla \cdot \Pi_2^{\pm} + \left(1 - \phi^{\pm}\right)\nabla p^{\pm} \tag{9}$$

On the left, are the dissipative stresses arising from inter-monolayer friction (coefficient b, taken to be independent of the lipid species) and the in-plane exchange friction associated with the exchange between the two species in the same leaflet (coefficient γ). Other sources of dissipation, such as in-plane viscous dissipation in the membrane and 3d dissipation into the surrounding viscous solvent, can be easily included. On the right, are contributions from reversible stresses – the osmotic pressure $\nabla \cdot \Pi_\alpha^{\pm} = \phi_\alpha^{\pm}\nabla \frac{\delta F}{\delta \phi_\alpha^{\pm}}$, the active stress Σ_1^{-} acting on the inner leaflet and the 2d hydrodynamic pressure p.

The 2d hydrodynamic pressure p can in principle be eliminated using the incompressibility condition on the hydrodynamic velocity $\mathbf{v}^{\pm}$, making the set of eqn (5)–(9) closed. For now, we set the barycentric hydrodynamic velocity $\mathbf{v}^{\pm}$ to zero, and eliminate the pressure p to obtain the final equation,

$$\frac{\partial \phi^{+}}{\partial t} = \nabla \cdot \left(M(\phi^{+})\left[\nabla \frac{\delta F}{\delta \phi^{+}}\right]\right) \tag{10}$$

$$\frac{\partial \phi^{-}}{\partial t} = \nabla \cdot \left(M(\phi^{-})\left[\nabla \frac{\delta F}{\delta \phi^{-}} - \nabla \cdot \Sigma^{-}\right]\right) \tag{11}$$

These equations are seen to be an extension of the usual Cahn–Hilliard equation that includes active stress Σ^{-}, with the current being a mobility times a force. The above calculation naturally leads to the appropriate mobility $M(1 - \phi^{\pm})(\phi^{\pm})$, the relative exchange mobility between the two components.[46,47] The form of the active stress Σ^{-} is determined from the hydrodynamic equations describing the active contractile flows that we derive below.

3.3 Active contractile stresses from the actomyosin cortex

The active hydrodynamic equations that describe the actomyosin fluid in terms of the concentration and polarization of actin filaments and myosin density, have been discussed in several of our earlier papers.[10,32,33] Here we will construct simpler effective two-variable active hydrodynamic equations in terms of actomyosin concentration $c(\mathbf{x}, t)$ and cortical velocity $\mathbf{v}(\mathbf{x}, t)$.

Local changes in actomyosin concentration are described by

$$\partial_t c = -\nabla \cdot \mathbf{J}_c + S(c, \phi^-) \tag{12}$$

where $S(c, \phi^-) = k_b(\phi^-) - k_u c$ represents the turnover of actomyosin from the inner leaflet membrane. The binding rate k_b, represents the local binding of actomyosin on the inner leaflet membrane enriched in saturated lipids, and so depends on ϕ^-. For instance, we take $k_b = k_{0b} + k_{1b} \tanh \phi^-$. The unbinding rate k_u would in general depend on the active stress (for instance, rapid unbinding when the local active stress is beyond a threshold), but here we take it to be constant. The actomyosin transport current is a sum of the diffusive and advective currents,

$$\mathbf{J}_c = -D\nabla(c + bc^3) + c\mathbf{v}, \tag{13}$$

where we include a nonlinear contribution to the diffusive flux to prevent unbounded growth of $c(\mathbf{x}, t)$. The cortical velocity $\mathbf{v}(\mathbf{x}, t)$ is determined by local force balance,

$$\gamma \mathbf{v} = \eta \nabla^2 \mathbf{v} + (\mu + \eta)\nabla(\nabla \cdot \mathbf{v}) + \nabla \cdot \sigma^a + \mathbf{f}, \tag{14}$$

where cortical friction is balanced against the bulk and shear viscous stresses, with viscosities μ and η, respectively, and the active stress σ^a. Since the actomyosin cortical fluid is isotropic, we take $\sigma^a = \zeta c\mathbf{I}$, where $\zeta > 0$ for contractile stresses. Apart from the active stress, myosin minifilaments drive an active self-propulsive current, which we represent as a non-conservative stochastic forcing $\mathbf{f}$ with zero mean and

$$\left\langle \mathbf{f}(\mathbf{x}, t) \cdot \mathbf{f}\left(\mathbf{x}', t'\right) \right\rangle = \frac{{f_0}^2}{\tau} \exp\left(-\frac{|t - t'|}{\tau} \right) \delta^{(2)}\left(\mathbf{x} - \mathbf{x}'\right), \tag{15}$$

where τ represents the lifetime of bound myosin minifilaments.

Eqn (12)–(15) describe the hydrodynamics of the actomyosin cortex, which we use to compute the active stress Σ^- on the inner leaflet membrane using the constitutive formula,

$$\Sigma^- = \frac{Kc}{1 + c}\mathbf{I} \tag{16}$$

where K represents the coupling between the actomyosin and the saturated lipids in the inner leaflet.

4 Results

We now solve eqn (10) and (11) together with eqn (12)–(16), for initial conditions where the two leaflets are in a mixed, homogeneous phase. It is convenient to make the equations dimensionless by scaling space and time by the length scale $l = \sqrt{\kappa/2a}$, where $a = \chi - \chi_c$ and the time scale $t^* = \kappa/(4Ma^2)$. For convenience, we take $\chi_c = 2$, the critical interaction parameter for a symmetric monolayer.

With this the non-dimensionalised equations read

$$\frac{\partial \phi^+}{\partial t} = \nabla \cdot \left(\frac{M(\phi^+)}{2} \nabla \left[-\frac{2\chi}{a}\phi^+ + \frac{1}{a}\ln\left(\frac{\phi^+}{1 - \phi^+}\right) - \nabla^2\phi^+ - W\phi^- \right] \right) \tag{17}$$

and

$$\frac{\partial \phi^-}{\partial t} = \nabla \cdot \left(\frac{M(\phi^-)}{2} \nabla \left[-\frac{2\chi}{a}\phi^- + \frac{1}{a}\ln\left(\frac{\phi^-}{1-\phi^-}\right) - \nabla^2\phi^- - W\phi^+ - \nabla\cdot\Sigma^- \right] \right) \quad (18)$$

for the membrane, and

$$\mathbf{v} = \eta\nabla^2\mathbf{v} + (\mu + \eta)\nabla(\nabla\cdot\mathbf{v}) + \nabla c + f \quad (19)$$

$$\partial_t c = D\nabla^2 c - \mathrm{Pe}\nabla\cdot(cv) + k_b(\phi^-) - k_u c \quad (20)$$

for the actomyosin cortex. In these dimensionless equations, the old parameters have been redefined as, $W/a \rightarrow W$, $K/a \rightarrow K$, and $\frac{2\eta a}{\gamma\kappa} \rightarrow \eta$, $\frac{2\mu a}{\gamma\kappa} \rightarrow \mu$, $\mathbf{v}/U \rightarrow \mathbf{v}$, with $U = \frac{\zeta}{\gamma l}$, $D/2Ma \rightarrow D$, $\zeta/(2\gamma Ma) \rightarrow \mathrm{Pe}$, $k_b t^* \rightarrow k_b$, $k_u t^* \rightarrow k_u$. The range that these scaled parameters take are displayed in Table1 in the Appendix.

Before presenting the results, we make some general observations. First, it is the combination of active stresses, transbilayer coupling and lateral interactions with saturated lipids that outcompetes the entropy of mixing and drives lateral and transverse organisation of lipids at temperatures below and above T_c. Note that T_c of the bilayer depends on the interaction parameter $\chi^{\pm}$, the transbilayer coupling W (eqn (4)), and the actomyosin–inner leaflet coupling K. Second, even for a symmetric quench (when the lo and ld concentrations are equal), the presence of the active stress Σ^- breaks the symmetry of interchange between lo and ld. This would imply that even for a symmetric quench, the domains will grow and saturate to a finite size. Third, from the structure of the active forces and reactive turnover in these equations, we should expect strong nonreciprocal effects to show up.

4.1 Active segregation in an asymmetric bilayer

For the most part we will deal with an asymmetric bilayer with the following composition – in the outer leaflet, 50% unsaturated lipids and 50% saturated lipids; while in the inner leaflet, 65% unsaturated lipids and 35% saturated lipids. We have also checked our qualitative results with other relative concentrations, keeping the bilayer asymmetric. These numbers closely resemble the lipid composition reported in ref. 15 and 24.

Starting from initial conditions where the two leaflets are in a mixed, homogeneous phase, we solve eqn (17)–(20) together with eqn (15) and (16), for a fixed net composition in the two leaflets and temperature. The numerical details are presented in the Appendix. We monitor the local lipid compositions in the two leaflets and the concentration of actomyosin over time. We characterise the distinct nonequilibrium phases and their dynamics and construct a phase diagram. In general, the phases reflect the interplay between the gradients of chemical potential and active stress. We first look at the regime of low contractility, Pe.

Symmetric bilayer, low contractility. We first study the case where the symmetric bilayer has equal net concentrations of saturated and unsaturated lipids. In the absence of any coupling to the actomyosin cortex (setting $K = 0$), the bilayer shows the usual spinodal instability[48] (Fig. 1B(i)) below T_c (eqn (4)), leading to growing bicontinuous patterns with a length scale (extracted from the

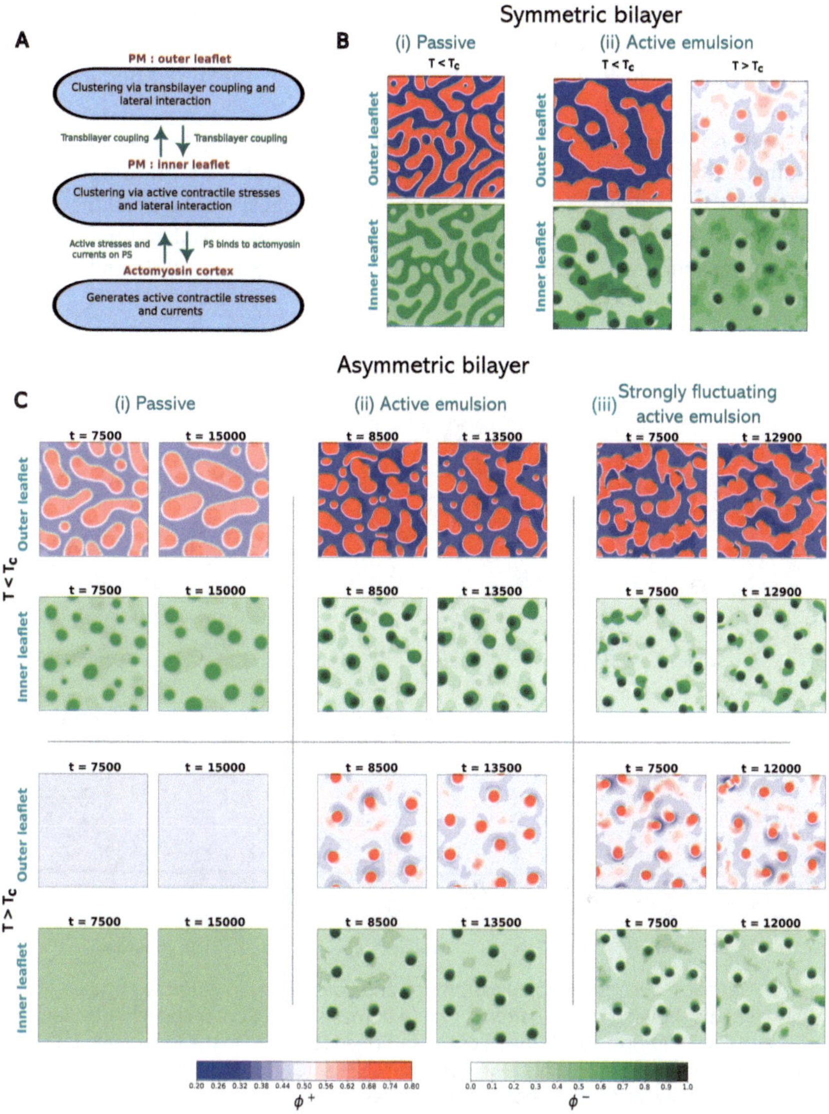

Fig. 1 Active contractile stresses, transbilayer coupling and lateral interactions drive mesoscale segregation in the bilayer. (A) Modular network of the main components and forces that act in the active composite membrane to drive lateral and transverse organisation of PM lipids. (B) Heat maps showing spatial patterning of lo–ld domains in the outer (ϕ^+, red–blue) and inner (ϕ^-, dark–light green) leaflets of a symmetric bilayer with $\langle\phi^+\rangle = \langle\phi^-\rangle = 0.5$ and $W = 0.5$, for $T < T_c$ and $T > T_c$. (i) Passive bilayer ($K = \mathrm{Pe} = 0$) showing the usual bicontinuous pattern at $T < T_c$. (ii) Active bilayer ($K = 10$, $\mathrm{Pe} = 24$) showing an active emulsion phase at both $T < T_c$ and $T > T_c$. Note the granules in the finite size lo-domains and the domain registry. (C) Same as the above for an asymmetric bilayer with $\langle\phi^+\rangle = 0.5$, $\langle\phi^-\rangle = 0.35$ with strong transbilayer coupling ($W = 2$) at $T = 0.94T_c < T_c$ and $T = 1.08T_c > T_c$. (i) Passive bilayer ($K = \mathrm{Pe} = 0$) showing finite size lo-domains at $T < T_c$ in both the leaflets and mixed phase at $T > T_c$. (ii) Active bilayer ($K = 5$, $\mathrm{Pe} = 17$) showing an active emulsion with granularity and registry at both $T < T_c$ and $T > T_c$. (iii) Strongly fluctuating active emulsion ($K = 10$, $\mathrm{Pe} = 27$), characterised by domain breakup, merger, and movement. The rest of the parameters are set at $\chi = 2.08$ ($T < T_c$), $\chi = 1.92$ ($T > T_c$), $k_u = 0.0032$, $c_0 = 0.2$, $k_{1b} = k_u c_0$, $\mu + \eta = 100$, $\eta = 25$, $D = 1$, $b = 0.5$, $f_0^2 = 0.0025$, $\tau = 1$.

real space correlation function) that grows as $R \sim t^{1/3}$ (data not shown). When the transbilayer coupling W is large, there is complete registry across the bilayer, which is lost when W is reduced. Of course, for $T > T_c$ the mixed homogeneous phase is stable.

With finite cortical coupling $K > 0$, and low contractility Pe, the phases are dramatically different. First, sufficiently high activity drives a finite wave-vector instability at temperatures below and above T_c. Because of the breaking of the symmetry of interchanging lo and ld, the domains grow as isolated patches and saturate at a finite size, both in the inner and outer leaflet (Fig. 1B(ii), and Movie 1,

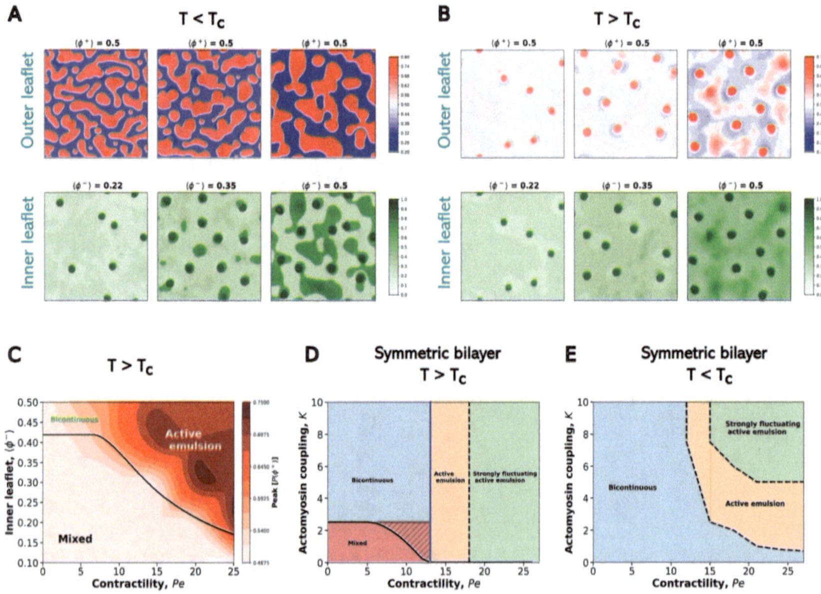

Fig. 2 Variation of compositional asymmetry and contractility in the active bilayer. (A and B) Heat maps showing spatial pattern of lo–ld domains on the outer and inner leaflets at $T < T_c$ and $T > T_c$, for increasing levels of the saturated lipid $\langle\phi^-\rangle$ in the inner leaflet. Overall there is an increase in domain size in both the leaflets. (C) Nonequilibrium phase diagram of the outer leaflet at $T > T_c$ in the contractility Pe–$\langle\phi^-\rangle$ plane (with $W = 2$), calculated from the peak of the $P(\phi^+)$, the probability distribution of ϕ^+ (colour bar). The line represents the curve eqn (21) obtained from a linear stability analysis about the mixed phase. The region marked bicontinuous refers to the equilibrium phase due to a shift in T_c coming from K, the coupling between actomyosin and the inner leaflet, with fluctuations which increase with Pe. (D) Nonequilibrium phase diagram for a symmetric bilayer with $\langle\phi^+\rangle = \langle\phi^-\rangle = 0.5$ and $W = 0.5$, at $T > T_c$ showing the different phases. The black line separating the mixed and segregated phases is obtained as before using eqn (21). The blue line separating the bicontinuous from the active emulsion is obtained from the linear instability of the homogeneous phase of actomyosin using eqn (19) and (20). The strongly fluctuating active emulsion is defined by having large temporal fluctuations in the Fourier components of ϕ^+. The hatched region corresponds to a mixed phase with large active fluctuations. (E) Nonequilibrium phase diagram for a symmetric bilayer with $\langle\phi^+\rangle = \langle\phi^-\rangle = 0.5$ and $W = 0.5$, at $T < T_c$ showing the different phases. The boundary between the bicontinuous and active emulsion is obtained from the steady state mean domain size. The strongly fluctuating active emulsion is defined by having large temporal fluctuations in the Fourier components of ϕ^+. The heat maps in (A) and (B) are obtained at $K = 10$, Pe $= 25$ with rest of the parameters as in Fig. 1.

Movie 2†). The domain size in the outer leaflet depends on the transbilayer coupling strength W, and recovers the bicontinuous pattern when W is small.

Another unusual feature of the segregation due to active contractile stress is that it drives significant density inhomogeneities that manifest as domain *granularity*, both below and above T_c (Fig. 1B(ii), and Movie 1, Movie 2†). At these low values of Pe and high W, the actomyosin asters in the cortex are in registry with the high density granules in the inner and outer leaflet, thus suggesting that these mesoscale domains, or active emulsions, are built from a collection of nanodomains, consistent with our experimental observations.[15]

Asymmetric bilayer, low contractility. We next study the asymmetric bilayer at the net composition mentioned at the start of this subsection. Here too, a threshold contractility Pe or actin binding affinity K, drives a finite wave-vector instability of the mixed phase at temperatures below and above T_c, with the domains growing and saturating at a finite size. As before, at strong transbilayer coupling, these active emulsions exhibit granularity, both in the inner and outer leaflet (Fig. 1C(ii), and Movie 3, Movie 5†).

A linear stability analysis of eqn (17) and (18), treating the actomyosin variables in eqn (19) and (20) as fast, shows that the transition from the mixed phase to the active emulsion phase is at (black line in Fig. 2C)

$$\mathrm{Pe} = \frac{2\sqrt{\left(A+(\alpha+1)^2\eta k_\mathrm{u}\right)\left(A+(\alpha+1)^2(D+\eta k_\mathrm{u})\right)}}{k_\mathrm{b}(\alpha+1)^2} + \frac{2A}{(\alpha+1)^2 k_\mathrm{b}} + \frac{D+2\eta k_\mathrm{u}}{k_\mathrm{b}} \tag{21}$$

where $A = Kk_\mathrm{u}\alpha(\phi_0^-)^2 \left.\frac{dk_\mathrm{b}}{d\phi^-}\right|_{\phi_0^-}$ and $\alpha = (\phi_0^- - 1)\phi_0^-\left(\chi - \frac{a^2W^2}{2-\chi}\right)$, with ϕ_0^- being the average inner leaflet lo area fraction.

Asymmetric bilayer, high contractility. As the contractility Pe or actin-binding affinity K increases, our theory predicts an unusual non-equilibrium steady state, a strongly fluctuating active emulsion. This state is characterised by continual domain breakdown and merger, including movement when the stochastic forcing f_0 is large (Fig. 1C(iii), and Movie 4, Movie 6†). The former is characterised by large fluctuations in the time series of the Fourier components of the segregation order parameter, similar to that reported earlier.[23,36] The granules are highly dynamic and appear to move restlessly across different mesoscale domains. We do not go into the technical details of this characterisation here, suffice to say this bears some resemblance to the phases observed in other nonreciprocal field theories.[49,50]

These intermittent strong fluctuations and domain movement are a consequence of the imbalance between the gradients of active stress and chemical potential which provides the driving force for the movement of the domain interface.

The local binding of actomyosin to the inner leaflet, and hence the active stress, depends on the local concentration of saturated lipids (eqn (12)). So we expect the dynamical patterning of composition in both the leaflets to be sensitive to the average lo composition in the inner leaflet, as corroborated in Fig. 2A and B. The phase diagram in Fig. 2C, for $T > T_c$, suggests that the inner leaflet should have a threshold concentration of saturated lipids to observe the active emulsion phase in the outer leaflet.

In Fig. 2D and E, we display nonequilibrium phase diagrams for $T > T_c$ and $T < T_c$ respectively for the symmetric bilayer, where all the above phases are represented.

Domain registry and slippage. We now ask whether the lo domains in the two leaflets are in registry, by defining the domain mismatch area[51–53] (Fig. 3A, see Appendix). The mismatch area decreases (registry increases) with increasing transbilayer coupling W, when the local lo order in the inner leaflet gets faithfully transmitted to the outer leaflet (Fig. 3A). The outer leaflet has a larger domain size owing to its higher net lo lipid concentration.

This opens up interesting possibilities for both inside-out and outside-in signalling.[11] In particular, the clustering of outer leaflet saturated lipids/proteins (*e.g.*, GPI-anchored proteins) could trigger local actomyosin activity immediately below it *via* transbilayer coupling, leading to downstream recruitment of signalling or effector molecules, as happens during local integrin activation and cell–substrate adhesion.[26]

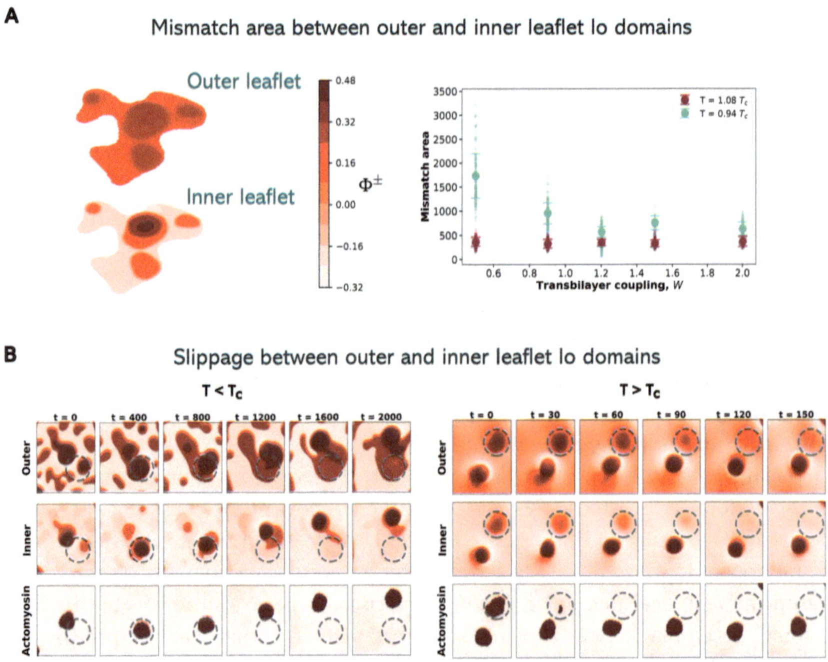

Fig. 3 Domain registry and slippage in the strongly fluctuating active emulsion in an asymmetric bilayer. (A, left) A single mesoscale domain on the two leaflets at $T < T_c$ acquired through image labeling (*binary thresholding*, and *python skimage* package), highlighting both registered and anti-registered areas across the leaflets. Colour-bar corresponds to the segregation parameter $\Phi^{\pm} = 2\phi^{\pm} - 1$, with positive values indicating saturated lipids. Using this, we calculate the mismatch area (the area where local saturated lipid at the outer leaflet is not matched with saturated lipid at the inner leaflet). (A, right) Variation of mismatch area with transbilayer coupling. (B) Segregated domains showing local slippage across the leaflets and across the inner leaflet and the cortex. This is a consequence of the strong active driving and is observed both below and above T_c. The parameters for this asymmetric bilayer with $\langle\phi^+\rangle = 0.5$, $\langle\phi^-\rangle = 0.35$ are $K = 10$ and $\mathrm{Pe} = 25$, and the rest of the parameters are as in Fig. 1.

Both in the strongly fluctuating active emulsion phase and when the stochastic forcing parameters f_0 and τ are large, we see the interesting phenomena of domain slippage (Fig. 3B, and Movie 7, Movie 8†), where the actomyosin domain slips past the inner leaflet lo domain which slips past the outer leaflet lo domain.

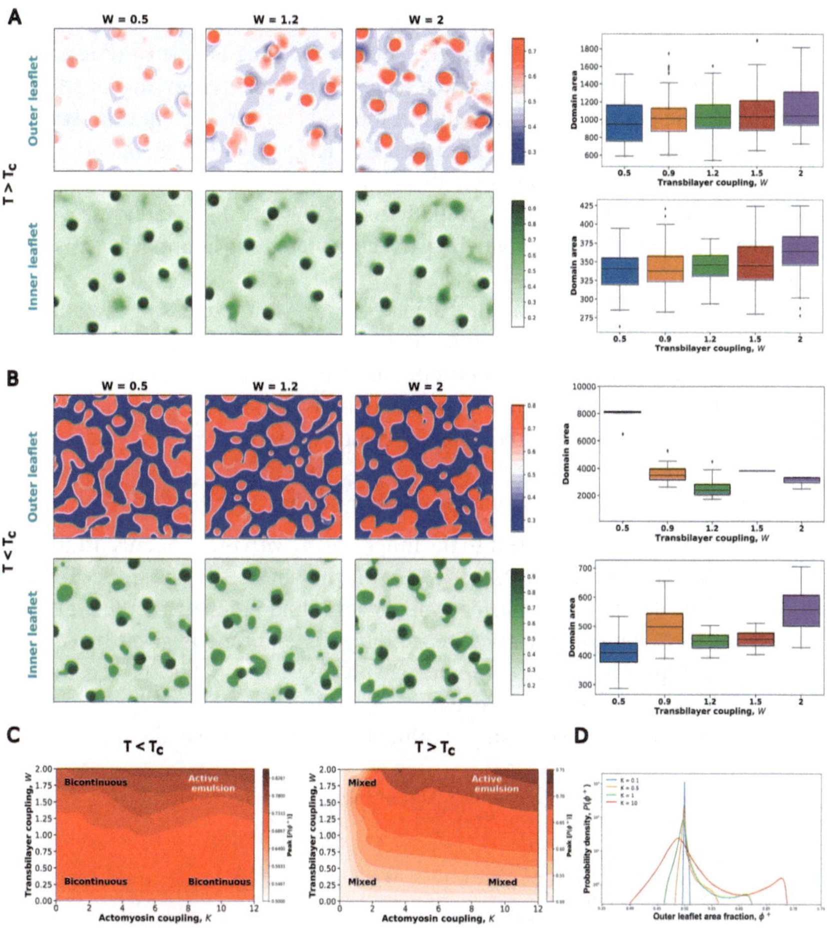

Fig. 4 Variation of transbilayer W and actomyosin coupling K in the asymmetric active bilayer. (A and B) Heat maps showing spatial patterning of lo–ld domains in the outer and inner leaflets at (A) $T > T_c$ and (B) $T < T_c$, in the strongly fluctuating emulsion phase at different values of W. The granularity is more prominent as W increases. (A and B right) Box plots, calculated through image processing tools (*python skimage*) at steady state, show the domain size as a function of W. The thickness of the box plot indicates the strong fluctuations in domain sizes at steady state. When $T < T_c$ (B, right) and W is small, the outer leaflet hardly feels the effects of non-equilibrium forces, and so the domain size is large. This will decrease as W increases. In contrast, domain size at the lower leaflet increases as transbilayer coupling increases. (C) Nonequilibrium phase diagram of the outer leaflet at $T < T_c$ and $T > T_c$ in the W–K plane, showing the transition between bicontinuous and active emulsion for $T < T_c$ and mixed to active emulsion for $T < T_c$. (D) Probability density function $P(\phi^+)$ for different values of actomyosin coupling K. Note that the width and tails of the distribution get broader with increasing K. The leaflets composition here is $\langle\phi^+\rangle = 0.5$ and $\langle\phi^-\rangle = 0.35$, with $K = 10$, Pe $= 25$, and the rest of the parameters as in Fig. 1.

Effect of transbilayer coupling. The effects of non-equilibrium active forces generated at the cortex are felt by the outer leaflet predominantly through the transbilayer coupling, which feeds back onto the inner leaflet and subsequently onto the active cortex, Fig. 1A.

Thus in Fig. 4A, we see that at $T > T_c$, the appearance of mesoscale domains at the outer leaflet is contingent on having strong transbilayer coupling. In Fig. 4B, we see that when $T < T_c$, the outer leaflet retains its bicontinuous pattern when W is small. However at larger W this pattern breaks up into finite size active emulsions. This implies that at large enough W, the transbilayer coupling is able to transmit the active stresses to the outer leaflet which represses the equilibrium in-plane forces.

Further, the granularity of domains in the outer leaflet, both at temperatures above and below T_c, is apparent only when W is large enough (Fig. 4A and B). The mean lo-domain size at steady state displays a non-monotonic relationship with W (Fig. 4B, upper leaflet box plot), initially decreasing and then increasing with W, together with improved domain registry and increasing granularity observed across the two leaflets.

Fig. 4D shows the probability density of ϕ^+ with increasing actomyosin coupling (and hence effective contractility). We note that as K increases, the width gets broader and shows a broad exponential tail at the lower end of ϕ^+. This is consistent with the broad exponential tails observed in the distribution of GPI-anchored proteins on the outer leaflet of cell membranes.[9,10,35]

4.2 Explicit PS – a 3-component asymmetric bilayer

We now explicitly introduce PS into the inner leaflet, with the net concentration of saturated lipids at 13% and PS at 22% (with the sum 35%, being the same as earlier). These levels are consistent with recent lipidomics data.[24] The FH free energy for the 3-component asymmetric bilayer membrane is given by eqn (2), where $\alpha = 3$ represents the PS component, with $\phi_2^+ = 1 - \phi_1^+$ in the outer leaflet and $\phi_2^- = 1 - \phi_1^- - \phi_3^-$ in the inner leaflet.

Following the sequence of steps outlined in Section 3.2, we obtain,

$$\frac{\partial \phi_1^+}{\partial t} = \nabla\cdot\left(M_1^+ \nabla \mu_{12}^+\right) \tag{22}$$

$$\frac{\partial \phi_1^-}{\partial t} = \nabla\cdot\left(M_1^- \nabla \mu_{12}^- - M_{31}^- \nabla\left(\mu_{32}^- - \nabla\cdot\Sigma_3^-\right)\right) \tag{23}$$

$$\frac{\partial \phi_3^-}{\partial t} = \nabla\cdot\left(M_3^-\left(\nabla \mu_{32}^- - \nabla\cdot\Sigma_3^-\right) - M_{31}^- \nabla \mu_{12}^-\right). \tag{24}$$

The mobilities $M_{ij}^{\pm} = M\phi_i^{\pm}\phi_j^{\pm}$ for $i \neq j$ and $M_i^{\pm} = M\phi_i^{\pm}(1 - \phi_i^{\pm})$, where for simplicity we have taken the coefficients M to be the same, and

$$\begin{bmatrix} \mu_{12}^+ \\ \mu_{12}^- \\ \mu_{32}^- \end{bmatrix} = \begin{bmatrix} \chi_{12}^+\left(\phi_2^+ - \phi_1^+\right) - W\left(\phi_1^- + \phi_3^-\right) + \ln\left(\phi_1^+/\phi_2^+\right) \\ \chi_{12}^-\left(\phi_2^- - \phi_1^-\right) + \left(\chi_{13}^- - \chi_{23}^-\right)\phi_3^- - W\phi_1^+ + \ln\left(\phi_1^-/\phi_2^-\right) \\ \chi_{23}^-\left(\phi_2^- - \phi_3^-\right) + \left(\chi_{13}^- - \chi_{12}^-\right)\phi_1^- - W\phi_1^+ + \ln\left(\phi_3^-/\phi_2^-\right) \end{bmatrix}$$

The active stress now acts only on PS in the inner leaflet,

$$\Sigma_3^- = \frac{K_{PS}c}{1 + c}. \tag{25}$$

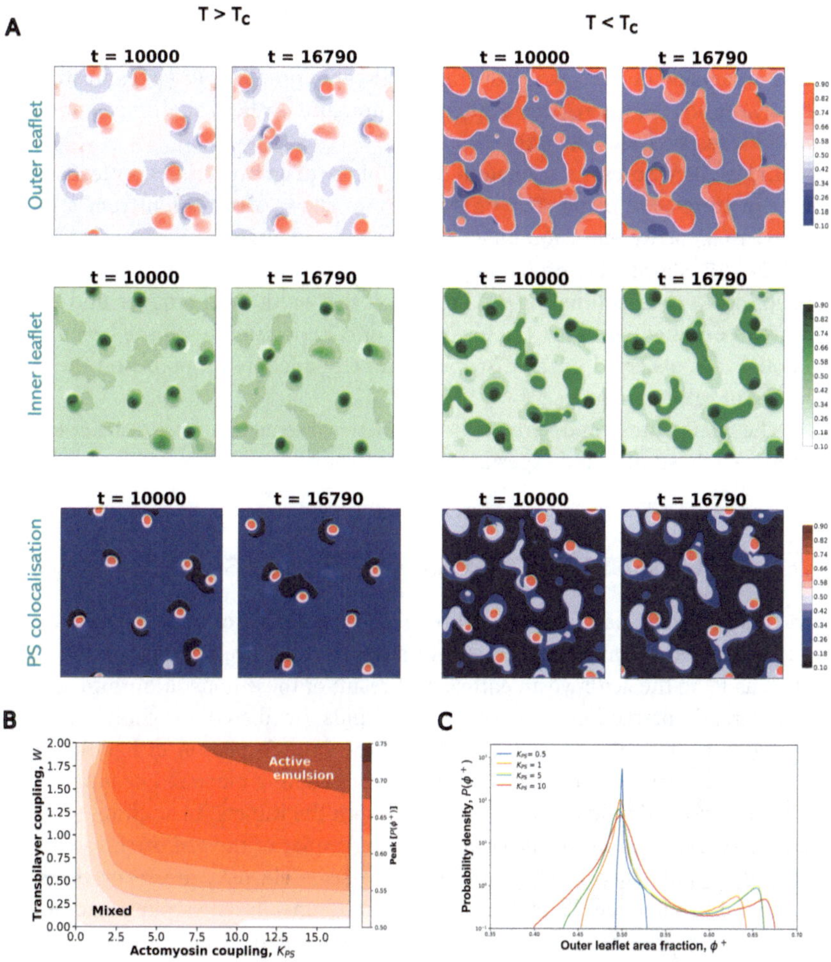

Fig. 5 Ternary mixture of saturated, unsaturated, and PS lipids at the inner leaflet of an asymmetric bilayer. (A) Heat maps showing spatial patterning of lo–ld domains in the outer and inner leaflets at $T > T_c$ and $T < T_c$. Both the lo domains and the inner leaflet PS show granularity. With the PS being at the centres of the lo-domain in the inner leaflet. The third panel highlights the colocalisation of PS with the saturated lipids in the outer leaflet. (B) Nonequilibrium phase diagram of the outer leaflet at $T > T_c$ in the W–K_{PS} plane, showing the transition from the mixed to active emulsion phase. (C) Probability density function $P(\phi^+)$ for different values of K_{PS}. As before, the width and tails of the distribution get broader with increasing K_{PS}. The lipid composition in the inner leaflet is $\langle\phi^+\rangle = 0.5$, $\langle\phi_1^-\rangle = 0.13$, and $\langle\phi_3^-\rangle = 0.22$, with $K = 10$, Pe $= 25$, $W = 2$, $\chi_{12}^- = \chi_{32}^- = \chi_{12}^+ = \chi$, $\chi_{13}^- = 0$, $\chi = 1.92$ ($T > T_c$), $\chi = 2.08$ ($T < T_c$), and the rest of the parameters as in Fig. 1.

The hydrodynamic equations describing the active cortex are the same as eqn (19) and (20), with the local binding rate of actomyosin on the inner leaflet being a function of PS, $k_b(\phi_3^-) = k_{0b} + \frac{1}{2}k_{1b}\tanh\phi_3^-$. The transbilayer coupling, W, now couples the saturated lipids in the outer leaflet and the saturated lipids and PS in the inner leaflet. The local lo content in the inner leaflet is obtained from $\phi_1^- + \phi_3^-$.

As mentioned earlier, the qualitative features of this 3-component asymmetric bilayer are the same as the previous 2-component asymmetric bilayer. However this model allows us to explicitly track the spatial profile of PS (ϕ_3^-) and study its colocalisation and coregistry with the lo domains in both the inner and outer leaflets, respectively (Fig. 5A).

In Fig. 5B, we present a nonequilibrium phase diagram of the outer leaflet for $T > T_c$ in the W–K_{PS} plane. The phase diagram shows a transition from a mixed phase to an active emulsion phase at large K_{PS} and large W.

Fig. 5C shows the probability density of ϕ^+ with increasing actomyosin coupling K_{PS}. As before, as K_{PS} increases, the width gets broader and shows a broad exponential tail at the lower end of ϕ^+, consistent with the broad exponential tails observed in the distribution of GPI-anchored proteins on the outer leaflet of cell membranes.[9,10]

The results presented here are in complete agreement with our earlier kinetic Monte Carlo studies and experimental observations.[15]

5 Discussion and future directions

In this paper we have discussed how the inner leaflet of the PM is inevitably subject to active contractile stresses arising from the coupling of specific lipids such as PS to the actomyosin cortex. As a result of these nonequilibrium stresses and lateral interactions with saturated lipids (mediated by cholesterol), lo-domains form in the inner leaflet. This drives the formation of lo-domains in the outer leaflet too, *via* a strong transbilayer coupling mediated by inner leaflet PS and saturated lipids and cholesterol in both the leaflets. In addition to lipids, lipid-tethered proteins localized to the outer leaflet of the PM, such as GPI-APs and glycolipids, also form nano and meso scale clusters, driven by the same nonequilibrium forces, lateral interactions and transbilayer coupling mediated by PS.[15,16] The finite-sized lo-domains generated this way are dynamic and occur both below and above the equilibrium phase transition temperature. We describe this as an active emulsion formed by local nonequilibrium stirring of the inner leaflet of the bilayer by stresses arising from the actomyosin cortex.[15] However, we note from Fig. 2C, that a reduction in the levels of inner leaflet PS, can drive the outer leaflet into a mixed phase. This implies that lo-domain formation, both in the inner and outer leaflets is contingent on adequate levels of PS in the inner leaflet.[15,16]

How are the levels of PS in the inner leaflet maintained in the face of inter-leaflet equilibration by lipid flip–flop?[54] The active maintenance of PS in the inner leaflet of the PM is regulated by nonequilibrium active flippases,[25] which provides an active control of an unstable configuration. We are currently working on how to incorporate the nonequilibrium effects of both actomyosin contractile forces and active flip–flop in an AFH theory for the asymmetric bilayer.

In order to obtain lo-domains, both the lateral and transbilayer interactions between the saturated lipids need to be mediated by adequate amounts of cholesterol in the two leaflets. Here we have taken the saturated lipids to be a complex of saturated lipids and cholesterol. It would be simple to include cholesterol as an explicit component; however this will not alter the qualitative results described here.

The results presented in Section 4 were for a specific choice of net lipid composition in the two leaflets.[24] This would be cell type dependent and may even be inhomogeneous as in polarised epithelial cells.[55] It would be interesting to apply our AFH formalism to study differences in lipid organisation in different cell types.

In summary, our work shows how sustained nonequilibrium processes operating at the plasma membrane, the ceaselessly restless living cell membrane, determine the dynamical lateral and transverse organisation of membrane components, their maintenance and control, under physiological conditions. As a consequence, the nonequilibrium force asymmetry arising from the actomyosin cortex is intimately linked to the compositional asymmetry of the cell membrane bilayer.

6 Conclusions

Our work shows how sustained nonequilibrium processes operating at the plasma membrane, the ceaselessly restless living cell membrane, might determine the dynamical organisation (both lateral and transverse) of lipids and proteins, their maintenance and control, under physiological conditions.

Data availability

All simulations were performed using the Python package Dedalus[61] (https://dedalus-project.org/). Data analysis packages are reported in the Appendix.

Author contributions

MR conceived and planned the project; AB and MR executed the project and jointly wrote the paper.

Conflicts of interest

There are no conflicts to declare.

7 Appendix

7.1 Shift in T_c due to transbilayer coupling

A passive equilibrated bilayer exhibits a diverging correlation length ξ and, consequently, a divergent structure factor $S(k = 0)$ at $k = 0$, as the critical temperature is approached, where $S(k)$ is the spatial Fourier transform of the equal time correlation function $\langle\phi(r, t)\phi(r', t)\rangle$.

The mean-field critical temperature is determined by expanding the free energy density to quadratic order and identifying the temperature at which $S(k = 0)$ diverges. The expansion of the free energy density around the homogeneous phase, where $\phi^{\pm} = 1/2$, is

$$f[\phi^{\pm}] = (-\chi^{\pm} + 2T)\phi^{\pm}(k)\phi^{\pm}(-k) + W\phi^{+}(k)\phi^{-}(-k) + \kappa^{\pm}k^2\phi^{\pm}(k)\phi^{\pm}(-k) \quad (26)$$

The mean field structure factor is given by,[45,56]

$$S_{\phi^\pm\phi^\pm}(k) = \frac{g_{\phi^\pm}}{4g_{\phi^+}g_{\phi^-} - W^2}, \quad S_{\phi^+\phi^-}(k) = \frac{W}{4g_{\phi^+}g_{\phi^-} - W^2} \tag{27}$$

where,

$$g_{\phi^\pm} = -\chi^\pm + 2T + K^\pm k^2 \tag{28}$$

The divergence of $S(k)$ at $k = 0$ occurs when $4(g_{\Phi^+}g_{\Phi^-})_c - W^2 = 0$, which sets the mean field critical temperature at

$$T_c = \frac{1}{4}\left(\chi^+ + \chi^- + \sqrt{(\chi^+ - \chi^-)^2 + W^2}\right) \tag{29}$$

We note that although we have taken values from the range above, the qualitative results remain robust to slight variation from the ranges.

7.2 Numerical methods

Eqn (17)–(20) are dimensionless by scaling space and time by,

(1) Unit of length, $l = \sqrt{\kappa/2a}$, which corresponds to $\approx$5–20 nm, where $a = \chi - \chi_c$.

(2) Unit of time, $t^* = \kappa/(4Ma^2) = l^2/D_m$. With $D_m = 1\ \mu m^2\ s^{-1}$,[15] this translates to $t^* = 0.0004$ s.

This scaling changes the parameters as shown in Table 1.

7.2.1 Simulation setup. The simulations were performed using the Python package Dedalus[61] (https://dedalus-project.org/), configured as follows:

• *De-aliasing factor*: set to 2 to accommodate third-order non-linearities in the actomyosin evolution equations.

Table 1 Scaled parameters (see Section 4, below eqn (20)) appear in the dimensionless eqn (17)–(20) outlined in the main text. Numerical values show the range of values explored, which correspond to the physical values obtained from earlier experimental or theoretical studies (references provided)

Parameters	Dimensionless values	Physical values
χ (interaction parameter)	2.15–1.85	χ/χ_c (= 2) sets the deviation from criticality
$\sqrt{\kappa/a}$ (correlation length)	1	$\approx$10–20 nm (ref. 15 and 57)
W (transbilayer coupling)	0.06–3.33	≈ 0.01–$0.5k_BT$ (ref.[57–59])
K (actomyosin coupling)	0–20 (ref. 15)	
Pe (Peclet no.)	0–60 (ref. 60)	
k_u (actomyosin–membrane unbinding rate)	0.00005–0.0032	0.125–8 s^{-1} (ref. 10 and 33)
k_b (actomyosin–membrane binding rate)	0.00625–0.425	2.5–160 $\mu m^2\ s^{-1}$ (ref. 10 and 33)
$\mu + \eta$ (shear + bulk cortical viscosity)	100	0.01 Pa s (ref. 10 and 33)
D (actomyosin diffusion coefficient)	1	1 $\mu m^2\ s^{-1}$

• *Boundary conditions*: periodic with a system size of $N_x = N_y = 256$ and $L_x = L_y = 256$.

• *Time-stepping parameters*: first-order time stepper SBDF1 with time steps of 0.001–0.005. Note that it is effective to use first-order time steppers owing to the stochastic forcing in the system.

• *CFL condition*: while not useful to calculate in most cases, we have observed that the current in the evolution of the area fraction of the inner leaflet (ϕ^-) is sufficient for satisfying the CFL condition.

7.2.2 Potential issues and troubleshooting. For troubleshooting, we strongly recommend utilizing the Dedalus Google group. Common issues include:

• *Gibbs phenomenon*: ringing artifacts in solutions caused by unresolved higher modes. Mitigated by reducing grid spacing or increasing the de-aliasing factor.

• *Singular solutions*: typically arise from accumulated errors. Resolved by reducing time steps spacing.

7.3 Data analysis: domain area, mismatch area, PDFs peak

Domain area. We binarize the area fraction (ϕ^+, ϕ^-) matrix using a threshold volume fraction of 0.5. Additionally, Otsu's thresholding method, implemented in the *skimage library*, owing to the inherent bimodal distribution of phase segregation provides robust results. To identify domains and extract their properties, we use the label and regionprops functions from *skimage*. These tools enable the computation and storage of various quantities, including the mean area fraction of the domains, label numbers, and domain areas. We then focus on regions with mean intensities greater than 0.5 as a secondary threshold, selecting their corresponding labels and areas. This method applied across time at distinct trans-bilayer coupling at steady state on the stored data of ϕ^+ and ϕ^- yields the box plots shown in Fig. 4.

Mismatch area. We extend the method described above, where we first analyze the upper leaflet and verify the existence of a corresponding domain in the lower leaflet. The mismatch area represents the area of the outer leaflet area fraction in the mesoscale domain absent in the inner leaflet, and *vice versa*. A representative plot of a sample domain obtained using this method is provided in the main text (Fig. 3). Here, each pixel length corresponds to the correlation length scale, and the mismatch area is expressed in units of l^2.

Probability density function peak. We first generate histograms from ϕ^+ time series data at steady state ($t \approx 10^4$), with 40 time points and 100 bins. From this histogram and bins we generate the probability density function curve and use *scipy.signal (find peaks)* with prominence of 0.2–0.4, to find peaks. We use this peak to plot the heat maps in (Fig. 2, 4 and 5).

7.4 Captions for supplementary movies†

• *Movie 1*: Strongly fluctuating active emulsions in a symmetric bilayer at $T < T_c$. Parameters are: $\chi = 2.08$, $W = 0.5$, $K = 10$, $\mathrm{Pe} = 24$, $k_u = 0.0032$, $c_0 = 0.2$, $k_{1b} = k_u c_0$, $\mu + \eta = 100$, $\eta = 25$, $D = 1$, $b = 0.5$, $\tau = 1$, ${f_0}^2 = 0.025$.

• *Movie 2*: Strongly fluctuating active emulsions in a symmetric bilayer at $T > T_c$. Parameters are: $\chi = 1.92$, $W = 0.5$, $K = 10$, $\mathrm{Pe} = 24$, $k_u = 0.0032$, $c_0 = 0.2$, $k_{1b} = k_u c_0$, $\mu + \eta = 100$, $\eta = 25$, $D = 1$, $b = 0.5$, $\tau = 1$, ${f_0}^2 = 0.025$.

• *Movie 3*: Active emulsions in an asymmetric bilayer at $T < T_c$. Parameters are: $\chi = 2.08$, $W = 2$, $K = 5$, $\mathrm{Pe} = 17$, $k_u = 0.0032$, $c_0 = 0.2$, $k_{1b} = k_u c_0$, $\mu + \eta = 100$, $\eta = 25$, $D = 1$, $b = 0.5$, $\tau = 1$, ${f_0}^2 = 0.025$.

• *Movie 4*: Strongly fluctuating active emulsions in an asymmetric bilayer at $T < T_c$. Parameters are: $\chi = 2.08$, $W = 2$, $K = 10$, $\mathrm{Pe} = 25$, $k_u = 0.0032$, $c_0 = 0.2$, $k_{1b} = k_u c_0$, $\mu + \eta = 100$, $\eta = 25$, $D = 1$, $b = 0.5$, $\tau = 1$, ${f_0}^2 = 0.025$.

• *Movie 5*: Active emulsions in an asymmetric bilayer at $T > T_c$. Parameters are: $\chi = 1.92$, $W = 2$, $K = 5$, $\mathrm{Pe} = 17$, $k_u = 0.0032$, $c_0 = 0.2$, $k_{1b} = k_u c_0$, $\mu + \eta = 100$, $\eta = 25$, $D = 1$, $b = 0.5$, $\tau = 1$, ${f_0}^2 = 0.025$.

• *Movie 6*: Strongly fluctuating active emulsions in an asymmetric bilayer at $T > T_c$. Parameters are: $\chi = 1.92$, $W = 2$, $K = 10$, $\mathrm{Pe} = 25$, $k_u = 0.0032$, $c_0 = 0.2$, $k_{1b} = k_u c_0$, $\mu + \eta = 100$, $\eta = 25$, $D = 1$, $b = 0.5$, $\tau = 1$, ${f_0}^2 = 0.025$.

• *Movie 7*: Slippage in an asymmetric bilayer at $T < T_c$. Parameters are: $\chi = 2.08$, $W = 2$, $K = 10$, $\mathrm{Pe} = 25$, $k_u = 0.0032$, $c_0 = 0.2$, $k_{1b} = k_u c_0$, $\mu + \eta = 100$, $\eta = 25$, $D = 1$, $b = 0.5$, $\tau = 1$, ${f_0}^2 = 0.025$.

• *Movie 8*: Slippage in an asymmetric bilayer at $T > T_c$. Parameters are: $\chi = 1.92$, $W = 2$, $K = 10$, $\mathrm{Pe} = 25$, $k_u = 0.0032$, $c_0 = 0.2$, $k_{1b} = k_u c_0$, $\mu + \eta = 100$, $\eta = 25$, $D = 1$, $b = 0.5$, $\tau = 1$, ${f_0}^2 = 0.025$. Actomyosin concentration c has been thresholded for better visualisation.

Acknowledgements

We thank S. Saha, A. Das, C. Patra and S. Mayor for earlier collaborations and useful discussions. MR acknowledges a J. C. Bose Fellowship (JCB/2018/00030), from the Department of Science and Technology, India, and support from the Department of Atomic Energy, India (under Project No. RTI 4006). AB and MR acknowledge support from Simons Foundation Grant 287975, and from the NCBS computing facilities. AB acknowledges the valuable help provided by the Dedalus community (Google group) and the free access to the Dedalus package provided by the developers.

Notes and references

1 M. Edidin, *Nat. Rev. Mol. Cell Biol.*, 2003, **4**, 414–418.
2 D. Lingwood and K. Simons, *Science*, 2010, **327**, 46–50.
3 E. Sezgin, I. Levental, S. Mayor and C. Eggeling, *Nat. Rev. Mol. Cell Biol.*, 2017, **18**, 361–374.
4 S. J. Singer and G. L. Nicolson, *Science*, 1972, **175**, 720–731.
5 K. Jacobson, E. D. Sheets and R. Simson, *Science*, 1995, **268**, 1441–1442.
6 K. Simons and E. Ikonen, *Nature*, 1997, **387**, 569–572.
7 T. Baumgart, A. T. Hammond, P. Sengupta, S. T. Hess, D. A. Holowka, B. A. Baird and W. W. Webb, *Proc. Natl. Acad. Sci. U. S. A.*, 2007, **104**, 3165–3170.
8 B. Alberts, R. Heald, A. Johnson, D. Morgan, M. Raff, K. Roberts and P. Walter, *Molecular Biology of the Cell: Seventh International Student Edition with Registration Card*, WW Norton & Company, 2022.
9 D. Goswami, K. Gowrishankar, S. Bilgrami, S. Ghosh, R. Raghupathy, R. Chadda, R. Vishwakarma, M. Rao and S. Mayor, *Cell*, 2008, **135**, 1085–1097.
10 K. Gowrishankar, S. Ghosh, S. Saha, C. Rumamol, S. Mayor and M. Rao, *Cell*, 2012, **149**, 1353–1367.

11 A. Chaudhuri, B. Bhattacharya, K. Gowrishankar, S. Mayor and M. Rao, *Proc. Natl. Acad. Sci. U. S. A.*, 2011, **108**, 14825–14830.
12 S. A. Freeman, A. Vega, M. Riedl, R. F. Collins, P. P. Ostrowski, E. C. Woods, C. R. Bertozzi, M. I. Tammi, D. S. Lidke, P. Johnson, *et al.*, *Cell*, 2018, **172**, 305–317.
13 P. Sil, N. Mateos, S. Nath, S. Buschow, C. Manzo, K. G. Suzuki, T. Fujiwara, A. Kusumi, M. F. Garcia-Parajo and S. Mayor, *Mol. Biol. Cell*, 2020, **31**, 561–579.
14 S. Arumugam, S. Schmieder, W. Pezeshkian, U. Becken, C. Wunder, D. Chinnapen, J. H. Ipsen, A. K. Kenworthy, W. Lencer, S. Mayor, *et al.*, *Nat. Commun.*, 2021, **12**, 3675.
15 S. Saha, A. Das, C. Patra, A. A. Anilkumar, P. Sil, S. Mayor and M. Rao, *Proc. Natl. Acad. Sci. U. S. A.*, 2022, **119**, e2123056119.
16 R. Raghupathy, A. A. Anilkumar, A. Polley, P. P. Singh, M. Yadav, C. Johnson, S. Suryawanshi, V. Saikam, S. D. Sawant, A. Panda, *et al.*, *Cell*, 2015, **161**, 581–594.
17 M. F. Garcia-Parajo, A. Cambi, J. A. Torreno-Pina, N. Thompson and K. Jacobson, *J. Cell Sci.*, 2014, **127**, 4995–5005.
18 T. S. van Zanten, A. Cambi, M. Koopman, B. Joosten, C. G. Figdor and M. F. Garcia-Parajo, *Proc. Natl. Acad. Sci. U. S. A.*, 2009, **106**, 18557–18562.
19 S. J. Plowman, C. Muncke, R. G. Parton and J. F. Hancock, *Proc. Natl. Acad. Sci. U. S. A.*, 2005, **102**, 15500–15505.
20 K. Jacobson, P. Liu and B. C. Lagerholm, *Cell*, 2019, **177**, 806–819.
21 S. Saha, A. A. Anilkumar and S. Mayor, *J. Lipid Res.*, 2016, **57**, 159–175.
22 M. C. Marchetti, J.-F. Joanny, S. Ramaswamy, T. B. Liverpool, J. Prost, M. Rao and R. A. Simha, *Rev. Mod. Phys.*, 2013, **85**, 1143.
23 A. Bansal, A. Das and M. Rao, *Indian J. Phys.*, 2022, **96**, 2567–2576.
24 J. Lorent, K. R. Levental, L. Ganesan, G. Rivera-Longsworth, E. Sezgin, M. Doktorova, E. Lyman and I. Levental, *Nat. Chem. Biol.*, 2020, **16**, 644–652.
25 H. M. Hankins, R. D. Baldridge, P. Xu and T. R. Graham, *Traffic*, 2015, **16**, 35–47.
26 J. M. Kalappurakkal, A. A. Anilkumar, C. Patra, T. S. van Zanten, M. P. Sheetz and S. Mayor, *Cell*, 2019, **177**, 1738–1756.
27 R. G. Fehon, A. I. McClatchey and A. Bretscher, *Nat. Rev. Mol. Cell Biol.*, 2010, **11**, 276–287.
28 I. Meenakshi, M. Rao, S. Mayor and R. Sowdhamini, *Front. Cell Dev. Biol.*, 2023, **11**, 1168050.
29 G. J. Doherty and H. T. McMahon, *Annu. Rev. Biophys.*, 2008, **37**, 65–95.
30 T. Lecuit and A. S. Yap, *Nat. Cell Biol.*, 2015, **17**, 533–539.
31 M. Rao and S. Mayor, *Curr. Opin. Cell Biol.*, 2014, **29**, 126–132.
32 K. Gowrishankar and M. Rao, *Soft Matter*, 2016, **12**, 2040–2046.
33 K. Husain and M. Rao, *Phys. Rev. Lett.*, 2017, **118**, 078104.
34 S. R. Hossein, R. Mandal and M. Rao, *arXiv*, 2021, preprint, arXiv:2105.11358, DOI: 10.48550/arXiv.2105.11358.
35 S. R. Hossein, R. Mandal and M. Rao, *Phys. Rev. E*, 2018, **98**, 052608.
36 A. Das, A. Polley and M. Rao, *Phys. Rev. Lett.*, 2016, **116**, 068306.
37 A. G. Clark, K. Dierkes and E. K. Paluch, *Biophys. J.*, 2013, **105**, 570–580.
38 A. G. Clark, O. Wartlick, G. Salbreux and E. K. Paluch, *Curr. Biol.*, 2014, **24**, R484–R494.

39 D. V. Köster, K. Husain, E. Iljazi, A. Bhat, P. Bieling, R. D. Mullins, M. Rao and S. Mayor, *Proc. Natl. Acad. Sci. U. S. A.*, 2016, **113**, E1645–E1654.
40 A. A. Bhat, A. Das, K. Husain, M. Rao, D. V. Koester and S. Mayor, *Biophys. J.*, 2020, **118**, 165a.
41 R. Chandran, G. Kale, J.-M. Philippe, T. Lecuit and S. Mayor, *Curr. Biol.*, 2021, **31**, 1726–1736.
42 P. Sharma, R. Varma, R. Sarasij, K. Gousset, G. Krishnamoorthy, M. Rao, S. Mayor, *et al.*, *Cell*, 2004, **116**, 577–589.
43 S. Saha, I.-H. Lee, A. Polley, J. T. Groves, M. Rao and S. Mayor, *Mol. Biol. Cell*, 2015, **26**, 4033–4045.
44 P. M. Chaikin, T. C. Lubensky and T. A. Witten, *Principles of Condensed Matter Physics*, Cambridge University Press, Cambridge, 1995, vol. 10.
45 R. Shlomovitz and M. Schick, *Biophys. J.*, 2013, **105**, 1406–1413.
46 H. Tanaka, *Faraday Discuss.*, 2012, **158**, 371–406.
47 M. Doi, *J. Phys.: Condens. Matter*, 2011, **23**, 284118.
48 A. J. Bray, *Adv. Phys.*, 2002, **51**, 481–587.
49 S. Saha, J. Agudo-Canalejo and R. Golestanian, *Phys. Rev. X*, 2020, **10**, 041009.
50 S. Saha and R. Golestanian, *arXiv*, 2022, preprint, arXiv:2208.14985, DOI: 10.48550/arXiv.2208.14985.
51 G. G. Putzel, M. J. Uline, I. Szleifer and M. Schick, *Biophys. J.*, 2011, **100**, 996–1004.
52 P. W. Fowler, J. J. Williamson, M. S. Sansom and P. D. Olmsted, *J. Am. Chem. Soc.*, 2016, **138**, 11633–11642.
53 A. Polley, S. Mayor and M. Rao, *J. Chem. Phys.*, 2014, **141**, 064903.
54 P. Bassereau and P. Sens, *Physics of Biological Membranes*, Springer, 2018, vol. 12.
55 S. C. van IJzendoorn, J. Agnetti and A. Gassama-Diagne, *Biochim. Biophys. Acta, Biomembr.*, 2020, **1862**, 183145.
56 Y. Hirose, S. Komura and D. Andelman, *Phys. Rev. E:Stat., Nonlinear, Soft Matter Phys.*, 2012, **86**, 021916.
57 S. L. Veatch, P. Cicuta, P. Sengupta, A. Honerkamp-Smith, D. Holowka and B. Baird, *ACS Chem. Biol.*, 2008, **3**, 287–293.
58 M. Schick, *Phys. Rev. E:Stat., Nonlinear, Soft Matter Phys.*, 2012, **85**, 031902.
59 S. May, *Soft Matter*, 2009, **5**, 3148–3156.
60 T. Moore, S. K. Wu, M. Michael, G. A. Gomez, Z. Neufeld, *et al.*, *Biophys. J.*, 2014, **107**, 2652–2661.
61 K. J. Burns, G. M. Vasil, J. S. Oishi, D. Lecoanet and B. P. Brown, *Phys. Rev. Res.*, 2020, **2**, 023068.

Nonequilibrium asymmetry in the living cell membrane

Ajay Kumar Bansal[a] and Madan Rao*[a]

Author affiliations

* Corresponding authors

[a] Simons Centre for the Study of Living Machines, National Centre for Biological Sciences, Bangalore 560 065, India

E-mail: rao.madan@gmail.com

Fax: +91 80 2363 6662

Tel: +91 80 2366 6270

Faraday Discuss., 2025, DOI: 10.1039/D4FD00207E. ***Amendment published 24 June 2025.***

The authors regret that the reference 57 was incorrect in the originating article, the correct details are shown here.

57. M. C. Blosser, A. R. Honerkamp-Smith, T. Han, M. Haataja and S. L. Keller, *Biophysical Journal*, 2015, **109**, 2317-2327.

The Royal Society of Chemistry apologises for these errors and any consequent inconvenience to authors and readers.

Faraday Discussions

DISCUSSIONS

Plasma membrane asymmetry and lipid homeostasis: general discussion

Francisco Barerra, Manuel Bibrowski, John Conboy, Niclas Paul Decker, Markus Deserno, Rumiana Dimova, Milka Doktorova, Thais Enoki, Oded Farago, Gerald W. Feigenson, Alexander P. Fellows, Gonen Golani, Carlo Guardiani, Heiko Heerklotz, Shinako Kakuda, Neha P. Kamat, Manpreet Kaur, Sarabjeet Kaur, Michael M. Kozlov, Kandice R. Levental, Ilya Levental, Reinhard Lipowsky, Erwin London, Joseph Lorent, Edward Lyman, Madan Rao, Tyler Reagle, Raya Sorkin, Malavika Varma and Axel Voigt

DOI: 10.1039/d5fd90018b

Michael M. Kozlov opened the discussion of the introductory Spiers Memorial Lecture by Gerald W. Feigenson: You mentioned the 2D pressure in one leaflet reaches a value of 34 mN m^{-1} (https://doi.org/10.1039/d5fd00041f). This implies that the second leaflet must develop a lateral tension of about the same value. The latter is an order of magnitude larger than the rupturing tension. How do you think a membrane could stand such stresses?

Gerald W. Feigenson replied: Sorry, but I have no information on the rupturing tension for this asymmetric bilayer. What I can tell you is that the van Deenen researchers were constantly watching out for rupture during their studies.[1]

1 B. Roelfesen, *et al*, *Biochim. Biophys. Acta*, 1971, **241**, 925–929.

Ilya Levental asked: You strongly claimed that pressure in the bilayer goes up due to enzyme treatment. Has that been directly measured and, if so, how?

Gerald W. Feigenson responded: In your papers, you never refer to the van Deenen group monolayer pressure measurements, which leads me to think that you did not realize that monolayer pressure is a variable. You have published papers that could have mentioned and criticized the van Deenen studies, and perhaps I have been remiss in my reading of your work, but I do not see such comments.[1]

1 B. Roelfesen, *et al*, *Biochim. Biophys. Acta*, 1971, **241**, 925–929.

Ilya Levental added: Do you have a reference that shows PLA2 enzyme activity in a monolayer increases the monolayer pressure? I have found papers saying the opposite.

Gerald W. Feigenson answered: Many references in my paper consider various aspects of your question. This is not a matter of thinking that Lyso-PC and FA have water solubility, but rather how intact is the monolayer of (mostly) cholesterol plus Lyso-PC + FA + ceramide. As you know, especially from Felix M. Goni's talk (https://doi.org/10.1039/d5fd00091b), the van Deenen groups have published many papers on this.[1]

1 R. B. Roelofsen and J. A. F. Op den Kamp, Techniques to determine transbilayer organization and dynamics of membrane phospholipids, *Membrane Biogenesis*, Springer, Berlin Heidelberg, 1988, pp. 1–13.

Edward Lyman said: A key piece of your argument is that as the pressure in a monolayer increases lipids are digested by PLA2, *etc.*, as products that result from enzymatic digestion accumulate. But a key assumption here is that the products (*e.g.* lysolipids) remain dissolved in the bilayer, but this runs counter to what we know about the solubility in the aqueous phase of many such products. Has this assumption been tested, and if so, where do I find the data?

Gerald W. Feigenson replied: There are perhaps too many publications from the van Deenen groups to know where to look, but one excellent summary is in ref. 1. As in my earlier response to Ilya Levental, do note that this is not a question of whether Lyso-PC and FA have some water solubility, but rather how intact is a monolayer that is mostly cholesterol, together with some Lyso-PC + FA + ceramide, which was continuously monitored and found to be intact.

1 B. Roelofsen and J. A. Op den Kamp, *Membrane Biogenesis*, Springer, Berlin Heidelberg, 1988, pp. 1–13.

Reinhard Lipowsky commented: In your talk and in Fig. 8 of your paper, you emphasized that the PLA2-catalyzed cleavage of the PC lipids into Lyso-PC lipids and fatty acids increases monolayer pressure, which is equivalent to a decrease in leaflet tension. This change in leaflet tension is based on the assumption that both reaction products, Lyso-PC and fatty acid, remain in the leaflet. It was not clear to me to what extent this assumption is corroborated by experimental data. In other words: is there experimental evidence that both reaction products remain in the leaflet? Which method was used to demonstrate this feature?

Monolayer pressure or leaflet tension is determined by the average lipid density within this leaflet. Using X-ray or neutron scattering methods, one should be able to measure these densities and, thus, the leaflet tensions for stacks of weakly curved bilayers. An even simpler system is provided by a PC monolayer at the air–water interface. For such a monolayer, it should be possible to study the PLA2-catalyzed cleavage of PC into Lyso-PC and fatty acid as well and measure the resulting density change within the monolayer.

Gerald W. Feigenson responded: I have no knowledge about interpreting X-ray diffraction results for asymmetric bilayers, and I am fairly certain that the van Deenen researchers did not make such a measurement. However, they did have an active research effort to create monolayers in a Langmuir trough, and to measure monolayer pressure during PC hydrolysis by PLA2. They found that catalysis of PLA2 hydrolysis stopped at a pressure of 34.8 mN m^{-1}. I do not know, but I will try to find out, if the van Deenen researchers have any direct measures of any sort, regarding reaction products of fatty acid and Lyso-PC.

Heiko Heerklotz said: It does not seem plausible to me to assume the same number of phospholipids to be present in the outer and inner leaflet: the preference of cholesterol for outer-leaflet lipid (particularly sphingomyelin) and its ability to equilibrate between the leaflets would then cause a significant asymmetry stress (higher lateral pressure in the outer leaflet).

Gerald W. Feigenson responded: I am trying to stay close to reliable data for asymmetric bilayers, and not so much to consider what is "reasonable". Especially given the unknown concentrations of cholesterol in the two leaflets, data for the phospholipids are most important. Very unfortunate that whoever in the Lorent *et al.* study[1] who was in charge of the PLA2 work did not understand more precisely the details of the van Deenen researchers. I want to know what you want to know, and so far, the most careful work has been done more than 50 years ago!

1 J. H. Lorent, *et al.*, *Nat. Chem. Biol.*, 2020, **16**, 644–652.

Ilya Levental commented: Your introductory lecture introduces the foundational work on membrane lipid asymmetry, focusing largely on observations reported by the van Deenen group in the 1970s. That work used phospholipase treatment of intact human red blood cells to infer the transbilayer distributions of the major phospholipid (PL) classes. Those data were most notably summarized by Verkleij *et al.*,[1] reporting that:

(a) almost all sphingomyelin (SM) is present in the outer plasma membrane (PM) leaflet, evidenced by almost complete digestion of SM by external SMase.

(b) all phosphatidylserine (PS) is in the inner leaflet, as none was accessible to external phospholipase A2 (PLA2).

(c) there is a robust enrichment of phosphatidylethanolamine (PE) in the inner leaflet.

(d) these effects combine to yield approximately equivalent distribution of overall phospholipids between leaflets.

The latter two claims were based on a combined treatment of cells with PLA2 and SMase, which led to increased digestion of PE and phosphatidylcholine (PC), compared to treatment with PLA2 alone.

In the introductory lecture these foundational findings were compared to those from a recent publication by our group,[2] a major claim of which is that phospholipid abundance may not be balanced between the two PM leaflets. Specifically, our paper claims that PLs are notably (by ~70%) enriched on the cytosolic leaflet and that these imbalances can be maintained by a combination of cholesterol asymmetry and differential leaflet stresses.

The validity of these findings and conclusions was challenged on the basis of specific methodological differences between the approaches of the van Deenen group and those from our group. The core of the criticism is that the phospholipase treatments must completely digest all available target lipids; however, when PLA2 is used alone, the products of its digestion (*i.e.* lysolipids and fatty acids), build up to produce pressure in the outer leaflet, which then inhibits PLA2 activity, thus preventing it from full digestion of outer leaflet PC and PE. Adding

SMase into this reaction (*i.e.* PLA2 + SMase) is then believed to reduce leaflet pressure by generating ceramide, allowing PLA2 to complete digestion. Below, I present evidence to show that these claims are either spurious or incorrect and that our report of PL imbalance in PMs is valid.

Choices of enzymes and complete digestion: it is indeed critical that enzymes complete digestion of all available target PLs. In our published work, we show evidence that PLA2 activity against PC in erythrocytes has an asymptotic behavior with time, saturating at ~15 min, at which time point no other PLs are digested (ESI Fig. 1D and E in Lorent *et al.*[3]). We also show that when erythrocytes are disrupted by sonication, essentially all glycerophospholipids are completely digested by PLA2 (ESI Fig. 1A–E) and all SM by SMase. To further ensure that digestion was complete, we increased treatment times to 2 h, increased osmotic tension on the membrane, and used higher treatment temperatures, none of which increased digestion of PLs by PLA2. Importantly, our findings with separate PLA2 and SMase treatments are completely consistent with those of the van Deenen group and others (see Fig. 1D and ESI Table S1 in Doktorova *et al.*[2]), implying that the outer leaflet has much less than 50% (we estimate 30–40%) of PM PLs.

Disagreement with van Deenen claims/approaches: our paper and updated model of the PM then explores the consequences of this potential imbalance. The van Deenen group appeared to believe that such an imbalance was impossible: "it is highly unlikely that the outer monolayer would contain only this relatively small amount (of phospholipids), as such a distribution would create a large gap with respect to the inner monolayer".[4] This assumption was incorrect, as it has become clear over the past several years that bilayers can indeed bear significant imbalances through cholesterol asymmetry and differential stress.[5–8] However, following on their incorrect assumption, the van Deenen group sought combinations of enzymes that would achieve 50% PL hydrolysis.

In all experiments of this type, it is essential that the PM remains intact, such that only outer leaflet lipids are accessible to the lipases. The van Deenen group claims that a sequential treatment of PLA2 followed by SMase allows digestion of only the outer leaflet, keeping the cells intact. Despite extensive efforts, we could not reproduce these observations under any condition: in our experiments, erythrocytes broke and leaked hemoglobin in any combination of PLA2 and SMase. It is also important to point out that hemoglobin is a large molecule (~65 kDa), and lack of frank hemolysis does not imply that the PM is intact. Indeed, we measured that erythrocytes become completely permeable to a 3 kD fluorescent dextran only after the combined sequential treatment with PLA2 then SMase (Fig. 1). The explanation for this leakage is simple: a bilayer where one leaflet is composed of ceramide, fatty acids, and lysolipids is unstable.

Two other significant issues with the van Deenen claims:

(a) in their experiments and ours, PLA2 treatment alone cleaves the majority of PC but does not access any PE,[1] despite PE being a preferred substrate for PLA2.[9] We (and others) also see no staining with a PE probe (duramycin) in intact cells (https://doi.org/10.103d/d4fd00205a). Further, Bretscher reported essentially no labelling of either PS or PE by an amine-reactive probe in intact erythrocytes.[10] These findings strongly imply that there is no PE on the outer leaflet prior to addition of SMase, and that PE only becomes accessible to PLA2 after SMase treatment because the PMs are broken, leaky, and/or scrambled.

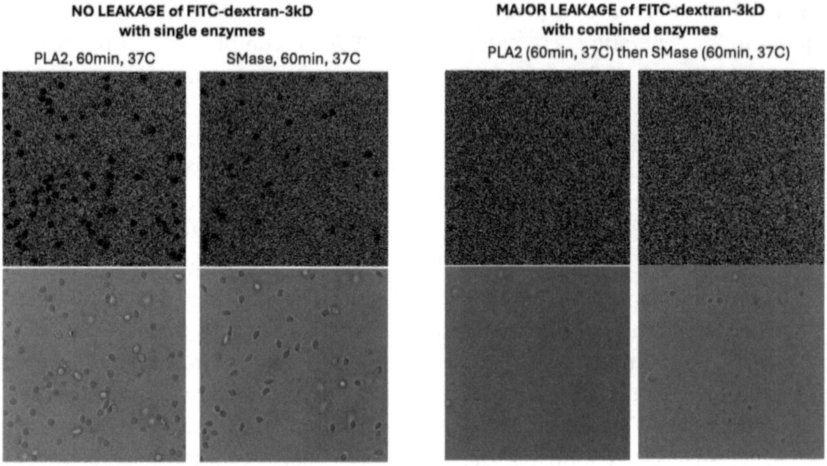

Fig. 1 (Right) Red blood cells become permeable to 3 kD FITC-labeled dextran after sequential PLA2 then SMase treatment, in contrast to either enzyme alone (left). This finding suggests that bilayers are destabilized by the enzyme combination.

(b) the combinations of PLA2 and SMase are reported to have strange, unexplained path dependence: PLA2 followed by SMase is supposedly non-lytic, while SMase followed by PLA2 produces complete hemolysis.[1] In our extensive experiments, a simpler conclusion emerges: combinations of SMase and PLA2 are lytic because they produce lipid combinations that are unable to maintain intact bilayers.

Claim for accumulated monolayer pressure inhibiting PLA2 activity: this claim is central to the criticism of our findings, as it underlies the supposed requirement of combining PLA2 and SMase, which is the only treatment that produces ~50% PL digestion. First, this claim—for which no references have been provided—is a conjecture, as monolayer pressure has not been directly measured in any bilayer, much less the PM of a living cell or under enzyme treatments. Second, there is significant evidence that monolayer pressure (in a Langmuir trough) is reduced by PLA2 activity[11] likely due to desorption of the relatively soluble reaction products (lysolipids) from the membrane. Similarly, it has been reported in several studies that when "lysolipids and fatty acids… accumulate in the lipid membrane, the hydrolysis rate of PLA2 increases by several orders of magnitude".[12] Third, if the claim were true, removal of the reaction products should reduce monolayer pressure and re-activate PLA2. However, papers from the van Deenen group reported that this does not occur. Instead they reported that "combination of PLA2 treatment with the removal of split products using serum albumin… (was) equally inefficient in that no further hydrolysis of phospholipids was obtained".[4]

In conclusion:

(a) we tested enzyme conditions to ensure complete digestion with no loss of bilayer integrity.

(b) combinations of SMase and PLA2 likely destabilize bilayers leading to digestion of inner leaflet PLs.

(c) there is no evidence for the conjecture that PLA2 activity is inhibited by the products of its reaction, rather there is abundant evidence to the contrary.

Thus, our claims of PL abundance imbalances in erythrocyte PMs are plausible and accurate, according to our data[2,3] and available evidence in literature.

1 A. J. Verkleij, *et al.*, The asymmetric distribution of phospholipids in the human red cell membrane. A combined study using phospholipases and freeze-etch electron microscopy, *Biochim. Biophys. Acta*, 1973, **323**, 178–193.
2 M. Doktorova, *et al.*, Cell membranes sustain phospholipid imbalance *via* cholesterol asymmetry, *Cell*, 2025, **188**, 2586–2602.
3 J. H. Lorent, *et al.*, Plasma membranes are asymmetric in lipid unsaturation, packing and protein shape, *Nat. Chem. Biol.*, 2020, **16**, 644–652, PMC7246138.
4 A. Rawyler, P. H. van der Schaft, B. Roelofsen and J. A. Op den Kamp, Phospholipid localization in the plasma membrane of friend erythroleukemic cells and mouse erythrocytes, *Biochemistry*, 1985, **24**, 1777–1783.
5 S. L. Foley, M. Varma, A. Hossein and M. Deserno, Elastic and thermodynamic consequences of lipid membrane asymmetry, *Emerging Top. Life Sci.*, 2023, **7**, 95–110.
6 M. Varma and M. Deserno, Distribution of cholesterol in asymmetric membranes driven by composition and differential stress, *Biophys. J.*, 2022, **121**, 4001–4018, PMC9674969.
7 T. Reagle, Y. Xie, Z. Li, W. Carnero and T. Baumgart, Methyl-beta-cyclodextrin asymmetrically extracts phospholipid from bilayers, granting tunable control over differential stress in lipid vesicles, *Soft Matter*, 2024, **20**, 4291–4307, PMC11135146.
8 M. P. K. Frewein, *et al.*, Distributing aminophospholipids asymmetrically across leaflets causes anomalous membrane stiffening, *Biophys. J.*, 2023, **122**, 2445–2455.
9 H. M. Verheij, A. J. Slotboom and G. H. de Haas, Structure and function of phospholipase A2, *Rev. Physiol., Biochem. Pharmacol.*, 1981, **91**, 91–203.
10 M. S. Bretscher, Asymmetrical lipid bilayer structure for biological membranes, *Nature*, 1972, **236**, 11–12.
11 X. Zhai, *et al.*, Direct observations of the cleavage reaction of an L-DPPC monolayer catalyzed by phospholipase A2 and inhibited by an indole inhibitor at the air/water interface, *ChemBioChem*, 2003, **4**, 299–305.
12 P. Hoyrup, J. Davidsen and K. Jorgensen, Lipid Membrane Partitioning of Lysolipids and Fatty Acids: Effects of Membrane Phase Structure and Detergent Chain Length, *J. Phys. Chem. B*, 2001, **105**, 2649–2657.

Reinhard Lipowsky addressed Gerald W. Feigenson and Ilya Levental: Gerald W. Feigenson and Ilya Levental express two contradictory opinions about the enzymatic cleavage of the PC lipids within one bilayer leaflet into Lyso-PC lipids and fatty acids. Gerald W. Feigenson believes that the two reaction products remain in the bilayer leaflet and increase the pressure in this leaflet, whereas Ilya Levental thinks that one or both reaction products are released from the leaflet into the surrounding aqueous solution. However, in both cases, the experimental evidence for these contradictory opinions remains unclear. So, which experiments support the view that the two reaction products stay in the bilayer leaflet and which experiments corroborate the opposite view that the reaction products are released in the surrounding solution?

Gerald W. Feigenson communicated in reply: Ref. 1 is one of the monolayer pressure studies on PLA2 action. Given that there are no direct ways to measure this pressure in an intact asymmetric bilayer, this paper addresses the questions of Ilya Levental about PLA2 behaviors, and by the controlled composition, addresses your question about PLA2 reaction products remaining in the asymmetric bilayer.

Ref. 2, specifically on p. 92, states that the exoplasmic leaflet is intact, in contrast to the claim that, being composed of Lyso-PC, FA, ceramide, and cholesterol, that it "must" fall apart.

1 R. A. Demel, *et al.*, *Biochim. Biophys. Acta*, 1975, **406**, 97–107.
2 R. F. Zwaal, *et al.*, *Biochim. Biophys. Acta*, 1975, **406**, 83–96.

Ilya Levental communicated in reply: Thanks for this important question Reinhard Lipowsky. Gerald W. Feigenson is correct that PLA2 activity is inhibited by high surface pressures (in monolayers). However, there is significant published evidence that PLA2 activity reduces surface pressure in monolayers and that lysolipids and FAs increase PLA2 activity in bilayers. I am unaware of any published evidence to the contrary.

Also, I am almost sure that there is no experimental evidence that PLA2 activity increases leaflet pressure in bilayers, mostly because I'm unaware of a method to directly measure this parameter (especially in cell membranes).

Raya Sorkin opened a general discussion of the paper by Erwin London: Regarding the Pro12A results with synthetic vesicles, Table 2 in your paper (https://doi.org/10.1039/d4fd00211c): when adding ionophore, for some compositions, you observe an increase in GP signal, but a decrease for others. Can you comment on this difference?

Erwin London answered: The effect of the ionophore may be a combination of its effect upon lipid packing, and its interaction with the fluorescent probe. The consistent increase in GP seen with Laurdan is suggestive of an ionophore effect on these parameters that is consistent for almost all lipid compositions. If an increase is indicative of interaction between probe and ionophore, in the case of Pro12A, interaction with the ionophore may be stronger in membranes with cholesterol. The decrease in the absence of cholesterol may reflect a lack of interaction with the ionophore along with a decrease in lipid packing.

Reinhard Lipowsky said: In your talk, you studied the effect of compositional asymmetry on the phase separation and domain formation within the individual leaflets of the bilayer. Let us label the two leaflets as 1 and 2. The lipids in leaflet 1 interact with the other lipids in leaflet 1 and with the lipids in leaflet 2. The latter interactions provide an external potential for the lipids in leaflet 1, which shifts the chemical potentials of these lipids and changes the two-phase coexistence region within leaflet 1.

Analogous shifts of chemical potentials and coexistence regions have been studied for lipid membranes exposed to adhesive surfaces or environments.[1,2] If the membrane is exposed to several such environments, the membrane is partitioned into several segments. One consequence of this ambience-induced segmentation is that phase separation is limited to one of these segments, which provides a generic mechanism for the suppression of phase separation and domain formation in cellular membranes, see Fig. 6 of ref. 2.

1 R. Lipowsky, *et al.*, *Soft Matter*, 2013, **9**, 8438, DOI: 10.1039/c3sm50712b.
2 R. Lipowsky, *Biol. Chem.*, 2014, **395**, 253, DOI: 10.1515/hsz-2013-0244.

Edward Lyman asked: I'm trying to wrap my head around the kinetics of scrambling. There are some numbers reported for rates of TMEM16F catalyzed flip-flop (about 10 000 lipids per second), but to build a kinetic model we also need additional information, like what are the copy numbers/expression levels of TMEM16F, and are they distributed uniformly?

Erwin London answered: These are important issues, but I do not know the expression level of scramblase proteins or how they are distributed within the plasma membrane.

Edward Lyman remarked: Is it known whether scrambling ever Goes to Completion? Or is it always a partial release of asymmetry?

Erwin London answered: We have not seen scrambling go to completion in our experiments. However, they involve scrambling of probes and a labelled lipid. What is needed for natural lipid scrambling is a reliable and fully quantitative asymmetry assay, and that is still not a fully solved problem. Our group is trying to determine how far we can progress towards a solution of this, using lipid exchange using cyclodextrins.

Edward Lyman added: I really think we need better/more ways to quantify lipid flip-flop in cells. Dithionite quenching has well known problems – requires fluorescent lipids, the reagent can diffuse across membranes. In particular, probe-free methods with improved resolution (in both space and time) have the potential to really advance the field.

Reinhard Lipowsky answered: It is important to note that the free energy barriers for lipid flip-flops depend strongly on the mechanical tensions, Sigma_1 and Sigma_2, within the two bilayer leaflets. For tensionless bilayers with Sigma_1 + Sigma_2 = 0, these free energy barriers depend on the stress asymmetry, Sigma_1 − Sigma_2, between the two leaflets. The associated flip-flop rates are displayed in Fig. 3 of ref. 1 for planar bilayers and in Fig. 6 of ref. 1 for nanovesicles (or small unilamellar vesicles). These leaflet tensions are difficult to determine and thus, to control in all-atom simulations. Therefore, the values of the leaflet tensions are usually not reported in those publications that use all-atom simulations to study lipid flip-flops, and the leaflet tensions then represent hidden variables with unknown numerical values. However, in order to scrutinize and elucidate the flip-flop rates as obtained from any molecular dynamics simulation, one should always include some estimates for the two leaflet tensions of the lipid bilayers.

1 A. Sreekumari and R. Lipowsky, *Soft Matter*, 2022, **18**, 6066, DOI: 10.1039/d2sm00618a.

Neha P. Kamat added: One way to monitor flip-flop of lipids is to use pH sensitive dyes. This is particularly useful in cases where the lipids flipping are negatively charged, in which case the protonated, non-polar lipids will be more predisposed to flip across the bilayer and release a proton that is detectable *via* a pH sensitive probe.

Milka Doktorova commented: Ionophores are toxic and kill cells, especially at higher concentrations. The plasma membrane of dead cells would be permeable to externally added fluorophores and that would interfere with many of the measurements. Assays for outer leaflet confinement of probes would also be compromised. Have you monitored for the presence of dead cells in the measured samples and verified that they do not affect the interpretation of the results?

Shinako Kakuda's response with: Yes, the ionophores can be toxic. Under our experimental conditions, using 8 micro M of 4-bromo A23187 at room temperature for 6 min, the viability of HEK 293T cells is approximately 80% as compared to 90-95% for untreated cells (as determined by Trypan Blue staining).

Erwin London added: The apparent FRET changes were seen within 2 minutes of ionophore addition. So those changes are unlikely to reflect cell death. 6NBD-PC became more inaccessible to extraction and inaccessible to dithionite reduction after ionophore treatment. This is not expected if there was increased membrane permeability upon ionophore treatment due to cell death. In addition, it is not clear that the behavior of cells that become permeable to Trypan Blue over very short times is not relevant to direct ionophore effects.

Heiko Heerklotz said: You showed that the outer leaflet of the unperturbed plasma membrane is completely or mostly in the ordered phase. Local perturbation (such as scrambling) may then give rise to the appearance or growth of liquid "lakes" in this ordered matrix. Given the high cholesterol content, this is very convincing. At the same time, it seems to contradict the traditional concept that so-called rafts are small ordered "islands" in a sea of disordered phase that act by co-accumulating "raft-associated" (*i.e.*, order-preferring) proteins to enhance their local concentration and speed up signalling between these proteins. Doesn't this mean we have to reconsider the very basic idea of raft function?

Erwin London answered: I think the idea that L_o domains could be the predominant phase in the outer leaflet, and that L_d domains could exist as lakes, has long been considered a possibility. Domain formation could still result in enhanced interactions when proteins segregate into the same type of domains, and inhibit interaction when they segregate into different domains. Plus, these models may be too simplistic. Certainly in the immune system strong clustering of L_o components during activation has been observed, although this probably involves the inner and outer leaflets.

Kandice R. Levental commented: You mention that the acceptor molecule in your FRET assay can have some fraction in the inner leaflet. We've seen that lipophilic dyes that access the cytosolic leaflet readily diffuse or are transported to other intracellular organellar membranes (*e.g.* ER). Does this confound your measurements, *e.g.* if a subset of the probe can desorb to the inner leaflet and therefore is lost to FRET? Is this seen your experiments?

Erwin London replied: The measurements are confounded if there is a change in the amount of acceptor in the outer leaflet upon whatever treatment is being studied. That is, if a probe scrambles to the inner leaflet, that is sufficient to reduce FRET by increasing the average distance between donor and acceptor. The reduction in FRET would be even more marked when probes move to internal membranes, which as you mention, often happens when a probe migrates from the inner leaflet to other organelles. In the experiments we carried out, the fraction of FRET acceptor probe in the outer leaflet did not change upon addition of scramblase activator (BrA23187). This suggests that this particular probe was not scrambled. Nor was the FRET donor scrambled. Other fluorescent probes were partly scrambled (*e.g.* Pro12A). This complicates interpretation of the fluorescence changes upon addition of scramblase inhibitor, as we noted, but we did not use them in the FRET assays. Nevertheless, as we pointed out in the talk and report, scramblase activator interactions with FRET probes already make interpretation in terms of domain changes, ambiguous at best.

Milka Doktorova queried: The protocol for preparation of large unilamellar vesicles in the paper consists only of hydration of the lipid film followed by freeze–thaw cycles. Without extrusion, this procedure has been shown to produce multilamellar liposomes (MLVs) with various techniques (*e.g.* Fig. 1 in ref. 1). The vesicles are also expected to be large and heterogeneous in size. Do you have data to support the unilamellarity and homogeneous size distribution of the liposomes that you prepared? It seems that MLVs are not going to affect the specific measurements in this study, however referring to them as LUVs (large unilamellar vesicles, a terminology typically employed for extruded liposomes with uniform size) can be misleading.

1 H. L. Scott, *et al.*, *Biophys. J.*, 2019, **117**, 1381.

Erwin London answered: With cholesterol containing membranes, extrusion has been shown to decrease cholesterol in membranes by publications from two different labs. When using vesicles containing cholesterol it should be avoided if unnecessary for the experiment being carried out, as in the studies we carried out. In my opinion the question of multilamellarity and size after freeze–thaw would need to be examined in many compositions and lipid concentrations before being considered a general finding. If more information is needed, dynamic light scattering is a facile method to establish average size and size distribution. An easy way to roughly estimate unilamellarity would be to include NBD-PE in vesicles and measure the extent of dithionite reduction. In any case, it would have been better to simply refer to the vesicles we used as large freeze–thaw vesicles.

Joseph Lorent asked: Might the ionophore provoking calcium influx not also activate cytoskeleton digesting enzymes at the investigated timescales and thereby increase the chances of domain formation, since the lateral restriction of the cytoskeleton is lost?

Erwin London answered: This is certainly possible. We did not investigate effects of calcium influx on the cytoskeleton. In fact, there is a large number of biological processes altered by calcium influx that have the potential to influence domain formation.

Alexander P. Fellows commented: The term "lipid order" is often used when discussing lipid membranes and their characterisation with GP measurements. When one thinks of "order", one naturally links it to entropy, but there are many different possible forms of order in such lipid systems. For example, one can think entirely intramolecularly in terms of the conformational order of the lipids (*i.e.*, the proportion of gauche defects or extent to which the tails are 'all-trans'). Alternatively, the order can be defined based on the width of the orientational distribution function of the ensemble of lipids, thus being more associated with relative crystallinity in the leaflet. This is then clearly linked to the length scale of orientational correlations within the molecular packing structure. Given that GP measurements are based on the ratio of the fluorescence intensities of a specific fluorescent tag that is intermolecularly interacting with the lipids in the membrane, to what extent do the different types of lipid order influence the observed values? In other words, what do GP measurements actually tell you about the complexities of the lipid order in the membrane?

Erwin London answered: GP is believed to measure how closely acyl chain tails of lipids are packed and the degree to which water is excluded. *i.e.* lack of local hydration. It is a very crude parameter, that has little if any orientation information and is used because GP is different in the different physical states of lipid bilayers. Low in the L_d and high in L_o and gel states.

Thais Enoki communicated: If I understand it correctly, your model suggests an asymmetric bilayer with one leaflet in the L_o phase and the other in the L_d phase. Could you provide some insights regarding the interactions occurring in the bilayer midplane? Do you expect the coupling between these leaflets to be negligible? Previous results indicate that an asymmetric membrane, with one leaflet containing domains (coexistence of two phases) and the other without domains (one phase), may either induce the formation of domains or suppress them, depending on the lipid composition.[1–4] These findings suggest that coupling impacts asymmetric bilayers.

In your previous papers (*e.g.* ref. 5), the discussions suggest models with strong coupling with ordered leaflet dominant, strong coupling with disordered leaflet dominant, strong coupling without a dominant leaflet, and not coupled. Taken all together, what are the expectations of how the L_o leaflet (rich in cholesterol and tightly packed) influences the L_d leaflet, and *vice versa*?

1 Q. Lin and E. London, Ordered raft domains induced by outer leaflet sphingomyelin in cholesterol-rich asymmetric vesicles, *Biophys. J.*, 2015, **108**, 2212–2222.
2 Q. Wang and E. London, Lipid Structure and Composition Control Consequences of Interleaflet Coupling in Asymmetric Vesicles, *Biophys. J.*, 2018, **115**, 664–678.
3 G. W. Feigenson, J. Huang and T. A. Enoki, An Unexpected Driving Force for Lipid Order Appears in Asymmetric Lipid Bilayers, *J. Am. Chem. Soc.*, 2023, **145**, 21717–21722.
4 T. A. Enoki and F. A. Heberle, Experimentally determined leaflet–leaflet phase diagram of an asymmetric lipid bilayer, *Proc. Natl. Acad. Sci. U. S. A.*, 2023, **120**, e2308723120.
5 E. London, Membrane Structure–Function Insights from Asymmetric Lipid Vesicles, *Acc. Chem. Res.*, 2019, **52**, 2382–2391.

Erwin London communicated in reply: Overall, as you note, there are several scenarios that have been observed, and when the exofacial (outer) leaflet has a composition that by itself in a symmetric bilayer would form ordered domains, or partly form ordered domains, and the cytofacial (inner) leaflet having a lipid composition that by itself would form a disordered state in a symmetric bilayer. One scenario is that ordered domains would form in both leaflets, which means due to interleaflet interactions the outer leaflet forced the inner leaflet to form ordered domains; the second is there would be no ordered domains in either leaflet, meaning that the disordered leaflet prevented the outer leaflet from forming ordered domains; and the third is that the outer leaflet would form ordered domains, and the inner leaflet would not. In other words, in this latter case the energy of interaction between the leaflets is insufficient to alter the physical state of either leaflets. All of these scenarios have been observed, and which one is observed depends upon lipid composition and the fraction of a leaflet that would form ordered domains. Therefore, the interactions between the leaflets can be strong enough to alter the physical state of a leaflet under some conditions. Models to explain the strength of interactions include acyl chain interdigitation, in which the acyl chain tails of one leaflet penetrate into the other leaflet, and the level of packing at the bilayer midplane. If I understand correctly, and there are other groups who have considered this issue much more carefully than I have, whether these interactions are sufficient to alter physical state/domain formation in one leaflet, involves their energetic cost or benefit relative to the overall cost of forming an ordered domain or preventing the formation of an ordered domain in a leaflet that would ordinarily not form, or form, an ordered domain, respectively. It has been proposed that this is related to the energetic cost of forming a boundary between the ordered and disordered domain in a leaflet relative to the cost of having a boundary between ordered and disordered physical states at the bilayer midplane.

Manpreet Kaur opened a general discussion of the paper by Kandice R. Levental: In your data, it is shown that PS scrambling is calcium dependent (https://doi.org/10.1039/d4fd00205a). Upon activation of T cells, specified regions exhibit Ca^{2+} dependent PS exposure. I would assume other lipids like phosphoinositides (PI) and PE would also be exposed? Is this supported by your data?

Kandice R. Levental replied: In mast cells, we see PE is also exposed at the same foci as PS. We have not yet looked at this in T cells. Additionally, we have not looked at PI in either cell type.

Manpreet Kaur asked: I would expect, sphingomyelin (SM) would flip to the inner leaflet on Ca^{2+} induced scrambling. On the inner leaflet, SM usually break-down to ceramide in the presence of SM-ase. If this is the case, would the lipid composition of the bilayer change when the cell is becoming active? What data support or oppose this in your study?

Kandice R. Levental replied: We are actively pursuing this direction. At this point, we have no evidence that SM flips to the inner leaflet upon Ca^{2+} induced scrambling. While it is often assumed that scrambling is bidirectional and fully non-selective, there is very little evidence supporting this assumption and almost none in cells.

However, while there may be no major net change to bilayer composition, the two leaflets are changing both in lipid headgroups and acyl chain composition. Upon scrambling, PS and PE become exposed on the exoplasmic leaflet, changing both the lipid headgroup and acyl chain composition of the two leaflets.

Francisco Barerra said: My question is regarding the reason why PS and PE lipids gets exposed in puncta; in some cases the puncta overlap with the localization of the IgE receptor, which would respond to receptor activation. However, many other puncta seem to have a different origin. Is it possible that those other puncta appear because scramblases are locally enriched?

Kandice R. Levental replied: Although there is occasional colocalization of PS/PE-positive foci with the IgE receptor, we do not believe this is specific co-enrichment, rather these are just random dots that happen to be close to each other. We are actively looking into the question of whether local enrichment of scramblases template the lateral localization of the foci, this seems a sensible and interesting hypothesis.

John Conboy commented: In addition to the punctate structures associated with PS exposure, did you observed positive membrane deformations (or budding) as a result of immune receptor-mediated activation of the mast cells? If so, what do you think the underlying mechanism of the budding is caused by?

Kandice R. Levental replied: We occasionally do see membrane deformations at the sites of larger foci (which are typically induced by calcium ionophores rather than cell activation). Occasionally these bend inward (*i.e.* endocytic), but generally the PS^+ puncta seem to bend outward (like small blebs). We do not know what causes the directionality nor the purpose of this bending (if there is one). That said, we and others have shown that PS scrambling is critical for outward membrane blebbing in the context of Giant Plasma Membrane Vesicles. We hypothesized that membrane scrambling 'softens' the bilayer, allowing intracellular forces (*e.g.* osmotic pressure) to push the plasma membrane outward.[1]

1 H.-Y. Wang, *et al.*, *Proc. Natl. Acad. Sci. U. S. A.*, 2025, 122, e2417145122, DOI: 10.1073/pnas.2417145122.

Rumiana Dimova said: You've shown that calcium influx is crucial for triggering localized scrambling. Given that localized lipid scrambling can alter the charge distribution across the membrane, have you considered how changes in membrane potential, or conversely, how pre-existing membrane potential gradients, might influence the spatial localization or kinetics of scrambling? This is particularly relevant in light of Sylvie Roke's presentation at this meeting, which highlighted the presence of potential charge 'hotspots' on cell membranes (https://doi.org/10.1039/d4fd00197d).

Furthermore, what are the key factors that determine the size and lifetime of these localized scrambled foci?

Kandice R. Levental responded: This is a great question and one that we are actively pursuing. Similar to a previous question in this discussion, we believe there are likely multiple mechanisms that determine the size and lifetime, that they're likely related to the (yet unknown) mechanism of why these scrambled lipids remain local. Likely possibilities include some combination of a (1) diffusion barrier, (2) the relative activities and rates of flippases/floppases/scramblases, (3) lipid-driven lateral organization *via* lipid demixing, and (4) induced curvature that may create a diffusion barrier of its own (or even a separate vesicle). The lifetime of these foci would then be regulated by the competing activity of the flippases/scramblases and the diffusion of the lipids within and outside of these foci.

We have not thought much about how potential gradients and inhomogeneities may regulate scrambling. We tend to think of changes in membrane potential as a consequence of scrambling (*i.e.*, loss of PS from the inner leaflet) rather than a regulator. But it is also likely that the cytosolic calcium that activates scrambling may be highly localized.

Tyler Reagle suggested: There are two hypotheses that could explain how these puncta of scrambled lipids remain local: that is, a diffusion barrier (such as from tight-binding of the revealed lipids or physical-barriers) or from active flippases/floppases outside of the puncta. Which hypothesis do you favor?

Kandice R. Levental replied: I would add two more possibilities to explain how these puncta of scrambled lipids remain local: (1) lipid-driven lateral organization *via* lipid demixing and (2) induced curvature (*i.e.* bending of the membrane somehow associated with local scrambling, that may create a diffusion barrier of its own, or even a separate vesicle). With the information we currently have, it is very difficult to 'pick' which is more important, but I would also guess that its very likely that more than one of these mechanisms are at play.

Erwin London commented: Controls to confirm Di4 is not flipped upon activation of lipid scrambling would be worthwhile.

Kandice R. Levental answered: Di4 scrambling appears to be temporally regulated. At the early stages of scrambling, we do not see Di4 being flipped across the bilayer. At later stages of ionomycin-induced scrambling, however, we do see Di4 on the cytosolic leaflet, as evidenced by its subsequent staining of internal organelle membranes.

Erwin London added: Di4 very likely partitions strongly into L_d domains, do you have any ideas for lifetime probes that would partition into L_o domains? These could provide additional information about the changes in bilayer physical properties at the site at which asymmetry is lost locally.

Kandice R. Levental replied: We should be a bit careful about 'partitioning' because the amount of signal (*i.e.*, number of photons) that we measure on a microscope, will be a function of both chemical partitioning (*i.e.*, where the probe is) and its quantum yield in that environment (*i.e.*, how much it emits). Di4 does report on L_o phases reasonably well.[1] We believe that FLIPPER-TR has a higher quantum yield in L_o domains compared to L_d domains, so it tends to report more on the L_o phase. We are developing approaches to covalently attach fluorescent probes to proteins that may allow us to target them to specific domains.

1 M. Amaro, *et al.*, *J. Phys. D: Appl. Phys.*, 2017, **50**, 134004, DOI: 10.1088/1361-6463/aa5dbc.

Sarabjeet Kaur asked: In what way does calcium influence membrane lipid scrambling? Have you explored whether other cations, such as sodium or potassium, can also trigger the membrane lipid scrambling process, and how their effects compare to that of calcium?

Kandice R. Levental answered: The major lipid scramblases are of the TMEM16 family which are calcium-activated ion channels and lipid scramblases. We hypothesize that downstream of receptor activation, IP3 production leads to calcium from ER stores, activates calcium-dependent calcium entry into the cell which activates TMEM16F to scramble the plasma membrane.[1]

1 J. Suzuki, *et al.*, *Nature*, 2010, **468**, 834–838, DOI: 10.1038/nature09583.

Gonen Golani queried: Are the scrambled domains fluid? Do they undergo coalescence or break over time?

Kandice R. Levental answered: We believe the scrambled domains are fluid as their biophysical properties (*e.g.* Di4 lifetime and Pro12A GP) reflect liquid-disordered structures. However, the domains appear mostly stable in that they're not moving laterally along the plasma membrane. Very occasionally we observe two domains that originate near each other, merging, though again this is diffraction limited, so they may just be too close to resolve but not merged.

Raya Sorkin asked: Can you speculate on the functional role of the observed puncta? Also, is there a critical PS concentration differentiating reversible *vs.* irreversible scrambling?

Kandice R. Levental answered: These are great questions to which we do not have very good answers.

First question: the functional effects of disrupting the major scramblase TMEM16F in T cells have been studied by others, and the data are confusing and contradictory. In some cases, it seems scrambling is essential to propagate T cell receptor activation. In other cases, scrambling 'limits' signalling to avoid exhaustion. We have no convincing evidence of our own in either direction, so anything I would add would be speculation. We are actively pursuing this line of research.

Second question: we also do not know much about this. Another version of this question is: if there is a concentration dependence that differentiates reversible and irreversible scrambling, is it a local concentration (*i.e.* the concentration of PS in the puncta) or the total amount of PS scrambled across the entirety of the plasma membranes (*i.e.* percent of the plasma membrane covered by PS)? Partly these questions are difficult to answer because our only toggles for asymmetry are (a) intracellular calcium *via* ionophores and (b) cell signalling activation. Both of these are extremely pleiotropic, making them difficult to control and interpret.

Malavika Varma suggested: Could looking at the timescale of formation and dissolution of these dots give an indication of whether the observed scrambling is signalling induced or related to another event such as apoptosis?

Kandice R. Levental replied: The timing of the PS^+ puncta following immune cell activation strongly suggest that this process is signalling induced downstream of receptor engagement and crosslinking.

Malavika Varma said: Is there a therapeutic angle to localized lipid scrambling? For example, could this be utilized in drug delivery to specific targets?

Kandice R. Levental answered: This is an interesting question. That would be great, but nothing specific comes to mind. One could perhaps target activated immune cells by their exposed PS. But this targeting isn't likely to be very specific, as other cells in the body expose PS for other reasons (*e.g.* cell–cell fusion). It is perhaps worth noting that cancer cells are thought to have more PS exposed than normal ones,[1] and there have been efforts to use this for targeting or detection.

1 B. Sharma and S. S. Kanwar, *Semin Cancer Biol.*, 2018, **52**, 17–25, DOI: 10.1016/j.semcancer.2017.08.012.

Reinhard Lipowsky remarked: In your talk, you emphasized several times that your optical images of the localized structures are diffraction-limited. As a consequence, your optical images are presumably limited to structures with linear dimensions that exceed about 300 nm. Do you have any estimate (or educated guess) how small the observed diffraction-limited structures really are? Have you tried to obtain a better resolution by applying some super resolution microscopy method such as STED?

Kandice R. Levental replied: We are working on using various super-resolution techniques (Minflux, STED, TEM) to better understand the ultrastructure of the scrambled foci. From TEM, we have observed PS^+ puncta that are on the order of 50 nm, but to be honest, this was in a single image, so may have been an artefact. At this point, we have no clue!

Markus Deserno said: We are all clearly wondering why the small puncta of scrambled domains do not diffusively spread into the surrounding plasma membrane. I understand we can see them in the first place because fluorescently labelled AnnexinV binds to scrambled PS lipids. It is not too hard to come up with plausible scenarios that would explain this—say, some diffusion barriers, whose construction might even be coordinated with the scrambling event itself—but I think your assay could not tell us. Once the puncta vanish, I presume that means the Annexins unbind because they have no PS to hold onto anymore. But what now happens to the lipids in that region? Do they remain confined by some diffusion barrier that may or may not be related to the initial scrambling event, or were there some temporary barriers in place that vanish after restoring asymmetry, so that lipids can now diffusively disperse? To tell these alternatives apart it seems to me we would need to label natively exoplasmic lipids in that domain and follow where they go after the Annexin fluorescence disappears. Is something like that even possible, or do I have an entirely wrong picture in my mind? I believe this would matter for the question whether immune cells have some specialized local "signalling regions" to which scrambling is confined and which have some additional meso-scale infrastructure in place that would then also locally suppress diffusion.

Kandice R. Levental answered: This is a big question. There's a small caveat worth correcting in the way we do the experiment: in the temporal experiments where puncta "disappear", AnxV is added at the particular time point we're assaying. So we are not watching a given cell and seeing AnxV unbind. Rather, AnxV does not bind to cells if added a few hours after activation. This was done intentionally to avoid the possibility of AnxV interfering with the native PS dynamics (*i.e.* 'holding on' to PS on the outer leaflet).

So where do the puncta go? I think your intuition is correct that it depends on the mechanism of what was holding them there in the first place. Our guess is either: (1) flippases finally 'win' after scramblases are turned off and slowly remove PE/PS from the outer leaflet and/or; (2) there's some trafficking event (endocytosis/ectocytosis) that removes the 'symmetrized patch'.

Manuel Bibrowski commented: Ca^{2+} can cause crosslinking and clustering of lipids. Did you have controls without Ca^{2+} in the buffer?

Kandice R. Levental responded: Ca^{2+} is necessary for the scrambling to occur as we believe these are downstream of calcium-activated scramblases. Calcium-mediated clustering of lipids likely doesn't explain the foci in activated cells as we don't see foci when the cells are fully scrambled and in the presence of calcium.

Carlo Guardiani opened a general discussion of the paper by Madan Rao: I am interested in the interaction between phosphatidyl-serin in the upper leaflet of the membrane and the saturated lipids of the upper leaflet (https://doi.org/10.1039/d4fd00207e). I would like to know whether this interaction can only be triggered by the contact with cytoskeletal elements or if it can also occur independently (possibly at a more basal level and with a decreased rate).

Madan Rao responded: I take it that you are referring to the cellular origins of the force asymmetry experienced by the inner leaflet compared to the outer leaflet. Indeed, apart from the coupling of the actomyosin cortex to the inner leaflet, there are other sources of force asymmetry, such as, (i) the differential forces arising from pH, ionic, and chemical differences in the solvents on either side of the bilayer; (ii) forces arising from the interaction of the outer leaflet with a dense extracellular matrix, glycans *etc.*; (iii) chemical forces arising from a variety of enzymatic and signalling reactions at the inner leaflet, such as enzymatic conversion of PIPs and other signalling lipids and proteins. We are very interested in studying the contributions from these other sources of force asymmetry.

Michael M. Kozlov asked: What are the characteristic values of the forces you used for your computations?

Madan Rao answered: The typical range of values of parameters used in our Active Flory–Huggins (AFH) theory has been obtained from experiments or previous theoretical estimates. In our simulations, we have varied the parameters over a range that brackets these estimated values. These estimates and the parameter range explored are listed in Table 1 in the manuscript.

Thus, for instance, experimental estimates of the transbilayer coupling W based on shear forces needed for domain sliding, are set at ≈ 0.01–$0.03\ k_BT$,[1] while earlier theoretical considerations explore the range $W \approx 0.1$–$0.5\ k_BT$.[2] In our article, we explore the range $W \in [0, 0.3]\ k_BT$. We see appreciable bilayer registry when $W = 0.15\ k_BT$ (Fig. 3A in our paper).

The active contractile stress density, parametrised by ζ is estimated from measured values of $\zeta/\gamma = 0$–$60\ \mu m^2\ s^{-1}$ at the cortex.[3] Similarly actin coupling parameter K (or KPS) is varied across the range 0–2.5 k_BT, following experimental estimates in ref. 4.

It is worth emphasizing that the interesting effects of active segregation for $T < T_c$ manifest when the magnitude of active contractile stress is comparable to the chemical potential (Fig. 1B and C in our paper). For $T > T_c$, the active stresses are always significant.

1 M. C. Blosser, A. R. Honerkamp-Smith, T. Han, M. Haataja and S. L. Keller, *Biophys. J.*, 2015, **109**, 2317–2327.
2 R. Shlomovitz and M. Schick, *Biophys. J.*, 2013, **105**, 1406–1413.
3 A. Saha, *et al.*, *Biophys. J.*, 2016, **110**, 6.
4 S. Saha, A. Das, C. Patra, A. A. Anilkumar, P. Sil, S. Mayor and M. Rao, *Proc. Natl. Acad. Sci. U. S. A.*, 2022, **119**, e2123056119.

Michael M. Kozlov commented: What is a qualitative difference between the action results of the active and passive contractile forces?

Madan Rao responded: I will take this to mean the differences between active actomyosin contractile forces and passive adhesion to the actomyosin cortex. The latter of course is a purely equilibrium contribution – passive adhesion to the cortex parametrised by K, leads to an upward shift in the equilibrium phase transition temperature T_c of the inner leaflet. This aspect will show up in both the passive and active theory, as shown in Fig. 2C and D in our paper. However, a purely passive theory would not be able to explain the experiments on the depletion of nano- and mesoscale clusters upon inhibiting myosin contractility using a variety of methods. Further, in a series of quantitative *in vivo* experiments,[1–5] we showed that purely equilibrium theories would not be able to describe the anomalous statistics of equal and unequal time correlations of the concentration of labelled GPI-anchored proteins (GPI-AP) or transmembrane proteins with an actin binding domain (TMABD), including large number fluctuations. In *in vitro* reconstitution experiments,[6] we observe the actomyosin-driven contractile flows that result in transient nanoclustering of GPI-APs and TMABD, exactly as predicted from the theory. None of these can be explained by a purely passive coupling of the membrane to the actomyosin cortex.

Over the last few years, we have formulated an AFH theory to describe the lateral mesoscale organisation of the plasma membrane that follows from our earlier theoretical picture of nanocluster formation. This is consistent with all of the above experiments at the nanoscale and makes precise predictions on the nature and dynamics of mesoscale segregation of membrane components, which are fundamentally different from their passive counterpart. This is discussed in detail in the article, and we list them here: first, the segregation is driven by the interplay of chemical potential gradients and active contactile stresses and thus segregation occurs even when $T > T_c$ (Fig. 1B, C, 2C and D). For $T > T_c$, any amount of activity drives the formation of finite mesoscale domains in both the inner and outer leaflets. For $T < T_c$, this occurs beyond a threshold activity; importantly, even for roughly equal concentration of L_o and L_d components, the active segregation would result in finite mesoscale domains, very different from the passive segregation (Fig. 1B, C and Fig. 2E in our paper).

Typically, active contractile stresses and turnover drive the system into a nonequilibrium steady state characterized by large fluctuations, both at the nano and meso scales. The AFH theory (and the earlier Kinetic Monte Carlo simulations[3,7,8]) predict macroscopic and intermittent fluctuations of domains with fragmentation and coalescence (Fig. 1B and C in our paper).

A large imbalance between local active stresses and chemical potential drives a finite current in the nonequilibrium steady state. The mesoscale domains have diffuse interfaces and significant density heterogeneities in the bulk, unlike conventional phase segregated domains. The latter gives rise to granular features within the mesoscale domains (Fig. 1B and C in our paper), whose size is correlated with the actomyosin aster size, allowing us to identify these granules with nanoclusters. There are other novel features that we predict, such as domain slippage between the two leaflets (Fig. 3B in our paper) and nonreciprocal sensing and coalescence of domains.[7]

1 D. Goswami, K. Gowrishankar, S. Bilgrami, S. Ghosh, R. Raghupathy, R. Chadda, R. Vishwakarma, M. Rao and S. Mayor, *Cell*, 2008, **135**, 1085–1097.
2 K. Gowrishankar, S. Ghosh, S. Saha, C. Rumamol, S. Mayor and M. Rao, *Cell*, 2012, **149**, 1353–1367.
3 S. Saha, A. Das, C. Patra, A. A. Anilkumar, P. Sil, S. Mayor and M. Rao, *Proc. Natl. Acad. Sci. U. S. A.*, 2022, **119**, e2123056119.
4 R. Raghupathy, A. A. Anilkumar, A. Polley, P. P. Singh, M. Yadav, C. Johnson, S. Suryawanshi, V. Saikam, S. D. Sawant and A. Panda, *et al.*, *Cell*, 2015, **161**, 581–594.
5 S. Saha, I.-H. Lee, A. Polley, J. T. Groves, M. Rao and S. Mayor, *Mol. Biol. Cell*, 2015, **26**, 4033–4045.
6 D. V. Köster, K. Husain, E. Iljazi, A. Bhat, P. Bieling, R. D. Mullins, M. Rao and S. Mayor, *Proc. Natl. Acad. Sci. U. S. A.*, 2016, **113**, E1645–E1654.
7 A. Bansal, A. Das and M. Rao, *Indian J. Phys.*, 2022, **96**, 2567–2576.
8 A. Das, A. Polley and M. Rao, *Phys. Rev. Lett.*, 2016, **116**, 068306.

Axel Voigt remarked: The cortex is modeled as an active but isotropic fluid. In the discussion of the results you argue that actomyosin asters in the cortex are in registry with the high density granules in the inner and outer leaflet. What are these asters? As the fluid is isotropic it cannot be a topological defect in an orientational field. Is it a defect in the velocity field?

Madan Rao replied: Our earlier models[1–3] treated the actomyosin cortex either as a two or three-component active fluid written in terms of actin and myosin densities, actin polarisation, and solvent velocity. Regardless, despite being in the isotropic phase, these equations support spatially localised topological defects such as asters (where the polar orientation of actin points radially inward) on account of strong density inhomogeneities, as explained in ref. 4. In the limit that the volume fraction of actomyosin is high, and assuming that the solvent flows quickly to balance any pressure gradients, one may simplify the equations and treat the cortex as a compressible fluid. Stresses generated by myosin then drive contractile flows in the cortical velocity, characterised by a large negative divergence and zero curl. A continuous supply of ATP and turnover of the active agents, ensure a nonequilibrium steady state.

1 K. Gowrishankar, S. Ghosh, S. Saha, C. Rumamol, S. Mayor and M. Rao, *Cell*, 2012, **149**, 1353–1367.
2 K. Gowrishankar and M. Rao, *Soft Matter*, 2016, **12**, 2040–2046.
3 K. Husain and M. Rao, *Phys. Rev. Lett.*, 2017, **118**, 078104.
4 M. C. Marchetti, J.-F. Joanny, S. Ramaswamy, T. B. Liverpool, J. Prost, M. Rao and R. A. Simha, *Rev. Mod. Phys.*, 2013, **85**, 1143.

Axel Voigt added: In deriving the model you consider velocity fields in the leaflets and interleaflet friction. These fields are incompressible. The cortex is modeled as a compressible fluid, but a direct coupling between the velocity in the cortex and the velocities in the leaflets is not discussed. The simulated model neglects the velocities in the leaflets but keeps the velocity in the cortex. Is there a physical justification for this approximation? Without this approximation, what would be a meaningful coupling between the incompressible velocity fields in the leaflets and the compressible velocity field in the cortex?

Madan Rao answered: Our calculation treats the cortex as a compressible active fluid that applies stresses on the inner leaflet. For simplicity, we have taken the volume fraction of the solvent to be small and hence ignored hydrodynamic stresses from the solvent. These active cortical stresses (here taken to be a scalar, but see ref. 1) in addition to the chemical potential enter into the force balance equations for the saturated and unsaturated lipid components at the inner leaflet (eqn (8) and (9) in our paper). Note that for simplicity, we have taken the membrane to be flat, and so the z-component of the velocities has been ignored. The velocities of the individual membrane components are compressible, while the barycentric velocities of the inner and outer leaflets are incompressible (below eqn (5) in our paper). This is valid since we have ignored lipid flip-flop. A more detailed derivation of these equations will appear in a later publication.

1 A. Bansal, A. Das and M. Rao, *Indian J. Phys.*, 2022, 96, 2567–2576

Oded Farago asked: Do patterns driven by active stresses of the cytoskeleton on the cytoplasmatic monolayer differ from patters appearing due to variations in thermodynamic variables like temperature, and if yes in which manner?

Madan Rao answered: Indeed, the effect of active stresses on the patterning of the inner leaflet are very different from the effects of varying temperature. To recall, the driving force for equilibrium segregation at temperatures below T_c is interfacial tension. The only effect of temperature on the dynamics of patterning is to vary the interfacial tension and interfacial width, *i.e.*, increasing temperature decreases the interfacial tension and increases the width. At temperatures above T_c, there is no phase segregation.

On the other hand, active stresses are always relevant and drive patterning both below and above T_c as a result of their interplay with chemical potential. This gives rise to all the distinctive features discussed in response to Michael M. Kozlov's earlier question on the action results of the active and passive contractile forces.

Gonen Golani said: In classical Flory–Huggins theory with no activity, the phase separated domain is the size of the system (macroscopic phase separation) at equilibrium. What determines the length scale of the 'active emulsions' in your model?

Madan Rao responded: As you say, in equilibrium two-component binary fluids at $T < T_c$, the phase segregated domains steadily grow without saturation. This is because the system attempts to reduce the interfacial energy by having two macroscopically large domains separated by a single flat interface.

The active emulsions, on the other hand, are driven by an interplay between the gradient of chemical potential (which drives the domain to grow when $T < T_c$) and contractile stresses (which draw flows inward), leading to finite-size domains. The contractile stresses depend on actomyosin parameters and their interaction with the membrane components in the inner leaflet. The length scale can be obtained from a linear stability analysis (eqn (21) in our paper). The domain size in the outer leaflet is dependent on the transbilayer coupling W. For small W and $T < T_c$, the size is set by the equilibrium properties of the outer leaflet, and hence one should expect a growing domain size without saturation (Fig. 4B in our paper). At larger W, the active stresses will start to contribute, leading to domain size saturation (Fig. 4B in our paper).

Markus Deserno remarked: Active forces do not just come with a *magnitude*; they come with a *time scale*, and that time scale usually also has a length scale associated with it. I am not very knowledgeable about driven or active systems, but my intuition tells me that active noise can only affect equilibrium modes if there is some sort of impedance matching going on, or maybe if some sort of resonance condition holds. For instance, if active forces are much slower than the equilibrium fluctuations in the system (say, membrane shape undulations), then I would expect them to just adiabatically move things around. Can you comment a bit more on that issue? Are there some simple rules of thumb, for instance, that tell me whether I have to worry about active forces, or which part of the frequency-resolved power spectrum of active noise is important?

Madan Rao answered: As you say, any effective force, be it passive or active, comes with a time scale. This includes van der Waals forces between fluctuating dipoles, exchange forces in magnetic solids, and active forces.

In the actomyosin system, this time scale has to do with the myosin contractile cycle, of the order of milliseconds, and involves an energy per ATP molecule of roughly 20 k_BT. A collection of myosins will thus give rise to non-thermal sources of noise that are exponentially correlated in time and violate the fluctuation–dissipation relation. This non-thermal noise will be significant in comparison with thermal noise at longer time scales.

In a macroscopic system with a finite concentration c of myosin aggregates, there will be a mean active stress proportional to c times the work done by each myosin motor per cycle. Gradients of this mean active stress drive systematic contractile flows in the cortex, resulting in the features we discuss in the article.

In addition, there are active stress fluctuations that contribute to say microrheological behaviour and show dramatic consequences of violation of the fluctuation–dissipation relation.

This is captured by the fluctuation–dissipation ratio (the ratio of the Fourier transform of the spatiotemporal fluctuation of the stochastic variable to the imaginary part of the response to an external force), whose deviation from unity describes how far the system is from equilibrium. This has been theoretically predicted and measured from the conformational fluctuations of a probe[1,2] embedded in an active medium. At high frequencies (short time scales), the mesoscopic fluctuations are dominated by thermal noise. At lower frequencies associated with the activity decorrelation rate, there is a "resonance-like" enhancement of the active stress fluctuations.

1 N. Kikuchi, *et al.*, *Proc. Natl. Acad. Sci. U. S. A.*, 2009, **106**, 47.
2 F. Gnesotto, *et al.*, *Rep. Prog. Phys.*, 2018, **81**, 6.

Kandice R. Levental said: In thinking about outside-in and inside-out signalling across the bilayer, what is the lateral distance that affects the opposite leaflet domain? I'm thinking specifically about the domain registry and slippage ideas. In a cell, would something like actomyosin connection to the inner leaflet necessarily be directly underneath (in complete registry) with the domain on the outer leaflet? Or could there be slippage such that the mesoscale organization in the outer leaflet can radiate outward further from the actomyosin connection?

Madan Rao responded: Our theoretical and experimental studies[1,2] show that the PS clustering at the inner leaflet is in co-registry with the actomyosin asters in the cortex. Simultaneously, the outer leaflet GPI-APS are in co-registry with the PS, owing to strong transbilayer coupling (Fig. 3A, Fig. 4A, B and Fig. 5A in our paper). As far as we can tell, this holds true for the mesoscale domains too (Fig. 4A, B and Fig. 5A in our paper). However, at large Peclet number, the domain movement and fluctuations would be large, leading to the prediction of slippage of the inner leaflet domains with the actomyosin clusters and the outer leaflet domains with the inner leaflet (Fig. 3B in our paper). As you say, this could have implications for outside-in or inside-out signalling, such as integrin or Ras signalling. In the context of integrin signalling,[3] triggering integrin activity *via* ECM ligand presentation activates the Rho-pathway, which locally induces the formation of actomyosin-driven GPI-AP nanoclusters that are enriched in an lo mesodomain on the outer leaflet. It would be interesting to measure the spatial and temporal response of such local integrin activation.

1 S. Saha, A. Das, C. Patra, A. A. Anilkumar, P. Sil, S. Mayor and M. Rao, *Proc. Natl. Acad. Sci. U. S. A.*, 2022, **119**, e2123056119.
2 R. Raghupathy, A. A. Anilkumar, A. Polley, P. P. Singh, M. Yadav, C. Johnson, S. Suryawanshi, V. Saikam, S. D. Sawant and A. Panda, *et al.*, *Cell*, 2015, **161**, 581–594.
3 J. M. Kalappurakkal, A. A. Anilkumar, C. Patra, T. S. van Zanten, M. P. Sheetz and S. Mayor, *Cell*, 2019, **177**, 1738–1756.

Niclas Paul Decker addressed Kandice R. Levental, Erwin London and Ilya Levental: Which physicochemical properties – including balanced liquid-ordered/disordered and *trans*-leaflet partitioning – should an ideal membrane probe have (or avoid) to minimise perturbation, and what checks will prove that the long-lived clusters seen in mast cells upon activation are genuine rather than probe-induced artefacts?

Kandice R. Levental responded: The 'best' probe really depends on what information you want to know. It is important to remember that because we're relying on diffraction limited microscopy, any single pixel contains hundreds of nm^2 of membrane. So there are likely to be many distinct regions contained therein. And if your probe has preference for one of those regions, then it will 'report' that environment preferentially, which can be particularly problematic if you're unaware of it.

So in that light, I would say either minimal preference between possible environments, or a well characterized preference that can be controlled for. Avoiding chemical reactivity is very important, as some fluorophores can change their properties or become oxidants under illumination.

Regarding the clusters, we strongly believe that they are not 'probe-induced artifacts'. We have observed them with many different probes (protein probes, small molecules, and synthetic fluorophores) under many different conditions (labelling before/after activation, using one *versus* multiple probes, *etc.*)

Erwin London added: Fluorescent probes used in low concentrations have negligible effects on lipid phase behavior. In terms of localization, one wants to have two probes with very different L_o and L_d localization properties. There are two explanations when a region of a membrane seems to be intermediate in properties relative to L_o and L_d. One is that it is a homogeneous intermediate state. The other is it contains coexisting L_o and L_d domains. In the former case both L_o and L_d preferring probes should sense an intermediate environment. In the latter case they should report different environments. The probe that localizes in the L_d domains should report an L_d environment, and the probe that prefers localizing in an L_o domain should report an L_o environment.

Ilya Levental added: A good question. So the "ideal" probe will of course depend on what you are trying to measure. I think the key point is to be aware that some are only in a single leaflet (usually charged or large heads) while others will be in both leaflets (if mostly hydrophobic). Another point is the preference for any potential L_o/L_d domains that might be present. In cells, there will be a lot going on in a single diffraction-limited pixel, so its important to be aware that your probe is likely preferentially reporting on some aspect (asymmetry, domain, *etc.*)

Ilya Levental addressed all: Consider a bilayer with phospholipids on one side, the other with ceramide, lysolipids, fatty acids *etc.*: would this be stable?

Conflicts of interest

There are no conflicts to declare.

Faraday Discussions

PAPER

Asymmetric phase transitions in lipid bilayers: coupling or bending?†

Mona Krompers,[a] Miriam Jaki,[a] Sinja Götz,[a] Jan Lembeck,[a] Laurine Kaul,[a] Martin Holzer[a] and Heiko Heerklotz*[abc]

Received 8th January 2025, Accepted 15th January 2025
DOI: 10.1039/d5fd00003c

Biomembranes show asymmetric lipid composition of their two leaflets. The phenomenon that ordered domains in one leaflet may affect the order of the other has been referred to as interleaflet coupling and discussed in terms of transmembrane signaling. Many coupling mechanisms have been proposed; one potential mechanism should arise from the fact that the isolated melting of an ordered, *e.g.*, gel phase gives rise to a significant expansion of this leaflet, resulting in a mismatch between the intrinsic areas of the leaflets. This asymmetry stress can be accommodated in a number of ways. One is interleaflet coupling – individually higher- and lower-melting leaflets melt together at intermediate melting temperature. Alternatively, the membrane may bend towards the larger-intrinsic-area leaflet, bud and release very small daughter vesicles (DVs). Here, we prepared lipid-asymmetric large unilamellar vesicles (aLUVs) with low-melting stearyl-oleyl-phosphatidylcholine (SOPC) in the inner and SOPC with ~20 mol% of high-melting dipalmitoyl phosphatidylglycerol (DPPG) in the outer leaflet. Phase transitions in aLUVs *versus* LUVs were recorded with pressure perturbation calorimetry; vesicle budding was monitored by asymmetric flow field-flow fractionation (AF4) and visualized by cryo-transmission electron microscopy. An HPLC protocol was established to quantify the total DPPG content; zeta potential was used to detect outer-leaflet DPPG. It turned out to be possible to prepare aLUVs at 5 and 15 °C, with the outer leaflet partially in the gel phase. The properties of the final aLUVs depended on the preparation temperature. aLUVs prepared at 5 and 15 °C caused the budding of exovesicles upon heating and only weak coupling of the phase transitions of the leaflets. aLUVs prepared at 30 °C with both leaflets in the fluid state showed stronger coupling upon asymmetric freezing/melting at lower temperature. This is in line with the hypotheses that (i) the exchange of lipid between close-to lipid-saturated cyclodextrin and acceptor vesicles at a given temperature results in largely stress-free bilayers and (ii) that outside budding and coupling are, to some extent, alternative responses of the bilayer to asymmetric expansion. These hypotheses help explaining

[a]*Department of Pharmaceutics, University of Freiburg, Institute of Pharmaceutical Sciences, Freiburg, Germany. E-mail: heiko.heerklotz@pharmazie.uni-freiburg.de*
[b]*University of Toronto, Leslie Dan Faculty of Pharmacy, Toronto, Canada*
[c]*BIOSS Center for Biological Signaling Studies, Freiburg, Germany*

† Electronic supplementary information (ESI) available. See DOI: https://doi.org/10.1039/d5fd00003c

our and some literature data; the overall understanding and prediction of coupling for any given aLUV system remains an urgent, open question.

Introduction

Recently, the effects of lipid asymmetry on the function of biological membranes and the roles of membrane intrinsic-area asymmetry stress in biological membrane remodeling phenomena have been attracting much interest. The study presented here considers the interplay between lipid species asymmetry and intrinsic-area asymmetry.

The intrinsic area of a lipid leaflet depends on the number of lipids and their optimal lateral area, at which repulsive and attractive interactions match at least in-plane.[1] Thereby, intrinsic areas of different molecules are not necessarily additive as illustrated impressively by the lipid-condensing effect of cholesterol.[2–4] Furthermore, even at the intrinsic area per molecule, *i.e.*, at zero lateral force within the interfacial plane, residual out-of-plane forces give rise to an intrinsic leaflet curvature.[5]

A mismatch between the intrinsic areas of two attached leaflets, also referred to as intrinsic-area asymmetry, asymmetry stress, differential stress, or bilayer spontaneous curvature, induces a tendency to bend the bilayer.[6] A straightforward way to study asymmetry stress effects in vesicular model systems and cells in detail is to insert or extract membrane-impermeant molecules to or from the accessible, outer leaflet. This can be achieved with membrane-partitioning yet impermeant lipopeptides,[7,8] lysolipids[9,10] and other surfactants.[11,12] Their addition to vesicles causes positive asymmetry stress, *i.e.*, overpopulation of the outer compared to the inner leaflet. Extraction from rather relaxed, surfactant-containing membranes, for example by dilution, induces negative asymmetry stress. The analogous approach has been demonstrated for phospholipids with the help of cyclodextrins able to solubilize lipids in the extravesicular solution.[13] Whereas "empty" cyclodextrin extracts lipid from the outer leaflet, oversaturated (supercooled) cyclodextrin–lipid complexes overpopulate it. The approaches using soluble and cyclodextrin-solubilized amphiphiles have revealed strong analogies.

A primary response to asymmetry stress is membrane bending, causing membrane shape changes and tubulation or budding and fission of daughter vesicles (DVs).[10,13–16] The extent to which these phenomena are possible can be increased by hypertonic conditions increasing the area-per-volume ratio of the membrane.[10,13] As bending effects reach their limit, one or more alternative asymmetry-stress responses kick in.[16] The fact that the activity of a lipid or surfactant in a membrane leaflet is increased in an overpopulated and reduced in an underpopulated leaflet[9,13,17] can, at some point, repel membrane-partitioning molecules into aqueous solution.[9,17,18] The linked leaflet may adjust by a gradual change of the area per lipid within a given phase[19,20] or by an enforced formation of a more condensed or ordered phase,[13] respectively. Lipid flip-flop, usually taking of the order of days, can be accelerated to occur within minutes.[21] Finally, asymmetry stress by surfactant insertion or lipid extraction may reach a threshold for transient membrane failure or even the collapse of the vesicle altogether.[13,17,22,23]

Asymmetry stress can also result from different thermal expansion of the leaflets in lipid-asymmetric membrane – the direct representation of Sheetz and Singer's[6] bimetallic couple analogy. Overpopulation, *i.e.*, increased lateral pressure can condense lipids into a gel phase well above its melting temperature, T_m, in symmetrical membranes.[13] Analogously, the melting-induced expansion in a low-T_m leaflet should stretch an opposed, high-T_m leaflet and may melt its lipid. In other words, asymmetric melting or formation of gel phase induces asymmetry stress and asymmetry stress should be a strong coupling mechanism of transitions in the two leaflets if it cannot be relaxed otherwise.

In fact, the coupling between transitions between ordered (*e.g.*, gel, liquid ordered) and fluid phase or state in one leaflet with the properties of the opposed leaflet of the bilayer has attracted much interest, also as a potential transmembrane signaling event. Evidence for such coupling has been found in large and giant liposomes[19,24–28] as well as in planar, supported or free-standing membranes.[29–31] Coupling effects have been found for some of the asymmetric membrane compositions investigated, mostly using scanning calorimetry or using fluorescent probes detecting membrane order. Several mechanisms have been proposed for coupling of membrane ordering transitions or changes and, in turn, the registration of ordered domains in the two leaflets. These include headgroup electrostatics, intrinsic lipid curvature, chain interdigitation, cholesterol flip-flop, or differential stress between two leaflets.[26,32–34] Interleaflet coupling is known to induce domain formation[35] and affect lateral lipid diffusion.[36] See Table S9 in ESI† for a more comprehensive overview of contributions. Overall, the effects governing interleaflet coupling appear to be so complex that in spite of a good basis of data, it seems still impossible to predict, for a chosen asymmetric vesicle, whether it shows strong, weak, or virtually no coupling of order-inducing or -reducing transitions between the leaflets.

The current study aims to elucidate whether asymmetric area changes cause coupling and or bending in vesicles of SOPC in the inner and 80 mol% SOPC + 20 mol% DPPG in the outer leaflet, in short, $SOPC^{in}|+20\% DPPG^{out}$ or without specification of the specific fraction of DPPG added, $SOPC^{in}|+DPPG^{out}$. This is done by detecting both bending (*i.e.*, outward budding of DVs) *via* AF4 and interleaflet coupling of the gel–fluid transition by differential scanning (DSC) and pressure perturbation calorimetry (PPC).

Furthermore, we will test the hypothesis that the lipid exchange between outer leaflet and largely saturated lipid-loaded cyclodextrin complexes produces rather relaxed, asymmetry-stress-free liposomes.[37,38] Of course, the range of suitable cyclodextrin and loaded-lipid concentrations must be limited – cyclodextrin with no or very little lipid will extract and lipid-supersaturated cyclodextrin will insert lipid, leading to negative and positive asymmetry stress, respectively.[13] However, there seems to be a substantial range in between these extremes where lipid exchange with cyclodextrin causes very little asymmetry stress.[37] This hypothesis would imply that the properties of aLUVs prepared at identical concentrations and, hence, with essentially equal outer and inner leaflet compositions would differ in their properties depending on the exchange temperature they have been prepared at. Tackling this question requires the preparation of aLUV also under exchange conditions where all or part of the membrane is in the gel phase. It remains to be shown whether this usually avoided procedure is feasible in the first place.

Specifically, we will test the suitability of the aLUV protocol that had been introduced for POPC^{in}|+20–50 mol% POPG^{out} [37,38] for vesicles of SOPC^{in}|+20 mol% DPPG^{out}. To this end, a novel HPLC protocol was set up to quantify DPPG. Transition temperatures for pure, symmetrical vesicles are 41.3 °C for DPPG[39] and about 5 °C for SOPC (see Marsh[40] for a compilation). The exchange will be done at 30 °C, where all lipids are fluid as well as at 15 °C and 5 °C, when the outer or both leaflets shall partially be in the gel phase. That means, at 15 °C, aLUVs will be disordered^{in}|ordered^{out}, also referred to as DI/OO in the literature. If the exchange really causes essentially asymmetry-stress-free liposomes, significant stress should build up as, after cyclodextrin removal, the temperature is changed and lipids melt and expand primarily in the outer leaflet. The phase transitions were monitored by DSC and PPC. The latter has the advantage to equilibrate and then probe heats of pressure increase and decrease at each temperature point instead of scanning, allowing for measurements at as little as 2 °C and for minimizing and revealing hysteresis effects.

Budding of exovesicles has been considered a possible response to asymmetry stress but not observed to accompany asymmetric phase transitions so far. The problem is that changes in vesicle size distribution have been monitored by dynamic light scattering which must fail to detect such effects. If, for example, "mother" vesicles are 130 nm in diameter and 20% of the lipid resides in DVs of 20 nm size, the latter will contribute only $(20/130)^4 = 0.3\%$ to the scattered light intensity. The fourth power describes the ratio between sphere surface, which correlates with the amount of lipid, and scattered light intensity. In other words, DLS is principally blind to vesicle budding. Here, budding of DVs will be tested by AF4, a technique detecting size distributions after separating the differently sized particles with incorporated fluorescent lipid (1 mol% NBD-DSPE). This way, it succeeds in characterizing vesicle budding as demonstrated upon addition of lysolipids recently.[10,16] Here, it reported the formation of small particles upon phase transitions that were confirmed to be very small DVs vesicles by cryo-TEM.

Materials and methods

Materials

1,2-Dipalmitoyl-*sn*-glycero-3-phospho-*rac*-glycerol sodium salt (DPPG-Na) was kindly contributed by Lipoid (Ludwigshafen, Germany). 1-Stearoyl-2-oleoyl-*sn*-glycero-3-phosphocholine (SOPC) and 1,2-distearoyl-*sn*-glycero-3-phosphoethanolamine-*N*-(7-nitro-2-1,3-benzoxadiazol-4-yl) ammonium salt (NBD-DSPE) were purchased from Avanti Polar Lipids (Alabaster, US). Randomly methylated β-cyclodextrin (mβCD) (>98%), sodium azide (NaN_3), sodium chloride (NaCl) and ammonium acetate (>99%) were purchased from Sigma-Aldrich (St. Louis, MO, US) and Vivaspin Turbo 4 centrifugal filters (4 mL; MWCO 100 kDa) were obtained from Sartorius (Stonehouse, UK). Ethylenediaminetetraacetic acid (EDTA; ≥99%) and tris(hydroxymethyl)aminomethane (Tris; ≥99.9%), and acetic acid (100%) were purchased from Carl Roth GmbH (Karlsruhe, Germany). Methanol (HPLC ULTRA LC-MS grade; ≥99.9%) was obtained from VWR International (Leuven, Belgium). Purified water was derived from an in-house arium pro water purification system from Sartorius AG (Göttingen, Germany). All other chemicals were purchased from Carl Roth (Karlsruhe, Germany) and were of analytical grade.

Vesicle preparation

Large unilamellar vesicles (LUVs) of SOPC were prepared according to standard procedures.[41,42] In brief, lipids or lipid mixtures dissolved in chloroform or chloroform/methanol (2 : 1) were dried using a rotational vacuum concentrator (Martin Christ, Osterode am Harz, DE) to a thin film and further under a high vacuum overnight. For AF4 experiments, 1 mol% of fluorescent lipid (NBD-DSPE) was incorporated into the film. The lipid film was rehydrated with buffer (110 mM NaCl, 10 mM Tris, and 0.5 mM EDTA at pH 7.4). After five freeze–thaw-cycles, the lipid dispersion was extruded 21 times through an 80 nm nucleopore track-etch polycarbonate membrane (Whatman, Maidstone, UK) at 25 °C using a LipoFast extruder (Avestin, Ottawa, Canada). Multi-lamellar vesicles (MLVs) composed of DPPG were prepared in the same way but were not extruded. The phospholipid concentration was determined by Bartlett Assay,[43] and the integrity of LUVs by dynamic light scattering (DLS).

Preparation of asymmetric liposomes

aLUVs were prepared as described previously.[37,38] Briefly, 1.0 mL of 5 mM SOPC liposomes with ∼20 mol% DPPG in the outer leaflet was prepared in two steps: first, 15 mM DPPG liposomes were completely disintegrated and the lipid complexed by adding 150 mM mβCD at 50 °C. Second, LUVs of ∼40 mM SOPC were prepared to serve as "acceptor" vesicles. Donor complexes and acceptor vesicles were brought to the desired exchange temperature (5, 15 or 30 °C) and then mixed to produce an incubation mixture containing a total of 5 mM SOPC, 0.63 mM DPPG and 50 mM mβCD. Standard incubation was carried out in a thermomixer at 1000 rpm for 20 minutes – spot checks for validation were done with longer incubation times as shown below.

After preparation, free mβCD and lipid–mβCD complexes were eliminated from the aLUV dispersion by centrifugation through Vivaspin Turbo 4 centrifugal filters, diluting the mβCD below a calculated value of <0.5 mM. The resulting DPPG content in the outer leaflet was quantified by ζ-potential measurements using a calibration curve provided as Fig. S3 in the ESI.† To prepare scrambled liposomes, aLUVs were dried under vacuum, the lipids re-dissolved in 0.5 mL chloroform/methanol (2 : 1) and dried again to obtain a lipid film, followed by rehydration and extrusion as described above.

Dynamic light scattering (DLS) and ζ

A Zetasizer Nano ZS (Malvern, Worcestershire, UK) equipped with a 633 nm He–Ne laser was used to measure particle size distributions as well as the ζ-potential of LUVs and aLUVs. Three measurements at 25 °C were taken for hydrodynamic diameter, polydispersity index (PDI), and ζ-potential for each sample. Effects of viscosity and refractive index of buffer components were taken into account. DLS was performed to confirm the intensity-weighted hydrodynamic diameter to be around 100 nm and the polydispersity index (PDI) to be ≤0.1. The sample was measured in a 4 mL polystyrene cuvette with a layer thickness of 10 mm (Sarstedt AG & Co. KG, Nümbrecht, Germany) at a detection angle of 173°. For ζ-potential measurements a flow-through, high-concentration ζ-potential cell (HCC; Malvern, UK) was used, as described before.[37] A transfer standard (ZTS1240, Malvern) with a specification of −40 ± 6 mV was measured at the beginning and end of each set of experiments to assess the quality of the setup. Liposomal dispersions were diluted to 1 mM prior to ζ-potential measurements.

Differential scanning calorimetry (DSC) and pressure perturbation calorimetry (PPC)

A VP-DSC (Malvern Instruments, Inc., Northampton, NC, USA) was used.[44] The sample cell was loaded with liposomes dispersed in buffer, while the reference cell was filled with plain buffer. Concerning DSC, samples were scanned from 2–45 °C at a scan rate of 20 K h^{-1}, and the heating power required for the desired temperature increase was measured. Data analysis with MicroCal Origin provided the apparent temperature-dependent molar heat capacity of the lipid vesicles, C_p. PPC measurements were performed as described previously.[45] In brief, the thermal volume expansion was determined by measuring the heat response to an isothermal pressure perturbation in a VP-DSC operating in isothermal mode (high gain, low noise mode). Pressure jumps between 5.3 bar and ambient were applied to a thermally equilibrated sample at each desired temperature point, typically within 2–50 °C. Control experiments with water *versus* water, buffer *versus* buffer and buffer *versus* water were performed for the evaluation of the absolute coefficient of thermal expansion, α_V, of the lipid phase.

High performance liquid chromatography (HPLC)

HPLC samples were analyzed utilizing an Ultimate 3000 HPLC System (Thermo Fisher Scientific GmbH, Dreieich, Germany) equipped with a quaternary pump (LPG-3400SD), an autosampler (WPS-3000TSL), a column compartment (TCC-3000SD), a diode array detector (DAD-3000), and a charged aerosol detector (Dionex Corona CAD; Dionex Softron GmbH, Germering, Germany). The flow rate was set to 1 mL min^{-1} and the autosampler was maintained at 20 °C. For all samples, a fixed injection volume of 10 μL was used. A reversed phase Luna C18(2) column (150 × 4.6 mm) with a particle size of 3 μm from Phenomenex Ltd (Aschaffenburg, Germany) was used for separation, which was kept at 45 °C. Separation was achieved using an isocratic eluent composition of 2% eluent A (4 mM ammonium acetate, pH 4.0) and 98% eluent B (4 mM ammonium acetate in methanol), with a total run duration of 15 minutes. The CAD detector employed for lipid quantification was operated with an inlet pressure of 35 psi, an evaporation temperature of 30 °C (factory fixed), a data acquisition rate of 10 Hz, and a maximum detection limit of 500 pA. The eluents were freshly prepared before usage. The HPLC system was equilibrated for a minimum of 20 minutes followed by two blank runs prior to sample analysis. Data evaluation was performed using Chromeleon® Software 7.2.1. The method was validated in accordance with the ICH guideline Q2(R2) addressing working range, precision, accuracy, detection limit, quantitation limit, and specificity.[46] Sample preparation was performed by diluting the original sample in ethanol. The samples were analyzed in triplicates. For further details, see ESI, Section 1.†

Asymmetric flow field flow fractionation (AF4) and multi-angle light scattering (MALS)

Daughter vesicles (DVs) and mother vesicles (MVs) were separated using AF4 as published elsewhere.[10] In brief, the AF4 setup consisted of a separation channel

(SC channel, regenerated cellulose membrane, cut-off 10 kDa, 490 μm spacer, wide type, Wyatt, Dernbach, Germany) and a flow controller (Eclipse AF4, Wyatt) connected online to MALS (DAWN Heleos II, Wyatt) and fluorescence (1260 Infinity G1321B, Agilent, Waldbronn, Germany) detectors. Separation buffer (110 mM NaCl, 10 mM Tris, 0.5 mM EDTA, and 0.02% NaN_3) was adjusted to pH 7.4 and filtered through 0.1 μm pores. Channel flow rate was set to 1.0 mL min^{-1} and 5–10 μg lipid sample was injected (0.2 mL min^{-1}). For separations at room temperature (RT; 23 ± 2 °C), the following AF4 sequence was used (cross flow rate $V_x = 0.08$ mL min^{-1} unless stated otherwise): (a) elution, 2 min; (b) focus, 2 min – focus + inject, 4 min – focus, 4 min; (c) elution, 20 min – elution, 23 min (linear V_x gradient: 0.80 to 0.05 mL min^{-1}) – elution, 15 min ($V_x = 0.05$ mL min^{-1}); (d) elution + inject, 5 min ($V_x = 0.0$ mL min^{-1}); (e) focus, 2 min – elution, 2 min. For separations at 4 °C, the same sequence was used with lower cross-flow rates (0.50 instead of 0.80 mL min^{-1} and 0.03 instead of 0.05 mL min^{-1}) to arrive at comparable retention times for a given sample at 23 °C and 4 °C. The geometric size of the vesicles was obtained from MALS data using the coated spheres model in ASTRA 8.0.1.21 (Wyatt). The fraction of fluorescence originating from the DVs was obtained from the fluorescence signal (λ_{ex} = 460 nm, λ_{ex} = 520 nm) as described earlier.[10]

Cryo-transmission electron microscopy (cryo-TEM)

Cryo-TEM samples were incubated 30 min at 35 °C before blotting. Then, 4 μL of liposomes at 6 mg mL^{-1} total lipid were deposited on a Quantifoil S 7/2 copper grid (Quantifoil, Jena, Germany). The grid was glow discharged using the PELCO easiGlow (Ted Pella, Redding, California, USA) immediately prior to use. The grid was blotted at 33 °C and 100% relative humidity and vitrified by plunge freezing into liquid ethane using a Vitrobot MK IV (ThermoFisher, Waltham, Massachusetts, USA). Imaging was performed under cryogenic conditions on a FEI Talos L120C at 120 kV operating voltage equipped with a Ceta 16-megapixel camera.

Results

Phase behavior in symmetrical LUVs of SOPC and DPPG

Usually, a thermodynamic assessment of lipid phase behavior is done with multilamellar vesicles, which fulfill the cooperativity requirements for true phases to a rather good approximation. The present study deals with aLUVs and, therefore, needs to record the state of the membranes in LUVs although their transitions are known to be broadened and deviate from the rules applying to true phases.

As a consequence, Fig. 1B obtained by DSC downscans for symmetric SOPC/DPPG LUVs (Fig. 1A) should not be referred to as a "phase diagram" but has been presented analogously. Fig. 1 represents an isomorphous melting behavior, *i.e.*, mixing of the lipids both in gel and fluid phase. Bold lines represent empirical approximations of effective solidus and liquidus boundaries, taking into account that phase boundaries must meet in a single T_m for pure components. Blue dash-dot lines illustrate the boundaries expected for ideal mixing of SOPC and DPPG in both phases, as calculated on the basis of the respective T_m and ΔH values. The deviations of the real boundaries can be explained by non-ideal mixing of the

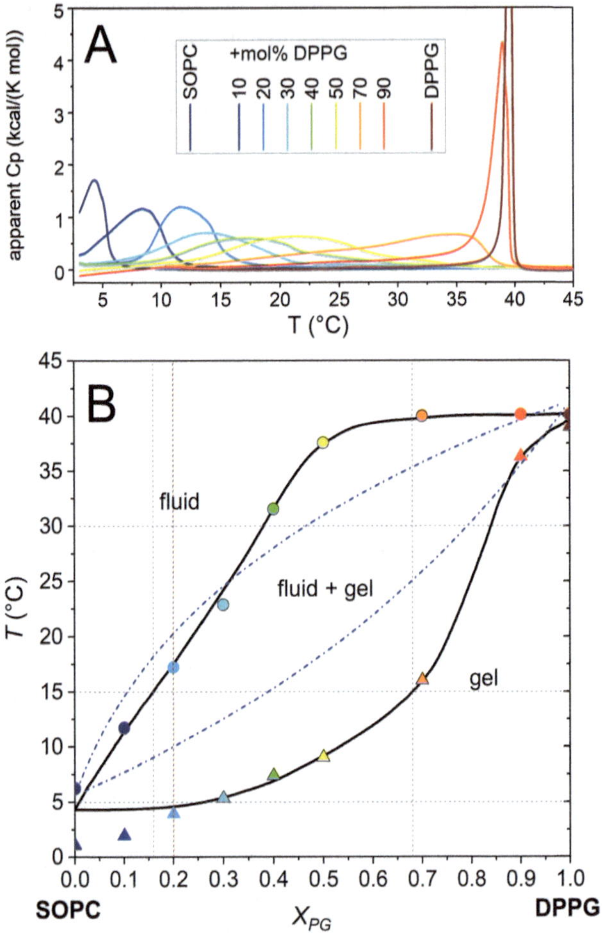

Fig. 1 Representation of the gel-to-fluid transition ranges in mixtures of SOPC and DPPG (panel A) presented analogously to a phase diagram (panel B). Temperatures at the onset (triangles) and completion (spheres) of transitions recorded by DSC at pH 7.4 are plotted as a function of the mole fraction of DPPG, X_{PG}, in symmetrical LUVs. The arbitrarily chosen, bold solid line would fulfill the requirements of a phase coexistence. Phase boundaries simulated for ideal mixing and high cooperativity (expected rather for MLV) are shown as dotted lines.

lipids (also seen in MLV phase diagrams) and compromised cooperativity and curvature effects in LUVs.

The calculations below were carried out interpreting the bold, arbitrary lines in Fig. 1B as approximations for true phase boundaries. Most important for the current study is the state of pure SOPC and SOPC with ~20 mol% DPPG at 5, 15, and 30 °C. The fraction of gel phase, x_{gel}, in pure SOPC is about 50% at 5 °C (close to T_m); at 15 and 30 °C, the membrane is all fluid. SOPC with 20 mol% DPPG (see vertical grid line) is fully fluid at 30 °C. At 15 °C, there is a coexistence of a gel phase with 68 mol% DPPG and a fluid phase with 16 mol% DPPG. The lever rule implies that the gel phase comprises $x_{gel} = (68 - 20)/(68 - 16) = 8$ mol% of the lipid; the remaining 92% is in the fluid phase. At 5 °C, the non-ideal behavior permits only a rather crude estimate. DPPG contents of gel and fluid phase of 28 and 2 mol%, respectively, suggest $x_{gel} \approx 70$ mol%. These data will be used below to assess the intrinsic states of the leaflets in asymmetric liposomes.

The exchange protocol yields aLUVs also in partial gel state, at 5 and 15 °C

The exchange protocol established to yield aLUVs of $POPC^{in}|+20$ mol% $POPG^{out}$ upon 20 minutes exchange at 28 °C[37,38] was applied to $SOPC^{in}|+DPPG^{out}$ at 30 °C, 15 °C, and 5 °C.

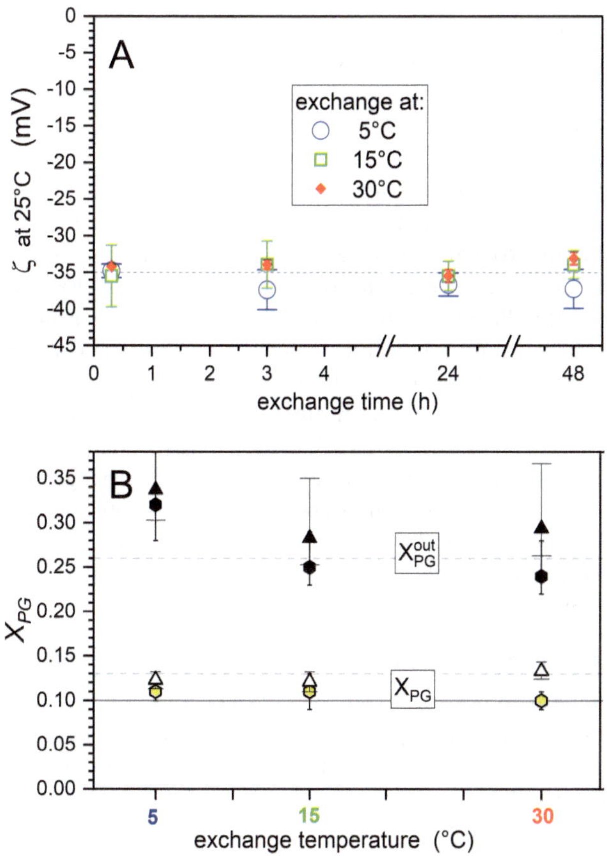

Fig. 2 Properties of aLUVs of $SOPC^{in}|+20$ mol% $DPPG^{out}$ prepared by exchange between SOPC "acceptor" vesicles with nearly DPPG-saturated mβCD solution as a function of exchange time (A) and exchange temperature (B). Panel A shows data of batch 5 (see Table S8† for detailed data) as a function of exchange temperature (see plot) and exchange time (abscissa). The zeta potential, ζ, is related to the DPPG-based charge density in the outer leaflet; it is always measured at 25 °C to eliminate effects of varying membrane area. Panel B displays total DPPG mole fractions, X_{PG}, obtained using Fig. S3† from ζ of scrambled aLUVs (open triangles, batch 4) and HPLC (yellow hexagons, batches 1–3). PG mole fractions in the outer leaflet of aLUVs, X_{PG}^{out}, are obtained from ζ for batch 4 (solid black triangles) and batches 1–3 (solid black hexagons). Errors of HPLC-based data are s.d. of triplicate measurements. Errors of ζ-based data reflect lower and upper limits of the 95% confidence interval of the calibration curve, Fig. S3.†

The zeta potential of batch 5 (see Table S8† for details) reached −35 mV after 20 minutes of exchange at all temperatures, with larger errors at lower exchange temperatures (see Fig. 2A). It is related to the surface charge density. Since all ζ measurements are carried out at 25 °C, where all lipids are fluid and share a similar surface area, the charge density should represent the mole fraction of the charged species, DPPG, to a good approximation. Longer incubation times up to 48 h did not lead to any further uptake of DPPG into the outer leaflet at 15 and 30 °C; a slight trend at 5 °C is possible but lies within the rather large error range.

The total amount of DPPG inserted into liposomes was obtained from HPLC as 10–11 mol% and from ζ measurements of scrambled aLUVs as ~12–14 mol% (Fig. 2B, open symbols); for raw data, see Table S8.† Conversion of the average ζ of 3 batches to X_{PG}^{out} was done by a calibration curve explained in Fig. S3.† The values obtained by the different methods are largely in line with each other and with the target value of 10% established for this protocol with POPC/POPG. The exchanged amount of DPPG did not significantly depend on exchange temperature. This is not trivial since cooling of mβCD–lipid complexes could induce supersaturation and the activity of DPPG in gel and fluid phase must be expected to be different.[13]

The ζ potential of aLUVs with all DPPG residing in the outer leaflet should correspond to twice the total DPPG content, $X_{PG}^{out} \approx 2X_{PG}$. Indeed, the experimental data (solid triangles for average of batches 1–3 and hexagons for batch 4 in Fig. 2B) are of the order of the upper dashed grid line at 26 mol%, twice that of the lower dashed grid line; again, the values are somewhat higher but mostly within error of the dashed line. As we will see below, samples prepared at 5 and 15 °C and then heated to room temperature show budding – we speculate that the coexistence of MV and DV with potentially different X_{PG}^{out} might cause minor deviations of ζ. In any case, the twofold higher X_{PG}^{out} compared to the average X_{PG} confirms the asymmetric insertion of DPPG, also at 5 °C and 15 °C exchange temperature.

Summarizing this section, the standard protocol established earlier applies also to prepare aLUVs of SOPC^{in}|+20% DPPG^{out}, even if the lipid is partially in the gel phase. The extent of DPPG insertion is quite similar to that of POPG.

Interleaflet coupling in aLUVs depends on exchange temperature

In order to gain information about interleaflet coupling in SOPC/DPPG liposomes, we studied the melting behavior in more detail using PPC, a calorimetric method that is capable of measuring the heat response to an isothermal pressure perturbation (see Fig. 3).

First, three replicate sets of aLUVs were prepared separately, one per set at an exchange temperature of 30 °C, followed by mβCD removal. This procedure is close to the standard one,[37,38] with both leaflets in the fluid phase. Each sample was cooled to 2 °C and several PPC runs carried out with temperature increasing from 2 °C to 45 °C, respectively. Unlike scanning calorimetry, in PPC, the calorimeter is equilibrated at each temperature point to be measured and then, down and up pressure jumps are applied to measure expansivity. Afterwards, the system heats to the next desired temperature. The PPC curves of the 3 replicate samples are presented as red lines in Fig. 3B, with their enclosed area filled with a red shading. A second plot of the data in a separate window and with detailed experimental points is provided as Fig. S5.† The width of the red area indicates some variability between the replicate aLUVs but overall, the curves agree in

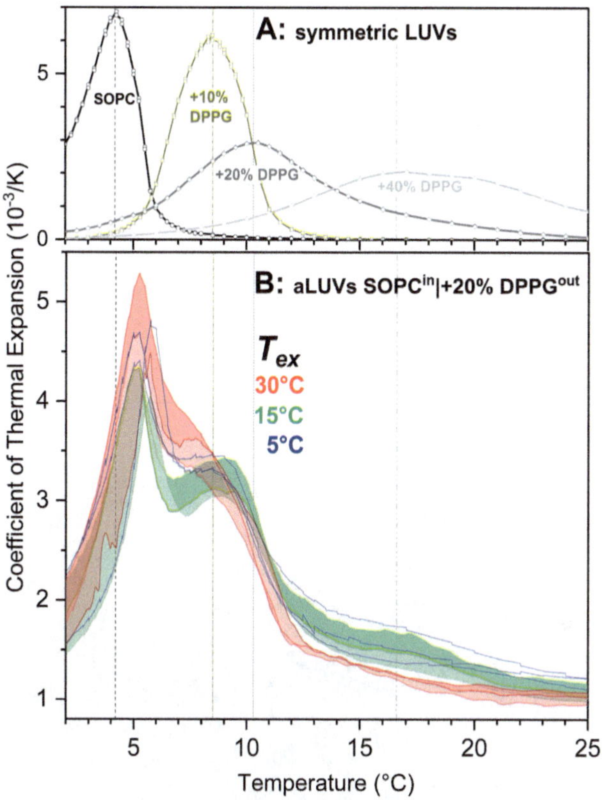

Fig. 3 PPC transition curves of aLUVs prepared at different exchange temperatures (B) compared to those of symmetric LUVs (A). For compositions, see plot (all percentages refer to mol%). In panel B, red curves enclosing a red area represent three replicate samples prepared at 30 °C exchange temperature. Green curves and green shading represent aLUVs prepared at 15 °C. aLUVs prepared at 5 °C are represented by blue curves (no shading). PPC transition curves can be interpreted analogously to DSC curves; here, the progress of the melting transition is indicated by a peak in thermal expansion instead of the – related – peak of apparent heat capacity detected by DSC. Panel B is focused on a general comparison – a repetition of all data with individual data points assigned to sample sets is provided as Fig. S5.†

showing a significant coupling. As discussed below, there is no standard way of quantifying interleaflet coupling from transition curves – our statement of "significant yet incomplete coupling" is based on two observations: first, the peak is considerable narrower and more overlapping than the superimposed transition curves of symmetric LUVs of SOPC (matching the composition of the inner leaflet) and SOPC + 20 mol% DPPG (as in outer leaflet) shown in Fig. 3A. This is taken as evidence for coupling. Second, the transition does not collapse into a single, homogeneous intermediate peak as, for example, that for symmetric LUVs with the average 10 mol% DPPG (Fig. 3A) but retains a sharp peak on the low-temperature side and a pronounced shoulder on the high-temperature side. This is interpreted in terms of incomplete coupling.

For each of the 3 sets of samples, there was also one sample with lipid exchange at 15 °C. Also these samples were measured in PPC starting at 2 °C,

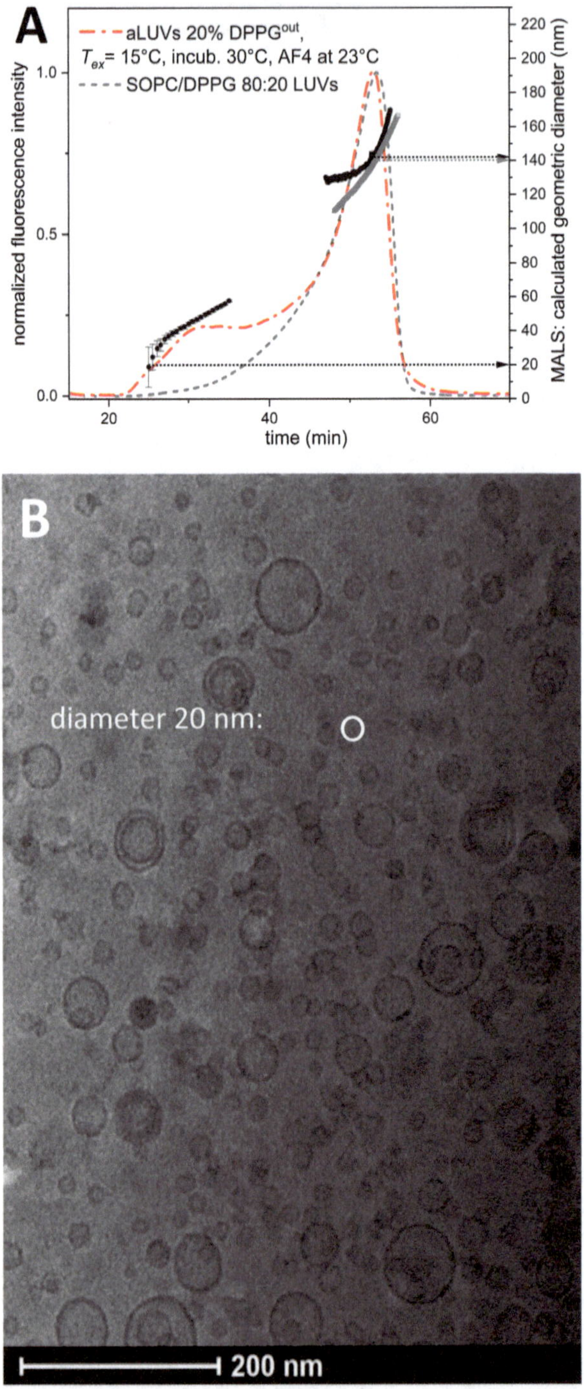

Fig. 4 Detection of small DVs in aLUV prepared at 15 °C after cyclodextrin removal and subsequent temperature increase as seen by AF4 (A) and cryo-transmission electron microscopy (cryo-TEM) (B). (A) The red dash–dot line (fluorescence signal, left axis) and black dots (MALS, right axis) represent a sample with 23 mol% DPPG in the outer leaflet after 30 min incubation at 35 °C, measured at 23 °C. The grey dashed curve (left axis) and

giving rise to the three green curves enclosing the green area in Fig. 3B (reproduced in detail in Fig. S5†). According to the criteria explained above, the transitions are rated to show significant but weaker coupling than that seen for the 30 °C-exchanged aLUVs. The two peaks of the green curves are farther apart and better separated than the shoulders of the red ones. Three samples prepared at 5 °C yielded the blue curves (see also Fig. S5†). For better visibility, no shading was done. The curves are largely within the green shading, indicating that there was no significant difference between the transitions of the aLUVs prepared at 5 and 15 °C.

Detecting budding of DVs after exposing aLUVs to temperature changes

Fig. 4 shows the typical AF4 elution profiles of symmetric LUVs (grey dashed curve) and of aLUVs produced at an exchange temperature of 15 °C (T_{ex} = 15 °C) and, after cyclodextrin removal, heated up to 35 °C and measured at 23 °C (red, dash–dot curve). The shoulder of the aLUV profile at short elution times implies the presence of a substantial number of much smaller particles. MALS data of the same run (black spheres, right axis) imply sizes of the order of 20 nm and somewhat larger at the shoulder. Symmetrical liposomes and remaining large aLUV show a peak size of ~140 nm.

The nature of the small particles created from aLUVs is clarified by cryo-TEM. Fig. 4B shows a large number of vesicles with diameters around 20 nm that illustrate and confirm the AF4 result. Note that the cryo-TEM picture displays DV sizes by number. The fluorescence signal (see red dash–dot line in Fig. 4A) corresponding to a given particle size scales with the fluorophore content of the vesicle, which increases with vesicle area, *i.e.*, the second power of size. Therefore, the local maximum of fluorescence seen for ~40 nm is not in conflict with a majority of DVs being in the 20 nm range.

The disagreement of the grey and black MALS curves in Fig. 4A illustrates that there is no strictly conserved correlation between size and elution time in the AF4 system. It is therefore important to consider MALS and fluorescence data in parallel for a given AF4 run.

Overall, Fig. 4B supports the interpretation of a short-time shoulder of AF4 profiles as shown here in terms of the presence of very small DVs, in line with previous results on lysolipid-induced budding of DVs.[10,16]

The question arises under which conditions such DVs appear and whether their formation is reversible. Fig. 5 shows AF4 elution profiles of aLUVs prepared at different exchange temperatures after cyclodextrin removal. The solid lines represent the first run, still at or somewhat below exchange temperature. These profiles are similar to those for symmetric liposomes (Fig. 4).

After incubating the aLUVs prepared at T_{ex} of 5 and 15 °C at 35 °C for 30 min, aliquots were taken again for AF4 carried out at 23 °C (red dash–dot lines in panels A and B). They show a pronounced shoulder assigned above to the appearance of small DVs with diameters of the order of 20 nm. The dashed curves in Fig. 5A and B were obtained by a third AF4 test of the samples after cooling

grey squares (right axis) were obtained with symmetrical LUVs of SOPC with 20% DPPG overall. (B) Cryo-TEM of sample treated analogously to aLUVs in (A), with specific DPPG content of 29 mol% in the outer leaflet according to ζ.

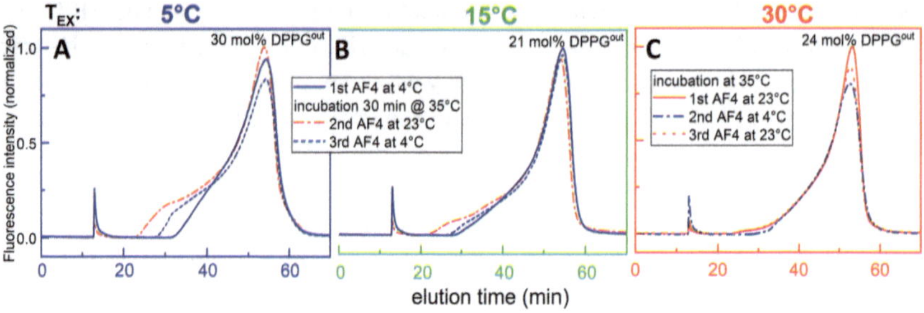

Fig. 5 AF4 elution profiles of aLUVs still at lipid-exchange temperature (solid lines), after incubation at different temperature (dash–dot) and back to T_{ex} (dashed) – line colors indicate AF4 temperatures of 4 °C (blue) or 23 °C (red). The different panels correspond to different exchange temperatures as marked in the plot. Zeta potential data yielded outer leaflet contents of DPPG of 21–30 mol% in SOPC vesicles, as indicated in the plots. Note that the first peak at 12 min is a technical artefact.

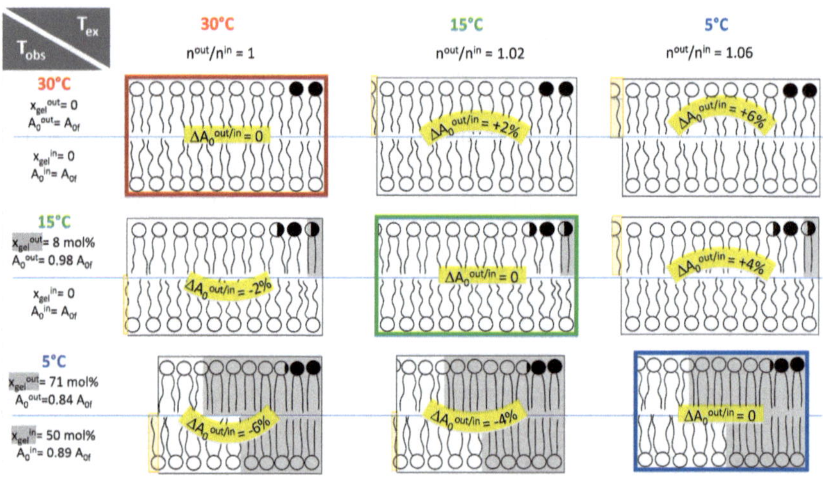

Fig. 6 Semiquantitative illustration of the model assuming negligible intrinsic area mismatch (*i.e.*, asymmetry stress), $\Delta A_0^{out/in} = 0$, at the exchange temperature, T_{ex}, at which aLUVs were prepared by equilibration with nearly lipid-saturated cyclodextrin. Columns of boxes represent selected exchange temperatures, rows temperatures of observation after cyclodextrin removal, T_{obs}. Note that the population number of the two leaflets, n^{out}/n^{in}, is conserved for a given T_{ex} (column) whereas the intrinsic fraction of gel phase, x_{gel}, and in turn, the average intrinsic area, A_0, for a given leaflet depends on observation temperature, T_{obs}. Temperature-induced intrinsic-area mismatch causes asymmetry stress and a tendency to bend the bilayer as indicated schematically.

them back down to 4 °C. The intermediate position of the shoulder between original (blue, first run) and budded (red, second run) state implies that DVs budding has been partially reversible; *i.e.*, some DVs seem to fuse upon cooling back.

The aLUVs produced at T_{ex} = 30 °C (Fig. 5C) do not show DVs, neither originally nor after cooling or re-heating. Summarizing, the appearance of small exovesicles is seen only for aLUVs prepared at 5 and 15 °C and then heated up to ≥23 °C.

Discussion

The hypothesis of "conserved intrinsic area upon exchange"

Baumgart and coworkers[13] showed for the exchange of lipid between lipid-loaded cyclodextrin and the outer leaflet of "acceptor" liposomes that there is a "crossing" lipid load of cyclodextrin. At lower loading, cyclodextrin tends to extract lipid from membranes. Lipid-supersaturated cyclodextrin inserts additional lipid into acceptor liposomes. To some extent, this consequence of an equilibrium distribution of the lipid is analogous to the cyclodextrin–vesicle partitioning of cholesterol.[47] The principal difference is that cholesterol equilibrates quickly between the leaflets and can be supplied or extracted quasi-symmetrically. Insertion *versus* removal of lipid from GUVs has been characterized by comparing the occurrence of invaginations or inward tubulation induced by outer-leaflet depletion with that of exvagination driven by overpopulation of the outer leaflet.[13]

Previous work[37,38] has shown that exchange of outer-leaflet POPC with POPG added in the form of cyclodextrin–lipid complexes replaces one POPC by one POPG as long as the added cyclodextrin is loaded with POPG close to saturation. There was no preference for one lipid over the other so that the PG : PC mole ratio in the outer leaflet after exchange agreed with that in the cyclodextrin complexes in solution and hence, with the total ratio between added DPPG and outer-leaflet POPC prior to exchange.

Importantly, the apparent saturation of cyclodextrin with lipid results from a high dissociation constant[13,37] and not from a full, stoichiometric binding; >90% of mβCD does not carry lipid. This renders the chemical potential of the complex little dependent on concentration. Over a finite range in cyclodextrin concentration, it simply absorbs the lipid that supersedes the intrinsic area of the outer leaflet and provides the lipid to fill the outer leaflet to match intrinsic and real area. Only well below saturation, cyclodextrin provides less lipid than extracting so that the outer leaflet becomes underpopulated[13] and ultimately, liposomes get dissolved.[13,37,48]

Defining a relative intrinsic-area mismatch, $\Delta A_0^{\text{out/in}}$:

$$\Delta A_0^{\text{out/in}} = \frac{2\left(n^{\text{out}} A_0^{\text{out}} - n^{\text{in}} A_0^{\text{in}}\right)}{n^{\text{out}} A_0^{\text{out}} + n^{\text{in}} A_0^{\text{in}}} \tag{1}$$

our hypothesis implies that upon proper exchange $\Delta A_0^{\text{out/in}} \approx 0$. Here, n stands for the number of lipid molecules in a given bilayer – setting $n^{\text{in}} = 1$, n^{out} simply becomes the number of outside lipids per inside lipid. A_0^{in} and A_0^{out} stand for the average intrinsic interfacial area per molecule in the inner and outer leaflet, respectively.

Let us, first, estimate A_0^{in} and A_0^{out} as a function of temperature. In the text concerning Fig. 1 above, we have estimated the intrinsic fraction of gel phase for SOPC forming the inner leaflet, $x_{\text{gel}}^{\text{in}}$, to be 50% at 5 °C and zero at ≥15 °C. For 20 mol% DPPG in SOPC, a typical composition of the outer leaflet, we obtained $x_{\text{gel}}^{\text{out}}$ of 71, 8 and 0 mol% at 5, 15 and 30 °C, respectively. Intrinsic fraction means here that the leaflet can form the phase that would be formed by a bilayer of the same composition and area per molecule, *i.e.*, it excludes coupling to a different leaflet.

The fraction x_{gel} governs A_0 because lipids shrink substantially upon freezing to the gel phase. For DPPG, $A_0 \approx 61$ Å^2 in the fluid[49] and 47 Å^2 in the gel[50,51] phase (for an overview, see Marsh[40]); *i.e.*, freezing reduces the area by a factor of 0.77. For simplicity and in lack of data for SOPC in the gel, we will assume for the semi-quantitative discussion here that both lipids have the same intrinsic areas of A_{0f} in the fluid and $A_{0gel} \approx 0.77A_{0f}$ in the gel phase. Then, A_0 becomes

$$\begin{aligned} A_0 &= A_{0f}\left(1 - x_{gel}\right) + A_{0gel}x_{gel} \\ &\approx A_{0f}\left(1 - 0.23x_{gel}\right) \end{aligned} \quad (2)$$

Lateral thermal expansion within the gel and fluid phase are neglected. With eqn (2) and the x_{gel} estimates given in the previous paragraph, one obtains the A_0 values for selected observation temperatures as provided in the left column of Fig. 6.

The upper left box in Fig. 6 illustrates aLUVs with 20 mol% DPPG^{out} that have been prepared at an exchange temperature of 30 °C, then separated from (partially lipid-loaded) cyclodextrin, incubated at 35 °C and finally characterized by AF4 at 23 °C. The DPPG content has been established maintaining a $\Delta A_0^{out/in}$ of zero – in line with our model – by exchanging 2 out of ten fluid SOPC with fluid DPPG. As the cyclodextrin has been removed, the lipids are fixed in their leaflets and the ratio $n^{out}/n^{in} = 1$ is conserved, independently of the observation temperature (first column in Fig. 6). If this sample is now cooled to 15 °C (one row down in Fig. 6), an intrinsic area mismatch is created by the freezing of 8 mol% of the outer leaflet lipid to the gel phase. Eqn (1) yields $\Delta A_0^{out/in} = 2 \times (0.98 - 1)/(0.98 + 1) = -2\%$, a negative stress that should favor a convex, inward bending. Cooling further to 5 °C yields, with the A_0^{out} and A_0^{in} shown on the left side, a mismatch of −6%.

The second column of boxes refers to aLUVs produced by exchange at 15 °C. As discussed in the previous paragraph, gel formation in the outer leaflet causes a shrinking of the average area per lipid by 2%. Then, the assumption of $\Delta A_0^{out/in}$ to be 0 as long as almost-lipid-saturated cyclodextrin is present implies that the outer leaflet contains a factor $n^{out}/n^{in} = 1.02$ times more lipid molecules. Again, this ratio is fixed as the cyclodextrin is removed and temperature-dependent changes in A_0^{out} and A_0^{in} cause intrinsic area mismatches that are positive, *i.e.*, outward-bending upon temperature increase and negative upon cooling. The same procedure yields a lipid number overpopulation of $n^{out}/n^{in} \approx 1.06$ for exchange at 5 °C and subsequent, positive intrinsic area mismatches at higher observation temperature.

Summarizing, the data support the concept that heating up aLUVs prepared under conditions of ordered outer and disordered inner (OO/DI) leaflets causes exovesiculation of small DVs upon OO melting. The underlying rationale predicts the same effect for aLUVs prepared in the all-fluid state and then cooled to a DO/OI scenario, *i.e.*, with lipid freezing in the inner leaflet.

In the literature, a number of such OO/DI systems have been studied without specifically addressing a potential DV budding. As discussed, spot checks with cryo-TEM can hardly provide a detailed assessment of budding but Fig. 3B of Eicher *et al.*[26] presents pictures of POPE^{in}|POPC^{out} that resemble our Fig. 4B. In this system, exovesiculation would be favored both by the shrinking of the inner leaflet upon POPE freezing and by the negative intrinsic curvature of the inner leaflet that would stabilize the strongly curved inner leaflet of a DV.

Budding as an alternative to asymmetry-stress-induced coupling?

In the introduction, we have mentioned a considerable number of interleaflet coupling mechanisms discussed in the literature. Here, we propose the interplay between intrinsic-area asymmetry and budding to be an additional coupling (or de-coupling) mechanism. Experimentally, the introduction of AF4 enabled us to sensitively detect exovesiculation (outward budding); we are not aware of information regarding inward budding from aLUVs. Whereas GUVs are free to bud cylindrical or other highly curved structures with microscopic dimensions to either side,[13] this may be different for LUVs. Scaling vesicles down by more than one order of magnitude brings sizes of DVs (or other budding structures) toward the order of membrane thickness. It should then matter whether budding proceeds towards or against the original curvature of the bilayer. Even if inward budding from LUVs can proceed, it would be limited to a smaller fraction of lipid than outward budding. The limit arises from the fact that small DVs have a larger surface-to-volume ratio than the original, large vesicle so that budding uses up excess area "stored" in deviations from spherical shape.[10,13] As mother vesicles become increasingly spherical, budding becomes increasingly opposed and stops. Inward DVs need even more membrane surface since their volume is enclosed by two bilayers.

Our data can, qualitatively, be explained on the basis of the "either coupling or (outward) budding" hypothesis. aLUVs prepared at 30 °C show no exovesiculation and increased interleaflet coupling of the melting transitions. Those prepared at 5 and 15 °C cause budding upon lipid melting and comparably weaker coupling. This is what one should expect if the other coupling mechanisms described play no major role here and inward budding is not or less possible.

aLUVs of POPC and the higher-melting POPE studied by Eicher *et al.*[26] do not support our simple "either coupling or budding" hypothesis. The fact that $\mathrm{POPE^{in}|{+}POPC^{out}}$ shows strong coupling of the melting transition, but $\mathrm{POPC^{in}|{+}POPE^{out}}$ does not, has been explained on the basis of the negative intrinsic curvature of POPE. From the perspective of intrinsic-area asymmetry, the question arises why the $\mathrm{POPE^{in}|{+}POPC^{out}}$ system, which might develop positive ΔA_0 and do outward budding, is coupling anyway. As mentioned, budding is limited by the available excess area.[10,13] The maximal intrinsic-area asymmetry in this system is quite high; at ~12–20 °C, the intrinsic state of the outer leaflet is virtually all fluid and the inner all gel,[26] causing a $\Delta A_0 \approx +26\%$. Hence, there might be some budding but this might not suffice to eliminate coupling. The reverse aLUVs, $\mathrm{POPC^{in}|{+}POPE^{out}}$, should develop a highly negative ΔA_0 which cannot be relaxed by outward budding. The finding that the transitions in the leaflets are hardly coupled points to some unknown pathway how this ΔA_0 is accommodated, with inward budding being a principal option to be explored.

A paper preparing $\mathrm{POPC^{in}|{+}DPPC^{out}}$ at room temperature and studying the sample at 55 °C, *i.e.*, likely undergoing $\mathrm{o} \rightarrow \mathrm{d^{in}|d^{out}}$ as in our study, does not seem to provide evidence suggesting or excluding budding.[19] It shows weak coupling. Other publications dealt with possible $\mathrm{d^{in}|d} \rightarrow \mathrm{o^{out}}$ scenarios[24,26,36,52–54] that are expected to induce negative ΔA_0. The respective papers report both strong and weak coupling for the different systems. A table compiling information about interleaflet coupling in aLUVs is provided as Table S9.†

Overall, there are sophisticated models describing coupling as such[30,55,56] but a prediction of the extent of coupling for a given combination of leaflet compositions and preparation conditions is still lacking. It is mechanistically convincing that asymmetry stress is a coupling mechanism that can be inhibited by budding. However, this neither excludes budding to happen in addition to coupling nor weak or no coupling to be found although budding is impossible or unlikely.

The problem of quantifying coupling

Another problem is that there is no clear-cut way to quantify the extent of coupling of phase transitions. Weak or no coupling is indicated by a thermotropic transition curve resembling a simple superposition of individual transitions. A good example[26] for weak coupling following this criterion is $POPC^{in}|$+$POPE^{out}$.

One model for a transition curve of aLUVs with strong coupling is that of the corresponding scrambled vesicles or of separately prepared, symmetric LUVs with the same overall composition. In Fig. 1B, we provide such a symmetrical analog curve measured with SOPC + 10 mol% DPPG symmetrically to model perfect coupling of aLUV with $SOPC^{in}|$+20 mol% $DPPG^{out}$. The agreement is poor, arguing against strong coupling. In contrast, aLUV data of $POPE^{in}|$+$POPC^{out}$ agreed quite well with their symmetrical analog,[26] indicating strong coupling.

It should, however, be noted that the model that a DSC curve of strong coupling resembles that of a scrambled sample is not unequivocal. Furthermore, the idea of "intermediate coupling" to give rise to transition curves with peaks in between those of averaged and scrambled-lipid curves is not trivial. Cheng and London[24] measured the fluorescence anisotropy of DPH and TMA-DPH in aLUVs of DOPE-$POPS^{in}|$+SM^{out}. The temperature-dependent steady-state anisotropy of DPH for these aLUVs was compared with those of symmetrical LUVs of the compositions of the inner, the outer, and scrambled leaflets. The resulting curves for averaged (not explicitly shown) and scrambled-like vesicles were, in fact, not much different from each other. Importantly, the experimental data for aLUVs were not at or in between these model curves but much closer to that for the SM-rich, outer leaflet. Assuming that DPH remained homogeneously distributed between the leaflets and showed about the same fluorescence lifetimes there, this argues for a strongly non-ideal mixing in the gel phase. In such a non-ideal system, it seems non-trivial to assess the extent of coupling in the first place.

Technical and biological implications of asymmetric transitions

Above, it has been demonstrated that the structural and thermodynamic properties of aLUVs depend on the specific conditions upon preparation such as the exchange temperature. Furthermore, potentially irreversible processes such as budding and fission of DVs render an aLUV sample dependent on its thermal history; storing an aLUV sample in the fridge might change it substantially.

The potential importance of interleaflet coupling for transmembrane signaling has long been recognized. An interesting aspect in this regard is the finding that the budding of DVs is partially reversible. That means that very small vesicles become fusogenic when their intrinsic-area asymmetry ΔA_0 decreases by a shrinking of the outer or expansion of the inner leaflet. This may also modulate the activity of protein-based remodeling machineries. Maiti and coworkers[57] have demonstrated that the SNARE-driven fusion of synaptic vesicles is promoted by a high serotonin content. Inner leaflet expansion by intercalating amphiphilic serotonin molecules should, indeed, cause a reduction of ΔA_0. Cell membranes with asymmetric order–disorder transitions have to regulate their intrinsic area asymmetry to avoid strong temperature dependencies of membrane remodeling processes.

Conclusions

The preparation of aLUVs *via* lipid exchange with cyclodextrin complexes works also in the gel phase. Mixing of DPPG and SOPC between cyclodextrin and aLUVs remained at least close to random.

The data are in line with the hypothesis that lipid exchange between aLUV outer leaflets and cyclodextrin loaded with another lipid close to saturation produces largely asymmetry-stress-free liposomes. This explains also the finding that aLUVs differ in their properties depending on the exchange temperature they have been prepared at.

Our data support the "coupling or budding" hypothesis that intrinsic-area asymmetry changes should impose interleaflet coupling of asymmetric melting or freezing transitions unless it can be accommodated by the budding and fission of daughter vesicles (DVs). However, published data are not in line with this simple idea – highlighting the complexity of coupling phenomena. A prediction of the extent of coupling based on leaflet compositions and preparation procedure remains impossible so far.

Partial reversibility of DV budding demonstrates that small vesicles can be rendered fusogenic as their intrinsic-area asymmetry is reduced.

Data availability

The data supporting this article are included as part of the supplementary information.†

Author contributions

H. H. conceived the study. M. K. coordinated the work and designed, carried out and analyzed DSC, PPC, DLS and most AF4 experiments. M. J. and M. K. established and applied the HPLC protocol. S. G. contributed some AF4 experiments; M. H. supported AF4 and established the AF4 setup at 5 °C. J. L. and L. K. did the cryo-TEM. H. H. supervised the experiments. M. K. wrote the first draft with assistance from other authors; all authors discussed and edited the draft and H. H. and M. K. finalized the manuscript. All authors approved the final version.

Conflicts of interest

The authors declare having no conflicts of interest.

Acknowledgements

Major instrumentation was funded by the Center for Biological Signaling Studies, BIOSS. S. G. acknowledges funding from the Deutsche Forschungsgemeinschaft (DFG, German Research Foundation) *via* the Research Training Group 2202 "Transport into and across membranes" (278002225/RTG 2202). We thank Lipoid GmbH (Ludwigshafen, Germany) for providing the lipids. The authors thank A. Meister, G. Pabst, T. Gutsmann, and D. Huster for constructive discussions.

References

1 J. Israelachvili, *Intermolecular and surface forces*, Academic Press, Amsterdam, 1984.
2 R. A. Demel and B. De Kruyff, *Biochim. Biophys. Acta, Rev. Biomembr.*, 1976, **457**, 109–132.
3 P. L. Yeagle, *Biochim. Biophys. Acta, Rev. Biomembr.*, 1985, **822**, 267–287.
4 H. Heerklotz and A. Tsamaloukas, *Biophys. J.*, 2006, **91**, 600–607.
5 S. M. Gruner, *Proc. Natl. Acad. Sci. U. S. A.*, 1985, **82**, 3665–3669.
6 M. P. Sheetz and S. J. Singer, *Proc. Natl. Acad. Sci. U. S. A.*, 1974, **71**, 4457–4461.
7 J. Steigenberger, Y. Verleysen, N. Geudens, J. C. Martins and H. Heerklotz, *Front. Microbiol.*, 2021, **12**, 2644.
8 H. Heerklotz and J. Seelig, *Eur. Biophys. J.*, 2007, **36**, 305–314.
9 H. Y. Fan, D. Das and H. Heerklotz, *Langmuir*, 2016, **32**, 11655–11663.
10 L. Hua, M. Kaiser, I. Carabadjac, A. Meister, G. Hause and H. Heerklotz, *Biophys. J.*, 2023, **122**, 4011–4022.
11 H. Y. Fan and H. Heerklotz, *J. Colloid Interface Sci.*, 2017, **504**, 283–293.
12 U. Kragh-Hansen, M. Le Maire and J. V. Møller, *Biophys. J.*, 1998, **75**, 2932–2946.
13 T. Reagle, Y. Xie, Z. Li, W. Carnero and T. Baumgart, *Soft Matter*, 2024, **20**, 4291–4307.
14 V. N. Georgiev, A. Grafmüller, D. Bléger, S. Hecht, S. Kunstmann, S. Barbirz, R. Lipowsky and R. Dimova, *Adv. Sci.*, 2018, **5**, 1800432.
15 S. Penič, L. Mesarec, M. Fošnarič, L. Mrówczyńska, H. Hägerstrand, V. Kralj-Iglič and A. Iglič, *Front. Phys.*, 2020, **8**, 342, DOI: 10.3389/fphy.2020.00342.
16 L. Hua, S. Akcesme, K. Müller and H. Heerklotz, *J. Phys. Chem. B*, 2025, **129**, 1260–1273.
17 H. Heerklotz, *Biophys. J.*, 2001, **81**, 184–195.
18 R. Jin and T. Baumgart, *Soft Matter*, 2021, **17**, 7506–7515.
19 F. A. Heberle, D. Marquardt, M. Doktorova, B. Geier, R. F. Standaert, P. Heftberger, B. Kollmitzer, J. D. Nickels, R. A. Dick, G. W. Feigenson, J. Katsaras, E. London and G. Pabst, *Langmuir*, 2016, **32**, 5195–5200.
20 B. Eicher, F. A. Heberle, D. Marquardt, G. N. Rechberger, J. Katsaras and G. Pabst, *J. Appl. Crystallogr.*, 2017, **50**, 419–429.
21 L. Dietel, L. Kalie and H. Heerklotz, *Biophys. J.*, 2020, **119**, 767–779.
22 J. Steigenberger, Y. Verleysen, N. Geudens, J. C. Martins and H. Heerklotz, *Front. Microbiol.*, 2021, **12**, 669709, DOI: 10.3389/fmicb.2021.669709.
23 U. Kragh-Hansen, M. Le Maire and J. V. Møller, *Biophys. J.*, 1998, **75**, 2932–2946.
24 H.-T. Cheng and E. London, *Biophys. J.*, 2011, **100**, 2671–2678.

25 G.-S. Yang, A. Wagenknecht-Wiesner, B. Yin, P. Suresh, E. London, B. A. Baird and N. Bag, *Biophys. J.*, 2024, **123**, 2256–2270.
26 B. Eicher, D. Marquardt, F. A. Heberle, I. Letofsky-Papst, G. N. Rechberger, M.-S. Appavou, J. Katsaras and G. Pabst, *Biophys. J.*, 2018, **114**, 146–157.
27 M. P. K. Frewein, P. Piller, E. F. Semeraro, O. Czakkel, Y. Gerelli, L. Porcar and G. Pabst, *Biophys. J.*, 2023, **122**, 2445–2455.
28 G. W. Feigenson, J. Huang and T. A. Enoki, *J. Am. Chem. Soc.*, 2023, **145**, 21717–21722.
29 V. Kiessling, C. Wan and L. K. Tamm, *Biochim. Biophys. Acta, Biomembr.*, 2009, **1788**, 64–71.
30 M. D. Collins and S. L. Keller, *Proc. Natl. Acad. Sci. U. S. A.*, 2008, **105**, 124–128.
31 T. R. Galimzyanov, R. J. Molotkovsky, M. E. Bozdaganyan, F. S. Cohen, P. Pohl and S. A. Akimov, *Phys. Rev. Lett.*, 2015, **115**, 088101, DOI: 10.1103/PhysRevLett.115.088101.
32 A. Hossein and M. Deserno, *Biophys. J.*, 2020, **118**, 624–642.
33 J. D. Nickels, J. C. Smith and X. Cheng, *Chem. Phys. Lipids*, 2015, **192**, 87–99.
34 M. P. K. Frewein, P. Piller, E. F. Semeraro, K. C. Batchu, F. A. Heberle, H. L. Scott, Y. Gerelli, L. Porcar and G. Pabst, *J. Membr. Biol.*, 2022, **255**, 407–421.
35 T. A. Enoki and G. W. Feigenson, *Biochim. Biophys. Acta, Biomembr.*, 2022, **1864**, 183995.
36 S. Chiantia and E. London, *Biophys. J.*, 2012, **103**, 2311–2319.
37 M. Markones, C. Drechsler, M. Kaiser, L. Kalie, H. Heerklotz and S. Fiedler, *Langmuir*, 2018, **34**, 1999–2005.
38 M. Markones, A. Fippel, M. Kaiser, C. Drechsler, C. Hunte and H. Heerklotz, *Biophys. J.*, 2020, **118**, 294–302.
39 Y.-P. Zhang, R. N. A. H. Lewis and R. N. McElhaney, *Biophys. J.*, 1997, **72**, 779–793.
40 D. Marsh, *Handbook of Lipid Bilayers*, 2013.
41 M. J. Hope, M. B. Bally, G. Webb and P. R. Cullis, *Biochim. Biophys. Acta, Biomembr.*, 1985, **812**, 55–65.
42 R. C. MacDonald, R. I. MacDonald, B. P. M. Menco, K. Takeshita, N. K. Subbarao and L. Rong Hu, *Biochim. Biophys. Acta, Biomembr.*, 1991, **1061**, 297–303.
43 G. R. Bartlett, *J. Biol. Chem.*, 1959, **234**, 466–468.
44 V. V Plotnikov, J. M. Brandts, L. N. Lin and J. F. Brandts, *Anal. Biochem.*, 1997, **250**, 237–244.
45 H. Heerklotz and J. Seelig, *Biophys. J.*, 2002, **82**, 1445–1452.
46 European Medicines Agency, Validation of Analytical Procedures Q2(R2), Committee for Medicinal Products for Human Use EMA/CHMP/ICH/82072/2006, 2023, https://database.ich.org/sites/default/files/ICH_Q2%28R2%29_Guideline_2023_1130.pdf.
47 A. Tsamaloukas, H. Szadkowska, P. J. Slotte and H. Heerklotz, *Biophys. J.*, 2005, **89**, 1109–1119.
48 T. G. Anderson, A. Tan, P. Ganz and J. Seelig, *Biochemistry*, 2004, **43**, 2251–2261.
49 G. Pabst, S. L. Grage, S. Danner-Pongratz, W. Jing, A. S. Ulrich, A. Watts, K. Lohner and A. Hickel, *Biophys. J.*, 2008, **95**, 5779–5788.

50 A. Watts, K. Harlos and D. Marsh, *Biochim. Biophys. Acta, Biomembr.*, 1981, **645**, 91–96.
51 G. Pabst, S. Danner, S. Karmakar, G. Deutsch and V. A. Raghunathan, *Biophys. J.*, 2007, **93**, 513–525.
52 H.-T. Cheng and E. London, *J. Biol. Chem.*, 2009, **284**, 6079–6092.
53 S. Chiantia, P. Schwille, A. S. Klymchenko and E. London, *Biophys. J.*, 2011, **100**, L1–L3.
54 Q. Lin and E. London, *Biophys. J.*, 2015, **108**, 2212–2222.
55 S. May, *Soft Matter*, 2009, **5**, 3148–3156.
56 A. J. Wagner, S. Loew and S. May, *Biophys. J.*, 2007, **93**, 4268–4277.
57 D. Saha Roy, A. Gupta, V. Vishvakarma, P. Krupa, M. S. Li and S. Maiti, *J. Phys. Chem. B*, 2024, **128**, 4975–4985.

Faraday Discussions

ROYAL SOCIETY OF CHEMISTRY

PAPER

Melting point matters: designing lipid nanocarriers for improved T cell activation†

Carina S. Fedosejevs,[ab] Lariana Cline [ab] and Neha P. Kamat*[abc]

Received 7th January 2025, Accepted 10th January 2025
DOI: 10.1039/d5fd00002e

Surface-modified lipid nanocarriers are increasingly used as artificial antigen-presenting cells for therapeutic applications in immunotherapy. Within these nanocarriers, the role of the lipids is typically limited to providing structure/stability of the particle, to anchoring a targeting moiety, and/or to altering the biodistribution of the nanocarriers *in vivo*. However, lipid membranes also possess special thermodynamic properties that impact their function. Here, we investigate the effect of the melting transition temperature of lipid nanocarriers on the activation efficiency of an immortalized line of T lymphocytes. Using an established *in vitro* activation assay and αCD3-functionalized lipid nanocarriers, we screened a variety of lipid nanocarriers with respect to their capacity to activate T cells. We observed a correlation between T cell activation efficiency and proximity of the melting transition temperature of the lipid nanocarrier to the temperature at which the activation study was conducted (37 °C). This relationship held across a variety of lipid compositions and appeared to be more important than the lipid headgroup or chain length. This trend was preserved when the activation temperature was shifted to 30 °C, supporting the role of the nanocarrier membrane state for target cell activation and the potential impact of phase-transition-related effects on nanocarrier activity. We conclude that lipid composition is indeed an important parameter for lipid-based nanocarrier design, not only for the more explored biochemical roles of the lipids but also for the thermodynamic properties the lipid mixtures generate. Our results provide a new consideration in therapeutic nanocarrier design that could significantly improve the efficacy of targeted nanocarrier formulations.

Introduction

The design of artificial antigen-presenting cells (aAPCs) that target and activate T cells, either *ex* or *in vivo*, has become an important approach for T cell

[a]*Department of Biomedical Engineering, Northwestern University, Evanston, IL 60208, USA. E-mail: nkamat@northwestern.edu*
[b]*Center for Synthetic Biology, Northwestern University, Evanston, IL 60208, USA*
[c]*Chemistry of Life Processes Institute, Northwestern University, Evanston, IL 60208, USA*

† Electronic supplementary information (ESI) available. See DOI: https://doi.org/10.1039/d5fd00002e

immunotherapy. Among the various nanoparticles used to design aAPCs, lipid-based nanoparticles are a popular choice due to their high biocompatibility and low systemic toxicity *in vivo*.[1–5] While several design principles have been identified that enhance the biodistribution and compatibility of lipid-based particles *in vivo*,[6–9] there remain a surprisingly low number of approved aAPC therapies for the clinic.[10] This dearth of approved therapies for immune-cell activation suggests a greater need for improved lipid nanocarriers that robustly and efficiently activate their target cells.

To better improve the design of aAPCs, it is useful to consider the sequence of events that T cells undergo during activation by an antigen-presenting cell. Specifically, for T lymphocytes, there have been a number of studies uncovering the complex processes that occur in the plasma membrane during activation.[11–21] Upon binding APC antigens, an immunological synapse is formed between the T cell and antigen-presenting cell, triggering a downstream signaling cascade activated by the influx of Ca^{2+} ions.[11,12] The formation of the immunological synapse is accompanied by a complex reorganization of the T cell membrane, referred to as "bullseye formation", during which the T cell receptors (TCRs) cluster in the center of the bullseye.[13,14] The asymmetry of the membrane in this region is reduced as phosphatidylserine (PS) lipids are relocated to the outer leaflet, creating a negative charge sink at the synapse and a local increase in lipid order.[15–19] Simultaneously, local deformations in the T cell membrane are necessary to bring antigens on the aAPC and receptors on the T cell close enough to bind and initiate this process, requiring local areas with increased elasticity.[20,21] Altogether, the activation of T cells requires rapid remodeling and reorganization of lipids and proteins on the T cell surface.

The dynamic rearrangements on the T cell membrane during synapse formation and activation suggest that the properties of the aAPCs facilitating this reorganization should enhance the efficiency of synapse formation and T cell activation. Along these lines, several properties of aAPCs have been shown to be important for T cell activation, including the biochemical identity and presentation of antigens and co-stimulatory molecules, as well as physical characteristics of the APCs and aAPCs themselves. Toward the former feature, the density of activating antibodies, such as the commonly used anti-CD3 (TCR stimulus) and anti-CD28 (costimulatory cue), has been shown to impact T cell activation.[22,23] Toward the latter feature, it has been found that the size,[24–26] shape,[27–29] and membrane fluidity[23,30] of the aAPC can impact aAPC efficacy. For example, the rigidity of the aAPC membrane is an important parameter for the activation process.[14,20,31] Since TCRs cluster during activation, a corresponding mobility of the activating antigen on the antigen-presenting cell is advantageous. Indeed, spatial clustering of the targeting moiety has been shown to enhance its binding efficiency to cellular targets.[32–34] Together, these studies suggest a critical role of the physical properties of functionalized aAPCs in effectively activating T cells.

While previous work has largely focused on the role of static physical properties of aAPCs in activating immune cells, perhaps a more important feature of aAPCs is their capacity to undergo dynamic changes in their properties. Lipid membranes and heterogenous cell membranes are able to undergo phase transitions, also known as melting transitions. Depending on their environment (temperature, pH, pressure, salt concentration, *etc.*), lipids adopt distinct orientations and phases within a bilayer, ranging from a less dense, disordered chain orientation (liquid disordered phase, L_D) to a more crystalline, highly ordered one (gel phase, L_O). Synthetic lipid membranes assembled from mixtures of lipids can undergo melting transitions between phase states. At these transition points, the membrane structure becomes highly dynamic, exhibiting large intrinsic fluctuations. The fluctuations can lead to unusual or unexpected events, as properties like compressibility, heat capacity, and area expansion coefficient diverge relative to points below or above the phase transition point.[35–38] While most studies on lipid membrane transitions have focused on synthetic lipid mixtures, recent studies have shown these transitions exist in biological membranes as well.[39–41] These phase transitions are expected to play an important role in cooperatively regulating various cellular functions such as transport processes, protein organization, and enzyme activities.[42,43] For example, a membrane might get very soft at the phase transition point with an associated increased permeability relative to points below and above the transition point, leading to enhanced signal transduction.[35–38,44] These prior results point to a potentially important but overlooked property in the design of nanocarrier membranes for activating signaling cascades in a target cell. Specifically, nanocarriers designed with a melting transition close to that of the target cell membrane's transition point might best support dynamic rearrangements of lipids and proteins on both membranes to efficiently drive cell synapse formation and activation.

Here, we set out to investigate the effect of the melting transition temperature of nanocarrier membranes on the activation of T cells. Using an *in vitro* platform to assemble aAPCs and assess T cell activation, we explored the effect of the nanocarrier melting transition temperature and its proximity to the cell growth temperature on T cell activation. Our results point to a new design consideration of surface functionalized, lipid-based aAPCs to effectively promote target cell activation.

Experimental

Materials

The following lipids were purchased from Avanti Polar Lipids: 1,2-dioleoyl-*sn*-*glycero*-3-phosphocholine (DOPC), cholesterol, 1,2-dioleoyl-*sn*-*glycero*-3-phosphoethanolamine-*N*-[benzylguanine(polyethylene glycol)-2000] (18:1 PE PEG BG), 1,2-distearoyl-*sn*-*glycero*-3-phosphoethanolamine-*N*-[methoxy(polyethylene glycol)-2000] (ammonium salt) (18:0 PE PEG 2000), 1,2-di-(9Z-octadecenoyl)-*sn*-*glycero*-3-phospho-L-serine (sodium salt) (18:1 PS; DOPS), 1,2-dilauroyl-*sn*-*glycero*-3-phospho-L-serine (sodium salt) (12:0 PS; DLPS), 1,2-dimyristoyl-*sn*-*glycero*-3-phospho-(1′-*rac*-glycerol) (sodium salt) (14:0 PG; DMPG), 1,2-dimyristoyl-*sn*-*glycero*-3-phosphocholine (14:0 PC; DMPC), 1-stearoyl-2-myristoyl-*sn*-*glycero*-3-phosphocholine (18:0-14:0 PC), 1-myristoyl-2-palmitoyl-*sn*-*glycero*-3-phosphocholine (14:0-16:0 PC), 1,2-dimyristoyl-*sn*-*glycero*-3-phospho-L-serine (sodium salt) (14:0 PS; DMPS), 1,2-dielaidoyl-*sn*-*glycero*-3-phosphoethanolamine (18:1 (Δ^9-*trans*)PE), 1-myristoyl-2-stearoyl-*sn*-*glycero*-3-phosphocholine (14:0-18:0 PC), 1,2-dipalmitoyl-*sn*-*glycero*-3-phospho-(1′-*rac*-glycerol) (sodium salt) (14:0 PG; DPPG), 1,2-dipalmitoyl-*sn*-*glycero*-3-phosphocholine (16:0 PC; DPPC), 1,2-diheptadecanoyl-*sn*-*glycero*-3-phosphocholine (17:0 PC), 1,2-dipalmitoyl-*sn*-*glycero*-3-phospho-L-serine (sodium salt) (16:0 PS; DPPS), 1,2-distearoyl-*sn*-*glycero*-3-phospho-(1′-*rac*-glycerol) (sodium salt) (18:0 PG, DSPG).

The fluorescent dyes, 2-dimethylamino-6-lauroylnaphthalene (LAURDAN) and 3-((3-((9-(diethylamino)-5-oxo-5*H*-benzo[*a*]phenoxazin-2-yl)oxy)propyl)(dodecyl) (methyl)ammonio)propane-1-sulfonatem (NR12S), were purchased from Fisher Scientific.

DNA preparation and protein expression

The αCD3 scFv-SNAP-tag plasmid was purchased from Twist Bioscience. The plasmid is designed with a T7 promoter and terminator, a ribosomal binding site and an N-terminal FLAG-tag, and a SNAP-tag based on a pJL1 vector. The DNA sequence is provided in Table S1.† DNA was amplified and purified with a PureLink™ HiPure Plasmid Midiprep Kit (Invitrogen). The protein was expressed cell-free with a PURExpress *In Vitro* Protein Synthesis kit (New England Biolabs) and additional PURExpress Disulfide Bond Enhancers (New England Biolabs). The reaction was assembled on ice with 3.3 nM plasmid DNA and run for 16 h at 30 °C. Afterwards, the protein was stored at 4 °C.

Liposome preparation

Liposomes were prepared *via* thin-film hydration: 70 mol% of a phospholipid of interest, 29 mol% cholesterol, 0.9 mol% 18:0 PE PEG 2000, and 0.1 mol% 18:1 PE PEG BG were mixed in a glass vial in chloroform and gently dried under a flow of nitrogen to create a thin film of the lipid mixture. The films were stored in a vacuum overnight to remove residual solvent. Films were rehydrated with phosphate-buffered saline (PBS) to a final lipid concentration of 4 mM at 60 °C for 30–60 minutes and vortexed in between. The vesicle suspensions were extruded through 100 nm filters (Whatman Nuclepore polycarbonate membranes) with a Mini-Extruder set (Avanti) with 11 passes through the membrane to form homogeneous large unilamellar vesicles. For functionalization, 4 μl of the αCD3 cell-free-expression crude mixture was incubated with 4 μl of liposomes for 2 h at room temperature to allow for conjugation of the αCD3-SNAP-tag to liposomes containing BG-lipids and was afterwards stored at 4 °C.

GUV preparation

Giant unilamellar vesicles (GUVs) were prepared *via* spontaneous swelling. Briefly, 60 μl of a 2.5 mM lipid solution was diluted in 200 μl chloroform in a 20 ml glass vial. Under a stream of nitrogen, the chloroform was evaporated and a thin film formed under constant rotation of the glass by hand. The vial was placed in a vacuum for 2 h to remove residual solvent. 2 ml sucrose solution (295 mOsm) was gently pipetted into the vial, which was then placed in a 60 °C oven overnight. Without further vortexing of the lipid suspension, 1 ml of the volume was taken out the next morning. To concentrate the GUVs, 200 μl of the suspension was carefully pipetted into 1 ml glucose (295 mOsm). After 20–30 min, 20 μl of the bottom fraction was collected and transferred to a new tube to either add the GUVs directly to an imaging dish for quality control or to further conjugate the GUVs with 4 μl of the αCD3 antibody fragment for 2 h at room temperature.

T cell activation

An in-house engineered Jurkat nuclear factor of activated T cells (NFAT) reporter cell line (described in detail in ref. 34) was used for this study. Cells were cultured in Roswell Park Memorial Institue (RPMI) 1640 medium (Gibco) with 10% fetal bovine serum (FBS) (Gibco) and 1% penicillin–streptomycin (Gibco) at 37 °C with 5% CO_2. For the activation study, 100 000 Jurkat cells in 100 µl full RPMI were seeded into a well of a 96 well plate, per each condition and replicate. 8 µl of the functionalized liposome suspension was added, and the plate was incubated overnight at 37 °C with 5% CO_2. As a positive control, phorbol 12-myristate 13-acetate (PMA) (0.1 ng ml^{-1} final) and ionomycin (2 ng ml^{-1} final) (both purchased from Sigma Aldrich) were added. For activation at 30 °C, the cells were incubated in an incubator at 30 °C and 5% CO_2. In total, for each lipid type at least 3 biological replicates have been measured at 37 °C with at least 2 technical replicates for each biological replicate. For studies conducted at 30 °C, 3 measurements with 2 technical replicates each were conducted.

Flow cytometry

Cells positive for green fluorescent protein (GFP) were detected *via* flow cytometry. Samples were diluted into 1 ml flow buffer (2 mM ethylenediamine tetraacetic acid (EDTA), 1% FBS) and washed (centrifugation at 500 g for 5 min, removal of supernatant, and resuspension in 300 µl flow buffer). Measurements were conducted on a BD64 LSRFortessa cell analyzer with sample excitation at 488 nm and GFP signal detection with a FITC filter. Events were gated for cells and single cells. The GFP-positive gate was based on the αCD3 only control (no liposomes) to account for background activation from unconjugated protein, although nearly no background activation has been observed with the soluble protein. Subsequent data analysis and creation of data plots was conducted in Python.

Laurdan GP spectroscopy

Liposomes containing 0.5 mol% Laurdan were diluted to a final concentration of 400 µM into the medium of choice (PBS or RPMI) and transferred into a cuvette (UV-cuvette, Brandtech). Measurements were conducted on a Cary Eclipse fluorescence spectrophotometer (Agilent technologies) equipped with a multicell holder, allowing up to 4 samples to be measured at the same time. Temperature ramps were conducted with 0.5 °C min^{-1} steps and data points were acquired in 0.5 °C steps with an integration time of 1 s. The temperature was detected with a sensor placed in a cuvette filled with water allowing a feedback loop for the spectrometer to adjust its temperature during the ramp. The Laurdan signal was excited at 380 nm and emission detected in two channels ($I_1 = 450$ nm and $I_2 = 500$ nm) in order to calculate the generalized polarization (GP) value: $GP = (I_1 - I_2)/(I_1 + I_2)$.

Dynamic light scattering

Liposome size and polydispersity were measured with dynamic light scattering (DLS) in a 100 µM dilution using a Malvern Zetasizer Nano. The results display the mean values of at least two biological replicates and each replicate is the average of 3 runs with 3 reads each.

Microscope activation study

Fluorescence images were recorded with confocal scanning microscopy on a Nikon Ti Eclipse microscope equipped with a confocal unit (Nikon C2 Si) and a 60× objective (Nikon Apo Lambda S 60× Oil). Activation was studied at 37 °C with a heated stage (PE100 Peltier system, Linkam). For imaging, cells were stained with 1 μM NR12S in RPMI and 1×10^5 cells in 130 μl full RPMI medium were transferred to a μ-Slide 18 Well imaging dish (#1.5, Ibidi). For activation, 20 μl of the GUV-αCD3 conjugation mixture was added to the cells. Imaging was started shortly after, as soon as the GUVs sank and started to attach to the Jurkat cells. GUVs were stained with free NR12S that was still in the cell solution. The background fluorescence of NR12S in the medium was not detectable. NR12S was excited at 488 nm and emission read in two channels (I_1 = 565–595 nm and I_2 = 600–630 nm). The emission spectrum is sensitive to changes in cholesterol and membrane state and accordingly spectral shifts were detected by measuring the ratio of two characteristic wavelengths (similarly to Laurdan). To quantify emission signals, the intensities were measured over regions of interest (ROIs) containing a GUV or a cell. Pixel-by-pixel analysis was not used for quantification as the slight movement of vesicles leads to small displacements and the edges of the membranes did not reflect correct ratio values. Ratio images and ROI analysis was conducted in the NIS-Elements software (Nikon).

Results and discussion

Lipid nanocarrier T_m affects the activation of Jurkat cells more so than lipid identity

To investigate the effect of nanocarrier melting transition temperature on T cell activation, we first assembled aAPCs that bind and activate T cells. aAPCs were assembled from 100 nm large unilamellar vesicles (LUVs) containing 70 mol% of a phospholipid of interest (18:1 PC as a reference), 29 mol% cholesterol, 0.9 mol% 18:0 PE PEG 2000, and 0.1 mol% 18:1 PE PEG BG. The latter presents the covalent binding partner of the SNAP-tag coupled to a PEGylated lipid anchor. We next used a simplified one-pot workflow to assemble aAPCs by conjugating a cell-free-expressed αCD3–SNAP-tag single-chain variable fragment (scFv) to LUVs, as we have previously described.[34] CD3 is an associated subunit of the TCR and presents a suitable target for this kind of study. An αCD3 scFv antibody fragment was designed as a fusion protein with a SNAP-tag and expressed in a bacterial cell-free expression system. Upon addition of the fusion protein to liposomes containing a benzylguanine (BG) lipid, the protein was covalently attached to liposomes and used for subsequent activation studies (Fig. 1A).

For one biological replicate, we expressed one batch of protein and divided the protein between different lipid LUV samples to ensure a similar protein concentration on each LUV and so that differences in activation efficiencies relate to the lipid composition and not to protein amount. Due to varying protein yields between batches, the absolute % activation of the cells varied between biological replicates. The activation levels generated by the different lipid/LUV types are noted as relative values compared to the activation of the reference sample 18:1 PC within each biological replicate. In the following, if not stated otherwise, relative values in % refer to the difference in activation between the sample and the reference (DOPC liposomes):

$$\left(100 \times \frac{\text{activation}_{\text{sample}}}{\text{activation}_{\text{DOPC}}}\right) - 100$$

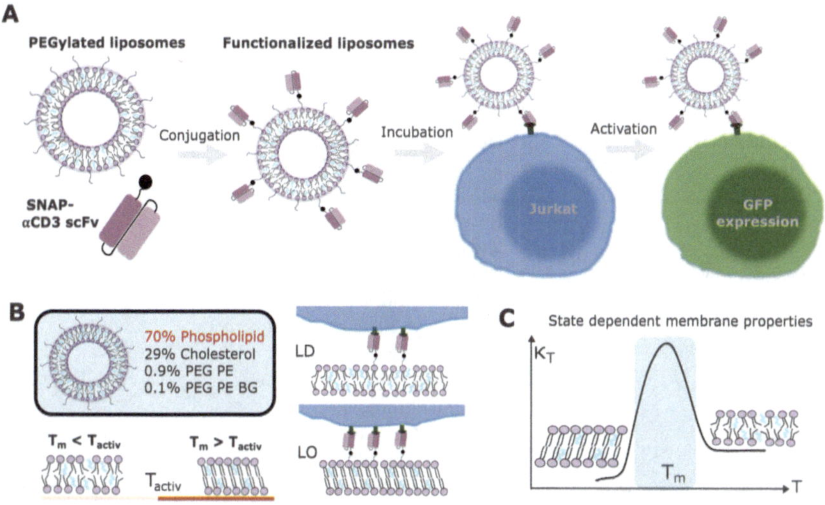

Fig. 1 Schematic overview of the assembly of lipid-based artificial antigen-presenting cells (aAPCs) and activation of Jurkat cells. (A) aAPCs are assembled from large unilamellar vesicles (LUVs) containing 70% phospholipids, 29% cholesterol, 0.9% PEG-lipids, and 0.1% benzylguanine (BG)-modified lipids, in which % refers to mol%. BG is a substrate for the self-labeling protein SNAP-tag and allows a SNAP-tag–αCD3 scFv fusion protein to be conjugated to LUVs *via* attachment of the SNAP-tag to the BG groups on the LUV surface. These particles are incubated overnight at 37 °C with an engineered Jurkat cell line expressing GFP upon activation. Activation is confirmed *via* microscopy and quantified with flow cytometry. This assay is based on the one-pot approach of LUV assembly introduced previously.[34] (B) The standard liposome formulation contains 18:1 PC (DOPC, $T_m = -17$ °C) keeping the liposome in the liquid disordered phase during interaction with the membrane of Jurkat cells. Phospholipids with $T_m > T_{activ}$ form more highly ordered membranes in the liquid ordered phase at T_{activ} when mixed with cholesterol. (C) Phase states in lipid bilayers are associated with distinct membrane properties, such as compressibility, heat capacity or diffusivity, density and permeability. More highly ordered states, such as the gel phase or cholesterol-induced liquid ordered phase, are more rigid and display lower diffusivity rates than the liquid disordered state. During the transition between two phases, membrane fluctuations are maximized and diverging physical properties and nonlinear behavior emerge. We hypothesize that this nonlinear behavior should promote membrane rearrangement and corresponding activation in target Jurkat cells.

The means of these relative activation values per lipid condition are presented in the following results. Data points represent the mean of at least 3 biological replicates.

All the lipid particles in this study were composed of the same amounts of cholesterol, PE PEG 2000, and PE PEG BG, with only the type of phospholipid changed, which represents 70 mol% of the total composition. Note that since the cholesterol, PEG and BG amounts are kept constant throughout the measurements, we will not explicitly refer to these components. Instead, we refer to the LUV type by the variable phospholipid type alone (*e.g.* PE LUV). An overview of the variable phospholipids tested in this study and their properties can be found in Table 1.

We first assessed the potential of different lipids to modify nanocarrier-mediated stimulation of an immortalized line of T lymphocytes. Lipids with

Table 1 Overview and characteristics of lipids used in this study

Phospholipid (PL) chain length	Abbreviation	PL name	Net charge	T_m (°C) of PL[a]
18:1 PC	DOPC	1,2-Dioleoyl-*sn-glycero*-3-phosphocholine	0	−17
14:0 PC	DMPC	1,2-Dimyristoyl-*sn-glycero*-3-phosphocholine	0	24
18:0-14:0 PC		1-Stearoyl-2-myristoyl-*sn-glycero*-3-phosphocholine	0	30
14:0-16:0 PC		1-Myristoyl-2-palmitoyl-*sn-glycero*-3-phosphocholine	0	35
14:0-18:0 PC		1-Myristoyl-2-stearoyl-*sn-glycero*-3-phosphocholine	0	39
16:0 PC	DPPC	1,2-Dipalmitoyl-*sn-glycero*-3-phosphocholine	0	41
17:0 PC		1,2-Diheptadecanoyl-*sn-glycero*-3-phosphocholine	0	50
18:1 PS	DOPS	1,2-Di-(9Z-octadecenoyl)-*sn-glycero*-3-phospho-L-serine	−1	−11
12:0 PS	DLPS	1,2-Dilauroyl-*sn-glycero*-3-phospho-L-serine	−1	14
14:0 PS	DMPS	1,2-Dimyristoyl-*sn-glycero*-3-phospho-L-serine	−1	35
16:0 PS	DPPS	1,2-Dipalmitoyl-*sn-glycero*-3-phospho-L-serine	−1	51
14:0 PG	DMPG	1,2-Dimyristoyl-*sn-glycero*-3-phospho-(1′-*rac*-glycerol)	−1	23
16:0 PG	DPPG	1,2-Dipalmitoyl-*sn-glycero*-3-phospho-(1′-*rac*-glycerol)	−1	41
18:0 PG	DSPG	1,2-Distearoyl-*sn-glycero*-3-phospho-(1′-*rac*-glycerol)	−1	54
18:1(Δ^9-*t*) PE		1,2-Dielaidoyl-*sn-glycero*-3-phosphoethanolamine	0	38

[a] Values taken from ref. 45.

different features according to headgroup, charge, chain length and saturation were tested. To measure lymphocyte activation, we used a Jurkat reporter cell line,[46] in which Jurkat cell activation is reported *via* GFP expression. Specifically, downstream activation of the NFAT transcription factor after antigen-mediated stimulation of Jurkat cells initiates GFP expression. The percentage of Jurkat cells that were activated was quantified *via* flow cytometry. After overnight incubation of liposome-conjugated protein with cells, the GFP signal was read out and quantified *via* flow cytometry. We confirmed that free SNAP-tag–αCD3 or blank liposomes without protein did not activate Jurkat cells in the investigated concentration regime of crude cell-free product and lipid vesicles, respectively (Fig. 2A). Treatment with ionomycin and PMA activates the cells without TCR stimulation and served as a positive control, yielding activation levels around 99% (Fig. 2A). However, due to varying yields of the αCD3–SNAP-tag protein in the cell-free-expression reactions, absolute activation levels for LUV-induced activation were found to differ between LUV batches and the variation is reflected in the error bars shown for each experiment. We first confirmed that Jurkat cells could be activated with our cell-free modified aAPCs by using 18:1 PC (DOPC) LUVs that

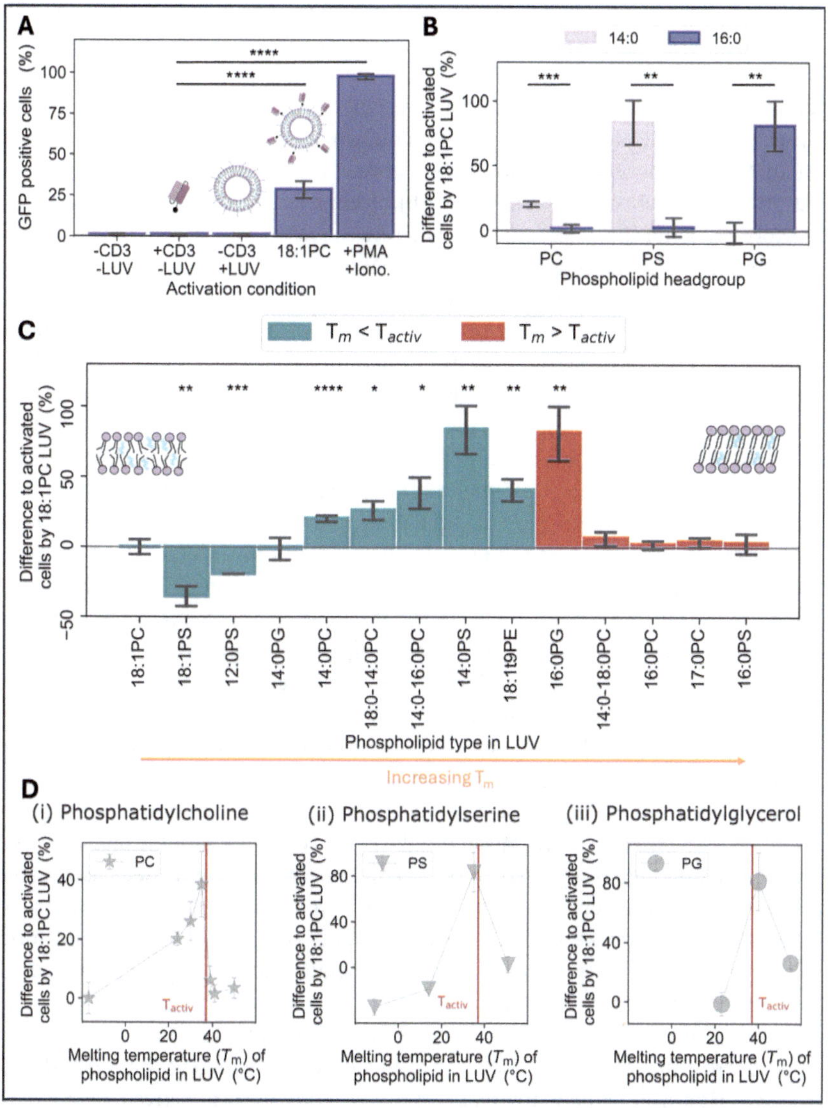

Fig. 2 Liposome composition determines the efficiency of Jurkat cell activation. (A) Flow cytometry analysis of activated Jurkat cells determined from the number of GFP-positive cells after incubation with LUVs. DOPC LUVs lacking CD3 (−CD3, +LUV) result in ~1% activation of Jurkat cells, similar to the % activation after incubation with unconjugated αCD3 only (+CD3, −LUV) or buffer (−CD3; −LUV). Addition of phorbol 12-myristate 13-acetate (PMA) and ionomycin activates the cells and is used as a positive control for the flow cytometry analysis. Functionalized liposomes (here: DOPC conjugated to αCD3) activate 20–50% of the cell population. GFP-positive cells are determined *via* flow cytometry. (B) Comparison of activation efficiencies for treatment with LUVs assembled from lipids with three different headgroups and two different chain lengths reveals no correlation of Jurkat cell activation with either parameter. LUVs based on PC lipids show lower activation potential for the investigated chain lengths compared to PS and PG LUVs. While LUVs with 14:0 PS perform far better than those with 16:0 PS (+70%), for PG LUVs this trend is reversed. (C) Screening of LUVs arranged according to ascending theoretical melting temperatures of the respective lipid type alone. Incubation temperature (T_{activ}) relative to T_m determines the state of the lipid membrane (neglecting the influence of

were subsequently included in each measurement. We measured the % of cells activated by DOPC LUVs to be between 20% and 50%.

We evaluated the impact of the lipid headgroup and chain length on T cell activation by assembling LUVs with a variable phospholipid containing one of three different headgroups (PC, PS, PG) and two different chain lengths (14:0 and 16:0). We did not observe a correlation between a given lipid headgroup or chain length and Jurkat cell activation (Fig. 2B). In contrast, for PS LUVs, the 14:0 lipid performed better than the 16:0 lipid, whereas the trend was reversed for PG LUVs. In comparison to the reference sample (DOPC LUVs), 14:0 PS and 16:0 PG LUVs increase the activation significantly by +70 to 85%. While 14:0 PS is expected to be in the L_D phase during activation, 16:0 PG is expected to be in the L_O phase. Interestingly, we noticed that both lipids that induced these enhanced activations have a melting transition temperature closer to the temperature at which activation studies were conducted (T_{activ} = 37 °C) with T_m (14:0 PS) = 35 °C and T_m (16:0 PG) = 41 °C compared to other samples (T_m (16:0 PS) = 54 °C and T_m (14:0 PG) = 23 °C). The theoretical melting temperatures reported (Table 1) for the variable phospholipid refer to pure lipid systems and do not account for either the amount of cholesterol, PEG and BG in the liposome, or the interaction of the LUVs with cells during activation studies. While these factors could shift the actual membrane transition point by some °C, this observation suggested a potential relationship between the lipid transition temperature and Jurkat cell activation.

To assess this potential relationship, we displayed our data as a function of increasing melting temperature of the variable phospholipid in our LUVs (from left to right in Fig. 2C). We observed that Jurkat cell activation relative to activation with DOPC LUVs displays a parabolic curve. LUVs containing the variable phospholipid with a T_m closest to the temperature of the activation studies resulted in the highest % of activated cells. For LUVs with a variable phospholipid T_m much lower or much higher than T_{activ}, the % activation decreased. This observation is independent of the headgroup type, chain length, saturation, or charge of the lipids we tested. In general, lipids with T_m < 37 °C should form disordered membranes during the activation studies at 37 °C, allowing increased mobility of the targeting moiety on the LUV surface, which could facilitate the clustering of the TCR receptor. Rigidity or stiffness of the substrate has been reported to enhance T cell activation before, but seems here to be limited to the L_D phase.[31] However, while many L_D LUVs in our study seem to activate better than L_O samples (T_m > 37 °C), membrane order alone does not appear to explain Jurkat cell activation. The LUV offering the highest mobility is the reference sample (DOPC or 18:1 PC) as it is composed of unsaturated lipids and has the furthest

cholesterol and medium). % Jurkat cells activated follows a parabolic curve as a function of LUV T_m with maximal activation occurring for LUVs that have a T_m close to T_{activ}. (D) Results for the activation efficiencies of Jurkat cells with different LUVs, as a function of the lipid T_m they are based on. Data is divided into three headgroups and different chain lengths yield different T_m values. Relative activation refers to the difference in activation efficiency compared to reference LUVs based on 18:1 PC. Data represents the mean of at least 3 biological replicates with error bars representing the standard error of the mean (SEM). P values are the results of one-way analysis of variance (ANOVA) comparing the % activation of the respective sample to the αCD3 only control (A), to a similar headgroup lipid (B) and to the 18:1 PC reference (C) (*p < 0.05; **p < 0.01; ***p < 0.001; ****p < 0.0001). Only significant changes are indicated.

distance between T_m (−17 °C) and T_{activ} (37 °C). Accordingly, it should remain in the fluid L_D phase during activation. Since nearly all LUVs containing a majority L_D phase showed enhanced activation compared to DOPC (except for negatively charged 18:1 PS and 12:0 PS), lipid fluidity alone does not dictate Jurkat cell activation. Instead, we observe that the LUV melting temperature appears to be a better predictor of Jurkat cell activation. When the data in Fig. 2C is plotted as a function of lipid T_m, we observe that LUVs with T_m values closest to the activation study temperature, T_{activ}, induced the most Jurkat cell activation (Fig. 2D). This trend holds regardless of lipid head group and suggests that LUV proximity to a melting transition plays an important role in Jurkat cell activation.

Jurkat cell activation correlates with the measured melting transitions in LUVs

To better confirm the trend between LUV T_m and Jurkat cell activation, we experimentally measured the T_m of the LUVs used in our studies. The structure and transition behavior of lipid bilayers can be affected by ions[47] as well as interactions with components in cell culture media. To evaluate the impact of these conditions on the LUV T_m, we determined LUV T_m values for LUVs expected to have a T_m close to 37 °C. To determine the T_m, we measured the Laurdan generalized polarization (GP) of LUVs as a function of temperature (Fig. 3). Laurdan is a fluorescent dye displaying a shift in emission spectrum between two phase states due to its solvatochromic character.[48] The GP value is determined by measuring Laurdan emission intensity at two characteristic wavelengths (I_1 = 450 nm and I_2 = 500 nm with GP = $(I_1 - I_2)/(I_1 + I_2)$). The GP shifts nonlinearly when a membrane in which it is integrated undergoes a phase transition, providing an experimental readout of the melting temperature of LUVs under our experimental conditions.

We measured the melting transition temperature of LUVs under three conditions: 100% phospholipid in PBS, 70% phospholipid/29% cholesterol/1% PEG in PBS and 70% phospholipid/29% cholesterol/1% PEG in full RPMI cell medium. Fig. 3 shows GP measurements as functions of temperature for 16:0 PG (A), 14:0 PS (B) and 16:0 PC (C) LUVs for these conditions. A nonlinearity in the GP signal as a function of temperature represents a transition between the ordered and fluid phase. This nonlinearity is visible as a maximum in the pointwise derivative of GP with respect to temperature and occurs at the temperature with the greatest GP shift. For pure phospholipid liposomes, the measured T_m is in good agreement with the literature values (16:0 PC: T_{meas} = 41.8 °C and T_{theory} = 41 °C; 14:0 PS: T_{meas} = 34.8 °C and T_{theory} = 35 °C; 16:0 PG: T_{meas} = 40.8 °C and T_{theory} = 41 °C). With the addition of cholesterol and PEGylated lipids to the LUVs, the transition gets less pronounced, which is typically observed with cholesterol addition; however, the melting temperature does not shift for all LUV types tested (Fig. 3 middle panel). In RPMI medium, the melting of 16:0 PC LUVs is not affected by the change in environment, whereas the 16:0 PG displays a strong increase in the transition temperature by ~6 °C. PG is highly susceptible to divalent ions, which probably interferes with its transition behavior and membrane properties.[49] For 14:0 PS LUVs, the transition is indicated as a point where the slope of the GP function changes significantly and a transition in the structural arrangement of the lipids happens. The inflection point of the derivative is at $T_{inflect}$ = 37.7 °C and is increased compared to the transition

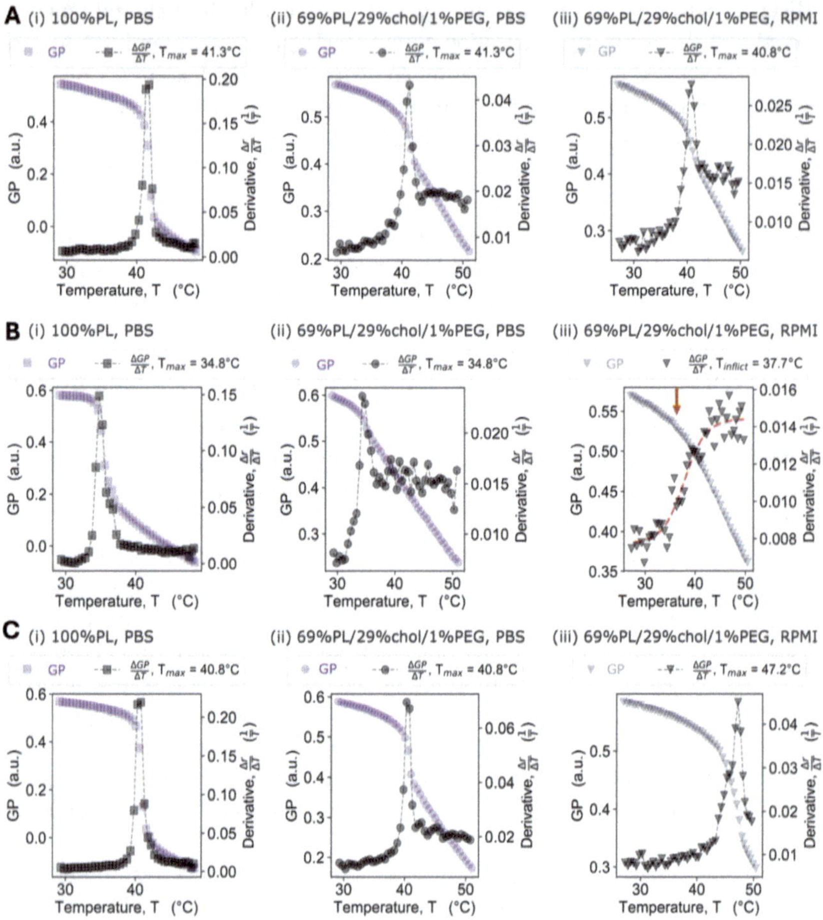

Fig. 3 Measurements of generalized polarization (GP) of Laurdan to determine the T_m of LUVs based on 16:0 PC (A), 14:0 PS (B) and 16:0 PG (C). From left to right: 100% PL in PBS, activation formulation in PBS, and activation formulation in full RPMI. For each condition, the GP values (purple) are plotted as a function of temperature and the phase transition temperature is identified as the temperature with the greatest GP change corresponding with a maximum in the pointwise derivative of the GP values with respect to the temperature (black). For pure phospholipid liposomes, the measured T_m is in good agreement with the literature values and similar holds for the addition of cholesterol and PEGylated lipids to the LUVs. In the RPMI medium in which the activation studies are conducted, the melting of PG- and PS-based LUVs is affected due to their interaction with divalent ions in the medium and the T_m differs from their literature value. For 14:0 PS LUVs, the shape of the GP function varies significantly as a nonlinear change in the slope is present, marked with a red arrow. An inflection point of the derivative is found at $T_{inflect} = 37.7$ °C and is increased compared to the transition temperature in PBS. Fluorescence intensity was measured with a spectrophotometer and GP values and their pointwise derivative were calculated with Python.

temperature in PBS. Overall, while we observed some variations in the actual transition temperature of LUVs relative to the transition temperature of the major variable phospholipid, most transition temperatures were close to the values reported in Table 1.

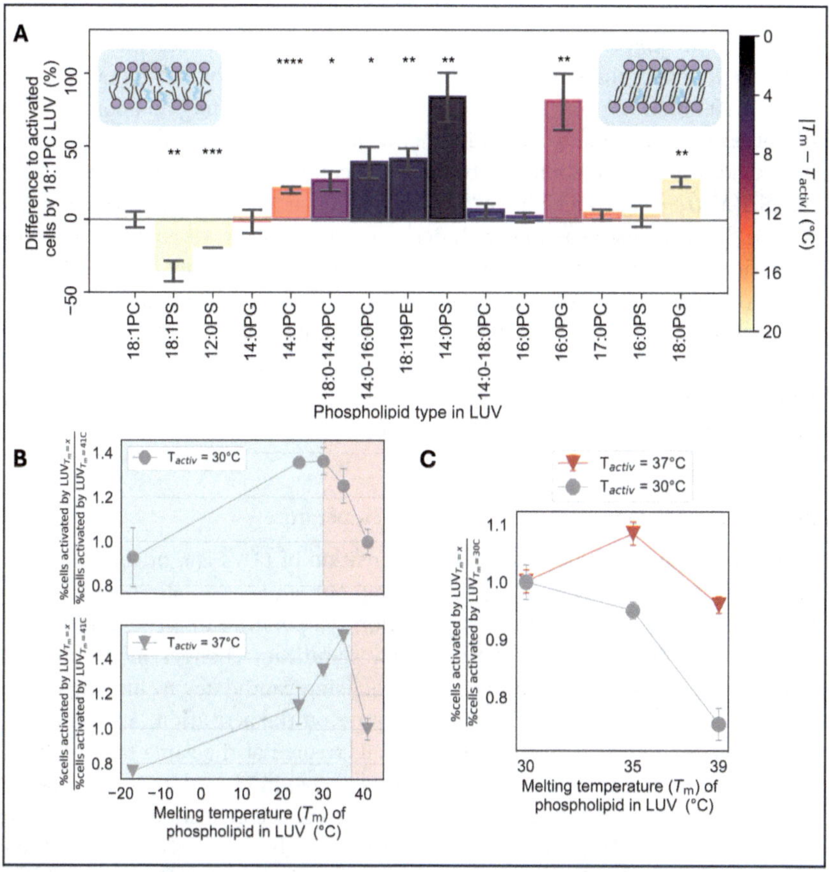

Fig. 4 Jurkat activation efficiency is a function of the proximity of LUV T_m to the temperature of activation studies, T_{activ}. (A) Jurkat activation data now plotted in ascending order of the corrected LUV T_m values from left to right using the T_m values determined in the previous section. Additional color scaling visualizes the proximity of the T_m to the temperature at which the activation was investigated (37 °C). Activation levels are enhanced when $|T_m - T_{activ}|$ is small and the LUVs are close to their melting transition. Except for 16:0 PG LUVs, the activation levels for LUVs in their L_O phase (T_m > 37 °C) are overall lower compared to those for lipids at a similar distance to 37 °C from the melting transition in the L_D phase. (B) Exemplary direct comparison of activation of Jurkat cells incubated with different PC-based LUVs as a function of LUV T_m measured at 30 °C (top) and 37 °C (bottom). For both measurements, the same batch of liposomes and conjugated antibody was used. At $T_{activ} = 30$ °C, 18:0-14:0 PC LUVs (data point at $T_m = 30$ °C) and 14:0 PC LUVs (data point at $T_m = 24$ °C) activated the most cells, while at $T_{activ} = 37$ °C the 14:0-16:0 PC LUVs ($T_m = 35$ °C) activated the most cells. This trend demonstrates the correlation between the activation efficacy of an LUV and the proximity of its T_m to the temperature of activation. Data points represent the mean of 2 technical replicates with error bars presenting the standard error of the mean (SEM). (C) Ratiometric comparison of LUVs with three different melting temperatures and their activation efficiencies of Jurkat cells at two different T_{activ} temperatures. Each data point represents the ratio of cells activated with lipid $LUV_{T_m=x}$ to cells activated with $LUV_{T_m=30°C}$ (% activation with $T_m = x$/% activation with $T_m = 30$ °C) at either 30 °C (grey) or 37 °C (red) to visualize temperature-dependent differences in performance. The ratio is at a maximum for LUVs with $T_m = 35$ °C (14:0-16:0 PC) at $T_{activ} = 37$ °C and at a maximum for LUVs with $T_m = 30$ °C (18:0-14:0 PC) at $T_{activ} = 30$ °C, demonstrating the increasing activation efficiency for LUVs with a T_m closer to T_{activ}. Measurements were performed n > 3 with error bars representing the

In summary, PS and PG headgroups show modified melting temperatures and shapes of the GP curve when measured in cell media, as they are especially sensitive to ions. While incubation in media shifts the T_m of 14:0 PS LUVs closer to 37 °C, it shifts the 16:0 PG LUV T_m further away, presenting an anomaly from our other data showing a correlation between LUV T_m and cell activation. As expected, PC lipids are not affected by the RPMI conditions and similar trends hold for 18:1(Δ^9-*trans*) PE, which has a T_m we measured at 35 °C in both PBS and RPMI (Fig. S1†). With these experimentally derived LUV T_m values, we could evaluate the activation data again and found that 14:0 PS LUVs have a higher T_m than 18:1 PE LUVs (reversed from when we presented T_m data according to the variable phospholipid alone in Fig. 2C). Hence, a reorganization of lipids when ordered according to their corrected transition points in the LUVs yields a cleaner correlation between the capacity to enhance activation and proximity of T_{activ} to LUV T_m, with DPPG as a clear exception (Fig. 4A).

Activation efficiencies at a different T_{activ} temperature

To test our hypothesis that the melting transition of LUVs is a driving factor for enhancing Jurkat cell activation, we repeated our study at an alternate temperature. Specifically, we reduced the incubation temperature to 30 °C during activation. Since PC headgroups do not show significant changes in the melting temperature, these lipids are the most promising candidates to investigate the direct influence of the incubation temperature on the activation. An exemplary, direct comparison of activation levels of cells treated at the same time with the same batch of PC LUVs at 30 °C and 37 °C is shown in Fig. 4B. Indeed, 18:0-14:0 PC LUVs (T_m = 30 °C) resulted in the highest Jurkat cell activation at 30 °C, while 14:0-16:0 PC LUVs (T_m = 35 °C) resulted in the highest activation at 37 °C. We applied statistical analysis to this data and activation levels of lipids relative to 14:0-18:0 PC LUVs (T_m = 39 °C) were plotted as a function of melting temperature at a given temperature. This ratio of both activation levels over $n = 3$ biological replicates underscores the trend observed in the previous analysis (Fig. 4B). The performance of different lipids at two different incubation temperatures shows a clear shift based on which lipid is closer to its transition at the given temperature (Fig. 4C). Adaption of the cells towards the 30 °C incubation can be excluded as they have been grown at 37 °C prior to the overnight activation at 30 °C and the time of incubation (16 h) is shorter than the doubling time of the Jurkat cells (20.7 ± 2.2 h).[50] These results clearly corroborate that transition-related effects in the membrane of the liposomes promote the activation process.

Due to the predicted anomalous behavior of the LUV in all properties during a melting transition, an exact mechanism responsible for the facilitated activation would be merely speculation.[35,36] However, nonlinear effects, like an increased amplitude of area-density fluctuations, as well as membrane undulations, are likely to increase the likelihood of antigen-receptor binding, to facilitate clustering and enhanced presentation of the antibody to the Jurkat cell surface, and to induce a mechanical signal stimulation.[14,20] During synapse formation, antigens and antibodies must overcome the distance of the immunological synapse (∼15

standard error of the mean (SEM). *P* values are the results of one-way analysis of variance (ANOVA) (*$p < 0.05$; **$p < 0.01$; ***$p < 0.001$; ****$p < 0.0001$).

nm) in order to bind. Thermal membrane undulations of the T cell membrane have been indeed proposed to initiate this binding.[21] Additional transition-induced undulations of the artificial antigen-presenting cell should further increase this interaction, especially for membranes in close proximity to their transition where they are expected to show maximal undulations.[36]

Besides PC-based liposomes, PG-based liposomes were also investigated at 30 °C. Surprisingly, in an overall comparison, 16:0 PG LUVs were still the best activating LUVs in absolute numbers compared to all tested lipids, even though the difference between T_m and T_{activ} has further increased (Fig. S2†). The origin of the special activation potential of 16:0 PG is not known and we therefore set out to more closely examine this composition *via* microscopy.

Visualization of activation with state-sensitive dye NR12S

To investigate the potential differences/conspicuous features of DPPG lipids and other intermediately activating lipid types in the cell activation process, we used microscopy to evaluate lipid-membrane–Jurkat-cell interactions. Instead of LUVs ($d = 100$–200 nm), we prepared giant unilamellar vesicles (GUVs) ($d = 5$–30 μm) and

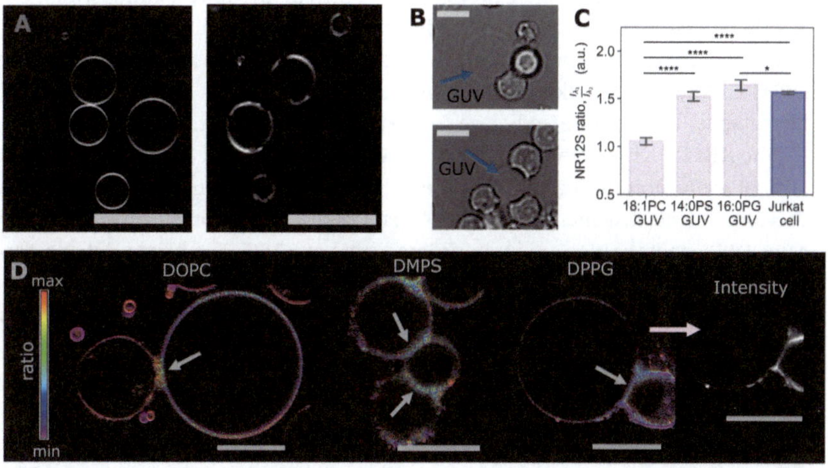

Fig. 5 Confocal microscopy images of Jurkat cell activation. (A) DPPG GUVs (right) display dark spots indicating phase separation in the membrane at room temperature, in contrast to DMPS (left). (B) Bright-field image of Jurkat cells accumulating around and adhering to GUVs modified with αCD3. (C) Analysis of NR12S emission ratio (I_1/I_2, where $I_1 = 565$–595 nm and $I_2 = 600$–630 nm) in three investigated GUV compositions and in Jurkat cells. An increased NR12S ratio corresponds to increased order in the membrane. Results are the mean of $n > 10$ GUVs/cells with error bars representing the standard error of the mean (SEM). (D) Exemplary ratiometric images for Jurkat cells incubated with DOPC, DMPS, and DPPG GUVs. Each pixel represents the ratio of the two wavelength channels (I_1/I_2). The color bar ranges from purple (lowest ratio) to red (highest ratio). The cutoff values are chosen individually for optimal visibility of ratio differences as the absolute numbers vary with the GUV compositions which are reported in (C). P values are the results of one-way analysis of variance (ANOVA) (*$p < 0.05$; **$p < 0.01$; ***$p < 0.001$; ****$p < 0.0001$). Areas of contact between GUVs and Jurkat cells are marked with an arrow indicating shifted NR12S intensity ratios. Next to the DPPG ratio image, an excerpt of the intensity weighted scan is shown to highlight bright areas on the GUV surface close to the interaction site with the Jurkat cell (immunological synapse). All scale bars display 10 μm.

attached the SNAP-tag–αCD3 protein similarly to the conjugation scheme used for LUVs (Fig. 5B). We confirmed that GUVs prepared in this way were able to activate Jurkat cells, as GFP fluorescence was observable after an overnight incubation of Jurkat cells with GUV-SNAP-tag–αCD3 conjugates (Fig. S3†). Cells and GUVs were stained with the state-sensitive membrane dye NR12S, and we measured the dye signal during the early activation period of the cells when the immunological synapse formed at 37 °C (30 to 60 minutes after addition of the GUVs to the cells). We measured the emission intensity ratio of NR12S at two wavelengths ($I_1 = 565$–595 nm and $I_2 = 600$–630 nm, with emission ratio parameter I_1/I_2). This ratio reflects the relative position of the emission spectrum, which is sensitive to cholesterol content and phase-state changes in the lipid and cell membranes.[51]

Three different LUVs were compared during their antigen-mediated stimulation of the Jurkat cells: DOPC LUVs, which serve as the standard reference in our study, DMPS LUVs, which have a melting temperature close to 37 °C, and DPPG LUVs, which displayed the highest activation potential in our previous experiments. The GUVs have a similar composition to the LUVs we used for the activation studies (70 mol% PL, 29 mol% cholesterol, 0.9 mol% 18:0 PE PEG 2000, and 0.1 mol% 18:1 PE PEG BG). The NR12S signals for the different GUV types differ because of the different levels of order in these membranes at 37 °C (Fig. 5C). DOPC GUVs, far in the L_D phase, display the lowest ratio, which can clearly be distinguished from the ratio of NR12S in the cell membrane at 37 °C (~1.5). DMPS is close to the optical ratio generated in the cell membrane and DPPG GUVs appear strikingly dark with a ratio around 1.6. Interestingly, small bright spots can be observed close to the cellular contact site (Fig. 5D).

The NR12S dye is known to have a reduced quantum yield (QY) in the L_O phase and an even more reduced QY in the gel phase. This feature helps explain the previous result, in which DPPG GUVs displayed a lower optical ratio, and indicates DPPG membranes are in an ordered state at 37 °C. The bright dots present in local areas of the membrane, on the other hand, indicate areas of less order and suggest these areas experience higher segregation of membrane components. Indeed, a low miscibility of cholesterol in PG membranes has been reported, which can lead to the appearance of a pure cholesterol phase.[52] With the likely segregation of cholesterol, the remaining membrane could indeed be closer to the gel phase than the L_O phase, leading to additional clustering of the unsaturated BG-conjugated PE anchored protein in the bright areas. Segregation of αCD3, which is attached to an unsaturated lipid, could better facilitate the Jurkat activation process. Interestingly, domains are already visible for the DPPG GUVs in 300 mOsm glucose before the dilution into RPMI medium (Fig. 5A), which was not observed for DMPS and DOPC GUVs. The presence of domains can also indicate coexistence of two phases in binary mixtures. NR12S segregation was observed at room temperature as well as 37 °C, and might provide an explanation for the observed enhanced activation induced by DPPG LUVs at both tested temperatures in our studies.

To further analyze differences in the interaction of cells and different GUVs, we investigated the optical signal of NR12S in the cellular contact sites, the potential immunological synapses (Fig. 5D). The ratiometric images present shifts of the color-coded NR12S emission ratio locally at the contact sites of the cells and DOPC and DMPS GUVs, which account for the local reorganization of the cell membrane in the synapse. For DOPC, the ratio of the interface between the GUV and cell is increased compared to the rest of the GUV and decreased compared to the cell. For DMPS, the immunological synapse has a decreased ratio compared to the cell and GUV, while DPPG shows no change compared to the cell. The interpretation of a decrease in the ratio of NR12S is not straight-forward, and either relates to a higher degree of disorder and a constant cholesterol amount or a decrease in the cholesterol amount itself.[51] For membranes in the solid gel phase, additional cholesterol decreases the order, while for membranes in the L_D phase, cholesterol increases order. For cell membranes, there are several examples showing that a depletion of cholesterol leads to an increase in membrane tension, which would lead to the observed local decrease in ratio,[31,53,54] supporting reports that the immunological synapse is more ordered compared to the rest of the cell.[15,17,18,55] A local accumulation of lipids with higher T_m and potentially local displacement of cholesterol in the Jurkat cell membrane at the site of the synapse could lead to the observed decrease in ratio and could explain the observed, local membrane state changes. At present, however, we do not have an explanation for the enhanced performance of DPPG-based vesicles.

Conclusion

Our results demonstrate that the interaction between T cells and substrates is not solely determined by the choice of targeting moiety, and neither is it determined by membrane fluidity alone. Rather, the choice of the phospholipid in nanocarrier design can either support the interaction efficiency of the targeting moiety with the target cell or even suppress it as a result of the modulation of properties of the functionalized membrane interface.

LUVs residing close to their transition regime during incubation tended to activate best in this study, independent of their individual features, such as headgroup, chain length, saturation, or even charge. Laurdan GP measurements indicate that there is indeed still a measurable melting transition present in the tested LUV compositions, despite their relatively high cholesterol content of 29 mol%. Our results suggest that phase-transition-related effects and the associated anomalous behavior can enhance the capacity of designed nanocarrier substrates to activate target cells. Specifically, we hypothesized that as the T_m of nanocarrier membranes comes in close proximity to the temperature at which activation studies are conducted, density fluctuations in both membranes should promote local accumulations of the antibody and increased membrane undulations should facilitate antigen–receptor binding.

An exception of the observed correlation with T_m was found for 16:0 PG LUVs, as this nanocarrier led to an increased Jurkat activation level compared to that induced by 14:0 PS LUVs, even though its T_m was measured to be ~10 °C above T_{activ}. Interestingly, 16:0 PG still performs best at 30 °C incubation temperature (Fig. S2†) and the activation capacity of this composition of LUV appears to be independent of the proximity of its T_m to T_{activ}. Domains observed in the GUVs made from the 16:0 PG lipids suggest cholesterol accumulations in which the PE anchor of the BG group would cluster too. Hence, segregation-related clustering could explain the improved activation capabilities for the temperatures of 30 °C and 37 °C, opening up another potential interesting feature for the design of nanocarriers. Our findings set the stage for future investigations exploring the design of LUVs and the role of both LUV T_m and domain-segregation of targeting molecules.

Data availability

The authors confirm that the data supporting the findings of this study are available within the article and its ESI.† Flow cytometry data can be requested from the authors.

Author contributions

The study was designed by C. S. F. and N. P. K. C. S. F and L. C. performed measurements. C. S. F. conducted data analysis and prepared the figures. C. S. F. and N. P. K. wrote the manuscript.

Conflicts of interest

C. S. F. and N. P. K. are inventors on a U.S. provisional patent (provisional US 63/746,157) submitted by Northwestern University that covers technology in this manuscript.

Acknowledgements

This work was supported by the National Science Foundation's MRSEC program (DMR-2308691) and made use of its Shared Facilities at the Materials Research Center of Northwestern University. This work was funded in part by the Chicago Biomedical Consortium with support from the Searle Funds at The Chicago Community Trust for C. S. F. and the International Institute for Nanotechnology (IIN) at Northwestern University (L. C.).

References

1 M. V. Maus, A. K. Thomas, D. G. B. Leonard, D. Allman, K. Addya, K. Schlienger, J. L. Riley and C. H. June, *Nat. Biotechnol.*, 2002, **20**, 143–148.
2 R. Zappasodi, M. Di Nicola, C. Carlo-Stella, R. Mortarini, A. Molla, C. Vegetti, S. Albani, A. Anichini and A. M. Gianni, *Haematologica*, 2008, **93**, 1523–1534.
3 K. Perica, J. G. Bieler, C. Schütz, J. C. Varela, J. Douglass, A. Skora, Y. L. Chiu, M. Oelke, K. Kinzler, S. Zhou, B. Vogelstein and J. P. Schneck, *ACS Nano*, 2015, **9**, 6861–6871.
4 S. Mandal, R. Hammink, J. Tel, Z. H. Eksteen-Akeroyd, A. E. Rowan, K. Blank and C. G. Figdor, *ACS Chem. Biol.*, 2015, **10**, 485–492.
5 T. R. Fadel, F. A. Sharp, N. Vudattu, R. Ragheb, J. Garyu, D. Kim, E. Hong, N. Li, G. L. Haller, L. D. Pfefferle, S. Justesen, K. C. Harold and T. M. Fahmy, *Nat. Nanotechnol.*, 2014, **9**, 639–647.
6 R. Nisini, N. Poerio, S. Mariotti, F. De Santis and M. Fraziano, *Front. Immunol.*, **9**, 155.
7 W. Zhang, C. E. Callmann and C. A. Mirkin, *ACS Appl. Mater. Interfaces*, 2021, **13**, 46325–46333.

8 D. Papahadjopoulos, S. Nir and S. Ohki, *Biochim. Biophys. Acta, Biomembr.*, 1972, **266**, 561–583.
9 J. Damen, J. Regts and G. Scherphof, *Biochim. Biophys. Acta, Lipids Lipid Metab.*, 1981, **665**, 538–545.
10 L. Sercombe, T. Veerati, F. Moheimani, S. Y. Wu, A. K. Sood and S. Hua, *Front. Pharmacol*, 2015, **6**, 286.
11 J. B. Huppa and M. M. Davis, *Nat. Rev. Immunol.*, 2003, **3**, 973–983.
12 M. L. Dustin, *Cancer Immunol. Res.*, 2014, **2**, 1023–1033.
13 M. J. P. Biggs, M. C. Milone, L. C. Santos, A. Gondarenko and S. J. Wind, *J. R. Soc. Interface*, 2011, **8**, 1462–1471.
14 M. Huse, *Nat. Rev. Immunol.*, 2017, **17**, 679–690.
15 W. Wu, X. Shi and C. Xu, *Nat. Rev. Immunol.*, 2016, **16**, 690–701.
16 J. A. Rudd-Schmidt, A. W. Hodel, T. Noori, J. A. Lopez, H. J. Cho, S. Verschoor, A. Ciccone, J. A. Trapani, B. W. Hoogenboom and I. Voskoboinik, *Nat. Commun.*, 2019, **10**, 5396.
17 K. Gaus, E. Chklovskaia, B. Fazekas De St. Groth, W. Jessup and T. Harder, *J. Cell Biol.*, 2005, **171**, 121–131.
18 T. Zech, C. S. Ejsing, K. Gaus, B. De Wet, A. Shevchenko, K. Simons and T. Harder, *EMBO J.*, 2009, **28**, 466–476.
19 A. MacDonald, B. Lam, J. Lin, L. Ferrall, Y. J. Kung, Y. C. Tsai, T. C. Wu and C. F. Hung, *Front. Immunol.*, 2021, **12**, 755995.
20 C. Hivroz and M. Saitakis, *Front. Immunol.*, 2016, **7**, 46.
21 M. A. Al-Aghbar, A. K. Jainarayanan, M. L. Dustin and S. R. Roffler, *Commun. Biol.*, 2022, **5**, 40.
22 D. Hollyman, J. Stefanski, M. Przybylowski, S. Bartido, O. Borquez-Ojeda, C. Taylor, R. Yeh, V. Capacio, M. Olszewska, J. Hosey, M. Sadelain, R. J. Brentjens and I. Rivière, *J. Immunother.*, 2009, **32**, 169–180.
23 A. S. Cheung, D. K. Y. Zhang, S. T. Koshy and D. J. Mooney, *Nat. Biotechnol.*, 2018, **36**, 160–169.
24 S. T. Reddy, A. Rehor, H. G. Schmoekel, J. A. Hubbell and M. A. Swartz, *J. Controlled Release*, 2006, **112**, 26–34.
25 K. Perica, A. De León Medero, M. Durai, Y. L. Chiu, J. G. Bieler, L. Sibener, M. Niemöller, M. Assenmacher, A. Richter, M. Edidin, M. Oelke and J. Schneck, *Nanomedicine*, 2014, **10**, 119–129.
26 L. E. Paulis, T. Geelen, M. T. Kuhlmann, B. F. Coolen, M. Schäfers, K. Nicolay and G. J. Strijkers, *J. Controlled Release*, 2012, **162**, 276–285.
27 R. A. Meyer, J. C. Sunshine, K. Perica, A. K. Kosmides, K. Aje, J. P. Schneck and J. J. Green, *Small*, 2015, **11**, 1519–1525.
28 J. C. Sunshine, K. Perica, J. P. Schneck and J. J. Green, *Biomaterials*, 2014, **35**, 269–277.
29 A. C. Wauters, J. F. Scheerstra, I. G. Vermeijlen, R. Hammink, M. Schluck, L. Woythe, H. Wu, L. Albertazzi, C. G. Figdor, J. Tel, L. K. E. A. Abdelmohsen and J. C. M. Van Hest, *ACS Nano*, 2022, **16**, 15072–15085.
30 Z. Liu, Y. R. Li, Y. Yang, Y. Zhu, W. Yuan, T. Hoffman, Y. Wu, E. Zhu, J. Zarubova, J. Shen, H. Nan, K. W. Yeh, M. M. Hasani-Sadrabadi, Y. Zhu, Y. Fang, X. Ge, Z. Li, J. Soto, T. Hsiai, L. Yang and S. Li, *Nat. Biomed. Eng.*, 2024, **8**, 1615–1633.

31 K. Lei, A. Kurum, M. Kaynak, L. Bonati, Y. Han, V. Cencen, M. Gao, Y. Q. Xie, Y. Guo, M. T. M. Hannebelle, Y. Wu, G. Zhou, M. Guo, G. E. Fantner, M. S. Sakar and L. Tang, *Nat. Biomed. Eng.*, 2021, **5**, 1411–1425.
32 T. Q. Vu, L. E. Sant'Anna and N. P. Kamat, *bioRxiv*, 2022, preprint, DOI: 10.1101/2022.10.18.512758.
33 T. Q. Vu, J. A. Peruzzi, L. E. Sant'Anna, E. W. Roth and N. P. Kamat, *Nano Lett.*, 2022, **22**, 2627–2634.
34 J. A. Peruzzi, T. Q. Vu, T. F. Gunnels and N. P. Kamat, *Small Methods*, 2023, **7**, 2201718.
35 E. Evans and R. Kwok, *Biochemistry*, 1982, **21**, 4874–4879.
36 A. Kuklin, D. Zabelskii, I. Gordeliy, J. Teixeira, A. Brûlet, V. Chupin, V. Cherezov and V. Gordeliy, *Sci. Rep.*, 2020, **10**, 5749.
37 B. Wunderlich, C. Leirer, A. Idzko, U. F. Keyser, A. Wixforth, V. M. Myles and T. Heimburg, *Biophys. J.*, 2009, **96**, 4592–4597.
38 T. Heimburg and A. D. Jackson, *Proc. Natl. Acad. Sci. U. S. A.*, 2005, **102**, 9790–9795.
39 T. Mužić, F. Tounsi, S. B. Madsen, D. Pollakowski, M. Konrad and T. Heimburg, *Biochim. Biophys. Acta, Biomembr.*, 2019, **1861**, 183026.
40 J. H. Crowe, F. A. Hoekstra, L. M. Crowe, T. J. Anchordoguy and E. Drobnis, *Cryobiology*, 1989, **26**, 76–84.
41 C. S. Fedosejevs and M. F. Schneider, *Proc. Natl. Acad. Sci. U. S. A.*, 2022, **119**, e2117521119.
42 L. Thilo, P. Overath and H. Träuble, *Biochemistry*, 1977, **16**, 1283–1290.
43 D. T. Hanisch and M. F. Schneider, *Langmuir*, 2024, **40**, 3369–3375.
44 W. V. Kraske and D. B. Mountcastle, *Biochim. Biophys. Acta, Biomembr.*, 2001, **1514**, 159–164.
45 D. Marsh, *Handbook of Lipid Bilayers*, CRC press, 2nd edn, 2013.
46 R. T. Abraham and A. Weiss, *Nat. Rev. Immunol.*, 2004, **4**, 301–308.
47 H. Träuble and H. Eibl, *Proc. Natl. Acad. Sci. U. S. A.*, 1974, **71**, 214–219.
48 T. Parasassi, E. K. Krasnowska, L. Bagatolli and E. Gratton, *J. Fluoresc.*, 1998, **8**, 365–373.
49 P. W. M. Van Dijck, P. H. J. T. Ververgaert, A. J. Verkleij, L. L. M. Van Deenen and J. De Gier, *Biochim. Biophys. Acta, Biomembr.*, 1975, **406**, 465–478.
50 N. W. Schoene and K. S. Kamara, *Free Radical Biol. Med.*, 1999, **27**, 364–369.
51 O. A. Kucherak, S. Oncul, Z. Darwich, D. A. Yushchenko, Y. Arntz, P. Didier, Y. Mély and A. S. Klymchenko, *J. Am. Chem. Soc.*, 2010, **132**, 4907–4916.
52 N. Borochov, E. J. Wachtel and D. Bach, *Chem. Phys. Lipids*, 1995, **76**, 85–92.
53 A. Biswas, P. Kashyap, S. Datta, T. Sengupta and B. Sinha, *Biophys. J.*, 2019, **116**, 1456–1468.
54 M. A. Ayee and I. Levitan, *Front. Biosci.*, 2016, 1245–1259.
55 D. M. Owen, S. Oddos, S. Kumar, D. M. Davis, M. A. A. Neil, P. M. W. French, M. L. Dustin, A. I. Magee and M. Cebecauer, *Mol. Membr. Biol.*, 2010, **27**, 178–189.

Faraday Discussions

PAPER

Fabrication and characterization of phosphoinositide containing asymmetric vesicles in physiological salt

Trevor A. Paratore, Alonzo H. Ross and Arne Gericke *

Received 9th December 2024, Accepted 30th January 2025
DOI: 10.1039/d4fd00191e

Phosphoinositide (PIPs) lipids mediate a broad range of physiological functions by attracting proteins at specific time points to distinct cellular sites. Many of these processes are associated with the local accumulation of PIPs and PIP/protein signaling platform formation. Studies aimed at determining the physicochemical underpinnings of PIP domain formation have been limited to model systems that exhibited the same lipid composition in both bilayer leaflets. However, biological membranes are asymmetric, and it is desirable to develop an experimental approach that allows for the fabrication of lipid model systems with a non-symmetric lipid bilayer, *i.e.*, a membrane mimic that exhibits a PIP containing lipid mixture in one leaflet and a different lipid composition in the opposing leaflet. We adapted the previously introduced hemifusion method for the fabrication of asymmetric Giant Unilamellar Vesicles (aGUVs) for the fabrication of aGUVs with phosphatidylinositol-(4,5)-bisphosphate (PI(4,5)P_2) in a physiological ionic strength buffer solution. The general method involved the bivalent cation-initiated fusion of a symmetric GUV (sGUV) with solid-supported lipid bilayers, which leads to the exchange of the outer leaflet of the sGUV. We find that initiating the hemifusion with 6 mM Ca^{2+} leads to a low yield and quality of the aGUVs. We attribute this to macroscopic Ca^{2+}/PI(4,5)P_2 domain formation of the solid support lipid bilayer (SLB), which leads to the interaction of the sGUVs with regions enriched in PI(4,5)P_2 (domain) and other areas that are void of the PIP lipid. Using 6 mM Mg^{2+} as the initiator instead led to an improvement in terms of yield and aGUV quality. The best results were obtained when using 1 mM Mg^{2+}. We are introducing several data analysis approaches that allow for the identification of aGUVs that exhibit high quality in terms of the outer leaflet exchange and composition of the two aGUV leaflets.

1 Introduction

Phosphoinositides (PIPs) are anionic phospholipids that mediate a broad range of cellular functions by attracting proteins to specific cellular sites at specific

Worcester Polytechnic Institute, Department of Chemistry and Biochemistry, 100 Institute Rd, Worcester, MA, 01609, USA. E-mail: agericke@wpi.edu; Tel: +1-5088315263

times. It has been shown for certain PIP moieties that the level of protein mediation they accomplish is associated with their enrichment in particular areas of the cell.[1,2] Such areas of enrichment, or domains, are distinct in their lipid and protein composition and provide an environment conducive to the signaling event taking place. Other examples of phosphoinositide distributions are the generation of PIP gradients during cell migration or the non-uniform distribution of phosphoinositides in polarized cells (*e.g.* epithelial cells).[3,4]

Biological membranes generally exhibit lateral and vertical asymmetry. Lateral asymmetry can manifest in the form of heterogeneous distributions of specific lipid and/or protein components, as is the case of domain formation. Vertical asymmetry refers to the difference in molecular composition between the two leaflets of a membrane bilayer. It has been a longstanding question to what extent these two types of asymmetries influence one another. For example, when a domain forms on one leaflet, what impact does it have on the physicochemical properties of the corresponding area directly across from it on the opposite leaflet? In other words, when two areas spanning the bilayer leaflets are coupled, how are their respective properties altered as a result? In the context of phosphoinositide signaling pathways, this is a major question because several PIP mediated signaling events have been associated with the formation of liquid-ordered domains (raft domains).[5–14] The notion that phosphoinositides could be enriched in raft domains has been controversial for decades[15,16] because the predominant acyl chain composition of PIPs – stearoyl-arachidonoyl – is considered too disordered for the lipid to accumulate in raft domains. In an elegant study, Myeong *et al.*[17] recently showed in cells that phosphatidylinositol-4-phosphate (PI(4)P) and -4,5-bisphosphate (PI(4,5)P_2) are present in both liquid-ordered (Lo) and liquid-disordered (Ld) phases, with no accumulation preference for either phase. Notably, they also found that PI(4,5)P_2 depletion by phospholipase C and the subsequent restoration of PI(4,5)P_2 levels was faster in the Lo than in the Ld phase. This finding suggests that the physical state of the phase where a phosphoinositide resides affects the activity of PIP modifying enzymes and potentially other proteins that interact with PIPs. Although the Myeong *et al.* study revealed how cholesterol depletion affects the global distribution of PI(4,5)P_2, several unresolved issues regarding the physicochemical mechanisms of raft-associated, PIP-mediated signaling events remain.

While the lipid composition of rafts in the outer leaflet of the plasma membrane has been thoroughly investigated,[18,19] we lack knowledge about inner leaflet raft lipid composition. Consequently, we have virtually no knowledge about the factors that affect PIP distribution between non-raft (Ld) and raft (Lo) regions. The physical properties of an inner leaflet raft domain containing PIPs have not been detailed and it is unclear to what extent PIP interaction with proteins is modulated by a raft *vs.* non-raft residency. While it has been shown for Langmuir films[20] and symmetric lipid vesicles[21–23] that calcium ions and/or cholesterol promote the formation of PIP enriched domains, it is currently unknown how this plays out in an asymmetric lipid bilayer environment exhibiting a lipid raft composition in one of the two leaflets.

To refine our understanding of interleaflet coupling, simplified model systems must be developed that exhibit vertical lipid asymmetry. One common method used to fabricate asymmetric small, large, and giant unilamellar vesicles is the cyclodextrin-catalyzed lipid exchange method.[24–27] However, it is important to note that the extension of this method to phosphoinositide-containing asymmetric lipid bilayer systems, while probably feasible, has limitations regarding the lipid composition of the inner and outer leaflets of the formed vesicle. More specifically, the extent of asymmetry in the formed vesicles is dependent on the level of favorability between the cyclodextrin pore and the lipid headgroups present.[28] Recently, Enoki and Feigenson introduced a hemifusion-based method to fabricate asymmetric giant unilamellar vesicles (aGUVs) and demonstrated its utility for studying raft phenomena.[29,30] This method involves the Ca^{2+} mediated fusion of a symmetric giant unilamellar vesicle (sGUV) with a solid-supported lipid bilayer (SLB) to obtain an aGUV. The addition of EDTA terminates the hemifusion process and leads to the fission of the aGUVs from the SLB. The inner leaflet of the aGUV has the lipid composition of the sGUV, while the outer leaflet exhibits the SLB lipid composition. One advantage of this method is that it does not rely on the selectivity of cyclodextrins thus generally expanding the breadth of lipid compositions that can be used to create asymmetric giant unilamellar vesicles. Our group recently broadened the scope of the hemifusion method to incorporate anionic lipids, enabling the creation of asymmetric giant unilamellar vesicles that contain phosphatidylserine and we did so using buffers of physiological ionic strength.[31]

The hemifusion process as described by Enoki and Feigenson[29] requires the addition of Ca^{2+} to initiate the SLB/sGUV hemifusion. For anionic lipid-containing SLBs, as in this study, the addition of Ca^{2+} may lead to lipid clustering and potentially macroscopic domain formation. It has been suggested that micron-sized phase separation in the SLB might impact the outer leaflet composition of the formed aGUVs.[30,31] This is because some sGUVs will settle during the hemifusion process on regions (domains) exhibiting one lipid composition, while other sGUVs will interact with areas of the SLB that have a different lipid makeup. Thus, selecting a system where the lipids in the SLB do not phase separate under the hemifusion conditions is desirable to obtain less compositional variation in the outer leaflet of the formed aGUVs. Previously, Ca^{2+} had been the reagent of choice in hemifusion experiments with charge-neutral lipids using low ionic strength buffers.[29,30,32,33] In our recent study where we incorporated phosphatidylserine (PS) into the outer leaflets of aGUVs,[31] we found that the addition of Ca^{2+} ions caused PS to form large, micron-sized clusters in the SLB over the course of a few hours in a physiological ionic strength buffer. Since the timescale of the hemifusion event is <25 minutes, we expected that this macroscopic domain formation would have no impact on the compositional quality of the aGUVs because the hemifusion process is completed before the macroscopic domains form. However, it is still possible that nanoscopic domain formation, which probably precedes macroscopic domain formation, impacts the compositional quality of the aGUVs. To test this and to evaluate whether there are alternative bivalent cations that can be used to initiate the hemifusion process, we investigated whether Mg^{2+} can be used instead of Ca^{2+} to obtain aGUVs. The rationale for this approach is that Mg^{2+} does not induce macroscopic PS clustering. We found that Mg^{2+} can indeed replace Ca^{2+} as the hemifusion initiator. In addition, we observed virtually no differences between the compositional quality of the aGUVs obtained *via* Ca^{2+} or Mg^{2+} initiated SLB/sGUV hemifusion, *i.e.*, even if PS/Ca^{2+} nanodomains exist within the hemifusion time window, they did not affect the compositional quality of the aGUVs.

The optimal system for studying phosphoinositide behavior in asymmetric lipid bilayers should provide for the following experimental freedoms. First, it should offer the opportunity to utilize a diverse range of lipid compositions for the PIP-containing leaflet. Second, the experimental approach should be able to accommodate lipid systems with phosphatidylinositol (PI) or any of the seven PIP species. Lastly, the system places PIP lipids in the outer leaflet of the vesicle to make studies aimed at investigating PIP interactions with cations and proteins possible. The hemifusion method is principally suitable for the fabrication of such systems. However, since the interaction of bivalent cations with a particular anionic lipid will depend on the chemical structure of the lipid's headgroup, it is possible that the optimal hemifusion conditions will deviate for particular anionic lipids from those identified for PS. $PI(4,5)P_2$ has an approximately −4 charge at physiological pH[34] and has been shown to react strongly with Ca^{2+} ions to form macroscopic domains.[20,35] These domains show remarkable thermal stability, persisting at temperatures as high as 80 °C.[23] On the other hand, Mg^{2+} does not induce micron-sized $PI(4,5)P_2$ domain formation (although it might form nano-sized domains). Also, the interaction between the $PI(4,5)P_2$ headgroup and Mg^{2+} is significantly weaker at physiological ionic strength than what is observed for buffer conditions with lower ion concentrations.[23,36,37]

In this paper, we present data from applying the hemifusion method, using our adapted approach for buffers with physiological ionic strength,[31] to a system containing phosphatidylinositol-4,5-bisphosphate ($PI(4,5)P_2$), a lipid with very complex physicochemistry[38–40] and of crucial biological importance.[41,42] For Ca^{2+} mediated hemifusion, we observed a broad compositional distribution in the formed aGUVs. This is attributed to Ca^{2+} induced clustering of $PI(4,5)P_2$, leading to macroscopic domain formation. Hemifusion of sGUVs with a laterally heterogeneous SLB results in some sGUVs interacting with $PI(4,5)P_2$ enriched regions (domains), while others settle on regions devoid of $PI(4,5)P_2$. Consequently, some of the formed aGUVs contain large quantities of $PI(4,5)P_2$, whereas others will exhibit only small amounts. To address this issue, we explored the possibility of using Mg^{2+} as the hemifusion initiator, as it is known to cluster $PI(4,5)P_2$ to a much lesser extent.[22] We find that the replacement of Ca^{2+} with Mg^{2+} as the hemifusion initiator leads indeed to a significantly improved compositional homogeneity throughout asymmetric bilayer leaflets.

2 Materials & methods

2.1. Materials

Porcine brain L-α-phosphatidylinositol-4,5-bisphosphate (brain-$PI(4,5)P_2$) ammonium salt, 1,2-dioleoyl-*sn*-glycero-3-phosphocholine (DOPC), and 1-oleoyl-2-[6-(4-(dipyrrometheneborondifluoride)butanoyl)amino]hexanoyl-*sn*-glycero-3-phosphoinositol-4,5-bisphosphate (TF-$PI(4,5)P_2$) ammonium salt were obtained as powders from Avanti Polar Lipids (Alabaster, Alabama) and used as received. 1,1′-Dioctadecyl-3,3,3′,3′-tetramethylindodicarbocyanine (DiD′) 4-chlorobenzene-sulfonate salt was obtained as a powder from Invitrogen™ (Waltham, MA). Nunc Lab-Tech II 8-well chambered coverglasses were purchased from Thermo Fisher (Waltham, MA, USA). Wide-orifice pipette tips were created by cutting the ends off standard plastic pipette tips. Bachmore© 50g artist-grade tracing paper was purchased from Amazon (Seattle, WA, USA). Liquinox was obtained from Alconox (White Plains, NY, USA). Sodium chloride, potassium chloride, calcium chloride, magnesium chloride, disodium ethylenediamine tetraacetic acid (Na_2EDTA), 4-(2-hydroxyethyl)-1-piperazineethanesulfonic acid (HEPES), sucrose, and glucose were obtained in high purity grades either through Fisher Scientific (Fairlawn, NJ, USA) or Sigma Aldrich (St. Louis, MO, USA). HPLC grade chloroform, ethanol, isopropanol, and methanol were also obtained through Fisher Scientific (Fairlawn, NJ, USA). Deionized water (18.2 MΩ cm^{-1}) was obtained using a RODI (Aztec, NM, USA) high-purity water system.

Phosphoinositide stock solutions were prepared by dissolving the lipid powder in a mixture of chloroform, methanol, and water in a ratio of 20 : 9 : 1 by volume. DOPC stocks were dissolved in a 2 : 1 mixture of chloroform and methanol by volume. The purity of lipid stock solutions was monitored continually by thin-layer chromatography. Concentrations of the phospholipid stock solutions were determined using an inorganic phosphate assay.[43] The DiD′ stock solution was prepared in HPLC grade ethanol. The fluorophore (DiD′ and TF-PI(4,5)P_2) stock solution concentrations were approximated using the Beer–Lambert relationship by taking absorbance measurements in methanol at 646 nm and 496 nm, respectively, and using the corresponding extinction coefficients (ε). The ε of DiD′ was obtained from the reagent lot's certificate of analysis ($\varepsilon = 250\,000$ $M^{-1}cm^{-1}$), while the ε of TF-PI(4,5)P_2 was obtained from Avanti Polar Lipids ($\varepsilon = 97\,000$ $M^{-1}cm^{-1}$). All lipid stock solutions and dyes were stored at −20 °C away from light.

2.2 Fabrication of asymmetric giant unilammellar vesicles

The procedure to fabricate asymmetric Giant Unilamellar Vesicles (aGUVs) using the hemifusion method requires three principal steps: Generation of symmetric Giant Unilamellar Vesicles (sGUVs), fabrication of solid supported lipid bilayers (SLB) from small unilamellar vesicles, and the hemifusion of the sGUVs with the SLB.

2.2.1 Generating symmetric giant unilamellar vesicles. Methods for obtaining a high yield of sGUVs will differ based on the desired lipid components. For the sGUVs representing the inner leaflet composition of the studied aGUVs, we used a molar composition of DOPC/DiD′ (99.02/0.08 mol%). For sGUVs representing the outer leaflet, we used a molar composition of DOPC/brain-PI(4,5)P_2/TF-PI(4,5)P_2 (95/4.94/0.06 mol%). A 0.5 mM total lipid stock solution of the desired inner leaflet composition was created. Separately, a 1 mM total lipid stock solution of the desired outer leaflet composition was prepared to a final volume of 600 μL prior to making the inner leaflet control sGUVs and the small unilamellar vesicles (SUVs) required to prepare the SLB. sGUVs were prepared using the PAPYRUS method with tracing paper as described by the Subramaniam group, modified for our purposes to exclude PEG lipids, incorporate a physiological concentration of salt, and include PI(4,5)P_2.[44,45]

In brief, a piece of tracing paper was first cleaned by submerging it in chloroform for 30 min and gently stirring every 10 min. This was then repeated with a fresh volume of chloroform. Afterwards, the chloroform was replaced with DI water and the previous process was repeated for another 30 min. Next, the paper was dried in a vacuum oven around 60 °C at 15 mmHg until dry. Once dry, two 6.35 mm diameter circular cutouts were removed from the paper using a single hole hole-puncher and coated with a 10 μL solution of either the outer leaflet lipid stock solution or the inner leaflet lipid stock solution. After applying this mixture, the paper cutout was placed back into the vacuum oven for 30 min. After drying, the cutout with the DOPC/DiD′ lipid mixture was placed in the bottom of a microwell plate at room temperature. The cutout with the DOPC/PI(4,5)P_2/TF-PI(4,5)P_2 mixture was placed in a 2 mL Eppendorf vial warmed to 49 °C. The following steps were taken to prepare both the inner and outer leaflet sGUVs, except all buffers used to make the PI(4,5)P_2 sGUVs were first warmed to 49 °C. 142.5 μL of 105 mM sucrose was added to the well containing the paper cutout, followed by 7.5 μL of a concentrated buffer solution containing 2.0 M NaCl, 1.0 M KCl, 0.020 M Na_2EDTA, 0.50 M HEPES pH = 7.4 10 min later to obtain an approximately 100 mM sucrose, 97 mM NaCl, 50 mM KCl, 1.0 mM Na_2EDTA, 25 mM HEPES, pH 7.4 buffer inside and outside of the formed sGUVs. The sGUVs were then collected after 110 min. To collect sGUVs after their formation, a pipette with a wide orifice tip was used to aspirate and dispense 150 μL of the solution 6 times before removing the entire solution. The osmolality of the sucrose + concentrated buffer solution was measured with an Advanced Instruments model 3300 Micro Osmometer (Norwood, MA, USA). All other buffers used in this study were adjusted to within ±2 mOsm kg^{-1} of this reading using either a 2 M NaCl solution (to adjust up) or DI water (to adjust down).

2.2.2 Preparing BSA-coated cover glasses. sGUV controls were imaged in chambered coverglasses coated with bovine serum albumin (BSA) to prevent the GUVs from rupturing upon contact with the glass. The coating procedure began with a thorough cleaning of the coverglass. The chambers were first rinsed three times in the following order: DI water, ethanol, isopropanol, ethanol, DI water. Next, the coverglass was sonicated for 15 min in dilute Liquinox detergent (around 2% (w/v) in DI water) at 69 °C in a Branson 1510 water bath sonicator (Brookfield, CT). Once done, the coverglass was rinsed again in the same order as before but finished with copious amounts of DI water. The coverglass was then dried under nitrogen gas and primed for BSA coating by filling each chamber with 0.01 M NaCl. After 40 min, 0.01 M NaCl was replaced with 1 mg per mL BSA in 0.01 M NaCl at pH = 5.0 to apply the BSA coating. After a minimum 3 h period, the coverglass was rinsed with copious DI water to remove excess BSA unbound from the glass surface. The DI water present in the BSA coated chambers was completely removed before the addition of sGUVs.

2.2.3 Fabrication of solid supported lipid bilayers. The solid-supported lipid bilayer composition is representative of the outer leaflet composition of the aGUVs formed using the hemifusion method. We used the same lipid stock solution to prepare both the outer leaflet sGUV controls and the small unilamellar vesicles (SUVs) used to create the SLB. The remaining 590 μL of the 1 mM total lipid stock solution containing PI(4,5)P_2 was dried under a steady stream of nitrogen gas at an elevated temperature. Following this, the vial containing the dried mixture is placed in a vacuum oven for 30 min to remove any residual solvent trapped between the layers of dried lipid. Once dry, the lipids are resuspended in a 97 mM NaCl, 100 mM KCl, 1 mM Na_2EDTA 25 mM HEPES, pH 7.4 buffer warmed to 49 °C and vortexed to create an opaque suspension of multilamellar vesicles (MLVs). The MLVs were then transferred from the vial into an Eppendorf tube and sonicated using a Sonics Vibra-Cell VCX130 Ultrasonic Processor tip sonicator (Newton, CT, USA), where an amplitude of 50% was applied to the vesicles in 15 s pulses (15 s on,

15 s off) for 30 min. After 30 min, a clear suspension of small unilamellar vesicles (SUVs) with an average diameter centered around 50–60 nm was retrieved. SUV diameters were confirmed using the dynamic light scattering function of a Malvern Zetasizer instrument (Malvern, Worcestershire, UK). These vesicles were then diluted fivefold using the resuspension buffer before use.

Before using the SUVs to form the SLBs, another chambered coverglass was first cleaned using the same method as described for the BSA-coated coverglass, but in this case, after drying the cleaned coverglass with nitrogen gas, it was plasma-cleaned under oxygen for 1.5 min using an LF-5 Plasma System (Mercator Control Systems, PVA TePla AG, Wettenberg, Germany). Immediately after plasma cleaning, a 0.5 mL aliquot of the diluted SUVs was added to two or three of the 8 chambers to begin SLB formation. In this process, the SUVs in each chamber settle to the bottom and fuse to form a planar bilayer. The SLBs were left to form, protected from light, for 20 min at room temperature. To remove residual unruptured vesicles and debris from the solid-supported lipid bilayer surface, 500 μL of resuspension buffer was added to the chamber and gently aspirated three times, discarding the buffer afterward. This process was repeated with fresh buffer two additional times, resulting in a total of nine aspirations. At the end, approximately 500 μL of buffer remained in the chamber. The quality (completeness) of the SLB was checked using fluorescence microscopy.

2.2.4. Hemifusion of symmetric giant unilamellar vesicles with supported lipid bilayers. Asymmetric GUVs were prepared using the hemifusion method we adapted in a recent paper for high salt concentrations.[31] In this procedure, 5 × 5 μL DOPC/DiD′ (99.9/0.1 mol%) sGUVs aliquots were added using a wide orifice pipette tip to a chamber containing a DOPC/PI(4,5)P_2/TF-PI(4,5)P_2 (95/5/0.06 mol %) SLB in 97 mM NaCl, 100 mM KCl, 1 mM Na_2EDTA, 25 mM HEPES, pH 7.4 (resuspension buffer). After waiting 10 min for the sGUVs to settle to the bottom of the chamber, hemifusion between the sGUVs and SLB was induced by adding 120 μL of 20 mM MCl_2 (M = Ca^{2+} or Mg^{2+}), 100 mM NaCl, 70 mM KCl, 25 mM HEPES, pH = 7.4 buffer. After waiting ∼23 min for hemifusion to progress, 250 μL of a 20 mM Na_2EDTA, 100 mM NaCl, 76 mM KCl, 25 mM HEPES, pH 7.4 buffer was added to chelate all of the bivalent cations and end the hemifusion process. A final 10 min period transpired before imaging to allow the system to equilibrate. Fresh sGUV controls were prepared and used on the same day as the hemifusion experiment. To ensure that any variations in results were directly attributable to the identity and concentration of the bivalent cations used to induce hemifusion, multiple experimental conditions were tested across different SLBs using the same control sGUV populations.

2.3 Fluorescence imaging

GUVs and SLBs images were captured on a Zeiss LSM510 confocal microscope (Oberkochen, Germany). SLBs and aGUVs were imaged in the chambers they were prepared in. sGUVs were imaged on BSA-coated coverglass by adding 20–30 μL of the sGUVs to the chamber, followed by 250 μL of a 100 mM glucose, 97 mM NaCl, 50 mM KCl, 1 mM Na_2EDTA, 25 mM HEPES, pH 7.4 buffer. Adding the glucose buffer after adding the GUVs helps the vesicles settle on the bottom and provides a better imaging contrast. All GUV images were obtained with plan-apochromat 63× magnification oil immersion objective. For the TF-channel, a 488 nm Ar laser with 505 nm long pass filter was used. For the DiD′ channel, a 633 nm HeNe laser with a 650 nm long pass filter was utilized. Unless otherwise specified, all GUV images were taken by capturing a single frame at the equator of the vesicle. Collected images were later processed in ImageJ by first applying the Gaussian blur filter with a radius of 1 μm, followed by an ImageJ code created by Dr Thais Enoki.[29] Fluorescence intensity data for each individual sGUV and aGUV were obtained by averaging the peak intensity profiles of 360 line scans in ImageJ and then plotted in Wolfram Mathematica©. Additional data plotting and statistics were performed in Excel and Origin Lab® 64 bit.

2.4 Analyzing asymmetric giant unilamellar vesicles

To determine the average fluorescence intensity of a vesicle, the ImageJ code generates 360 rays that radiate from the center of the vesicle image. These rays trace the perimeter of the vesicle, capturing the peak fluorescence intensity at 360 points along the edge. These peak intensities are then averaged together to give an average fluorescence intensity of a given vesicle. The average of all fluorescence intensities for each control population (TF-PI(4,5)P_2 controls and DiD′ controls) and aGUV population were taken. Any vesicles that fell outside two times the standard deviation of the channel average were discarded as outliers. A vesicle's DiD′ or TF-PI(4,5)P_2 intensity is chiefly dependent on the fraction of DiD′ or TF-PI(4,5)P_2 present in it. Theoretically, aGUVs that experienced complete outer leaflet exchange with the SLB should have half the fraction of DiD′ and TF-PI(4,5)P_2 as their sGUV counterparts, meaning their DiD′ and TF-PI(4,5)P_2 intensities should be half that of the sGUVs and with half the distribution width of the sGUVs. We term this region the ‘passing range’.

We also used two other parameters to assess the quality of the formed aGUVs: leaflet exchange (P^{ex}) and percentage agreement ($P^{a\%}$). P^{ex} is defined as how much DiD′ diffused out of or TF-PI(4,5)P_2 incorporated into the aGUV relative to the control GUVs and can be calculated as follows:

$$\text{Exchange (TF-PI(4,5)P}_2) = P^{ex}_{TF} = \frac{I_a}{I_c/2} \tag{1}$$

$$\text{Exchange (DID)} = P^{ex}_{DID} = 2\left(1 - \frac{I_a}{I_c}\right) \tag{2}$$

where I_a represents the average intensity of an aGUV channel and I_c represents the average intensity of all the symmetric control GUVs in the corresponding channel. $P^{a\%}$ is a simple parameter used to estimate how well the exchange percent values for each channel are aligned. $P^{a\%}$ can be calculated as shown below:

$$\text{Percentage agreement} = P^{a\%} = 100\% - [|P^{ex}_{TF} - P^{ex}_{DiD'}| \times 100].$$

A $P^{a\%}$ value of 100% for a given aGUV would mean that the amount of DiD′ exchanged with TF-PI(4,5)P_2 is equal, while a value below 100% indicates incomplete exchanges for one or both leaflets.

To summarize, an aGUV was considered to have undergone complete exchange with the proximal SLB leaflet only if the following conditions were met:

(1) Both the green (TF) and the red (DiD′) channel average fluorescence intensity values of the aGUV fell within the passing range.

(2) The leaflet exchange ratio for each channel of an aGUV fell within ±0.30 of complete outer exchange.

(3) The percentage agreement between the red and green channels in the aGUV fell within ±30%.

3 Results and discussion

3.1 Evaluation of the quality of solid supported $PI(4,5)P_2$/PC bilayers upon addition of Ca^{2+} or Mg^{2+}

The morphology and integrity of the solid-supported lipid bilayer used for the hemifusion and the time course of domain development, if it occurs, upon the addition of bivalent cations is an important factor for the quality of the aGUVs. Fig. 1 displays images of $PI(4,5)P_2$/PC supported lipid bilayers before and after either Ca^{2+} or Mg^{2+} was added. The images were taken less than five minutes after adjusting the bivalent cation concentration to the indicated amount. As seen in Fig. 1b, within moments, Ca^{2+} causes $PI(4,5)P_2$ to form long "finger-like" domains consisting of what appears to be multiple smaller, circular areas. Meanwhile, no domains appear to form in the SLB containing Mg^{2+}. Longer time scales result in the domains becoming larger and more circular.

3.2 Comparison of aGUVs obtained through hemifusion initiated with either 6 mM Ca^{2+} or Mg^{2+}

Fig. 2 shows DOPC and DOPC/brain $PI(4,5)P_2$ sGUVs in the left column, while the right hand columns display aGUVs that contain $PI(4,5)P_2$ on the outer leaflet and PC on the inner leaflet using either 6 mM Ca^{2+} or 6 mM Mg^{2+} as the initiator for hemifusion. We selected these concentrations to serve as a baseline of comparison to the method we developed for PS/PC aGUVs at physiological salt concentrations.[31] While these bivalent cation concentrations significantly exceed their physiological levels in cells, the addition of EDTA (hemifusion termination step) effectively masks these cations after the hemifusion is completed.

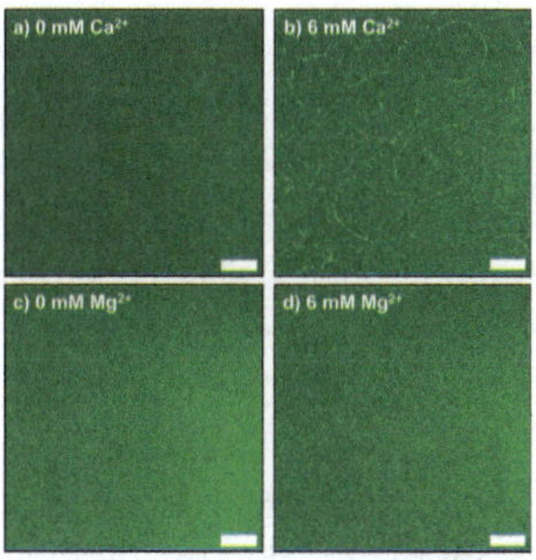

Fig. 1 Images of 95/4.94/0.06 mol% DOPC/brain-$PI(4,5)P_2$/TF-$PI(4,5)P_2$ solid-supported lipid bilayers. The left column displays SLBs with a buffer composition of 97 mM NaCl, 100 mM KCl, 1 mM Na_2EDTA 25 mM HEPES, pH 7.4 (resuspension buffer) prior to adding bivalent cations. The right column displays the same area of the SLB from the left column with either 6 mM Ca^{2+} or 6 mM Mg^{2+} present. (a) 0 mM Ca^{2+} (b) 6 mM Ca^{2+} (c) 0 mM Mg^{2+} (d) 6 mM Mg^{2+}. We accounted for the presence of EDTA by assuming 1 : 1 stoichiometry between the bivalent cation and EDTA. Scale bar = 20 µm.

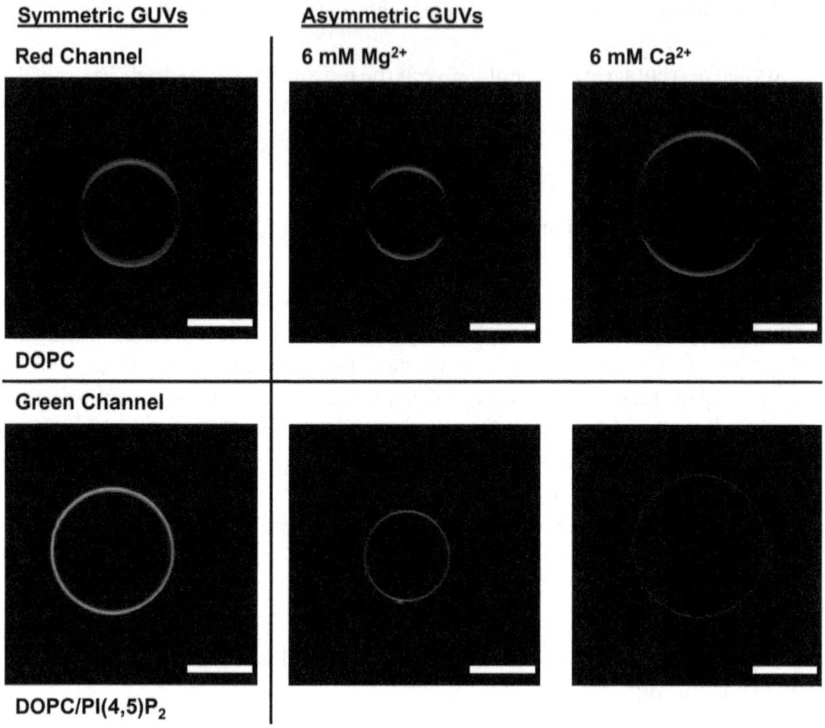

Fig. 2 Images of symmetric control GUVs (left column) and asymmetric GUVs (right column) prepared with either 6 mM Mg^{2+} or 6 mM Ca^{2+} as indicated. Scale bar = 10 μm. sGUV controls were prepared in 100 mM sucrose but were imaged in 100 mM glucose for a total buffer composition of 100 mM saccharide, 97 mM NaCl, 50 mM KCl, 1 mM Na_2-EDTA, 20 mM HEPES, pH = 7.4 on either side of the vesicle bilayer. aGUVs were imaged above the SLB after hemifusion in the resuspension buffer plus the indicated amount of bivalent cation and ~7 mM Na_2EDTA. Vesicles prepared with DiD′ appear dimmer at the equator due to a well-known polarization artifact arising from the properties of the fluorophore and are not due to phase separation in a pure DOPC bilayer.

Fig. 3 shows the processed intensity data (see methods section for details) for the sGUV controls and the aGUVs obtained through Ca^{2+} initiated hemifusion. Even after removing outliers, we found that the sGUV controls (Fig. 3a, left) containing PI(4,5)P_2 (green controls) have a wider distribution in fluorescence intensities compared to the pure DOPC sGUVs (red control). While the general fragility of GUVs with high (>5 mol%) levels of PI(4,5)P_2 (ref. 46) might be one of the reasons for the broader intensity distribution, it cannot be ruled out that interactions between PI(4,5)P_2 and the cellulose of the PAPYRUS during the swelling process may also be a contributing factor for the varying levels of PI(4,5)P_2 in the sGUVs. It is important to note that the PI(4,5)P_2/PC sGUVs are solely used to obtain fluorescence intensity values for a lipid bilayer so that the degree of exchange for the aGUVs can be evaluated.

The right side of Fig. 3a shows the normalized intensity distribution for DOPC/PI(4,5)P_2/TF-PI(4,5)P_2 (outer leaflet/green channel) and DOPC/DiD′ (inner leaflet/red channel) aGUVs. Since the aGUVs contain each lipid mixture only in one

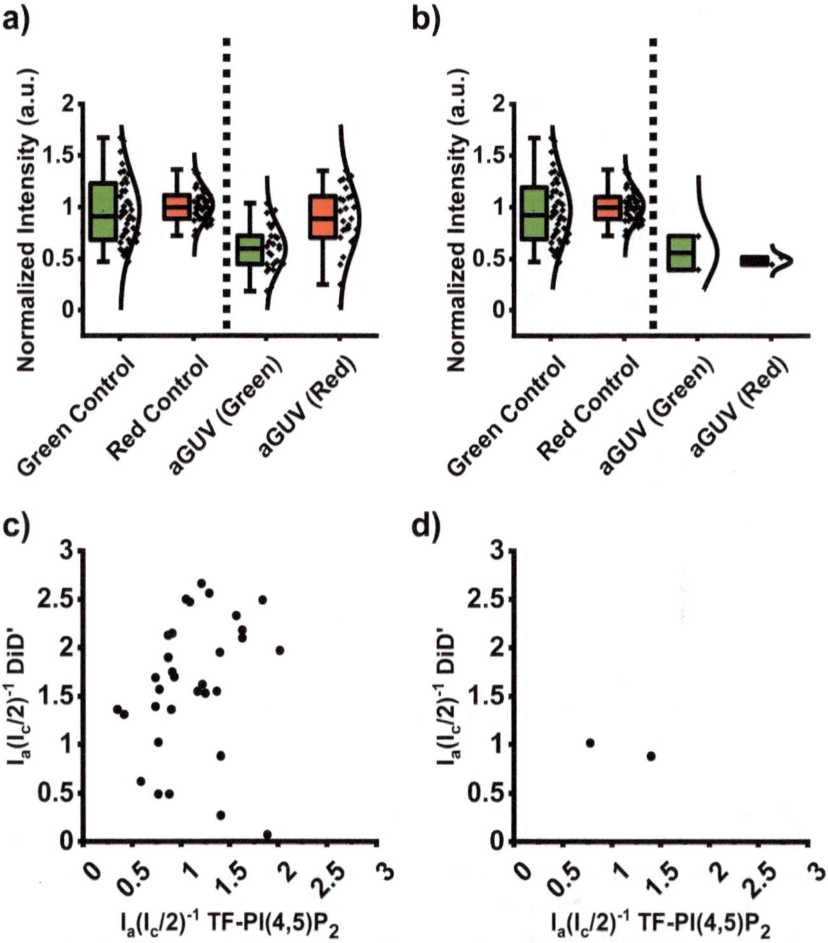

Fig. 3 DOPC/DiD (inner leaflet, red channel): DOPC/PI(4,5)P_2/TF-PI(4,5)P_2 (outer leaflet, green channel) asymmetric giant unilamellar vesicles prepared using 6 mM Ca^{2+} as the hemifusion initiator. (a) Comparison of sGUV controls for the green (DOPC/PI(4,5)P_2/TF-PI(4,5)P_2; $N = 51$) and red (DOPC/DiD′; $N = 47$) channels to the corresponding fluorescence channels of imaged aGUVs ($N = 32$) across 6 separate experiments. (b) The range-passing aGUVs (see text) from the samples presented in panel A. (c) Cross-correlation of the two channels for individual aGUVs. A "perfect" aGUV would be found at the data point (1,1) which corresponds to a completely exchanged outer leaflet and no artifacts like multi-lamellarity. (d) Cross-correlation of the two channels for the range-passing aGUVs. sGUV buffer: 97 mM NaCl, 100 mM KCl, 1 mM Na_2EDTA, 25 mM HEPES pH = 7.4 buffer. aGUV buffer after hemifusion: 97 mM NaCl, 100 mM KCl, 7 mM Na_2EDTA, 25 mM HEPES pH = 7.4 buffer. The controls displayed in (a) and (b) are the same experiments and are reproduced for comparison.

leaflet (as opposed to both leaflets in sGUVs), the intensities for the aGUVs are expected to be half of the intensity values for the corresponding sGUVs. While the median for the aGUV TF-PI(4,5)P_2 (outer leaflet) intensities is near the expected 0.5 intensity value, the DiD′ median intensity is slightly above 0.5 of the corresponding sGUV intensity and the distribution is broader than that observed for the TF-PI(4,5)P_2 (green) channel.

To identify those aGUVs that are artifact-free and exhibit a near complete DOPC/DiD′ inner leaflet, we analyzed the scatter plot in Fig. 3c, which correlates the fluorescence intensities of DiD′ and TF-PI(4,5)P_2 in aGUVs. Please note that the *x*-axis in Fig. 3c and d represents the fraction of the sGUV leaflet exchanged for the TF-PI(4,5)P_2 containing lipid mixture, while the *y*-axis represents the fraction of DiD′ leaflet remaining in the formed aGUV (the same applies to Fig. 4, 5c and d). By dividing the aGUV fluorescence intensities (I_a) for each channel by half of the corresponding sGUV intensities ($0.5I_c$), we obtain a value of "1" for a complete leaflet. This normalization places the "near perfect" aGUVs (outer sGUV leaflet completely hemifusion exchanged) around the (1,1) data mark on the scatter plot. It is clear from the scatter plot that many of the aGUVs don't fulfill the requirement of being close to the (1,1) mark. Furthermore, even for an incomplete exchange of the outer leaflet, the sum of the TF-PI(4,5)P_2 and DiD′ values should yield "2" since this corresponds to a unilamellar vesicle (single lipid bilayer). However, for several aGUVs the sum of the two values is significantly above "2", suggesting that the quality of these aGUVs has been compromised.

Vesicles can deviate from expected hemifusion values for multiple reasons. The most obvious reason is multi-lamellarity. Multilamellar vesicles present in the aGUV population can distort the observed distribution of fluorescence intensities since the presence of multiple bilayers intrinsically results in unpredictable fusion properties. While the algorithm used to determine the GUV fluorescence intensity values filters multilamellar symmetric giant vesicles out, it may fail to identify asymmetric giant vesicles that have partially exchanged their outermost leaflet. Another important aspect to consider is that in particular, TF-PI(4,5)P_2 is subject to quenching when the (local) fluorophore density changes, *e.g.*, due to the formation of domains (even though macroscopic domains are not discernible, nanoscopic domains might exist). This will obscure any quantitative analysis that is based on fluorescence intensities. Another reason is the presence of domains in the solid-supported lipid bilayer as can be seen in Fig. 1 for the Ca^{2+} initiated hemifusion. As the added symmetric GUVs make contact with the SLB, some vesicles will settle on regions enriched in PI(4,5)P_2, while others will interact with regions deficient in PI(4,5)P_2. Since the fluorescent probe used to monitor the extent of exchange for the SLB was a chain-labeled PI(4,5)P_2 molecule, this will lead to aGUVs that are either enriched or depleted in TF-PI(4,5)P_2, depending on whether the sGUV interacted with a domain or not.

To filter for the aGUVs that show fluorescence intensities for both channels within an acceptable range around half the corresponding intensity of the sGUVs, we utilized a data analysis method introduced in our recent paper on PS containing aGUVs.[31] Asymmetric GUVs with a fluorescence intensity in each channel that fall within one-half the average and standard deviation of their corresponding controls are considered 'range-passing', *i.e.*, in both channels, the acceptable intensity spread is one-half the intensity spread observed for the corresponding sGUV. Out of 32 imaged aGUVs across 6 repeat experiments, only two aGUVs prepared with 6 mM Ca^{2+} as the hemifusion initiator were range-passing (Fig. 3b). Consequently, those two aGUVs exhibit data points in the scatter plot near the (1,1) mark, however, overall, the yield of high quality aGUVs when Ca^{2+} is used as the hemifusion initiator, is unsatisfactory.

To test whether other bivalent cations yield better hemifusion results when PI(4,5)P_2 is present in the SLB, we expanded our study to Mg^{2+}. Fig. 4 shows the

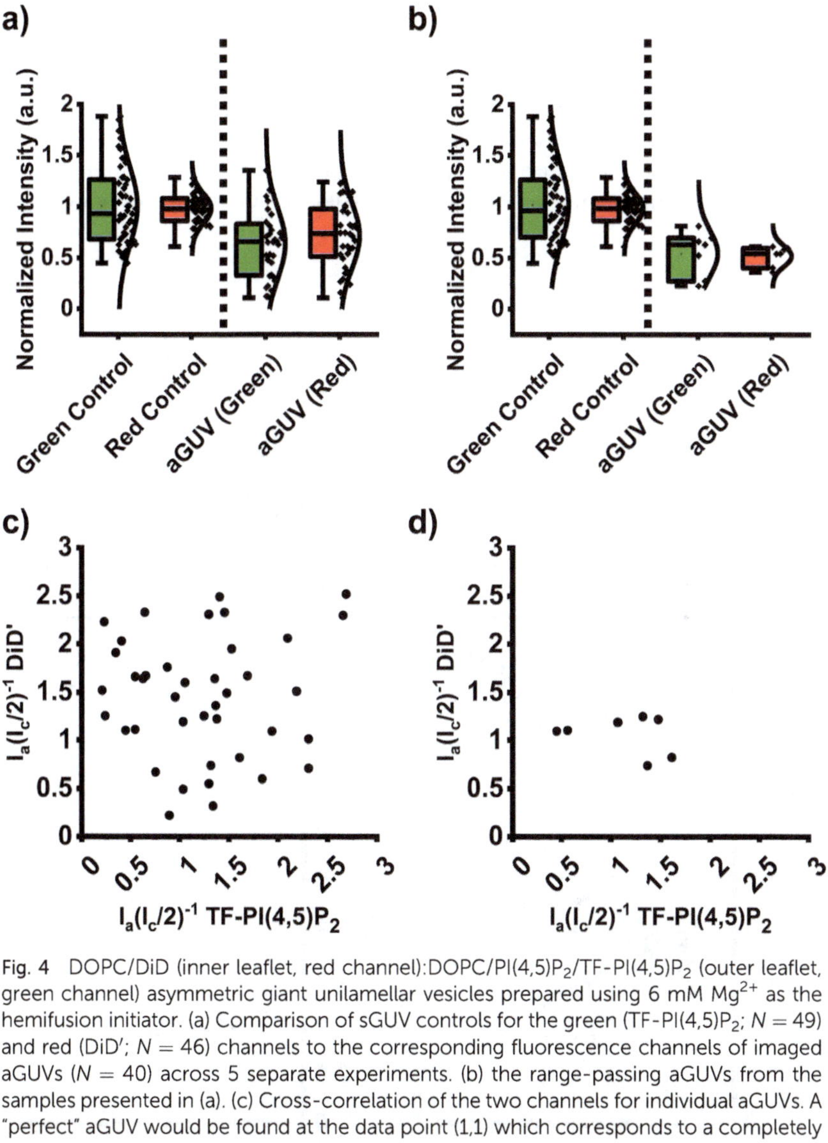

Fig. 4 DOPC/DiD (inner leaflet, red channel):DOPC/PI(4,5)P_2/TF-PI(4,5)P_2 (outer leaflet, green channel) asymmetric giant unilamellar vesicles prepared using 6 mM Mg^{2+} as the hemifusion initiator. (a) Comparison of sGUV controls for the green (TF-PI(4,5)P_2; $N = 49$) and red (DiD′; $N = 46$) channels to the corresponding fluorescence channels of imaged aGUVs ($N = 40$) across 5 separate experiments. (b) the range-passing aGUVs from the samples presented in (a). (c) Cross-correlation of the two channels for individual aGUVs. A "perfect" aGUV would be found at the data point (1,1) which corresponds to a completely exchanged outer leaflet and no artifacts like multi-lamellarity. (d) Cross-correlation of the two channels for the range-passing aGUVs. sGUV buffer: 97 mM NaCl, 100 mM KCl, 1 mM Na_2EDTA, 25 mM HEPES pH = 7.4 buffer. aGUV buffer after hemifusion: 97 mM NaCl, 100 mM KCl, 7 mM Na_2EDTA, 25 mM HEPES pH = 7.4 buffer. The controls displayed in (a) and (b) are the same experiments and are reproduced for comparison.

results for hemifusion initiated with 6 mM Mg^{2+}. Fig. 4a displays the fluorescence intensities for both channels, while Fig. 4c displays the corresponding scatter plot. Also in this case, many aGUVs are distant from the high-quality (1,1) exchange value. Fig. 4b and d show the data after the range-passing algorithm (described above) was applied. The number of vesicles that were range-passing was 7 out of 40 imaged vesicles giving a yield of approximately 18%. For the TF-PI(4,5)P_2 channel, some of the data points that are range passing are distant from the (1,1) data point. Since the distribution of the PI(4,5)P_2 containing sGUVs is quite broad, more aGUVs become range-passing even though only 50% of the outer leaflet has been exchanged. Overall, the quality of the aGUVs when using 6 mM Mg^{2+} instead of Ca^{2+} is improved, however, a greater yield is still desirable.

We were interested to see whether a lower concentration of Mg^{2+} would be sufficient to induce hemifusion in a system that contains PI(4,5)P_2 and whether this would result in a higher yield of high-quality aGUVs. There are two reasons

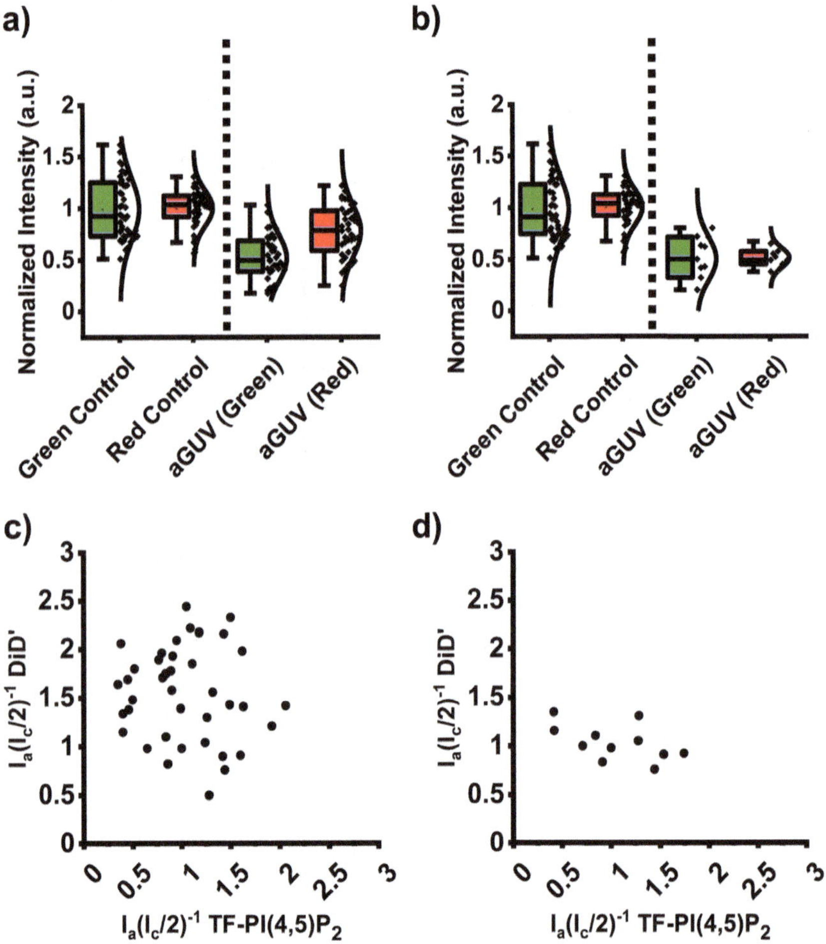

Fig. 5 DOPC/DiD (inner leaflet, red channel): DOPC/PI(4,5)P_2/TF-PI(4,5)P_2 (outer leaflet, green channel) asymmetric giant unilamellar vesicles prepared using 1 mM Mg^{2+} as hemifusion initiator. (a) Comparison of sGUV controls for the green (TF-PI(4,5)P2; $N = 47$) and red (DiD′; $N = 47$) channels to the corresponding fluorescence channels of imaged aGUVs ($N = 40$) across 5 separate experiments. (b) The range-passing aGUVs from the samples presented in (a). (c) Cross-correlation of the two channels for individual aGUVs. A "perfect" aGUV would be found at the data point (1,1) which corresponds to a completely exchanged outer leaflet and no artifacts like multi-lamellarity. (d) Cross-correlation of the two channels for the range-passing aGUVs. sGUV buffer: 97 mM NaCl, 100 mM KCl, 1 mM Na_2EDTA, 25 mM HEPES pH = 7.4 buffer. aGUV buffer after hemifusion: 97 mM NaCl, 100 mM KCl, 7 mM Na_2EDTA, 25 mM HEPES pH = 7.4 buffer. The controls displayed in (a) and (b) are the same experiments and are reproduced for comparison.

for this: First, a lower Mg^{2+} concentration limits the interference from $PI(4,5)P_2$/Mg^{2+} domains that may exist at the nanoscopic level. Second, to anticipate experimental difficulties that will arise when the aGUVs are used in experiments to model bivalent cation-induced domain formation of anionic lipids. To stop hemifusion, EDTA is added to the suspension, which binds the Mg^{2+} and leads to the detachment of the aGUVs from the SLB. For experiments aimed at studying the effect of Ca^{2+} on $PI(4,5)P_2$ domain formation in asymmetric lipid bilayers, the addition of Ca^{2+} will set free the EDTA bound Mg^{2+} since Ca^{2+} binds EDTA more strongly than Mg^{2+}. To use the aGUVs containing $PI(4,5)P_2$ as inner leaflet plasma membrane models, the ideal scenario would be one where the salt concentration more closely resembles that of the cytoplasm post-hemifusion. Therefore, using 1 mM Mg^{2+} instead of 6 mM will yield, after the addition of Ca^{2+} (over titration of EDTA), a Mg^{2+} concentration that more closely resembles cellular concentrations (0.5–1 mM).

Fig. 5 shows the data for aGUVs with an outer leaflet composition of DOPC/brain-$PI(4,5)P_2$/TF-$PI(4,5)P_2$ (95/4.94/0.06 mol%) using 1 mM Mg^{2+} to initiate hemifusion. Fig. 5a and b show that the yield of aGUVs is generally very high. Furthermore, nine aGUVs were considered range-passing out of 40 total for an approximately 23% yield, which is notably higher than in either of the other conditions. The distribution of P^{ex} values of both the green and red channels appear tighter than in either of the 6 mM Ca^{2+} or 6 mM Mg^{2+} conditions, as shown in Fig. 5c. Of the nine aGUVs that are range-passing, six fit our criteria of P^{ex} and $P^{a\%}$ values of less than 30%.

3.3 Evaluating aGUV quality using DiD′/TF-$PI(4,5)P_2$ exchange agreement value

In Fig. 3–5, we we are utilizing the formula $(I_a/(I_c/2)^{-1}$ for both DiD and TF-$PI(4,5)P_2$ to evaluate the quality of the exchange. In the case of the DiD containing leaflet this

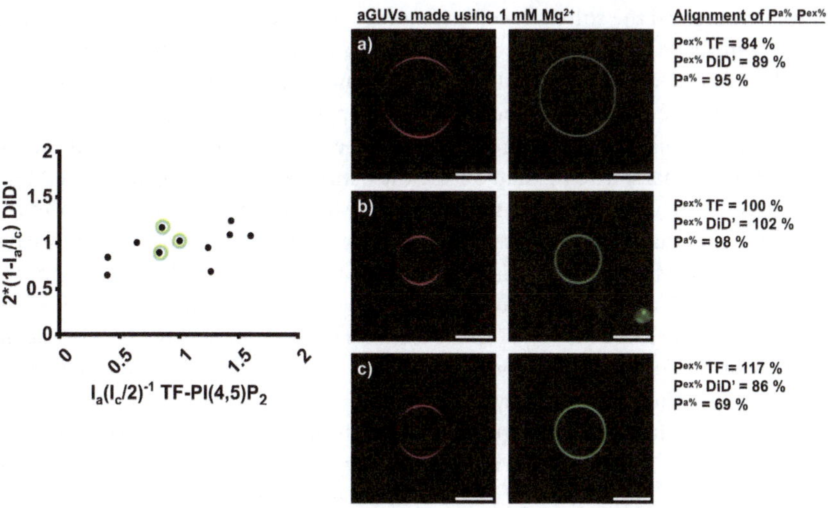

Fig. 6 Identification of the aGUVs with both, DiD′ and TF-$PI(4,5)P_2$, fluorescence intensity exchange P^{ex} are within 30% of each other (green circle identifies the selected aGUVs). Exchange agreement = $P^{a\%} = 100\% - [(P^{ex}(TF) - P^{ex}(DiD')) \times 100\%]$.

calculation determines the fraction of the DiD leaflet remaining rather than the fraction of DiD leaflet exchanged. In Fig. 6 eqn (2) is utilized to calculate exchange DiD. A major strength of using fluorescence microscopy of GUVs is that one can select for further studies the aGUVs that meet certain quality criteria, namely unilamellarity and a near perfect exchange of the outer leaflet. In Fig. 3–5, we based our yield calculation on how well each leaflet matches one-half the average and standard deviation of its corresponding symmetric control population. Doing so identifies a number of aGUVs that fall within a range that matches the theoretical leaflet composition of an ideal aGUV prepared from those symmetric lipid bilayers. In Fig. 6, we reproduce Fig. 5d (please note that the *y*-axis is changed to represent the fraction of exchanged DiD′ leaflet) and highlight those data points that fulfill the requirement that the data for both channels are within ±30% of each other and show the images for three of these aGUVs. The $P^{a\%}$ aGUV quality value shows excellent exchange and bilayer properties. Our desired outer leaflet composition, as determined by the lipid composition of the SLB used, was DOPC/brain-PI(4,5)P_2 (95/4.94/0.06 mol%). Thus, we can be reasonably confident that we obtained four vesicles that approximate the desired inner and outer leaflet asymmetric GUV composition of 100 mol% DOPC and 95/4.95 DOPC/PI(4,5)P_2, respectively. By multiplying the desired leaflet compositions by the P^{ex} values for each channel one can approximate the asymmetric vesicle composition, however, this assumes that the TF-PI(4,5)P_2 amount accurately reflects the overall PI(4,5)P_2 concentration. For the aGUVs prepared in 1 mM Mg^{2+}, the variation in PI(4,5)P_2 spans from ~3 mol% to ~5 mol% for the leaflet. This 2 mol% difference is equal to the range of PI(4,5)P_2 concentrations believed to exist on the plasma membrane (1–2 mol% across the plasma membrane bilayer). We chose a higher concentration of PI(4,5)P_2 to avoid potential quenching artifacts from TF-PI(4,5)P_2 which occur as a result of interactions with bivalent cations at plasma membrane concentrations of PI(4,5)P_2.[22,23]

4 Conclusions

We investigated the suitability of Ca^{2+} or Mg^{2+} initiated hemifusion of a PC sGUV with a mixed PC/PI(4,5)P_2 SLB to obtain aGUVs with the phosphoinositide species in the outer leaflet. We found that using 6 mM Ca^{2+} to initiate hemifusion in a system where PI(4,5)P_2 was present in the supported lipid bilayer resulted in the formation of micron-sized domains which affected the quality of aGUVs produced by creating many aGUVs that were either enriched or lacking in TF-PI(4,5)P_2 content. By relying on a known tendency of Mg^{2+} to cluster PI(4,5)P_2 to a much lesser extent than Ca^{2+}, we were able to slightly improve our yield of usable aGUVs using 6 mM Mg^{2+} as the initiator. However, when applying the most stringent quality selection, the number of aGUVs considered of the highest quality was still limited. The use of 1 mM Mg^{2+} as the initiator yielded not only the best results in terms of the aGUV quality and yield, it also provides the additional benefit of enabling studies aimed at investigating the interaction of Ca^{2+} with PI(4,5)P_2 without releasing unphysiologically high levels of Mg^{2+} from the EDTA upon the addition of Ca^{2+}. Compared to an earlier study aimed at fabricating PS containing aGUVs,[31] obtaining PI(4,5)P_2 containing sGUVs and aGUVs in physiological salt conditions turns out to be significantly more challenging. Overall, however, this study has shown that PI(4,5)P_2 containing aGUVs can be obtained, which will provide the foundation for future studies aimed at studying lipid raft/PI(4,5)P_2

domain interactions. Expanding the work to other PIPs or anionic lipids will require in all likelihood careful adjustments to the experimental conditions.

Data availability

Original data are available at https://asymvesiclepip2.wpi.edu/

Author contributions

TAP, AHR and AG were responsible for conceptualization, methodology, writing (original draft and review), TAP was additionally responsible for the investigation, validation and formal data analysis. AG, AHR supervised the project and acquired funding.

Conflicts of interest

The authors report no conflict of interest.

Note added in proof

Fig. 6 was revised following a question from Dr Thais Enoki, see DOI: 10.1039/D5FD90019K.

Acknowledgements

This work was funded by the National Science Foundation CHE-1904886.

References

1 Y. Wen, V. M. Vogt and G. W. Feigenson, *Annu. Rev. Biochem.*, 2021, **90**, 681–707.
2 Y. Wen, G. W. Feigenson, V. M. Vogt and R. A. Dick, *J. Mol. Biol.*, 2020, **432**, 5343–5364.
3 X. Li, Y. Miao, D. S. Pal and P. N. Devreotes, *Semin. Cell Dev. Biol.*, 2020, **100**, 133–142.
4 M. P. Krahn, *Front. Cell Dev. Biol.*, 2020, **8**, 277.
5 M. G. Waugh, S. Minogue, D. Chotai, F. Berditchevski and J. J. Hsuan, *J. Biol. Chem.*, 2006, **281**, 3757–3763.
6 R. M. Adam, N. K. Mukhopadhyay, J. Kim, D. Di Vizio, B. Cinar, K. Boucher, K. R. Solomon and M. R. Freeman, *Cancer Res.*, 2007, **67**, 6238–6246.
7 A. Arcaro, M. Aubert, M. E. Espinosa del Hierro, U. K. Khanzada, S. Angelidou, T. D. Tetley, A. G. Bittermann, M. C. Frame and M. J. Seckl, *Cell Signal*, 2007, **19**, 1081–1092.
8 B. Cinar, N. K. Mukhopadhyay, G. Meng and M. R. Freeman, *J. Biol. Chem.*, 2007, **282**, 29584–29593.
9 C. M. Johnson, G. R. Chichili and W. Rodgers, *J. Biol. Chem.*, 2008, **283**, 29920–29928.
10 S. Das, S. Chakraborty and A. Basu, *J. Neurochem.*, 2010, **115**, 537–549.

11 F. Furt, S. Konig, J. J. Bessoule, F. Sargueil, R. Zallot, T. Stanislas, E. Noirot, J. Lherminier, F. Simon-Plas, I. Heilmann and S. Mongrand, *Plant Physiol.*, 2010, **152**, 2173–2187.
12 X. Gao, P. R. Lowry, X. Zhou, C. Depry, Z. Wei, G. W. Wong and J. Zhang, *Proc. Natl. Acad. Sci. U. S. A.*, 2011, **108**, 14509–14514.
13 L. Zhang, J. Li, P. Zhang, Z. Gao, Y. Zhao, X. Qiao and C. Chen, *Biophys. Rep.*, 2018, **4**, 25–38.
14 O. Cizmecioglu, J. Ni, S. Xie, J. J. Zhao and T. M. Roberts, *eLife*, 2016, **5**, e17635.
15 J. van Rheenen, E. M. Achame, H. Janssen, J. Calafat and K. Jalink, *EMBO J.*, 2005, **24**, 1664–1673.
16 I. Parmryd, J. Adler, R. Patel and A. I. Magee, *Exp. Cell Res.*, 2003, **285**, 27–38.
17 J. Myeong, C. G. Park, B. C. Suh and B. Hille, *Proc. Natl. Acad. Sci. U. S. A.*, 2021, **118**, e2025343118.
18 Y. Nagatsuka and Y. Hirabayashi, *Biochim. Biophys. Acta*, 2008, **1780**, 405–409.
19 A. S. Klymchenko and R. Kreder, *Chem. Biol.*, 2014, **21**, 97–113.
20 I. Levental, D. A. Christian, Y. H. Wang, J. J. Madara, D. E. Discher and P. A. Janmey, *Biochemistry*, 2009, **48**, 8241–8248.
21 Z. Jiang, R. E. Redfern, Y. Isler, A. H. Ross and A. Gericke, *Chem. Phys. Lipids*, 2014, **182**, 52–61.
22 Y. Wen, V. M. Vogt and G. W. Feigenson, *Biophys. J.*, 2018, **114**, 2630–2639.
23 T. A. Paratore, G. E. Schmidt, A. H. Ross and A. Gericke, *Chem. Phys. Lipids*, 2024, **264**, 105424.
24 H. T. Cheng, Megha and E. London, *J. Biol. Chem.*, 2009, **284**, 6079–6092.
25 H. T. Cheng and E. London, *Biophys. J.*, 2011, **100**, 2671–2678.
26 S. Chiantia, P. Schwille, A. S. Klymchenko and E. London, *Biophys. J.*, 2011, **100**, L1–L3.
27 F. A. Heberle, D. Marquardt, M. Doktorova, B. Geier, R. F. Standaert, P. Heftberger, B. Kollmitzer, J. D. Nickels, R. A. Dick, G. W. Feigenson, J. Katsaras, E. London and G. Pabst, *Langmuir*, 2016, **32**, 5195–5200.
28 M. Son and E. London, *J. Lipid Res.*, 2013, **54**, 3385–3393.
29 T. A. Enoki and G. W. Feigenson, *Biophys. J.*, 2019, **117**, 1037–1050.
30 T. A. Enoki and G. W. Feigenson, *Biochim. Biophys. Acta, Biomembr.*, 2022, **1864**, 183995.
31 J. McDonough, T. A. Paratore, H. M. Ketelhohn, B. C. DeCilio, A. H. Ross and A. Gericke, *Membranes*, 2024, **14**, 181.
32 T. A. Enoki, J. Wu, F. A. Heberle and G. W. Feigenson, *Biochim. Biophys. Acta, Biomembr.*, 2021, **1863**, 183586.
33 T. A. Enoki, J. Wu, F. A. Heberle and G. W. Feigenson, *Data Brief*, 2021, **35**, 106927.
34 E. E. Kooijman, K. E. King, M. Gangoda and A. Gericke, *Biochemistry*, 2009, **48**, 9360–9371.
35 Y. H. Wang, A. Collins, L. Guo, K. B. Smith-Dupont, F. Gai, T. Svitkina and P. A. Janmey, *J. Am. Chem. Soc.*, 2012, **134**, 3387–3395.
36 K. Han, A. Gericke and R. W. Pastor, *J. Phys. Chem. B*, 2020, **124**, 1183–1196.
37 K. Han, S. H. Kim, R. M. Venable and R. W. Pastor, *Proc. Natl. Acad. Sci. U. S. A.*, 2022, **119**, e2202647119.
38 T. Balla, *Physiol. Rev.*, 2013, **93**, 1019–1137.
39 J. O. De Craene, D. L. Bertazzi, S. Bar and S. Friant, *Int. J. Mol. Sci.*, 2017, **18**, 634.

40 Y. Posor, W. Jang and V. Haucke, *Nat. Rev. Mol. Cell Biol.*, 2022, **23**, 797–816.
41 V. S. Yerramilli, A. H. Ross, S. Scarlata and A. Gericke, *Biophys. J.*, 2022, **121**, 793–807.
42 T. E. Rusten and H. Stenmark, *Nat. Methods*, 2006, **3**, 251–258.
43 G. Rouser, S. Fkeischer and A. Yamamoto, *Lipids*, 1970, **5**, 494–496.
44 J. Pazzi and A. B. Subramaniam, *ACS Appl. Mater. Interfaces*, 2020, **12**, 56549–56561.
45 A. Cooper, V. Girish and A. B. Subramaniam, *Langmuir*, 2023, **39**, 5579–5590.
46 K. Carvalho, L. Ramos, C. Roy and C. Picart, *Biophys. J.*, 2008, **95**, 4348–4360.

Faraday Discussions

PAPER

Engineering of lipid membranes asymmetrically functionalized with chondroitin sulfate†

Teresa Rodríguez-García, [a] Loretta Akakpo,[b] Sadie L. Nickles,[c] Ryan J. Schuck, [a] Daiane S. Alves,[a] Katherine G. Schaefer,[c] Frederick A. Heberle, [b] Gavin M. King[cd] and Francisco N. Barrera *[a]

Received 9th December 2024, Accepted 29th January 2025
DOI: 10.1039/d4fd00195h

One of the defining properties of the eukaryotic plasma membrane is the glycocalyx, which is formed by glycosylated lipids and proteins. The glycocalyx is arranged asymmetrically, as it is exclusive to the extracellular side of the membrane. Membrane asymmetry therefore includes both lipid and carbohydrate asymmetry, whereby the latter has the most skewed trans-leaflet imbalance. The glycocalyx plays a structural role that protects cell integrity and it also participates in mechanosensing and other cellular processes. However, our understanding of glycocalyx function is hampered by the lack of suitable model systems to perform biophysical investigation. Here, we describe the engineering of lipid bilayers that are chemically conjugated at the outer surface with one of the most abundant glycocalyx components, chondroitin sulfate (CS). Membranes were doped with a reactive phospholipid, which allowed thiol–maleimide conjugation of thiol-modified CS at the lipid headgroup. Our data show that we achieved CS conjugation of large unilamellar vesicles, supported lipid bilayers, and giant unilamellar vesicles. CS conjugation of vesicles allowed electrostatic recruitment of poly-L-lysine, which could recruit other CS-coated vesicles or CS in solution. Overall, we describe a simple and robust method for polysaccharide functionalization of vesicles which can be applied to gain new mechanistic understanding of the pathophysiological role of the glycocalyx.

Introduction

The plasma membrane of human cells is a highly complex and heterogeneous bilayer formed primarily by three families of molecules: lipids, proteins, and

[a]*Department of Biochemistry & Cellular and Molecular Biology, University of Tennessee, Knoxville, USA. E-mail: fbarrera@utk.edu*
[b]*Department of Chemistry, University of Tennessee, Knoxville, USA*
[c]*Department of Physics and Astronomy, University of Missouri, Columbia, USA*
[d]*Department of Biochemistry, University of Missouri, Columbia, USA*

† Electronic supplementary information (ESI) available. See DOI: https://doi.org/10.1039/d4fd00195h

carbohydrates. The arrangement of these molecules in a specific pattern allows the membrane to be a multi-functional entity that is able to respond dynamically to changes in the cell microenvironment. A fundamental property of the plasma membrane is asymmetry, as the two lipid bilayer leaflets have drastically different molecular composition. Plasma membrane asymmetry primarily entails a strong gradient of two different molecule classes: (1) transversal lipid asymmetry refers to the large differences in lipid composition between the inner (cytoplasmic) and outer membrane leaflets,[1,2] while (2) carbohydrate asymmetry occurs because glycosylated lipids and proteins are oriented in such a way that the sugar moieties face the cell exterior, constituting what is generally known as the glycocalyx.[3] The glycocalyx forms a thick and dense matrix composed of a diverse array of glycosylated molecules: glycolipids, proteoglycans, mucins and glycoproteins.[4,5] Far from merely working as a protective layer to the plasma membrane, the glycocalyx plays an active role that regulates a range of physiological and pathological processes.[4] The cellular processes where the glycocalyx is involved include regulation of cell morphology,[6] membrane protein diffusion,[7] and the immune system.[8] From a therapeutic standpoint, the glycocalyx also participates in target recognition by viruses,[9,10] fungi[11] and bacteria,[12,13] and controls cancer cell development[14] and can regulate cancer immunotherapy (checkpoint inhibition).[15] Despite its importance, we have a limited understanding of glycocalyx functions, in part due to the lack of adequate model systems that allow quantitative investigation of the interplay between lipid and saccharide components.

Chondroitin sulfate (CS) belongs to the glycosaminoglycan family, and is one of the major components of the glycocalyx and the extracellular matrix. CS is a long (∼10–100 kDa) linear polysaccharide formed by repetition of a disaccharide unit:[4] a D-glucuronic acid and a sulfated *N*-acetylgalactosamine.[16,17] The different types of CS (A, C, D and E), vary on the position of the sulfation and epimerization.[17] CS chains are linked to a core protein unit to form chondroitin sulfate proteoglycans (CSPG). CSPG are linked to the plasma membrane, as part of the glycocalyx, or linked at the extracellular matrix, where they participate in tissue formation. CSPG also regulate signaling by binding to growth factors, as they control their access to their target receptor.[18]

Two recent publications report on approaches to decorate vesicle with CS. Jahnke and co-workers chemically conjugated CS to cholesterol,[19] and the soluble CS-cholesterol hybrid molecule is able to partition into the surface of vesicles. However, this strategy leads to cholesterol asymmetry in the plasma membrane, which is expected to strongly impact the physical properties of the membrane.[20,21] Shioiri *et al.* used a synthetic approach to conjugate CS to the phospholipid phosphatidylethanolamine.[22] While this is an elegant chemical approach, it requires precipitation and purification steps. An additional potential drawback of the two previous approaches, where the lipid-CS partitions into already formed vesicles, is the poor control over the final membrane density of the CS.

Here, we report a facile method to form lipid bilayers that are coated with CS (Fig. 1). Our data show that this approach generates CS-functionalized bilayers for the three most common types of model membranes: large unilamellar vesicles (LUVs), supported membranes, and giant unilamellar vesicles (GUVs).

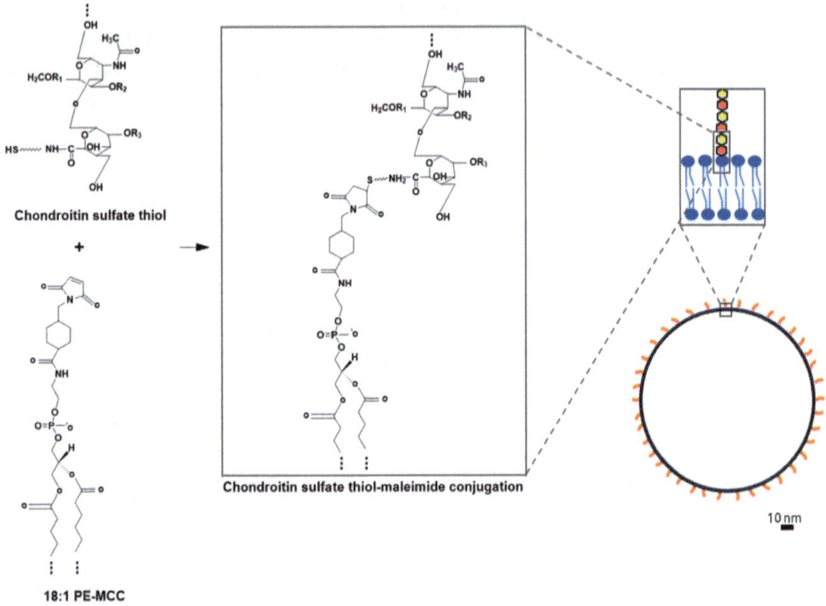

Fig. 1 Conjugation of chondroitin sulfate thiol (CS) to a maleimide-modified lipid to form CS-functionalized large unilamellar vesicles. The chemical structure of CS and 18 : 1 PE-MCC is shown before and after thiol–maleimide bond formation. In the CS chemical structure, R1, R2, R3 = H or SO_3H, depending on the type of CS subtype (A, C, D, or E) and the squiggly line indicates a spacer linker. Only part of the fatty acid chains and CS polymer are shown (marked with ellipsis). D-glucuronic acid and sulfated *N*-acetylgalactosamine are shown as red and yellow hexagons. The figure is approximately to scale.

Methods

Reagents

The lipids POPC (1-palmitoyl-2-oleoyl-*glycero*-3-phosphocholine) (Catalog No. 850457), 18 : 1 PE-MCC (1,2-dioleoyl-*sn*-glycero-3-phosphoethanolamine-*N*-[4-(*p*-maleimidomethyl) cyclohexane-carboxamide] (sodium salt)) (Catalog No. 780201), DOPC (1,2-dioleoyl-*sn*-glycero-3-phosphocholine) (Catalog No. 850375), 18 : 1 PE-TopFluorAF594 (1,2-dioleoyl-*sn*-glycero-3-phosphoethanolamine-*N*-(TopFluorAF594) (ammonium salt)) (Catalog No. 810387), and 18 : 1 NBD PE (1,2-dioleoyl-*sn*-glycero-3-phosphoethanolamine-*N*-(7-nitro-2-1,3-benzoxadiazol-4-yl)) (ammonium salt) (Catalog No. 810145) were purchased from Avanti Polar Lipids, Alabaster, AL. Tris base (2-amino-2-(hydroxymethyl)-1,3-propanediol) (Catalog No. BP152-500) and Wisteria Floribunda Lectin-FITC Labeled (WFA) (Catalog No. L32481) were purchased from Fisher Scientific. We used a mixture of chondroitin sulfate A, C, D and E that was functionalized with a thiol group from Haworks LLC, Bedminster, NJ (CS-Thiol-25k), which is abbreviated simply as CS. Poly-L-lysine (PLL) labeled with the fluorophore FITC (Catalog No. P3543-10MG) was purchased from Millipore Sigma, Burlington, MA.

CS conjugation to large unilamellar vesicles

POPC and 18 : 1 PE-MCC stocks were prepared in chloroform. Aliquots of lipids with 99.5 mol% POPC and 0.5% 18 : 1 PE-MCC were dried under argon gas and then placed in a vacuum overnight. Lipid films were resuspended in 10 mM Tris, 350 mM NaCl buffer, pH 7.5. Large unilamellar vesicles (LUVs) were formed by extrusion with a Mini-Extruder (Avanti Polar Lipids, Alabaster, AL) through a 100 nm pore size membrane (Whatman, United Kingdom) at room temperature. Functionalization of the LUVs was performed with CS prepared in the same buffer at a 2 : 1 CS : lipid molar ratio. The conjugation reaction with CS was incubated for 2 hours with constant shaking at room temperature.

NBD head group environment fluorescence assay

LUVs conjugated with chondroitin sulfate were prepared as described above, but in this assay the lipid composition of the LUVs was 98.5 mol% POPC, 1% 18 : 1 NBD PE, and 0.5% 18 : 1 PE-MCC. Samples were loaded into a black 96-well plate (Corning, Kennebunk, ME) to measure the fluorescence spectra on a Cytation 5 plate reader (BioTek, Winooski, VT), using an excitation wavelength of 460 (±10) nm and an emission wavelength of 535 (±10) nm.

Dynamic light scattering (DLS)

We used a Dynapro Nanostar I DLS instrument (WYATT Technology, Santa Barbara, CA), which uses a 658 nm laser at a controlled temperature of 25 °C. The scattered light was measured at an angle of 90°. For the analysis, the LUVs were prepared as described in the previous paragraph. A titration with FITC-conjugated poly-L-lysine (PLL-FITC) was performed after a 30 minutes incubation. CS and PLL controls were also prepared. All samples were diluted with 10 mM Tris, 350 mM NaCl buffer, pH 7.5 for the corresponding final concentrations, and to a final volume of 100 µL. The samples were measured into a cyclic olefin copolymers cuvette (WYATT Technology, NC0616299). All DLS data were processed using the DYNAMICS software (WYATT Technology, Santa Barbara, CA), and the size distribution is shown considering the mass particle size distribution.

Removal of unconjugated chondroitin sulfate

To remove the unconjugated chondroitin sulfate of the glycoliposome samples, five washings of the PCMal-CS *via* centricon (Vivaspin 6, 100 kDa; Millipore Sigma, Catalog No. GE28-9323-19) were performed. Glycoliposomes were prepared as stated in the "CS conjugation to large unilamellar vesicles" section. After the conjugation with CS, five centrifugation steps (3260×*g*, 10 minutes, 25 °C) were done. Flow through samples of every washing were collected in order to measure the concentration of CS after each one. To perform this measurement, we used the lectin WFA labeled with FITC. WFA binding to CS causes a fluorescence increase, allowing to create a calibration curve where the fluorescence intensity of the FITC labeled in WFA increases as the [CS] increases. Samples were loaded into a black 96-well plate (Corning, Kennebunk, ME) to measure the fluorescence spectra on a Cytation 5 plate reader (BioTek, Winooski, VT), using an excitation wavelength of 495 (±9) nm and an emission wavelength of 515 (±9) nm. In each replicate, a linear calibration curve was created at the same time as the flow through samples were measured. After the centrifugation, the PCMal-CS sample was measured using DLS.

Lipid preparation for AFM

DOPC and 18 : 1 PE-MCC at a 99.5 : 0.5 molar ratio, were dried in glass culture tubes, back-filled with argon, sealed, and stored at −20 °C. At the time of extrusion, room-temperature phosphate-buffered saline (PBS) was added to swell the lipids. The lipid solution was extruded (Liposofast, Avestin) through 100 nm membranes 51 times, to form unilamellar vesicles. The solution was aliquoted and stored at −80 °C until the time of the AFM experiments. On the day of experiments, CS was incubated with the DOPC PE-MCC lipid in a 2 : 1 CS : lipid molar ratio for 1 hour with constant shaking.

Supported lipid bilayer formation and AFM imaging

80 μL of the sample solution was deposited on a freshly cleaved mica disc and incubated for 10 min at room temperature to allow for adhesion to the surface. Loosely bound material was washed away *via* buffer exchange (80 μL of PBS, 5–6 times). All images were acquired in PBS using biolever mini tips (Bruker, $k \sim 0.1$ N m^{-1}, $F_0 \sim 25$ kHz in fluid) on a Cypher, Asylum Research AFM in tapping mode. Care was taken to keep the tip-sample force <100 pN. Analysis was performed using commercial software (Asylum Research, Inc).

GUVs formation and functionalization

GUVs were electroformed in the presence of 100 mM sucrose using an analog heat block connected to a function generator. 250 nanomoles of lipid (97 mol% POPC, 2% 18 : 1 PE-MCC and 1% PE-TopFluorAF594) dissolved in 200 μL of chloroform, were spread homogeneously on the conductive site of indium-tin-oxide (ITO) coated glass slides (Delta Technologies, Loveland, CO). Subsequently, the lipid-film-coated ITO slides were desiccated for 2 hours by placing the slides in a heated chamber (55 °C) attached to a vacuum pump, to remove the remaining chloroform. An O-ring spacer was placed on top of the lipid-coated ITO slide, which was then filled up with 600 μL of 100 mM sucrose, 10 mM Tris, pH 7.5 solution. A sealed chamber was created by placing a second ITO slide on top. Electrodes were connected with the conductive sides of the ITO slides and the preinstalled standard program was run, generating an AC field of 2 V and 10 Hz for 2 h at 30 °C.

After electroformation, GUVs were functionalized with CS. For the conjugation, 70 μL of GUVs were incubated for 1 hour with 70 μL of CS 25 μM, in 10 mM Tris, 100 mM sucrose, pH 7.5 buffer. GUVs were then incubated with 0.5 μM PLL-FITC for 30 minutes. GUVs conjugated with chondroitin in the presence of PLL-FITC were harvested by sedimentation after 20 minutes in 1 mL of 100 mM glucose, 10 mM Tris, pH 7.5 solution. Subsequently, they were placed on a slide with a spacer between slide and coverslip for confocal imaging.

Confocal microscopy imaging of GUVs

Confocal microscopy was performed on an inverted Zeiss LSM 900/Airyscan laser scanning confocal microscope (Carl Zeiss AG). The pinhole was set to 1 airy unit, and all experiments were conducted at room temperature. For image acquisition, a 63X oil immersion objective (HCPL APO 63X/1,40 OIL CS2) was used. The rAF594 channel used an excitation wavelength of 561 nm and the FITC channel of 488 nm. Images were analyzed and adjusted in ZEISS ZEN 3.11 software.

Results

We sought to develop a straightforward approach for covalent conjugation of CS to lipid membranes. We employed the thiol–maleimide reaction, commonly used for protein labeling, and applied it in this case to create a hybrid lipid molecule with a long and negatively charged CS chain at the headgroup (Fig. 1, left). To this end, we employed two commercially available reagents: a 25 kDa chain of CS that was modified with a thiol moiety, and the lipid 18 : 1 PE-MCC, which we will refer to as PCMal, where a maleimide group is incorporated at the phosphatidylethanolamine headgroup. Incubation at room temperature and neutral pH leads to formation of a thiol–maleimide covalent bond.

We expected that the presence of the negatively charged CS chain would increase the polarity of the lipid headgroup layer. We devised an assay to test the success of the conjugation, by incorporating an environmentally-sensitive reporter at the lipid headgroup. We chose an NBD fluorophore incorporated into a phospholipid headgroup, as the fluorescence of NBD decreases in a more polar medium.[23–27] To perform this assay, we formed POPC large unilamellar vesicles (LUVs) that contained 0.5 mol% of PCMal and 1 mol% of an NBD-labeled phospholipid. We allowed these vesicles to react with thiol-modified CS (for simplicity we will refer to this reagent simply as CS hereafter). We observed a large change in the NBD fluorescence spectra (Fig. 2A), characterized by an important decrease in fluorescence intensity (Fig. 2C). To ensure that this fluorescence change resulted from CS conjugation to the LUVs, we performed control experiments where PE-MCC was not present in the vesicles. In this case we observed only a small change in fluorescence (Fig. 2B and C), which might result from non-covalent interaction of CS with the bilayer. The larger fluorescence intensity

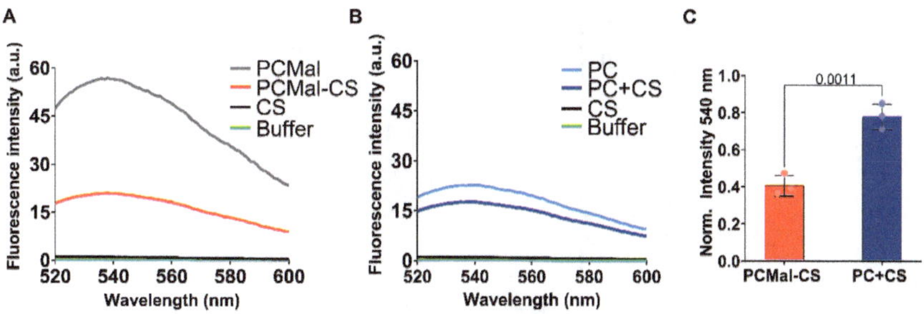

Fig. 2 NBD headgroup environmental assay. Representative fluorescence spectra for CS incubation with PCMal (A) or non-reactive PC (B) LUVs. Control fluorescence for buffer and CS in buffer are shown. (C) Quantification of NBD fluorescence intensity at 540 nm are shown for PCMal-CS and for PC + CS, normalized to conditions in the absence of CS. The error bars are the S.D., and the p value was obtained from a one-tailed Student's t-test ($n = 3$).

change observed with PCMal suggests that there is covalent attachment of CS to PCMal LUVs.

The 25 kDa CS chains that we employed contain more than 50 disaccharide units. Therefore, after they are conjugated to the LUVs, they are expected to cause a detectable increase in vesicle hydrodynamic radius (Fig. 1, right). We employed dynamic light scattering (DLS) to investigate vesicle size. Control DLS experiments with free CS showed that the polysaccharide particle has an effective diameter of 7.5 ± 5.3 nm (mean ± S.D.) (Fig. 3 and ESI Fig. 1†), suggesting that the CS chains are partially elongated in solution. We performed DLS before and after conjugation, and observed an increase in vesicle size of ~15 nm (Fig. 3), compatible with CS-coated LUVs (Fig. 1). This result supports that the conjugation was successful.

Since we performed the maleimide-thiol conjugation with a molar excess of CS, we expect free CS to be found outside the vesicles. We performed serial centrifugal concentration steps using a Centricon with a 100 kDa bilayer, which is expected to allow free CS to go into the flow-through (FT), while CS-coated LUVs remain inside the centrifugal device. After each step, fresh buffer was added and the process was repeated a total of five times, where we expect a serial dilution to occur in each FT. We determined the CS levels in the FTs by quantification with a lectin that binds CS, conjugated to a fluorophore. Specifically, we used the *Wisteria floribunda* (WFA) lectin labelled with fluorescein. As expected, we observed a sequential decrease is free CS in each centrifugal step, and that five steps were enough to remove most CS (ESI Fig. 2†). The vesicle size did not change significantly after CS removal. However, we observed the appearance of an unexplained particle population of ~25 nm. Therefore, we performed the rest of the experiments without the centrifugal step. Next, we studied the colloidal stability of the PCMal-CS vesicles. After the conjugation we stored the samples at 4 °C, and performed DLS experiments over two weeks. The data showed no major change in vesicle diameter over time (ESI Fig. 3†), suggesting that the CS-conjugated LUVs are stable long-term under refrigerated storage. Finally, we tested the effect of a −80 °C freeze/thaw

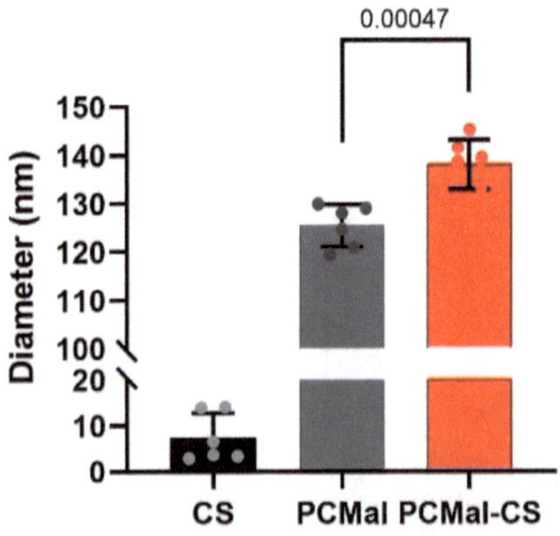

Fig. 3 DLS study of LUV size. DLS data are shown for thiol-modified chondroitin sulfate (CS) in solution, LUVs containing maleimide-reactive PE-MCC lipid (PCMal), and for PCMal LUVs after conjugation with CS (PCMal-CS). Error bars are the S.D., and the p value was obtained from a one-tailed Student's t-test ($n = 6$).

cycle. DLS was performed after thawing, and we observed no diameter change (ESI Fig. 3†). These DLS results suggest that our protocol for coating LUVs with CS yields robust vesicles that can be stored at 4 and −80 °C. These results underscore the ease of experimentation with these vesicles, which do not require daily preparation and can be reused after −80 °C storage.

The use of bilayers deposited on solid supports, forming supported lipid bilayers (SLB), is a popular method that allows the use of different types of microscopy to study lipid membranes. We applied our protocol for CS functionalization to SLB, which were imaged using atomic force microscopy (AFM). We employed membranes of DOPC, the lipid of choice for AFM due to the ease of spreading onto the underlying mica surface,[28,29] doped with PCMal. Fig. 4 displays AFM images showing that conjugation with CS causes changes in membrane morphology. As expected, pristine bilayers were largely homogeneous and flat (Fig. 4A), while bilayers that were subjected to CS conjugation followed by extensive rinsing, showed heterogeneities on top of the bilayer plane (Fig. 4B and ESI Fig. 4†).

We estimated surface homogeneity by quantification of the root mean square roughness in 4 non-overlapping 100 nm × 100 nm regions. The lipid-only bilayer surface was smooth, with a roughness of less than 1 Å. However, after conjugation with CS, the roughness approximately doubled (Fig. 4 and ESI Fig. 4†). Line analysis illustrates local increases in height of 5–10 Å that are compatible with conjugation of CS molecules that lay mostly flat on the lipid membrane. Under these conditions, we expect the CS chains not to be perpendicular to the SLB surface, as they are highly flexible and could be deflected downward by the AFM tip, unlike the DLS experiments, where they are expected to be more extended on the surface of the LUVs (Fig. 1, right), hence the significantly larger dimensions observed by DLS (compare Fig. 3 and 4B). Finally, we performed a control experiment where CS was deposited directly on the mica, in the absence of lipid.

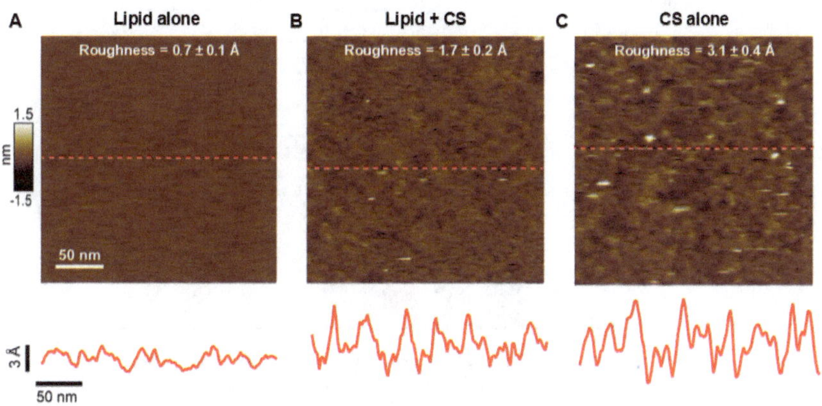

Fig. 4 Chondroitin sulfate conjugation increases the roughness of the lipid bilayer surface. (A) AFM image of DOPC in the absence of CS. The root mean square roughness (mean ± standard deviation) is displayed on top. The dashed red line identifies a line scan that is displayed below the image. (B) Data for DOPC PCMal lipid conjugated with CS. A greater than 2-fold increase in bilayer surface roughness is observed over bare DOPC. (C) Chondroitin sulfate deposited directly on mica (in the absence of lipid). All data was acquired in aqueous buffer solution at room temperature.

As expected, these data show roughness (Fig. 4C) due to the polysaccharide. This control experiment supports the notion that membrane roughness originates from CS functionalization. Taken all together, the AFM results indicate that conjugation of CS to the membrane results in a molecularly rough surface. This result is expected for the low (0.5 mol%) maleimide lipid density that we employed. In these conditions the anchor points of CS to the membrane, if one considers that they are distributed randomly in the membrane, would be separated by ~10 nm (Fig. 1, right).

CS coating of the lipid is expected to change the physical properties of the vesicles, and allow for the introduction of new functionalities. As a simple test, we studied the interaction with poly-L-lysine (PLL). PLL is a synthetic polypeptide that is often used in cell culture experiments to provide a favorable substrate for human cells to grow on a plastic surface. PLL favors cell spreading and attachment as it establishes favorable electrostatic interactions with negative charges in the glycocalyx.[16,30] We chose a FITC-labeled and long PLL chain (24 kDa), reasoning that after binding to the surface of a CS-coated LUV, the PLL polymer might protrude and interact with a second LUV, in practice non-covalently crosslinking different vesicles. We tested this hypothesis incubating PCMal LUVs with PLL. Control DLS experiments showed that PLL interacted with free CS (Fig. 5), forming a complex that was larger (20.7 ± 3.0 nm; mean ± S.D.) than the addition of the sizes of the PLL and CS particles. This result indicates that PLL interacts electrostatically with sulfate groups on CS. When we added PLL to the PCMal-CS LUVs, we observed a magnified effect. Fig. 5 shows DLS data for CS vesicles incubated with increasing concentrations of PLL (ESI Fig. 5†). The diameter of the particles increased several fold, up to a size of ~3 µm (note that the vertical axis uses a logarithmic scale). These results suggest agglomeration of vesicles, showing that CS is able to endow vesicles with new functionality.

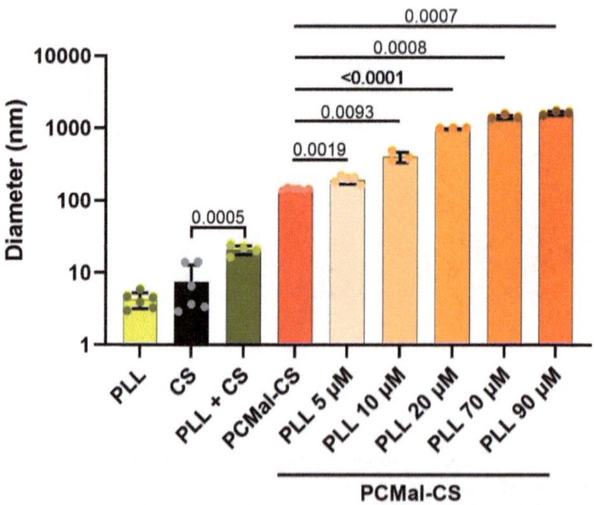

Fig. 5 Poly-L-lysine crosslinks PCMal-CS LUVs. Control DLS data are shown for Poly-L-lysine-FITC (PLL), CS, and CS incubated with PLL. PCMal-CS LUVs were incubated with increasing concentrations of PLL and DLS measurements performed. The error bars are S.D., and p values were obtained from a one-tailed Student's t-test ($n = 3–6$).

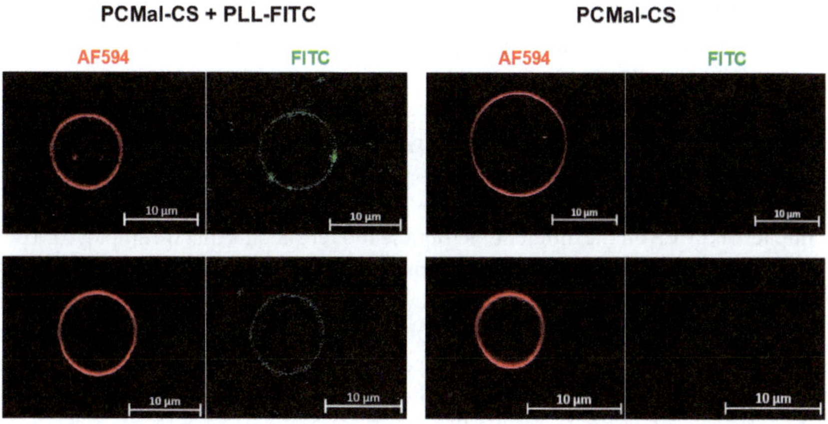

Fig. 6 Giant unilamellar vesicles functionalization with CS. Left panel, PCMal-CS GUVs were incubated with PLL-FITC. The red channel shows the AF594 lipid, and the green channel PLL-FITC. Right panel, control experiment in the absence of PLL-FITC. Two GUVs are shown for each experimental condition. The equatorial variation in fluorescence intensity of AF594 is likely due to a polarization effect. Scale bars are 10 μm.

Giant unilamellar vesicles (GUVs) are cell-sized liposomes that have become the standard membrane model system for confocal microscopy investigation of lipid membranes.[31] We tested if our thiol–maleimide conjugation scheme was able to achieve CS conjugation of GUVs. We used a standard electro-swelling protocol to form GUVs, which was followed by incubation with CS. GUV visualization was achieved by vesicle doping with the fluorescent lipid TopFluor-A594. Since the DLS data indicated binding of PLL-FITC to CS-conjugated vesicles, we used this reagent for fluorescence detection of CS on the GUVs. However, since GUVs have low mechanical resistance, we used different experimental conditions that were geared towards avoiding potential GUV disruption due to vesicle agglomeration. Therefore, we used a low GUV density, where the probability of two vesicles interacting with the same PLL-FITC molecule are expected to be low. Fig. 6 shows data collected after CS conjugation in the presence (left) or absence (right) of PLL-FITC. As expected, we observed GUVs larger than 5 μm in diameter, with a range of sizes. We observed that PCMal-CS GUVs were covered with CS, as reported by green labeling due to PLL-FITC recruitment. We observed that GUVs were frequently decorated with bright green spots. Since we use a molar excess of CS, we expect CS to be found in solution. Our images contained green particles on the vesicles (Fig. 6), which we attribute to free CS that is aggregated by PLL-FITC. We propose that the green spots observed on the vesicle surface result from PLL recruitment of these particles into the GUVs, analogous of how CS crosslinked LUVs (Fig. 5).

Conclusions

This work describes a facile new method to functionalize model membranes with long CS polymers. Our approach uses the 18 : 1 PE-MCC (PCMal) reactive lipid, which contains a maleimide group in the headgroup that is able to form a covalent bond with thiol-modified CS. Both reagents are commercially available, as well as for alternate lipid and glycosaminoglycan options. Such conjugation created a synthetic glycolipid that remained anchored to the membrane due to the hydrophobicity of the acyl chains. We demonstrate the success of the approach in the three most common types of reconstituted membranes: LUVs, GUVs and SLBs.

We performed the covalent linkage after membrane formation to minimize impact of the CS on the different membrane model systems that we employed. This approach is expected to yield strongly asymmetric membranes with a skewed CS distribution. The SLBs we formed will have no lipid acyl chain asymmetry, since the acyl chains of the DOPC and PCMal lipids are identical. However, the LUVs were formed with POPC, and therefore we expect some asymmetry in the distribution of the palmitoyl acyl chain between the inner and outer membrane leaflet. We expect that the rate of trans-bilayer flip-flop of asymmetric glycolipid to be extremely slow,[32] due to the large size (25 kDa) and multiple negative charges of the CS chain. Therefore, we expect CS asymmetry to be maintained for long periods of time.

The strategy that we present for functionalization of lipid membranes with a polysaccharide is straightforward and can be easily carried out in most biochemistry laboratories. Unlike previous methods for lipid modification with polysaccharides,[14,19,22,33] our approach does not require complex chemical synthesis or purification of modified lipid species. We observed that our CS-liposomes were stable at least over a period of two weeks, and were not significantly disrupted by a freeze–thaw cycle, underscoring their ease of use and robustness as a new model system.

Functionalization of lipid vesicles with glycosaminoglycans like CS allows building membrane complexity in a systematic manner and provides a pathway towards creating reconstituted glycocalyx model systems with near-native characteristics. The glycocalyx is a dense medium where glycoproteins and glycopolymers are heavily entangled. As a first step towards reproducing this scenario, we used PLL for non-covalent crosslinking of CS. Our data show that our method allows association between vesicles (Fig. 5) and non-covalent recruitment of CS chains into the membrane (Fig. 6). Potential future applications for this technology include the development of membrane-based biosensors, biophysical investigations of the glycocalyx, stealth drug delivery, and the creation of more chemically complex artificial cells.

Data availability

Data are available upon request from the authors.

Conflicts of interest

There are no conflicts to declare.

Acknowledgements

This work was supported by NIH grant R35GM140846 (to FNB), NSF grant 2122027 (to GMK), and NSF Grant CHE-2204126 (to FAH). We are thankful to Prof. Amit Joshi (University of Tennessee) for use of his Zeiss LSM Airyscan microscope.

References

1 J. H. Lorent, K. R. Levental, L. Ganesan, G. Rivera-Longsworth, E. Sezgin, M. Doktorova, E. Lyman and I. Levental, Plasma membranes are asymmetric in lipid unsaturation, packing and protein shape, *Nat. Chem. Biol.*, 2020, **16**(6), 644–652, DOI: 10.1038/s41589-020-0529-6.

2 A. J. Verkleij, R. F. Zwaal, B. Roelofsen, P. Comfurius, D. Kastelijn and L. L. van Deenen, The asymmetric distribution of phospholipids in the human red cell membrane. A combined study using phospholipases and freeze-etch electron microscopy, *Biochim. Biophys. Acta, Biomembr.*, 1973, **323**(2), 178–193, DOI: 10.1016/0005-2736(73)90143-0.

3 L. Fu, H. N. Kim, J. D. Sterling, S. M. Baker and M. S. Lord, The role of the cell surface glycocalyx in drug delivery to and through the endothelium, *Adv. Drug Delivery Rev.*, 2022, **184**, 114195, DOI: 10.1016/j.addr.2022.114195.

4 L. Möckl, The Emerging Role of the Mammalian Glycocalyx in Functional Membrane Organization and Immune System Regulation, *Front. Cell Dev. Biol.*, 2020, **8**, 253, DOI: 10.3389/fcell.2020.00253.

5 X. Cui, X. F. Zhang and A. Jagota, Penetration of Cell Surface Glycocalyx by Enveloped Viruses Is Aided by Weak Multivalent Adhesive Interaction, *J. Phys. Chem. B*, 2023, **127**(2), 486–494, DOI: 10.1021/acs.jpcb.2c06662.

6 C. R. Shurer, J. C. Kuo, L. M. Roberts, J. G. Gandhi, M. J. Colville, T. A. Enoki, H. Pan, J. Su, J. M. Noble, M. J. Hollander, J. P. O'Donnell, R. Yin, K. Pedram, L. Möckl, L. F. Kourkoutis, W. E. Moerner, C. R. Bertozzi, G. W. Feigenson, H. L. Reesink and M. J. Paszek, Physical Principles of Membrane Shape Regulation by the Glycocalyx, *Cell*, 2019, **177**(7), 1757–1770, DOI: 10.1016/j.cell.2019.04.017.

7 S. A. Freeman, A. Vega, M. Riedl, R. F. Collins, P. P. Ostrowski, E. C. Woods, C. R. Bertozzi, M. I. Tammi, D. S. Lidke, P. Johnson, S. Mayor, K. Jaqaman and S. Grinstein, Transmembrane Pickets Connect Cyto- and Pericellular Skeletons Forming Barriers to Receptor Engagement, *Cell*, 2018, **172**(1–2), 305–317, DOI: 10.1016/j.cell.2017.12.023.

8 S. M. Haslam, S. Julien, J. M. Burchell, C. R. Monk, A. Ceroni, O. A. Garden and A. Dell, Characterizing the glycome of the mammalian immune system, *Immunol. Cell Biol.*, 2008, **86**(7), 564–573, DOI: 10.1038/icb.2008.54.

9 F. Broszeit, N. Tzarum, X. Zhu, N. Nemanichvili, D. Eggink, T. Leenders, Z. Li, L. Liu, M. A. Wolfert, A. Papanikolaou, C. Martínez-Romero, I. A. Gagarinov, W. Yu, A. García-Sastre, T. Wennekes, M. Okamatsu, M. H. Verheije, I. A. Wilson, G. J. Boons and R. P. de Vries, N-Glycolylneuraminic Acid as a Receptor for Influenza A Viruses, *Cell Rep.*, 2019, **27**(11), 3284–3294, DOI: 10.1016/j.celrep.2019.05.048.

10 L. Byrd-Leotis, N. Jia, S. Dutta, J. F. Trost, C. Gao, S. F. Cummings, T. Braulke, S. Müller-Loennies, J. Heimburg-Molinaro, D. A. Steinhauer and R. D. Cummings, Influenza binds phosphorylated glycans from human lung, *Sci. Adv.*, 2019, **5**(2), eaav2554, DOI: 10.1126/sciadv.aav2554.

11 J. Lin, J. Miao, K. G. Schaefer, C. M. Russell, R. J. Pyron, F. Zhang, Q. T. Phan, N. V. Solis, H. Liu, M. Tashiro, J. S. Dordick, R. J. Linhardt, M. R. Yeaman, G. M. King, F. N. Barrera, B. M. Peters and S. G. Filler, Sulfated glycosaminoglycans are host epithelial cell targets of the Candida albicans

toxin candidalysin, *Nat. Microbiol.*, 2024, **9**(10), 2553–2569, DOI: 10.1038/s41564-024-01794-8.

12 A. L. Lewis, C. M. Szymanski, R. L. Schnaar, and M. Aebi, Bacterial and Viral Infections, in *Essentials of Glycobiology*, ed. A. Varki, R. D. Cummings, J. D. Esko, P. Stanley, G. W. Hart, M. Aebi, D. Mohnen, T. Kinoshita, N. H. Packer, J. H. Prestegard, et al., The Consortium of Glycobiology Editors, La Jolla, California, Cold Spring Harbor Laboratory Press, Cold Spring Harbor (NY), 2022, pp. 555–568.

13 M. A. McGuckin, S. K. Lindén, P. Sutton and T. H. Florin, Mucin dynamics and enteric pathogens, *Nat. Rev. Microbiol.*, 2011, **9**(4), 265–278, DOI: 10.1038/nrmicro2538.

14 E. C. Woods, N. A. Yee, J. Shen and C. R. Bertozzi, Glycocalyx Engineering with a Recycling Glycopolymer that Increases Cell Survival In Vivo, *Angew. Chem., Int. Ed.*, 2015, **54**(52), 15782–15788, DOI: 10.1002/anie.201508783.

15 H. Xiao, E. C. Woods, P. Vukojicic and C. R. Bertozzi, Precision glycocalyx editing as a strategy for cancer immunotherapy, *Proc. Natl. Acad. Sci. U. S. A.*, 2016, **113**(37), 10304–10309, DOI: 10.1073/pnas.1608069113.

16 N. Kanyo, K. D. Kovacs, A. Saftics, I. Szekacs, B. Peter, A. R. Santa-Maria, F. R. Walter, A. Dér, M. A. Deli and R. Horvath, Glycocalyx regulates the strength and kinetics of cancer cell adhesion revealed by biophysical models based on high resolution label-free optical data, *Sci. Rep.*, 2020, **10**(1), 22422, DOI: 10.1038/s41598-020-80033-6.

17 L. Djerbal, H. Lortat-Jacob and J. Kwok, Chondroitin sulfates and their binding molecules in the central nervous system, *Glycoconjugate J.*, 2017, **34**(3), 363–376, DOI: 10.1007/s10719-017-9761-z.

18 K. Forsten-Williams, C. L. Chu, M. Fannon, J. A. Buczek-Thomas and M. A. Nugent, Control of growth factor networks by heparan sulfate proteoglycans, *Ann. Biomed. Eng.*, 2008, **36**(12), 2134–2148, DOI: 10.1007/s10439-008-9575-z.

19 K. Jahnke, M. Pavlovic, W. Xu, A. Chen, T. P. J. Knowles, L. R. Arriaga and D. A. Weitz, Polysaccharide functionalization reduces lipid vesicle stiffness, *Proc. Natl. Acad. Sci. U. S. A.*, 2024, **121**(22), e2317227121, DOI: 10.1073/pnas.2317227121.

20 S. Chakraborty, M. Doktorova, T. R. Molugu, F. A. Heberle, H. L. Scott, B. Dzikovski, M. Nagao, L. R. Stingaciu, R. F. Standaert, F. N. Barrera, J. Katsaras, G. Khelashvili, M. F. Brown and R. Ashkar, How cholesterol stiffens unsaturated lipid membranes, *Proc. Natl. Acad. Sci. U. S. A.*, 2020, **117**(36), 21896–21905, DOI: 10.1073/pnas.2004807117.

21 J. T. Groves, Bending mechanics and molecular organization in biological membranes, *Annu. Rev. Phys. Chem.*, 2007, **58**, 697–717.

22 T. Shioiri, J. Tsuchimoto, K. Fukushige, T. Takeuchi, M. Naito, H. Watanabe and N. Sugiura, Chondroitin sulfate liposome: clustering toward high functional efficiency, *J. Biochem.*, 2024, **176**(3), 229–236, DOI: 10.1093/jb/mvae041.

23 J. M. Westerfield, A. R. Sahoo, D. S. Alves, B. Grau, A. Cameron, M. Maxwell, J. A. Schuster, P. C. T. Souza, I. Mingarro, M. Buck and F. N. Barrera, Conformational Clamping by a Membrane Ligand Activates the EphA2 Receptor, *J. Mol. Biol.*, 2021, **433**(18), 167144, DOI: 10.1016/j.jmb.2021.167144.

24 V. P. Nguyen, L. Palanikumar, S. J. Kennel, D. S. Alves, Y. Ye, J. S. Wall, M. Magzoub and F. N. Barrera, Mechanistic insights into the pH-dependent membrane peptide ATRAM, *J. Controlled Release*, 2019, **298**, 142–153.

25 T. J. Pucadyil, S. Mukherjee and A. Chattopadhyay, Organization and dynamics of NBD-labeled lipids in membranes analyzed by fluorescence recovery after photobleaching, *J. Phys. Chem. B*, 2007, **111**(8), 1975–1983.

26 H. L. Scott, J. M. Westerfield and F. N. Barrera, Determination of the Membrane Translocation pK of the pH-Low Insertion Peptide, *Biophys. J.*, 2017, **113**(4), 869–879.

27 V. P. Nguyen, A. C. Dixson and F. N. Barrera, The Effect of Phosphatidylserine on a pH-Responsive Peptide Is Defined by Its Noninserting End, *Biophys. J.*, 2019, **117**(4), 659–667.

28 K. G. Schaefer, A. E. Pittman, F. N. Barrera and G. M. King, Atomic force microscopy for quantitative understanding of peptide-induced lipid bilayer remodeling, *Methods*, 2022, **197**, 20, DOI: 10.1016/j.ymeth.2020.10.014.

29 S. J. Attwood, Y. Choi and Z. Leonenko, Preparation of DOPC and DPPC Supported Planar Lipid Bilayers for Atomic Force Microscopy and Atomic Force Spectroscopy, *Int. J. Mol. Sci.*, 2013, **14**(2), 3514–3539.

30 M. Chighizola, T. Dini, S. Marcotti, M. D'Urso, C. Piazzoni, F. Borghi, A. Previdi, L. Ceriani, C. Folliero, B. Stramer, C. Lenardi, P. Milani, A. Podestà and C. Schulte, The glycocalyx affects the mechanotransductive perception of the topographical microenvironment, *J. Nanobiotechnol.*, 2022, **20**(1), 418, DOI: 10.1186/s12951-022-01585-5.

31 K. S. Nair and H. Bajaj, Advances in giant unilamellar vesicle preparation techniques and applications, *Adv. Colloid Interface Sci.*, 2023, **318**, 102935, DOI: 10.1016/j.cis.2023.102935.

32 M. H. L. Nguyen, M. DiPasquale, B. W. Rickeard, M. Doktorova, F. A. Heberle, H. L. Scott, F. N. Barrera, G. Taylor, C. P. Collier, C. B. Stanley, J. Katsaras and D. Marquardt, Peptide-Induced Lipid Flip-Flop in Asymmetric Liposomes Measured by Small Angle Neutron Scattering, *Langmuir*, 2019, **35**(36), 11735–11744.

33 D. Rabuka, M. B. Forstner, J. T. Groves and C. R. Bertozzi, Noncovalent cell surface engineering: incorporation of bioactive synthetic glycopolymers into cellular membranes, *J. Am. Chem. Soc.*, 2008, **130**(18), 5947–5953, DOI: 10.1021/ja710644g.

Faraday Discussions

ROYAL SOCIETY OF CHEMISTRY

DISCUSSIONS

Engineering plasma membrane mimics: general discussion

Francisco Barrera, Manuel Bibrowski, John C. Conboy, Aileen Cooney, Markus Deserno, Rumiana Dimova, Milka Doktorova, Thais A. Enoki, Oded Farago, Alexander P. Fellows, Arne Gericke, Alberto Giacomello, Felix M. Goni, Heiko Heerklotz, Ben John, Neha P. Kamat, Michael M. Kozlov, Reinhard Lipowsky, Erwin London, Joseph H. Lorent, Ioanna Mela, Walter Nickel, Georg Pabst, Tyler Reagle, Karin Riske, Raya Sorkin, Tsu-Wang Sun and Martin E. Villanueva

DOI: 10.1039/d5fd90019k

Milka Doktorova opened a discussion of the paper by Heiko Heerklotz: The idea that changes in the phase state of one leaflet can drive interleaflet coupling and morphological transitions due to the generated imbalance in leaflet areas is novel and exciting! Here the main observation is that performing the exchange at low temperature (where domains are present in the outer leaflet) followed by heating (and consequently, melting of these domains) results in budding of daughter vesicles due to the increased area of the outer leaflet relative to the inner. In that case, can you detect differences in the extent of budding when you vary the amount of lipid exchange in the asymmetric liposomes? In other words, if you start with a small gel domain *vs.* a large one, its melting should produce different amounts of area imbalance and thus, logically, different amounts of the response (*i.e.*, budded vesicles).

Related to that, when vesicles bud outward or inward, they lose some of their area and it seems natural that they would also become smaller. Do you see changes in the diameter of mother vesicles upon formation of daughter vesicles? It is possible that such changes may be difficult to observe in bulk measurements, but asymmetric flow field-flow fractionation (AF4) should be more sensitive than dynamic light scattering (DLS).

Heiko Heerklotz answered: Yes, this is as you are saying. As qualitatively seen in Fig. 5 of our paper (https://doi.org/10.1039/d5fd00003c), the budded fraction is larger (integration yields a factor of 2 – not shown) in panel A than in B. Generally though, it should be noted that budding has been shown to be limited (by excess area) so that more intrinsic-area asymmetry does not necessarily mean more budding (but here it does).

When it comes to the size change of mother vesicles upon budding, yes, they should shrink, but the decrease is quite small. For the maximal budding of some 20–30% seen upon surfactant insertion, we used to see slight down-shifts of mother vesicle (MV) elution time and size.[1,2] Here, the budding shown in Fig. 5 of our paper is maximally ~12% and a shift for daughter vesicle (DV) size is hardly detectable. Note that, for spheres, a 12% area change would reduce the diameter from 100 nm to 94 nm only (radius related to the square root of the area). Second, in order to bud, the mothers must NOT be spheres. I don't know how deflation changes the hydrodynamic size, but in any case, one should not expect to resolve size changes resulting from a 10% area change. That is also, as you mention, why there is no chance to detect budding *via* DLS (which detects, practically exclusively, the MVs).

1 L. Hua, S. Akcesme, K. Müller and H. Heerklotz, *J. Phys. Chem. B*, 2025, **129**, 1260–1273, DOI: 10.1021/acs.jpcb.4c05868.
2 L. Hua, M. Kaiser, I. Carabadjac, A. Meister, G. Hause and H. Heerklotz, *Biophys. J.*, 2023, **122**, 4011, DOI: 10.1016/j.bpj.2023.08.023.

Tyler Reagle commented: It is thought that the curvature-partitioning of lipids with small spontaneous curvatures is weak if the mixture is far from a phase boundary.[1,2] Since these buds seem to form from vesicles in response to melting of ordered, phase-separated domains in their outer leaflets, it would be expected that the composition of the released vesicles should match that of the bulk lipid, which was initially complexed by the cyclodextrins during the exchange procedure. Is this the case or is some enrichment of one of the outer leaflet species observed?

1 A. Tian and T. Baumgart, *Biophys. J.*, 2009, **96**, 2676.
2 B. Sorre, *Proc. Natl. Acad. Sci. U. S. A.*, **106**, 5622.

Heiko Heerklotz responded: Preliminary data suggest a modest enrichment of DVs by dipalmitoyl phosphatidylglycerol (DPPG) by about a factor of two. This is very little compared to lysolipid-induced budding (factor of ≥12).[1] In light of the papers you are referring to, the fact that there seems to be at least some sorting of DPPG might be related to the apparently quite non-ideal mixing of the two lipids, as suggested by Fig. 1 of our paper (https://doi.org/10.1039/d5fd00003c) – with all due caution since it refers to large unilamellar vesicles (LUVs), not multi-lamellar vesicles (MLVs).

1 L. Hua, M. Kaiser, I. Carabadjac, A. Meister, G. Hause and H. Heerklotz, *Biophys. J.*, 2023, **122**, 4011, DOI: 10.1016/j.bpj.2023.08.023.

Tyler Reagle said: At the end of your presentation, you asked about other effects that could determine the different outcomes of vesicle bending *versus* coupling of leaflet phase behaviors in asymmetric vesicles prepared *via* cyclodextrin-mediated exchange. The osmotic conditions that the vesicles are exposed to as their leaflet surface-area differences are adjusted is probably an important determinant.

A recent paper provides some experimental evidence supporting this idea.[1] When supersaturated cyclodextrins (CDs) were used to overfill the leaflets of giant unilamellar vesicles (GUVs) composed of POPC at room temperature (where the symmetric vesicles began in a fluid state), the outcomes of outward bending or formation of ordered domains seem to depend on the solution conditions during lipid insertion. When the GUVs were exposed to supersaturated CDs in the bulk solution, outward bending resulted; when the GUVs were exposed to supersaturated CDs by local delivery (*via* micropipette), ordered domains formed. Because the CD solutions were slightly hypertonic compared to the solutions inside the GUVs (see the ESI of the referenced study[1]), the GUVs were presumably deflated during lipid delivery. During local-delivery of CDs, however, the extent of osmotic deflation experienced by the GUVs was likely smaller than during bulk administration (because the hypertonicity of the locally-administered CD solution was presumably buffered by the rest of the osmotically-balanced medium). Since vesicle shape theories indicate that outward bending is facilitated by osmotic deflation,[2] the larger osmotic deflation experienced in the bulk-delivery scheme might explain why bending was the outcome. In the local-delivery scheme, ordered-domain formation might have resulted because outward bending was insufficiently facilitated and phase transformation provided an alternative pathway through which the bilayers could alleviate their leaflet surface-area differences.[3]

1 T. Reagle, Y. Xie, Z. Li, W. Carnero and T. Baumgart, *Soft Matter*, 2024, **20**, 4291–4307.
2 H.-G. Döbereiner, E. Evans, M. Kraus, U. Seifert and M. Wortis, *Phys. Rev. E: Stat. Phys., Plasmas, Fluids, Relat. Interdiscip. Top.*, 1997, **55**, 4458.
3 S. Foley, A. Hossein and M. Deserno, *Biophys. J.*, 2022, **121**, 2997–3009.

Heiko Heerklotz replied: This is fascinating. One should repeat the coupling experiments at different outside osmolarities and check whether inflated vesicles show more coupling…! That inflated vesicles show, as to be expected, less budding has been supported for lysolipid-induced budding.[1]

1 L. Hua, M. Kaiser, I. Carabadjac, A. Meister, G. Hause and H. Heerklotz, *Biophys. J.*, 2023, 122, 4011, DOI: 10.1016/j.bpj.2023.08.023.

Reinhard Lipowsky remarked: In your talk and in your written contribution to the Faraday Discussion (https://doi.org/10.1039/d5fd00003c), you describe the budding and subsequent creation of daughter vesicles, induced by the expansion of one bilayer leaflet. Very similar processes have been observed in coarse-grained molecular dynamics simulations. In these simulations, unilamellar nanovesicles with a diameter of up to 40 nm have been studied.[1,2] The stress asymmetry in the vesicle bilayers can be directly controlled by the assembly of different lipid numbers in the two leaflets. Starting from a spherical vesicle, the reduction of the vesicle volume, which mimics osmotic deflation, can lead to out-budding or in-budding of the vesicle. Out-budding is observed when the outer leaflet is compressed or the inner leaflet is stretched. In-budding is induced by a compressed inner leaflet or a stretched outer leaflet. In both cases, the vesicle membranes form a closed membrane neck. For in-budded vesicles, the fission of the membrane neck was also observed, leading to the division of the nanovesicle into two daughter vesicles; see Fig. 22 of ref. 2. We also studied the fusion of two nanovesicles and observed rather fast fusion when the outer leaflets of the two vesicles were stretched; see Fig. 44 in ref. 2.

1 R. Ghosh, V. Satarifard, A. Grafmüller and R. Lipowsky, *Nano Lett.*, 2019, 19, 7703–7711, DOI: 10.1021/acs.nanolett.9b02646.
2 R. Lipowsky, R. Ghosh, V. Satarifard, A. Sreekumari, M. Zamaletdinov, B. Różycki, M. Miettinen and A. Grafmüller, *Biomolecules*, 2023, **13**, 926, DOI: 10.3390/biom13060926.

Heiko Heerklotz answered: Thanks for these very valuable comments and references supporting some of our hypotheses. We will have a detailed look to connect our data with the theoretical results as closely as possible. A particularly interesting point is also the general possibility of inward budding in asymmetric LUVs (aLUVs) – a phenomenon we have not been able to resolve experimentally so far. Possibly, the stronger coupling in the system that would require inward budding for avoiding intrinsic-area mismatch does not result from a principal impossibility of inward budding but from its lower limit, given its larger "consumption" of excess area per daughter vesicle.

Oded Farago asked: Can you comment on the interplay between the inward/outward budding effect and the spontaneous curvature of the asymmetric membrane, and to what extent budding depends on the size of the vesicle?

Heiko Heerklotz responded: We see outward budding from vesicles showing positive spontaneous curvature – as one would expect. We are not able to detect inward budding in the first place – so our vesicles developing negative spontaneous curvature might undergo some inward budding. In any case, they show a stronger coupling, suggesting that the negative spontaneous curvature cannot as effectively be balanced as the positive one – maybe due to inward budding being more restricted (limited to a smaller fraction of the lipid) than outward budding. We haven't systematically studied the effects of vesicle size. We assume that size might be a possible reason for budding and fission to occur from LUVs whereas GUVs can also develop tubules that remain connected to the mother GUV. For a size-dependent study of aLUVs, it is crucial to keep in mind that excess area plays a key role for budding and different preparation techniques and conditions might yield different excess areas. Such effects would confuse a correlation between the size of mother vesicles and the extent of budding.

Martin E. Villanueva queried: Could you clarify the method used to determine the thermal expansion coefficient?

Heiko Heerklotz replied: The progress of the transitions was monitored *via* pressure perturbation calorimetry (PPC). It measures the heat response of a sample to small pressure jumps of ∼5 bar at (many) constant temperature(s) and converts it into the coefficient of thermal expansion of the solute using a Maxwell relation.[1,2] The shape of the curve should agree with that of heat capacity *versus* temperature seen *via* differential scanning calorimetry (DSC), but PPC has a number of technical advantages. Mainly, since it is not scanning temperature, one gets stable measurements right above 0 °C (we started at 2 °C). Furthermore, it applies both up- and down-jumps in temperature and hence, tests for the absence of hysteresis effects at each single point. So the bottom line: PPC was better, but DSC should work as well and the curves can be interpreted here in terms of the progress of the melting transition, just like DSC curves.

1 L.-N. Lin, J. F. Brandts, J. M. Brandts and V. Plotnikov, *Anal. Biochem.*, 2002, **302**, 144–160.
2 H. Heerklotz, *Methods Mol. Biol.*, 2002, **400**, 197–206.

Martin E. Villanueva enquired: Have you considered performing signal deconvolution analysis to achieve a more accurate quantification and identification of the coupled transitions?

Heiko Heerklotz answered: We agree that it would be great to have a model that predicts a progress (*e.g.*, DSC or PPC) curve of two partially coupled transitions (such as the inner and outer leaflet) based on those of their symmetrical counterparts (here: symmetrical stearyl-oleyl-phosphatidylcholine (SOPC) and symmetrical SOPC|+20% DPPG, respectively) and one or more coupling parameters. Unfortunately, we did not find such a model and it seemed unrealistic to establish such a model here. A simple linear combination of the two symmetrical curves can be considered a model for "zero coupling" and clearly did not fit the data.

Tsu-Wang Sun asked: In cases where asymmetric vesicles (aLUVs) undergo budding instead of interleaflet coupling, how is the required excess area for budding achieved on the inner leaflet? This may also relate to your question about inward budding in aLUVs, specifically regarding the possible structure of the buds.

Heiko Heerklotz responded: The area of the inner leaflet cannot change in the budding process so the possible extent of budding is limited by the excess area – *i.e.*, deviation from sphericity – of the original mother vesicles. Indeed, this area-enforced budding limit should be lower for (hypothetical) inward budding since the volume of intravesicles has to be enclosed in two bilayers (daughter and surrounding mother) so that more excess area is "used up" per daughter vesicle. If inward budding from aLUVs is principally possible, such a lower limit might explain the stronger coupling in liposomes that would require inward budding to avoid intrinsic-area mismatch.

Rumiana Dimova commented: You observe a suppression of budding under certain conditions. Could this be attributed to an increase in the membrane bending rigidity, which would resist the necessary curvature changes for bud formation?

Heiko Heerklotz replied: I agree that this option should be considered. For example, unpublished experiments of lysolipid-induced budding with DPPC in the gel phase showed no budding (very much in contrast to fluid POPC and similar liposomes). Maybe this effect was due to a stress-induced drop in lysolipid membrane insertion ("staying out" in our terminology) but this is a hen-and-egg question: if the membrane had been able to compensate intrinsic-area mismatch by budding, the lysolipid would likely have inserted. For the differential melting presented in the Faraday Discussion, we always found some budding when we expected positive spontaneous curvature (heating outside-gel liposomes) and we had no means to check whether intra-vesiculation happened in the reverse case. The question remains whether bending rigidity affects (along with the loss of excess area) the budding limit or the size and composition of the daughter vesicles. Much work is left to be done.

John C. Conboy queried: With regards to the observation of positive curvature but the lack of negative curvature for the appropriate asymmetric lipid systems (vesicles), respectively, what role do you think the presence of methylated β-cyclodextrin (mβCD) has on these observations? We have observed a 5 Å^2 per molecule increase in the mean molecular area (MMA) of DPPC when pre-equilibrated with mβCD for Langmuir isotherms at the air/water interface (see Fig. 1). Do you think the possible interaction of mβCD at concentrations less than those required for lipid extraction and dissolution might be contributing to the observed behavior? That is, if mβCD is only present on the exterior leaflet of the vesicle, this might explain the lack of negative curvature.

Heiko Heerklotz answered: I agree that possible remnants of cyclodextrin remaining attached to aLUVs after exchange deserve more thorough attention in general. We can also consider the possibility of recording effects on order parameter profiles, and your Langmuir balance data are very interesting in this context as well. It would be helpful for me to better understand the actual concentration of cyclodextrin at your monolayer interface to compare it with the amounts that might be present in our vesicle samples. Also, is the transition you see at ~48 mN m^{-1} the solubilization of DPPC into the subphase by cyclodextrin?

In our studies, we carried out the respective controls suggesting that residual cyclodextrin did not pose any problem. For example, cyclodextrin removal

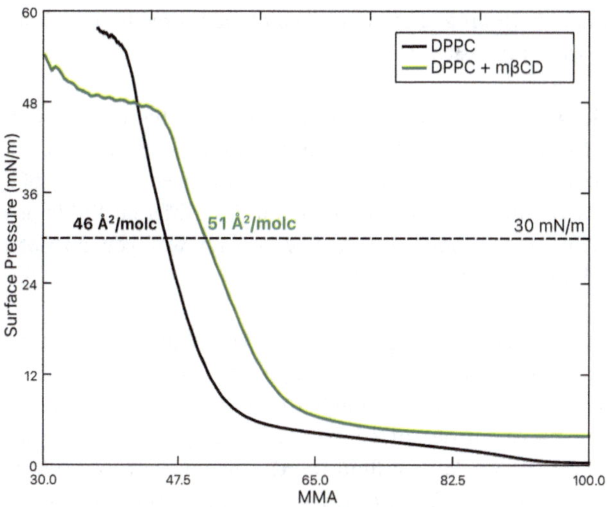

Fig. 1 Langmuir isotherms of DPPC (black) and DPPC + mβCD (green) were taken in triplicate on a pure water subphase. For the DPPC + mβCD, DPPC was prepared at a 1 : 8 (DPPC : mβCD) ratio in chloroform (1 mg mL^{-1} DPPC) and warmed above the phase transition (65 °C) for 30 minutes. The incubated solution was then cooled to room temperature and spread dropwise at the air/water interface. At least 10 minutes was allowed for equilibration prior to compression. Compression for both data sets was performed at 4 mm min^{-1}.

(starting at the very high concentration used for exchange!) did not change the zeta potential (reflecting the surface charge density)[1] and cyclodextrin-free LUVs and the corresponding aLUV (after cyclodextrin removal) did not differ in zeta potential, detergent-induced leakage and detergent solubilization properties.[2] Of course, this does not exclude potential residual cyclodextrin affecting other studies that might be more susceptible. Regarding negative curvature – we interpret the enhanced coupling in the case of cooling aLUVs prepared at 30 °C as a negative-spontaneous-curvature effect that can less effectively be relaxed than a positive spontaneous curvature in 5 °C-prepared aLUVs. A significant intercalation of cyclodextrin to expand the outer leaflet would – according to this view – reduce rather than increase coupling in this case.

1 M. Markones, C. Drechsler, M. Kaiser, L. Kalie, H. Heerklotz and S. Fiedler, *Langmuir*, 2018, **34**, 1999–2005, DOI: 10.1021/acs.langmuir.7b03189.
2 L. Dietel, L. Kalie and H. Heerklotz, *Biophys. J.*, 2020, **119**, 767–779, DOI: 10.1016/j.bpj.2020.07.004.

Raya Sorkin said: I think you mentioned that daughter vesicles can fuse with the mother vesicles. Under which conditions and why does that occur?

Heiko Heerklotz responded: Small daughter vesicles (DVs) had budded from mother vesicles (MVs) equilibrated with cyclodextrin–DPPG at 5 °C upon heating to 30 °C. This was detected by taking a little aliquot from the liposome dispersion and running it on the AF4. This finding was explained by the stronger expansion of the outer leaflet, which had contained a larger fraction of gel phase at 5 °C. Cooling the sample containing the DVs and MVs back to 5 °C and taking another aliquot for AF4 revealed that the amount of DVs had decreased. We speculate that this resulted from fusion between DVs (or maybe DVs and MVs) that was induced by the stronger shrinking of the outer leaflet, now stretching its lipids to an area larger than their intrinsic area.

Alexander P. Fellows remarked: You mentioned that there is very little or no preference for specific lipids in the different budding geometries/curvatures (*i.e.*, inward or outward). This is somewhat surprising given the range of spontaneous curvatures and conical shapes/volumes that the different lipids possess. Could you comment on this and suggest a reason for why you do not observe any preference based on these effects?

Heiko Heerklotz replied: I agree with your view that positive intrinsic curvature of a molecule favors localization in daughter vesicles. For lysoPC[1] and similarly other surfactants[2] this accumulation in the outer leaflet of DVs starts at a factor of ~12 and increases further in subsequent budding events as the excess area decreases. Preliminary data for fluid DPPG/SOPC suggest some DV-accumulation of DPPG on the order of a factor of 2. This is relatively very little, but, I agree, more than nothing. So yes, some more positive intrinsic curvature of DPPG compared to SOPC seems effective here as well.

1 L. Hua, M. Kaiser, I. Carabadjac, A. Meister, G. Hause and H. Heerklotz, *Biophys. J.*, 2023, **122**, 4011, 10.1016/j.bpj.2023.08.023.
2 L. Hua, S. Akcesme, K. Müller and H. Heerklotz, *J. Phys. Chem. B*, 2025, **129**, 1260–1273, 10.1021/acs.jpcb.4c05868.

Milka Doktorova commented: The ability of cyclodextrins (CDs) to dissolve liposomes has been shown to be composition-dependent.[1] This observation indicates that CDs have different affinities for different types of lipids. When preparing asymmetric liposomes, CDs have to pull some lipids from the outer leaflet of the vesicles and "offload" different ones into it. The amount of lipid "extraction" and "delivery" should be determined by equilibration of the chemical potentials of the two types of lipids between the membrane and CD environments. If the chemical potentials of the lipids are different (which seems likely due to the differential affinities for CD), the exchange may not be 1 : 1 (*i.e.*, one lipid being delivered for every lipid being extracted). Because of that, it seems difficult to predict what the 'net' extraction *vs.* delivery would be as it may vary depending on lipid compositions. Moreover, differential and curvature stresses may be balanced in a way that allows the membrane to remain flat (no excessive curvature or budding) while still being differentially stressed.[2,3]

It is therefore hard to conceptualize why CD exchange would always produce stress-free liposomes as stated in the text (https://doi.org/10.1039/d5fd00003c), and how this can be verified in general. Can you elaborate a little bit more on why you believe this to be the case and what types of observations you are basing this on?

1 Z. Huang and E. London, *Langmuir*, 2013, **29**, 14631.
2 A. Hossein and M. Deserno, *Biophys. J.*, 2020, **118**, 624–642.
3 M. Doktorova, J. L. Symons, X. Zhang, H.-Y. Wang, J. Schlegel, J. H. Lorent, F. A. Heberle, E. Sezgin, E. Lyman, K. R. Levental and I. Levental, *Cell*, 2025, **188**, 2586–2602, DOI: 10.1016/j.cell.2025.02.034.

Heiko Heerklotz answered: Our data so far have supported the working hypothesis that cyclodextrin-based exchange creates largely relaxed "acceptor" vesicles over a broad range of cyclodextrin saturations by the donor lipid and cyclodextrin concentrations and with a variety of lipid combinations. This had been counterintuitive and surprising to us too. And, of course, the fact that in several different studies we did not find evidence to falsify this hypothesis does not prove it. A key question is what exactly "largely relaxed" means for a selected observable in the context of a given study. Let me list a few findings. Before producing lipid-asymmetric aLUVs, we found that partitioning of (membrane impermeant) dodecyl maltoside is stronger into underpopulated than into overpopulated leaflets (in our example, by a factor of 2).[1] When we established our assay to produce aLUVs without donor vesicles,[2] we found the exchange to be independent of the cyclodextrin concentration over a wide range. For example, dissolving 0.5 mM POPG in cyclodextrin and adding the complexes to vesicles of 5 mM POPC would result in a PG content of the outer leaflet of about 0.5 mM PG/(0.5 mM PG + 2.5 mM outer-leaflet PC) = 17% over a wide range at cyclodextrin concentrations of 40–60 mM, corresponding to 75–25% saturation of the cyclodextrin. The same was observed for other PG contents, up to 30% (Fig. 3 of ref. 2). This was stable over time, indicating no substantially enhanced lipid flip-flop. Some content leakage was observed only for extremely "hungry" cyclodextrin at 70 mM, saturated with 25% lipid only, and none for up to 60 mM and 25–100% saturation. From today's perspective, one might expect that the stress had been relaxed by budding (at the time, there was no means to tell), but in the paper presented here (https://doi.org/10.1039/d5fd00003c; Fig. 5), we showed that exchange (at 5, 15, or 30 °C, which changes the degree of saturation of the cyclodextrin) alone did not induce budding – only subsequent temperature changes did for 5 and 15 °C exchange. The temperature-induced budding seen here implies – in line with the hypothesis – that more lipid has been inserted into the membrane when the area per lipid was lower. So, yes, it is counterintuitive and might not hold forever and for everything but so far, this hypothesis has worked for us to design and interpret our experiments.

1 H. Heerklotz, *Biophys. J.*, 2002, **81**, 184–195, DOI: 10.1016/S0006-3495(01)75690-1.
2 M. Markones, C. Drechsler, M. Kaiser, L. Kalie, H. Heerklotz and S. Fiedler, *Langmuir*, 2018, **34**, 1999, DOI: 10.1021/acs.langmuir.7b03189.

Michael M. Kozlov enquired: Could it be that a fully budded vesicle remains connected to the mother vesicle by a thin membrane tether?

Heiko Heerklotz responded: The quantification of the amount of lipid in daughter vesicles (DVs) *via* AF4 works with their macroscopic separation from mother vesicles (MVs) – this excludes the presence of tethers at this point in time. Of course, this does not strictly exclude tethers being present after budding and then breaking upon separation. However, the cryo-TEM pictures of non-separated samples never seem to show any tethers. So it seems important to understand how fission could be achieved in these systems but at the moment, nothing can strictly be ruled out.

Felix M. Goni opened a discussion of the paper by Neha P. Kamat: You observe a higher degree of activation when the lipid is near the transition temperature. This could be related to the fact that, near the transition, the vesicle is a mosaic of gel and fluid nanodomains. For example, 1,2-dipalmitoyl-*sn-glycero*-3-phosphocholine (DPPC) vesicles become totally leaky at the transition temperature. Do you think this phenomenon is related to your observations?

Neha P. Kamat replied: Spatial arrangement may be critical for optimizing the multivalent interactions required for effective T cell engagement and it is likely that the presence of nanodomains indeed facilitates this. Clustering of antigens could enhance the local antigen density, thereby improving the avidity of the liposome–T cell interactions and potentially promoting a more robust immunological response.

Felix M. Goni said: You mentioned that when vesicle–vesicle contacts occur, a flat surface is observed. I have seen something similar, *i.e.*, flat surfaces rather than mere contact points, in many vesicle aggregation/tethering events. Do you think this is a general phenomenon?

Neha P. Kamat answered: Our interpretation has been that the observed flat areas arise primarily due to interactions within the immunological synapse, which is facilitated by localized deformation of the interacting cells. This flattening has been previously observed during the engagement of T cells with antigen-presenting cells, supporting this view. However, we agree that the possibility of this being a more general phenomenon is indeed intriguing and merits further consideration.

Manuel Bibrowski remarked: In your small unilamellar vesicle (SUV) formulation, you had 1% polyethylene glycol (PEG) and 0.1% substrate. PEG prevents membrane interactions, so why does the membrane composition have an influence?

Neha P. Kamat responded: 1% PEG is typically included in vesicle formulations made for cellular studies since it helps prevent nonspecific interactions with cell membranes. Our vesicles also contained a targeting protein, anti-CD3, which was essential for the activation response observed in the cells. In the absence of anti-CD3, we did not see activation of Jurkat cells. So a combination of both the membrane properties and the presence of a targeting molecule that interacts with a Jurkat cell receptor was important for cell activation.

Oded Farago asked: In your approach to the problem of organization of the immunological synapse – do you consider the membrane as an effective medium through which the dynamical process takes place, or do you make further assumptions about the role played by the lipids?

Neha P. Kamat replied: We believe that the properties of the lipid bilayer membrane play a critical role in the activation process. The cooperative interactions among the lipids confer the membrane with the ability to support dynamic processes and undergo self-reorganization. While the lipids contribute specific properties to the membrane, derived from their charge, structure, and transition temperature, individual lipids alone cannot establish distinct material properties. It is the collective behavior of the lipids that governs the membrane's characteristics, including its rigidity and density, thus driving these essential processes.

Georg Pabst commented: The samples you are studying contain 29 mol% cholesterol, which is close to – if not above – the concentration at which the melting transition disappears. Is it then reasonable to expect a simple correlation with the proximity to the melting transition? In other words, the melting transition is a first-order transition. Are you certain that a first-order transition is actually occurring in your systems?

Neha P. Kamat answered: Upon observing a general correlation between the melting temperature (T_m) of the primary phospholipid in the liposome formulation and the temperature of incubation of T cells with liposomes ($T_{activ.} = 37$ °C), we further investigated the extent to which a transition is still detectable in whole liposomes. By measuring Laurdan generalized polarization (GP) temperature profiles, we observed that liposomes containing 29% cholesterol in full media exhibited a broadened transition compared to the pure phospholipid sample in PBS. Nevertheless, since the nonlinearity in Laurdan GP remains evident in most cases, albeit less pronounced than in the pure samples, we interpret this as a cooperative transition still occurring within the liposomes.

That being said, we cannot definitively determine the underlying mechanism responsible for the observed correlation, nor can we completely exclude the possibility that it is unrelated to the transition itself. However, given that this trend persists across lipids with varying head groups, chain lengths, and charges of the primary phospholipid, and is also observed when $T_{activ.}$ is shifted to 30 °C, with probes possessing a T_m closer to 30 °C activating more efficiently than those measured at 37 °C, it is unlikely that the transition does not play a role in the observed behavior, or at least in the specific membrane properties near the transition point.

Joseph H. Lorent queried: Might the charge of the headgroups not play a role in T cell receptor (TCR) activation?

Neha P. Kamat responded: Given that negatively charged phosphatidylserine (PS) lipids undergo reorganization during the early stages of TCR activation in T cells, being transferred to the outer leaflet of the plasma membrane, it is reasonable to assume that a charged membrane may interact more strongly with the T cell during activation. Our results support the idea that charged liposomes can activate T cells more efficiently, but this effect is observed only when activation occurs near the T_m of the liposomes. For instance, incubation with charged 1,2-di-(9Z-octadecenoyl)-*sn-glycero*-3-phospho-L-serine (DOPS) (18:1 PS, $T_m = -11$ °C) or 1,2-dilauroyl-*sn-glycero*-3-phospho-L-serine (DLPS) (12:0 PS, $T_m = 14$ °C) liposomes does not result in enhanced T cell activation compared to zwitterionic samples with a T_m similarly distant from 37 °C. Therefore, charge alone cannot account for the enhanced activation. Rather, it appears that a combination of charge and T_m plays a crucial role. While we have not yet systematically investigated this interplay, it is an area worth exploring in future studies.

Karin Riske said: In your paper (https://doi.org/10.1039/d5fd00002e), it is said that the headgroup charge of the phospholipid of interest in the nanocarrier liposome has no effect on the activation response of the T cells. However, the magnitude of the response is about double for the anionic lipids phosphatidyl glycerol (PG) and phosphatidyl serine (PS) close to their phase transition as compared to that of the zwitterionic phosphatidyl choline (PC). So, probably, headgroup charge does also play a significant role in the activation response.

Neha P. Kamat replied: Our data reveal a clear correlation between activation efficiency and the proximity of the transition temperature (T_m) of the liposomes to the temperature where the liposomes are incubated with T cells and activation is induced (37 °C). When examining activation as a function of lipid T_m by headgroup type, we observe similar trends for both charged and zwitterionic head groups. The effect of charge on this correlation appears minimal, as charged liposomes, with T_m values farther from 37 °C, do not exhibit improved activation compared to zwitterionic liposomes. Only for samples with a T_m closer to 37 °C is enhanced activation observed with charged liposomes. Thus, while charge may play a significant role in the activation response, no substantial difference in activation is observed between charged and neutral liposomes for probes with a T_m significantly lower than 37 °C.

Thais A. Enoki opened a discussion of the paper by Arne Gericke: I want to mention a point that you did not discuss in your paper (https://doi.org/10.1039/d4fd00191e). Eqn (1) in your paper describes the percentage of lipid exchange for dye initially present in the supported lipid bilayer (SLB) that enters the aGUV outer leaflet.

$$P^{\mathrm{TF}}_{\mathrm{exchange}} = \frac{I^{\mathrm{TF}}_{\mathrm{aGUV}}}{I^{\mathrm{TF}}_{\mathrm{GUV}}/2}, \tag{1}$$

where $I^{\mathrm{TF}}_{\mathrm{aGUV}}$ is the intensity of the TopFluor dye, the dye originally in the SLB, measured in the aGUV after lipid exchange, and $I^{\mathrm{TF}}_{\mathrm{GUV}}$ is the intensity of the TopFluor dye measured in the symmetric GUVs.

However, we must use a different equation (eqn (2)) to report the percentage of lipid exchange for the dye originally on the symmetric GUV.

$$P^{\mathrm{DiD}}_{\mathrm{exchange}} = 2\left(1 - \frac{I^{\mathrm{DiD}}_{\mathrm{aGUV}}}{I^{\mathrm{DiD}}_{\mathrm{GUV}}}\right), \tag{2}$$

where $I^{\mathrm{DiD}}_{\mathrm{aGUV}}$ is the intensity of the DiD dye, the dye originally in the symmetric GUV, measured in the aGUV after lipid exchange, and $I^{\mathrm{DiD}}_{\mathrm{GUV}}$ is the intensity of the DiD dye measured in the symmetric GUVs; see for instance ref. 1.

It is important to note that this dye (which is called the host dye in a previous paper, and here is the dye DiD) exits the outer leaflet of the initially symmetric GUV. Still, the inner leaflet of the newly formed aGUV does not exchange lipids/dyes with lipids of the SLB during hemifusion.

The intensities measured in the aGUVs and reported in your paper (Fig 3a, 4a, and 5a – the second box plot colored in red, aGUVs) show the *remaining dye* in the aGUVs, which includes the inner leaflet of the aGUV (not available to exchange) and that remaining in the aGUV outer leaflet (for incomplete exchange).

I have included an example here.

When $I^{\mathrm{DiD}}_{\mathrm{aGUV}} = I^{\mathrm{DiD}}_{\mathrm{GUV}}$, it indicates that no dye has transferred from the vesicle outer leaflet to the SLB. Consequently, applying eqn (2) leads us to conclude that $P = 0$, which aligns with our expectations.

The use of eqn (1) instead suggests that when $I^{\mathrm{DiD}}_{\mathrm{aGUV}} = I^{\mathrm{DiD}}_{\mathrm{GUV}}$, $P = 2$, which implies that both leaflets were exchanged entirely; this is incorrect because no dye has transferred from the GUV outer leaflet to the SLB.

In addition, the second equation of the paper, written here as eqn (3), calculates the percentage of agreement. However, that will not give a correct number if one uses eqn (1) to calculate the percentage of lipid exchange of DiD (the dye initially in the symmetric GUV).

$$P^{a\%} = 100\% - [P^{\mathrm{ex}}_{\mathrm{TF}} - P^{\mathrm{ex}}_{\mathrm{DiD}}] \times 100\% \tag{3}$$

In Fig. 3c, 4c, and 5c of your paper, the *y*-axis actually illustrates the fraction of lipids that were not exchanged, *i.e.*, the fraction that did not exit the aGUV. It would be more beneficial to compare the fraction of lipids escaping the vesicle's outer leaflet to the fraction of lipids from the SLB that were transferred to the aGUVs' outer leaflets. The comparisons presented in the plots (Fig. 3c, 4c, and 5c) appear to be inconsistent and may require further examination for clarity and accuracy.

1 T. A. Enoki and F. A. Heberle, *Proc. Natl. Acad. Sci. U. S. A.*, 2023, **120**, e2308723120.

Arne Gericke answered: We have adjusted our paper (https://doi.org/10.1039/d4fd00191e) to correct the mistake in the calculation of the leaflet exchange degree. Thank you for pointing this out.

Rumiana Dimova said: You've elegantly used fluorescence intensity to map the lipid distribution in your asymmetric GUVs. However, it's well-known that the fluorescence quantum yield of dyes can be sensitive to their microenvironment. Factors such as the presence of divalent cations like calcium or magnesium, and variations in lipid packing density between the symmetric GUVs, the SLB, and the resulting asymmetric GUVs, could all potentially affect the quantum yield of your fluorescent probes. This would complicate the quantitative interpretation of your intensity measurements. Have you explicitly considered and addressed this potential variability in quantum yield? And do you think that control experiments performed on symmetric GUVs with lipid compositions matching those of the individual leaflets of your asymmetric GUVs could provide a useful baseline for comparison and help to deconvolute changes in intensity due to lipid distribution from those due to quantum yield effects?

Thais A. Enoki responded: We have observed that the fluorescence intensity of TopFluor PC could be slightly brighter in the L_d phase compared to the L_o phase. The difference in the fluorescence quantum yield depending on the packing environment must be taken into account when analyzing fluorescent intensities to estimate the percentage of lipid exchange. We discuss that in the supporting material of ref. 1.

1 T. A. Enoki and G. W. Feigenson, *Biophys. J.*, 2019, **117**, 1037–1050.

Arne Gericke replied: I agree that the microenvironment is important to consider when using fluorescence intensities to evaluate the quality of the aGUVs. In a previous study[1] we did not see an effect of calcium or magnesium on fluorescence intensity (outside of effects due to clustering). Changes in lipid packing density (*e.g.*, due to domain formation) are affecting the fluorescence intensities, which is particularly true for TopFluor-labeled lipids (see the paper cited above). Therefore, GUVs with lateral inhomogeneities (domains) need to be evaluated very carefully. Keeping the fluorophore concentrations as low as feasible might be a remedy if fluorescence quenching is observed. The sGUVs that are used to baseline the fluorescence intensities exhibit the same lipid compositions as the respective aGUV leaflets. In addition, we have conducted many experiments with sGUVs that exhibited varying lipid compositions to assess the impact on fluorescence intensities. In all of these experiments we have not seen any concerning results. However, it is paramount that proper control experiments for each aGUV lipid composition are carried out, in particular, when phase-separated GUV model systems are involved.

1 T. A. Paratore, G. E. Schmidt, A. H. Ross and A. Gericke, *Chem. Phys. Lipids*, 2024, **264**, 105424.

Walter Nickel enquired: Comparing GUVs with either an asymmetric or symmetric transbilayer distribution of phosphatidylinositol-(4,5)-bisphosphate ($PI(4,5)P_2$), did you identify a biological process that depends on an asymmetric distribution of $PI(4,5)P_2$? An example would be fibroblast growth factor 2 (FGF2) membrane translocation as part of its unconventional secretory pathway, a process that has been shown to be facilitated by $PI(4,5)P_2$ transbilayer asymmetry of plasma membranes.[1]

1 M. Kaur, F. Lolicato and W. Nickel, *bioRxiv*, 2025, preprint, DOI: 10.1101/2025.04.19.649664.

Arne Gericke answered: Your paper is very interesting since its shows how the alteration of membrane physical properties due to the local accumulation of $PI(4,5)P_2$ results in a distinct physiological response that requires asymmetry. I am not aware of a similar case in the literature. Obviously, local accumulation of phosphoinositides (PIPs) leads to a higher charge density and hence electrostatic binding of cationic proteins; however, this is not dependent on asymmetry like it is in your paper.

Karin Riske asked: Have you done any experiments to check how the presence of asymmetric PIP_2 in GUVs impacts membrane properties? And did you try to obtain aGUVs with other molar fractions of PIP_2?

Arne Gericke responded: We have not done such experiments yet, but it is our goal to proceed to experiments like those you are outlining.

Manuel Bibrowski queried: Have you had other verification methods for the abundance of PIP_2 in the membrane, that do not rely on chemical modifications of the PIP_2 lipid?

Arne Gericke replied: Unfortunately, we don't have another method that would examine the quality of the aGUVs on a single-GUV level.

Ioanna Mela opened a discussion of the paper by Francisco N. Barrera: What drives the higher roughness of the sample with chondroitin sulphate alone in Fig. 4 of your paper (https://doi.org/10.1039/d4fd00195h)? What drives the high roughness observed, taking into account that CS is a small molecule?

Francisco N. Barrera answered: CS can aggregate, likely through hydrogen bonding between chains. In solution, aggregation is reduced by physiological concentrations of NaCl. Maybe the roughness results from a similar process.

Ioanna Mela enquired: How is the negatively charged CS immobilised on the negatively charged mica surface? Could insufficient immobilisation be giving the appearance of "roughness"?

Francisco N. Barrera responded: We expect that CS attaches non-specifically to the mica surface, potentially *via* the aminated region of the monosaccharide.

Ioanna Mela asked: What is the theoretical length of CS and how does that correspond to the observed increase in diameter of the LUVs?

Francisco N. Barrera replied: The expected polymer length is 5–10 nm, which agrees with the observed DLS diameter of chondroitin sulfate: 7 nm. There is a good match with the diameter increase after grafting to the vesicles, of approximately 15 nm, given the expected continuous CS layer around the vesicle.

Markus Deserno remarked: In your experimental protocol you are – in soft matter lingo – grafting polymers onto a surface. From your paper I estimate that the individual polymers are roughly 10 nm apart and between 5 nm and 10 nm long. This, I think, might still be in the so-called "mushroom" regime, where you have isolated polymer coils, but you are not too far away from the brush regime, where the entropic repulsion between neighboring grafted polymers would stretch them out. This in turn would lead to a sizable lateral repulsion between them and an associated torque that might create outward growing membrane tubules. Did you ever see that?

Francisco N. Barrera answered: Markus, I appreciate you doing the calculation. Indeed, we expect the chondroitin sulfate (CS) chains to be in the mushroom regime due to us grafting them to the membrane with only 0.5 mol% of the maleimide lipid. So far, we have not observed outward membrane tubules, but we recently started doing experiments with 5 mol% of the reactive lipid in order to favor the brush regime. However, it is still to be determined if attaining this high coverage is feasible; one has to consider that access of the CS-SH molecule to the maleimide groups at the membrane surface could be hampered once a high level of CS coating of the membrane exists. Therefore, it seems likely that electrostatic repulsion between the CS chains might limit completion of the maleimide–SH conjugation. In that case, we would employ higher levels of salt in the solution to minimize such repulsion.

Karin Riske commented: It would be interesting to follow in real time the conjugation of chondroitin sulfate to the GUVs containing the reactive lipid, so that possible changes in spontaneous curvature, for instance, could be observed.

Francisco N. Barrera responded: This is an excellent suggestion. However, our current experimental approach is far from ideal to perform this experiment, since the chondroitin sulfate chains are not fluorescent, and we can only detect them after anchoring to the membrane by using fluorescently labeled molecules such as fluorescein isothiocyanate (FITC)-PLL or FITC-*Wisteria floribunda* lectin (WFA). To perform this experiment, which is a good idea, it would be probably better to work with a fluorescently-labeled polysaccharide.

Ben John queried: In the atomic force microscopy (AFM) image (Fig 4 in the paper; https://doi.org/10.1039/d4fd00195h), there are small features, possibly nanoclusters; do you think these arise from complete demixing? Could these clusters be formed by the polymers, the lipids, or both? And have you investigated how varying the mixing ratios influences domain or nanocluster formation?

Francisco N. Barrera replied: It is our interpretation that the roughness on the surface corresponds to several chains of chondroitin sulfate laying on top of the bilayer. A hydrogen-bonding network could be established between the chondroitin sulfate chains under these conditions. We are in the process of testing different levels of chondroitin sulfate grafting by titrating the maleimide reactive lipid (1,2-dioleoyl-*sn-glycero*-3-phosphoethanolamine-*N*-[4-(*p*-maleimidomethyl) cyclohexane-carboxamide] (PE-MCC).

Erwin London enquired: Was there any spectral shift of the fluorescence of the nitrobenzoxadiazole (NBD)-labeled lipid that would give additional clues as to the change in its local environment after attachment of carbohydrate to the lipid vesicles?

Francisco N. Barrera answered: The conjugation of chondroitin sulfate chains to membranes doped with NBD-labeled lipid affect the fluorescence intensity of the dye, but we did not observe a significant spectral shift. This result suggests that the polysaccharide chains do not lie flat on the surface of the membrane, where they would strongly alter the NBD environment, but protrude away from the membrane, as expected in the mushroom regime.

Martin E. Villanueva asked: Regarding the AFM measurements, have you considered using an alternative substrate to mica? Employing silicon oxide (SiO_2), which presents slightly higher intrinsic surface roughness (R_q), could still support lipid-bilayer formation while potentially enhancing the detection of surface modifications following chondroitin sulfate (CS) grafting. The use of SiO_2 might promote the formation of bilayers with a higher density of defects, thereby facilitating CS attachment and leading to a more pronounced increase in surface roughness, as reflected in the R_q parameter.

Francisco N. Barrera responded: We have formed supported bilayers on silicon oxide supports, and under these conditions the grafting with chondroitin sulfate is successful as well. However, we have not yet done a side-by-side comparison on the efficiency of grafting.

Joseph H. Lorent queried: How long are the poly-L-lysine polymers and could the cross-linking be tuned by modifying the polymer size ?

Francisco N. Barrera replied: The poly-L-lysine (PLL) polymers that we employed were long (15–30 kDa), as we intended them to be able to interact with and bridge several chondroitin sulfate chains. This effect could be potentially magnified with longer poly-L-lysine chains that are also commercially available.

Neha P. Kamat enquired: Do you imagine that your vesicles are predisposed to grow?

Francisco N. Barrera answered: The DLS results of vesicles grafted with CS revealed large particles in the presence of PLL. While at the moment we have a poor understanding of the structural makeup of the particles, it is conceivable that vesicle fusion occurs and that indeed vesicle growth has occurred.

Aileen Cooney asked: Did you notice any impact of the maleimide lipid on the yield of vesicle formation, especially above 1 mol%?

Francisco N. Barrera responded: We were initially worried that the reactive lipid would impact membrane properties, and hinder GUV formation or vesicle fusion for the spreading and formation of supported lipid bilayers. However, so far we have detected no impact of the maleimide-conjugated lipid, up to 5% molar.

Alberto Giacomello opened a general discussion: I was wondering how the plasma membrane mimics discussed in this session can help settle methodological problems in the asymmetry of plasma membranes. Preparing an artificial membrane has the advantage of controlling more precisely the composition of the leaflets and thus being able to test the merits and limitations of each technique. For instance, in the opening lecture (https://doi.org/10.1039/d5fd00041f), potential artefacts in experiments with PLA2 have been discussed, which could perhaps be settled with mimics of known composition.

Joseph H. Lorent replied: It would certainly be a very useful study to systematically investigate the enzymatic digestion of phospholipids in asymmetric model membranes with different compositions and leaflet properties. I further think that the used digestion protocol probably has to be adapted to the outer leaflet phospholipid packing of the plasma membrane of the investigated cell. Further controls have to be performed to control the possible loss of membrane asymmetry, as described in ref. 1.

1 J. H. Lorent, K. R. Levental, L. Ganesan, G. Rivera-Longsworth, E. Sezgin, M. Doktorova, E. Lyman and I. Levental, *Nat. Chem. Biol.*, 2020, **16**, 644–652.

Reinhard Lipowsky answered: It is always useful to compare the complex behavior of a biological membrane with the simpler behavior of a biomimetic model membrane. In fact, I strongly believe that the physical mechanisms that govern a certain membrane behavior can only be understood by studying such model membranes, using a combination of experimental methods, theoretical considerations, and computer simulations. This integrated approach is now widely recognized in the context of bottom-up synthetic biology. The asymmetry of bilayer membranes has been intensely studied *via* computer simulations, which revealed the importance of leaflet tensions and stress asymmetry.[1,2] From an experimental point of view, bilayer asymmetry and spontaneous curvature can be precisely controlled by binding His-tagged and fluorescent proteins, such as His-tagged green fluorescent protein (GFP), to vesicle membranes doped with anchor lipids.[3] Furthermore, a new experimental method has been recently introduced, by which one can create giant unilamellar vesicles with different intramembrane domains that possess distinct bilayer asymmetries.[4]

1 R. Lipowsky, R. Ghosh, V. Satarifard, A. Sreekumari, M. Zamaletdinov, B. Różycki, M. Miettinen and A. Grafmüller, *Biomolecules*, 2023, **13**, 926, DOI: 10.3390/biom13060926.
2 R. Lipowsky, *Faraday Discuss.*, 2025, DOI: 10.1039/d4fd00184b.
3 J. Steinkühler, R. L. Knorr, Z. Zhao, T. Bhatia, S. M. Bartelt, S. Wegner, R. Dimova and R. Lipowsky, *Nat. Commun.*, 2020, **11**, 905, DOI: 10.1038/s41467-020-14696-0.
4 N. Yandrapalli, T. Seemann, R. Lipowsky and T. Robinson, *Nat. Commun.*, under review.

Georg Pabst responded: I believe this would be a worthwhile approach to pursue. However, directly translating findings from such simplified systems remains challenging. It reminds me of the difficulties encountered when extrapolating the properties of L_o/L_d domains in lipid-only vesicles to live cells ("rafts"), which led to considerable controversy. Somewhat unexpectedly related: we previously investigated the activity of sphingomyelinase on large unilamellar vesicles composed of palmitoyl-oleoyl phosphatidylcholine (POPC) and egg sphingomyelin (ESM).[1] Our data suggests that the enzyme could access lipids in the inner leaflet, at least after sufficient incubation time. At that point, we did not quantify any induced asymmetry, as it was not the primary focus of our study. Nevertheless, reaction timing is certainly a critical factor in cell-based experiments. It is also important to emphasize that our vesicle system lacked cholesterol and was initially symmetric.

1 B. Boulgaropoulos, H. Amenitsch, P. Laggner and G. Pabst, *Biophys. J.*, 2010, **99**, 499–506, DOI: 10.1016/j.bpj.2010.04.028.

Conflicts of interest

There are no conflicts to declare.

Faraday Discussions

ROYAL SOCIETY OF CHEMISTRY

PAPER

The interplay of composition and mechanics in the thermodynamics of asymmetric ternary lipid membranes

Malavika Varma and Markus Deserno *

Received 9th December 2024, Accepted 13th January 2025

DOI: 10.1039/d4fd00196f

Eukaryotic lipid membranes are both compositionally complex and strongly asymmetric. Preferential lipid interactions enable coexistence between two fluid phases and an associated critical point, while bilayer asymmetry leads to leaflet-specific values for many observables—most saliently composition, but also a difference in leaflet tensions, for which we introduced the term "differential stress." Lipid mixing thermodynamics has been extensively studied, notably in idealized ternary model systems, and interest in asymmetry has grown significantly in the past decade, but their interplay remains poorly understood. Here we propose a conceptual framework for the thermodynamics of asymmetric ternary lipid membranes. Cholesterol emerges as an essential actor playing two different roles: first, it controls lipid mixing; second, it couples the compositional phase points of the two leaflets by achieving chemical equilibrium between them. Since differential stress can squeeze cholesterol from one leaflet into the other, this couples mechanical properties such as lateral stresses and curvature torques directly to mixing thermodynamics. Using coarse-grained simulations, we explore implications for leaflet coexistence, mechanical stability of giant vesicles, and differential stress driven phase segregation in a single leaflet. We hope this framework enables a fresh look at some persistent puzzles in this field, most notably the elusive nature of lipid rafts.

1 Introduction

Our overarching goal in this contribution is to develop a theoretical framework, largely based on thermodynamics, within which we can describe lipid membranes that are both compositionally complex and asymmetric. The scope for the composition degree of freedom will remain ostensibly modest: we do not aim to grapple with the enormous richness of biological lipid mixtures, and we will eschew proteins entirely. Instead, we will restrict our study to the widely studied case of ternary systems comprising a saturated lipid species, an unsaturated one, and cholesterol. These are the simplest known lipid mixtures that capture a phenomenon widely believed to be of physiological significance: the

Department of Physics, Carnegie Mellon University, Pittsburgh, PA 15213, USA

possibility of coexistence between two fluid phases exhibiting different degrees of order. Considering bilayer asymmetry we will go beyond the usual discussion of leaflet-specific lipidomes and emphasize the importance of other thermodynamic variables that can differ between the two sides of the membrane, with an emphasis on stresses and torques.

This introduction serves to very briefly set the stage for these two main players—reminding our readers of a few key facts of complex biomembranes, and how these tend to be captured in simplified model systems.

1.1 Compositional complexity in nature

Biomembranes feature an astoundingly diverse lipidome:[1,2] thousands of chemically distinct lipids are synthesized in the endoplasmic reticulum (ER) of a typical eukaryotic cell, whose individual membrane organelles are kept at unique compositions while still hosting hundreds of individual species.[3] This complexity is challenging to even categorize,[4] but it is anything but random, even though its purpose is not well understood.

Since chemically distinct lipids interact differently with one another, we should expect that they do not mix ideally. If the interactions are sufficiently dissimilar, phase separation might occur. The by far most extensively studied instances of this possibility are "lipid rafts",[5–9] whose 2006 consensus definition describes them as "small (10–200 nm), heterogeneous, highly dynamic, sterol- and sphingolipid-enriched domains that compartmentalize cellular processes."[10] While having many putative effects on a wide spectrum of biomembrane functions, their precise nature (and even existence) has been famously controversial.[11,12] All the same, decades of data on highly non-ideal mixing need to be explained. Recent reviews[13,14] suggest that rafts will remain protagonists in this story, even though almost surely as more subtle actors than originally envisioned.

Of course, in addition to lipids, biomembranes also contain a host of peripheral or integral membrane proteins, which add further complexity, but for the sake of the present discussion we will summarily ignore them. This smacks of impermissible simplification, so let us offer a few thoughts to ease the discomfort:

(1) We will subsequently also ignore more than 99% of all lipid species, aiming for a model system that captures some emergent phase behavior, not an intricate lipidomic fingerprint.

(2) If the lipid matrix becomes laterally inhomogeneous, this will subsequently affect the proteins, but in a first step we can imagine them simply adjusting their distribution in response to the spatially nontrivial lipid background.

(3) In a second step, uneven protein partitioning between different lipid phases will in turn change the properties of these phases, further tweaking the distribution of lipids.

We can think of points 2 and 3 as the beginnings of an iterative process that adjusts the lipids, then the proteins, then again the lipids, then again the proteins—and so on. Mathematically, this constitutes a perturbation expansion in some lipid–protein coupling parameter. We will proceed on the hope that its first terms capture the main physics reasonably well. This is rigorously true if that coupling parameter is (in some sense) small—but, of course, this almost surely depends on the situation, and so we have to remain vigilant.

1.2 Ternary model systems

Mixed membranes comprising hundreds of distinct lipid species cannot be studied with any expectation of experimental reproducibility and precise theoretical understanding. Model systems are clearly needed, but from the outset it is not obvious how far we may simplify. Remarkably, more than two decades of research strongly suggest that ternary systems consisting of a saturated lipid species, an unsaturated species, and cholesterol capture many of the key physical properties at the heart of biomembrane phase behavior.[15–31] Specifically, these systems exhibit the phenomenon of liquid–liquid phase separation, which offers tantalizing connections to raft-like domains. A particularly intriguing aspect is that the liquid–liquid coexistence region terminates in a critical point,[32–38] which has also been observed in cellular membranes (including its Ising-like critical exponents!).[39] This underscores the presence of universality—which often justifies the use of even highly simplified models.

1.2.1 Ternary phase diagrams. Phase diagrams of (symmetric) ternary lipid mixtures consisting of a saturated lipid, an unsaturated lipid, and cholesterol have been studied extensively. Since they will feature prominently in the rest of our discussion, we shall remind the reader of their main "topography", which will double as an opportunity to revisit the "Gibbs triangle" representation of ternary mixtures.

The constraint that the three mole fractions $\{\phi_u, \phi_s, \phi_c\}$ must sum to 100% reduces the independent degrees of freedom from 3 to 2. An elegant way to turn this into a "plotting strategy" is to exploit Viviani's theorem: the sum of the shortest distances from any interior point of an equilateral triangle to its sides is constant and equal to the triangle's height. As a corollary, the three symmetric lines radiating out from this point parallel to the triangle's sides can be used to set up coordinates which add up to a fixed number, conveniently chosen as 100%. This is illustrated in Fig. 1 using the point $\phi_u : \phi_s : \phi_c = 20 : 50 : 30$. We can hence represent ternary mixtures as follows: pure components correspond to the corners of the triangle, binary mixtures live on the sides connecting the respective two pure corners, and ternary mixtures are located anywhere in the interior of the triangle.

1.2.2 Typical phase diagram of ternary (and "raft like") lipid mixtures. Let us return to the particular case of ternary mixtures comprising saturated, unsaturated, and cholesterol lipids. Their phase diagram depends of course on the chosen lipids, but at a temperature at which the (pure) saturated lipid is in a gel phase and the (pure) unsaturated lipid in a fluid phase, these diagrams typically take the general form shown in Fig. 1: the *u*- and *s*-lipids exhibit 2-phase coexistence along the binary triangle edge (conventionally picked to be the lower baseline, almost always with "*s*" on the right hand side).

Adding cholesterol (*i.e.*, "moving up"), opens a second 2-phase region near the saturated corner (tie-lines are shown in green) with a cholesterol-poor gel-phase coexisting with a more cholesterol-rich so-called "liquid ordered" (ℓ_o) fluid phase. Both 2-phase regions border at a 3-phase triangle, on whose third side a third 2-phase region is attached. Unlike the other two, it does not extend to its nearest triangle side (namely, *uc*), since at the chosen temperature cholesterol and the unsaturated species mix well (except that cholesterol precipitates at sufficiently high concentration—see below). Instead, the two phases coexisting

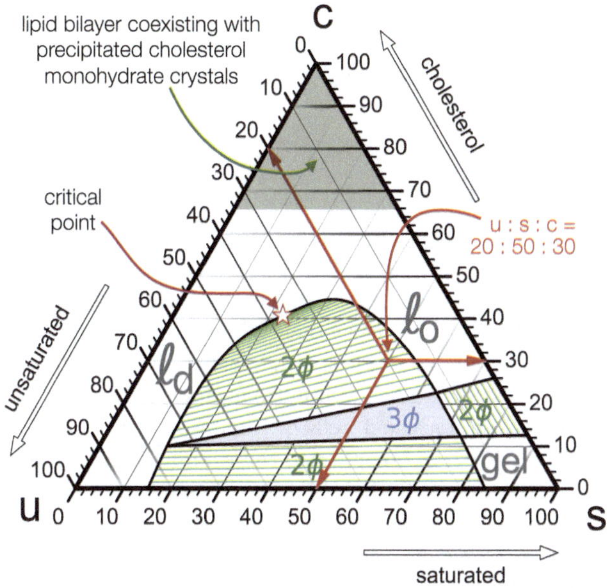

Fig. 1 Gibbs triangle representation of a typical ternary phase diagram of a lipid mixture of saturated, unsaturated, and cholesterol lipids. This image is merely a schematic, but it qualitatively reflects the basic topography measured in many experimental studies. White regions correspond to pure phases, hatched green regions exhibit two-phase coexistence, with the green lines actually being the tie-lines between coexisting phase points. The central blue triangle is a region of three-phase coexistence, in which the pure phases at the triangle's corners coexist. The three red arrows illustrate how to read off the composition of some given point. The sum of their lengths is the same no matter which point in the triangle we pick, which is the key geometric fact that enables this representation in the first place.

across this region—the ℓ_o phase and its "partner", a less ordered (since more unsaturated) "liquid disordered" (ℓ_d) phase, become more similar and merge at a critical point. "Beyond" this point there is no meaningful distinction between a fluid ordered or disordered phase, any more than a meaningful distinction between liquid water and steam can be made past water's critical point—both merely being fluid phases of the same symmetry.

The solubility limit of cholesterol in PC-phospholipids is typically around 66%,[40] above which it precipitates from the bilayer into crystals of cholesterol monohydrate. Once this happens, a point in the Gibbs phase triangle no longer reflects the membrane composition, but the resulting states still feature a stable membrane. In fact, they are experimentally very useful, because they anchor cholesterol's chemical potential to that of its monohydrate, permitting quantitative calibration between different membrane systems.[41]

If the temperature is such that even the saturated lipid is in a fluid phase, then the two gel-fluid coexistence regions vanish from the phase diagram, and with it also the 3-phase triangle. They are effectively "pushed down" to the *us*-binary line and all that survives is the ℓ_o/ℓ_d coexistence. This may look like a very different phase diagram, but recall that gel phases are rarely physiologically relevant. What matters is the fluid–fluid coexistence region and its critical point, which remains intact, not gel-phase physics created by the pure-*s*-corner. We will subsequently discuss simpler "gel-free" versions of the ternary phase diagram and argue that for the purpose of understanding physiologically interesting ternary lipid mixtures these will suffice.

1.3 Membrane asymmetry

Lipid membrane asymmetry is not a recent addition to a biomembrane's complexity list. It was discovered at the very dawn of modern biomembrane science: Mark Bretscher's seminal paper[42] on the subject appeared just 12 days after Singer and Nicholson published their "Fluid Mosaic Model."[43] The following year, Verkleij *et al.* proposed the first rough leaflet distribution in a human red blood cell,[44] followed by similar results from other authors for platelets,[45] the plasma membrane of nucleated cells,[46–48] and some membrane organelles.[49] Recent work by Lorent *et al.*[50] confirmed these early findings and added highly fine-grained detail. Moreover, their bioinformatics analysis of transmembrane proteins anchored by a single-pass α-helix (in particular, that anchor's surface area imbalance) strongly suggests that plasma membrane asymmetry is evolutionarily conserved across all eukarya.

It is worth highlighting that leaflet-specific lipidomes can exist only because spontaneous transitions of phospholipids between leaflets ("flip-flop" events) happen so slowly—between hours and days[51]—that active but slow cellular transmembrane sorting mechanisms can successfully counter the decay into a fully scrambled state. An important exception is cholesterol, whose flip-flop time is believed to be somewhere in the sub-microsecond to millisecond range.[51–54] Hence, its concentration is thermodynamically equilibrated between leaflets on most experimentally and biophysically relevant timescales.

1.3.1 Biological model systems. The most straightforward way to do experiments on asymmetric membranes is to work with those given to us by nature—using live cells, or at least membranes derived from them. A particularly convenient and frequently used model system are so-called giant plasma membrane vesicles (GPMV), which one can obtain by chemically inducing cells to vesiculate (or "bleb").[55–58] In fact, ℓ_o/ℓ_d-like critical fluctuations in biomembranes were first observed in GPMVs.[39]

However, such bio-derived systems as exemplars for asymmetry have a number of drawbacks. The most obvious one is that they are very complicated: they consist of far more than just a few types of lipids, besides also containing many proteins.[58–60] This not only makes it difficult to know (let alone control) their composition; if the goal is to specifically probe asymmetry, their substantial complexity adds numerous confounding factors. Furthermore, it appears that whatever asymmetric composition the plasma membrane has, GPMVs have lost at least some of it (they appear to be at least partially scrambled).[60,61]

1.3.2 Artificially created asymmetric membranes. Clean asymmetric model bilayers are preferable if the goal is to specifically examine asymmetry, but making them has been tricky. This is likely one of the two primary reasons (the other one being the challenge to measure leaflet-resolved properties) why our knowledge of asymmetric membranes lags behind that of symmetric ones, even though we have known basically from day one that the biologically relevant situation is asymmetric.

Luckily, this situation has changed dramatically over the past decade: by now more than 70 protocols for synthesizing asymmetric bilayers have been published, which Krompers and Heerklotz have recently reviewed, classified into four major categories, and analyzed in terms of advantages and drawbacks.[62] We believe that the availability of such clean model systems is a key driver of asymmetry's renaissance. And yet, the community has only just begun to explore the exciting opportunities this affords. This includes many now feasible research questions that are waiting to be realized, likely offering consequential insights into the biological situation.

1.3.3 Types of asymmetry. A leaflet-specific lipid content is the most salient aspect of membrane asymmetry, but it is by no means the only one. In fact, if we break the symmetry of one specific observable (here: composition), other symmetries are prone to break in the process. The generic assumption should therefore be that any observable that can be defined at the leaflet level is symmetry broken once one such observable is symmetry broken. This could be structural properties (*e.g.*, area per lipid, thickness, charge density), mechanical properties (*e.g.*, elastic moduli, spontaneous curvature, leaflet tensions and torques), or dynamical ones (*e.g.*, diffusion constant, viscosity, relaxation rates). In this paper, we will focus—besides the obvious compositional degree of freedom—on two mechanical observables: differential stress, $\Delta\Sigma$, and bilayer torque, $\mathscr{T}$.

Differential stress is the difference between the two individual mechanical leaflet tensions,

$$\Delta\Sigma = \Sigma_{+} - \Sigma_{-}, \tag{1}$$

where we use "+" and "−" to distinguish the two leaflets, which we henceforth will refer to as "leaf$_{+}$" and "leaf$_{-}$" (think of "+" as the "upper" or "outer" one, if you wish). Differential stress $\Delta\Sigma$ is the "orthogonal partner" to the more common total tension $\Sigma = \Sigma_{+} + \Sigma_{-}$, in the sense that these two observables define a new set of orthogonal axes in the leaflet-resolved tension space $\{\Sigma_{+}, \Sigma_{-}\}$. Our group has pointed out that differential stress must be included when discussing membrane asymmetry; in fact, since often $\Delta\Sigma \gg \Sigma$, it can be the more important variable.[63–67]

Bilayer torque $\mathscr{T}$ is the thermodynamic observable conjugate to a membrane's extrinsic curvature J, *i.e.*, the (generalized) force that drives bending,

$$\mathscr{T} = \left(\frac{\partial f}{\partial J}\right)_{T,\Sigma,\ldots}, \tag{2}$$

where f is the free energy per area and the subscripts remind us which other observables are meant to remain fixed.

Differential stress can be rephrased as the existence of a preferred bilayer curvature $J_{0,\mathrm{s}}$ at which this stress vanishes, because at fixed lipid content bending changes the leaflet reference areas measured some distance z_0 away from the bilayer midsurface, to lowest order linear in the curvature. (The relevant reference surface at this distance z_0 is the so-called "neutral surface," at which bending and stretching energies decouple.) This implies that we should be able to write[63]

$$\Delta\Sigma = \frac{\kappa_{\mathrm{nl}}}{z_0}(J - J_{0,\mathrm{s}}), \tag{3}$$

where $\kappa_{nl} \simeq K_A z_0^2$ and K_A is a membrane's area expansion modulus. We can then combine bending and differential stress into a single curvature-elastic expression of the form[63–67]

$$f = \frac{1}{2}\kappa(J - J_{0,b})^2 + \frac{1}{2}\kappa_{nl}(J - J_{0,s})^2, \tag{4}$$

where κ is the ordinary bending modulus and $J_{0,b}$ the more common lipid-shape based spontaneous bilayer curvature. The torque is hence

$$\mathcal{T} = \left(\frac{\partial f}{\partial J}\right)_{T,\Sigma,\ldots} = \underbrace{\kappa(J - J_{0,b})}_{\mathcal{T}_\kappa} + \underbrace{\kappa_{nl}(J - J_{0,s})}_{\mathcal{T}_\Sigma} \tag{5a}$$

$$\overset{(3)}{=} \kappa(J - J_{0,b}) + z_0\Delta\Sigma. \tag{5b}$$

Fig. 2 illustrates the sign conventions for these two terms.

Demanding that a flat membrane is torque free (*i.e.*, it will voluntarily stay flat) yields

$$0 = \mathcal{T}(J = 0) \overset{(5b)}{=} -\kappa J_{0,b} + z_0\Delta\Sigma \quad \Rightarrow \quad \Delta\Sigma = \frac{\kappa J_{0,b}}{z_0}. \tag{6}$$

Using common values on the right hand side (such as $\kappa \simeq 30k_BT$, $J_{0,b} \simeq$ few $\times 10^{-2}$ mN m^{-1}, and $z_0 \simeq 1$ nm) shows that we should expect $\Delta\Sigma$ to be on the order of a few mN m^{-1}, which is between one and two orders of magnitude larger than typical cellular membrane tensions.[68]

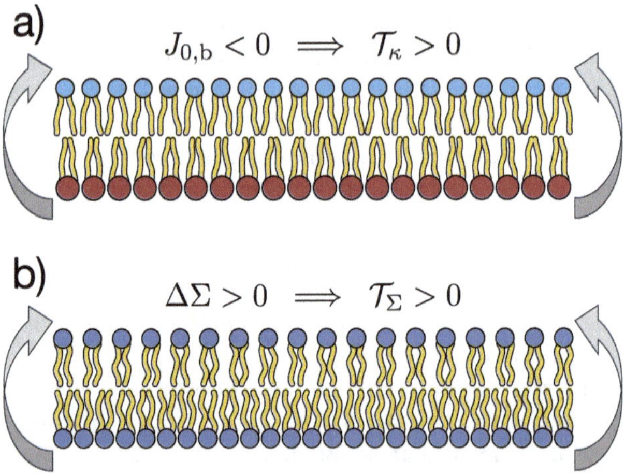

Fig. 2 Clarifying the sign of the torque, $\mathcal{T} = \mathcal{T}_\kappa + \mathcal{T}_\Sigma$. (a) The torque due to lipid curvature, $\mathcal{T}_\kappa$, strives to bend the membrane up when the spontaneous bilayer curvature is negative, $J_{0,b} < 0$, *i.e.* when the upper leaflet has a more negative spontaneous curvature than the lower one. (b) The torque due to differential stress, $\mathcal{T}_\Sigma$, strives to bend the membrane up when the differential stress is positive, $\Delta\Sigma > 0$, *i.e.* when the upper leaflet is under excess tension.

2 Towards ternary asymmetric phase diagrams

Whatever the precise physics of asymmetric ternary lipid mixtures will be, it is clear that these systems have more degrees of freedom than their symmetric counterparts—most obviously since we now have to deal with two nontrivial leaflet compositions. This means that we will not merely be asking how the phases in a diagram such as the one in Fig. 1 might change. The more profound question is: what type of phase diagrams should we be drawing in the first place? What are good thermodynamic variables to put on the axes of what kind of plot? How many degrees of freedom do we have, and what are they?

2.1 Effective binary modeling

So far experiments offer little guidance, since leaflet-resolved phase diagrams are practically nonexistent. The first (and so far only) measurement of a leaflet-resolved *binary* phase diagram (a mixture of two phospholipids, DOPC and DPPC) was only published in 2023.[69] While experimentally challenging, it is conceptually straightforward: the mole-fraction on each side is a single number, which yields a straightforward two-dimensional diagram for which theoretical predictions exist.[70–72] Ternary lipid systems have also been studied theoretically, but the cholesterol component has been treated implicitly, essentially by considering a single order parameter that distinguishes ℓ_o from ℓ_d (rendering the problem effectively that of a binary mixture), to which a rapidly flip-flopping cholesterol component is locked.[70–75] At any rate, differential stress or considerations regarding torque have not been part of any analysis.

2.2 Counting degrees of freedom

2.2.1 Simplistic counting gives 4 degrees of freedom. Let us count the degrees of freedom to establish the dimensionality of the problem. Each leaflet is a ternary mixture, which has a two-dimensional phase-space—one Gibbs triangle for each side. A convenient way to parametrize points in those is *via* the following two coordinates. First, define leaflet-specific saturation ratios

$$r_\pm = \frac{S_\pm}{U_\pm + S_\pm} \equiv \frac{S_\pm}{L_\pm}, \tag{7}$$

which measure the relative percentage of saturated phospholipids $S_\pm$ among all phospholipids $L_\pm = U_\pm + S_\pm$. States of constant $r_\pm$ form lines that pass through the cholesterol corner and divide the opposite *us*-side at the respective ratio of *s*-lipids. Second, to single out a point on that line, we may specify its cholesterol mole fraction, which we will henceforth denote with the symbol $\phi_\pm$. These coordinates are illustrated in Fig. 3. Since both Gibbs triangles parametrize two-dimensional phase spaces, the total problem appears to have 4 degrees of freedom.

2.2.2 More careful counting also gives 4 degrees of freedom. Four degrees of freedom is in fact the correct answer, but to properly understand why, we need to account for two additional and very important effects—one which will remove a degree of freedom, and another one which will add a degree of freedom.

First the removal. Since cholesterol rapidly flip-flops between the leaflets, we need to satisfy the condition of equal chemical potential,

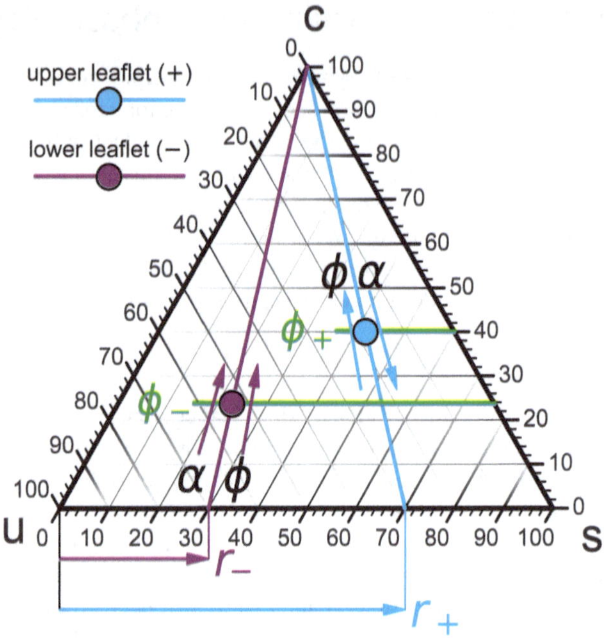

Fig. 3 Parametrizing points in a ternary lipid mixture *via* the saturation ratios $r_\pm$ and the cholesterol mole fractions $\phi_\pm$. The two colors correspond to compositions in the two leaflets—cyan being leaf$_+$, magenta leaf$_-$. The arrows next to the two leaflet compositions indicate how each would respond to an increase in either the total cholesterol content ϕ or the phospholipid leaflet abundance α.

$$\mu^{+}_{\text{chol}} = \mu^{-}_{\text{chol}}. \tag{8}$$

This constraint reduces the number of degrees of freedom from 4 to 3. The cholesterol fractions ϕ_+ and ϕ_- in the two leaflets cannot be set to whatever values we desire; instead, we can only decide on the total mole fraction ϕ of cholesterol, defined *via*

$$\phi = \frac{C_+ + C_-}{L_+ + L_- + C_+ + C_-} \equiv \frac{C}{L + C} \equiv \frac{N_+\phi_+ + N_-\phi_-}{N_+ + N_-}, \tag{9}$$

where $N_\pm$ are the total lipid numbers in each leaflet. As cholesterol equalizes its chemical potential on both sides *via* flip-flop, the individual coexisting leaflet fractions $\phi_\pm$ arise.

Now the addition. Compositional ternary diagrams such as the one in Fig. 1 only record mole fractions; they know nothing about the absolute number of lipids on the two different sides. It proves useful to include this missing thermodynamic information not *via* the individual lipid numbers $N_\pm$ but in terms of total lipid content, $N_+ + N_-$, and number difference, $N_+ - N_-$. The former captures potentially interesting physics pertaining to system size (which we will sidestep in this paper), while the latter homes in more directly on asymmetry. Considering that cholesterol rapidly flip-flops, it is more convenient to specify the abundance asymmetry *via* the phospholipid contingent, which does not change under cholesterol redistribution; let us hence define the phospholipid abundance asymmetry

$$\alpha = \frac{L_+ - L_-}{L_+ + L_-}. \tag{10}$$

In the absence of cholesterol, changing the abundance by $\Delta\alpha$ changes the differential stress by $\Delta\Delta\Sigma = -K_A\Delta\alpha$ (assuming area additivity, since abundance then couples directly to area difference, the observable conjugate to differential stress). Shifting α by just a few percentage points would then change differential stress by several mN m^{-1}, rendering α a fairly "stiff" degree of freedom.

We will see that once cholesterol is present, its ability to offset a phospholipid imbalance by redistributing into the less crowded leaflet permits much larger changes of α without incurring huge differential stress. Of note, recent experiments have claimed that the cytosolic leaflet of the human red blood cell contains about twice as many phospholipids as the exoplasmic one (*i.e.*, $\alpha \sim -33\%$),[76] in line with the raw data of no less than five previous studies (see Table S1 in that reference). Such an enormous abundance asymmetry would profoundly affect cholesterol distribution and leaflet-specific phase behavior, both fields rife with their own controversial claims. We expect that further work dedicated to this complex interplay has the potential to resolve numerous contentious issues. Our goal here is to develop a theoretical framework that accounts for phospholipid abundance asymmetry, however large it may prove to be, such that its thermodynamic implications can be quantitatively examined.

2.2.3 Using ϕ and α to independently tune ϕ_+ and ϕ_-. Having leaflet abundance as a new tuning knob, a fascinating possibility emerges: despite the fact that we cannot pick two values ϕ_+ and ϕ_- for the cholesterol leaflet mole fractions and expect them to automatically stay that way, we might be able to make them coexist by a suitable choice of cholesterol fraction ϕ and phospholipid abundance α. The reason is that these two tuning parameters drive linearly independent responses in the cholesterol leaflet fractions $\phi_\pm$. Consider again Fig. 3: if we increase ϕ, we expect both leaflet fractions $\phi_\pm$ to also increase; likewise, a decrease in ϕ entails decreases in $\phi_\pm$. But the response to changes in α is very different: if we increase α (at fixed total tension), then we put leaf$_+$ under compression and leaf$_-$ under tension. As a result, we will "squeeze" cholesterol from leaf$_+$ to leaf$_-$ and hence decrease ϕ_+ while simultaneously increasing ϕ_-.

Small changes in ϕ and α trigger small changes in the $\phi_\pm$, which are presumably linear. This shows that

$$\begin{pmatrix} \Delta\phi_+ \\ \Delta\phi_- \end{pmatrix} = \begin{pmatrix} a_{11} & -a_{12} \\ a_{21} & a_{22} \end{pmatrix} \begin{pmatrix} \Delta\phi \\ \Delta\alpha \end{pmatrix} \tag{11}$$

with positive numbers a_{ij}. These depend on ϕ and α (and other parameters of the problem), and they are not entirely independent, since conservation of the total cholesterol content linearly couples $\Delta\phi_+$ and $\Delta\phi_-$ (in some cumbersome way). All the same, the "parallel" and "antiparallel" responses to changes in ϕ and α add the minus sign to the upper right entry of the matrix in eqn (11), which guarantees that it is invertible. Hence, if we wish to make a specific (small) change in the cholesterol leaflet fractions, the required adjustments of cholesterol content and abundance asymmetry are

$$\begin{pmatrix} \Delta\phi \\ \Delta\alpha \end{pmatrix} = \frac{1}{a_{11}a_{22} + a_{12}a_{21}} \begin{pmatrix} a_{22} & a_{12} \\ -a_{21} & a_{11} \end{pmatrix} \begin{pmatrix} \Delta\phi_+ \\ \Delta\phi_- \end{pmatrix}. \tag{12}$$

To be clear: we are not guaranteed that we can always make every conceivable pair $\{\phi_+, \phi_-\}$ coexist, since this local argument does not clarify global reach. Besides the fact that cholesterol does not dissolve in a membrane beyond ~66%,[40,41] permissible $\Delta\phi$ and especially $\Delta\alpha$ will have feasibility limits of their own. Additionally, edge cases are likely problematic. For instance, it will be impossible to make a nonzero ϕ_+ coexist with a vanishing ϕ_-, because mixing entropy terms of the form $\phi \log \phi$ will create an infinite driving force at $\phi \to 0$ that cannot be balanced by any finite enthalpic terms.

To summarize the present discussion: we have argued that the thermodynamic phase space of our system is four-dimensional—already disregarding "obvious" variables such as the total number of lipids $N = N_+ + N_-$ (*i.e.*, the system size) and the temperature T, both of which we imagine fixed once and for all. Possible independent degrees of freedom are $\{r_+, r_-, \phi_+, \phi_-\}$ or alternatively $\{r_+, r_-, \phi, \alpha\}$. The former can be easily visualized as two points in a ternary phase diagram, and realizing them requires a judicious choice of ϕ and α. The latter may be more easily tunable experimentally and yield leaflet-specific cholesterol fractions ϕ_+ and ϕ_- that arise after flip-flop assisted chemical potential equilibration.

2.3 Some important new linear response functions

The matrix elements in eqn (11) describe the local linear relation for how $\{\phi_+, \phi_-\}$ vary with $\{\phi, \alpha\}$. In other words, they quantify the *linear response* of the leaflet cholesterol fractions to the external control parameters of cholesterol content and phospholipid abundance.

More generally, we can define susceptibilities that measure the response to changes in external control variables—as is routinely done in thermodynamics. For now, let us focus on responses to changing $\{\phi, \alpha\}$, which are most closely related to the question of asymmetry and ternary mixtures, even though we could also include $\{r_+, r_-\}$ into the mix.

The observables whose perturbation we will discuss in a bit more detail here are differential stress $\Delta\Sigma$ and bilayer torque $\mathscr{T}$, so let us define their α-related susceptibilities

$$\chi^{\Delta\Sigma}_{\alpha|\phi}(\phi,\ \alpha,\ r_+,\ r_-,\ T,\ N,\ J,\ \ldots) = -\left(\frac{\partial\Delta\Sigma}{\partial\alpha}\right)_{\phi,r_+,r_-,T,N,J,\ldots} \tag{13a}$$

$$\chi^{\mathscr{T}}_{\alpha|\phi}(\phi,\ \alpha,\ r_+,\ r_-,\ T,\ N,\ J,\ \ldots) = -\left(\frac{\partial\mathscr{T}}{\partial\alpha}\right)_{\phi,r_+,r_-,T,N,J,\ldots}, \tag{13b}$$

as well as their ϕ-related partners

$$\chi^{\Delta\Sigma}_{\phi|\alpha}(\phi,\ \alpha,\ r_+,\ r_-,\ T,\ N,\ J,\ \ldots) = \left(\frac{\partial\Delta\Sigma}{\partial\phi}\right)_{\alpha,r_+,r_-,T,N,J,\ldots} \tag{13c}$$

$$\chi^{\mathscr{T}}_{\phi|\alpha}(\phi,\ \alpha,\ r_+,\ r_-,\ T,\ N,\ J,\ \ldots) = \left(\frac{\partial \mathscr{T}}{\partial \phi}\right)_{\alpha,r_+,r_-,T,N,J,\ldots}. \tag{13d}$$

These would almost surely benefit from easier notations (at least this one is informative), as well as some intuitive names, since cumbersome attempts like "iso-compositional differential stressability" for $\chi^{\Delta\Sigma}_{\alpha|\phi}$ are not likely to catch on. Nevertheless, as linear response functions they satisfy the joint differential relation

$$\begin{pmatrix} \mathrm{d}\mathscr{T} \\ \mathrm{d}\Delta\Sigma \end{pmatrix} = \begin{pmatrix} -\chi^{\mathscr{T}}_{\alpha|\phi} & \chi^{\mathscr{T}}_{\phi|\alpha} \\ -\chi^{\Delta\Sigma}_{\alpha|\phi} & \chi^{\Delta\Sigma}_{\phi|\alpha} \end{pmatrix} \begin{pmatrix} \mathrm{d}\alpha \\ \mathrm{d}\phi \end{pmatrix}, \tag{14}$$

which also implies that they obey the obvious Maxwell relations

$$\frac{\partial}{\partial \alpha}\chi^{\mathscr{T}}_{\phi|\alpha} = -\frac{\partial}{\partial \phi}\chi^{\mathscr{T}}_{\alpha|\phi} \quad \text{and} \quad \frac{\partial}{\partial \alpha}\chi^{\Delta\Sigma}_{\phi|\alpha} = -\frac{\partial}{\partial \phi}\chi^{\Delta\Sigma}_{\alpha|\phi}. \tag{15}$$

While it might be difficult to determine these susceptibilities quantitatively in experiment, they will all have a noticeable impact on the response of vesicles, especially micron-scale giant unilamellar vesicles (GUVs), to the respective changes (see Section 3 below). At the very least, their signs would be straightforward to observe, and possibly even whether the response is "strong" or "mild". Of course, in simulations they are very accessible, and we will discuss some examples below.

Let us elaborate on the sign. We expect the susceptibilities with respect to changes in α to have a definite sign, because α itself has a definite sign baked into it. Recall that increasing α means that we will increase the phospholipid content in leaf$_+$ relative to leaf$_-$. This will put leaf$_+$ under compression and leaf$_-$ under tension. Since stress is the negative of pressure, this change will make the derivative $\partial\Delta\Sigma/\partial\alpha$ negative, and for that reason we propose the extra minus sign in the definition of $\chi^{\Delta\Sigma}_{\alpha|\phi}$, which would then render it positive. A similar argument explains the extra minus sign in the definition of $\chi^{\mathscr{T}}_{\alpha|\phi}$: pushing more phospholipids into leaf$_+$ gives rise to a torque that would drive "down-bending" (*i.e.*, leaf$_+$ will bulge "out"), and from eqn (5b) or (6) it is clear that this corresponds to a negative torque, which the extra minus sign then turns into a positive susceptibility.

The situation is slightly more subtle, though, because in the presence of cholesterol the membrane can relax the differential stress induced by a change in α by cholesterol relocation from the compressed to the tense leaflets. The same holds for the torque: its change, too, will be curtailed by cholesterol flip-flop. Unless this cholesterol relocation triggers a compositional instability (which might happen if we push one side into a phase coexistence region), we would expect it to reduce the changes in $\Delta\Sigma$ or $\mathscr{T}$ (compared to the cholesterol-free case), but not to actually flip the sign. This is not a rigorous argument, though.

Importantly, the response functions with respect to cholesterol, $\chi^{\Delta\Sigma}_{\phi|\alpha}$ and $\chi^{\mathscr{T}}_{\phi|\alpha}$, behave very differently. If we increase ϕ, this neither means that cholesterol will increase in a specific leaflet, nor does it distribute in equal amounts between the two leaflets (even though it will distribute such that ${\mu_{\mathrm{chol}}}^+ = {\mu_{\mathrm{chol}}}^-$ remains true). Consider the following illustrative scenarios:

(1) A cholesterol-free compositionally symmetric membrane is under some lipid abundance asymmetry $\alpha > 0$. If we add cholesterol, the chemical partitioning forces are equal, but the differential stress $\Delta\Sigma$ due to α will guide cholesterol preferentially into the leaflet whose tension is larger, thus reducing the magnitude of $\Delta\Sigma$. This will render $\chi_{\phi|\alpha}^{\Delta\Sigma}$ negative if the differential stress is positive and positive if the differential stress is negative.

(2) A compositionally asymmetric membrane is under vanishing differential stress. If we add cholesterol, there is no driving force coming from the stress, but the compositional difference will now bias cholesterol—away from the phase that already contains a higher mole fraction of cholesterol but also towards the side into which cholesterol partitions better (say, the more saturated one). The resulting change in differential stress—and hence the sign of $\chi_{\phi|\alpha}^{\Delta\Sigma}$—can be either positive or negative, depending on how these drivers pan out.

We have previously observed both of these cases in simpler binary systems.[77] Notice that the undetermined sign of $\chi_{\phi|\alpha}^{\Delta\Sigma}$, rather than being annoying, means that the mere *direction* of an effect will be informative about some conceivably difficult to ascertain membrane observables—such as, is there a cholesterol imbalance between the leaflets, or is there pre-existing differential stress. Experiments that change ϕ might thus be very informative.

We expect the situation for $\chi_{\phi|\alpha}^{\mathscr{T}}$ to be similar: the sign is not pre-determined and instead depends on the specifics of the underlying situation. It is unfortunately more difficult to make analogous arguments, because the response of cholesterol's inter-leaflet distribution to changes in torque are more subtle. The differential stress contribution to the torque, $\mathscr{T}_\Sigma = z_0\Delta\Sigma$ is of course the same, but the part of the torque due to lipid shape, $\mathscr{T}_J = -\kappa J_{0,\mathrm{b}}$, requires an answer to the question how $J_{0,\mathrm{b}}$ changes with cholesterol concentration—a famously tricky problem that might not be simply captured by a linear combination of some "bare" intrinsic lipid curvatures.[78–80] It seems to us that a workable rule of thumb is that at sufficiently large cholesterol content, any further increase will tend to reduce the magnitude of torque. One way of seeing this is that in the "theorist's limit" of $\phi \to 1$ we reach a perfectly symmetric membrane, which hence has zero torque (and also zero differential stress). The "ultimate" direction into which $\mathscr{T}$ must change as ϕ increases is hence clear, even though at intermediate concentrations additional drivers may well complicate matters. (Of course, the same argument can be made for the differential stress.)

2.4 Zero torque foliations

After identifying a set of variables that specify the thermodynamic state, all other observables will be functions of those. For instance, if we initially pick two desired leaflet compositions $\{r_-, \phi_-\}$ and $\{r_+, \phi_+\}$, then the $\{\phi, \alpha\}$ values needed to realize them are functions of those. Likewise, observables such as differential stress, $\Delta\Sigma$, or bilayer torque, $\mathscr{T}$, are now set, and neither of them is generally zero. For the differential stress that is intuitively easy to see: two arbitrarily chosen leaflet compositions generally do not coexist in their cholesterol content across the leaflets. To prevent a net flux from one leaflet to the other, we need to set up a stress difference that opposes such a translocation.

Let us follow this idea a bit further. Pick a composition $\{r_-, \phi_-\}$ for leaf$_-$ (the magenta point in Fig. 4) and try to arrange for chemical coexistence with

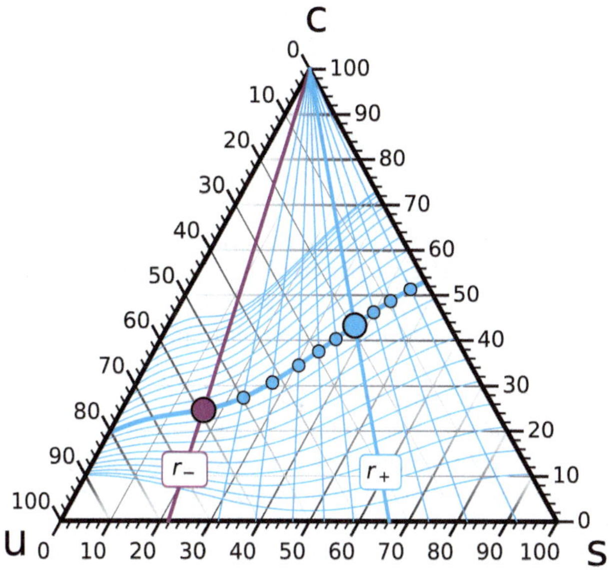

Fig. 4 Schematic illustration of a zero torque foliation. A given composition $\{r_-, \phi_-\}$ in leaf$_-$ (magenta point) can be made to coexist with a wide range of $\{r_+, \phi_+\}$ compositions in leaf$_+$, but at a specified r_+ value (say, the bold cyan line) only a small number of ϕ_+ values (likely only a single one, the big cyan point) will additionally coexist at zero torque $\mathscr{T}$. The collection of all such points over a range of r_+ values forms a one-dimensional slice of zero-torque states through the ternary diagram. The family of all such curves, parametrized by different ϕ_- values on the r_- line of leaf$_-$, constitute a foliation of the ternary phase diagram. (The specific one shown here is merely for illustration; it is not based on a particular free energy model.)

compositions in leaf$_+$ that have a fixed saturation ratio r_+ but different values of ϕ_+ (points on the bold cyan r_+ line in Fig. 4). Each such point will require some specific $\{\phi, \alpha\}$ combination, and once we have found it, it will result in a bilayer with some differential stress, and some torque, whose value depends on ϕ_+:

$$\Delta\Sigma(\phi_+) = \Delta\Sigma(\phi_+ | r_-, \phi_-, r_+), \tag{16a}$$

$$\mathscr{T}(\phi_+) = \mathscr{T}(\phi_+ | r_-, \phi_-, r_+). \tag{16b}$$

Consider for instance the torque: the function $\mathscr{T}(\phi_+)$ varies with ϕ_+ in some conceivably complicated and yet-to-be-determined way, and for some value of ϕ_+ it might vanish (the cyan point on the bold cyan r_+ line in Fig. 4). At this special point the two ternary compositions in the two leaflets do not just chemically coexist; they coexist at zero torque, which constitutes an additional mechanical condition. Constraint counting of this type does not answer the question how many solutions $\mathscr{T}(\phi_+) = 0$ has. However, in light of the underlying physical situation we expect this equation to identify a small number of points along the r_+ curve, likely just a single one, that coexist with the composition $\{r_-, \phi_-\}$ at vanishing bilayer torque.

Let us now also vary the r_+ line, and for each one find the special zero-torque-point(s) on it (a few are illustrated as smaller cyan points in Fig. 4). Their collection forms a one-dimensional curve in the ternary phase diagram, namely, the locus of all leaflet compositions in leaf$_+$ that can coexist at vanishing torque with the given point $\{r_-, \phi_-\}$ in leaf$_-$. This curve has to pass through the $\{r_-, \phi_-\}$ point, because this gives rise to a symmetric bilayer, and we know that such a system would also have zero torque.

Finally, we can construct such zero-torque-curves for any cholesterol fraction ϕ_- on the r_- line. This yields a foliation of the ternary phase diagram into zero torque coexistence curves: a specified composition $\{r_-, \phi_-\}$ in leaf$_-$ can in principle coexist with any composition $\{r_+, \phi_+\}$ in leaf$_+$ (a set of dimension two), but only a one-dimensional subset $\phi_+(r_+|r_-, \phi_-)$ coexists at zero torque.

We hasten to clarify that the zero-torque-curves we constructed depend on the points $\{r_-, \phi_-\}$ in leaf$_-$ to which we pinned them. We are not claiming that any two points selected on a given curve will coexist with one another at zero torque—this is a stronger condition that does not follow from the simple counting argument we have presented.

3 Zero torque states and their perturbations

We concluded the previous section by specifically discussing a coexistence between the two leaflets that ensures a vanishing torque. Why is this interesting?

3.1 Stability of giant unilamellar vesicles

There is no reason why compositionally asymmetric membranes should be flat: we have broken the reflection symmetry of a membrane (with respect to its midplane) in terms of chemical identity, so we should expect that membrane shape itself also fails to obey that symmetry: it will generally be curved—meaning, it will be driven into a curved state by a nonzero torque $\mathscr{T}$. Let us estimate how much curvature we should expect. If the membrane is free of any additional stresses (especially differential stress) then its spontaneous bilayer curvature should be the difference of the individual spontaneous leaflet curvatures $J_{0,\mathrm{m}\pm}$, weighted by their respective monolayer bending rigidities $\kappa_{\mathrm{m}\pm}$:

$$J_{0,\mathrm{b}} = \frac{\kappa_{\mathrm{m}+}J_{0,\mathrm{m}+} - \kappa_{\mathrm{m}-}J_{0,\mathrm{m}-}}{\kappa_{\mathrm{m}+} + \kappa_{\mathrm{m}-}} \approx \frac{1}{2}(J_{0,\mathrm{m}+} - J_{0,\mathrm{m}-}). \tag{17}$$

Considering the $J_{0,\mathrm{m}}$ values of typical lipids,[81] equilibrium radii $R_{0,\mathrm{b}} = 2/J_{0,\mathrm{b}}$ of asymmetric vesicles are usually on the order of a few tens of nanometers (the size of small unilamellar vesicles, SUVs), unless we pick two lipid species that just happen to have very similar spontaneous curvatures. However, many techniques have been developed to create asymmetric giant unilamellar vesicles (GUVs),[62] whose curvature radii are easily two orders of magnitude larger than those of SUVs. Why are they stable against submicroscopic tubulation?

A possible answer is that despite the large spontaneous bilayer curvature $J_{0,\mathrm{b}}$ originating from the lipid shape asymmetry, the net torque is actually very small, because a counter-torque arising from differential stress cancels the spontaneous curvature torque.[63,64,66,67] As we argued following eqn (6), this would mean that an asymmetric flat membrane—and hence, essentially any asymmetric GUV—has to experience a differential stress of a few mN m^{-1}.

Conversely, that asymmetric GUVs should have $\mathscr{T} \simeq 0$ implies that the zero torque foliations discussed in Section 2.4 become constraints on the compositions that may coexist across a GUV's leaflet: once the composition on one side is fixed, the composition on the other is limited to a one-dimensional slice through the ternary phase diagram.

As an example: we propose that the $\mathscr{T} = 0$ constraint will fix the abundance asymmetry of GUVs during their creation. The precise mechanism is likely complicated and will depend on the protocol for making asymmetric GUVs in the first place, but our stability argument sidesteps such details and simply notes the following: if the saturation ratios $\{r_+, r_-\}$ are set, then all that needs to be determined are the $\{\phi_+, \phi_-\}$ values, *via* suitable choices of $\{\phi, \alpha\}$. However, the condition $\mathscr{T} = 0$ selects a one-dimensional subset from those. For instance, if the asymmetric creation protocol somehow fixes ϕ, then the condition $\mathscr{T} = 0$ sets α, because any other choice of α would not produce a GUV that is mechanically stable against tubulation. Almost the same argument holds if the creation process instead fixes ϕ_-. The very existence of a stable asymmetric GUV means that we must have achieved torque balance, and whatever compositional arrangements materialized, there is only one corresponding α that will do so.

3.2 Perturbing cholesterol content and abundance

Since perturbations of a vesicle's thermodynamic state need not keep the two leaflet compositions on the same zero-torque-foliation, we expect that they trigger shape deformations—anywhere from mild ones such as sphere → prolate → oblate → stomatocyte,[82,83] up to the formation of membrane tubules.[84–87] Let us discuss two specific examples, related to changes of ϕ and α.

3.2.1 Changing cholesterol content. Consider two leaflet compositions $\{r_-, \phi_-\}$ and $\{r_+, \phi_+\}$ coexisting in a mechanically stable asymmetric GUV—such as the filled magenta and cyan circles sharing the same bold cyan zero-torque foliation curve in Fig. 5. Let us now change the cholesterol content of the membrane. Experimentally, this is commonly achieved with the help of an exchange agent such as methyl-β-cyclodextrin (MβCD).[88–90] Briefly, cyclodextrins are water soluble cyclic oligosaccharides which posses hydrophobic cavities that can transport small hydrophobic molecules across aqueous environments. If we for instance expose GUVs to “empty” MβCD, it will extract cholesterol from the outer leaflet it has access to, but since cholesterol molecules in both leaflets are in chemical equilibrium, we effectively remove them from both leaflets—meaning, we lower ϕ, to an extent dependent on the details of MβCD exposure. In practice, the situation is a bit more complicated, though, because MβCD can also exchange phospholipids. For instance, Rahimi *et al.*[91] have conducted experiments like those we propose here, aiming to add cholesterol to vesicles *via* pre-loaded MβCD, and found that discharged MβCD can in turn remove phospholipids from the outer leaflets of their GUVs. To avoid this, one should restrict to MβCD concentrations too low to noticeably extract phospholipids.[92]

Lowering ϕ will reduce the cholesterol leaflet concentrations $\phi_\pm$, even though not necessarily by the same amount. More importantly, we generally have no guarantee that the two new points will still share a zero-torque foliation curve.

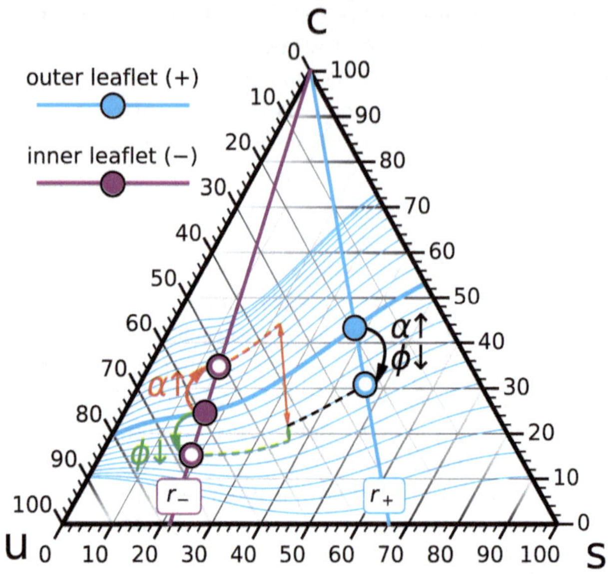

Fig. 5 Perturbation of a ternary asymmetric GUV initially set up such that its inner composition $\{r_-, \phi_-\}$ (big magenta dot) and outer composition $\{r_+, \phi_+\}$ (big cyan dot) chemically coexist at $\mathscr{T} = 0$. The two independent perturbations which either decrease ϕ or increase α will have a similar effect on leaf$_+$: a reduction of ϕ_+. But their effects on leaf$_-$ are opposite: depleting cholesterol will reduce ϕ_- while increasing α will increase ϕ_-. The extent to which this moves points that originally shared the same zero-torque foliation onto different foliation curves is very different.

Stated in terms of the response functions defined in Section 2.3, there is no reason to believe the susceptibility $\chi^{\mathscr{T}}_{\phi|\alpha}$ vanishes (we will see it might, but only for very special conditions). As a consequence, we expect that the GUV would want to deform—if it can. Deflated GUVs should therefore have a tendency to assume new shapes, and since the shape diagrams for deflated vesicles are well understood,[82,83] the nature of these deformations will alert us at the very least to the sign of $\chi^{\mathscr{T}}_{\phi|\alpha}$ under the present conditions.

3.2.2 Changing phospholipid abundance. Consider the same two coexisting points we just studied in Section 3.2.1, but instead of depleting the system of cholesterol, we now add a small amount of phospholipids to the outer side (to be specific: leaf$_+$) of the GUV. This can also be accomplished with MβCD, namely, by pre-loading it with the lipids we wish to deliver.[87] However, since our vesicles contain both phospholipids and cholesterol, and MβCD can transport both, it is advantageous to change to α-cyclodextrins: their smaller hydrophobic cavity (6 glucopyranoside units instead of 7) is too narrow to fit cholesterol but can still host individual lipid tails. Specifically, hydroxylpropyl-α-cyclodextrin (HPαCD)[93,94] and methyl-α-cyclodextrin (MαCD)[95–97] have proven to be very suitable in exchanging phospholipid content without touching cholesterol, and MαCD has been shown to be "lipid-loadable".[95]

To simplify the discussion, we will assume that the saturation ratio r_+ on the outer leaflet of the GUV remains fixed during the lipid addition process. This can be done by loading the cyclodextrins with a ratio of u- and s-lipids that reflects r_+

(possibly adjusted to account for different complexation strengths). Alternatively, we could appeal to the fact that percent-level changes in α require percent-level changes in L_+ (in fact, $\Delta L_+/L = \Delta\alpha/(1 - \alpha)$), but these do generally not change the saturation ratio r_+ significantly, unless it is close to 0.

As Fig. 5 indicates, the change in abundance α has opposite effects on the leaflets. Adding phospholipids to the outer leaflet will put it under compression and hence squeeze cholesterol into the inner leaflet, which had been put under tension due to the change of α. Unlike in the cholesterol depletion scenario from Section 3.2.1, where the compositions moved into the same directions and so could potentially remain quite close to $\mathcal{T} = 0$, this is not an option when we tune α: ϕ_+ moves down while ϕ_- moves up, and so the membrane will definitely develop a net torque. While it is difficult to directly compare the magnitudes of effects driven by ϕ and α, since these two variables (albeit dimensionless) measure different things, we could compare the change in $\mathcal{T}$ under conditions that give rise to the same change in ϕ_+, as is illustrated in Fig. 5. It appears evident that such a change in abundance would perturb the torque balance more strongly. Rather loosely speaking, this might translate to the statement that ternary GUVs are easier to mechanically perturb with phospholipids than with cholesterol.

4 Computational illustrations

Many of the connections we have described so far cannot (yet) be checked experimentally, but we are convinced that the rapidly growing sophistication of experimental techniques will change this in the near future. Until then, it will be useful to employ computational techniques to not just check the theoretical picture but also identify unexpected phenomena that warrant further study.

4.1 Coarse-graining as the arena for modeling

The reality of finitely available computing time puts limits on how big a system we can simulate, and for how long. Coarse-grained (CG) simulations are a popular way to overcome this constraint: develop a force field at a lower level of resolution that strives to capture physics at longer time scales and larger length scales by summarily encoding finer-scale physics in effective interactions between CG degrees of freedom. While pragmatically expedient, we would like to add that this is also good physics: a universally intriguing property of nature is that large-scale physics almost always depends on only a small number of effective parameters, with much of the intricate finer-scale detail having been rendered irrelevant. If so, it behooves us to leave such irrelevant detail out of our models, too. Notice that if we happen to ignore too much and our model no longer works, we thereby learn a deep lesson about what constitutes relevant physics.

4.1.1 The unbearable slowness of diffusion. Simulating ternary mixtures is challenging for a number of reasons, one being statistical sampling. Curvature elastic membranes are known to sample phase space slowly: their bending mode relaxation rate is $r_\kappa = \kappa q^3/4\eta$, with q being the wave vector of the mode and η the viscosity of the embedding solvent—usually water, but we could also picture a more viscous intracellular environment.

Recall, though, that (i) in particle-based simulations we are (for better or worse) not interested in the long scales but the short ones and (ii) we will also need to sample compositional fluctuations, which relax on the diffusive scale $r_D = Dq^2$, with D being the lipid diffusion constant. Comparing q^3 *vs.* q^2 shows that diffusive modes will actually relax even slower than curvature modes at sufficiently small scales. Where is the cross-over? From $\kappa q_\times{}^3/4\eta = Dq_\times{}^2$ we get $\lambda_\times = 2\pi/q_\times = \pi\kappa/2D\eta$, and using typical values $\kappa \simeq 30k_BT$, $D \simeq 5\ \mu m^2\ s^{-1}$ and $\eta = 10^{-3}$ Pa s yields $\lambda_\times \simeq 40\ \mu m$. Hence, at any computational scale that bothers to actually represent lipids, we are deeply in the regime where compositional relaxation is the bottleneck.

Since the largest wavelength at which we need to sample in a simulation is the box length L, the characteristic relaxation time is $r_D{}^{-1} = L^2/4\pi^2D$. Taking $L \simeq 50$ nm (chosen to host raft-like heterogeneities at the few tens of nanometer scale), we find $r_D{}^{-1} \simeq 13\ \mu s$. While doable at the atomistic level with modern high performance computing, this remains a serious challenge and is not a feasible means to scan parameter space.

4.1.2 The course grained model we employ. We will use a highly coarse-grained solvent-free model recently developed by us.[98] It fine-tunes some key control parameters of an earlier model,[99] which itself is based on a widely-applied CG model that, however, cannot represent some of the key physics needed once asymmetry enters the stage.[100–102] Our CG model captures the notion of saturated and unsaturated lipids that differ in their specific area and the usual order parameters, a smaller rapidly flipping species of cholesterol, stretching- and bending-elasticity, spontaneous lipid curvature, and lipid diffusion; for details, see ref. 98. Crucially, this model can represent ℓ_o/ℓ_d phase coexistence in ternary mixtures, as illustrated in Fig. 6.

4.1.3 Simulation details. The CG model has its own intrinsic length-, energy-, and time-scales: σ, ε, and τ, respectively. The length scale follows

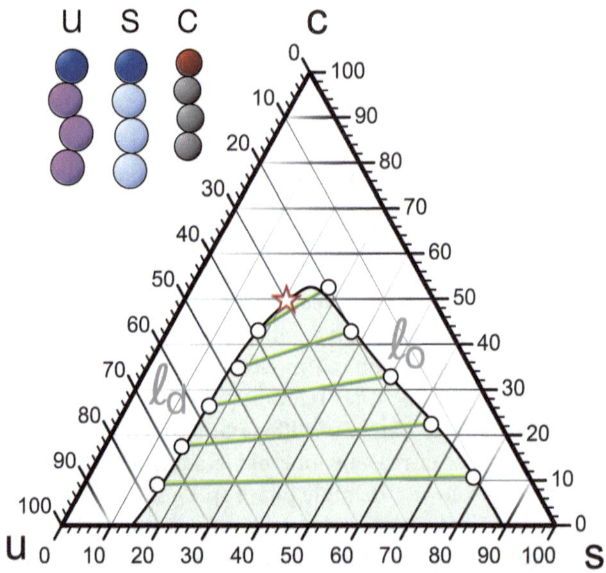

Fig. 6 Phase diagram of a ternary mixture in the Cooke lipid model of a lipid mixture, showing ℓ_o/ℓ_d coexistence and (the approximate location of) its critical point (adapted from ref. 98).

straightforwardly by matching physical dimensions and leads to $\sigma \simeq 0.75$ nm. The energy scale is fixed by the temperature: we run our simulations at $k_BT = 1.4\varepsilon$, which sets ε if we assume T corresponds to the temperature at which the experiment is run, say 310 K. This finalizes the translation for various units: with $k_BT = 310\ \text{K} \times 1.38 \times 10^{-23}\ \text{J K}^{-1} = 4.28$ pN nm we get $\varepsilon = 3.06$ pN nm, $\varepsilon/\sigma = 4.08$ pN, and $\varepsilon/\sigma^2 = 5.43$ pN nm^{-1} for the units of energy, torque density, and stress (or surface tension), respectively.

The time scale is more subtle. In principle, CG degrees of freedom also have a mass m, and this defines a time scale $\tau = \tau_{\text{bare}} = \sigma\sqrt{m/\varepsilon}$, but this scale only matters for the calculation of instantaneous dynamical quantities, such as the kinetic energy. Longer time scale dynamics—diffusion, bending mode or chain relaxations, lipid flip-flop, *etc.*—is not well described by τ_{bare}, because CG models are virtually always tuned to reproduce thermodynamic equilibrium properties, not to also rescue the dynamics. In fact, the strongly sped-up dynamics resulting from a much smoother free energy landscape is a major redeeming quality of coarse-graining.

The way to translate CG dynamics into real world units is then to agree on a specific dynamical process—say, diffusion—and interpret the CG unit τ such that CG simulations of that process quantitatively map to those in the real world. For example, when a CG lipid diffusion constant is $D = 0.01\sigma^2/\tau$ (as measured for the ℓ_d phase in our model[98]) and $D = 5\ \mu\text{m}^2\ \text{s}^{-1}$ in the lab,[103] setting those equal (and recalling $\sigma = 0.75$ nm) defines $\tau \approx 1$ ns. This for instance shows that diffusive relaxation over a 50 nm $= 67\sigma$ scale happens over the time scale 11 000τ, which in our case takes about 20 h on a 32 core node.

We ran our simulations using the ESPResSo package,[104] using an integration time step $\delta t = 0.005\tau$. We reach the canonical ensemble *via* a standard Langevin thermostat,[105] with a friction constant $\gamma = 1m/\tau$. To realize membranes under zero lateral tension we employed semi-anisotropic boundary conditions using a barostat of Kolb/Dünweg[106] type with a box mass $Q = 0.01m/\sigma^4$ and a friction constant $\gamma_Q = 2 \times 10^{-4}m/\sigma^4$. Our simulations typically contained 2048 lipids and ran between 80 000τ and 100 000τ, with the first 20 000τ being used for thermalization.

4.2 Coexisting leaflets at very different saturation ratios

We will start by exploring coexisting leaflets that differ markedly in their saturation ratio, which creates sizable driving forces for cholesterol from one into the other. Specifically, we picked

$$r_- = 1/8 = 0.125, \tag{18a}$$

$$r_+ = 5/6 \approx 0.833. \tag{18b}$$

The low-saturation r_- line in leaf$_-$ bypasses the coexistence region, while the more ordered high-saturation r_+ line in leaf$_+$ stays outside it for $\phi \gtrsim 20\%$. Given how far the coexistence region reaches on the *s*-side for low cholesterol content, it is difficult to entirely avoid it, unless we move very close to the triangle's *sc*-side, which might bring us uncomfortably close to some not fully resolved gel complications[98] in the *s*-corner of the diagram. Observe that with this choice, cholesterol chemically prefers to partition into leaf$_+$.

4.2.1 The torque surface. We have simulated systems with $N = 2048$ lipids, picking a combination of $\{\phi, \alpha\}$ states within the range $\phi \in [0, 50\%]$ and $\alpha \in [-5\%, 5\%]$, let the systems find equilibrium during approximately $20\,000\tau$, and sampled for typically another $80\,000\tau$ during which we measured a variety of observables, in particular the cholesterol distribution and asymmetry, differential stress, and torque. We found that within this domain the torque $\mathcal{T}(\phi,\alpha)$ can be represented remarkably well by a quadratic (reduced $\chi^2 = 0.11$), so we empirically fit it to

$$\mathcal{T}(\phi,\alpha) = c_0 + c_{1,\phi}\phi + c_{1,\alpha}\alpha + c_{2,\phi\phi}\phi^2 + c_{2,\alpha\alpha}\alpha^2 + c_{2,\phi\alpha}\phi\alpha. \tag{19}$$

We will refer to this as the "torque surface".

Fig. 7 shows a contour plot of that surface. For sufficiently negative values of the abundance α, *i.e.* when leaf$_+$ becomes increasingly depleted of phospholipids, the torque is positive (*i.e.*, the membrane would want to "curl up" if not prevented by the periodic boundary conditions; see Fig. 2 for a clarification of the torque's sign). Conversely, if leaf$_+$ is sufficiently overcrowded, the torque becomes negative. In between the torque crosses zero at a location dependent on the overall cholesterol content ϕ. This zero-torque-curve describes the possible states of mechanically stable GUVs, which due to their essential flatness cannot harbor any significant torque before deforming or even tubulating.

In the absence of cholesterol, $\phi = 0$, when the abundance asymmetry vanishes as well, $\alpha = 0$, but the torque is nevertheless not zero but slightly positive. This happens because the system is still not symmetric; most notably, leaf$_+$ contains an approximately six times larger fraction of saturated lipids, which have (at least in pure phases) an approximately 25% smaller area per lipid.[98] Since for $\alpha = 0$ the number of phospholipids is the same in both leaflets, the relaxed leaflet area in leaf$_+$ is smaller than in leaf$_-$, giving rise to a positive differential stress, whose torque contribution $\mathcal{T}_\Sigma = z_0 \Delta\Sigma$ wants to bend the bilayer up (see Fig. 2b).

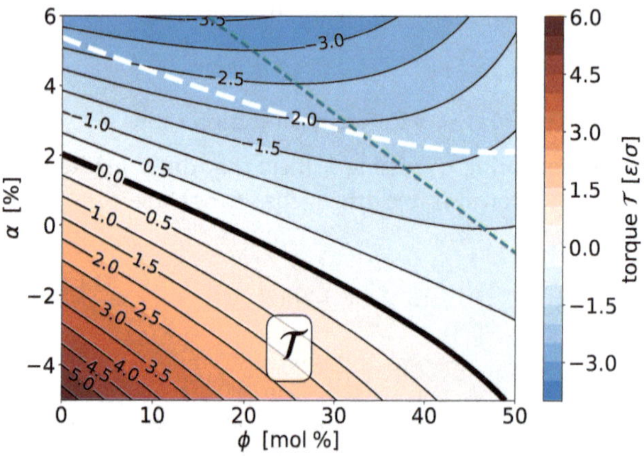

Fig. 7 Contour plot of the (empirically fitted) torque surface $\mathcal{T}(\phi,\alpha)$ for the saturation ratios $r_- = 0.125$ and $r_+ = 0.833$. The bold solid curve is the location where the torque vanishes and hence GUVs would be mechanically stable, the thin teal dashed curve is the nullcline at which $\partial\mathcal{T}/\partial\phi = \chi^{\mathcal{T}}_{\phi|\alpha} = 0$, *i.e.*, where the torque does not change upon small variations of overall cholesterol content. The bold white dashed curve is the locus of states for which $\Delta\Sigma = 0$ (see also Fig. 10).

Adding cholesterol to this state will preferentially recruit it into leaf$_+$ to release the area strain, an effect that will become even stronger when we lower α. Independent of these mechanical considerations, the higher saturation ratio r_+ in leaf$_+$ will also favor the partitioning of cholesterol into it. Hence, there are two reasons that favor ϕ_+ growing faster than ϕ_-, and as a consequence, the torque decreases. Notice, though, that as we increase the abundance asymmetry α, the stress-derived torque $\mathscr{T}_\Sigma$ weakens and at some point reverses, as leaf$_-$ becomes depleted. Where will the cholesterol go, now that the two drivers compete? As the contour lines in Fig. 7 show, the trend is indeed non-monotonic: initially, $\mathscr{T}$ still decreases, since chemical partitioning bias still drives more cholesterol into leaf$_+$. But at some point it reverses, as the increasing ϕ_+ value weakens further recruitment of even more cholesterol and entropy favors a more even distribution, which together ends up reducing the magnitude of the torque. The teal dashed line in Fig. 7 marks the location of that reversal: at it, the contour lines are horizontal and $\chi^{\mathscr{T}}_{\phi|\alpha} = 0$ (which is exactly the nullcline of $\nabla\mathscr{T}$ for the variable α).

4.2.2 The response function $\chi^{\mathscr{T}}_{\phi|\alpha}$. If we cut the torque surface at some fixed value of α, we can single out the dependence of torque on ϕ: $\mathscr{T}(\phi|\alpha)$. The derivative of this function with respect to ϕ is exactly the susceptibility $\chi^{\mathscr{T}}_{\phi|\alpha}$ defined in eqn (13d), which measures how the torque changes with cholesterol content at a fixed value of α. This response function is shown in Fig. 8—itself as a function of ϕ and parametrized for a set of α-values. Due to our simple quadratic representation (19) of $\mathscr{T}(\phi,\alpha)$, we find simple lines. For most of the values the response function is negative, showing that addition of cholesterol tends to reduce the torque. However, this reduction weakens as cholesterol increases, and the $\chi^{\mathscr{T}}_{\phi|\alpha}(\phi)$ lines cross zero and become positive. This happens earlier when the abundance asymmetry is already larger. These zero crossings reflect the "trend reversal" we discussed in the previous section, *i.e.*, the nullcline included in Fig. 7.

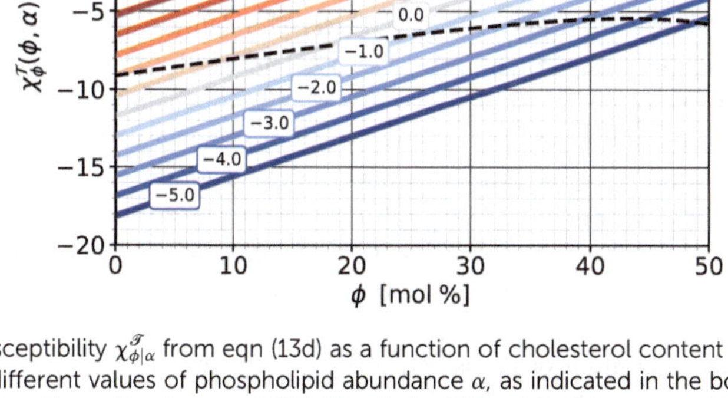

Fig. 8 Susceptibility $\chi^{\mathscr{T}}_{\phi|\alpha}$ from eqn (13d) as a function of cholesterol content ϕ, parametrized by different values of phospholipid abundance α, as indicated in the boxed labels, and the saturation ratios from eqn (18). The dashed line shows the susceptibility for the special ϕ-dependent α values at which $\mathscr{T} = 0$ (*i.e.*, on the zero-torque contour in Fig. 7).

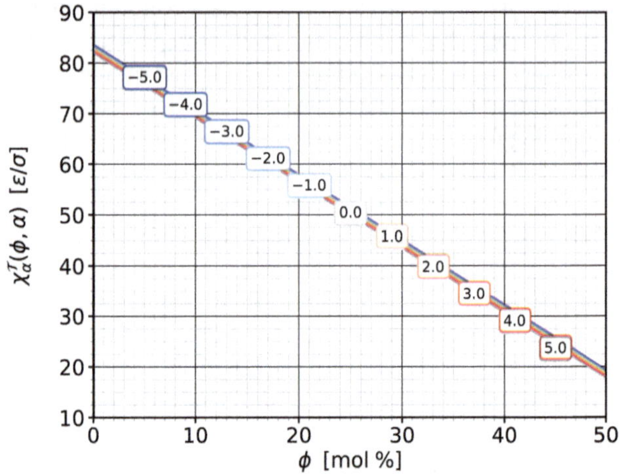

Fig. 9 Susceptibility $\chi^{\mathcal{T}}_{\alpha|\phi}$ from eqn (13b) as a function of cholesterol content ϕ, parametrized by different values of phospholipid abundance α, as indicated in the boxed labels. The lines almost coincide, with smaller α values having ever so slightly larger susceptibilities.

4.2.3 The response function $\chi^{\mathcal{T}}_{\alpha|\phi}$. We can also cut the torque surface along constant-ϕ-slices, which gives us the complementary function $\mathcal{T}(\alpha|\phi)$, whose derivative with respect to α leads to the other torque-related response function, $\chi^{\mathcal{T}}_{\alpha|\phi}$ from eqn (13b). This susceptibility—again as a function of ϕ and for a set of different α is shown in Fig. 9.

Several observations are notable here. First, the susceptibilities hardly depend on α. This means that $\partial\chi^{\mathcal{T}}_{\alpha|\phi}/\partial\alpha = -\partial^2\mathcal{T}/\partial\alpha^2 \approx 0$: the cuts of the torque surface along constant ϕ are essentially straight lines. In Fig. 7 this can be recognized by the fact that the contour curves intersect any line with a fixed ϕ value at very evenly spaced points. These lie closer together for smaller ϕ, leading to larger slopes and higher $\chi^{\mathcal{T}}_{\alpha|\phi}$ values. Physically this means that a change $\Delta\mathcal{T}$ in membrane torque due to a change $\Delta\alpha$ in phospholipid abundance does not depend on the pre-existing abundance α. A change $\Delta\Delta A$ of the area excess ΔA between the two leaflets always changes the curvature in the same way—and the parallel surface theorem[107,108] agrees: $\Delta\Delta A = 2Az_0\Delta J$, provided that adding or removing phospholipids always adds or removes the same area.

Second, the susceptibility $\chi^{\mathcal{T}}_{\alpha|\phi}$ is positive. The definite sign (unlike what we have seen for the complementary partner $\chi^{\mathcal{T}}_{\phi|\alpha}$) derives from the expectation that adding phospholipids on one side invariably bends the membrane away from that side. Since adding lipids to leaf$_+$ increases α, but a downward bending counts as a negative torque (*cf.* again Fig. 2), we have added an additional minus sign to the definition (13b) of $\chi^{\mathcal{T}}_{\alpha|\phi}$ to arrange for a convenient positive sign. This mimics definitions such as $\kappa_T = -(\partial V/\partial P)_T/V$ for the isothermal compressibility (whose sign is fixed by a rigorous thermodynamic argument, though, not merely a strong expectation).

Third, the magnitude of $\chi^{\mathcal{T}}_{\alpha|\phi}$ is noticeably larger than that of $\chi^{\mathcal{T}}_{\phi|\alpha}$. Small changes of the phospholipid abundance α change the torque more strongly than comparably small changes of the cholesterol content ϕ. As mentioned above, the reason is that changes in α have a sign built into it: all phospholipids are added or removed from the same leaflet, while cholesterol addition or removal is shared between the two leaflets (besides the fact that cholesterol molecules are also smaller).

4.2.4 The differential stress surface. Just as we can measure the torque $\mathscr{T}$ as a function of the two thermodynamic variables ϕ and α, we can do the same with the differential stress $\Delta\Sigma$. The procedure mirrors the one for the torque surface from Section 4.2.1; in particular, we again find that a quadratic fit captures the simulated data very well. Fig. 10 shows the result *via* a contour plot, amended again by the nullcline at which $\partial\Delta\Sigma/\partial\phi = \chi^{\Delta\Sigma}_{\phi|\alpha} = 0$, a bold contour curve to highlight the place where the differential stress vanishes, and for comparison also the location of the curve where the torque vanishes. Let us summarize several notable points:

(1) As we deplete phospholipids from leaf$_+$ (*i.e.*, reduce α), the differential stress increases.

(2) The variation with ϕ is again a bit more subtle, since for sufficiently large α an initial reduction in $\Delta\Sigma$, driven by preferential partitioning, can reverse direction for sufficiently large ϕ, when stress and entropy take over.

(3) Torque and differential stress never vanish at the same time: zero torque states have a positive differential stress, while zero differential stress states have a negative torque.

(4) The $\mathscr{T} = 0$ curve lies very close to the $\Delta\Sigma = 1.0\varepsilon/\sigma^2$ contour line, which shows that all possible zero torque states have (within about ±7%) the same differential stress.

(5) Just as for the torque, we could also plot the two susceptibilities $\chi^{\Delta\Sigma}_{\phi|\alpha}$ and $\chi^{\Delta\Sigma}_{\alpha|\phi}$, but they behave qualitatively very similarly to their torque counterparts. Briefly, the $\chi^{\Delta\Sigma}_{\phi|\alpha}$ are lines with positive slope that at the nullcline transition from negative to positive values, and this happens earlier for larger α. The $\chi^{\Delta\Sigma}_{\alpha|\phi}$ are

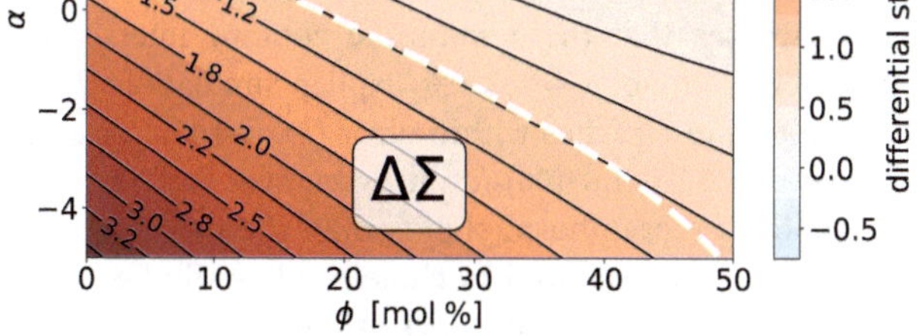

Fig. 10 Contour plot of the (empirically fitted) differential stress surface $\Delta\Sigma(\phi, \alpha)$ for the saturation ratios $r_- = 0.125$ and $r_+ = 0.833$. The bold solid curve is the location where the differential stress vanishes, the thin teal dashed curve is the nullcline at which $\partial\Delta\Sigma/\partial\phi = \chi^{\Delta\Sigma}_{\phi|\alpha} = 0$, *i.e.*, where the differential stress does not change upon small variations of overall cholesterol content. The bold white dashed curve is the locus of states for which $\mathscr{T} = 0$ (see also Fig. 7).

positive, generally larger in magnitude, decreasing with ϕ, and again hardly dependent on α.

4.2.5 Leaflet spontaneous curvature and its torque. Let us specialize eqn (5b), which links torque and stress, to a flat state—as relevant to our simulations:

$$\kappa J_{0,\mathrm{b}} = z_0 \Delta\Sigma - \mathscr{T} \text{ (flat membrane).} \tag{20}$$

Since we now have both the torque and the differential stress surface available, we can obtain the lipid-shape affiliated spontaneous bilayer curvature torque $\kappa J_{0,\mathrm{b}}$ for a wide range of conditions, assuming we know z_0. Taking $z_0 \simeq 2\sigma$ for our model, we find that $\kappa J_{0,\mathrm{b}}$ varies fairly little over the explored parameter range (mostly between $1.5\varepsilon/\sigma$ and $2.5\varepsilon/\sigma$). This is surprising, since we would expect that changing cholesterol content or compressing/stretching a leaflet affects the conformational ensemble of lipids, and hence their preferred curvature. Clearly, the extent to which this happens will also depend on details of the lipid model, and this question should be revisited with different models, especially those at a more refined resolution.

Let us map this finding to realistic units: recalling that $\sigma \approx 0.75$ nm and $k_{\mathrm{B}}T = 1.4\varepsilon$, we find $z_0 \approx 1.5$ nm and $\kappa J_{0,\mathrm{b}} \approx 2\varepsilon/\sigma \approx 1.9k_{\mathrm{B}}T/\mathrm{nm}$. If we recall that the bending rigidity of our CG membranes is around $\kappa \approx 30k_{\mathrm{B}}T$,[99] which is similar to real lipid membranes, the lipid torque translates to a spontaneous bilayer curvature of $J_{0,\mathrm{b}} \approx 0.063\ \mathrm{nm}^{-1}$. This corresponds to an equilibrium vesicle radius $R_0 = 2/J_{0,\mathrm{b}} \approx 32$ nm, in line with the typical expectations we have outlined in Section 3.1. This reiterates that unless differential stress cancels the torque associated with lipid shape, these systems cannot exist as stable GUVs. But it also shows that our fairly simple CG model reproduces the orders of magnitude of some of these effects quite well.

4.2.6 Tentative application to biological systems. The exoplasmic leaflet of cell membranes is significantly more saturated than the cytosolic one; for instance, Lorent *et al.*[50] show that cytosolic phospholipids in the human red blood cell membrane have about twice as many double bonds as their exoplasmic counterparts. Their data and models also suggest $r_+/r_- \approx 5.5$—not too far off from our current example $r_- = 0.125$ and $r_+ = 0.833$, which yields the slightly larger contrast $r_+/r_- \approx 6.7$. Let us hence take these saturation ratios as a crude proxy for a plasma membrane (with leaf$_+$ being the outer one) and add 40 mol% cholesterol to match the physiological situation. What type of membrane do we get, if we insist on an overall torque-free state?

From Fig. 7 we see that the zero-torque contour intersects $\phi = 40\%$ at a phospholipid abundance of $\alpha \approx -3.1\%$. Explicit simulations at this state point confirm that the torque vanishes within error ($\mathscr{T} = 0.12(12)\varepsilon/\sigma$) but the differential stress does not: $\Delta\Sigma = 1.01(06)\varepsilon/\sigma^2$, in agreement with Fig. 10. The cytosolic leaflet hence contains more phospholipids than the exoplasmic one, an abundance asymmetry partially balanced by cholesterol: we find $\phi_+ = 45\%$ and $\phi_- = 34\%$, showing that about 57% of all cholesterol is in leaf$_+$.

That the exoplasmic leaflet contains fewer phospholipids is a subtle balance between several competing factors: at first one might think that a leaflet richer in more saturated lipids, which have a smaller specific area, should contain more of those lipids. This is indeed true when we remove all cholesterol: as we have seen, the zero-torque contour intersects $\phi = 0\%$ at $\alpha \simeq 2\%$, *i.e.*, at a slightly positive phospholipid abundance. But we have two competing effects: first, the excess of saturated lipids in the outer leaflet renders $J_{0,\mathrm{b}}$ slightly positive. This creates a negative bilayer torque $\mathscr{T}_\kappa$ which we must "undo" by a positive differential stress to stay at $\mathscr{T} = 0$ (see eqn (5b)); this favors reducing the phospholipid contingent in leaf$_+$ even in the absence of cholesterol. And second, the saturated lipids in the outer leaflet recruit cholesterol more avidly. As we increase ϕ, the cholesterol mole fraction ϕ_+ will hence grow more strongly than ϕ_-.

Taking everything together, the initial exoplasmic abundance "flips" beyond $\phi \approx 18\%$: torque-free membranes now have more phospholipids in their cytosolic leaflet. This is qualitatively in line with experimental observations, but the quantitative comparison is far off: in our case, at 40% cholesterol the cytosolic leaflet contains about 6% more phospholipids than the exoplasmic one, while recent experiments argue that the excess can be 100% or even more—a factor of 2.[76] Of course, we must be careful with predictions based on coarse-grained models as simplified as ours: while we have tried to capture many important characteristics of this system when we developed our force-field (such as lipid area, cholesterol partitioning, and the overall phase behavior),[98] more subtle phenomena (*e.g.*, how does lipid spontaneous curvature depend on saturation and cholesterol content) need to be further examined. That being said, this large discrepancy serves to remind us how extraordinary the experimental claims are, and how difficult it would be to achieve a torque balanced state with an acceptable differential stress at a much larger abundance asymmetry (assuming, of course, that torque balance is relevant to begin with).

4.3 Demixing driven by differential stress

As a final illustration we show how a system of fixed overall lipid content can be driven to phase segregate in one leaflet by depleting its cholesterol content *via* a suitably induced differential stress.

4.3.1 System setup. For equilibration reasons it is easier to create ℓ_o domains in an ℓ_d background, and so we will pick the ℓ_d-leaflet (which for consistency we make leaf$_-$) to be the one to phase separate. Specifically, let us pick a saturation ratio $r_- = 0.2$, which results in a line that intersects the $\ell_\mathrm{o}/\ell_\mathrm{d}$ coexistence region in the vicinity of $\phi_- = 20\%$ (see Fig. 11). To keep matters simple, we will choose the composition in leaf$_+$ to be as far away from coexistence as possible, namely, on the binary *sc*-side of the triangle (*i.e.*, at $r_+ = 1$).

With this choice of $r_\pm$, we ran a set of simulations at $\alpha = 0$ over a range of overall cholesterol concentrations $\phi \in \{10\%, 15\%, 20\%, \ldots, 50\%\}$ and let these relax until both sides found their equilibrium cholesterol content $\phi_\pm$. The purpose is to find the ϕ-value that lets ϕ_- sit as close as possible to the binodal of the coexistence region, so that subsequent changes in differential stress, which raise or lower ϕ_-, will move leaf$_-$ further away or more deeply into the coexistence region.

Since the specific area of *s*-lipids is about 25% smaller than that of *u*-lipids, we expect a system with the same number of phospholipids on both sides (*i.e.*, $\alpha = 0$) to be under negative differential stress (*i.e.*, leaf$_+$ is under tension while leaf$_-$ is compressed). The presumably cleanest way to run the simulations is to compensate for this and increase the abundance until $\Delta\Sigma = 0$. This is technically challenging, though, since the necessary increase is itself ϕ-dependent (recall the

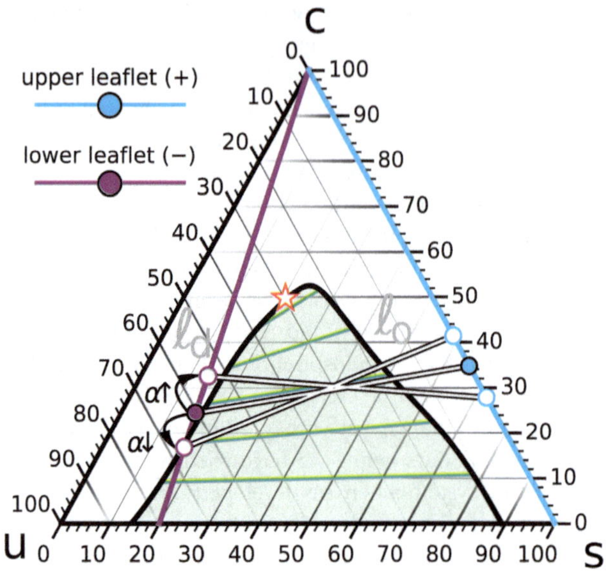

Fig. 11 The set of saturation ratios $r_+ = 1$ for leaf$_+$ (cyan line) and $r_- = 0.2$ for leaf$_-$ (magenta line), combined with an overall cholesterol content $\phi = 30\%$, yields a pair of coexisting points (solid cyan and magenta symbols) of which the leaf$_-$ composition resides very close to the ℓ_d-side of the binodal. Increasing the abundance asymmetry $\alpha \rightarrow +10\%$ squeezes cholesterol from leaf$_+$ to leaf$_-$ and pushes the latter further away from the two phase region. Reducing $\alpha \rightarrow -10\%$ instead removes cholesterol from leaf$_-$ and plunges it into ℓ_o/ℓ_d coexistence (see Fig. 12 for illustrations). The compositions in leaf$_+$ pivot oppositely but always stay in a homogeneous ℓ_o phase.

nontrivial ϕ–α relation on the $\Delta\Sigma = 0$ contour of the differential stress surface shown in Fig. 10). To avoid an extra round of iterations, we decided to forgo this ambition and instead select an overall cholesterol content ϕ that results in a ϕ_- slightly above the coexistence region, as we expect the slight net compression in leaf$_-$ to assist the formation of ordered domains. With this in mind, we selected $\phi = 30\%$, which resulted in $\phi_- \approx 24.5\%$, about 5 percentage points above the local cholesterol content of the binodal. Fortuitously, the resulting system has an almost vanishing torque, $\mathscr{T} = -0.33(17)\varepsilon/\sigma \simeq -1.3(7)$ pN, meaning, it would be voluntarily (close to) flat.

Observe that the r_- line and the binodal intersect at a relatively small angle, such that small movements of the binodal to the left or right would shift the intersection by a fairly large amount. The binodal is indeed not known very precisely, as its location is not merely dependent on sampling (slow) statistical fluctuations in the compositions of coexisting ℓ_o/ℓ_d phases (see ref. 98 for details) but also on difficult to quantify systematic errors inherent in the Hidden Markov Model's phase identification.[98,109] With these complications in mind, we chose to not over-engineer the precise location of ϕ_-.

4.3.2 Driving the leaflet into the coexistence region. While the chosen system with $\phi = 30\%$ has leaf$_-$ close to the ℓ_d-side of the binodal, leaf$_+$ is far from coexistence and manifestly in a homogeneous ℓ_o phase at $r_+ = 1$ and $\phi_+ = 34.8\%$, see again the phase diagram in Fig. 11. We now create two new systems in which we change the abundance asymmetry to $\alpha = +10\%$ and $\alpha = -10\%$. The former increases the number of phospholipids in leaf$_+$ relative to leaf$_-$ and hence expels some of the cholesterol into leaf$_-$, where its concentration increases to about $\phi_- = 32.5\%$, moving it further away from the binodal. In the other case we instead deplete leaf$_+$ relative to leaf$_-$ and thus draw additional cholesterol from leaf$_-$, where its concentration hence drops even more, to about $\phi_- = 16.7\%$, thereby plunging this leaflet into the coexistence region.

Fig. 12 shows stylized snapshots of the leaf$_-$ lipid configuration for the three systems with $\alpha \in \{-10\%, 0\%, +10\%\}$. As the abundance changes from negative to positive, and as a consequence the leaflet's cholesterol content from a small to a larger value, we observe that very distinct stable ℓ_o domains visible at $\alpha = -10\%$ melt away, with merely some remnant transient non-ideally mixed "flakes" remaining, which dynamically fluctuate in and out of existence.

4.3.3 Associated mechanical variables. The differential stress driving cholesterol translocation is quite substantial: as the values in Fig. 12 show, $\Delta\Sigma$ changes by almost 25 pN nm^{-1}, while the torque changes by about 37 pN. Since these changes appear linear with abundance, we get simple estimates for the α-driven susceptibilities:

$$\chi^{\Delta\Sigma}_{\alpha|\phi} \approx -\frac{\Delta\Delta\Sigma}{\Delta\alpha} \approx 23\varepsilon/\sigma^2 \simeq 124 \text{ pN nm}^{-1}, \tag{21a}$$

$$\chi^{\mathscr{T}}_{\alpha|\phi} \approx -\frac{\Delta\mathscr{T}}{\Delta\alpha} \approx 46\varepsilon/\sigma \simeq 187 \text{ pN}. \tag{21b}$$

Interestingly, the value for $\chi^{\mathscr{T}}_{\alpha|\phi}$ is essentially the same we found in Section 4.2.3 (see Fig. 9 at $\phi = 30\%$), even though the $r_\pm$ values were different. This suggests that the response to a change in lipid abundance is dominated by mechanics. We

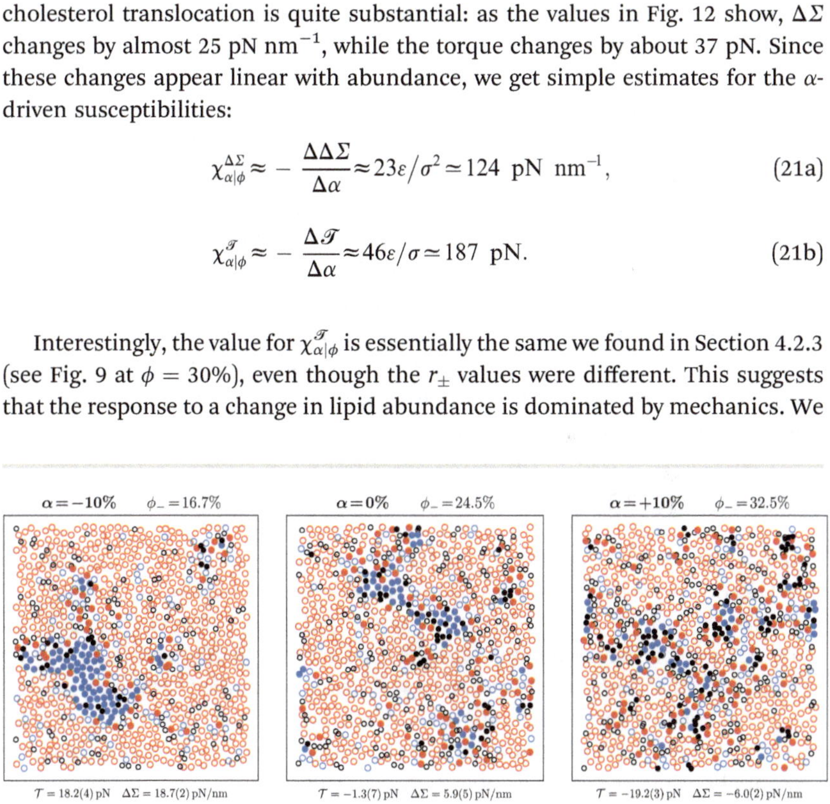

Fig. 12 Stylized representative lipid configurations in leaf$_-$ of the three systems discussed in Section 4.3 and represented in the phase diagram of Fig. 11. Colors indicate lipid type—red: unsaturated, blue: saturated, black: cholesterol—while style indicates the phase state—open circles: ℓ_d, filled circles: ℓ_o—as identified by a Hidden Markov Model analysis.[98,109] At zero abundance, the phase state in leaf$_-$ is very close to the ℓ_d-side of the coexistence binodal and we see only fleeting occurrences of small transient ℓ_o regions. At $\alpha = -10\%$ cholesterol is drawn out of leaf$_-$ which pushes the phase state into the coexistence region and we get a small but persistent ℓ_o domain. Conversely, at $\alpha = +10\%$ cholesterol is pushed from leaf$_+$ into leaf$_-$, increasing the state's distance from the coexistence region and further melting any remaining ordered domains. All systems contain 2048 lipids, and the box length is approximately $L \approx 32\sigma \approx 24$ nm.

already saw that $\chi^{\mathscr{T}}_{\alpha|\phi}$ itself hardly depends on α; here we get additional support from the fact that the change in torque is almost exclusively driven by the change in differential stress.

Using eqn (20) to calculate the intrinsic torque $\kappa J_{0,\mathrm{b}}$, and taking again $z_0 = 2\sigma \simeq 1.5$ nm, we find that $\kappa J_{0,\mathrm{b}} \simeq 10$ pN within error for all three systems. The value itself is quite reasonable (using again $\kappa \approx 30k_\mathrm{B}T$ we get $J_{0,\mathrm{b}} \simeq 0.08\ \mathrm{nm}^{-1}$), but its ϕ-independence appears surprising, given that the $\alpha = +10\%$ system has about twice as much cholesterol in leaf$_-$ as the $\alpha = -10\%$ system, with a slightly smaller but opposite effect in leaf$_+$, and so we would expect some effect at the spontaneous curvature level. Of course, the precise answer depends on exactly what value we use for z_0, and that value itself can change as leaflets become more or less ordered and hence lipids stretch or shrink. Furthermore, the mechanism by which cholesterol affects intrinsic lipid curvature is notoriously subtle,[78,80] and we should not expect this to be fully captured by a coarse-grained model as simple as ours.

The magnitude of these stresses and torques raises concerns about whether under experimental conditions these systems would remain stable. Unbalanced torques of order 18 pN may be associated with characteristic curvature radii $R \sim 2\kappa/\mathscr{T} \sim 14$ nm. Our box length $L \approx 24$ nm is not much larger than this, and so curvature deformations such as tubulation are not an option, but they would be for macroscopic systems at the micron scale. To still observe a single leaflet phase transition we have to relocate enough cholesterol with less stress. We suspect the experimental situation is easier, though, as we will not need to change a leaflet's cholesterol content by as much as 15% to affect a very noticeable difference in its phase state. Macroscopic systems respond much more sharply when crossing first order phase boundaries. However, considering that transitions from nanoscopic domains to macroscopic phase separation appear to be another characteristic of these systems,[110–112] we must be careful not to over-interpret findings obtained from small simulations.

Conclusions

We have proposed a thermodynamic framework that describes the conditions for coexistence between the two leaflets of an asymmetric ternary lipid membrane comprising a saturated lipid, an unsaturated lipid, and cholesterol. Our goal was not to make specific predictions—the discussion remains fully agnostic about the detailed form of the system's free energy. Instead, we clarified the dimensionality of the underlying thermodynamic state space, arguing that even though cholesterol will transition between leaflets to equilibrate its chemical potential, two essentially arbitrary ternary compositions can still coexist. However, doing so will generally require a nonzero differential stress $\Delta\Sigma$ that creates a mechanical counter-pressure to act against a generally non-zero chemical driving force. This implies that $\Delta\Sigma$ is an essential thermodynamic variable that must be part of any description of the asymmetric ternary system, for otherwise not the entire available state space is also accessible.

Differential stress not only helps achieve a certain desired cholesterol imbalance; since individual leaflet tensions act some distances $\pm z_0$ displaced from the membrane midplane, $\Delta\Sigma$ creates a torque. This torque, in turn, will combine with the generally nonzero intrinsic torque due to lipid compositional asymmetry into an overall torque $\mathscr{T}$ that will try to bend the membrane. We argue that large membranes such as GUVs are therefore only stable against small-scale tubulation if the overall torque indeed vanishes.

Keeping track of the accessible degrees of freedom is hence rather subtle: two two-dimensional composition spaces (*i.e.*, two Gibbs triangles per leaflet) combine to 4 degrees of freedom, but the equilibrium condition $\mu_{\rm chol}{}^{+} = \mu_{\rm chol}{}^{-}$ removes one, leaving only 3. However, differential stress can help balance cholesterol, so including it we bounce back to 4. Except, if the resulting torque is not also close to zero, the membrane is unstable against tubulation, so we again drop down to 3.

All these complications can be traced back to cholesterol, a remarkable actor that plays two entirely different roles here: on the one hand it co-determines the phase behavior as one of the compositional axes in the Gibbs triangles. On the other hand it can transition between leaflets and hence change the inter-leaflet stresses. Recall now that cholesterol also affects the intrinsic curvature of mixtures, usually not additively, and that its effect on leaflet area is not just non-additive but maybe also non-positive (because under certain conditions adding cholesterol will condense the membrane). This shows that writing down an actual free energy, or equations of state, for these coexisting asymmetric ternary systems is going to be a significant challenge, which we probably have to approach by adding complications one step at a time. This was not the goal of this paper, but we hope that the conceptual framework we have provided here will make it easier to progress on this difficult journey.

Data availability

The code used to set up the simulations, along with various data analysis scripts used in this article, are available on GitHub: https://github.com/m-varma-phys/asymmetric_ternary_mixtures.

Author contributions

Malavika Varma: conceptualization (supporting); formal analysis (lead); investigation (equal); methodology (lead); software (lead); visualization (supporting); writing—review & editing (equal). Markus Deserno: conceptualization (lead); investigation (equal); funding acquisition (lead); visualization (lead); project administration (lead); supervision (lead); writing—original draft (lead); writing—review & editing (equal).

Conflicts of interest

There are no conflicts to declare.

Acknowledgements

The authors are grateful for many stimulating discussions on the subject with Ilya Levental, Kandice Levental, Sarah Veatch, Milka Doktorova, Fred Heberle, Thais Enoki, Alex Sodt, Gerald Feigenson, Samuel Foley, Seamus Gallagher, and Tyler Reagle. This material is based upon work supported by the National Science Foundation under Award No. NSF/CHE 2102316.

Notes and references

1 G. van Meer, *EMBO J.*, 2005, **24**, 3159–3165.
2 M. R. Wenk, *Nat. Rev. Drug Discovery*, 2005, **4**, 594–610.
3 G. van Meer, *Annu. Rev. Cell Biol.*, 1989, **5**, 247–275.
4 J. E. Kyle, L. Aimo, A. J. Bridge, G. Clair, M. Fedorova, J. B. Helms, M. R. Molenaar, Z. Ni, M. Orešič, D. Slenter, E. Willighagen and B.-J. M. Webb-Robertson, *Metabolomics*, 2021, **17**, 55.
5 K. Simons and E. Ikonen, *Nature*, 1997, **387**, 569–572.
6 D. A. Brown and E. London, *Annu. Rev. Cell Dev. Biol.*, 1998, **14**, 111–136.
7 D. A. Brown and E. London, *J. Biol. Chem.*, 2000, **275**, 17221–17224.
8 K. Simons and D. Toomre, *Nat. Rev. Mol. Cell Biol.*, 2000, **1**, 31–39.
9 K. Simons and M. J. Gerl, *Nat. Rev. Mol. Cell Biol.*, 2010, **11**, 688.
10 L. J. Pike, *J. Lipid Res.*, 2006, **47**, 1597–1598.
11 S. Munro, *Cell*, 2003, **115**, 377–388.
12 I. Levental and S. L. Veatch, *J. Mol. Biol.*, 2016, **428**, 4749–4764.
13 F. M. Goñi, *Chem. Phys. Lipids*, 2019, **218**, 34–39.
14 I. Levental, K. R. Levental and F. A. Heberle, *Trends Cell Biol.*, 2020, **30**, 341–353.
15 G. W. Feigenson and J. T. Buboltz, *Biophys. J.*, 2001, **80**, 2775–2788.
16 S. L. Veatch and S. L. Keller, *Phys. Rev. Lett.*, 2002, **89**, 268101.
17 S. L. Veatch and S. L. Keller, *Biophys. J.*, 2003, **85**, 3074–3083.
18 D. Scherfeld, N. Kahya and P. Schwille, *Biophys. J.*, 2003, **85**, 3758–3768.
19 S. L. Veatch, I. V. Polozov, K. Gawrisch and S. L. Keller, *Biophys. J.*, 2004, **86**, 2910–2922.
20 S. L. Veatch and S. L. Keller, *Biochim. Biophys. Acta, Biomembr.*, 2005, **1746**, 172–185.
21 S. L. Veatch and S. L. Keller, *Phys. Rev. Lett.*, 2005, **94**, 148101.
22 H. McConnell, *Biophys. J.*, 2005, **88**, L23–L25.
23 S. L. Veatch, K. Gawrisch and S. L. Keller, *Biophys. J.*, 2006, **90**, 4428–4436.
24 G. W. Feigenson, *Annu. Rev. Biophys. Biomol. Struct.*, 2007, **36**, 63–77.
25 T. Baumgart, G. Hunt, E. R. Farkas, W. W. Webb and G. W. Feigenson, *Biochim. Biophys. Acta, Biomembr.*, 2007, **1768**, 2182–2194.
26 A. Bunge, P. Müller, M. Stöckl, A. Herrmann and D. Huster, *Biophys. J.*, 2008, **94**, 2680–2690.
27 D. Marsh, *Biochim. Biophys. Acta, Biomembr.*, 2009, **1788**, 2114–2123.
28 Y. Z. Yoon, J. P. Hale, P. G. Petrov and P. Cicuta, *J. Phys.: Condens. Matter*, 2010, **22**, 062101.
29 F. A. Heberle, J. Wu, S. L. Goh, R. S. Petruzielo and G. W. Feigenson, *Biophys. J.*, 2010, **99**, 3309–3318.
30 F. A. Heberle and G. W. Feigenson, *Cold Spring Harbor Perspect. Biol.*, 2011, **3**, a004630.
31 M. C. Blosser, J. B. Starr, C. W. Turtle, J. Ashcraft and S. L. Keller, *Biophys. J.*, 2013, **104**, 2629–2638.
32 S. L. Veatch, O. Soubias, S. L. Keller and K. Gawrisch, *Proc. Natl. Acad. Sci. U. S. A.*, 2007, **104**, 17650–17655.
33 A. R. Honerkamp-Smith, P. Cicuta, M. D. Collins, S. L. Veatch, M. Den Nijs, M. Schick and S. L. Keller, *Biophys. J.*, 2008, **95**, 236–246.

34 A. R. Honerkamp-Smith, S. L. Veatch and S. L. Keller, *Biochim. Biophys. Acta, Biomembr.*, 2009, **1788**, 53–63.

35 L. S. Hirst, P. Uppamoochikkal and C. Lor, *Liq. Cryst.*, 2011, **38**, 1735–1747.

36 S. L. Veatch and P. Cicuta, *Physics of Biological Membranes*, Springer, 2018, pp. 141–168.

37 E. Cammarota, C. Soriani, R. Taub, F. Morgan, J. Sakai, S. L. Veatch, C. E. Bryant and P. Cicuta, *J. R. Soc. Interface*, 2020, **17**, 20190803.

38 T. R. Shaw, S. Ghosh and S. L. Veatch, *Annu. Rev. Phys. Chem.*, 2021, **72**, 51–72.

39 S. L. Veatch, P. Cicuta, P. Sengupta, A. Honerkamp-Smith, D. Holowka and B. Baird, *ACS Chem. Biol.*, 2008, **3**, 287–293.

40 J. Huang, J. T. Buboltz and G. W. Feigenson, *Biochim. Biophys. Acta, Biomembr.*, 1999, **1417**, 89–100.

41 J. Huang and G. W. Feigenson, *Biophys. J.*, 1999, **76**, 2142–2157.

42 M. S. Bretscher, *Nature (London), New Biol.*, 1972, **236**, 11.

43 S. J. Singer and G. L. Nicolson, *Science*, 1972, **175**, 720–731.

44 A. J. Verkleij, R. F. A. Zwaal, B. Roelofsen, P. Comfurius, D. Kastelijn and L. L. M. Van Deenen, *Biochim. Biophys. Acta, Biomembr.*, 1973, **323**, 178–193.

45 P. K. Schick, K. B. Kurica and G. K. Chacko, *J. Clin. Invest.*, 1976, **57**, 1221–1226.

46 A. Sandra and R. E. Pagano, *Biochemistry*, 1978, **17**, 332–338.

47 J. A. F. Op den Kamp, *Annu. Rev. Biochem.*, 1979, **48**, 47–71.

48 P. F. Devaux, *Biochemistry*, 1991, **30**, 1163–1173.

49 M. G. P. Vale, *Biochim. Biophys. Acta, Biomembr.*, 1977, **471**, 39–48.

50 J. H. Lorent, K. R. Levental, L. Ganesan, G. Rivera-Longsworth, E. Sezgin, M. Doktorova, E. Lyman and I. Levental, *Nat. Chem. Biol.*, 2020, **16**, 644–652.

51 M. M. Sperotto and A. Ferrarini, *The Biophysics of Cell Membranes: Biological Consequences*, Springer Singapore, Singapore, 2017, pp. 29–60.

52 W. F. D. Bennett, J. L. MacCallum, M. J. Hinner, S. J. Marrink and D. P. Tieleman, *J. Am. Chem. Soc.*, 2009, **131**, 12714–12720.

53 S. Jo, H. Rui, J. B. Lim, J. B. Klauda and W. Im, *J. Phys. Chem. B*, 2010, **114**, 13342–13348.

54 R.-X. Gu, S. Baoukina and D. P. Tieleman, *J. Chem. Theory Comput.*, 2019, **15**, 2064–2070.

55 R. E. Scott, *Science*, 1976, **194**, 743–745.

56 R. E. Scott and P. B. Maercklein, *J. Cell Sci.*, 1979, **35**, 245–252.

57 D. Holowka and B. Baird, *Biochemistry*, 1983, **22**, 3466–3474.

58 E. Sezgin, H.-J. Kaiser, T. Baumgart, P. Schwille, K. Simons and I. Levental, *Nat. Protoc.*, 2012, **7**, 1042–1051.

59 D. M. Engelman, *Nature*, 2005, **438**, 578–580.

60 T. Baumgart, A. T. Hammond, P. Sengupta, S. T. Hess, D. A. Holowka, B. A. Baird and W. W. Webb, *Proc. Natl. Acad. Sci. U. S. A.*, 2007, **104**, 3165–3170.

61 H. Keller, M. Lorizate and P. Schwille, *ChemPhysChem*, 2009, **10**, 2805–2812.

62 M. Krompers and H. Heerklotz, *Membranes*, 2023, **13**, 267.

63 A. Hossein and M. Deserno, *Biophys. J.*, 2020, **118**, 624–642.

64 A. Hossein and M. Deserno, *J. Chem. Phys.*, 2021, **154**, 014704.

65 S. L. Foley, A. Hossein and M. Deserno, *Biophys. J.*, 2022, **121**, 2997–3009.

66 S. L. Foley, M. Varma, A. Hossein and M. Deserno, *Emerging Top. Life Sci.*, 2023, **7**, 95–110.

67 M. Deserno, *Curr. Opin. Struct. Biol.*, 2024, **87**, 102832.
68 C. E. Morris and U. Homann, *J. Membr. Biol.*, 2001, **179**, 79–102.
69 T. A. Enoki and F. A. Heberle, *Proc. Natl. Acad. Sci. U. S. A.*, 2023, **120**, e2308723120.
70 A. J. Wagner, S. Loew and S. May, *Biophys. J.*, 2007, **93**, 4268–4277.
71 G. G. Putzel and M. Schick, *Biophys. J.*, 2008, **94**, 869–877.
72 S. May, *Soft Matter*, 2009, **5**, 3148–3156.
73 J. J. Williamson and P. D. Olmsted, *Biophys. J.*, 2015, **108**, 1963–1976.
74 J. J. Williamson and P. D. Olmsted, *Soft Matter*, 2015, **11**, 8948–8959.
75 J. J. Williamson and P. D. Olmsted, *Biophys. J.*, 2018, **115**, 1956–1965.
76 M. Doktorova, J. L. Symons, X. Zhang, H.-Y. Wang, J. Schlegel, J. H. Lorent, F. A. Heberle, E. Sezgin, E. Lyman, K. R. Levental and I. Levental, *Cell*, 2025, **188**, 1–17.
77 M. Varma and M. Deserno, *Biophys. J.*, 2022, **121**, 4001–4018.
78 A. J. Sodt, R. M. Venable, E. Lyman and R. W. Pastor, *Phys. Rev. Lett.*, 2016, **117**, 138104.
79 D. W. Allender, A. J. Sodt and M. Schick, *Biophys. J.*, 2019, **116**, 2356–2366.
80 A. H. Beaven, K. Sapp and A. J. Sodt, *Biophys. J.*, 2023, **122**, 2162–2175.
81 R. M. Venable, F. L. Brown and R. W. Pastor, *Chem. Phys. Lipids*, 2015, **192**, 60–74.
82 U. Seifert, K. Berndl and R. Lipowsky, *Phys. Rev. A: At., Mol., Opt. Phys.*, 1991, **44**, 1182–1202.
83 U. Seifert, *Adv. Phys.*, 1997, **46**, 13–137.
84 R. Lipowsky, *Faraday Discuss.*, 2013, **161**, 305–331.
85 Y. Liu, J. Agudo-Canalejo, A. Grafmüller, R. Dimova and R. Lipowsky, *ACS Nano*, 2016, **10**, 463–474.
86 R. Lipowsky, *Adv. Biol.*, 2022, **6**, 2101020.
87 T. Reagle, Y. Xie, Z. Li, W. Carnero and T. Baumgart, *Soft Matter*, 2024, **20**, 4291–4307.
88 R. Zidovetzki and I. Levitan, *Biochim. Biophys. Acta, Biomembr.*, 2007, **1768**, 1311–1324.
89 S. Mahammad and I. Parmryd, *Methods in membrane lipids*, 2015, pp. 91–102.
90 L. Szente and É. Fenyvesi, *Struct. Chem.*, 2017, **28**, 479–492.
91 M. Rahimi, D. Regan, M. Arroyo, A. B. Subramaniam, H. A. Stone and M. Staykova, *Biophys. J.*, 2016, **111**, 2651–2657.
92 E. London, *Acc. Chem. Res.*, 2019, **52**, 2382–2391.
93 Z. Huang and E. London, *Langmuir*, 2013, **29**, 14631–14638.
94 Q. Lin and E. London, *PLoS One*, 2014, **9**, e87903.
95 G. Li, J. Kim, Z. Huang, J. R. St. Clair, D. A. Brown and E. London, *Proc. Natl. Acad. Sci. U. S. A.*, 2016, **113**, 14025–14030.
96 G. Li, S. Kakuda, P. Suresh, D. Canals, S. Salamone and E. London, *PLoS One*, 2019, **14**, e0223572.
97 P. Suresh and E. London, *Biochim. Biophys. Acta, Biomembr.*, 2022, **1864**, 183774.
98 M. Varma, F. Khuri-Makdisi and M. Deserno, *J. Chem. Phys.*, 2024, **161**, 114103.
99 S. Foley and M. Deserno, *J. Chem. Theory Comput.*, 2020, **16**, 7195–7206.
100 I. R. Cooke, K. Kremer and M. Deserno, *Phys. Rev. E: Stat., Nonlinear, Soft Matter Phys.*, 2005, **72**, 011506.

101 I. R. Cooke and M. Deserno, *J. Chem. Phys.*, 2005, **123**, 224710.
102 M. Deserno, *Macromol. Rapid Commun.*, 2009, **30**, 752–771.
103 G. Orädd, P. W. Westerman and G. Lindblom, *Biophys. J.*, 2005, **89**, 315–320.
104 F. Weik, R. Weeber, K. Szuttor, K. Breitsprecher, J. de Graaf, M. Kuron, J. Landsgesell, H. Menke, D. Sean and C. Holm, *Eur. Phys. J.: Spec. Top.*, 2019, **227**, 1789–1816.
105 G. S. Grest and K. Kremer, *Phys. Rev. A: At., Mol., Opt. Phys.*, 1986, **33**, 3628.
106 A. Kolb and B. Dünweg, *J. Chem. Phys.*, 1999, **111**, 4453–4459.
107 M. Deserno, *Chem. Phys. Lipids*, 2015, **185**, 11–45.
108 M. P. Do Carmo, *Differential geometry of curves and surfaces: revised and updated*, Courier Dover Publications, 2nd edn, 2016.
109 A. J. Sodt, M. L. Sandar, K. Gawrisch, R. W. Pastor and E. Lyman, *J. Am. Chem. Soc.*, 2014, **136**, 725–732.
110 T. M. Konyakhina, S. L. Goh, J. Amazon, F. A. Heberle, J. Wu and G. W. Feigenson, *Biophys. J.*, 2011, **101**, L8–L10.
111 F. A. Heberle, R. S. Petruzielo, J. Pan, P. Drazba, N. Kucerka, R. F. Standaert, G. W. Feigenson and J. Katsaras, *J. Am. Chem. Soc.*, 2013, **135**, 6853–6859.
112 R. D. Usery, T. A. Enoki, S. P. Wickramasinghe, M. D. Weiner, W.-C. Tsai, M. B. Kim, S. Wang, T. L. Torng, D. G. Ackerman, F. A. Heberle, et al., *Biophys. J.*, 2017, **112**, 1431–1443.

Faraday Discussions

PAPER

The many faces of membrane tension for biomembranes and vesicles†

Reinhard Lipowsky *

Received 27th November 2024, Accepted 13th December 2024
DOI: 10.1039/d4fd00184b

Membrane tension is generated by external forces and constraints that affect the membrane's surface area. Many studies have been pursued to elucidate this intuitive concept. Here, we focus on biomimetic model systems, for which the notion of membrane tension can be elaborated in a quantitative manner. Bottom-up, these membrane systems are planar bilayers, unilamellar nanovesicles, and giant unilamellar vesicles (GUVs). For planar bilayers and nanovesicles, we need to distinguish the mechanical bilayer tension within the whole bilayer from the mechanical leaflet tensions within the individual bilayer leaflets. In addition, the fluctuation spectrum of the bilayer defines the bilayer's fluctuation tension. These different tensions can be determined by molecular dynamics simulations. On the micron-scale, GUV membranes as observed by light microscopy, experience both a mechanical and a curvature-elastic tension, as follows from the theory of curvature elasticity. The curvature-elastic tension represents a material parameter, whereas the mechanical tension is a kind of hidden variable that depends on the size and shape of the membrane as demonstrated for multispherical shapes of GUVs. Furthermore, the excess membrane area stored in the shape fluctuations is again governed by a fluctuation tension. Many membrane systems exhibit characteristic tension thresholds. Examples include, the shapes of tubular membranes and the morphological responses of vesicle membranes to condensate droplets.

1 Introduction

Experimental procedures that generate membrane tension include osmotic inflation, adhesion, and micropipette aspiration of vesicles. Membrane tension is intimately related to membrane area but difficult to study because the membrane ruptures when its area is increased by a few percent. In order to elucidate the concept of membrane tension, it is very useful to consider biomimetic model

Max Planck Institute of Colloids and Interfaces, 14424 Potsdam, Germany. E-mail: lipowsky@mpikg.mpg.de; Tel: +49 331 5679600

† Electronic supplementary information (ESI) available: Sections S1–S9, Fig. S1–S18, and Table S1. See DOI: https://doi.org/10.1039/d4fd00184b

membranes. Here, we will study three such systems: planar lipid bilayers, unilamellar nanovesicles, and giant unilamellar vesicles (GUVs).

On the nanometer scale, the mechanical tension of the membrane is provided by the bilayer tension as determined by molecular dynamics simulations. This bilayer tension can be decomposed into two leaflet tensions, which control the spatio-temporal remodelling of lipid bilayers and nanovesicles.[1] On the micrometer scale, the total membrane tension of a biomembrane with transbilayer asymmetry is equal to the sum of its mechanical and its spontaneous or curvature-elastic tension.[2,3]

This paper is organized as follows. The next Section 2 emphasizes three fundamental aspects of biomembranes as provided by their molecular bilayer structure, their fluidity, and their shape fluctuations. In Section 2.3, we remind the reader about the important differences between membrane tension and interfacial tension. The subsequent Sections 3–5 elucidate the different notions of membrane tension for planar bilayers, unilamellar nanovesicles, and giant unilamellar vesicles (GUVs), respectively. In Section 6, we show that the remodelling of GUVs by condensate droplets is governed by the competition between the curvature-elastic tensions of the membrane segments and the interfacial tension of the liquid–liquid interface. The last Section 7 describes the most important conclusions.

2 Fundamental aspects of biomembranes

2.1 Symmetric and asymmetric bilayer membranes

All biomembranes are based on lipid bilayers, consisting of two leaflets (or monolayers) of lipid molecules. In general, the two leaflets may differ in their lipid composition, in the packing densities of these lipids – as described by area per lipid or volume per lipid – and in the molecular composition of the two aqueous solutions adjacent to the two leaflets. All of these differences contribute to the transbilayer asymmetry and the associated spontaneous curvature.

For the closed membrane of a vesicle, the inner leaflet is in contact with the interior aqueous solution, the outer leaflet with the exterior aqueous solution. Furthermore, biomembranes typically contain membrane proteins which often exhibit an average orientation towards the interior or towards the exterior solution. Therefore, a symmetric bilayer membrane is obtained if both leaflets have the same lipid composition and exhibit the same packing densities for all lipid components. Furthermore, a symmetric bilayer also requires that the exterior solution has the same molecular composition as the interior solution. Such a symmetric membrane represents a highly idealized bilayer state but also provides a reference state that is useful from a conceptual point of view. Furthermore, such a symmetric bilayer is easy to prepare in molecular dynamics simulations, provided one uses a planar bilayer with periodic boundary conditions. Experimentally, a planar membrane geometry is obtained for hole-spanning or black lipid membranes[4] and for pore-spanning membranes.[5]

2.2 Fluidity of membranes

The most fundamental property of a biomembrane is its fluidity. On the molecular scale, the lipid molecules undergo fast lateral diffusion with a diffusion

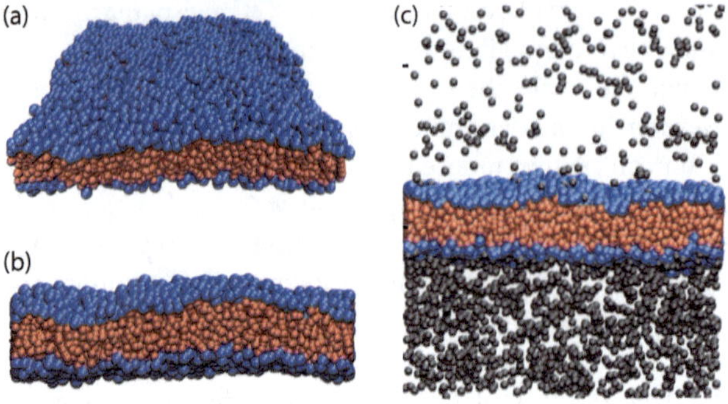

Fig. 1 Two examples for one-component lipid bilayers with transbilayer asymmetry: (a and b) planar bilayer consisting of 920 lipids in its upper and 738 lipids in its lower leaflet. The lipids have blue headgroups and red chains; and (c) planar bilayer with 832 lipids in the upper and 832 lipids in the lower leaflet but exposed to two different concentrations of solutes (grey beads) in the upper and lower solution. The solutes adsorb onto the bilayer and form different adsorption layers at the two leaflets. The bilayers have a thickness of about 4 nm.[7]

constant of the order of a few μm^2 per second. Thus, if we consider two lipids that are in contact with each other, it takes in the order of one or a few nanoseconds for them to swap their spatial positions. On the nanoscale, fluid membranes can exhibit two types of elastic deformations. First, they can be stretched and compressed, which leads to different areas per lipid and different volumes per lipid. Second, they can attain bent or curved states, which deviate from the essentially planar states in Fig. 1. These two elastic deformations follow from classical elasticity theory[6] for a liquid material with vanishing shear modulus.

2.2.1 Stretching and compression of membranes. First, let us consider the stretching and compression of a membrane in response to a mechanical tension, Σ, acting on this membrane. The fluidity of the membrane implies that this mechanical tension must be spatially uniform and isotropic for any (meta)stable state of the membrane. Indeed, tension gradients generate lipid flows within the membrane but these flows decay when the membrane relaxes towards a stable state.

If the membrane is tensionless, it attains a certain optimal area $A = A_0$. In general, the membrane area A differs from this optimal value and is then subject to the mechanical membrane tension

$$\Sigma = K_A(A - A_0)/A_0 \equiv K_A \Delta a, \tag{1}$$

which is proportional to the area compressibility modulus K_A and to the relative area dilation $(A - A_0)/A_0 = \Delta a$. A positive mechanical tension, $\Sigma > 0$, leads to membrane stretching, a negative tension, $\Sigma < 0$, to membrane compression. The Hooke-type relationship between Σ and $(A - A_0)/A_0$ has been used, *e.g.*, to analyse micropipette aspiration experiments.[8]

The corresponding elastic energy is obtained from eqn (1) by integration over the membrane area A, which leads to the elastic stretching energy

$$E_{st} = \int_{A_0}^{A} dA' \Sigma\left(A'\right) = \frac{1}{2} K_A \frac{(A - A_0)^2}{A_0}. \tag{2}$$

This energy represents the reversible work that we have to perform in order to stretch or compress the membrane from its optimal value A_0 at zero tension to its actual value A. Note that the elastic energy E_{st} is always positive, both for stretching with $A > A_0$ and for compression with $A < A_0$.

2.2.2 Bending of membranes. So far, we have not considered bent or curved states of the membrane and the associated bending energies. On the nanoscale, the curvature of the membrane is described by its principal curvatures, C_1 and C_2, or equivalently by the mean and Gaussian curvatures $M \equiv \frac{1}{2}(C_1 + C_2)$ and $G \equiv C_1 C_2$. As explained in Section S1,† the shape of a vesicle with volume V and membrane area A is computed by minimizing the elastic shape functional

$$F_{el} = -\Delta PV + \Sigma' A + 2\kappa \int dA(M - m)^2 \tag{3}$$

which depends on the two Lagrange multipliers ΔP and Σ' as well as on the area integral over the bending energy density $2\kappa(M - m)^2$. This energy density involves two curvature-elastic parameters, the bending rigidity κ and the spontaneous curvature m.

The Lagrange multiplier Σ' represents a tension. It can be shown[9] that the Lagrange multiplier tension Σ' is equal to the mechanical tension Σ in eqn (1), that is,

$$\Sigma' = \Sigma = K_A \Delta a. \tag{4}$$

This equality applies to all (meta)stable states of the vesicle membrane as obtained by the constrained minimization of the elastic shape functional F_{el} in eqn (3). This minimization of F_{el} leads to the local shape equation, which has the general form[2]

$$\Delta P \approx 2\Sigma_{tot} M \text{ with } \Sigma_{tot} \equiv \Sigma + 2\kappa m^2 \tag{5}$$

to first order in the mean curvature M and defines the total membrane tension Σ_{tot} and the spontaneous or curvature-elastic tension $2\kappa m^2$. The curvature-elastic tension can be directly measured by micropipette aspiration of tubulated GUVs see Fig. S1.† [10]

2.3 Shape fluctuations of fluid membranes

In the low tension regime, GUVs exhibit shape fluctuations in the form of bending undulations that are clearly visible with an optical microscope. Likewise, these undulations can be observed in molecular dynamics simulations, see Fig. 2. The influence of membrane tension on the membrane's shape fluctuations led to a long and controversial debate. This debate started in the early 1980s with several studies,[11–13] which considered both liquid–liquid interfaces with low interfacial tension[11,13] and fluid membranes with low membrane tension.[12,13] These two tensions must be distinguished, however, because they have fundamentally different properties.

2.3.1 Interfacial tension *versus* mechanical membrane tension. First, the interfacial tension is always positive as follows from the thermodynamic stability of the corresponding liquid–liquid interface,[15] whereas the mechanical

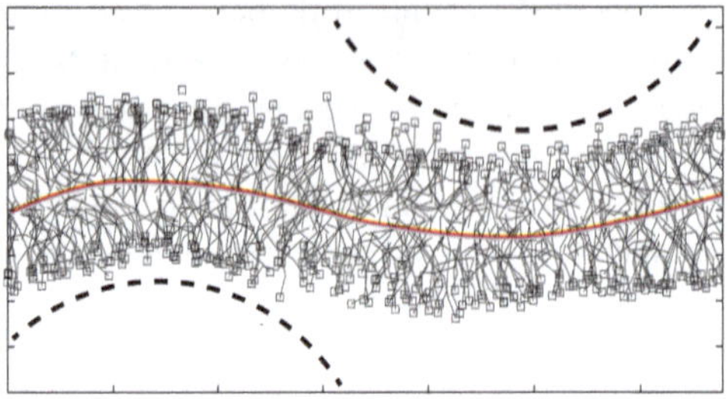

Fig. 2 Simulation snapshot of a symmetric and tensionless bilayer. The lipid bilayer has a thickness of about 4 nm. The smallest curvature radius of its midsurface (red) is about 6 nm, the radius of the dashed circles. Reused with permission from ref. 14, American Physical Society, 1999.

membrane tension Σ in eqn (1) can be positive, zero, or negative. Second, the mechanical membrane tension depends on the size and shape of the membrane, as demonstrated for multispherical vesicle shapes in Section 5.1 below. This shape dependence of Σ also follows when we apply the identity $\Sigma' = \Sigma$ in eqn (4) to the curvature-elastic theory of vesicle adhesion.[16] In contrast, the interfacial tension is independent of the size and shape of the liquid–liquid interface. Third, the shape fluctuations of membranes and interfaces have very different scaling properties as described in Section S2.† In particular, in contrast to interfaces, fluid membranes have a finite persistence length ξ_p for the correlations of their normal vectors, with different types of shape fluctuations on length scales below and above the persistence length see Fig. S2 and S3.†

2.3.2 Persistence length of fluid membranes. Below their persistence length, fluid membranes undergo fluctuations with relatively small amplitudes around their average shape, as in Fig. 2. In contrast, the membranes attain crumpled states with no average orientation of their normal vectors when their sizes are large compared to the persistence length ξ_p. Likewise, boundary conditions imposed on the membrane have no effect on those membrane segments that have a large spatial separation $\Delta L \gg \xi_p$ from this boundary.

The persistence length ξ_p of a fluid membrane grows exponentially with the ratio of the bending rigidity κ to the thermal energy $k_B T$ according to[12,13,17]

$$\xi_p = \ell \exp\left[c_p \kappa / k_B T\right] \text{with a molecular length scale } \ell \tag{6}$$

and the dimensionless coefficient $c_p = 4\pi/3$ as determined by Monte Carlo simulations of discretized fluid membranes.[17]

From a conceptual point of view, the persistence length as given by eqn (6) provides an important cross-over length between two different regimes of shape fluctuations. In practise, this length scale is not relevant for lipid bilayers. Indeed, using the bending rigidity $\kappa \simeq 20\ k_B T$ for lipid bilayers at room temperature and the molecular length scale $\ell \simeq 6$ nm, corresponding to the smallest curvature radius in Fig. 2, we obtain the estimate that ξ_p is of the order of 10^{23} km, which is astronomical compared to the size of the largest biomembranes. In fact, a fluid membrane is likely to form a pore across the bilayer when the lateral size of the membrane is still much smaller than the persistence length. Indeed, as shown in Section 3.3 below, any small but finite mechanical tension $\Sigma > 0$ leads to poration for a sufficiently large membrane.

3 Membrane tensions of planar bilayers

We now model the membranes as molecular bilayers assembled from one or several lipid components. These bilayers can be studied by molecular dynamics simulations. Using such simulations, one can determine several types of tension: the mechanical bilayer tension, the mechanical leaflet tensions, and the fluctuation tension. The mechanical tensions are computed *via* the stress profile across the bilayer. The fluctuation tension is deduced from the spectrum of bending undulations on length scales that are small compared to the persistence length.

In order to reduce boundary effects, all simulation results for planar bilayers as described here have been obtained for periodic boundary conditions. Thus, we will not consider any "frame tension" as sometimes discussed in the literature. In fact, if we fixed the edge of the membrane by a frame, we would also have to prescribe the normal vectors of the membrane along this frame.

3.1 Mechanical tensions of planar bilayers and leaflets

Each planar bilayer consists of two leaflets, the lower and the upper leaflet labelled ll and ul, respectively. Both leaflets represent two-dimensional liquids, which are subject to the leaflet tensions $\Sigma_{\rm ll}$ and $\Sigma_{\rm ul}$. In mechanical equilibrium, each of these leaflet tensions must be laterally uniform. As a consequence, the bilayer tension Σ can be decomposed according to

$$\Sigma = \Sigma_{\rm ll} + \Sigma_{\rm ul}. \tag{7}$$

The computation of these tensions from the stress profile across the bilayer is described in Section S3.†

3.2 Leaflet tension space for planar bilayers

Planar bilayers with a fixed total number of lipids, $N_{\rm ll} + N_{\rm ul}$, define a two-dimensional leaflet tension space, as depicted in Fig. 3 for $N_{\rm ll} + N_{\rm ul} = 1682$. The origin of this leaflet tension space as given by $\Sigma_{\rm il} = \Sigma_{\rm ul} = 0$, defines the relaxed reference state with two tensionless leaflets. This reference state of the planar bilayers can be obtained by focusing on symmetric bilayers with $N_{\rm ll} = N_{\rm ul} = 841$ and identifying the unique bilayer with vanishing bilayer tension $\Sigma = 0$. To characterize the elastic response of the reference state, it is useful to distinguish two types of elastic deformation, corresponding to the red and black data in Fig. 3.

The red data represent elastic deformations of the reference state with equal leaflet tensions (ELTs), that is, with

$$\Sigma_{\rm ll} = \Sigma_{\rm ul} \text{ (equal leaflet tensions).} \tag{8}$$

The ELT states are located along the main diagonal of the leaflet tension space in Fig. 3. These ELT states are described in more detail in Section S4 and Fig. S5–S7.† On the other hand, the black data in Fig. 3 correspond to elastic deformations of the reference state with opposite leaflet tensions (OLTs), that is, with

$$\Sigma_{\rm ll} = -\Sigma_{\rm ul} \text{ (opposite leaflet tensions).} \tag{9}$$

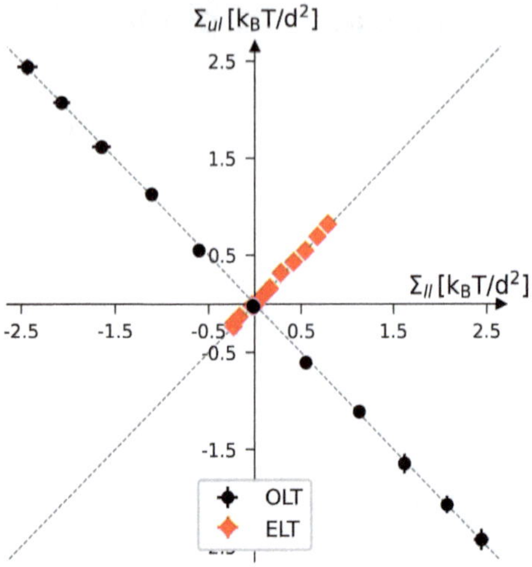

Fig. 3 Leaflet tension space for planar bilayers that contain a total number of $N_{ll} + N_{ul} = 1682$ lipids.[18] The two coordinates are the leaflet tensions Σ_{ll} and Σ_{ul} in the lower and upper leaflets. Negative and positive leaflet tensions describe compressed and stretched leaflets. The reference state with tensionless leaflets, corresponding to $\Sigma_{ll} = \Sigma_{ul} = 0$, is obtained for the symmetric bilayer with $N_{ll} = N_{ul} = 841$ lipids. The red data points describe elastic deformations with equal leaflet tensions (ELTs), $\Sigma_{ll} = \Sigma_{ul}$. The black data represent bilayers with opposite leaflet tensions (OLTs), $\Sigma_{ll} = -\Sigma_{ul}$ and $\Sigma = 0$. All OLT states can be obtained from the reference state by reshuffling lipids from one leaflet to the other and adjusting the base area of the simulation box to obtain tensionless bilayers.

The OLT states are located on the diagonal which is orthogonal to the main diagonal in Fig. 3. All OLT states can be obtained from the reference state by reshuffling lipids from one leaflet to the other, keeping the total lipid number $N_{ll} + N_{ul}$ constant and imposing the constraint of vanishing bilayer tension $\Sigma = \Sigma_{ll} + \Sigma_{ul} = 0$. As a consequence, one leaflet becomes compressed by a negative leaflet tension whereas the other leaflet becomes stretched by a positive leaflet tension.

The different OLT states of the planar bilayer can be distinguished by their stress asymmetry[7,19]

$$\Delta\Sigma \equiv \Sigma_{ul} - \Sigma_{ll}. \tag{10}$$

The stress asymmetry $\Delta\Sigma$ vanishes for all ELT states of the planar bilayer, for which both leaflets contain the same number of lipids, *i.e.*, for which $N_{ul} = N_{ll}$. In contrast, all OLT states of the planar bilayer exhibit a nonzero stress asymmetry, which increases in magnitude as we move away from the reference state with $\Sigma_{il} = \Sigma_{ul} = 0$. Furthermore, all bilayer states with the same nonzero value of the stress asymmetry, $\Delta\Sigma$ = constant, are located on a line parallel to the main diagonal. Stress asymmetry in planar bilayers has also been studied by Deserno and coworkers,[20,21] who used the term "differential stress" instead of "stress asymmetry". The elastic response of the planar bilayer leaflets to the leaflet tensions can be described both in terms of area per lipid and volume per lipid (Section S4, Fig. S5 and S6†), where volume per lipid is obtained *via* Voronoi tessellation (Fig. S7†).

3.3 Formation of stable bilayer pores

For a sufficiently large and positive bilayer tension Σ, the bilayer will rupture and form a pore. To illustrate this poration process, we now consider a planar and symmetric bilayer, which contains $N/2$ identical lipid molecules in each leaflet. We assume that the membrane area stored in the shape fluctuations of the bilayer can be ignored. As a consequence, the surface area A of the bilayer membrane is close to its projected area $A_{\|}$, which can be directly controlled in molecular dynamics simulations by the geometry of the simulation box. Indeed, for a planar bilayer that spans a cubic simulation box, the projected area of the bilayer is equal to the base area of this box. Furthermore, the bilayer has an optimal projected area $A_{\|} = A_0$ for which the mechanical tension $\Sigma = (A_{\|} - A_0)/A_0$ as given by eqn (1), vanishes. When $A_{\|}$ is larger than A_0, the bilayer is stretched by a positive mechanical tension $\Sigma > 0$.

Now, let us compare, for a given projected area $A_{\|} > A_0$, two different bilayer states: State (Un) corresponds to the unporated state with constant area per lipid, $a = 2A_{\|}/N$, state (Po) to a porated bilayer with a single circular pore of radius R. The stretching free energy of state (Un) is equal to

$$F_{\mathrm{un}} = \frac{1}{2}K_{\mathrm{A}}\left(A_{\|} - A_0\right)^2 A_0^{-1}(\mathrm{State(Un)}) \tag{11}$$

as follows from eqn (2). The free energy F_{po} of the porated bilayer state (Po) has two contributions: (i) the stretching energy of the remaining bilayer with area $A_{\|} - A_0 - \pi R^2$ and (ii) the edge energy of the pore with circumference $2\pi R$ and edge tension λ, which implies[22,23]

$$F_{\mathrm{po}} = \frac{1}{2}K_{\mathrm{A}}\left(A_{\|} - A_0 - \pi R^2\right)^2 A_0^{-1} + 2\pi R\lambda \quad (\mathrm{State(Po)}). \tag{12}$$

The free energy difference ΔF between (Po) and (Un) is

$$\Delta F(R) = F_{\mathrm{po}} - F_{\mathrm{un}} = 2\pi R\lambda - \pi\Sigma R^2 + \frac{1}{2}\pi^2 K_{\mathrm{A}} R^4 A_0^{-1} \tag{13}$$

with the mechanical tension $\Sigma = K_{\mathrm{A}}(A_{\|} - A_0)/A_0 = K_{\mathrm{A}}\Delta a$ as in eqn (1). The free energy difference ΔF depends on the pore radius R and involves two material parameters, the area compressibility modulus K_{A} and the edge tension λ. In addition, ΔF depends on the mechanical tension Σ, which is proportional to the relative area dilation $\Delta a = (A_{\|} - A_0)/A_0$ and, thus, represents a geometric control parameter that can be varied by changing the base area $A_{\|}$ of the simulation box.

When the bilayer is compressed by a negative value of Σ, all terms on the right hand side of eqn (13) are positive and $\Delta F(R)$ is a monotonically increasing function of R, which implies that the unporated bilayer state (Un) is always the state of lower free energy. On the other hand, when the bilayer is stretched by a positive value of Σ, the free energy difference $\Delta F(R)$ in eqn (13) develops an inflection point for mechanical tension

$$\Sigma_1 = c_1 K_{\mathrm{A}}^{1/3}\lambda^{2/3}/A_0^{1/3} \tag{14}$$

with $c_1 = 3\pi^{1/3}/2^{2/3} = 2.77$. For $\Sigma > \Sigma_1$, the inflection point splits up into a maximum and a minimum of $\Delta F(R)$. The minimum at $R = R_2$ describes a bilayer with a stable pore, which becomes more stable with increasing Σ until it attains the same free energy as the unporated state. This coexistence of the porated and unporated state is obtained for the mechanical tension

$$\Sigma_2 = c_2 K_A^{1/3} \lambda^{2/3} / A_0^{1/3} \tag{15}$$

with $c_2 = 2^{1/3} c_1 = 3.49$. For $\Sigma > \Sigma_2$, the porated bilayer provides the state of lower free energy.

Therefore, as we vary the mechanical bilayer tension Σ within a symmetric bilayer, the stretched bilayer has no pore for $\Sigma \le \Sigma_1$, forms a metastable pore for $\Sigma_1 < \Sigma < \Sigma_2$, and a globally stable pore for $\Sigma > \Sigma_2$. The threshold values Σ_1 and Σ_2 as given by eqn (14) and (15) are in good agreement with the results of molecular dynamics simulations.[23] It is important to note that both tension thresholds Σ_1 and Σ_2 are proportional to $1/A_0^{1/3}$ and thus go to zero for large bilayer area A_0. As a consequence, sufficiently large bilayers can always lower their free energy by forming a stable pore.

3.4 Leaflet tensions of tensionless planar bilayers

In order to avoid pore formation and bilayer rupture, we will now focus on tensionless bilayers with $\Sigma = \Sigma_{ll} + \Sigma_{ul} = 0$ or $\Sigma_{ll} = -\Sigma_{ul}$, corresponding to OLT states of the bilayer. For bilayers with a total lipid number $N_{ll} + N_{ul} = 1682$ as in Fig. 3, the tensionless bilayers or OLT states are given by the black data points in this figure. The corresponding leaflet tensions are displayed in Fig. 4 as a function of the lipid number N_{ul} assembled in the upper leaflet. As shown in this figure, tensionless leaflets are obtained for the symmetric bilayer with $N_{ll} = N_{ul} = 841$.

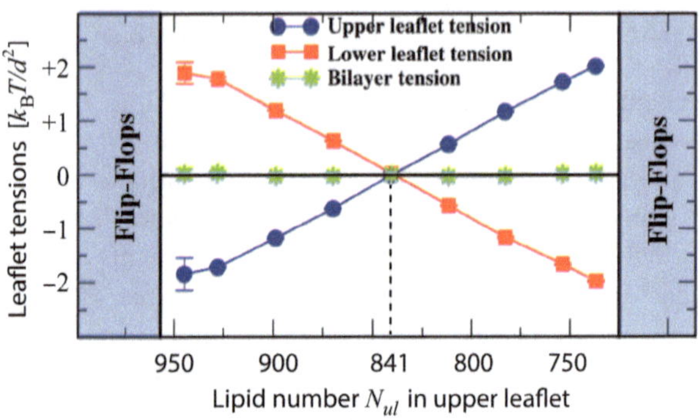

Fig. 4 Leaflet tensions of tensionless planar bilayers: upper leaflet tension Σ_{ul} (blue) and lower leaflet tension Σ_{ll} (red) *versus* lipid number N_{ul} for constant total lipid number $N_{ul} + N_{ll} = 1682$.[19] The lipid number N_{ul} is reduced by moving lipids from the upper to the lower leaflet, thereby increasing the upper leaflet tension and decreasing the lower one. Both leaflet tensions vanish for $N_{ul} = 841$ (vertical dashed line) which defines the relaxed reference state of the planar bilayers. The green data display the bilayer tension $\Sigma = \Sigma_{ul} + \Sigma_{ll}$, which is close to zero. For $945 \ge N_{ul} \ge 737$, the bilayer remains stable but becomes unstable (i) for $N_{ul} \ge 957$ *via* flip-flops from the compressed upper to the stretched lower leaflet and (ii) for $N_{ul} \le 725$ *via* flip-flops from the compressed lower to the stretched upper leaflet.

The stability regime for planar tensionless bilayers as displayed in Fig. 4, is bounded by two instability lines: the left instability line at $N_{ul} = 957$ and $N_{ll} = 725$, as well as the right instability line at $N_{ul} = 725$ and $N_{ll} = 957$. For $N_{ul} > 957$, the lipids in the compressed upper leaflet start to undergo flip-flops into the stretched lower leaflet. For $N_{ul} < 725$, the lipids in the compressed lower leaflet start to flip-flop into the stretched upper leaflet. In addition to flip-flops of lipid molecules, tensionless planar bilayers undergo structural instabilities outside of the stability regime in Fig. 4. Data for lipid flip-flops and structural instabilities are displayed in Fig. S8 and S9.† On the other hand, if a multicomponent and tensionless lipid bilayer contains a lipid species such as cholesterol that undergoes frequent flip-flops, these flip-flops act to decrease the stress asymmetry between the bilayer leaflets (Section S5 and Fig. S10†).[24]

3.5 Fluctuation tension of planar bilayers

In this last subsection on planar bilayers, we briefly look at the effect of the membrane's shape fluctuations. These fluctuations can be decomposed into Fourier modes, which leads to the fluctuation spectrum

$$S(q) \approx \frac{1}{A_{\|}} \frac{k_B T}{\Sigma_{fl} q^2 + \kappa q^4} \tag{16}$$

for small wavenumber q. The prefactor of the q^2-term defines the fluctuation tension Σ_{fl}. For large q-values, the spectrum is changed by molecular protrusions.

One interesting and open question is the relation between the fluctuation tension Σ_{fl} in eqn (16) and the mechanical bilayer tension Σ. In our original molecular dynamics study of the fluctuation spectrum,[14] we assumed that vanishing mechanical tension, $\Sigma = 0$, implies vanishing fluctuation tension, $\Sigma_{fl} = 0$. This assumption has been confirmed by the fluctuation spectra of planar bilayers with two lipid components, see simulation data in Fig. S11.†

On the other hand, several molecular dynamics simulations found small differences between the numerical values of the two tensions Σ and Σ_{fl},[25–27] in agreement with the theory described in ref. 28. However, the most recent study,[29] which addressed this issue using Monte Carlo simulations, came to the conclusion that the two tensions are, in fact, identical, that is, that $\Sigma_{fl} = \Sigma$, in agreement with our original assumption.[14]

4 Membrane tensions of nanovesicles

For a closed vesicle, we distinguish the inner from the outer leaflet by the subscripts il and ol and denote the corresponding leaflet tensions by Σ_{il} and Σ_{ol}. The bilayer tension Σ is again equal to the sum of the two leaflet tensions, that is, $\Sigma = \Sigma_{il} + \Sigma_{ol}$.

4.1 Mechanical tensions of spherical bilayers and leaflets

For a spherical bilayer, the stress profile depends on the radial coordinate r, which measures the distance from the center of the sphere. The corresponding mechanical tension Σ can be obtained from the stress profile $s = s(r)$ *via*

$$\Sigma = \int_0^{+\infty} dr\, s(r). \quad (17)$$

The bilayer tension Σ of a spherical vesicle depends on the vesicle volume in close analogy to the experimental procedures of osmotic deflation and inflation. Osmotic deflation leads to a reduction of the bilayer tension whereas osmotic inflation increases this tension.

The inner leaflet occupies a certain range of r-values, which lie below the r-values for the outer leaflet. The molecular interface between these two leaflets defines the midsurface of the spherical bilayer at $r = r_{mid}$.[1] The leaflet tensions are then obtained from

$$\Sigma_{il} = \int_0^{r_{mid}} dr\, s(r) \text{ and } \Sigma_{ol} = \int_{r_{mid}}^{+\infty} dr\, s(r). \quad (18)$$

4.2 Leaflet tension space for spherical bilayers

Now, consider an ensemble of spherical vesicles, which all have the same total lipid number $N_{il} + N_{ol}$. The corresponding leaflet tension space is displayed in Fig. 5 for total lipid number $N_{il} + N_{ol} = 2525$. The origin of this space, corresponding to the relaxed reference state with tensionless leaflets $\Sigma_{il} = \Sigma_{ol} = 0$, is obtained for $N_{il} = 840$ lipids in the inner leaflet and $N_{ol} = 1685$ lipids in the outer leaflet. Therefore, the bilayer state with two tensionless leaflets is obtained when the outer lipid number N_{ol} is about twice as large as the inner lipid number N_{il}.

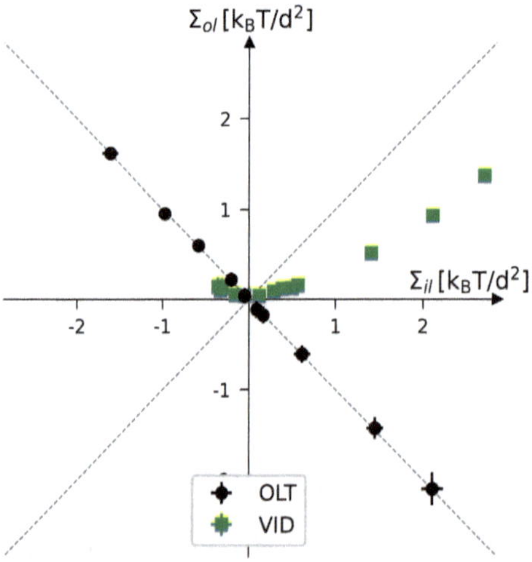

Fig. 5 Leaflet tension space for vesicle bilayers with a total number of $N_{il} + N_{ol} = 2525$ lipids in both leaflets. The two coordinates are the leaflet tensions Σ_{il} and Σ_{ol} in the inner and outer leaflets. The reference state with tensionless leaflets, corresponding to $\Sigma_{il} = \Sigma_{ol} = 0$, is obtained for a vesicle bilayer with $N_{il} = 840$ lipids in the inner leaflet and $N_{ol} = 1685$ lipids in the outer one.[18] Negative and positive leaflet tensions describe compressed and stretched leaflets. The black data correspond to elastic OLT deformations obtained from the reference state by reshuffling lipids from one leaflet to the other and adjusting the vesicle volume to obtain tensionless bilayers with $\Sigma = \Sigma_{il} + \Sigma_{ol} = 0$. The green data represent the elastic deformations arising from changes in vesicle volume, corresponding to vesicle inflation or deflation (VID).

Because the reference state has a large lipid number asymmetry, it is difficult to identify the ELT states with $\Sigma_{ol} = \Sigma_{il}$ for spherical bilayers. In contrast, it is relatively simple to generate the OLT states with $\Sigma_{ol} = -\Sigma_{il}$ from the relaxed reference state with $\Sigma_{ol} = \Sigma_{il} = 0$ by: (i) reshuffling lipids between the two leaflets and (ii) adjusting the vesicle volume to obtain a tensionless bilayer.[18,19,30] The resulting OLT states correspond to the black data points in Fig. 5. The different OLT states are distinguished by the stress asymmetry

$$\Delta\Sigma_{ve} \equiv \Sigma_{ol} - \Sigma_{il} \tag{19}$$

which increases in magnitude as we move away from the reference state with $\Sigma_{il} = \Sigma_{ol} = 0$ along the black data points in Fig. 5.

To generate the green data points for vesicle inflation or deflation (VID) in Fig. 5, we start again from the reference state with $\Sigma_{il} = \Sigma_{ol} = 0$ but now change the vesicle volume in order to increase or decrease the bilayer tension, thereby mimicking the experimental procedure of osmotic inflation or deflation. Note that the green data do not follow the main diagonal, in contrast to the equal leaflet tension (ELT) data in Fig. 3, but become parallel to the main diagonal for large leaflet tensions, which implies a constant stress asymmetry, $\Delta\Sigma_{ve}$.

4.3 Leaflet tensions of tensionless vesicle bilayers

The leaflet tensions of tensionless vesicle bilayers, which are assembled from a total number of $N_{il} + N_{ol} = 2875$ lipids, are displayed in Fig. 6. For these vesicle bilayers, the relaxed reference state with tensionless leaflets is obtained for $N_{ol} = 1921$ lipids in the outer leaflet and $N_{il} = 954$ lipids in the inner leaflet, corresponding to the dashed vertical line in Fig. 6. Thus, when both leaflet tensions vanish, the lipid number N_{ol} in the tensionless outer leaflet is more than twice as large as the lipid number N_{il} in the tensionless inner leaflet.

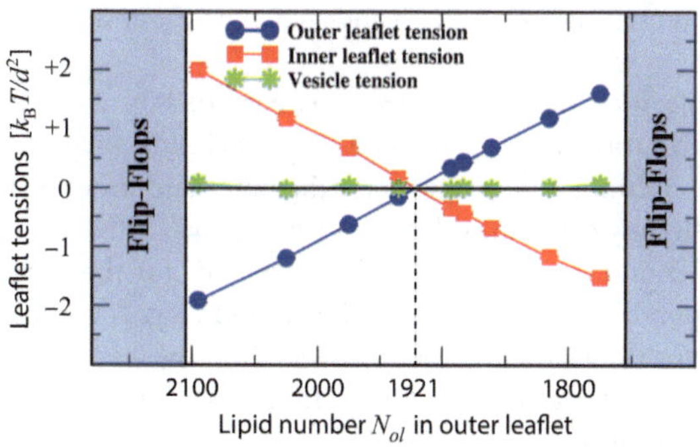

Fig. 6 Leaflet tensions for tensionless vesicle bilayers with a total number of $N_{il} + N_{ol} = 2875$ lipids. The vesicles have a diameter of 23.8 d or 19 nm. The blue and red data represent the outer and inner leaflet tensions.[19] The green data correspond to the bilayer tension $\Sigma = \Sigma_{ol} + \Sigma_{il}$, which is close to zero. During the run time of the simulations, we observed no flip-flops within the stability regime (white), corresponding to $1775 \leq N_{ol} \leq 2095$. The left vertical line at $N_{ol} = 2105$ represents the instability line at which the lipids start to undergo flip-flops from the compressed outer to the stretched inner leaflet. The right vertical line at $N_{ol} = 1755$ represents the instability line at which the lipids start to undergo flip-flops from the compressed inner to the stretched outer leaflet.

The white stability regime in Fig. 6 is bounded by two instability lines. When we reach the left instability line at $N_{ol} = 2105$, lipids start to flip-flop from the compressed outer leaflet to the stretched inner leaflet. Likewise, at the right stability line with $N_{ol} = 1755$, the compressed inner leaflet becomes instable with respect to flip-flops from the inner to the outer leaflet. The cumulative distribution function for the first flip-flop is displayed in Fig. S12.† In addition to flip-flops, tensionless vesicle bilayers undergo structural instabilities outside of their stability regime as illustrated in Fig. S13.†

4.4 Volume per lipid and lateral volume compressibility

The elastic response of a planar bilayer to mechanical tension can be described in terms of area per lipid or in terms of volume per lipid. In order to determine area per lipid for nanovesicles, we need to introduce some projection onto the curved surfaces of the vesicle bilayers, which represents an ambiguous procedure. Such an ambiguity can be avoided by using the volume per lipid, ν, to characterize the lipid bilayer.[18]

The volume per lipid can be obtained from the following procedure based on Voronoi tessellation. In the coarse-grained molecular model, the molecules are built up from different types of spherical beads. Voronoi tessellation assigns a polyhedral cell to each bead, such that all points in this cell are closer to the center of the chosen bead than to the center of any other bead. In Fig. S7,† we illustrate the tessellation method for a planar bilayer and for an individual lipid molecule. Next, the volumes V_{il} and V_{ol} of the inner and outer leaflets are computed by summing up the volumes of all polyhedral cells that belong to the respective leaflet. Finally, the volumes per lipid, ν_{il} and ν_{ol}, of the inner and outer leaflets are then obtained by dividing the leaflet volumes by the lipid numbers, N_{il} and N_{ol}, which leads to

$$\nu_{il} = \frac{V_{il}}{N_{il}} \text{ and } \nu_{ol} = \frac{V_{ol}}{N_{ol}}. \tag{20}$$

These lipid volumes vary with the leaflet tensions Σ_{il} and Σ_{ol} in the inner and outer leaflets.

One example for this dependence is shown in Fig. 7 for a vesicle bilayer assembled from a total number of $N_{il} + N_{ol} = 1500$ lipids. Inspection of this figure shows that the optimal lipid volumes for the outer and inner leaflet have almost identical but slightly different values as given by[18] $\nu_{il,0} = (3.5790 \pm 0.0004)d^3$ for the inner leaflet and by $\nu_{ol,0} = (3.5612 \pm 0.0003)d^3$ for the outer, leaflet of the vesicle, corresponding to the vertical dashed lines in Fig. 7. Therefore, the optimal lipid volumes for the inner and outer leaflets differ by less than one percent.

The deviations of the lipid volumes from their optimal values define the relative volume dilations

$$\Delta\nu_{il} \equiv \frac{\nu_{il} - \nu_{il}^0}{\nu_{il}^0} \text{ and } \Delta\nu_{ol} \equiv \frac{\nu_{ol} - \nu_{ol}^0}{\nu_{ol}^0} \tag{21}$$

for the inner and outer leaflets. The elastic response of the two leaflets to the leaflet tensions Σ_{il} and Σ_{ol} is then described by the tension–volume relations

$$\Sigma_{il} \approx B_{il}\Delta\nu_{il} = B_{il}\frac{\nu_{il} - \nu_{il}^0}{\nu_{il}^0} \tag{22}$$

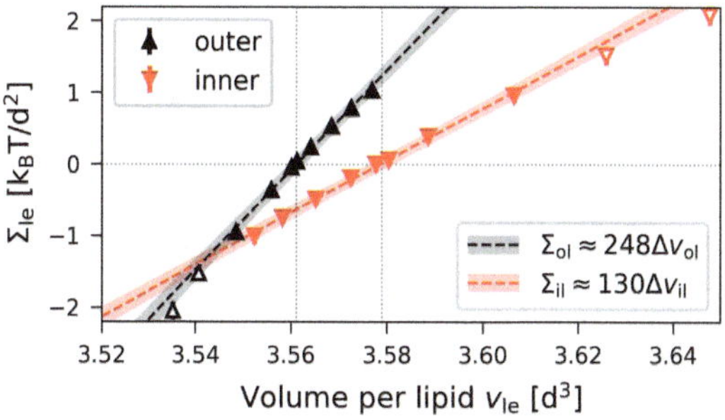

Fig. 7 Leaflet tensions *versus* volume per lipid for nanovesicles with a total number of $N_{il} + N_{ol} = 1500$ lipids.[18] Leaflet tensions $\Sigma_{le} = \Sigma_{ol}$ and $\Sigma_{le} = \Sigma_{il}$ of the outer (black triangles) and inner leaflets (red triangles) *versus* volumes per lipid, $v_{le} = v_{il}$ and $v_{le} = v_{ol}$, in the two leaflets of the tensionless bilayers. The two dotted vertical lines are located at the optimal lipid volumes $v = v^0_{ol} \simeq 3.56d^3$ and $v = v^0_{il} \simeq 3.58d^3$ of the outer and inner leaflet.

and

$$\Sigma_{ol} \approx B_{ol}\Delta v_{ol} = B_{ol}\frac{v_{ol} - v^0_{ol}}{v^0_{ol}} \tag{23}$$

to first order in the relative volume dilations, which define the lateral volume compressibilities B_{il} and B_{ol}. These two elastic moduli are significantly different for the two leaflets, as follows from the data in Fig. 7.[18]

4.5 Tension-induced fusion of nanovesicles

Membrane fusion of two nanovesicles can be induced by their bilayer tensions or by their leaflet tensions. A particularly interesting fusion process is observed for tensionless bilayers with $\Sigma = 0$ and $\Sigma_{ol} = -\Sigma_{il}$, which avoids bilayer rupture as an alternative pathway to fusion.

4.5.1 Membrane fusion induced by bilayer tension. Experimentally, tension-induced fusion was originally studied by osmotic swelling of vesicles that adhere to a planar bilayer.[31] The molecular mechanisms underlying tension-induced fusion have been visualized and elucidated by molecular dynamics simulations.[32–34] More recently, several experimental studies provided additional evidence for tension-induced fusion,[35–38] which can be strongly enhanced by electrostatic interactions between oppositely charged membranes.[36] In the simulations, the close proximity of two lipid bilayers can lead to a variety of alternative outcomes such as stable hemifused bilayers and bilayer rupture, but fusion events are always observed within a certain range of positive bilayer tensions.

One example for membrane fusion as observed in molecular dynamics simulations, is depicted in Fig. 8 for a nanovesicle in contact with a planar bilayer.[33,39] In order to visualize the fusion pathway, the lipids in the planar bilayer have red head groups and green tails whereas the lipids in the vesicle have orange head groups and yellow tails.

The observed fusion pathway proceeds *via* three distinct substeps. After the vesicle and the planar bilayer have come into contact at time t = 78.5 ns, the vesicle starts to spread onto the bilayer, thereby forming a highly curved

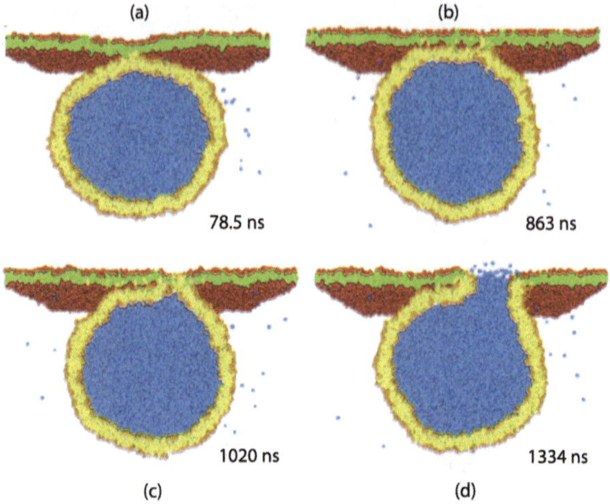

Fig. 8 Fusion of nanovesicle with planar bilayer: The vesicle bilayer consists of 6869 lipids (orange heads, yellow chains) and has a diameter of about 30 nm, while the planar bilayer contains 6911 lipids (red heads, green chains). The water beads originally inside the vesicle are blue, the water beads in the exterior compartment are not shown for clarity. The figure displays four time lapse snapshots which illustrate the development of the fusion process from 78.5 ns, when the first contact between the two membranes occurs and the lipid tails start to undergo interbilayer flip-flops, until the opening of the fusion pore after 1334 ns. Reused with permission from ref. 33, American Physical Society, 2007.

membrane segment along the contact line with the planar bilayer. The lipids along the contact line then undergo interbilayer flip-flops from the vesicle to the planar bilayer, see the yellow lipid tails that appear in the lower leaflet of the planar bilayer. These flip-flops lead to intermixing and partial fusion of the two bilayers, followed by the nucleation of a bean-shaped hemifusion diaphragm at about $t = 1177$ ns. The bean-shape of the diaphragm can be clearly seen when one looks at cross-sections parallel to the planar bilayer, which are perpendicular to the side views in Fig. 8. Finally, this diaphragm ruptured close to the contact line at $t = 1334$ ns, thereby opening up a strongly asymmetric fusion pore that provides a water channel between the interior vesicle solution (blue) and the water compartment (white) above the planar bilayer.

The fusion process displayed in Fig. 8 was induced by increasing the area per lipid, a, of the planar bilayer, thereby increasing the mechanical tension Σ, experienced by this bilayer. The resulting fusion process competes with several alternative pathways such as membrane adhesion, hemifusion, and rupture. Because of these alternative pathways, the fraction of successful fusion events, which defines the fusion probability, depends strongly on the area per lipid, a, and on the mechanical tension Σ within the planar bilayer. Indeed, as shown in Fig. S14,† the fusion probability increases linearly with increasing $a \sim\Sigma$, reaches a maximum, and then decreases again for larger values of $a \sim\Sigma$. The decay of the fusion probability for large a is caused by bilayer rupture.

4.5.2 Vesicle fusion driven by leaflet tensions. Bilayer rupture is strongly suppressed when we consider tensionless bilayers with vanishing mechanical tension, $\Sigma = 0$, corresponding to one stretched and one compressed leaflet. Somewhat surprisingly, two nanovesicles with $\Sigma = 0$ also undergo fusion,

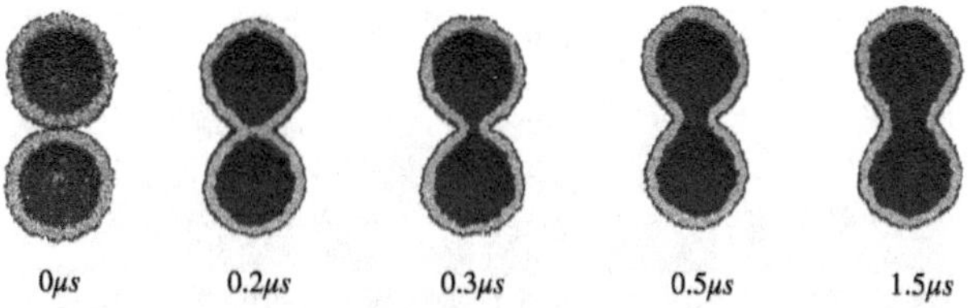

Fig. 9 Two identical vesicles with tensionless bilayers undergo fusion, when their outer leaflets are stretched by a sufficiently large leaflet tension Σ_{ol} and their bilayers experience a sufficiently large stress asymmetry. Each vesicle contains $N_{il} = 4500$ lipids in its inner and $N_{ol} = 5600$ lipids in its outer leaflet, corresponding to the outer leaflet tension $\Sigma_{ol} = +1.02\ k_BT/d^2$ and the inner leaflet tension $\Sigma_{il} = -1.02\ k_BT/d^2$. The vesicles come into contact at time $t = 0$ and undergo fusion within 0.3 μs.[1]

provided their outer leaflets are stretched by a sufficiently large leaflet tension Σ_{ol}. One example for such a fusion process is displayed in Fig. 9.

The fusion of nanovesicles with tensionless bilayers but stretched outer leaflets can be understood as follows. First, within the emerging contact area, the shielding of the hydrophobic cores of the two bilayers is reduced by stretching the two outer leaflets and increasing their areas and volumes per lipid. Second, both leaflet tensions drive the subsequent opening of the fusion pore. Indeed, the stretched outer leaflet tries to reduce its area, thereby pulling on the emerging fusion pore, whereas the compressed inner leaflet tries to increase its area, thereby curving the rim of the pore. Furthermore, recent molecular dynamics simulation show that this fusion process proceeds without bilayer rupture and without leakage of vesicle contents.[40]

5 Membrane tensions of giant vesicles

In this section, we will address membrane tensions of giant vesicles and will focus on multispherical and two-sphere shapes as well as on different shapes of tubular membranes. All of these shapes involve closed membrane necks, which also represent the necessary initial step for vesicle division. Additional aspects of multispherical shapes are discussed in Section S6†, more details on positive two-sphere shapes are provided in Section S7,† which also describes the experimentally observed division of positive two-sphere shapes.[41] Negative two-sphere shapes are further examined in Section S8.†

5.1 Multispherical shapes of giant vesicles

Multispherical shapes of giant vesicles can be obtained by exposing the two leaflets of the vesicle membranes to two different sugar solutions which generate a spontaneous curvature within the membranes.[42,43] If the inner leaflet is in contact with sucrose and the outer leaflet exposed to glucose, the spontaneous curvature m is positive. On the other hand, the GUV membrane acquires a negative spontaneous curvature, $m < 0$, when the interior leaflet is exposed to glucose and the outer leaflet is in contact with sucrose. Some examples for the resulting multispheres are displayed in Fig. 10.

5.1.1 Multispheres consist of large and small spheres. Inspection of Fig. 10 reveals that each multisphere involves only two types of sphere: a large sphere with

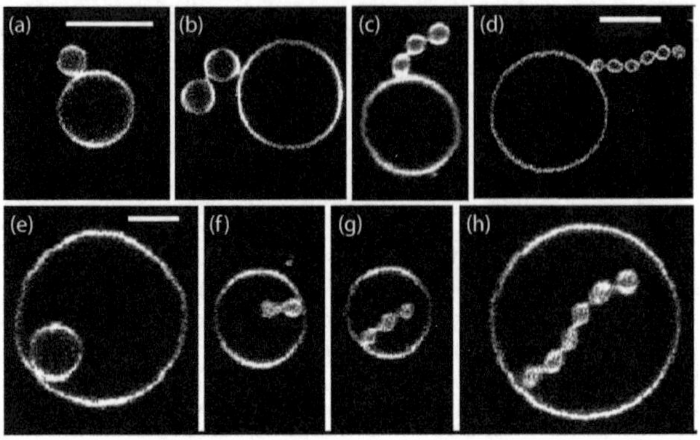

Fig. 10 (a–d) (1 + N_s)-spheres for positive spontaneous curvatures, consisting of one large sphere and N_s small spheres, all with positive mean curvature. The scale bars in (a) and (d) are 10 μm, the scale bar in (a) also applies to (b) and (c). (e–h) (1 + N_s)-spheres for negative spontaneous curvatures, consisting of one large sphere with positive mean curvature and N_s small spheres with negative mean curvature. The membrane necks of the positive and negative multispheres have a positive and negative effective mean curvature, respectively, as follows from eqn (24). The scale bar in (e) is 5 μm and applies to (e–h).[42]

radius R_l and positive mean curvature $M_l = +1/R_l$, as well as small spheres with identical radius R_s and identical mean curvature $M_s = \pm 1/R_s$. This general property of multispheres is a direct consequence of the local shape equation for spherical membrane segments as given by eqn (S24),† which is quadratic in the constant mean curvature of the spherical segment. For positive spontaneous curvature, multispheres can also consist of several equally sized spheres, corresponding to a degenerate (double) root of the quadratic shape equation, see the example in Fig. 12c below.

5.1.2 Limit shapes *versus* persistent shapes. Each multispherical shape can be characterized by its stability regime, that is, by a certain subregion of the morphology diagram, which depends on two dimensionless shape parameters, the volume-to-area ratio $\bar{v}$ and the rescaled spontaneous curvature $\bar{m}$ as defined by eqn (S9) and (S10).† The stability regime of each multisphere is bounded by two lines of limit shapes. When we leave the stability regime by crossing a line of limit shapes, at least one closed membrane neck opens up. On the other hand, away from these limit shapes, the multispherical shapes represent *persistent shapes* that depend only on $\bar{v}$ but are independent of $\bar{m}$.[44]

5.1.3 Positive and negative multispheres. Each closed neck between two connected spheres i and j with mean curvature M_i and M_j can be characterized by its effective mean curvature as given by[44]

$$M_{ij}^{\text{eff}} \equiv \frac{1}{2}(M_i + M_j) = \frac{1}{2}\left(\pm\frac{1}{R_i} \pm \frac{1}{R_j}\right). \tag{24}$$

This effective curvature represents a purely geometric quantity that can be directly obtained from microscopy images as in Fig. 10. Therefore, we can distinguish positive necks with $M_{ij}^{\text{eff}} > 0$ from negative necks with $M_{ij}^{\text{eff}} < 0$.

In fact, for each multisphere, all necks have the same sign. Indeed, all membrane necks are positive for each multisphere in Fig. 10a–d and negative for each multisphere in Fig. 10e–h. As a consequence, we can distinguish positive multispheres with positive membrane necks from negative multispheres with negative membrane necks.

5.1.4 Mechanical membrane tension of multispheres. The constant mean curvature M_{sp} of a spherical membrane segment is obtained by solving the local shape equation as given by eqn (S24).† This equation is quadratic in M_{sp} and depends on ΔP and Σ. The two solutions of this equation are equal to the mean curvatures of the large and the small spheres, M_l and M_s, which build up the multispheres as in Fig. 10. It is possible to invert these solutions to obtain ΔP and Σ, which leads to the mechanical tension[44]

$$\Sigma = 2\kappa m(M_l + M_s - m) = 2\kappa m\left(\frac{1}{R_l} \pm \frac{1}{R_s} - m\right). \tag{25}$$

The $\pm$ sign applies to positive and negative multispheres, which are characterized by small spheres with positive and negative mean curvature, respectively. The relationship in eqn (25) remains valid for the degenerate case of a single solution with $M_l = M_s = M_*$ and $R_l = R_s = R_*$.

The mechanical tension as given by eqn (25) involves both the curvature-elastic parameters κ and m as well as the mean curvatures M_l and M_s (or M_*) of the large and small spheres. Therefore, eqn (25) directly demonstrates that the mechanical tension Σ depends on the size and shape of the multispherical vesicle.

5.2 Two-sphere shapes of giant vesicles

The simplest multispheres are provided by two spheres connected by a closed membrane neck as shown in panels a and e of Fig. 10. The positive two-sphere shape in Fig. 10a consists of one large sphere and one smaller spherical out-bud, and the negative two-sphere shape in Fig. 10e of one large sphere and one smaller spherical in-bud. The stability regime of the positive two-spheres is described in the next paragraph and displayed in Fig. 11, the stability regime of the negative two-spheres in Section S8 and Fig. S16.†

5.2.1 Stability regime for positive two-sphere shapes. For positive spontaneous curvature, the two-sphere shape in Fig. 10a involves a positive ls-neck between the large and the small sphere. The condition for this ls-neck to be stably closed is given by the inequality[44]

$$m \geq M_{ls}^{eff} = \frac{1}{2}(M_l + M_s) = \frac{1}{2}\left(\frac{1}{R_l} + \frac{1}{R_s}\right) > 0. \tag{26}$$

In the limiting case, in which this inequality becomes an equality, we obtain the neck closure condition that determines the line of limit shapes L_{1+1}^{pos}, see the upper boundary line of the stability regime in Fig. 11.

The lower boundary line of the stability regime in Fig. 11 is provided by another line of limit shapes, denoted by L_{2*} and consisting of two equally sized spheres with radius $R_l = R_s = R_* = R_{ve}/\sqrt{2}$. In the morphology diagram, the L_{2*}-line is located at $\bar{v} = \bar{v}_{2*} = 1/\sqrt{2}$ and $\bar{m} \geq \bar{m}_{2*} = \sqrt{2}$.

The two lines of limit shapes, L_{1+1}^{pos} and L_{2*}^{pos}, meet at the corner point with $\bar{v} = \bar{v}_{2*} = 1/\sqrt{2}$ and $\bar{m} = \bar{m}_{2*} = \sqrt{2}$. At this corner point, the limit shape consists of two equally sized spheres with mean curvature $M_* = 1/R_* = m_{2*}$, which implies zero bending energy, $E_{be} = 0$, as follows from eqn (S3).†

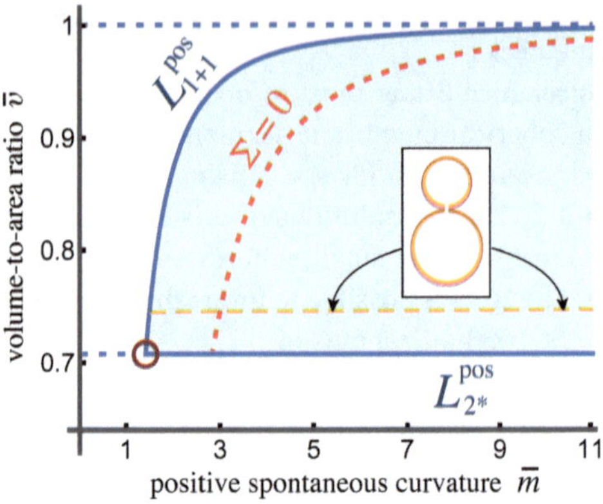

Fig. 11 Stability regime (light blue) for positive (1 + 1)-spheres with positive spontaneous curvature $\bar{m}$. The stability regime is bounded by two lines (dark blue) of limit shapes, L^{pos}_{1+1} and L^{pos}_{2*}, which meet at the corner point (purple circle) with $\bar{m} = \sqrt{2}$ and $\bar{v} = 1/\sqrt{2}$. The persistent (1 + 1)-spheres along the red dashed line are characterized by tensionless membranes with mechanical tension $\Sigma = 0$; this tension is positive and negative to the left and to the right of the red dashed line. The inset displays a persistent (1 + 1)-shape that remains unchanged when we vary the spontaneous curvature along the horizontal dashed line (orange). This shape invariance is a direct consequence of the area and volume relations in eqn (S27) and (S28).†

5.2.2 Mechanical membrane tension of positive two-spheres. Along the line of limit shapes L^{pos}_{1+1} in Fig. 11, the neck closure condition $M_l + M_s = 2M^{eff}_{ls} = 2m$ implies the mechanical tension

$$\Sigma = 2\kappa m^2 = 2\kappa (M^{eff}_{ls})^2 \text{ for all limit shapes } L^{pos}_{1+1} \tag{27}$$

as follows from eqn (25). The geometry of the limit shape L^{pos}_{1+1} changes with the spontaneous curvature m because the effective mean curvature $M^{eff}_{ls} = m$ of the ls-neck increases with increasing m. The relation in eqn (25) also implies that

$$\Sigma = 0 \text{ for } m = M_l + M_s, \tag{28}$$

which is valid for the persistent shapes along the red dashed line in Fig. 11.

Along the line of limit shapes L^{pos}_{2*} with $\bar{v} = \bar{v}_{2*} = 1/\sqrt{2}$ and $\bar{m} \geq \bar{m}_{2*} = \sqrt{2}$, the mean curvature M_* of the equally sized spheres is equal to $M_* = m_{2*} = \sqrt{2}/R_{ve}$. It then follows from eqn (25) with $M_l = M_s = M_*$ that the associated membrane mechanical tension is given by

$$\Sigma = 2\kappa m\left(\frac{2\sqrt{2}}{R_{ve}} - m\right) \text{ for the limit shapes } L^{pos}_{2*}. \tag{29}$$

This tension vanishes at $m = 2m_{2*} = 2\sqrt{2}/R_{ve}$, which represents the lower end point of the dashed red line in Fig. 11. Furthermore, the mechanical tension Σ as given by eqn (29) behaves as $\Sigma \approx -2\kappa m^2$ for large m, that is, the mechanical tension Σ becomes large and negative for large and positive spontaneous curvature m.

As shown in Fig. 11, the L_{1+1}-line meets the L_{2*}-line of two equally sized spheres at the corner point with $\overline{m} = \overline{m}_{2*} = \sqrt{2}$ and $\overline{v} = \overline{v}_{2*} = 1/\sqrt{2}$ where the mechanical tension Σ has the value

$$\Sigma = 2\kappa m_{2*}{}^2 = 4\kappa/R_{\text{ve}}{}^2 \text{ (corner point).} \tag{30}$$

5.2.3 Division of positive two-sphere vesicles. Closed membrane necks of multispherical vesicles also play a pivotal role for the division of these vesicles. As an example, consider the positive two-sphere vesicle in the inset of Fig. 11. This shape remains unchanged when we increase the spontaneous curvature m of the vesicle membrane. However, as explained in Section S7.2† such an increase of m also increases the constriction force f^{pos} acting against the positive closed neck of the two-sphere vesicle, as described by eqn (S30).† A sufficiently large constriction force leads to the cleavage of the closed neck and to the division of the two-sphere vesicle into two daughter vesicles, as observed experimentally for GUVs.[41] Because the constriction force is a local neck property, such a force is present at each closed neck of all multispheres.

5.3 Different shapes of tubular membranes

Tubular membranes and vesicles can exhibit different shapes, corresponding to multispherical, unduloidal, and cylindrical shapes.

5.3.1 Multispherical shapes of membrane tubes. We first consider multispherical shapes of membrane tubes as shown in Fig. 12. In Fig. 12b, the tube is part of a (1 + 14)-sphere and consists of 14 small spheres emanating from a somewhat larger sphere. In Fig. 12c, the multispherical tube is built up from 15 equally sized spheres, corresponding to a (15*)-sphere. The stability regime for the (1 + 14)-spheres is displayed in Fig. 12a as the light blue subregion of the morphology diagram. As for (1 + 1)-spheres, the stability regime for the (1 + 14)-spheres is bounded by two lines of limit shapes, denoted by L^{pos}_{1+14} and L^{pos}_{15*}.

The line of limit shapes L^{pos}_{1+14} in Fig. 12a is given by

$$\overline{v} = \overline{v}_{1+14}(\overline{m}) = \left(1 - \frac{14}{\overline{m}^2}\right)^{3/2} + \frac{14}{\overline{m}^3}, \tag{31}$$

and the line of limit shapes L^{pos}_{15*} by

$$\bar{m} \geq \bar{m}_{15*} \equiv \sqrt{15} \text{ and } \bar{v} = \bar{v}_{15*} \equiv 1/\sqrt{15} \tag{32}$$

as follows from the area and volume relations in eqn (S27) and (S28)† with small-sphere radius $r_s = 1/\bar{m}$. The red dashed line in Fig. 12a provides the location for vesicle membranes with vanishing mechanical tension, $\Sigma = 0$. Along this line, the spontaneous curvature m is equal to $2M^{\text{eff}}_{\text{ls}}$, which can be combined with eqn (S27) and (S28)† to obtain a parametric plot.[43]

The small spheres of the (1 + 14)-spheres in Fig. 12b transform into small prolates along an instability line as given by the relationship

$$r_s - \frac{r_s{}^2}{\sqrt{1 - 14r_s{}^2}} = \frac{3}{\overline{m}} \tag{33}$$

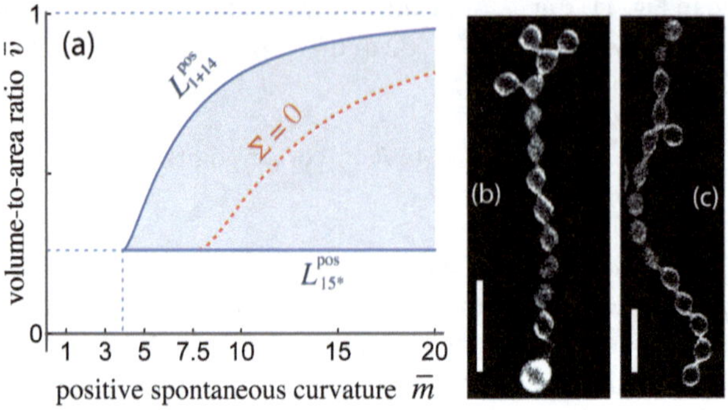

Fig. 12 (a) Stability regime (light blue) for positive (1 + 14)-spheres. The stability regime is bounded by two lines of limit shapes (dark blue), L^{pos}_{1+14} and L^{pos}_{15*}. These two lines meet at the corner point with $\overline{m} = \overline{m}_{15*} = \sqrt{15} = 3.87$ and $\overline{v} = \overline{v}_{15*} = 1/\sqrt{15} = 0.258$. The persistent (1 + 14)-shapes along the red dashed line are enclosed by membranes with vanishing mechanical tension, $\Sigma = 0$; (b) example for a GUV that forms a (1 + 14)-sphere consisting of one somewhat larger sphere at the lower end and a multispherical tube of 14 small spheres; and (c) example for a GUV forming a (15*)-sphere with 15 equally sized spheres. The images in (b) and (c) are from ref. 42, both scale bars are 10 μm.

between the small-sphere radius r_s and the spontaneous curvature $\bar{m}$. This equation has no solution for $\bar{m} < \bar{m}_{ss} \equiv 21.47$, one solution for $\bar{m} = \bar{m}_{ss}$, and two solutions for $\bar{m} > \bar{m}_{ss}$. Therefore, the small spheres of the persistent (1 + 14)-spheres are stable for $\bar{m} < \bar{m}_{ss} = 21.47$ but transform into small prolates for $\bar{m} > \bar{m}_{ss}$ and a certain $\bar{m}$-dependent range of $\bar{v}$-values. At $\bar{m} = \bar{m}_{ss}$, the instability line emerges from the single point $(\bar{m}_{ss},\bar{v}_{ss}) = (21.47,0.4035)$, which opens up into a parabola-like line for $\bar{m} > \bar{m}_{ss}$. For large $\bar{m}$, the upper and the lower branch of this parabola-like line approach the two lines of limit shapes, L^{pos}_{1+14} and L^{pos}_{15}, respectively. This bifurcation is located outside of the morphology diagram displayed in Fig. 12a.

5.3.2 Multispherical tubes *via* fusion of proteoliposomes. Membrane nanotubes have been prepared by the fusion of proteoliposomes that contain certain membrane proteins, which hydrolyse GTP.[45,46] Some insight into this process of nanotube formation from proteoliposomes can be obtained from the following thought experiment. Thus, let us consider N_* proteoliposomes that undergo fusion to form an (N_*)-multisphere. In the absence of GTP, the individual proteoliposome are taken to be relaxed in the sense that the mean curvature of the vesicle membranes are close to the membrane's spontaneous curvature m, which implies that the radius R of the proteoliposomes is close to $1/m$ and that their bending energy as given by eqn (S3)† is close to zero. The dimensionless shape parameters $\bar{v}$ and $\bar{m}$ in eqn (S9) and (S10)† are then equal to $\bar{v} = \bar{v}_1 = 1$ and $\bar{m} = \bar{m}_1 = 1$ for each proteoliposome.

When we add GTP, the proteoliposomes start to fuse and to form multispherical tubes. More precisely, the fusion of N_* proteoliposomes leads to an (N_*)-multisphere with the shape parameters $\overline{v} = \overline{v}_{N*} \equiv 1/\sqrt{N^*}$ and $\overline{m} = \overline{m}_{N*} \equiv \sqrt{N_*}$, provided both the total membrane area and the total vesicle volume are conserved during the fusion process. For $N_* = 2$, the shape parameters $\overline{v} = \overline{v}_{2*} = 1/\sqrt{2}$ and $\overline{m} = \overline{m}_{2*} = \sqrt{2}$ represent the corner point of the stability regime in Fig. 11, at which the two lines of limit shapes, L^{pos}_{1+1} and L^{pos}_{2*}, meet. For $N_* = 15$, the shape parameters $\bar{\nu} = \bar{\nu}_{15*} \equiv 1/\sqrt{15}$ and $\bar{m} = \bar{m}_{15*} \equiv \sqrt{15}$ define the location of the corner point in Fig. 12a.

5.3.3 Unduloidal shapes of membrane tubes. So far, super-resolution microscopy has not been used to resolve the shapes of nanotubes reconstituted from proteoliposomes *in vitro*. However, structured illumination microscopy has been applied to the nanotubular networks of the endoplasmic reticulum *in vivo*,[47,48] which revealed that these shapes resemble unduloids[49] as depicted in Fig. 13. Within the morphology diagram in the $(\bar{m},\bar{\nu})$-plane, long unduloids with mean curvature $M = m$ and zero bending energy are located between two hyperbolae with $\bar{\nu} = \bar{\nu}_1(\bar{m}) \equiv 1/\bar{m}$ and $\bar{\nu} = \bar{\nu}_2(\bar{m}) \equiv 3/(4\ \bar{m})$. Such unduloid shapes will form *via* the fusion of proteoliposomes when some membrane proteins get lost during the fusion process, thereby reducing the spontaneous curvature $\bar{m}$, or when the fusion process is leaky and the interior volume of the proteoliposomes is reduced during the process.

If one of the membrane necks as indicated by the blue arrow heads in Fig. 13, became closed for an extended period of time, the neck can undergo fission, thereby breaking the nanotubes into several fragments. For reticular networks, such a fragmentation of the nanotubes has indeed been induced experimentally, both *in vivo* and *in vitro*, by downregulating the biosynthesis of GTP-hydrolyzing membrane proteins[50] and by interrupting the supply of GTP.[45] As a consequence, any mechanism that keeps the membrane necks open also prevents nanotube fragmentation.

5.3.4 Peristaltic modes of cylindrical nanotubes. The unduloidal tubes displayed in Fig. 13 can be viewed as peristaltic modes of cylindrical nanotubes. Thus, consider a cylindrical tube of length L and radius R_{cy}. The shape fluctuations of such a cylindrical tube can be decomposed into Fourier modes. We will focus on peristaltic modes which preserve the axisymmetry of the cylinder and include the most unstable mode. It is convenient to introduce a fluctuation tension Σ_{fl} that governs the excess area stored in these shape fluctuations. An analogous tension term has been considered for planar membrane segments[51] and for quasi-spherical vesicles.[52,53]

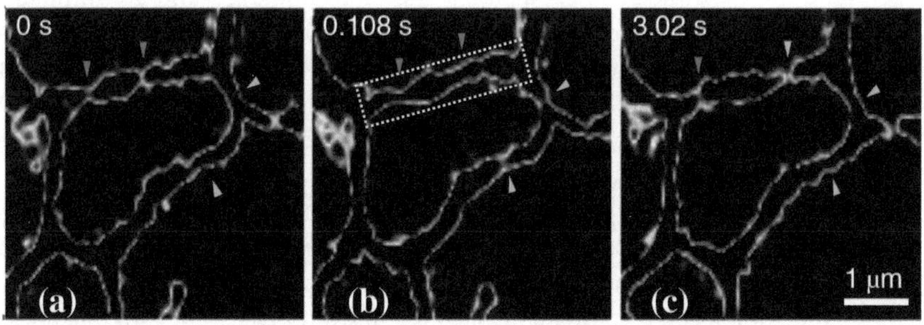

Fig. 13 Time-lapse snapshots of an irregular pentagon formed by membrane nanotubes and five three-way junctions as observed by structured illumination microscopy. Reused with permission from ref. 47, Springer-Nature, 2018. Time points in seconds (upper left corners); scale bar in (c) also applies to (a) and (b). The pentagon consists of five three-way junctions connected by five tubular segments. One tube segment is marked by the white dotted rectangle in (b). The four blue arrow heads point to transient membrane necks (or constrictions) formed along three tubular segments, which resemble peristaltic shape deformations of these tubules.[49]

For a cylindrical tube with radius R_{cy} and length L, the most unstable mode, denoted by l_{n*}, is characterized by wavenumber $p = p_* = 1/R_{cy}$, period $2\pi R_{cy}$, and mean-squared amplitude[49]

$$\langle |l_{n*}| \rangle^2 = \frac{k_B T}{2\pi R_{cy} L \Sigma_{fl} p_*^2} = \frac{k_B T}{2\pi (L/R_{cy}) \Sigma_{fl}}, \tag{34}$$

which grows as $1/\Sigma_{fl}$ for small Σ_{fl}. Thus, the tube should become fragmented for sufficiently small tensions, for which the amplitude of the most unstable mode exceeds the tube radius, that is, for

$$\left\langle |l_{n*}|^2 \right\rangle \geq R_{cy}{}^2 \text{ or } \Sigma_{fl} \leq \Sigma_{th,1} \equiv \frac{k_B T}{2\pi R_{cy} L}. \tag{35}$$

The tension threshold $\Sigma_{th,1}$ is very low. As an example, consider the tube segment within the dotted white rectangle in Fig. 13b, which has a length of $L = 2.5$ μm and a radius of $R_{cy} = 100$ nm. For this tube, the tension $\Sigma_{th,1}$ has the value 4×10^{-6} mN m^{-1} at room temperature.

In Fig. 13, the peristaltic modes do not lead to tube fragmentation. Inspection of this figure indicates that the most unstable modes have amplitudes of about half the cylinder radius which defines another tension threshold $\Sigma_{th,2}$ *via*

$$\left\langle |l_{n*}|^2 \right\rangle = (R_{cy}/2)^2 \text{ or } \Sigma_{th,2} \equiv \frac{4k_B T}{2\pi R_{cy} L} = 4\Sigma_{th,1}. \tag{36}$$

Thus, the fragmentation of the membrane nanotubes can be avoided by increasing the fluctuation tension Σ_{fl} from $\Sigma_{th,1}$ to $\Sigma_{th,2} = 4\Sigma_{th,1}$.

6 Membranes in contact with condensate droplets

The term 'condensate droplet' implies that the droplet is enclosed by a liquid–liquid rather than by a liquid–gas interface. Aqueous two-phase (or biphasic) systems based on biopolymers such as PEG and dextran have been applied for several decades in biochemical analysis and biotechnology[54] and are intimately related to water-in-water emulsions.[55] The aqueous phase separation leads to the formation of two coexisting liquid phases, denoted here by α and β as in Fig. 14. The two liquid phases α and β are separated by a liquid–liquid interface with interfacial tension $\Sigma_{\alpha\beta}$, which represents a new tension scale for the vesicle-droplet systems.

The aqueous phase separation of PEG-dextran solutions provides an example for segregative phase separation, in which one phase is enriched in one macromolecular component such as PEG whereas the other phase is enriched in the other macromolecular component such as dextran. The segregative behavior implies that the different species of macromolecules effectively repel each other. The phase diagram and the interfacial tension of aqueous PEG-dextran solutions is described in Section S9 and Fig. S17.† Another type of aqueous two-phase system is created by associative phase separation, for which one phase is enriched in the macromolecular components whereas the other phase represents a dilute solution of the macromolecules.[57–60] The associative behavior implies that the different macromolecular species effectively attract each other. Associative

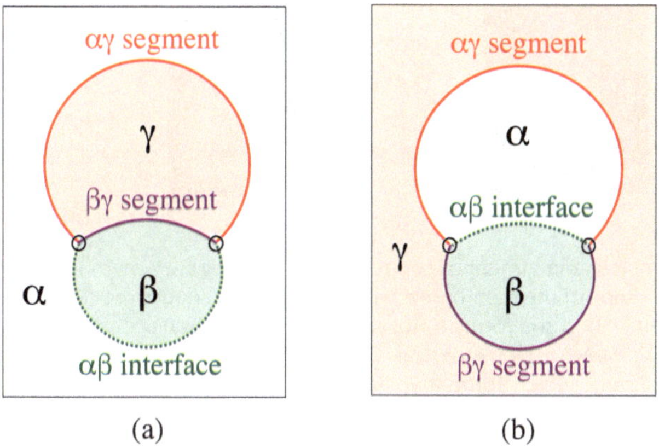

Fig. 14 Partial wetting geometry of vesicle-droplet systems, which involve three liquid phases α (white), β (green), and γ (light red). The phases α and β represent two coexisting phases that arise *via* segregative or associative liquid–liquid phase separation; the γ phase is an inert spectator phase: (a) phase separation of the exterior solution and adhesion of an exterior β droplet to the outer leaflet of the vesicle membrane; and (b) phase separation of the interior solution creating one interior α and one interior β droplet, both of which are in contact with the inner leaflet of the membrane.[56]

phase separation is observed, for instance, in solutions of two, oppositely charged polyelectrolytes,[59,60]

Condensate droplets have also been observed in living cells where they provide separate liquid compartments or organelles which are not enclosed by intracellular membranes. These biomolecular condensates are believed to form *via* liquid–liquid phase separation in the cytoplasm[61,62] and can be reconstituted *in vitro*.[63–66] They are enriched in certain types of proteins that have intrinsically disordered domains and interact *via* multivalent macromolecular interactions.[62,65–68]

6.1 Geometry of vesicle-droplet systems

When a condensate droplet comes into contact with a vesicle membrane, attractive interactions between droplet and membrane lead to the adhesion of the droplet to the membrane. The geometry of these vesicle-droplet systems involves three liquid phases α, β, and γ as shown in Fig. 14. The two phases α and β are formed by segregative or associative liquid–liquid phase separation and are separated by the $\alpha\beta$ interface. When the droplet partially wets the membrane, the $\alpha\beta$ interface forms a contact line with the membrane, which divides the membrane up into two segments: the $\alpha\gamma$ segment exposed to the α and γ phases as well as the $\beta\gamma$ segment in contact with the β and γ phases. In Fig. 14a and b, the coexisting phases α and β are located outside and inside the vesicle, respectively. Therefore, Fig. 14a displays one exterior β droplet at the outer leaflet of the bilayer membrane, whereas Fig. 14b shows one interior α and one interior β droplet in contact with the inner leaflet of the membrane.

6.2 Remodelling of GUV membranes by condensate droplets

The adhesion of condensate droplets to vesicles generates strong morphological responses of the vesicle membranes. One example for this response is provided by

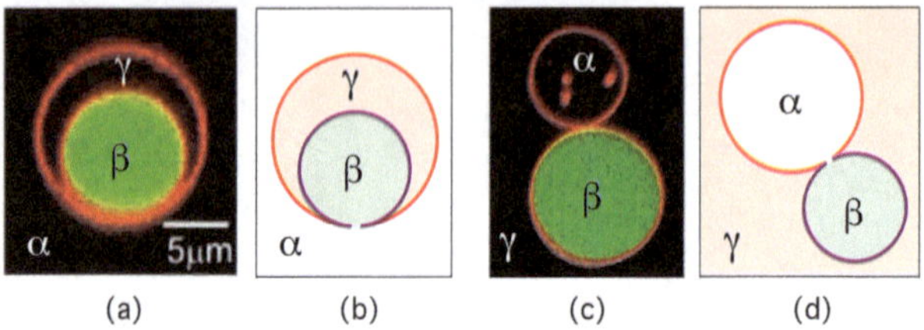

Fig. 15 Complete engulfment of condensate droplets by GUV membranes (red): (a and b) two-sphere shape of the membrane, which completely engulfs the dextran-rich β droplet (green) *via* an in-bud and forms a closed membrane neck adjacent to the droplet; reused with permission from ref. 71, American Chemical Society, 2012 (c and d) two-sphere shape of the membrane, which forms two spherical segments around both the dextran-rich β (green) and the PEG-rich α (black) droplet, connected by a closed membrane neck[72] In both cases, the systems's free energy is strongly reduced by eliminating the αβ interface.

the apparent kinks, which are observed in the optical microscope along the contact line between the adhering droplet and the vesicle membrane, see Fig. 14. On the nanoscopic scale, these kinks represent membrane segments with a very high curvature, caused by the capillary forces that the αβ interface exerts onto the membrane at the contact line.[69,70] For PEG-dextran systems, the particularly interesting processes are the complete engulfment of the droplets by the membranes[71,72] and the formation of membrane nanotubes,[73] as shown in Fig. 15 and 16, respectively. The different patterns of nanotubes displayed in Fig. 16a and b reveal complete and partial wetting of the GUV membrane by the aqueous two-phase system. More details on this wetting behavior are provided in Section S9 and Fig. S18.†

6.3 Interfacial tension *versus* curvature-elastic tensions

The two morphological pathways displayed in Fig. 15 and 16 arise from the competition between the interfacial tension of the liquid–liquid interface with the curvature-elastic membrane tensions of the two membrane segments αγ and βγ.

When the area $A_{\alpha\beta}$ of the αβ interface is reduced by $\Delta A_{\alpha\beta}$, the interfacial free energy of the vesicle-droplet system decreases by

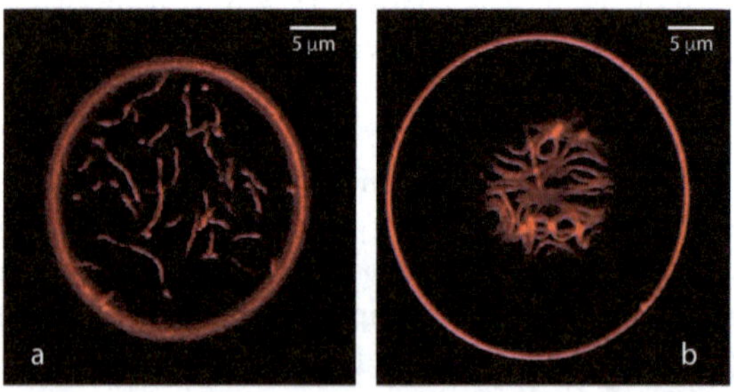

Fig. 16 Spontaneous tubulation of the αγ membrane segment in contact with the PEG-rich phase α leads to different patterns of nanotubes:[73] (a) disordered pattern corresponding to a vesicle membrane that is completely wetted by the PEG-rich phase; and (b) layer of densely packed tubes corresponding to a membrane that is partially (de)wetted by both aqueous phases. All tubes are connected to the outer vesicle membranes (big circles). In both images, the diameter of the fluorescence-labelled tubes is below the diffraction limit of the light microscope.

$$\Delta E_{\text{int}} = -\Sigma_{\alpha\beta}\Delta A_{\alpha\beta} \tag{37}$$

which is proportional to the interfacial tension $\Sigma_{\alpha\beta}$. The latter tension is a material parameter, which is determined by the molecular composition of the α and β phases but is independent of the size and shape of the interface. When the β droplet is completely engulfed by the membrane as in Fig. 15, the $\alpha\beta$ interface is completely eliminated from the vesicle-dropet system, thereby reducing its free energy by $\Sigma_{\alpha\beta}A_{\alpha\beta}$ with the total interfacial area $A_{\alpha\beta}$. To better visualize the shape of the vesicle-droplet systems for completely engulfed droplets, the optical images in panels a and c of Fig. 15 are compared to schematic drawings in panels b and d of the same figure.

For GUVs filled with PEG-dextran solutions, inward-pointing nanotubes are formed by the membrane segment $\alpha\gamma$ exposed to the PEG-rich phase α, as displayed in Fig. 16. The associated spontaneous curvature $m_{\alpha\gamma}$ is generated by the adsorption of PEG molecules onto the membranes.[73] The free energy reduction by the nanotube formation competes with the free energy reduction by the complete engulfment of the droplets. Which process is energetically more favorable depends on three parameters, the interfacial tension $\Sigma_{\alpha\beta}$ and the curvature-elastic tensions

$$\sigma_{\alpha\gamma} \equiv 2\kappa_{\alpha\gamma}{m_{\alpha\gamma}}^2 \text{ and } \sigma_{\beta\gamma} \equiv 2\kappa_{\beta\gamma}{m_{\beta\gamma}}^2 \tag{38}$$

of the two membrane segments.[70]

When the $\alpha\gamma$ membrane segment has a large spontaneous curvature $m_{\alpha\gamma} \gg 1/R_{\text{ve}}$ but a small mean curvature, this segment has a large bending energy of the order $\sigma_{\alpha\gamma}\, A_{\alpha\gamma}$, which is proportional to the curvature-elastic tension $\sigma_{\alpha\gamma}$ of the $\alpha\gamma$ segment and to the surface area $A_{\alpha\gamma}$ of this segment. Now, when the membrane area ΔA_{me} of the weakly curved mother vesicle is transferred to form a nanotube of mean curvature $M \simeq m_{\alpha\gamma}$, we gain the bending energy

$$\Delta E_{\text{be}} = -\sigma_{\alpha\gamma}\Delta A_{\text{me}} = -2\kappa_{\alpha\gamma}{m_{\alpha\gamma}}^2\Delta A_{\text{me}}. \tag{39}$$

We now repeat this argument for the membrane segment $\beta\gamma$ in contact with the dextran-rich phase β, which generates the spontaneous curvature $m_{\beta\gamma}$ of the $\beta\gamma$ segment. When the membrane area ΔA_{me} of this segment forms a nanotube with mean curvature $M \simeq m_{\beta\gamma}$, we gain the bending energy

$$\Delta E_{\text{be}} = -\sigma_{\beta\gamma}\Delta A_{\text{me}} = -2\kappa_{\beta\gamma}{m_{\beta\gamma}}^2\Delta A_{\text{me}}. \tag{40}$$

Because the membrane segment $\beta\gamma$ in contact with the dextran-rich phase is not observed to form nanotubes, we can directly conclude from eqn (39) and (40) that the curvature-elastic tension $\sigma_{\beta\gamma}$ of the $\beta\gamma$ segment is smaller than the curvature-elastic tension $\sigma_{\alpha\gamma}$ of the $\alpha\gamma$ segment.

The reduction of the membrane's bending energy by nanotube formation as given by eqn (39) remains to be compared with the reduction of the interfacial free energy by the engulfment of the droplet as described by eqn (37). We then conclude that complete droplet engulfment should be observed for[70]

$$\Sigma_{\alpha\beta} \gg \sigma_{\alpha\gamma} = 2\kappa_{\alpha\gamma}{m_{\alpha\gamma}}^2 \quad \text{(complete engulfment)}, \tag{41}$$

that is, when the interfacial tension $\Sigma_{\alpha\beta}$ is large compared to the curvature-elastic tension $\sigma_{\alpha\gamma}$ of the membrane segment $\alpha\gamma$ exposed to the PEG-rich α phase. On the other hand, the $\alpha\gamma$ membrane segment will form nanotubes if

$$\sigma_{\alpha\gamma} \gg \Sigma_{\alpha\beta} \text{ (nanotube formation)}, \tag{42}$$

that is, if the curvature-elastic tension $\sigma_{\alpha\gamma}$ of the $\alpha\gamma$ segment is large compared to the interfacial tension $\Sigma_{\alpha\beta}$. For the PEG-dextran two-phase system, the interfacial tension $\Sigma_{\alpha\beta}$ varies by several orders of magnitude (Fig. S17b†). As a consequence, this two-phase system can lead to both $\Sigma_{\alpha\beta} \gg \sigma_{\alpha\gamma}$ and $\sigma_{\alpha\gamma} \gg \Sigma_{\alpha\beta}$, that is, to both complete engulfment of the condensate droplet as in Fig. 15 and to the formation of membrane nanotubes as in Fig. 16.

7 Conclusions

In this paper, we elucidate the notion of membrane tension for three biomimetic membrane systems as provided by planar lipid bilayers, unilamellar nanovesicles, and GUVs. For planar bilayers and nanovesicles, we distinguish the bilayer tension Σ from the two leaflet tensions, Σ_1 and Σ_2, with $\Sigma = \Sigma_1 + \Sigma_2$. A global view of the leaflet tensions is provided by the two-dimensional parameter spaces in Fig. 3 and 5 for planar and vesicle bilayers, respectively. Particularly interesting bilayer states are tensionless but have opposite leaflet tensions (OLTs) as in Fig. 4 for planar bilayers and in Fig. 6 for nanovesicles. Outside their stability regimes, planar and vesicle bilayers exhibit lipid flip-flops (Fig. S8 and S12†) and structural instabilities (Fig. S9 and S13†).

For GUV membranes, the total membrane tension Σ_{tot} is equal to the mechanical membrane tension Σ plus the curvature-elastic tension $2\kappa m^2$. The mechanical membrane tension depends on the size and shape of the membrane as demonstrated for multispheres (Fig. 10) in Section 5.1 and for positive two-sphere shapes (Fig. 11) in Section 5.2. We then addressed the different shapes of tubular membranes, which can be multispherical (Fig. 12), unduloidal (Fig. 13), or cylindrical. Unduloidal shapes represent peristaltic deformations of cylindrical shapes. The fluctuation tension Σ_{fl} for the excess membrane stored in the peristaltic modes exhibits two tension thresholds as given by eqn (35) and (36).

The adhesion of condensate droplets to GUVs is considered in Section 6. The geometry of the vesicle-droplet systems (Fig. 14) involves three surfaces, the liquid–liquid interface with interfacial tension $\Sigma_{\alpha\beta}$ as well as two membrane segments $\alpha\gamma$ and $\beta\gamma$ with curvature-elastic tensions $\sigma_{\alpha\gamma}$ and $\sigma_{\beta\gamma}$ as defined in eqn (38). The relative magnitude of these tensions determines the remodelling of the GUV membranes: for large interfacial tension, the condensate droplets are completely engulfed by the vesicle membranes as in Fig. 15, whereas one of the membrane segments forms membrane nanotubes as in Fig. 16 for small interfacial tension.

Finally, it follows from the finite persistence length of tensionless fluid membranes as given by eqn (6) and from the poration of sufficiently large membranes in Section 3.3, that the thermodynamic route to membrane tension is ill-defined, in contrast to the thermodynamic definition of interfacial tension by J. Willard Gibbs.

Data availability

This paper provides a theoretical analysis of simulation data and experimental data that are available in the published and cited literature.

Conflicts of interest

There are no conflicts to declare.

Acknowledgements

I thank Bartosz Różycki, Udo Seifert, Gerhard Gompper, and Alberto Imperato for useful correspondence, and all my collaborators for enjoyable and fruitful interactions. This research was conducted within the Max Planck School "Matter to Life" supported by the German Federal Ministry of Education and Research (BMBF) in collaboration with the Max Planck Society and the Max Planck Institute of Colloids and Interfaces. Open Access funding provided by the Max Planck Society.

References

1 R. Lipowsky, R. Ghosh, V. Satarifard, A. Sreekumari, M. Zamaletdinov, B. Różycki, M. Miettinen and A. Grafmüller, *Biomolecules*, 2023, **13**, 926.
2 R. Lipowsky, *Faraday Discuss.*, 2013, **161**, 305–331.
3 R. Lipowsky, *Adv. Biol.*, 2022, **6**, 2101020.
4 M. D. Collins and S. L. Keller, *Proc. Natl. Acad. Sci. U.S.A.*, 2008, **105**, 124–128.
5 A. Orth, L. Johannes, W. Römer and C. Steinem, *ChemPhysChem*, 2012, **13**, 108–114.
6 L. D. Landau and E. M. Lifshitz, *Theory of Elasticity*, Pergamon Press, New York, 1986.
7 B. Różycki and R. Lipowsky, *J. Chem. Phys.*, 2015, **142**, 054101.
8 E. Evans and R. Hochmuth, Current Topics in Membranes and Transport, *Elsevier*, 1978, **10**, 1–64.
9 R. Lipowsky, *Adv. Colloid Interface Sci.*, 2014, **208**, 14–24.
10 T. Bhatia, J. Agudo-Canalejo, R. Dimova and R. Lipowsky, *ACS Nano*, 2018, **12**, 4478–4485.
11 P.-G. De Gennes and C. Taupin, *J. Phys. Chem.*, 1982, **86**, 2294–2304.
12 W. Helfrich and W. Harbich, *Chem. Scripta*, 1985, **25**, 32–36.
13 L. Peliti and S. Leibler, *Phys. Rev. Lett.*, 1985, **54**, 1690–1693.
14 R. Goetz, G. Gompper and R. Lipowsky, *Phys. Rev. Lett.*, 1999, **82**, 221–224.
15 J. Rowlinson and B. Widom, *Molecular Theory of Capillarity*, Oxford University Press, Oxford, 1989.
16 U. Seifert and R. Lipowsky, *Phys. Rev. A*, 1990, **42**, 4768–4771.
17 G. Gompper and D. Kroll, *Phys. Rev. E*, 1995, **51**, 514–525.
18 M. Zamaletdinov, M. Miettinen and R. Lipowsky, *Soft Matter*, 2023, **19**, 6929–6944.
19 A. Sreekumari and R. Lipowsky, *Soft Matter*, 2022, **18**, 6066–6078.
20 A. Hossein and M. Deserno, *Biophys. J.*, 2020, **118**, 624–642.
21 M. Varma and M. Deserno, *Biophys. J.*, 2022, **121**, 4001–4018.

22 O. Farago, *J. Chem. Phys.*, 2003, **119**, 596–605.
23 T. V. Tolpekina, W. K. den Otter and W. J. Briels, *J. Chem. Phys.*, 2004, **121**, 8014–8020.
24 M. Miettinen and R. Lipowsky, *Nano Lett.*, 2019, **19**, 5011–5016.
25 A. Imparato, J. Shillcock and R. Lipowsky, *Europhys. Lett.*, 2005, **69**, 650–656.
26 A. Imparato, *J. Chem. Phys.*, 2006, **124**, 154714.
27 J. Stecki, *J. Chem. Phys.*, 2006, **125**, 154902.
28 J.-B. Fournier and C. Barbetta, *Phys. Rev. Lett.*, 2008, **100**, 078103.
29 H. Shiba, H. Noguchi and J.-B. Fournier, *Soft Matter*, 2016, **12**, 2373–2380.
30 R. Ghosh, V. Satarifard, A. Grafmüller and R. Lipowsky, *Nano Lett.*, 2019, **19**, 7703–7711.
31 J. Z. A. Finkelstein and F. Cohen, *Annu. Rev. Physiol.*, 1986, **48**, 163–174.
32 J. Shillcock and R. Lipowsky, *Nat. Mater.*, 2005, **4**, 225–228.
33 A. Grafmüller, J. Shillcock and R. Lipowsky, *Phys. Rev. Lett.*, 2007, **98**, 218101.
34 L. Gao, R. Lipowsky and J. C. Shillcock, *Soft Matter*, 2008, **4**, 1208–1214.
35 T.-T. Kliesch, J. Dietz, L. Turco, P. Halder, E. Polo, M. Tarantola, R. Jahn and A. Janshoff, *Sci. Rep.*, 2017, **7**, 12070.
36 R. B. Lira, T. Robinson, R. Dimova and K. A. Riske, *Biophys. J.*, 2019, **116**, 79–91.
37 I. Ivanov, R. B. Lira, T.-Y. D. Tang, T. Franzmann, A. Klosin, L. C. da Silva, A. A. Hyman, K. Landfester, R. Lipowsky, K. Sundmacher and R. Dimova, *Adv. Biosyst.*, 2019, **3**, 1800314.
38 S. Deshpande, S. Wunnava, D. Hueting and C. Dekker, *Small*, 2019, **15**, 1902898.
39 A. Grafmüller, J. C. Shillcock and R. Lipowsky, *Biophys. J.*, 2009, **96**, 2658–2675.
40 B. Rozycki, R. Ghosh and R. Lipowsky, Fusion of nanovesicles driven by the stress asymmetry between leaflettensions, in preparation.
41 J. Steinkühler, R. L. Knorr, Z. Zhao, T. Bhatia, S. Bartelt, S. Wegner, R. Dimova and R. Lipowsky, *Nat. Commun.*, 2020, **11**, 905.
42 T. Bhatia, S. Christ, J. Steinkühler, R. Dimova and R. Lipowsky, *Soft Matter*, 2020, **16**, 1246–1258.
43 R. Lipowsky, *Adv. Colloid Interface Sci.*, 2022, **301**, 102613.
44 R. Lipowsky, *The Giant Vesicle Book*, Taylor & Francis, 2019, ch. 5, pp. 73–168.
45 R. E. Powers, S. Wang, T. Y. Liu and T. A. Rapoport, *Nature*, 2017, **543**, 257–272.
46 M. A. Betancourt-Solis, T. Desai and J. A. McNew, *J. Biol. Chem.*, 2018, **293**, 18514–18524.
47 D. Holcman, P. Parutto, J. E. Chambers, M. Fantham, L. J. Young, S. J. Marciniak, C. F. Kaminski, D. Ron and E. Avezov, *Nat. Cell Biol.*, 2018, **20**, 1118–1125.
48 C. J. Obara, A. S. Moore and J. Lippincott-Schwartz, *Cold Spring Harbor Perspect. Biol.*, 2023, **15**, a041259.
49 R. Lipowsky, S. Pramanik, A. S. Benk, M. Tarnawski, J. P. Spatz and R. Dimova, *ACS Nano*, 2023, **17**, 11957–11968.
50 G. Orso, D. Pendin, S. Liu, J. Tosetto, T. J. Moss, J. E. Faust, M. Micaroni, A. Egorova, A. Martinuzzi, J. A. McNew and A. Daga, *Nature*, 2009, **460**, 978–983.
51 W. Helfrich and R.-M. Servuss, *Il Nuovo Cimento D*, 1984, **3**, 137–151.
52 S. Milner and S. Safran, *Phys. Rev. A*, 1987, **36**, 4371–4379.
53 U. Seifert, *Z. Physik B*, 1995, **97**, 299–309.

54 P. A. Albertsson, Partition of Cell Particles and Macromolecules: Separation and Purification of Biomolecules, Cell Organelles Membranes, and Cells, in *Aqueous Polymer Two-phase Systems and Their Use in Biochemical Analysis and Biotechnology*, Wiley, 3rd edn, 1986.
55 J. Esquena, *Curr. Opin. Colloid Interface Sci.*, 2016, **25**, 109–119.
56 R. Lipowsky, *Membranes*, 2023, **13**, 223.
57 C. D. Crowe and C. D. Keating, *Interface Focus*, 2018, **8**, 20180032.
58 H. G. B. de Jong and H. R. Kruyt, *Proc. K. Ned. Akad. Wet.*, 1929, **32**, 849–856.
59 C. E. Sing, *Adv. Colloid Interface Sci.*, 2017, **239**, 2–16.
60 L. Li, S. Srivastava, M. Andreev, A. B. Marciel, J. J. de Pablo and M. V. Tirrell, *Macromolecules*, 2018, **51**, 2988–2995.
61 C. P. Brangwynne, C. R. Eckmann, D. S. Courson, A. Rybarska, C. Hoege, J. Gharakhani, F. Jülicher and A. A. Hyman, *Science*, 2009, **324**, 1729–1732.
62 S. F. Banani, H. O. Lee, A. A. Hyman and M. K. Rosen, *Nat. Rev. Mol. Cell Biol.*, 2017, **18**, 285–295.
63 P. Li, S. Banjade, H.-C. Cheng, S. Kim, B. Chen, L. Guo, M. Llaguno, J. V. Hollingsworth, D. S. King, S. F. Banani, P. S. Russo, Q.-X. Jiang, B. T. Nixon and M. K. Rosen, *Nature*, 2012, **483**, 336–341.
64 A. Patel, H. O. Lee, L. Jawerth, S. Maharana, M. Jahnel, M. Y. Hein, S. Stoynov, J. Mahamid, S. Saha, T. M. Franzmann, S. Grill, D. Drechsel, A. A. Hyman and S. Alberti, *Cell*, 2015, **162**, 1066–1077.
65 Y. Lin, D. S. W. Protter, M. K. Rosen and R. Parker, *Mol. Cell*, 2015, **60**, 208–219.
66 A. Molliex, J. Temirov, J. Lee, H. J. Kim, T. Mittag and J. P. Taylor, *Cell*, 2015, **163**, 123–133.
67 T. W. Han, M. Kato, S. Xie, L. C. Wu, H. Mirzaei, J. Pei, M. Chen, Y. Xie, J. Allen, G. Xiao and S. L. McKnight, *Cell*, 2012, **149**, 768–779.
68 L. Guo and J. Shorter, *Mol. Cell*, 2015, **60**, 189–192.
69 H. Kusumaatmaja, Y. Li, R. Dimova and R. Lipowsky, *Phys. Rev. Lett.*, 2009, **103**, 238103.
70 R. Lipowsky, *J. Phys. Chem. B*, 2018, **122**, 3572–3586.
71 Y. Li, H. Kusumaatmaja, R. Lipowsky and R. Dimova, *J. Phys. Chem. B*, 2012, **116**, 1819–1823.
72 R. Dimova and R. Lipowsky, *Adv. Mater. Interfaces*, 2017, **4**, 1600451.
73 Y. Liu, J. Agudo-Canalejo, A. Grafmüller, R. Dimova and R. Lipowsky, *ACS Nano*, 2016, **10**, 463–474.

Faraday Discussions

PAPER

Interaction of lipid domains originating from differential domain–monolayer contact energy

Avishai Barnoy[ab] and Michael M. Kozlov *[a]

Received 29th November 2024, Accepted 3rd January 2025
DOI: 10.1039/d4fd00186a

We consider a flat membrane containing pure lipid domains located in the membrane monolayers and separated in the membrane plane. We assume the energy of contact along the membrane mid-surface between a domain and the underlying monolayer to be different from that between the two monolayers. We theoretically analyse the effect of the differential contact energy on the elastic deformations of tilt and splay in the membrane monolayers and the resulting interaction between two domains situated in the apposed monolayers. We demonstrate that the character of this interaction depends on the ratio, η, between the domain rigidity and that of a regular membrane monolayer. For the rigidity ratio smaller than a critical value, $\eta < \eta^* \approx 3$, the domain interaction is predicted to be attractive for all inter-monolayer distances. For the super-critical values of the rigidity ratio, $\eta > \eta^*$, the interaction is repulsive for small distances and attractive for large distances with a certain equilibrium inter-domain separation corresponding to a vanishing interaction force. The predicted attractive interaction is proposed to favor the registration in the membrane plane of apposed domains as observed in most domain-containing membranes.

Introduction

Studies of the formation and organization of membrane domains whose physicochemical properties such as the lipid and protein composition and the lipid phase state differ from those of the regular membrane monolayers represent an extensive and, practically, self-standing field of membrane biophysics.[1–4] In the context of cell membranes, the most familiar and extensively discussed are the domains which consist mainly of glycosphingolipids and cholesterol, contain specific proteins, and are referred to as the lipid rafts.[5] In multicomponent pure lipid membranes, the domain formation results from the in-plane segregation of

[a]Department of Physiology and Pharmacology, Faculty of Medical and Health Sciences, Tel Aviv University, 69978 Tel Aviv, Israel. E-mail: michk@tauex.tay.ac.il
[b]School of Chemistry, Faculty of Exact Sciences, Tel Aviv University, Israel

coexisting lipid phases, most commonly, the regions of liquid-ordered (L_o)[1] or gel (L_β)[6,7] phase within the bulk of a liquid-disordered (L_d) (ref. 8 and references therein) phase.

Theoretical modelling of domain formation in lipid bilayers was performed by various approaches ranging from sophisticated methods of soft matter physics[9,10] to state-of-the-art numerical simulations.[11] This led to understanding the molecular interactions in the lipid monolayer plane driving the domain formation.[12,13] Yet, one aspect of the domain arrangement within membranes remained largely open, namely, the forces responsible for the commonly observed strong spatial correlation between domains formed in the apposed membrane monolayers. In most of the experimental studies, the domains in one membrane monolayer were observed to be in register with those in the apposed monolayer so that both the lipid rafts in cell membranes and the regions of ordered lipid phases in pure lipid membranes were mostly seen to span the whole membrane thickness (see ref. 14 for review). The origin of the trans-monolayer interactions mediating a cross-talk between domains in the apposed monolayers remained hypothetical.

The major proposal for explaining this phenomenon is that the lipid phase separation and the resulting domain formation in one membrane monolayer trigger the same process right underneath in the second monolayer due to a hypothetical trans-membrane coupling interaction.[9,15,16] At the same time, it was possible to experimentally separate the trans-bilayer domains into unregistered monolayer domains by applying tangential shear forces to the top and bottom membrane surfaces.[17] This implied that the trans-membrane registration was not crucial for the monolayer domain stability and that an attractive interaction must exist between laterally separated domains located in apposed monolayers favoring their mutual approach and ultimate overlapping. While an effective interaction related to the domain line tension was suggested to be responsible for fine-tuning the relative positioning of the already overlaying domains,[18] an understanding of interactions between separated non-overlapping domains has remained missing.

Here we propose a mechanism leading to an attractive interaction between domains formed in the apposed membrane monolayers and stemming from the membrane deformations. We suggest that the origin of the membrane deformations and, hence, the domain interaction is the difference in the energies of contact between a monolayer and a domain and that between two regular membrane monolayers. The physical essence of the differential contact energy is equivalent to that of the domain's surface tension proposed in ref. 15 and 19 and discussed in ref. 14. The same concept of differential contact energy was recently considered in the context of the membrane shaping by caveolin discs.[20] Our theoretical analysis of flat membranes shows that the propensity of a domain-containing membrane to minimize the overall differential contact energy generates the deformations of tilt and splay of the membrane monolayers, which propagate along the membrane plane and mediate an interaction of separated membrane domains. We determine the conditions under which this interaction is attractive and demonstrate the feasibility of these conditions for realistic lipid membranes.

Qualitative essence of the model

Single domain

To qualitatively explain the model's main idea, we first consider a lipid bilayer containing a single lipid domain in one of its monolayers (Fig. 1). We use the following terminology illustrated in Fig. 1A. The monolayer lipid domain will be called simply the domain. The membrane monolayer containing the domain will be referred to as the proximal monolayer. The second monolayer apposed to the proximal one will be called the distal monolayer. The fragment of the distal monolayer whose lipid molecules touch by their hydrocarbon chains the hydrophobic surface of the domain will be called the contact zone. For simplicity, we assume that the physical properties of the distal monolayer, including the contact zone, are identical to those of the proximal monolayer but differ from those of the domain. For the bilayer with multiple domains (Fig. 2), the bottom and top monolayers will be referred to as the proximal and the distal ones, respectively.

The central hypothesis of the model is that the contact interaction between the lipid molecules of the distal and proximal monolayers along the bilayer mid-plane is different from that between the contact zone and the hydrophobic plane of the domain. This differential contact interaction is the origin of the domain surface

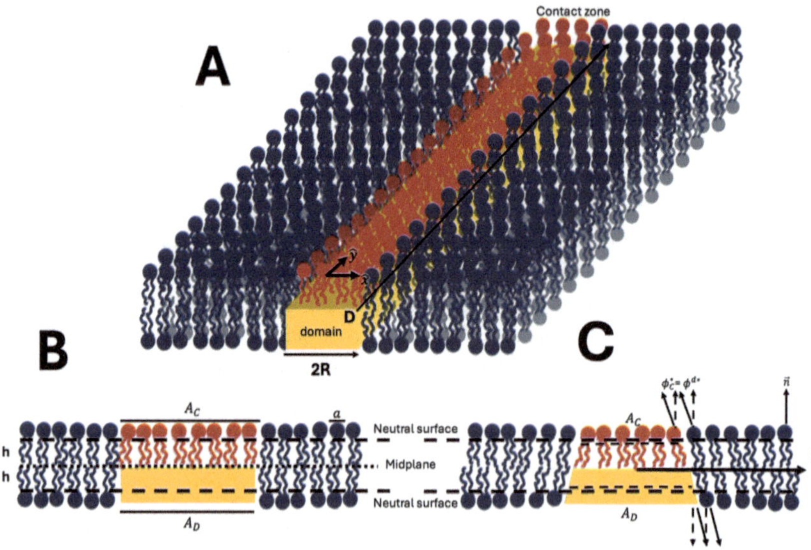

Fig. 1 A one-dimensional model of a flat membrane containing one lipid domain. The domain is shown in yellow. The lipid molecules in the contact zone are shown in red. (A) 3D representation of the initial state of the system. The dimension of the domain in the x-direction is $2R$ and in the y-direction is D. (B) The cross-section in the x-direction of the system in the initial state. The dashed lines represent the monolayer neutral surfaces, the dotted line represents the bilayer midplane; h is the distance between the neutral surface and a monolayer midplane; A_c and A_D are the areas of the contact zone and the domain respectively, determined at the corresponding neutral surface; a is the lipid molecular area at the neutral surface. (C) The tilt and splay deformations in the distal and proximal monolayers following the relaxation of the contact energy. Dashed arrows represent the normal to the neutral surface, solid arrows represent the lipid director.

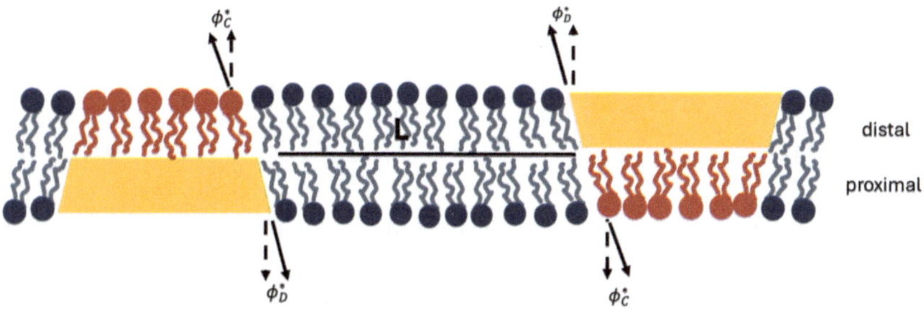

Fig. 2 Cross-section of a membrane with two domains located in apposed monolayers. Due to symmetry, the tilt angles in equivalent sections of the apposed monolayers are identical. The color scheme follows that of Fig. 1. The inter-domain distance along the midplane is L. Dashed arrows represent the normal to the neutral surface, solid arrows represent the lipid director.

tension introduced in ref. 15 and 19. To be specific, we consider the contact between the distal and proximal monolayers to be energetically more favorable than that between the distal monolayer and the domain. In the following, the energy of the contact interaction will be called the contact energy, for brevity.

To keep the computations simple, we consider the one-dimensional version of the model in which a domain has the shape of a strip of finite width, $2R$, and infinite length (Fig. 1A). We choose the planar coordinates system in the bilayer mid-plane with the axes x and y directed perpendicularly and parallel to the domain's length, respectively (Fig. 1A). For the one-domain system the origin of the x coordinate, $x = 0$, is chosen in the middle of the domain (Fig. 1A). For the two-domain system the domains are oriented parallel to the axis y and, hence, to each other, the distance between the domain edges is denoted by L (Fig. 2). The monolayer deformations depend only on the x-coordinate, simplifying the analysis. Despite this simplification, the predictions must also be qualitatively valid for the realistic two-dimensional distributions of deformations.

For the following, we need to briefly recall a few basic notions used to describe a lipid monolayer as an elastic surface (see for review ref. 21). While, generally, a lipid monolayer, as any interface, can be described by any surface parallel to the monolayer plane called the dividing surface,[22] most conveniently it is represented by a specific dividing surface referred to as the monolayer neutral surface.[23–25] The deformations of the monolayer stretching–compression determined at the neutral surface are energetically decoupled from the deformations of the monolayer bending[23–25] and, hence, from the deformation of splaying the hydrocarbon chains of lipid molecules.[26] According to the analysis of experimental data on lipid monolayers of different compositions, the monolayer neutral surface underlies the effective boundary between the region of the lipid polar heads and that of the hydrocarbon moieties of the lipid molecule.

The premise of our model is that the in-plane area per one lipid molecule, a, determined at the neutral surface of any of the membrane monolayers does not change upon the splaying and/or tilting of lipid molecules and/or generation of the monolayer tension. The constancy of a upon splaying is the consequence of the above-mentioned property of the neutral surface. The invariance of a upon tilting follows from the essence of the lipid tilt deformation as the transverse shear of the lipid material, which keeps constant the molecular volume and in-plane area.[26,27] Finally, the background for the constancy of a upon the tension generation is the large stretching–compression modulus of a lipid monolayer whose typical value of about 100 mN m^{-1} (ref. 28) exceeds by a few orders of magnitude the experimentally feasible tension values of less than 1 mN m^{-1} (see *e.g.* ref. 29).

In addition, we will assume that the lipid molecular area, a, is equal for all lipid molecules of the system including those composing the regular monolayers and the domains. In real systems, the lipid molecular area at the neutral surface can vary within the range of about 30–40% depending on the lipid species and phase state.[25,30–32] We neglect this variation meaning that the predictions of our modelling have qualitative rather than quantitative character.

We consider, for simplicity, the bilayer to be flat, which implies that it is subjected to a lateral tension preventing its deviations from the planar shape. We denote by A_D and A_C the areas of, respectively, the domain and the contact zone determined at the neutral surfaces of the corresponding monolayers (Fig. 1B). The numbers of lipid molecules constituting the domain and the contact zone are, respectively, $N_D = \frac{A_D}{a}$ and $N_C = \frac{A_C}{a}$.

We consider the initial state of the bilayer to exhibit no tilt of the lipid molecules in any part of the system (Fig. 1A and B). In this state, the areas and the molecular numbers are equal for the contact zone and the domain, $A_C = A_D$, and $N_C = N_D$ (Fig. 1B).

The system tends to reduce the contact energy by decreasing the number of lipid molecules, N_C, and, hence, the area, A_C, of the contact zone, while keeping constant the domain's parameters, A_D and N_D. This can be achieved by tilting the lipid molecules[26,27,33] at the domain's boundary (Fig. 1C). A simple geometrical consideration shows that the tilting of the lipid molecules of the contact zone inward and those of the domain outward from the domain's center reduces A_C and, hence, decreases N_C (Fig. 1C). The molecular tilt induced at the domain boundary propagates along the monolayer plains decaying with the distance from the boundary, which generates the splay deformation (Fig. 1C).[26,27,33]

Hence, the reduction of the number of molecules in the contact zone, N_C, and, thus, the contact energy is accompanied by the cost of the elastic energy of the tilt and splay.[26,27,33] The energetically most favorable and, therefore, equilibrium configuration of the system is characterized by a certain extent of the lipid tilt-splay that is set by the interplay between the contact energy driving the deformations and the elastic energy resisting it.

Multiple domains

In the case where the bilayer contains two (Fig. 2) or more lipid domains, the factors driving the system's deformations are the same as those described for a single domain. Yet, an additional important factor to be considered in this case is the overlap of membrane deformations generated by neighboring domains. This overlap results in a membrane-mediated interaction between the domains, which can be either repulsive or attractive depending on the system's parameters.

Main definitions and equations

We consider a flat lipid bilayer whose monolayers contain domains. We analyse two systems: a lipid bilayer with one domain (Fig. 1), and a bilayer with two domains located in apposed monolayers (Fig. 2).

For clarity, we introduce the main notions and equations for the system with one domain (Fig. 1). They can be straightforwardly used in the two-domain case (Fig. 2). In a few cases where such a direct usage is not possible, we provide additional definitions.

The system will be described by three mutually parallel flat surfaces: the bilayer's mid-plane and the neutral surfaces of the proximal and distal monolayers separated from the bilayer's midplane by a distance, h, approximately equal to the monolayer thickness (Fig. 1B).

The physical values of the domain and the contact zone will be denoted by the subscripts D and C, respectively, whereas those describing the distal or the proximal monolayers will be indicated by the superscripts d and p. We will use no indices by introducing notations and relationships common for all parts of the system.

The system is characterized at every point of its neutral surface by the tilt, t, and splay, $\bar{J}$, of the constituent lipid molecules. The notions of tilt and splay are introduced and discussed in ref. 26, 27 and 33. In brief, for the 1D system considered here, the lipid tilt, t, is equal to $t = \tan\phi$, where ϕ is the angle between the unit vector, $\vec{n}$, describing the average orientation of the lipid hydrocarbon chains and the unit normal, $\vec{N}$, of the monolayer's neutral surface (Fig. 3). We define the tilt angle, ϕ, to be positive if $\vec{n}$ leans away from $\vec{N}$ in the clockwise direction and negative otherwise. The values of the tilt angle in the domain, the contact zone, the proximal monolayer outside the domain, and the distal monolayer outside the contact zone will be denoted by ϕ_D, ϕ_C, ϕ^p, and ϕ^d, respectively.

The lipid splay, $\bar{J}$, quantifies the variation of the lipid chain orientation along the monolayer neutral plane (Fig. 3). For the flat 1D bilayers considered here, the splay is defined by $\bar{J} = \frac{dt}{dx}$.

For simplicity, we assume the monolayer deformations in all parts of the system to be weak leading to small tilt angles, $\phi < 1$, so that tilt and splay can be approximated as $t = \phi$ and $\bar{J} = \frac{d\phi}{dx}$. The smallness of the splay is expressed by $\left|\frac{d\phi}{dx}\right| h < 1$.

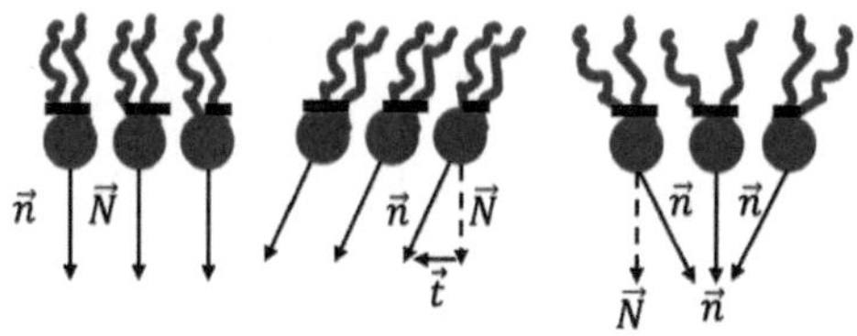

Fig. 3 Tilt-splay deformations of a lipid monolayer. Left: initial undeformed state. Middle: tilt deformation. Right: splay deformation. The dashed arrows represent the normal to the neutral surface, the solid arrows represent the lipid director, and the dashed line shows the monolayer's neutral surface.

Energy of the system

Our goal is to find the equilibrium configuration of the system corresponding to the minimum of the system's energy, F. The energy will be computed with respect to the initial state in which the tilt and, hence, the splay vanish, $\phi = 0, \frac{d\phi}{dx} = 0$, in all parts of the system (Fig. 1A and B).

We consider the energy F to consist of the contact energy, F_{cont}, and the elastic energy of tilt-splay, F_t.

$$F = F_{cont} + \sum_i F_t{}^i. \tag{1}$$

The meaning of $\sum_i$ in eqn (1) is the summation of the contributions of all parts of the system.

To introduce the contact energy, F_{cont}, we first define the partial contact energies, ε_D and ε_L. The value ε_D is the energy of contact between the contact zone and the domain related to the unit area of the contact zone's neutral plane. The value ε_L is the energy of contact between the distal and proximal monolayers related to the unit area of the neutral surface of the former. The contact energy, F_{cont}, is proportional to the variation in the number of lipid molecules in the contact zone, N_C, resulting from the emerging tilt-splay deformations. Taking into account that in the initial state $N_C = N_D$ and using the relationship $N = \frac{A}{a}$, the contact energy can be presented as

$$F_{cont} = (A_C - A_D)\Delta\varepsilon, \tag{2}$$

where $\Delta\varepsilon = \varepsilon_D - \varepsilon_L$ is assumed to be positive, $\Delta\varepsilon > 0$. We are not aware of any experimental data enabling evaluation of the differential contact energy, $\Delta\varepsilon$. Taking it to be at least one order of magnitude smaller than the surface tension of the oil–water interface, this value can be estimated as $\Delta\varepsilon \leq 1k_BT$ nm^{-2}, which is supported by the phenomenological and computational estimations providing values in the range (0.1–1) k_BT nm^{-2} [14,15,19] (where $k_BT \approx 4 \times 10^{-21}$ joules is the product of the Boltzmann constant and the absolute temperature).

For the one-dimensional model considered here, we will present the domain area as $A_D = 2 \cdot D \cdot R$, where $2R$ is the domain width along the x axis, as already mentioned, and D is a constant multiplier whose meaning is the length along the direction of the system's uniformity (Fig. 1A). For estimations, we will use the minimal relevant value of the domain's half-width, $R \approx 100$ nm.

The sum of the tilt and splay energies of a monolayer related to the unit area of its neutral surface, $f_b + f_t$, can be expressed in the case considered here of small deformations as,[26]

$$f_b + f_t = \frac{1}{2}\kappa\left(\frac{d\phi}{dx}\right)^2 + \frac{1}{2}\kappa_t\phi^2, \tag{3}$$

where κ is the monolayer splay (bending) modulus, and κ_t is the monolayer tilt modulus. The typical splay modulus of a monolayer is $\kappa \approx 10k_BT \approx 4 \times 10^{-20}$ joules.[34,35] The value of the monolayer's tilt modulus constitutes a few tens of mN m^{-1} (ref. 26 and 36) and will be taken as $\kappa_t \approx 30$ mN m^{-1}. This equation (eqn (3)) implies that the monolayer spontaneous splay (spontaneous curvature) vanishes and there is no contribution of the energy of saddle splay (Gaussian curvature). The total energy of a monolayer tilt-splay, $F_b + F_t$, is obtained by integration of eqn (3) over the area of the neutral surface,

$$\frac{1}{D}(F_b + F_t) = \int \left[\frac{1}{2}\kappa\left(\frac{d\phi}{dx}\right)^2 + \frac{1}{2}\kappa_t\phi^2\right]dx. \quad (4)$$

The elastic moduli of splay and tilt of the domain denoted by κ_D and κ_{tD} must be, generally, larger than the corresponding moduli, κ and κ_t, of the regular parts of the system's monolayers. To keep the calculations simple, we assume the splay and tilt moduli of the domains to differ from those of the regular monolayers by the same factor, η,

$$\kappa_D = \eta\cdot\kappa,\ \kappa_{tD} = \eta\cdot\kappa_t \quad (5)$$

The factor η will be referred to below as the relative rigidity of the domain and assumed to be $\eta > 1$.

Main equations

The distributions of the tilt angle, $\phi(x)$, and, hence, of the splay, $\frac{d\phi(x)}{dx}$, can be found either by minimizing the full energy (eqn (1)) or by solving the related Lagrange equation, which for a flat monolayer and a small tilt-splay deformation is

$$\frac{d^2\phi}{dx^2} = \frac{1}{\lambda^2}\phi, \quad (6)$$

where λ is the characteristic relaxation length determined by

$$\lambda = \sqrt{\frac{\kappa}{\kappa_t}}. \quad (7)$$

Based on the cited-above values of the elastic moduli, the relaxation length λ has a value in the range of 1–2 nm.

Due to the simplifying assumption (eqn (5)), the relaxation length (eqn (7)) is the same for the domain, the contact zone and all other parts of the system.

The solutions of Lagrange equations (eqn (6)) must satisfy certain boundary conditions. For simplicity, we explicitly present here the boundary conditions for the system with one domain (Fig. 1C). For the system with two domains, the boundary conditions can be found analogously. For symmetry reasons, in the middle of the domain, $x = 0$, the tilt angles of the domain and the contact zone must vanish,

$$\phi_D(x = 0) = 0,\ \phi_C(x = 0) = 0. \quad (8)$$

At large distances, $x = \infty$, the tilt angles of the proximal and distal monolayers must vanish,

$$\phi^p(x = \infty) = 0, \text{ and } \phi^d(x = \infty) = 0, \quad (9)$$

to avoid an infinite energy of tilt (eqn (4)).

Finally, a set of conditions has to be satisfied at the domain boundary, $x = R$. The geometrical characteristics at this boundary will be denoted by asterisks and referred to as the boundary values (Fig. 1C). First, the orientation of the hydrocarbon chains, $\vec{n}$, in each leaflet must be continuous. The continuity of $\vec{n}$ can be expressed through the boundary tilt angles of the domain, $\phi_D(x = R) = \phi_D^*$, the contact zone, $\phi_C(x = R) = \phi_C^*$, the proximal monolayer $\phi^p(x = R) = \phi^{p*}$, and $\phi^d(x = R) = \phi^{d*}$, by

$$\phi^{p*} = \phi_D^*, \ \phi^{d*} = \phi_C^*. \tag{10}$$

Using the tilt boundary values, the contact energy (eqn (2)) related to the unit length of the y axis can be expressed as

$$\frac{1}{D}F_{\text{cont}} = h\left(\phi_C^* + \phi_D^*\right)\Delta\varepsilon. \tag{11}$$

Way of computations

In the following, we find the distribution of tilt and splay in each part of the system by solving the Lagrange equation (eqn (6)) with the boundary conditions (eqn (8) and (10)). Inserting the resulting functions $\phi(x)$ into eqn (4), performing the integration over the area of each part of the system, inserting the results into eqn (1) and summing them along with eqn (11) we obtain the full energy of the system as a function of the boundary tilt angles, ϕ_C^* and ϕ_D^*. Minimizing the resulting energy with respect to ϕ_C^* and ϕ_D^* we find the equilibrium values of these angles and the energy of the equilibrium state. In the case of the two-domain system, the dependence of the equilibrium energy on the distance between the domains will enable the analysis of the domain interaction mediated by the membrane deformations.

Because of the structure of the Lagrange equation (eqn (6)), all the results depend on the dimensionless form of the system's variables and parameters having units of length. Specifically, the dimensionless coordinate, domain half-width, and distance between the domains are, respectively,

$$\xi = \frac{x}{\lambda}, \ \rho = \frac{R}{\lambda}, \ l = \frac{L}{\lambda}, \tag{12}$$

where λ is the relaxation length (eqn (7)).

Results

One domain

Due to the system's symmetry with respect to the middle of the domain, $x = 0$, we analyse half of the system, $x \geq 0$ (Fig. 1C). The distribution of the tilt angle in the domain, the contact zone, and the proximal and distal monolayers resulting from our computations are given by

$$\begin{aligned} \phi_D(\xi) &= \phi_D^* \cdot \frac{\sinh(\xi)}{\sinh(\rho)}; \quad & \phi_C(\xi) &= \phi_C^* \cdot \frac{\sinh(\xi)}{\sinh(\rho)}; \\ \phi^p(\xi) &= \phi_D^* \cdot e^{-(\xi-\rho)}; \quad & \phi^d(\xi) &= \phi_C^* \cdot e^{-(\xi-\rho)}. \end{aligned} \tag{13}$$

The equilibrium boundary angles in the contact zone and the domain are, respectively,

$$\phi_{\mathrm{C}}^{*} = -\frac{h\cdot\Delta\varepsilon}{\sqrt{\kappa\kappa_{\mathrm{t}}}}\cdot\frac{1}{\mathrm{cth}(\rho)+1}, \tag{14}$$

and

$$\phi_{\mathrm{D}}^{*} = -\frac{h\cdot\Delta\varepsilon}{\sqrt{\kappa\kappa_{\mathrm{t}}}}\cdot\frac{1}{\eta\cdot\mathrm{cth}(\rho)+1}. \tag{15}$$

In eqn (14) and below, $\mathrm{cth}(\square) = 1/\tanh(\square)$ is a hyperbolic cotangent function.

The energy of the equilibrium state computed with respect to the initial state is

$$\frac{1}{D}F^{*} = -\frac{1}{2}\cdot\frac{(h\cdot\Delta\varepsilon)^{2}}{\sqrt{\kappa\kappa_{\mathrm{t}}}}\cdot\frac{2+(1+\eta)\mathrm{cth}(\rho)}{[\mathrm{cth}(\rho)+1]\cdot[\eta\cdot\mathrm{cth}(\rho)+1]}. \tag{16}$$

Taking into account that a realistic value of the dimensionless half-width of the domain is large compared to the relaxation length, $\rho = \frac{R}{\lambda} \gg 1$, so that $\mathrm{cth}(\rho) \approx 1$, the expressions eqn (14)–(16) can be simplified. In particular, the energy is given by

$$\frac{1}{D}F^{*} = -\frac{1}{4}\cdot\left(\frac{\eta+3}{\eta+1}\right)\cdot\frac{(h\cdot\Delta\varepsilon)^{2}}{\sqrt{\kappa\kappa_{\mathrm{t}}}}. \tag{17}$$

The dependence of the equilibrium energy on the relative rigidity is presented in Fig. 4A, and the distribution along the x-axis of the tilt angles for both the distal and proximal monolayers is shown in Fig. 4B for different values of the relative domain rigidity. The extent of tilting in the distal monolayer is independent of the relative domain rigidity. The extent of tilting in the proximal monolayer including the domain decreases with growing domain rigidity. When the rigidity of the domain is equal to the rigidity of the membrane, $\eta = 1$, the tilt angles in the distal and proximal monolayers are equal (Fig. 4B, orange and purple dashed lines). The energy relaxation is most effective if the domain is as flexible as the regular monolayers, $\eta = 1$ (Fig. 4A), but even if the domain is infinitely rigid, $\eta \rightarrow \infty$, the energy relaxes to a substantial extent due to the tilting in the distal monolayer.

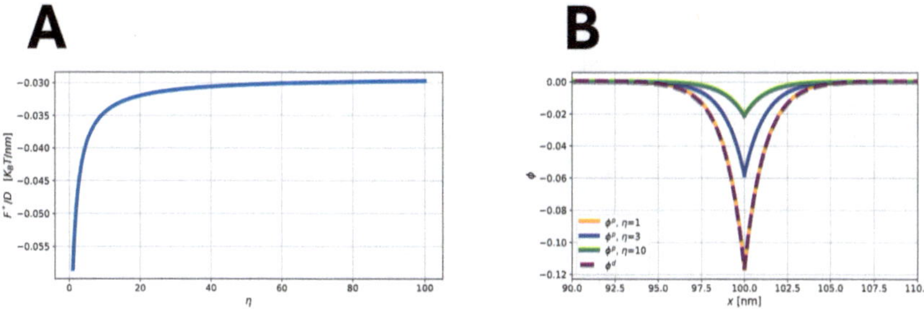

Fig. 4 Results for a single-domain system. (A) The energy of the equilibrium state as a function of the relative domain rigidity, η. (B) The distribution of the tilt angles in the proximal (continuous lines) and distal (dashed line) monolayers for different values of the relative domain rigidity. Parameters used for all curves: $h = 2$ nm, $\Delta\varepsilon = 0.5\ k_{\mathrm{B}}T\ \mathrm{nm}^{-2}$, $\kappa = 10k_{\mathrm{B}}T$, $\kappa_{\mathrm{t}} = 30\ \mathrm{mN\ m}^{-1}$, $R = 100$ nm.

Two domains in apposed monolayers

Now we consider a bilayer containing two domains. If the domains are inserted into the same membrane leaflet, they induce similar tilt-splay deformations in the proximal and distal monolayers. Based on the previous work on membrane deformations by caveolin discs[20] and cap-like inclusions,[37] the overlap of the membrane deformations generated by the two similar domains located in the same monolayer must generate a repulsive interaction between the domains.

Here we consider a bilayer containing two domains located in the apposed monolayers and separated by a distance L between the domains' edges (Fig. 2).

Since the main goal of this section is to analyse the membrane-mediated interaction between the domains, we describe the monolayer deformations and the energy only of the mid-part of the system located between the domains' centers and dependent on the inter-domain distance, L. Since the relevant domain size, R, substantially exceeds the nanometer-large relaxation length, $R \gg \lambda$, the deformations and energies of the left and right peripheral parts of the system do not depend on the separation, L, and, therefore, do not contribute to the domain interaction.

Due to the system's symmetry, the boundary tilt angles for the domain located in the distal, $\phi_D^{d^*}$, and proximal, $\phi_D^{p^*}$, monolayers must be equal, $\phi_D^{d^*} = \phi_D^{p^*} = \phi_D^*$ (Fig. 2). The same is true for the boundary angles of the two contact zones, $\phi_C^{d^*} = \phi_C^{p^*} = \phi_C^*$.

For the distal monolayer of the system, the computed distributions of the tilt angle in the contact zone, the domain, and the monolayer between them (Fig. 2) are given by

$$\begin{aligned}
\phi_C^d(\xi) &= \phi_C^* \cdot \frac{\sinh\left(\rho + \frac{l}{2} + \xi\right)}{\sinh(\rho)}, \\
\phi_D^d(\xi) &= \phi_D^* \cdot \frac{\sinh\left(\rho + \frac{l}{2} - \xi\right)}{\sinh(\rho)}, \\
\phi^d(\xi) &= \frac{1}{2} \cdot \left[(\phi_D^* + \phi_C^*) \cdot \frac{\cosh(\xi)}{\cosh\left(\frac{l}{2}\right)} + (\phi_D^* - \phi_C^*) \cdot \frac{\sinh(\xi)}{\sinh\left(\frac{l}{2}\right)} \right].
\end{aligned} \tag{18}$$

For the proximal monolayer, the distributions of the tilt angle are obtained by inverting the sign of the dimensionless coordinate, ξ, in eqn (18).

To present the computed expressions for the boundary tilt angles and the energy, we first define two auxiliary functions of the dimensionless inter-domain distance $l = L/\lambda$,

$$B1(l) = 1 + \frac{\left[\eta \cdot \mathrm{cth}(\rho) + \tanh\left(\frac{l}{2}\right)\right] \cdot [\mathrm{cth}(\rho) + \mathrm{cth}(l/2)]}{\left[\eta \cdot \mathrm{cth}(\rho) + \mathrm{cth}\left(\frac{l}{2}\right)\right] \cdot [\mathrm{cth}(\rho) + \tanh(l/2)]}, \tag{19}$$

and

$$B2(l) = \frac{[\mathrm{cth}(\rho) + \mathrm{cth}(l/2)]}{\left[\eta \cdot \mathrm{cth}(\rho) + \mathrm{cth}\left(\frac{l}{2}\right)\right]}. \tag{20}$$

The dependences of the boundary tilt angles of the contact zone and the domain on the dimensionless distance, l, are given by

$$\phi_{\mathrm{C}}^{*}(l) = -\frac{2}{B1(l)} \cdot \left[\frac{1}{\mathrm{cth}(\rho) + \tanh(l/2)}\right] \cdot \frac{h \cdot \Delta\varepsilon}{\sqrt{\kappa\kappa_{\mathrm{t}}}}, \tag{21}$$

$$\phi_{\mathrm{D}}^{*}(l) = B2(l) \cdot \phi_{\mathrm{C}}^{*}(l). \tag{22}$$

The energy of the middle part of the system is expressed through the boundary angles by

$$\frac{1}{D}F^{*}(l) = \sqrt{\kappa\kappa_{\mathrm{t}}}\left[\mathrm{cth}(\rho) \cdot \phi_{\mathrm{C}}^{*}(l)^{2} + \eta \cdot \mathrm{cth}(\rho) \cdot \phi_{\mathrm{D}}^{*}(l)^{2} + \frac{1}{2} \cdot \left(\phi_{\mathrm{C}}^{*}(l) + \phi_{\mathrm{D}}^{*}(l)\right)^{2} \tanh\left(\frac{l}{2}\right) + \frac{1}{2} \cdot \left(\phi_{\mathrm{C}}^{*}(l) - \phi_{\mathrm{D}}^{*}(l)\right)^{2} \mathrm{cth}\left(\frac{l}{2}\right)\right] + 2 \cdot h \cdot \Delta\varepsilon \cdot \left[\phi_{\mathrm{C}}^{*}(l) + \phi_{\mathrm{D}}^{*}(l)\right]. \tag{23}$$

The dependence of the energy on the inter-domain distance, $F^{*}(l)$, determined by eqn (23) along with eqn (19)–(22) is presented in Fig. 5 for different values of the relative domain's rigidity η.

There are two regimes of interaction. For values, η, larger than a critical value, η^{*}, the interaction is repulsive for small inter-domain distances, and attractive for large inter-domain distances. This regime is characterized by an equilibrium distance, l^{*}, between the domains, corresponding to the minimum of the energy

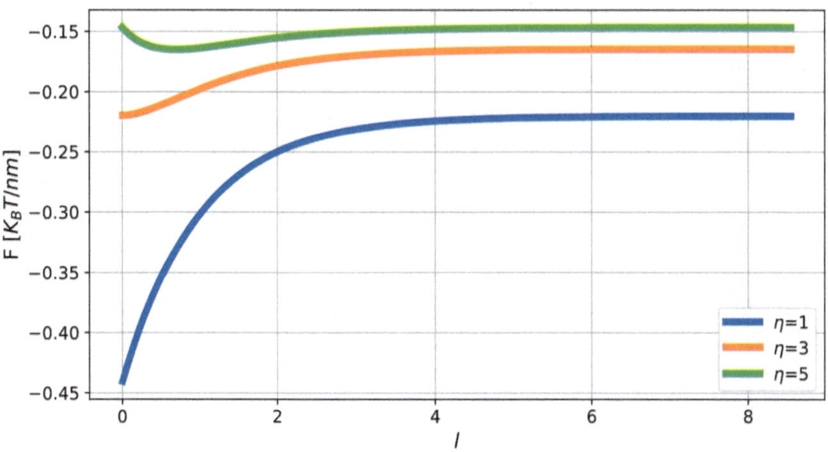

Fig. 5 The membrane-mediated interaction between domains. The energy of the middle part of the system as a function of the dimensionless distance between the domains, l, for different values of the relative domain rigidity, η. Parameters used for all curves: $h = 2$ nm, $\Delta\varepsilon = 0.5k_{\mathrm{B}}T\ \mathrm{nm}^{-2}$, $\kappa = 10k_{\mathrm{B}}T$, $\kappa_{\mathrm{t}} = 30\ \mathrm{mN\ m}^{-1}$, $R = 100$ nm.

(Fig. 5). For the values of the relative rigidity smaller than the critical value, $\eta < \eta^*$, the interaction is attractive for all distances. The critical relative rigidity, η^*, can be found from the approximation of the energy function (eqn (23)) valid for small distances, $l \ll 1$, which is

$$\frac{1}{D}F^* = -\frac{1}{2}\cdot\frac{(h\cdot\Delta\varepsilon)^2}{\sqrt{\kappa K_t}}\left\{\frac{12}{(\eta+1)} + \frac{\left[(\eta-1)^2\cdot \mathrm{cth}(\rho)^2 - 4\right]}{(\eta+1)^2\cdot \mathrm{cth}(\rho)^2}\cdot l\right\}. \tag{24}$$

The corresponding interaction force, $\mathcal{F} = -\frac{\mathrm{d}F^*}{\mathrm{d}L}$, at a vanishing distance is given by

$$\frac{1}{D}\mathcal{F} = \frac{1}{2}\cdot\frac{(h\cdot\Delta\varepsilon)^2}{\kappa}\cdot\frac{\left[(\eta-1)^2\cdot \mathrm{cth}(\rho)^2 - 4\right]}{(\eta+1)^2\cdot \mathrm{cth}(\rho)^2}. \tag{25}$$

The critical value of the relative domain rigidity, η^*, separating the two regimes of the domain interaction is that for which the force (eqn (25)) vanishes,

$$\eta^*(\rho) = 1 + \frac{2}{\mathrm{cth}(\rho)}. \tag{26}$$

For the realistic values of the dimensionless half-width of the domain, $\rho \gg 1$, the critical value is $\eta^* \approx 3$. The dependence of the critical relative rigidity, η^*, on the dimensionless domain size, ρ, is presented in the phase diagram in Fig. 6.

The dependence of the equilibrium inter-domain distance, l^*, on the relative rigidity, η, is presented in Fig. 7.

The maximal inter-domain attractive force corresponds to the case in which the domain rigidity is similar to that of the regular lipid monolayer, $\eta = 1$, and is equal $\frac{1}{D}\mathcal{F}_{\max} = \frac{1}{2}\cdot\frac{(h\cdot\Delta\varepsilon)^2}{\kappa}$.

Discussion

We considered a flat membrane containing lipid domains in its monolayers. We hypothesized that the partial energy of contact between a monolayer and a domain along the membrane midplane differs from that between the two membrane monolayers. Such a differential contact energy is the origin of the domain's surface tension proposed in ref. 15 and 19. We demonstrated by computations that the differential contact energy generates the intra-membrane deformations of lipid tilt and splay which in turn give rise to a membrane-mediated interaction between the domains. The central finding of our analysis is that the character of the interaction between domains located in apposed membrane monolayers depends on the ratio between the domains' and the monolayer's rigidities. Assuming that the domains are more rigid than the regular monolayers we predict that domains whose rigidity is less than threefold that of a regular lipid monolayer exhibit an attractive interaction for any inter-domain separation (Fig. 6 and 7). More rigid domains are predicted to mutually repel at small and attract at large separations, and infinitely rigid domains exhibit

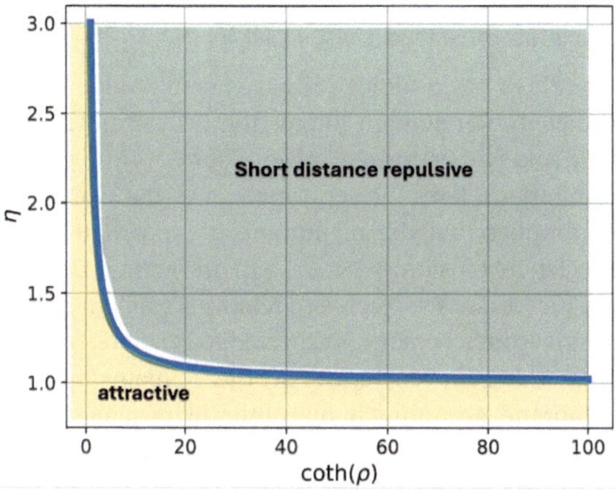

Fig. 6 Phase diagram of the regimes of interaction between domains located in apposed monolayers. The yellow-colored area represents the regime of attractive interaction for all inter-domain distances. The green-colored area represents the regime where the interaction is repulsive at short and attractive at large distances. The phase boundary corresponds to the critical relative domain rigidity, η^*.Parameters used for all curves: $h = 2$ nm, $\Delta\varepsilon = 0.5k_BT$ nm^{-2}, $\kappa = 10k_BT$, $\kappa_t = 30$ mN m^{-1}, $R = 100$ nm.

a repulsion for all distances. The scale of the interaction energy, F_0, is set, according to (eqn (24)), by the combination of the above-defined parameters of the system, $F_0 = \frac{(h\cdot\Delta\varepsilon)^2}{\sqrt{\kappa\kappa_t}}D$. For the differential partial contact energy of $\Delta\varepsilon = 0.5k_BT\,\text{nm}^{-2}$, and the common values of the monolayer thickness, $h = 2$ nm, the monolayer splay (bending) modulus $\kappa = 10k_BT$, the monolayer tilt modulus, $\kappa_t = 30$ mN m^{-1}, and the domain size, $D = 100$ nm, the energy scale is $F_0 \approx 10k_BT$. The

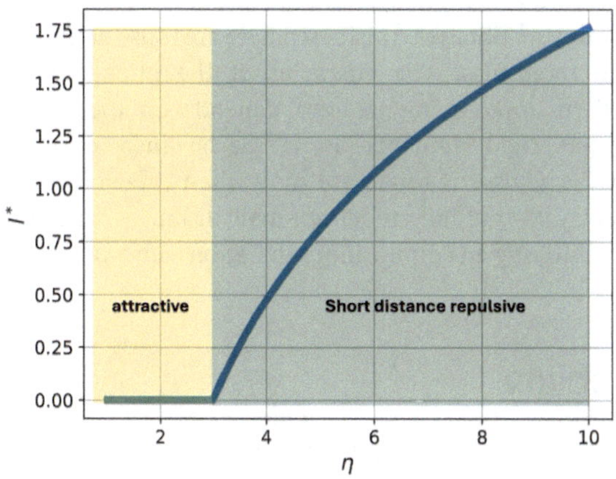

Fig. 7 The equilibrium separation between the domains for different values of the relative domain rigidity, η. For low relative rigidity, $\eta < \eta^*$, the interaction is attractive for all separations. For $\eta > \eta^*$ the interaction is repulsive at small and attractive at large distances creating a non-vanishing equilibrium separation. Parameters used for all curves: $h = 2$ nm, $\Delta\varepsilon = 0.5k_BT$ nm^{-2}, $\kappa = 10k_BT$, $\kappa_t = 30$ mN m^{-1}, $R = 100$ nm.

scale of the interaction force (eqn (25)), is set by $\mathcal{F}_0 = \frac{(h \cdot \Delta\varepsilon)^2}{\kappa} D$, which for the above parameter values has a biologically reasonable value of $\mathcal{F}_0 \approx 40$ pN. The range of the predicted inter-domain interaction is determined by the relaxation length, λ (eqn (7)), and is therefore, relatively short.

The major limitation of the present analysis is the consideration of a flat membrane. This implies that the membrane is exposed to a stretching force generating a large lateral tension, $\gamma \gg \Delta\varepsilon$, and preventing, therefore, membrane bending. In loose membranes subjected to relatively small membrane tensions, $\gamma \ll \Delta\varepsilon$, the differential contact energy is expected to generate, in addition to the tilt and splay deformation, the curvature of the domains and, hence, of the surrounding membrane. According to our preliminary analysis, the development of the domain's curvature considerably reinforces the domain interaction and increases the interaction range to $\lambda_\gamma = \sqrt{\frac{\kappa}{\gamma}}$, which can be substantially larger than that given by eqn (7) for biologically relevant tensions, $\gamma \leq 0.1$ mN m^{-1}, much smaller than the tilt modulus, $\gamma \ll \kappa_t \approx 10$ mN m^{-1}. A detailed analysis of the domain interaction in loose membranes will be presented in a separate article.

Finally, our model does not account for the effects of proteins present in cell membranes in general and cell membrane domains in particular. Membrane proteins can substantially influence the domain interaction in case they are localized to the boundaries of the domains and/or contact zones and modify the degree of the lipid tilting. Therefore, our model is directly applicable to the domain interaction in purely lipid membranes but may miss some protein effects as far as lipid rafts are concerned.

Conclusions

According to our model, the differential energy of the intra-membrane contact between a lipid domain and the underlying lipid monolayers generates deformations of the membrane lipid matrix and the resulting domain interactions. The deformations for flat membranes considered here are those of tilt and splay of the constituent lipid molecules. The character of the interaction between two domains located in apposed monolayers depends on the domain's relative rigidity. Moderately rigid domains attract each other at all inter-domain distances. Domains whose rigidities are more than threefold those of regular membrane monolayers mutually repel at small distances, and attract at large distances, thus, exhibiting an equilibrium separation with a vanishing interaction force.

Data availability

This study was carried out using publicly available data from the published articles (see references).

Conflicts of interest

The authors declare no conflicts of interest.

Acknowledgements

MMK acknowledges support from the Israel Science Foundation (Grant No. 1994/22) and holds the Joseph Klafter Chair in Biophysics.

References

1 T. Baumgart, S. T. Hess and W. W. Webb, Imaging coexisting fluid domains in biomembrane models coupling curvature and line tension, *Nature*, 2003, **425**(6960), 821–824.
2 H. A. Rinia and B. de Kruijff, Imaging domains in model membranes with atomic force microscopy, *FEBS Lett.*, 2001, **504**(3), 194–199.
3 H. A. Rinia, M. M. Snel, J. P. van der Eerden and B. de Kruijff, Visualizing detergent resistant domains in model membranes with atomic force microscopy, *FEBS Lett.*, 2001, **501**(1), 92–96.
4 E. Sezgin, I. Levental, S. Mayor and C. Eggeling, The mystery of membrane organization: composition, regulation and roles of lipid rafts, *Nat. Rev. Mol. Cell Biol.*, 2017, **18**(6), 361–374.
5 K. Simons and E. Ikonen, Functional rafts in cell membranes, *Nature*, 1997, **387**(6633), 569–572.
6 J. H. Ipsen, G. Karlstrom, O. G. Mouritsen, H. Wennerstrom and M. J. Zuckermann, Phase equilibria in the phosphatidylcholine-cholesterol system, *Biochim. Biophys. Acta, Biomembr.*, 1987, **905**(1), 162–172.
7 B. D. Ladbrooke and D. Chapman, Thermal analysis of lipids, proteins and biological membranes. A review and summary of some recent studies, *Chem. Phys. Lipids*, 1969, **3**(4), 304–356.
8 G. W. Feigenson, Phase diagrams and lipid domains in multicomponent lipid bilayer mixtures, *Biochim. Biophys. Acta, Biomembr.*, 2009, **1788**(1), 47–52.
9 G. G. Putzel, M. J. Uline, I. Szleifer and M. Schick, Interleaflet coupling and domain registry in phase-separated lipid bilayers, *Biophys. J.*, 2011, **100**(4), 996–1004.
10 S. Komura and D. Andelman, Physical aspects of heterogeneities in multi-component lipid membranes, *Adv. Colloid Interface Sci.*, 2014, **208**, 34–46.
11 W. F. Bennett and D. P. Tieleman, Computer simulations of lipid membrane domains, *Biochim. Biophys. Acta, Biomembr.*, 2013, **1828**(8), 1765–1776.
12 S. Garg, J. Ruhe, K. Ludtke, R. Jordan and C. A. Naumann, Domain registration in raft-mimicking lipid mixtures studied using polymer-tethered lipid bilayers, *Biophys. J.*, 2007, **92**(4), 1263–1270.
13 M. D. Collins and S. L. Keller, Tuning lipid mixtures to induce or suppress domain formation across leaflets of unsupported asymmetric bilayers, *Proc. Natl. Acad. Sci. U. S. A.*, 2008, **105**(1), 124–128.
14 S. May, Trans-monolayer coupling of fluid domains in lipid bilayers, *Soft Matter*, 2009, **5**(17), 3148–3156.
15 M. D. Collins, Interleaflet coupling mechanisms in bilayers of lipids and cholesterol, *Biophys. J.*, 2008, **94**(5), L32–L34.
16 J. J. Williamson and P. D. Olmsted, Comment on "Elastic Membrane Deformations Govern Interleaflet Coupling of Lipid-Ordered Domains", *Phys. Rev. Lett.*, 2016, **116**(7), 079801.

17 M. C. Blosser, A. R. Honerkamp-Smith, T. Han, M. Haataja and S. L. Keller, Transbilayer Colocalization of Lipid Domains Explained via Measurement of Strong Coupling Parameters, *Biophys. J.*, 2015, **109**(11), 2317–2327.

18 T. R. Galimzyanov, R. J. Molotkovsky, M. E. Bozdaganyan, F. S. Cohen, P. Pohl and S. A. Akimov, Elastic Membrane Deformations Govern Interleaflet Coupling of Lipid-Ordered Domains, *Phys. Rev. Lett.*, 2015, **115**(8), 088101.

19 H. J. Risselada and S. J. Marrink, The molecular face of lipid rafts in model membranes, *Proc. Natl. Acad. Sci. U. S. A.*, 2008, **105**(45), 17367–17372.

20 A. Barnoy, R. G. Parton and M. M. Kozlov, Mechanism of membrane curvature generation and caveola formation by flat disc-like complexes of caveolin, *bioRxiv*, 2024, preprint, DOI: 10.1101/2024.07.09.602763.

21 F. Campelo, C. Arnarez, S. J. Marrink and M. M. Kozlov, Helfrich model of membrane bending: from Gibbs theory of liquid interfaces to membranes as thick anisotropic elastic layers, *Adv. Colloid Interface Sci.*, 2014, **208**, 25–33.

22 J. W. Gibbs, *The Scientific Papers*, New York, Dover, 1961.

23 M. M. Kozlov, S. L. Leikin and V. S. Markin, Elastic properties of interfaces - elasticity moduli and spontaneous geometric characteristics, *J. Chem. Soc., Faraday Trans. 2*, 1989, **85**, 277–292.

24 M. M. Kozlov and M. Winterhalter, Elastic-Moduli and Neutral Surface for Strongly Curved Monolayers - Analysis of Experimental Results, *J. Phys. II*, 1991, **1**(9), 1085–1100.

25 S. Leikin, M. M. Kozlov, N. L. Fuller and R. P. Rand, Measured effects of diacylglycerol on structural and elastic properties of phospholipid membranes, *Biophys. J.*, 1996, **71**(5), 2623–2632.

26 M. Hamm and M. Kozlov, Elastic energy of tilt and bending of fluid membranes, *Eur. Phys. J. E*, 2000, **3**(4), 323–335.

27 M. Hamm and M. Kozlov, Tilt model of inverted amphiphilic mesophases, *Eur. Phys. J. B*, 1998, **6**(4), 519–528.

28 E. Evans and R. Skalak, *Mechanics and Thermodynamics of Biomembranes*, CRC, Boca Raton, Florida, 1980.

29 B. Pontes, P. Monzo and N. C. Gauthier, Membrane tension: A challenging but universal physical parameter in cell biology, *Semin. Cell Dev. Biol.*, 2017, **71**, 30–41.

30 J. F. Nagle and S. Tristram-Nagle, Structure of lipid bilayers, *Biochim. Biophys. Acta, Rev. Biomembr.*, 2000, **1469**(3), 159–195.

31 N. Fuller and R. P. Rand, The influence of lysolipids on the spontaneous curvature and bending elasticity of phospholipid membranes, *Biophys. J.*, 2001, **81**(1), 243–254.

32 J. A. Szule, N. L. Fuller and R. P. Rand, The effects of acyl chain length and saturation of diacylglycerols and phosphatidylcholines on membrane monolayer curvature, *Biophys. J.*, 2002, **83**(2), 977–984.

33 M. M. Terzi and M. Deserno, Novel tilt-curvature coupling in lipid membranes, *J. Chem. Phys.*, 2017, **147**(8), 084702.

34 R. Dimova, Recent developments in the field of bending rigidity measurements on membranes, *Adv. Colloid Interface Sci.*, 2014, **208**, 225–234.

35 W. Helfrich, Elastic properties of lipid bilayers: theory and possible experiments, *Z. Naturforsch. C*, 1973, **28**(11–12), 693–703.

36 M. Doktorova, D. Harries and G. Khelashvili, Determination of bending rigidity and tilt modulus of lipid membranes from real-space fluctuation

analysis of molecular dynamics simulations, *Phys. Chem. Chem. Phys.*, 2017, **19**(25), 16806–16818.

37 T. R. Weikl, M. M. Kozlov and W. Helfrich, Interaction of conical membrane inclusions: Effect of lateral tension, *Phys. Rev. E*, 1998, **57**(6), 6988–6995.

Faraday Discussions

PAPER

Characterization of lipid chain order and dynamics in asymmetric membranes by solid-state NMR spectroscopy†

Oskar Engberg, Viola Döbel, Kathrin M. Engel and Daniel Huster*

Received 9th December 2024, Accepted 16th January 2025
DOI: 10.1039/d4fd00192c

We studied the structure and dynamics of asymmetric $POPC^{out}/(POPE/POPG)^{in}$ and $POPS^{out}/(POPE/POPG)^{in}$ lipid membranes. To this end, the outer layer of multilamellar POPE/POPG (molar ratio 9 : 1) vesicles was exchanged (using methyl-β-cyclodextrin) by either chain deuterated POPC-d_{31} or POPS-d_{31}, for which ^{2}H NMR order parameters were measured. As controls, we prepared symmetric POPC-d_{31}/POPE/POPG and POPS-d_{31}/POPE/POPG membranes of the composition of just the outer membrane of the asymmetric multilamellar vesicles and pure POPC-d_{31} or POPS-d_{31} multilamellar vesicles. Compared to symmetric membranes of the same lipid composition, chain order parameters (S) of the asymmetric preparations were higher in the upper half of the chain and lower in the lower half. This reshuffling of acyl chain order is also expressed in higher ^{2}H NMR Zeeman order relaxation rates (R_{1Z}) of the chain segments in asymmetric membranes indicating alterations in the elastic properties of asymmetric bilayers as inferred from plots of R_{1Z} *vs.* S^2. Asymmetric membranes showed increased stiffness and rigidity although the lipid acyl chain composition between the inner and outer leaflets were identical. There were no indications for chain interdigitation between the two leaflets in the NMR spectra, which led us to speculate that the interleaflet coupling could be accomplished by sensing the differences in lipid packing densities between the two leaflets. These alterations in leaflet properties should have consequences for lipid protein interaction and ultimately protein function.

1 Introduction

The overwhelming lipid variety and asymmetry of the plasma membrane remains one of the most enigmatic characteristics of biological cells.[1,2] The energetic cost of synthesizing hundreds of different lipid species and maintaining an entropically unfavorable leaflet asymmetry of the plasma membrane is enormous and

Institute of Medical Physics and Biophysics, University of Leipzig, Härtelstr. 16/18, D-4107 Leipzig, Germany. E-mail: daniel.huster@medizin.uni-leipzig.de

† Electronic supplementary information (ESI) available. See DOI: https://doi.org/10.1039/d4fd00192c

consumes significant resources of the cell. One would expect evolution to have optimized (*i.e.* reduced) the lipid variety, composition and structure of the most important barrier for the cell unless there are important reasons to maintain this diversity. Till this day, our understanding of how specific lipids control biological function is limited to very few species such as cholesterol, phosphatidylinositolbisphosphate (PIP_2), diacylglycerol (DAG), docosahexaenoic acid (DHA), and phosphatidylserine (PS).[3–6] For the majority of lipids, however, the biological reason for the tremendous lipid variety is still unknown. Various examples of lipid function as second messengers, "force from lipid", or specific lipid–protein interactions are known,[3–6] however, most of our knowledge has been achieved from membrane models with symmetric lipid distribution,[6–8] ignoring the most prevailing feature of all plasma membranes. But an increasing body of studies suggests that numerous core functions of the plasma membrane are related to membrane leaflet asymmetry.[9]

Recent work has demonstrated that (i) preparation of asymmetric membrane models is possible with a reasonable effort,[10] and (ii) both membrane and protein properties are sensitive to lipid asymmetry,[11–13] which provides an imperative to represent this important aspect in our membrane models for biophysical studies. The limited amount of data that is available for asymmetric membrane preparations describes numerous unique findings.[9] One such aspect is intraleaflet coupling. Differential scanning calorimetry work has revealed that asymmetric $POPC^{out}/POPE^{in}$ and $POPE^{out}/POPC^{in}$ membranes have different melting characteristics.[11] A coupled phase transition was only observed for the $POPC^{out}/POPE^{in}$ vesicles, while the $POPE^{out}/POPC^{in}$ vesicles showed the individual melting temperatures of each lipid species. Another study of symmetric and asymmetric DPPC/DPPS vesicles also showed quite different DSC thermograms.[14] Intraleaflet coupling is thought to be related to interdigitation of the lipid chains.[15] Cholesterol, although it is known for its very fast flip flop between the two membrane leaflets, has also been found to be highly asymmetrically distributed between the leaflets, where the exoplasmic leaflet contains significantly more cholesterol.[16,17] Accordingly, lipid leaflet asymmetry also influences lateral domain formation.[18] Most importantly, first studies demonstrate that membrane asymmetry influences membrane protein function.[13,19]

NMR methods along with X-ray and neutron scattering as well as FTIR spectroscopy have shaped our understanding of lipid membranes as highly dynamic entities that show a transient lateral heterogeneity.[20–24] Lipid molecules undergo multiple different motions with varying amplitudes involving timescales from ps to ms.[24] This leads to a high degree of disorder and membrane elastic properties that are designed to optimize protein function,[25–28] but also other functions like permeability, fusion, small molecule binding *etc.* As the plasma membrane is enriched in saturated lipids and cholesterol in the outer and unsaturated lipids in the inner leaflet, it has been suggested that the two leaflets have different mechanical properties.[2] Indeed, higher bending moduli were measured in asymmetric egg $sphingomyelin^{out}/POPE^{in}$ vesicles.[29] The high lipid dynamics is also very well represented in modern all atom molecular dynamics (MD) simulations, which can also be done for asymmetric systems and easily span trajectories of tens of microseconds.[30] Nevertheless, similar to the situation in experimental work, there are much fewer MD simulations on asymmetric membranes in comparison to the huge amount of MD data on symmetric membranes.[31–33]

While the dynamic nature of lipid-induced effects on protein function has long been a matter of discussion,[25–28,34] considering the aspect of membrane asymmetry has to be taken into account in the next step. While especially scattering and some FTIR data is available for asymmetric membranes,[12,14,29,35] the powerful arsenal of NMR methods to characterize membrane properties has so far only rarely been utilized, essentially only to characterize the degree of asymmetry.[36,37] In this discussion, we present ^{2}H NMR spectroscopy data on two asymmetric membrane preparations that compare (i) acyl chain order parameters, (ii) lipid packing properties, and (iii) dynamic/elastic properties of the two leaflets of asymmetric membranes composed of monounsaturated POPE/POPG (9 : 1) in the inner leaflet and either POPC or POPS in the outer. For the asymmetric preparations, we used multilamellar vesicles (MLVs) whose outer shell is exchanged with either POPC or POPS. This allows for studying membrane properties of the outer leaflet only as all inner membranes of the MLVs are symmetric. However, using acyl chain perdeuterated lipids for the exchange, the isotopic enrichment of the outer leaflet is sufficient to study the structure and dynamics of this lipid leaflet with atomic resolution.

2 Materials and methods

1-Palmitoyl-2-oleoyl-*sn*-glycero-3-phosphocholine (POPC), 1-palmitoyl-2-oleoyl-*sn*-glycero-3-phosphoethanolamine (POPE), 1-palmitoyl-2-oleoyl-*sn*-glycero-3-phospho-(1′-*rac*-glycerol) (POPG), and 1-palmitoyl-2-oleoyl-*sn*-glycero-3-phospho-L-serine (POPS) were purchased from Avanti Polar Lipids (Alabaster, AL, USA). POPC-d_{31}, and POPS-d_{31} were perdeuterated in the palmitoyl *sn*-1 chain and also purchased from Avanti. All other chemicals purchased were of the highest purity if not stated otherwise. Solvents were obtained in the highest commercially available purity from Biosolve BV (Valkenswaard, Netherlands). The chemicals 2,5-dihydroxybenzoic acid (DHB) and 9-aminoacridine (9-AA) were purchased from Sigma Aldrich (Taufkirchen, Germany) and Acros Organics (Geel, Belgium), respectively.

2.1 Asymmetric MLV sample preparation

For preparation of asymmetric membranes, we by and large followed the protocol of Doktorova *et al.*[10] with some essential modifications. Most importantly, there was no need to use sucrose as the acceptor lipid structures were MLVs, and SUVs as donor vesicles. For separation of donor and acceptor vesicles, simple centrifugation worked well. Therefore, there was no worry that sucrose could interfere with lipid structure and dynamics. An extra centrifugation step was added for easier handling of the final pellet to be transferred into an NMR rotor.

Lipids were dissolved in $CHCl_3$/MeOH (1 : 1 v/v) and evaporated at 40 °C using a rotary evaporator. The donor vesicles (composed of either POPC-d_{31} or POPS-d_{31}) were prepared at a concentration of 10 mg ml^{-1} in mqH_2O containing 35 mM NaCl and probe sonicated on ice (duty cycle 50%) until transparent (~30 min) to form small unilamellar vesicles (SUV). The acceptor vesicles were prepared from POPE/POPG (9 : 1 molar ratio) lipid mixtures suspended at a concentration of

10 mg ml^{-1} in mqH_2O containing 35 mM NaCl and freeze–thawed between liquid nitrogen and a 50 °C water bath for 10 times to form multilamellar vesicles (MLV). The 10 mol% POPG helped decreasing the large aggregates formed when pure POPE is dispersed in aqueous solution.[38] The acceptor MLV and the donor SUV dispersions were centrifuged at 20 000 × *g* for 30 min at 4 °C to remove small vesicles from the acceptor MLV and large vesicles from the donor SUV solutions. The remaining pellet for the acceptor MLVs and the SUV supernatant was kept.

The SUVs dispersion was diluted with aliquots of a 70 mM mβCD solution and mqH_2O to reach a final methyl-β-cyclodextrin (mβCD) concentration of 35 mM and a molar ratio of 8 : 1 mβCD to lipid. This dispersion was incubated for 2 h at 37 °C under magnetic stirring to preload the mβCD with donor lipids. After 2 h, the dispersion was used to dilute the MLV pellet at 2 : 1 donor : acceptor lipid molar ratio and incubated for 30 min at 37 °C under magnetic stirring to exchange the outer leaflet lipids on the MLV acceptor preparation. Subsequently, the samples were diluted with a 35 mM NaCl solution to 50 ml and transferred to 50 ml Falcon tubes, which were centrifuged at 20 000 × *g* for 30 min at 4 °C. The pellet including ~0.5 ml of the remaining supernatant on the bottom of the tube was redissolved with mqH_2O containing 35 mM NaCl to a volume of 1 ml and transferred to 1.5 ml Eppendorf tubes. The tubes were centrifuged at 20 000 × *g* for 30 min at 4 °C to form a pellet. Finally, the lipid pellets were transferred into 4 mm MAS NMR rotors and immediately measured.

For symmetric control MLV samples, lipids were dissolved in 2 ml cyclohexane after solvent evaporation and lyophilized overnight to form a fluffy powder that is easy to hydrate. The powders were hydrated to 50 wt% using mqH_2O containing 35 mM NaCl and freeze–thawed between 50 °C water bath and LN_2 to form MLVs. Samples were transferred into 4 mm MAS NMR rotors and stored at −20 °C until measured.

2.2 Solid state NMR measurements

The samples were measured on a Bruker 750 Avance I NMR spectrometer using a resonance frequency of 115.1 MHz for ^{2}H. A 4 mm triple gradient MAS probe was used but without spinning to acquire static ^{2}H NMR spectra using a phase-cycled quadrupolar echo sequence[39] with a 30 μs delay between the two π/2 pulses of ~3 μs length each. The spectral width was 500 kHz and the recycling delay 1 s. The order parameters[40] and chain extension profiles were calculated using Mathcad programs for dePakeing and previously described[41] methods using the mean torque model.[42] For measuring the T_{1Z} relaxation times an inversion recovery pulse sequence with quadrupole echo and a delay list of 0.001, 0.007, 0.015, 0.03, 0.07, 0.12, 0.18, 0.25, 0.5, 1, 2.5 s with a relaxation delay of 2.5 s was used. The R_{Z1} relaxation rates of the ^{2}H NMR powder spectra were calculated using a program written in Mathcad as previously described.[43] Briefly, the ^{2}H NMR line shapes were simulated using 15 Pake doublets and one isotropic peak for the asymmetric samples. From the fit of the inversion recovery data, the spin lattice relaxation time of each peak was determined.

2.3 Preparation of asymmetric LUV for testing lipid exchange efficiency

For the preparation of asymmetric LUVs, we followed the protocol of Doktorova *et al.*[10] without significant modifications. For both donor and acceptor vesicle preparation, lipids were dissolved in $CHCl_3$/MeOH (1 : 1 v/v) and evaporated as above. Both acceptor and donor vesicles were dissolved in 35 mM NaCl H_2O to reach a concentration of 10 mg ml^{-1}. Donor MLVs were prepared from the respective lipid mixes by freeze thawing. Acceptor vesicles were prepared by extrusion using two 100 nm Teflon filters 25 times to obtain large unilamellar vesicles (LUV). The donor MLV dispersions were centrifuged at 20 000 × *g* for 30 min at 4 °C to separate the small vesicles from the donor MLV. The donor MLV pellet was dissolved and diluted in 70 mM mβCD solution and further diluted with mqH_2O to reach a final mβCD concentration of 35 mM (at an 8 : 1 mβCD to lipid molar ratio) and incubated at 37 °C for 2 h under magnetic stirring to preload the mβCD with donor lipids. After 2 h, the LUV acceptor solution was added at 2 : 1 donor : acceptor molar ratio incubated for 30 min at 37 °C under magnetic stirring for the exchange of the outer leaflet lipids on the LUV acceptor. Subsequently, the samples were diluted with 35 mM NaCl H_2O to 50 ml and transferred to a 50 ml Falcon tube and centrifuged at 20 000 × *g* for 30 min at 4 °C. The supernatant was carefully removed with a glass Pasteur pipet until ~2 ml of the pellet was left. To remove mβCD and mβCD–lipid complexes from the supernatant, Amicon® Ultra Filters (Merck KGaA, Darmstadt, Germany) with a 100 kDa MWCO were used in a swing bucket centrifuge (4000 × *g* at room temperature) until the volume was reduced to a few ml. Before use the centrifugation filters were wetted with 10 ml of mqH_2O and centrifuged at 4000 × *g* for 10 min at room temperature. Subsequently, the samples were extracted and lipid concentration was quantified using TLC.

2.4 Thin layer chromatography

The lipids from the LUV dispersions were extracted following the method described by Bligh and Dyer.[44] Specifically, 200 μl of the LUV dispersion were combined with 400 μl of $CHCl_3$/MeOH (1 : 1 [v/v]) in a glass vial and mixed for 30 s. A 5 min centrifugation step at 10 000 × *g* and 20 °C provided phase separation. The lipid-containing $CHCl_3$ phase was transferred to a clean glass vial and for the remaining aqueous phase, the procedure was repeated twice using 200 μl $CHCl_3$ for each step and organic phases were combined. $CHCl_3$ from the combined phases was evaporated using a vacuum rotation concentrator (RVC 2-25, Martin Christ Gefriertrocknungsanlagen GmbH, Osterode am Harz, Germany). The resulting lipid film was redissolved in 200 μl $CHCl_3$. Triplicates of the mixture were applied in 1.5, 2, 2.5 μl spots for POPS, POPC and POPE and in 6, 7, 8 μl spots for POPG to a high-performance thin-layer chromatography (HPTLC) silica gel 60 plate (Merck KGaA) by a CAMAG Linomat 5 sample applicator (CAMAG, Berlin, Germany). POPS, POPE, and POPC (0.5 mg ml^{-1} in $CHCl_3$) and POPG (1 mg ml^{-1} in $CHCl_3$) were applied as lipid calibration standards to the same HPTLC plate. Chromatographic separation was carried out in a CAMAG glass chamber using a mobile phase consisting of chloroform/ethanol/water/triethylamine (30 : 35 : 7 : 35 [v/v/v/v]) for 45 min.[45] Lipid spots were visualized by staining with primuline (Sigma-Aldrich, Taufkirchen, Germany), 100 mg l^{-1} in acetone/H_2O (4 : 1 [v/v]), followed by excitation at 366 nm. The picture was recorded in the black and white mode and the densitometric analysis (band volume) of each band was calculated using totallab quant 2.2 software (totallab, Newcastle, UK). The calibration curve was fitted with Origin software (version 9.60, Originlab corporation,

Northampton, MA, USA), using a single exponential model equation, from which the sample lipid concentrations were derived.

2.5 Matrix-assisted laser desorption/ionization time-of-flight mass spectrometry to detect mβCD

The asymmetric MLVs were analyzed by matrix-assisted laser desorption/ionization time-of-flight mass spectrometry (MALDI-TOF MS) to check for remaining mβCD. One μl of each sample (~5% of the sample) was filled to 10 μl with methanol and subsequently mixed with 10 μl of the respective matrix. DHB[46] (0.5 M in methanol) was used as the matrix to record mass spectra in the positive ion mode, 9-AA (10 mg ml^{-1} in 2-propanol/acetonitrile (3 : 2, v/v)) served as the matrix for negative ion spectra. The diluted asymmetric MLV samples were mixed 1 : 1 (v/v) with the respective matrix and vortexed for good homogeneity. Subsequently, 0.75 μl of the respective solution were transferred onto an aluminum-coated MALDI target (Bruker Daltonics GmbH, Bremen, Germany). Instantly after air drying, positive and negative ion mass spectra were recorded on an Autoflex Speed mass spectrometer (Bruker Daltonics) utilizing a Neodym-YAG laser emitting at 355 nm. The extraction voltage was 20 kV and gated matrix suppression was applied to prevent the saturation of the detector by matrix ions. For each mass spectrum, 10 000 single laser shots were averaged. The laser fluency was kept about 10% above threshold (*i.e.* the minimum laser fluency required to detect any signals) to obtain an optimum signal-to-noise ratio. All spectra were acquired in the reflector mode using delayed extraction conditions. Raw data were processed using the software "Flex Analysis" version 3.0 (Bruker Daltonics).

3 Results

We prepared two samples of asymmetric membranes. These systems consisted of POPE/POPG (molar ratio 9 : 1) multilamellar vesicles (MLVs) in which only the outer bilayer was prepared asymmetrically. To this end, SUVs of either POPC-d_{31} or POPS-d_{31}, both perdeuterated in the *sn*-1 chain, served as donor vesicles. After mβCD-catalyzed exchange, the MLVs were centrifuged and the pellet was used as sample for the ^{2}H NMR experiments. As control, we used four symmetric MLV preparations composed of POPC-d_{31}, POPS-d_{31}, POPC-d_{31}/POPE/POPG (molar ratio 10 : 9 : 1) and POPS-d_{31}/POPE/POPG (molar ratio 10 : 9 : 1). The lipid mixing ratio of the symmetric vesicles was chosen assuming that all lipids are homogeneously distributed between the two leaflets.

Typical ^{2}H solid-state NMR spectra are shown in Fig. 1. The NMR spectra of both pure POPC and POPS as well as the symmetric vesicles show powder patterns of well-resolved Pake doublets assignable to the individual methylene and methyl segments of the chain. The NMR spectra of the asymmetric membranes are characterized by a larger line width and an isotropic peak indicating somewhat slower motions (*i.e.* shorter T_2 times) and a minor fraction (~7–9%) of lipids in highly curved geometry.

In the series where deuterated POPC-d_{31} was used, pure POPC gives the narrowest spectrum. Both the symmetric and the asymmetric membranes provide broader ^{2}H NMR spectra, indicating higher chain order parameters. In contrast,

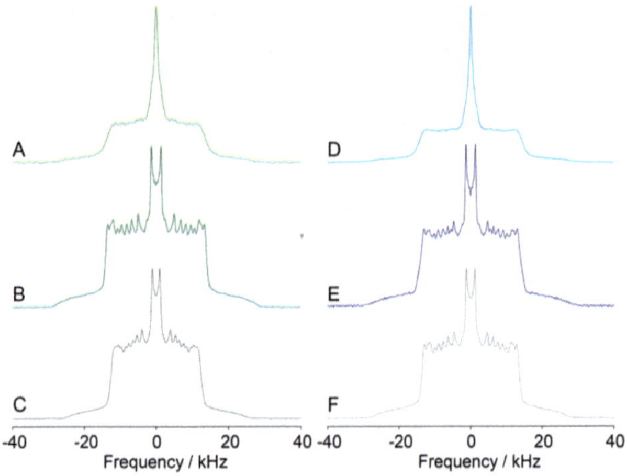

Fig. 1 Solid-state ^{2}H NMR spectra of multilamellar vesicle preparations with symmetric and asymmetric lipid distribution. (Left) ^{2}H NMR spectra of asymmetric POPC-d_{31}^{out}/(POPE/POPG)^{in} membranes (A), symmetric POPC-d_{31}/POPE/POPG bilayers (B), and pure POPC-d_{31} membranes (C). (Right) ^{2}H NMR spectra of asymmetric POPS-d_{31}^{out}/(POPE/POPG)^{in} membranes (D), symmetric POPS-d_{31}/POPE/POPG bilayers (E), and pure POPS-d_{31} membranes (F).

the NMR spectra of the POPS-d_{31}-containing samples are of relatively similar width. These differences can be discussed more quantitatively after dePakeing of the spectra and calculating the order parameters (S) along the chains for all samples.

Fig. 2A compares the order parameters (S) of the POPC-d_{31}-containing samples. For pure POPC, the typical order parameter plateau of the upper chain and the strongly decreasing order parameters in the lower chain are observed. The symmetric POPC-d_{31}/POPE/POPG membranes show the same features, but for the entire chain, the order parameters are about 10–20% larger. This is caused by the presence of POPE, which is characterized by a negative intrinsic curvature that induces tighter lipid packing in PE-containing mixtures.[8,47] The order parameters

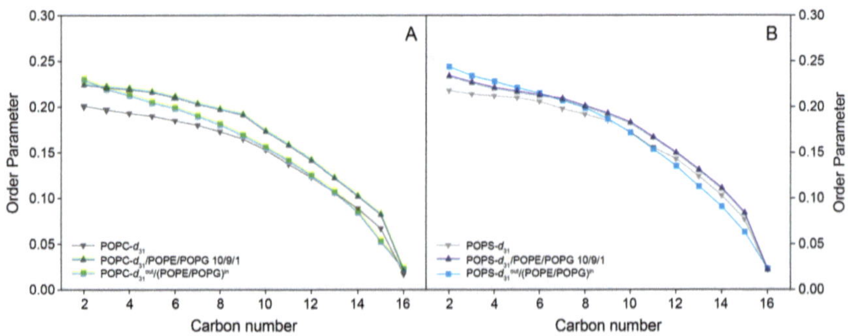

Fig. 2 Smoothed ^{2}H NMR order parameter plots of symmetric and asymmetric MLVs determined from the ^{2}H NMR spectra shown in Fig. 1. (A) Order parameters of POPC-d_{31} and (B) order parameters of POPS-d_{31} are plotted. The temperature of the measurement was 37 °C.

of the asymmetric $POPC^{out}/(POPE/POPG)^{in}$ membranes show a distinctly different shape. While the upper ~3 carbons show slightly higher or similar order as the POPC in symmetric membranes, order parameters drop rather steeply, fall even below those of pure POPC starting at carbon position 8, and remain lower for the entire lower chain. Such a behavior of the chain order parameters has for instance been observed for the lipid modifications of Ras proteins that were inserted into membranes of lower hydrophobic thickness than the length of the Ras lipid modification.[48]

In Fig. 2B, the order parameters of pure POPS-d_{31}, symmetric POPS-d_{31}/POPE/POPG, and asymmetric POPS-${d_{31}}^{out}/(POPE/POPG)^{in}$ are compared. PS membranes are also characterized by higher order parameters than observed for PC bilayers.[47] Therefore, the addition of POPE as in the symmetric POPS-d_{31}/POPE/POPG bilayers does not lead to a very significant increase in chain order. In the asymmetric $POPS^{out}/(POPE/POPG)^{in}$, we observe a similar tendency as seen for asymmetric $POPC^{out}/(POPE/POPG)^{in}$ bilayers: the upper chain segments of the POPS chain show higher order parameters, while the lower segments are characterized by lower order compared to pure POPS and the symmetric POPS-d_{31}/POPE/POPG bilayers.

The differences in the order parameters are also reflected in the length of the lipid's acyl chains, where lower order generally corresponds to shorter projected chain lengths.[42,49] We calculated these geometric parameters for the lipid's acyl chains using Brown's mean torque model.[42] For pure POPC, we measure an average chain length of 12.6 Å in agreement with previous data.[50] Due to the presence of POPE that is characterized by a high negative intrinsic curvature as discussed above, chain length of POPC in the presence of POPE increases to 13.3 Å. Interestingly, the chain length of the POPC in the outer leaflet of the asymmetric membranes is found to be 12.7 Å, which corresponds well to the pure POPC membranes.

For POPS, we had seen relatively similar order parameter profiles and the chain lengths calculated are also very similar (13.2 Å for pure POPS), 13.3 Å for symmetric POPS/POPE/POPG membranes, and 13.2 Å for asymmetric $POPS^{out}/(POPE/POPG)^{in}$ membranes. This is also reflected in the chain extension plots, which are shown in ESI Fig. S1.†

Obviously, the packing and dynamic properties of POPC and POPS in asymmetric membranes are quite different from symmetric bilayers. To further characterize the elastic properties of the asymmetric preparations, we measured ^{2}H T_{1Z} relaxation times and plotted the relaxation rate R_{1Z} as a function of the square of the order parameter (S^2) for each lipid segment. These "square law plots" have empirically been shown to describe the elastic properties of membranes.[51–53] For saturated lipid membranes, a straight line is observed suggesting a $R_{1Z} \propto S^2$ proportionality.[51–53] In the presence of detergents, which make membranes softer, the R_{1Z} *vs.* S^2 plot is curved and characterized by a steeper slope and a $R_{1Z} \propto S^n$ proportionality where $n < 2$.[51] Such behavior has also been observed for monounsaturated lipids as POPC.[54] In the presence of cholesterol, which makes membranes stiffer, the slope of the R_{1Z} *vs.* S^2 plot is shallower but linear. These plots represent a useful measure for a qualitative evaluation of the membrane's elastic properties. Alterations in membrane structure that lead to a rigidization are accompanied by a decrease in the slope of the R_{1Z} *vs.* S^2 plot; alterations that make membranes more elastic are manifested in a convexly shaped curve in the R_{1Z} *vs.* $S^{n<2}$ plot with a steeper slope.

Fig. 3 shows these square law plots for both asymmetric membrane preparations and the respective controls. For pure POPC-d_{31} MLVs, a similar plot is observed as reported before.[54] The data can be fitted with the function $R_{1Z} = y_0 + a \times S^n$, where n should be 2 in an ideal square law plot. For the pure POPC membranes, we determined $n = 1.02$, indicating a significant deviation from the square law in agreement with soft POPC membranes characterized by a $R_{1Z} \propto \sqrt{S^2}$ proportionality. For the symmetric POPC-d_{31}/POPE/POPG membranes, the plot becomes shallower and the exponent determined from the fit increased to $n = 1.52$. For the asymmetric POPC^{out}/(POPE/POPG)^{in} membranes, the exponent further increased to $n = 1.62$.

The empiric relation between R_{1Z} and S^2 can be used to compare the elastic properties of the different membranes. Monounsaturated membranes are generally more elastic than saturated ones, which leads to a convexly curved R_{1Z} *vs.* S^2 plot. Mathematically, this is manifested by a $R_{1Z} \propto S^n$ plot, where $n < 2$. If the

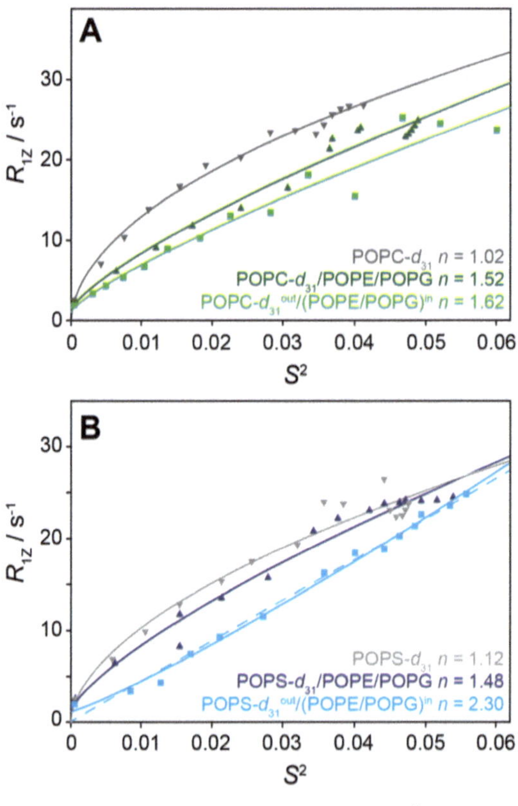

Fig. 3 (A) Square law plots of POPC-d_{31}^{out}/(POPE/POPG)^{in} membranes (light green squares), symmetric POPC-d_{31}/POPE/POPG bilayers (dark green triangles), and pure POPC-d_{31} membranes (gray upside down triangles). (B) Square law plots of POPS-d_{31}^{out}/(POPE/POPG)^{in} membranes (light blue squares), symmetric POPS-d_{31}/POPE/POPG bilayers (dark blue triangles), and pure POPS-d_{31} membranes (light gray upside down triangles). The lines represent best fits to the function $R_{1Z} = y_0 + a \times S^n$, and the dashed solid line represents a linear regression.

membrane becomes more rigid, n also increases, which is what is observed here; n increases from 1.02 for pure POPC membranes to 1.52 for symmetric and 1.62 to asymmetric bilayers. This suggests that the symmetric POPC-d_{31}/POPE/POPG membranes are more rigid than pure POPC, which is explained by the high negative intrinsic curvature of POPE. The outer leaflet of asymmetric POPC^{out}/(POPE/POPG)^{in} is slightly more rigid, although it is composed of a relatively pure POPC layer. This suggests that the higher rigidity of the inner PE-containing leaflet dominates the elastic properties of the entire membrane.

An important parameter for discussing this interesting behavior is the efficiency of the lipid exchange of the outer leaflet of the MLVs. For instance, if the outer layer still contained a high proportion of POPE, the difference in the R_{1Z} *vs.* S^2 plots of pure POPC and asymmetric POPC^{out}/(POPE/POPG)^{in} membranes could be explained by the presence of POPE, which has a negative intrinsic curvature. Lipid analysis of the MLVs, where only the outer leaflet was exchanged, is difficult as all lipids of the inner (symmetric) layers are also measured. Therefore, we prepared POPE/POPG (molar ratio 9 : 1) LUVs and exchanged the outer layer with POPC-d_{31} using a similar procedure as for the asymmetric MLVs. By quantitative thin layer chromatography (ESI Fig. S2†), we determined the amount of POPC to be 48.7% and the POPE to be 51.3%. For this preparation, the POPG spot was too weak for quantification. This suggests that the exchange of the outer layer was almost complete. For comparison, we analysed the symmetric POPC-d_{31}/POPE/POPG membranes and determined a molar lipid ratio of 50.3% POPC, 46.0% POPE, and 3.7% POPG, which is very close to the original mixing ratio. We also performed MALDI-TOF mass spectrometry to check for remaining mβCD in the MLV samples used for ^{2}H NMR (ESI Fig. S3†). Although quantification of the mass spectra is not possible, residual mβCD was clearly identified.

Next, we measured the relaxation rates for pure POPS-d_{31} membranes and plot them as a function of S^2 (Fig. 3B). Again, a curved plot is obtained that can be fitted with an exponent of $n = 1.12$. For the symmetric POPS-d_{31}/POPE/POPG membranes, the plot is similar but has a slightly higher exponent of $n = 1.48$ most likely because of the presence of POPE. For the asymmetric POPS^{out}/(POPE/POPG)^{in} membranes, an almost perfectly linear square law plot is obtained (dashed line). However, using the same fit function as above, an exponent of $n = 2.30$ is found.

Finally, we also determined the relative composition of asymmetric LUVs and the symmetric preparations using TLC (ESI Fig. S2†). For the asymmetric POPS^{out}/(POPE/POPG)^{in} LUVs, we measured 65.9% POPS, 30.5% POPE, and 3.6% POPG, and for the symmetric POPE/POPG/POPS membranes, the ratio between the three lipids was 74% POPS, 24.4% POPE, and 1.6% POPG. The MALDI-TOF spectra of the preparations also showed remaining mβCD (ESI Fig. S3†).

4 Discussion

Asymmetric plasma membranes are a unique feature of pro- and eukaryotic organisms. As maintaining a broad lipid variety and the asymmetry between the two membrane leaflets consumes a considerable amount of energy, it is obvious that such a high effort is only maintained if absolutely required. Yet, our understanding of this intriguing biological situation is in its infancy. Now that robust procedures have been developed to prepare and characterize such systems,[10] biophysical studies of the properties of asymmetric membranes and the impact on biological questions such as permeation, protein function, fusion, or lateral domain formation can be addressed.

Here, we discuss two asymmetric membrane preparations in comparison to symmetric bilayers. The first represents a model for the plasma membrane of mammalian cells with phosphatidylcholine in the outer and phosphatidylethanolamine and a negative lipid (here phosphatidylglycerol) in the inner leaflet. A methodological reason to use a small fraction of POPG was because pure POPE MLVs produce too large aggregates that are difficult to handle, which is remedied by 10 mol% POPG, which makes the MLV slightly smaller and prevents sedimentation of the acceptor vesicles. In the other system, the acidic lipid phosphatidylserine is in the outer leaflet, which is rather artificial as PS is normally found in the cytoplasmic leaflet unless cells undergo apoptosis.[55] Our choice of system is limited by the fact that for ^{2}H solid-state NMR applications, multilamellar vesicles need to be prepared as anisotropic NMR interactions that provide the structural and dynamic information are largely averaged in LUV systems.[56] This comes with the disadvantage that in an MLV system, we can exchange only the very outer leaflet of the vesicle, here by deuterated lipids, providing a unique probe to selectively detect the lipids in the outer leaflet. The drawback of this procedure is that the lipids that are naturally found in the cytoplasmic leaflet are not accessible in this preparation. But the outer bilayer of a MLV is a very good model for the plasma membrane as its dimensions compare much better to a cell than the LUVs.

A first question that we studied concerns the completeness of the lipid exchange. Given the characteristics of the MLV system, a quantitative measurement of the composition of the outer layer is difficult because the inner bilayers of the MLVs are symmetric and not accessible by the mβCD-catalyzed exchange. Therefore, we prepared LUVs by extrusion and exchanged the outer leaflet using the same strategy as in the MLV preparation. In the POPC^{out}/(POPE/POPG)^{in} LUVs, we found complete exchange of POPE/POPG with POPC rendering the resulting outer membrane of the MLVs completely asymmetric. In the POPS^{out}/(POPE/POPG)^{in} LUVs, we measured a slightly higher amount of POPS than what could be accommodated in the outer leaflet. It is not clear at this point if PS would possibly be subject to faster flip flop or if the cyclodextrin exchange could have resulted in a higher deposition of PS also in the inner leaflet. Another study also found donor lipids in the inner leaflet that should not have exchanged.[11] Nevertheless, it is quite obvious that highly asymmetric membranes were prepared and we assume that the exchange procedure was equally successful for the outer leaflet of the two MLV systems. We also checked if the NMR samples contained remaining mβCD and found the sugar to be present in both preparations, but a quantification of the mass spectra was not possible. If remaining mβCD may have an influence on the membrane is not clear at this point.

Our study shows that the order parameter profiles of the lipid molecules in symmetric and asymmetric membranes are different. In comparison to symmetric membranes, we find that lipid chain order is higher in the upper and lower in the lower half of the acyl chains for both POPC and POPS when exclusively located in the outer layer. This indicates different packing constraints in the asymmetric membranes. Remarkably, the acyl chain lengths of POPC in the outer leaflet only or in symmetric vesicles were identical within experimental error

(±0.1 Å); the same is observed for POPS, where the chain lengths are identical in both preparations. This supports the results from lipid analytics, which revealed that our preparations showed a very high degree of asymmetry. If POPE was prominently found in the outer leaflet, it would have increased the chain lengths in the outer leaflet. More interestingly, there seems to be a preference for higher order in the upper and lower order in the lower acyl chain when membranes show an asymmetric lipid distribution. This could possibly indicate an increased negative intrinsic curvature in the outer leaflet of asymmetric membranes. Such a steeper decrease in order along the acyl chains will likely modify the properties of the lipid/water interface and potentially alter the partitioning of small molecules, drugs, or peptides. MD simulations have also identified chain order parameter changes and altered lateral stress profiles between the leaflets in asymmetric membranes.[14,57] While the MD of asymmetric DPPC/DPPS membranes revealed a decrease in PC order and an increase in PS order when making membranes asymmetric,[14] our ^{2}H NMR results indicated a distinctly different chain order parameter profile.

From this, one may speculate about the mechanism of interleaflet coupling. The most invoked mechanism for this phenomenon is acyl chain interdigitation. For the two asymmetric systems (POPC^{out}/(POPE/POPG)^{in} and POPS^{out}/(POPE/POPG)^{in}), there is no neutron scattering data available that would most directly detect possible acyl chain interdigitation. ^{2}H NMR spectra and order parameters can also indicate chain interdigitation, although very clear information has only been obtained for lipids with large differences in acyl chain lengths.[58,59] The order parameters measured for the two systems studied here do not suggest acyl chain interdigitation, but it cannot safely be excluded either. Supportive of this notion is the systematic study by Frewein *et al.*,[15] showing that interdigitation was not observed if POPC was used as donor lipid (with DPPC on the inner membrane leaflet) but only for systems with a larger *sn*-1/*sn*-2 chain length difference. Altogether, it is fair to say that acyl chain interdigitation is unlikely to occur in the two systems.

Even though chain interdigitation is unlikely, the two membrane leaflets are not uncoupled from one another. Clearly, the inner membrane leaflet influenced the order parameters and elastic properties of the outer leaflet. An obvious difference between outer and inner leaflet is in lipid packing density. As the inner leaflet is rich in POPE, which is known for its negative intrinsic curvature, one could assume that the tighter packing of the inner leaflet is transmitted into the outer leaflet. One can speculate that these packing density alterations may very well be transmitted from one leaflet to the other. But with the current very limited data set, such conclusions remain fairly speculative.

We turned our attention to dynamic aspects of the lipids in the outer membrane leaflet. NMR relaxation rates are great parameters to study the molecular motions of the lipids in detail.[24,60] From first inspection of the ^{2}H NMR spectra of asymmetric membranes, it is clear that the line widths of the individual Pake doublets is increased, which indicates shorter T_2 values usually associated with slower lipid motions. Very straightforward is the measurement of ^{2}H T_{1Z} relaxation times of the deuterated chains. Relaxation rates ($R_{1Z} = 1/T_{1Z}$) decrease for asymmetric membranes compared to pure or symmetric preparations. A very convenient way of addressing the dynamic behavior of lipid membranes and assess their elastic properties are the square law plots, where R_{1Z} is plotted against the square of the order parameter (S^2) of the respective segment.[60] In the pioneering work from the Brown laboratory, a square law dependence ($R_{1Z} \propto S^2$) was obtained for saturated (symmetric) membranes of varying chain lengths,[60] which is interpreted as a dominant contribution of low frequency motions in agreement with a collective model for slow motions on the mesoscopic scale (1–100 nm). In the presence of cholesterol, which makes membranes stiffer, R_{1Z} is still proportional to S^2, but the slope of the curves decreases with increasing cholesterol content.[61] In the presence of detergents, such as $C_{12}E_8$,[51] unsaturated lipid chains,[62,63] or the lipid modifications of a lipidated peptide[48] are found to follow a $R_{1Z} \propto S^n$ proportionality, where $n < 2$ and in some cases even a linear $R_{1Z} \propto S$ relationship has been observed.[54]

Here, we found rather strong deviations from the $n = 2$ power law as expected for monounsaturated membranes in agreement with relatively soft lipid membranes.[54] Pure POPC or POPS membranes showed highest elasticity, while the addition of POPE in symmetric membranes slightly rigidified the membranes. Plotting R_{1Z} *vs.* S^2 for the outer leaflet of POPC^{out}/(POPE/POPG)^{in} and POPS^{out}/(POPE/POPG)^{in} revealed an additional increase in stiffness where especially the POPS-rich leaflet behaved quite like a fully saturated membrane. These results are in agreement with studies that observed that cholesterol-free asymmetric bilayers feature increased bending rigidities.[12,29,64] However, the increased bending rigidity of asymmetric membranes has been discussed in relation to the enrichment of saturated sphingolipids and cholesterol in the outer and (poly-)unsaturated lipids in the inner leaflet. Here, all the membranes were composed of monounsaturated lipids only. It is well known that POPE increases the packing density of membranes leading to increased stiffness.[25] In agreement with our results, Frewein *et al.*[15] also observed an increase in bending rigidity in asymmetric POPC^{out}/POPE^{in} bilayers compared to scrambled POPC/POPE membranes. This clearly indicates that the higher bending rigidity of the inner POPC layer dominates the elastic properties of the asymmetric membranes. This, along with the other recent results on asymmetric membranes, provides an imperative to consider leaflet asymmetry as a crucial determinant of membrane properties and likely protein functions as shown in first studies.

5 Conclusions

We investigated biophysical properties of lipid membranes with leaflet asymmetry in the distribution of lipid species featuring POPE/POPG (9 : 1) lipids in the inner and either POPC or POPS in the outer leaflet. ^{2}H NMR chain order parameters revealed an intriguing reshuffling of order from the lower to the upper chain segments of the outer leaflet of asymmetric compared to symmetric membranes. Interestingly, the overall chain lengths remained constant compared to symmetric bilayers of the same lipid species. This suggests that the lateral stress and lateral pressure profile of asymmetric bilayers is considerably different, leading to alteration in the lipid–protein interactions with likely consequences for protein function. This is also implied from the increase in stiffness of asymmetric membranes differing in the headgroup but not in the chain composition. An interleaflet coupling is the likely mechanism for this interesting behavior, but from the results of this work, no indications for interdigitation of the acyl chains between the lipids of the individual leaflets could be observed. One could speculate that the packing density of one leaflet is sensed by the other and that these parameters are correlated between the leaflets. This would suggest that lipid dynamics is also correlated between the molecules of the two leaflets. While the tools to investigate such correlations in lipid dynamics within specific correlation time windows have been developed,[24,65] a current limitation for NMR applications is the necessity to rely on anisotropic NMR interactions that are largely averaged in unilamellar vesicles. But solutions to this problem are underway, which will allow also detecting NMR parameters of the lipids in the inner leaflet.

Data availability

Tables with all order parameters and R_{1Z} relaxation rates for the various membrane preparations determined from the experimental NMR spectra are available in the supplementary information file.†

Author contributions

D. H. designed the study. O. E. made all sample preparations, and carried out the NMR experiments. O. E. and D. H. analysed the ^{2}H NMR spectra. V. D. determined the lipid concentration. K. M. E. carried out the mass spectrometry and analyzed the data. D. H. and O. E. wrote the paper with contributions from all coauthors.

Conflicts of interest

There are no conflicts to declare.

Acknowledgements

The study was funded by the Deutsche Forschungsgemeinschaft (DFG, German Research Foundation), project numbers 546534953 and 511424302.

References

1 J. H. Lorent, K. R. Levental, L. Ganesan, G. Rivera-Longsworth, E. Sezgin, M. Doktorova, E. Lyman and I. Levental, Plasma membranes are asymmetric in lipid unsaturation, packing and protein shape, *Nat. Chem. Biol.*, 2020, **16**, 644–652.

2 G. van Meer, D. R. Voelker and G. W. Feigenson, Membrane lipids: where they are and how they behave, *Nat. Rev. Mol. Cell Biol.*, 2008, **9**, 112–124.

3 B. J. Litman and D. C. Mitchell, A role for phospholipid polyunsaturation in modulating membrane protein function, *Lipids*, 1996, **31**, 193–197.

4 M. Liscovitch and L. C. Cantley, Lipid second messengers, *Cell*, 1994, **77**, 329–334.

5 R. Baccouch, E. Rascol, K. Stoklosa and I. D. Alves, The role of the lipid environment in the activity of G protein coupled receptors, *Biophys. Chem.*, 2022, **285**, 106794.

6 N. Thakur, A. P. Ray, L. Sharp, B. Jin, A. Duong, N. G. Pour, S. Obeng, A. V. Wijesekara, Z. G. Gao, C. R. McCurdy, K. A. Jacobson, E. Lyman and

M. T. Eddy, Anionic phospholipids control mechanisms of GPCR-G protein recognition, *Nat. Commun.*, 2023, **14**, 794.

7 M. Damian, M. Louet, A. A. S. Gomes, C. M'Kadmi, S. Denoyelle, S. Cantel, S. Mary, P. M. Bisch, J. A. Fehrentz, L. J. Catoire, N. Floquet and J.-L. Banères, Allosteric modulation of ghrelin receptor signaling by lipids, *Nat. Commun.*, 2021, **12**, 3938.

8 O. Engberg, D. Ulbricht, V. Dobel, V. Siebert, C. Frie, A. Penk, M. K. Lemberg and D. Huster, Rhomboid-catalyzed intramembrane proteolysis requires hydrophobic matching with the surrounding lipid bilayer, *Sci. Adv.*, 2022, **8**, eabq8303.

9 G. Pabst and S. Keller, Exploring membrane asymmetry and its effects on membrane proteins, *Trends Biochem. Sci.*, 2024, **49**, 333–345.

10 M. Doktorova, F. A. Heberle, B. Eicher, R. F. Standaert, J. Katsaras, E. London, G. Pabst and D. Marquardt, Preparation of asymmetric phospholipid vesicles for use as cell membrane models, *Nat. Protoc.*, 2018, **13**, 2086–2101.

11 B. Eicher, D. Marquardt, F. A. Heberle, I. Letofsky-Papst, G. N. Rechberger, M. S. Appavou, J. Katsaras and G. Pabst, Intrinsic curvature-mediated transbilayer coupling in asymmetric lipid vesicles, *Biophys. J.*, 2018, **114**, 146–157.

12 M. P. K. Frewein, P. Piller, E. F. Semeraro, O. Czakkel, Y. Gerelli, L. Porcar and G. Pabst, Distributing aminophospholipids asymmetrically across leaflets causes anomalous membrane stiffening, *Biophys. J.*, 2023, **122**, 2445–2455.

13 P. Piller, E. F. Semeraro, G. N. Rechberger, S. Keller and G. Pabst, Allosteric modulation of integral protein activity by differential stress in asymmetric membranes, *PNAS Nexus*, 2023, **2**, pgad126.

14 L. Pasalic, P. Males, A. Cikos, B. Pem and D. Bakaric, The rise of FTIR spectroscopy in the characterization of asymmetric lipid membranes, *Spectrochim. Acta, Part A*, 2024, **305**, 123488.

15 M. P. K. Frewein, P. Piller, E. F. Semeraro, K. C. Batchu, F. A. Heberle, H. L. Scott, Y. Gerelli, L. Porcar and G. Pabst, Interdigitation-induced order and disorder in asymmetric membranes, *J. Membr. Biol.*, 2022, **255**, 407–421.

16 M. Doktorova, J. L. Symons, X. Zhang, H.-Y. Wang, J. Schlegel, J. H. Lorent, F. A. Heberle, E. Sezgin, E. Lyman, K. R. Levental and I. Levental, Cell membranes sustain phospholipid imbalance *via* cholesterol asymmetry, *bioRxiv*, 2024, preprint, DOI: 10.1101/2023.07.30.551157.

17 Y. Zhu, L. Porcar, T. Ravula, K. C. Batchu, T. L. Lavoie, Y. Liu and U. Perez-Salas, Unexpected asymmetric distribution of cholesterol and phospholipids in equilibrium model membranes, *Biophys. J.*, 2024, **123**, 3923–3934.

18 Q. Lin and E. London, The influence of natural lipid asymmetry upon the conformation of a membrane-inserted protein (perfringolysin O), *J. Biol. Chem.*, 2014, **289**, 5467–5478.

19 E. Perozo, A. Kloda, D. M. Cortes and B. Martinac, Physical principles underlying the transduction of bilayer deformation forces during mechanosensitive channel gating, *Nat. Struct. Biol.*, 2002, **9**, 696–703.

20 H. I. Petrache, N. Gouliaev, S. Tristram-Nagle, R. Zhang, R. M. Suter and J. F. Nagle, Interbilayer interactions from high resolution x-ray scattering, *Phys. Rev. E: Stat. Phys., Plasmas, Fluids, Relat. Interdiscip. Top.*, 1998, **57**, 7014–7024.

21 M. C. Wiener and S. H. White, Structure of a fluid dioleoylphosphatidylcholine bilayer determined by joint refinement of x-ray and neutron diffraction data III. Complete structure, *Biophys. J.*, 1992, **61**, 434–447.
22 S. L. Veatch, O. Soubias, S. L. Keller and K. Gawrisch, Critical fluctuations in domain-forming lipid mixtures, *Proc. Natl. Acad. Sci. U. S. A.*, 2007, **104**, 17650–17655.
23 D. Huster, K. Arnold and K. Gawrisch, Investigation of lipid organization in biological membranes by two-dimensional nuclear Overhauser enhancement spectroscopy, *J. Phys. Chem. B*, 1999, **103**, 243–251.
24 A. A. Smith, A. Vogel, O. Engberg, P. W. Hildebrand and D. Huster, A method to construct the dynamic landscape of a bio-membrane with experiment and simulation, *Nat. Commun.*, 2022, **13**, 108.
25 A. V. Botelho, N. J. Gibson, R. L. Thurmond, Y. Wang and M. F. Brown, Conformational energetics of rhodopsin modulated by nonlamellar-forming lipids, *Biochemistry*, 2002, **41**, 6354–6368.
26 M. F. Brown, Soft matter in lipid-protein interactions, *Annu. Rev. Biophys.*, 2017, **46**, 379–410.
27 W. E. Teague, O. Soubias, H. Petrache, N. Fuller, K. G. Hines, R. P. Rand and K. Gawrisch, Elastic properties of polyunsaturated phosphatidylethanolamines influence rhodopsin function, *Faraday Discuss.*, 2013, **161**, 383–395.
28 O. Soubias, W. E. Teague Jr, K. G. Hines, D. C. Mitchell and K. Gawrisch, Contribution of membrane elastic energy to rhodopsin function, *Biophys. J.*, 2010, **99**, 817–824.
29 B. W. Rickeard, M. H. L. Nguyen, M. DiPasquale, C. G. Yip, H. Baker, F. A. Heberle, X. B. Zuo, E. G. Kelley, M. Nagao and D. Marquardt, Transverse lipid organization dictates bending fluctuations in model plasma membranes, *Nanoscale*, 2020, **12**, 1438–1447.
30 C. M. Brown and S. J. Marrink, Modeling membranes in situ, *Curr. Opin. Struct. Biol.*, 2024, **87**, 102837.
31 M. D. Weiner and G. W. Feigenson, Molecular dynamics simulations reveal leaflet coupling in compositionally asymmetric phase-separated lipid membranes, *J. Phys. Chem. B*, 2019, **123**, 3968–3975.
32 R. Vacha, M. L. Berkowitz and P. Jungwirth, Molecular model of a cell plasma membrane with an asymmetric multicomponent composition: water permeation and ion effects, *Biophys. J.*, 2009, **96**, 4493–4501.
33 E. H. Chaisson, F. A. Heberle and M. Doktorova, Building asymmetric lipid bilayers for molecular dynamics simulations: what methods exist and how to choose one?, *Membranes*, 2023, **13**, 629.
34 D. Huster, S. Maiti and A. Herrmann, Phospholipid membranes as chemically and functionally tunable materials, *Adv. Mater.*, 2024, **36**, e2312898.
35 F. A. Heberle, D. Marquardt, M. Doktorova, B. Geier, R. F. Standaert, P. Heftberger, B. Kollmitzer, J. D. Nickels, R. A. Dick, G. W. Feigenson, J. Katsaras, E. London and G. Pabst, Subnanometer structure of an asymmetric model membrane: interleaflet coupling influences domain properties, *Langmuir*, 2016, **32**, 5195–5200.
36 D. Marquardt, F. A. Heberle, T. Miti, B. Eicher, E. London, J. Katsaras and G. Pabst, ^{1}H NMR Shows Slow Phospholipid Flip-Flop in Gel and Fluid Bilayers, *Langmuir*, 2017, **33**, 3731–3741.

37 V. V. Kumar and W. J. Baumann, Bilayer asymmetry in lysophosphatidylcholine/cholesterol (1:1) vesicles. A phosphorus-31 NMR study, *Biochem. Biophys. Res. Commun.*, 1986, **139**, 25–30.
38 N. Kucerka, J. Pencer, J. N. Sachs, J. F. Nagle and J. Katsaras, Curvature effect on the structure of phospholipid bilayers, *Langmuir*, 2007, **23**, 1292–1299.
39 J. H. Davis, K. R. Jeffrey, M. Bloom, M. I. Valic and T. P. Higgs, Quadrupolar echo deuteron magnetic resonance spectroscopy in ordered hydrocarbon chains, *Chem. Phys. Lett.*, 1976, **42**, 390–394.
40 M. Lafleur, B. Fine, E. Sternin, P. R. Cullis and M. Bloom, Smoothed orientational order profile of lipid bilayers by ^{2}H-nuclear magnetic resonance, *Biophys. J.*, 1989, **56**, 1037–1041.
41 D. Huster, K. Arnold and K. Gawrisch, Influence of docosahexaenoic acid and cholesterol on lateral lipid organization in phospholipid membranes, *Biochemistry*, 1998, **37**, 17299–17308.
42 H. I. Petrache, S. W. Dodd and M. F. Brown, Area per lipid and acyl length distributions in fluid phosphatidylcholines determined by ^{2}H NMR spectroscopy, *Biophys. J.*, 2000, **79**, 3172–3192.
43 A. Vogel, K.-T. Tan, H. Waldmann, S. E. Feller, M. F. Brown and D. Huster, Flexibility of ras lipid modifications studied by ^{2}H solid-state NMR and molecular dynamics simulations, *Biophys. J.*, 2007, **93**, 2697–2712.
44 E. G. Bligh and W. J. Dyer, A rapid method of total lipid extraction and purification, *Can. J. Biochem. Physiol.*, 1959, **37**, 911–917.
45 M. Leopold, P. B. Mass-Sanchez, M. Krizanac, P. Stancl, R. Karlic, P. Prabutzki, V. Parafianczuk, J. Schiller, A. Asimakopoulos, K. M. Engel and R. Weiskirchen, How the liver transcriptome and lipid composition influence the progression of nonalcoholic fatty liver disease to hepatocellular carcinoma in a murine model, *Biochim. Biophys. Acta*, 2025, **1870**, 159574.
46 J. Schiller, J. Arnhold, S. Benard, M. Müller, S. Reichl and K. Arnold, Lipid analysis by matrix-assisted laser desorption and ionization mass spectrometry: a methodological approach, *Anal. Biochem.*, 1999, **267**, 46–56.
47 D. Huster, K. Arnold and K. Gawrisch, Strength of Ca^{2+} binding to retinal lipid membranes: consequences for lipid organization, *Biophys. J.*, 2000, **78**, 3011–3018.
48 A. Vogel, C. P. Katzka, H. Waldmann, K. Arnold, M. F. Brown and D. Huster, Lipid modifications of a ras peptide exhibit altered packing and mobility *versus* host membrane as detected by ^{2}H solid-state NMR, *J. Am. Chem. Soc.*, 2005, **127**, 12263–12272.
49 J. F. Nagle, Area/lipid of bilayers from NMR, *Biophys. J.*, 1993, **64**, 1476–1481.
50 T. Meyer, D. J. Baek, R. Bittman, I. Haralampiev, P. Muller, A. Herrmann, D. Huster and H. A. Scheidt, Membrane properties of cholesterol analogs with an unbranched aliphatic side chain, *Chem. Phys. Lipids*, 2014, **184**, 1–6.
51 D. Otten, M. F. Brown and K. Beyer, Softening of membrane bilayers by detergents elucidated by deuterium NMR spectroscopy, *J. Phys. Chem. B*, 2000, **104**, 12119–12129.
52 M. F. Brown, R. L. Thurmond, S. W. Dodd, D. Otten and K. Beyer, Composite membrane deformation on the mesoscopic length scale, *Phys. Rev. E: Stat. Phys., Plasmas, Fluids, Relat. Interdiscip. Top.*, 2001, **64**, 010901.

53 M. F. Brown, R. L. Thurmond, S. W. Dodd, D. Otten and K. Beyer, Elastic deformation of membrane bilayers probed by deuterium NMR relaxation, *J. Am. Chem. Soc.*, 2002, **124**, 8471–8484.

54 H. A. Scheidt, R. M. Badeau and D. Huster, Investigating the membrane orientation and transversal distribution of 17β-estradiol in lipid membranes by solid-state NMR, *Chem. Phys. Lipids*, 2010, **163**, 356–361.

55 S. Nagata, Apoptosis and clearance of apoptotic cells, *Annu. Rev. Immunol.*, 2018, **36**, 489–517.

56 D. B. Fenske and P. R. Cullis, Acyl chain orientational order in large unilamellar vesicles: comparison with multilamellar liposomes: a ^{2}H and ^{31}P nuclear magnetic resonance study, *Biophys. J.*, 1993, **64**, 1482–1491.

57 T. L. Mai, P. Derreumaux and P. H. Nguyen, Structure and elasticity of mitochondrial membranes: a molecular dynamics simulation study, *J. Phys. Chem. B*, 2023, **127**, 10778–10791.

58 R. N. Lewis, R. N. McElhaney, M. A. Monck and P. R. Cullis, Studies of highly asymmetric mixed-chain diacyl phosphatidylcholines that form mixed-interdigitated gel phases: Fourier transform infrared and 2H NMR spectroscopic studies of hydrocarbon chain conformation and orientational order in the liquid-crystalline state, *Biophys. J.*, 1994, **67**, 197–207.

59 M. Bosse, J. Sibold, H. A. Scheidt, L. J. Patalag, K. Kettelhoit, A. Ries, D. B. Werz, C. Steinem and D. Huster, Shiga toxin binding alters lipid packing and the domain structure of Gb3-containing membranes: a solid-state NMR study, *Phys. Chem. Chem. Phys.*, 2019, **21**, 15630–15638.

60 M. F. Brown, Theory of spin-lattice relaxation in lipid bilayers and biological membranes. ^{2}H and ^{14}N quadrupolar relaxation, *J. Chem. Phys.*, 1982, **77**, 1576–1799.

61 G. V. Martinez, E. M. Dykstra, S. Lope-Piedrafita, C. Job and M. F. Brown, NMR elastometry of fluid membranes in the mesoscopic regime, *Phys. Rev. E*, 2002, **66**, 050902.

62 K. Rajamoorthi, H. I. Petrache, T. J. McIntosh and M. F. Brown, Packing and viscoelasticity of polyunsaturated omega-3 and omega-6 lipid bilayers as seen by ^{2}H NMR and X-ray diffraction, *J. Am. Chem. Soc.*, 2005, **127**, 1576–1588.

63 A. Vogel, T. Schröder, C. Lange and D. Huster, Characterization of the myristoyl lipid modification of membrane-bound GCAP-2 by ^{2}H solid-state NMR spectroscopy, *Biochim. Biophys. Acta*, 2007, **1768**, 3171–3181.

64 Y. Elani, S. Purushothaman, P. J. Booth, J. M. Seddon, N. J. Brooks, R. V. Law and O. Ces, Measurements of the effect of membrane asymmetry on the mechanical properties of lipid bilayers, *Chem. Commun.*, 2015, **51**, 6976–6979.

65 A. A. Smith, E. M. Pacull, S. Stecher, P. W. Hildebrand, A. Vogel and D. Huster, Analysis of the dynamics of the human growth hormone secretagogue receptor reveals insights into the energy landscape of the molecule, *Angew. Chem., Int. Ed.*, 2023, **62**, e202302003.

Faraday Discussions

ROYAL SOCIETY OF CHEMISTRY

PAPER

Exploring the sensitivities of experimental techniques to various types of membrane asymmetry using atomistic simulations

Frederick A. Heberle [a] and Milka Doktorova *[b]

Received 16th December 2024, Accepted 14th January 2025
DOI: 10.1039/d4fd00200h

Biological membranes have two leaflets that can differ in both lipid composition and total lipid abundance. These different types of asymmetries play a major role in determining the biophysical properties of the membrane; however, they have proven challenging to assay experimentally even in simpler model systems. Molecular dynamics simulations offer the means for detailed computational investigation of systematically varied interleaflet lipid distributions, but opportunities for critical validation with wet lab experiments are scarce. To help address this problem, here we use atomistic simulations of asymmetric bilayers to generate synthetic experimental data and thus investigate the sensitivity of various approaches to changes in relative lipid composition, number, and cholesterol distribution. Contrary to trends in symmetric bilayers, the simulations showed a decrease in lipid packing with increasing cholesterol in differentially stressed asymmetric bilayers, with more pronounced changes in the more loosely packed leaflet. Representative experimental data computed from the simulation trajectories indicated that the detection of asymmetry-induced changes in leaflet properties should be possible with environment-sensitive fluorescent probes and NMR observables, but may require optimization of sample preparation conditions. On the other hand, small-angle scattering data are already experimentally accessible and can reveal differential leaflet packing densities through a model-free analysis. We further show that computationally generated cryo-EM intensity profiles are highly sensitive to phospholipid imbalance between membrane leaflets. Together, these findings provide a roadmap for developing targeted applications of the *in vitro* techniques and obtaining experimental data critical for validating computationally derived principles related to membrane asymmetry.

Introduction

As biological interfaces, lipid bilayers have evolved to support a large number of functions and processes. This is achieved by a rich palette of biophysical

[a]*Department of Chemistry, University of Tennessee Knoxville, Knoxville, TN 37916, USA*
[b]*Department of Biochemistry and Biophysics, Stockholm University, Science for Life Laboratory, Solna 171 65, Sweden. E-mail: milka.doktorova@dbb.su.se*

properties created in part by the bilayer lipid composition.[1] Extensive studies of the physio-chemical behavior of single component lipid bilayers have linked lipid structure (*e.g.*, chain length, head group type) to membrane permeability, fluidity and elasticity.[2–6] Further analyses of the thermodynamics of lipid mixing have revealed the ability of membranes to compartmentalize their protein constituents *via* lateral heterogeneities characterized by in-plane (within leaflet) co-existence of regions with different lipid compositions and biophysical profiles.[7–9] These versatile structure–property relationships have been investigated with a large suite of experimental and computational techniques,[10–13] and formulate the basis for current models of membrane structure and function.

In addition to their in-plane complexity, lipid bilayers can exhibit non-random organization with respect to the two leaflets. For example, the existence and prevalence of compositional (and by extension biophysical) asymmetry in biological membranes has been recognized since the 1970s.[14–16] Cells use valuable resources to actively maintain gradients in lipid saturation and charge across the bilayer midplane, which lead to contrasting fluidity and electrostatic surface potentials in the two leaflets.[16–20] Lipids also have different spontaneous curvatures, and their interleaflet distributions may induce curvature stress in the bilayer.[2,21–23] Cholesterol, with its unique structure and dynamics, further contributes to these effects *via* its (re)distribution.[21,24] The overall compositional asymmetry of model and cell membranes has been assayed with various approaches including fluorescence quenching, peptide–membrane interactions, and NMR and mass spectrometry, although in many cases the exact quantification of the lipid make-up of both leaflets has not been possible.[16,25–29]

Lipid composition however, may not be the only (or even the main) determinant of the properties of a membrane. Lipid bilayers can tolerate a range of imbalances in the relative abundance of lipids between their leaflets.[22,30,31] This is realized by adjustments in the lipid packing in each monolayer to conform to the area imposed by the closed membrane surface while minimizing hydrocarbon–water contacts. The result is a build-up of tension in one leaflet and pressure in the other (termed differential, or asymmetry, stress) with concomitant opposite effects on the structure and dynamics of the two leaflets.[32] Analogously to externally applied net bilayer tension, differential stress can be tolerated up to a certain threshold before the membrane ruptures or changes morphology.[33] For example, the addition of lipids to one leaflet of pre-formed membranes has been used to induce excessive asymmetry stress and budding or membrane shape changes in both synthetic and living systems.[34–37] At the same time, theory and experiment indicate that substantial resting differential stress is likely present in giant synthetic vesicles with engineered compositional asymmetry as well as unperturbed cell plasma membranes, 'silently' affecting their measurable properties without any clear morphological signatures.[22,38–40]

Differential stress can be quantified from the lateral pressure distribution of simulated lipid bilayers.[22,30,41] However, the ability to unambiguously detect and measure it in experiments remains a formidable challenge.[42] To begin to address this problem, here we examine the sensitivities of various experimental techniques to different types of interleaflet membrane asymmetries including in phospholipid composition, number and cholesterol distribution. To this end, we employ molecular dynamics simulations to analyze the properties and dynamics of detailed atomistic models, then use them to generate representative experimental data from liposomes with the corresponding bilayer structures. The experiments we focus on include fluorescence lifetime measurements of environment-sensitive probes, NMR observables, small-angle scattering form factors, and cryo-EM images and intensity profiles. Our goal is to examine the differences in biophysical properties induced by the interplay of the various types of asymmetries and evaluate the capabilities of the *in vitro* approaches to detect them. Our results present a framework for identifying the conditions and controls for each technique that are best suited for analysis of the multifaceted asymmetry of lipid membranes, thus providing a roadmap to help guide future research efforts in these directions.

Methods

Atomistic bilayer models

Solvated lipid bilayer models were constructed with CHARMM-GUI[43–45] and were composed of 1,2-dipalmitoyl-*sn*-glycero-3-phosphocholine (DPPC), 1,2-diarachidonoyl-*sn-glycero*-3-phosphocholine (DAPC) and cholesterol (Chol). In particular, eight symmetric bilayers had either DPPC or DAPC with 0, 10, 30 and 50 mol% Chol, and one symmetric bilayer (referred to as 'scramble') was composed of DPPC/DAPC/Chol 35/35/30 mol%. All bilayers contained 100 lipids per leaflet and were hydrated with 50 waters per lipid and no ions. Two additional symmetric bilayers with DPPC, 1,2-dioleoyl-*sn-glycero*-3-phosphocholine (DOPC) and Chol, were constructed, with DPPC/DOPC/Chol in molar percents of 56/9/35 (liquid-ordered, Lo) and 23/67/10 (liquid-disordered, Ld).

In addition, six asymmetric bilayers were taken from ref. 46 for further analysis. The bilayers had 70 DPPC lipids in one leaflet, a varying number of DAPC lipids in the opposite leaflet, and Chol at 30 mol% of all lipids distributed according to its chemical potential (as characterized from initial coarse-grained simulations, see ref. 46). Table 1 lists the number of lipid and cholesterol molecules in each leaflet. All bilayers were similarly constructed with CHARMM-GUI and hydrated with 50 waters per lipid and no ions.

Molecular dynamics simulations

All bilayers were simulated with NAMD[47] and the CHARMM36 force field for lipids,[48] as described in ref. 46. Briefly, the systems were first equilibrated with CHARMM-GUI's 6-step equilibration protocol, then simulated with a 2 fs timestep and the length of all hydrogen bonds constrained (*rigidbonds* parameter set to *all*). Nonbonded interactions were modeled with a 10–12 Å Lennard-Jones potential function using force switching *via* the *vdwforceswitching* option in NAMD. The particle mesh Ewald method with a grid spacing of 1 Å was used for electrostatic interactions. The simulations were performed with semi-isotropic pressure coupling at a constant temperature of 40 °C (or 25 °C for the Lo and Ld bilayers) and pressure at 1 atm using Langevin thermostat (damping coefficient of 5 ps^{-1}) and barostat (period 200 fs and decay 50 fs). The symmetric bilayers were run for 600–1180 ns. All asymmetric bilayers were run for over 1 μs with the last 400 ns used for analysis, as detailed in Table S4 in ref. 46.

Calculation of bilayer structural parameters

Simulations were first centered so that the average position of all terminal methyl carbons was at $(x,y,z) = (0,0,0)$ in every frame. Average area per lipid (APL) was

Table 1 Composition and structure of simulated asymmetric bilayers. Shown are the name of each bilayer with the corresponding phospholipid imbalance (N_{DPPC}/N_{DAPC} lipids in parentheses); the corresponding numbers of DPPC, DAPC and Chol molecules in each leaflet; the Chol leaflet mol%; area per lipid; leaflet thickness (distance between average z position of the phosphate atoms and bilayer center at $z = 0$); and tension (differential stress) in each leaflet with their corresponding errors in parenthesis

Bilayer (PL imb.)	Composition, # lipids (DPPC/DAPC/Chol)		Chol leaflet (mol%)		Area per lipid (Å^2)		Leaflet thickness (Å), phosphate to midplane		Leaflet tension (mN m^{-1})	
	Top	Bot	Top	Bot	Top	Bot	Top	Bot	Top	Bot
A.1 (0.91)	70/0/56	0/77/7	44	8	42.0 (0.36)	63.0 (0.54)	23.5 (0.26)	20.2 (0.29)	14.8 (0.9)	−14.8 (0.9)
A.2 (1.00)	70/0/52	0/70/8	43	10	41.3 (0.30)	64.7 (0.46)	23.7 (0.22)	19.8 (0.28)	7.2 (0.9)	−7.2 (0.9)
A.3 (1.11)	70/0/47	0/63/10	40	14	40.8 (0.28)	65.4 (0.45)	23.9 (0.21)	19.5 (0.27)	−0.1 (1.0)	0.1 (1.0)
A.4 (1.25)	70/0/43	0/56/11	38	16	40.3 (0.19)	67.9 (0.33)	24.1 (0.18)	18.9 (0.26)	−11.1 (1.0)	11.1 (1.0)
A.5 (1.43)	70/0/37	0/49/14	35	22	40.1 (0.23)	68.1 (0.39)	24.1 (0.20)	18.7 (0.28)	−14.5 (1.0)	14.5 (1.0)
A.6 (2.00)	70/0/29	0/35/16	29	31	39.7 (0.20)	77.0 (0.39)	24.1 (0.17)	16.8 (0.28)	−31.7 (1.0)	31.7 (1.0)

calculated by dividing the lateral area of the simulation box by the number of lipids (including Chol) in each leaflet. Lateral pressure profiles were obtained with NAMD as described in detail in ref. 30. The differential stress and bilayer torque density were calculated from the integral and first moment of the pressure profiles, respectively, as outlined in ref. 46. The order parameters and relaxation rates of the carbon–hydrogen bonds were calculated following the protocols in ref. 49. Errors on APL and order parameters were obtained by block averaging, while errors on differential stress and bilayer torque density were calculated with bootstrapping analysis. Number and charge density profiles were computed for the whole systems in thin slabs of thickness 0.2 Å with the density profile plugin in VMD.[50] The bilayer dipole potential was calculated from the charge density distribution following the approach in ref. 51.

Calculation of small-angle scattering intensity

Scattering intensity was calculated from the time-averaged atomic number density distributions projected onto the bilayer normal, $n_i(z)$, where z denotes position along the bilayer normal and i indexes the individual lipid and water atoms. Each distribution was scaled by the appropriate atomic scattering factor (*i.e.*, the X-ray atomic form factor or the coherent neutron scattering length) and then summed to produce a scattering length density profile, $\rho(z)$. The scattering intensity as a function of scattering vector q is given by:[52]

$$I(q) = \frac{1}{q^2}\left|\int_{-D/2}^{D/2} \Delta\rho(z)(\cos qz + i\sin qz)\mathrm{d}z\right|^2, \quad (1)$$

where $\Delta\rho(z) = \rho(z) - \rho_s$ with ρ_s being the scattering length density of the aqueous solvent, and the integral extends from the bulk solvent (*i.e.*, where $\Delta\rho(z) = 0$) on one side of the bilayer to the other (D is the length of the simulation box). The scattering form factor was then computed from the intensity profile as $F(q) = \sqrt{I(q)q^2}$. The center-of-mass of the form factor was calculated as:

$$\overline{q} = \int_{q_{\min}}^{q_{\max}} qF(q)\mathrm{d}q \Big/ \int_{q_{\min}}^{q_{\max}} F(q)\mathrm{d}q, \quad (2a)$$

$$\overline{F} = \frac{1}{2}\int_{q_{\min}}^{q_{\max}} F(q)^2\mathrm{d}q \Big/ \int_{q_{\min}}^{q_{\max}} F(q)\mathrm{d}q, \quad (2b)$$

where the q-range ($q_{\min}$ to $q_{\max}$) was 0.03–0.37 Å^{-1} for SAXS data and 0–0.30 Å^{-1} for SANS data.

Calculation of cryo-EM intensity profiles

In a projection image of a liposome, the azimuthally averaged intensity profile (IP) contains information about the bilayer structure. IPs were calculated as previously described[53] and briefly summarized here. First, the electrostatic potential within the lipid bilayer, $\Phi(z)$, was approximated as a sum of contributions from individual lipid and water atoms with an additional contribution from the membrane dipole potential Φ_d, *i.e.*,

$$\Phi(z) = \sum_i V_i n_i(z) + \Phi_d(z). \quad (3)$$

In eqn (3), the sum is over individual lipid atoms, $n_i(z)$ is the atom's number density profile described in the previous section, and V_i is the spatially integrated, shielded coulomb potential for an isolated neutral atom ($V_i = 25, 130, 108, 97$, and 267 V Å^3 for H, C, N, O and P, respectively). The phase shift experienced by an electron wave passing through the bilayer is given by:

$$g(z) = \sigma_e \Phi(z), \tag{4}$$

where σ_e accounts for the dependence of the electron phase on the projected potential and is equal to 0.65 mrad V^{-1} Å^{-1} for 300 keV electrons.[54] A cryo-EM image of a liposome corresponds to a projection of the vesicle's spherical scattering profile, $\gamma(r,\theta,\phi)$, onto a plane, followed by convolution with a contrast transfer function (CTF). For a spherically symmetric liposome of radius R where each leaflet is uniformly mixed, γ has no angular dependence and can be approximated with the flat bilayer profile $g(z)$ using the coordinate transformation $r = z + R$, such that $\gamma(r,\theta,\phi) = \gamma(r) = g(r - R)$. In this case, the projected density $\Gamma(\boldsymbol{r})$ is given by the Abel transform:

$$\Gamma(\boldsymbol{r}) = 2\int_r^{\infty} \frac{g(a - R)}{\sqrt{1 - (r/a)^2}} \mathrm{d}a. \tag{5}$$

To mimic experimental images, $\Gamma(\boldsymbol{r})$ is convolved with a CTF, $c(\boldsymbol{s})$, and associated phase perturbation factor, $\chi(\boldsymbol{s})$:

$$c(\boldsymbol{s}) = [\sin \chi(\boldsymbol{s}) - Q \cos \chi(\boldsymbol{s})]\exp(-B|\boldsymbol{s}|^2), \tag{6}$$

$$\chi(\boldsymbol{s}) \approx -\pi\lambda\Delta Z|\boldsymbol{s}|^2 + \frac{\pi}{2}C_s\lambda^3|\boldsymbol{s}|^4. \tag{7}$$

In eqn (6) and (7), $\boldsymbol{s}$ is the spatial frequency, B is the amplitude decay factor, Q is the dimensionless amplitude contrast factor, λ is the electron wavelength, ΔZ is the defocus length (defined such that positive values indicate underfocus), and C_s is the spherical aberration coefficient. The reciprocal space image is calculated as:

$$I(\boldsymbol{s}) \propto 1 + mc(\boldsymbol{s})\mathcal{F}_s[\Gamma(\boldsymbol{r})](\boldsymbol{s}), \tag{8}$$

where $\mathcal{F}_s$ is the Fourier transform of the projected 2D phase shift image $\Gamma(\boldsymbol{r})$ and m is an arbitrary scale factor that adjusts the intensity contrast in the spatial domain. The corresponding real-space image $I(\boldsymbol{r})$ is the inverse Fourier transform of $I(\boldsymbol{s})$,

$$I(\boldsymbol{r}) = \mathcal{F}_r^{-1}[I(\boldsymbol{s})](\boldsymbol{r}). \tag{9}$$

For the calculations in this work, we used the following cryo-EM parameter values: $R = 50$ nm, $B = 300$ Å^2, $Q = 0.075$, $\lambda = 1.97$ pm (corresponding to 300 keV electrons), $\Delta Z = 2.5$ μm, $C_s = 2.0$ mm, $m = 400$.

Results and discussion

Asymmetry-mediated changes in leaflet composition and properties

To investigate the effects of biologically relevant asymmetries on membrane properties, we analyzed a series of molecular dynamics (MD) simulation trajectories of all-atom bilayers where one leaflet contained the saturated lipid DPPC, the other contained the unsaturated lipid DAPC, and cholesterol was at a fixed overall concentration and free to redistribute between the leaflets. This simplified lipid composition aims to capture the asymmetry in lipid saturation reported for red blood cell membranes,[16] which also plays a key role in determining the chemical potential of cholesterol in the leaflets.[24] The bilayers had a fixed number of DPPC lipids (N_{DPPC}) in the top leaflet but varying numbers of DAPC lipids (N_{DAPC}) in the bottom leaflet (Table 1). Initial coarse-grained simulations showed that while such bilayers with more DPPC than DAPC lipids were stable up to relatively large imbalances, deviations in the opposite direction (more DAPC lipids) were not tolerated as well and produced unphysical bilayer morphologies even at smaller imbalances.[46] The phospholipid (PL) imbalance, N_{DPPC}/N_{DAPC}, in the systems we studied here thus ranged from 0.91 to 2.0.

All systems had cholesterol at 30 mol% of all PLs. In real membranes, cholesterol can quickly flip between leaflets and redistribute according to its chemical potential. Since proper sampling of this movement is challenging in all-atom representation, the equilibrium cholesterol distribution in the models was determined from initial long coarse-grained trajectories, and did not change during the all-atom simulations (Table 1).[46] In all but one of the bilayers, cholesterol showed a strong preference for the saturated DPPC leaflet (Fig. 1A). Its strong pairwise interactions with DPPC competed with other factors including the overabundance of the saturated lipid (relative to DAPC), as predicted from theory.[24] Cholesterol's enrichment in the top leaflet thus decreased gradually across the systems until it completely disappeared in the A.6 bilayer which had twice as many DPPC than DAPC lipids (Fig. 1A and Table 1).

Cholesterol has a well characterized condensing effect on lipid packing in both saturated and unsaturated membranes.[55–57] We confirmed this trend in symmetric DPPC and DAPC bilayers which showed a systematic decrease in the average area per lipid (APL) with increasing amounts of cholesterol (Fig. 1B, open symbols). Since the individual leaflets of the asymmetric bilayers are binary

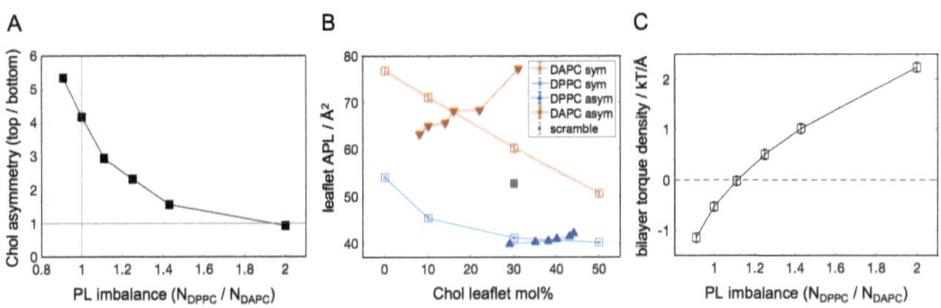

Fig. 1 Biophysical properties of simulated asymmetric bilayers. (A) Cholesterol asymmetry (cholesterol mol% in the top *vs.* the bottom leaflet) as a function of interleaflet PL imbalance. Solid gray lines indicate the conditions of no PL imbalance and no cholesterol asymmetry. (B) Leaflet area per lipid (APL) as a function of cholesterol mol% in the individual leaflets of symmetric DPPC or DAPC bilayers (open squares, '*sym*') and the asymmetric DPPC/DAPC bilayers from Table 1 (filled triangles, '*asym*'). Shown for comparison is a symmetric 'scramble' bilayer with 1 : 1 DPPC : DAPC and 30 mol% cholesterol (gray square). (C) Bilayer torque density as a function of PL imbalance in the asymmetric bilayers from Table 1.

mixtures with varying concentrations of cholesterol, we can similarly analyze the effect of the sterol on their APLs. In contrast to the expected trend observed in the symmetric membranes, cholesterol had an exactly opposite effect in the asymmetric bilayers, increasing rather than decreasing the leaflet APL (Fig. 1B, closed symbols). This counter-intuitive trend can be related to the internal stresses in the leaflets, *i.e.* differential stress, resulting from the interleaflet PL and cholesterol distributions (Table 1). In most bilayers, one leaflet had tension while the lipids in the other leaflet were compressed (*i.e.*, a negative tension). These stresses constrained the effects of cholesterol, making them deviate from the trends observed in symmetric bilayers which by definition have zero leaflet tension.[30] Only the A.3 bilayer, which had a PL imbalance of 1.11 and about 3-fold more cholesterol in the DPPC leaflet, had no differential stress. Interestingly, in that bilayer, the APL of the DAPC leaflet containing 14 mol% cholesterol (Table 1) was lower than the APL of a symmetric DAPC bilayer with similar cholesterol concentration, while the condensing effect of cholesterol in the DPPC leaflet was similar to its symmetric counterpart (Fig. 1B). These results reveal that even in the absence of internal stresses, compositional asymmetry can influence bilayer properties in ways that cannot be fully recapitulated in symmetric bilayers, which is a manifestation of interleaflet coupling.[58–60] How the specific lipid compositions of the leaflets affect the observed relationships between PL imbalance and differential stress, and APL and cholesterol concentration, remains to be investigated.

The high resolution of the simulated bilayers also enables analysis of the propensity of the bilayers to curve, which in experiments can be monitored *via* changes in membrane morphology.[34,61,62] Depending on its lipid composition, each leaflet has a spontaneous curvature that determines the curvature stress experienced by the leaflet in a given configuration (*e.g.*, when flat).[63,64] The balance between curvature stress and differential stress governs the energy cost of changes in the bilayer morphology.[22] In particular, the first moment of the lateral pressure profile (whose integral gives the leaflet stresses) is the bilayer torque density.[41] Its deviation from zero provides an estimate of the bilayer tendency to curve, with positive and negative values indicating curvature towards the top and bottom leaflets, respectively. Analysis of the asymmetric bilayers shows that this parameter is highly sensitive to their molecular organization, with a consistent preference for curvature towards the compressed leaflet (Fig. 1C). Only in the tensionless A.3 bilayer is a flat morphology the most energetically favorable. Since this bilayer has no area strain, the result implies that it has vanishing curvature stress (*i.e.*, a spontaneous curvature of zero)[41] despite its highly asymmetric profile (Table 1). This is likely due to the modulation of the leaflets' spontaneous curvature by cholesterol.[21] However, vanishing differential stress is not a prerequisite for a stable flat configuration, as the latter can exist even when the leaflets experience significant tension and compression that counterbalances curvature stress.[46] These findings demonstrate that the absence of curvature in experimental asymmetric membranes cannot be used to infer the presence or absence of both differential and curvature stresses.

Sensitivity of environment-sensitive probes

One commonly applied approach for analyzing the structural properties of lipid membranes in both model systems and cells, is measuring changes in their lipid packing *via* environment-sensitive probes.[65] A typical experiment relies on incorporation of the probe in one or both bilayer leaflets and recording its fluorescence lifetime (for *e.g.* Di4-ANEPPDHQ,[66] Flipper[67]) or generalized polarization (laurdan,[68] NR12S[69]). While each of these probes likely senses an interplay of multiple leaflet properties,[70] they are often used as reporters of lipid packing. This has been shown to be a good approximation for Di4, whose lifetime in symmetric giant unilamellar vesicles is very strongly correlated with the average area per lipid calculated from corresponding simulations of flat bilayer patches for a wide range of simple 1- and 2-component lipid mixtures.[46] To verify that the relationship holds for 3-component bilayers, we simulated the two timeline endpoints of DPPC/DOPC/Chol 40/40/20 representing its coexisting Lo and Ld compositions (see Methods). The bilayers were thus symmetric and uniform, and we compared the APLs from the simulations to the experimentally measured Di4 lifetimes in the mixtures from ref. 16. To account for any differences in the experimental conditions between ref. 16 and 46, we used the Di4 lifetime of DOPC reported in both studies to normalize the results. The Lo and Ld compositions both fell precisely on the calibration curve as shown in Fig. 2A, confirming that even in more complex multi-component mixtures, Di4 lifetime is highly correlated with simulated APLs.

Using the experimentally derived linear relationship between the two parameters, we then converted the leaflet APLs in our other simulated bilayers to Di4 lifetimes and compared the sensitivity of this parameter to the asymmetries in the membranes (Fig. 2). Since Di4 is charged and does not easily flip between leaflets,[16] we analyzed the expected lifetime measured when the probe is added to the bilayer from the outside, labeling only the top leaflet, or selectively incorporated exclusively in the bottom leaflet. Thus, while the simulations do not contain the

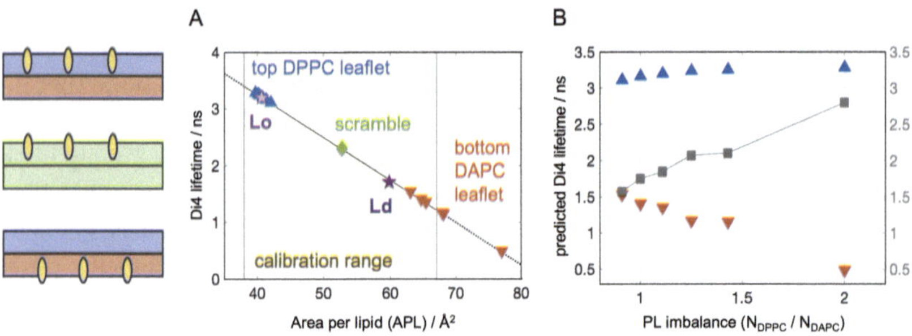

Fig. 2 Predicted sensitivity of Di4 lifetime to membrane asymmetries. (A) Fluorescence lifetime of Di4 calculated from leaflet APLs in simulated bilayers using the experimentally derived calibration curve from ref. 46. Shown is data for the simulated top DPPC leaflets (blue triangles pointing up), bottom DAPC leaflets (red triangles pointing down) and symmetric scrambled bilayer (green diamond). Also shown are the results for symmetric 3-component Lo and Ld compositions of DPPC/DOPC/Chol as explained in the text (pink and purple stars). The solid black curve shows the linear relationship between simulation and experiment derived in ref. 46 from model membranes, while the dotted lines indicate regions outside of the calibration range. (B) Calculated Di4 lifetime as a function of PL imbalance for the top and bottom leaflets in the asymmetric bilayers (same symbols as in (A)). Also shown is the difference between the two (gray squares, right *y*-axis) which shows the largest sensitivity to the transverse lipid organization in the membranes.

probe itself, the structural properties of the leaflets are used to estimate the leaflet-specific fluorescence lifetime read-out from Di4 in an experiment. In experiments, adding Di4 to outer membrane leaflets is straightforward since in aqueous solutions the probe partitions strongly into lipid bilayers due to its hydrophobic groups. Adding Di4 to selectively label the inner (cytosolic) leaflet is more challenging but has been accomplished experimentally by microinjection of cells.[16] This was confirmed by the difference in Di4 lifetime of the cytoplasmic leaflet relative to the exoplasmic leaflet, and the lack of BSA back-extraction from the PM (see Fig. S7 in ref. 16). In model membranes, selective labeling of the inner leaflet of liposomes has been accomplished using cyclodextrin to exchange the outer leaflet probe with lipid from unlabelled donor vesicles,[71] although this has not been specifically demonstrated for Di4. The selective incorporation of a lipid dye in the inner leaflet can be verified with externally added quenchers, and other environment-sensitive probes can in principle be used in analogous ways (see Introduction).

The first noticeable result from our computational analysis was that two of the data points—corresponding to two DAPC-rich leaflets—fell outside of the calibration range between simulation and experiment (Fig. 2A).[46] One of them is a leaflet in bilayer A.6 that has the most extreme PL imbalance of 2.0 and substantial differential stress (Table 1) and bilayer torque density (Fig. 1C). This highly stressed bilayer state is tolerated in simulations due to the relatively small system size and applied periodic boundary conditions but may exceed the tolerance threshold of liposomes *in vitro*, where the stress can be relieved instead by mechanisms such as extreme curvature and budding.[34,72] The calibration range could be extended by analyzing more disordered symmetric bilayers under net bilayer tension or at higher temperatures (both of which can be applied in simulations and experiments to increase leaflet APL). However, the tolerance threshold for bilayer stress in the experiments, as well as any potential changes to the correspondence between the simulations and Di4 measurements from external perturbations, must be investigated and accounted for.

Comparison between the calculated Di4 lifetimes of the outer and inner leaflets across the series of simulated bilayers reveals differences in the sensitivity of this parameter to the properties of the respective lipid compositions (Fig. 2B). The relatively tightly packed DPPC leaflet, whose APL varies only from 39.7 ± 0.2 $Å^2$ to 42 ± 0.36 $Å^2$ between the two extremes (bilayers A.1 and A.6), shows marginal variation in Di4 lifetime. In contrast, the predicted changes in the DAPC leaflet, whose APL increases by 22% (Table 1), are more pronounced. Since the trends in the leaflet APLs are inversely correlated, where the smaller APL becomes smaller while the larger APL increases, most sensitive to the asymmetries is the difference between outer and inner leaflet Di4 lifetimes (Fig. 2B, gray squares). Measuring this difference *in vitro* would require concurrent experiments of membranes with either their top or bottom leaflets labelled with Di4, or calibrating and incorporating in the two leaflets different environment-sensitive probes with non-overlapping fluorescence spectra that can be imaged and analyzed simultaneously.

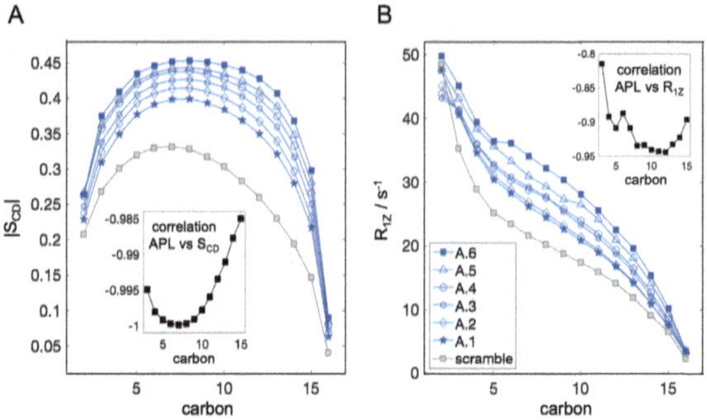

Fig. 3 Variation in simulation parameters corresponding to properties measured with NMR. (A) Acyl chain order parameter profiles for carbons 2 through 16 averaged over the *sn*-1 and *sn*-2 chains of DPPC in the asymmetric bilayers A.1 to A.6 from Table 1, and the scramble analog with DPPC/DAPC 1 : 1 and 30 mol% cholesterol. Errors are less than 10^{-2} and are omitted for clarity. (B) Spin-lattice relaxation rates of the corresponding CH bond fluctuations analyzed in (A). Insets in both (A) and (B) show the correlation between S_{CD} or R_{1Z} and leaflet APL (Table 1) calculated from the plotted data for the asymmetric bilayers.

Sensitivity of NMR structural parameters

In addition to APL, bilayer structure is often characterized by the degree of order of the lipid acyl chains. This property is quantified from the dynamic reorientation of the carbon–hydrogen (CH) bonds along the lipid chains relative to the bilayer normal. In particular, the average order parameter (S_{CD}) characterizing the mean of the fluctuations, as well as their relaxation rate (R_{1Z}), can be measured with NMR and also calculated from simulations (*e.g.* ref. 49 and 73). The experiments are often done with deuterium NMR which allows for better spectral resolution and identification of chain segments.[74] The measurements are typically performed on oriented lipid stacks or multilamellar lipid dispersions,[75] but can also be done on unilamellar liposomes,[76,77] thus making them suitable for asymmetry studies using model membranes. Since deuterated lipids with saturated chains are in general more readily available from commercial vendors, we analyzed the CH bond dynamics for DPPC in the simulated bilayers as representative experimental data (Fig. 3). Similar comparisons of the order at specific carbon segments can also be made with electron spin resonance (ESR) experiments *via* spin probes selectively incorporated into one of the leaflets.[71,78]

The resulting order parameter profiles show an overall increase in chain order (*i.e.*, higher $|S_{CD}|$) with a decrease in leaflet tension and APL (Fig. 3A and Table 1), as expected. There is a strong negative correlation between APL and S_{CD} for all carbons (inset in Fig. 3A) but the best sensitivity to the asymmetry-mediated properties (*i.e.*, the largest separation between all conditions) is achieved in the bottom parts of the chains, at carbons 12 through 15 (Fig. 3A). This is not surprising since this is the part of the leaflet in closest contact with the opposing leaflet. In contrast, the order parameter profiles for some bilayers (*e.g.* A.5 and A.4) overlap in the plateau region (carbons 5 through 10). Interestingly, in the same bilayers, these carbon segments have detectable differences in their relaxation rates even though the correlation between APL and R_{1Z} is slightly weaker (Fig. 3B), indicating that the structural information from S_{CD} and R_{1Z} is complementary.

Thus, if available, data from both parameters can maximize the ability to distinguish the leaflet states in experiments. Most challenging in this endeavour would be optimizing the size and hydration level of the asymmetric vesicles for the measurements. While the simulations revealed overall strong correlation between the NMR parameters and APL (Fig. 3), experimental conditions may be key for determining the consistency of the measured bilayer structure with other techniques.[73] The relationship between the order parameters and relaxation rates in the region of leaflet–leaflet interaction has also been connected to bilayer elasticity *via* the membrane thickness in symmetric membranes.[49] Further research is needed to test and develop this dependence in asymmetric bilayers.

Sensitivity of small-angle scattering data

Small-angle X-ray and neutron scattering (SAXS and SANS, respectively) are robust and well-established techniques for the precise measurement of bilayer structural parameters including area per lipid and bilayer thickness.[79,80] Using atomic number density profiles from simulated bilayers as input, we calculated form factors for SAXS (Fig. 4A) and SANS (Fig. 4B–D), the latter using different combinations of protiated and perdeuterated lipids to explore the effects of internal bilayer contrast. Whereas symmetric bilayers produce clearly delineated scattering lobes separated by minima that approach zero intensity, asymmetry results in a characteristic "liftoff" of the minima that is particularly pronounced in cases of extreme transbilayer scattering length density contrast (*e.g.*, protiated DAPC in the inner leaflet and DPPC-d62 in the outer leaflet as in Fig. 4C and D).[81–83] Such liftoff is also visible (albeit to a smaller extent) in the first minimum following the first peak at the lowest q in the SAXS form factors (Fig. 4A). In principle, structural parameters can be extracted from the data by fitting to a suitable structural model; however, this is a daunting task for an asymmetric membrane where each leaflet has its own variables thus requiring a multitude of free parameters in the fitting routine.[84,85] Furthermore, as illustrated in the fully protiated condition in Fig. 4B, bilayers with very different PL and cholesterol asymmetries can lead to almost indistinguishable scattering curves, making it difficult to unambiguously determine the underlying lipid organization.

An alternative approach to analyzing such data uses model-free quantities that correlate with structural features. For example, increasing phospholipid imbalance results in systematic changes in the intensity profiles that are particularly prominent in the first two scattering lobes for SAXS (Fig. 4A) and internally contrasted SANS (Fig. 4C and D) data. These changes are readily apparent in the centers-of-mass of the $F(q)$ curves, calculated as described in the Methods and plotted as small filled symbols (highlighted by solid arrows) in Fig. 4A–D. Fig. 4E shows the corresponding trends and reveals that the q-coordinate of the center-of-mass increases nearly linearly with phospholipid imbalance in all but the fully protiated SANS data. A useful aspect of this analysis is that it is insensitive to sample-to-sample differences in lipid concentration. Furthermore, the q-range over which it is calculated can be adjusted to include only data with the highest signal-to-noise in the experiments. Importantly, this quantification demonstrates that both SAXS and SANS with at least one asymmetrically distributed perdeuterated lipid are sensitive to compositional and number imbalances, in contrast to SANS of fully protiated samples. Joint analysis of these various conditions may

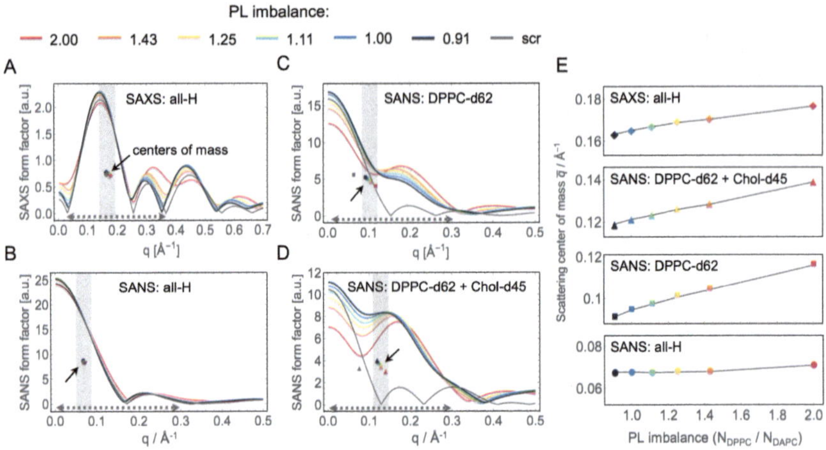

Fig. 4 Small-angle scattering data from simulated bilayers. (A) Small-angle X-ray scattering (SAXS) form factors for the asymmetric and scramble bilayers calculated from the simulation trajectories as a function of q. Also shown are the corresponding center-of-mass for each scattering form factor (filled symbols) as explained in the text. (B–D) Small-angle neutron scattering (SANS) form factors for the simulated bilayers for three isotopic conditions: fully protiated (hydrogenated) sample with either none (all-H), only DPPC chains (DPPC-d62) or both DPPC chains and Chol (DPPC-d62 + Chol-d45) perdeuterated. In all samples, water molecules were 100% deuterated. The corresponding scattering centers-of-mass are also shown (filled symbols). (E) Plots of the q-values of the scattering centers-of-mass as a function of PL imbalance for each of the data sets in A–D. All plots span q-regions of the same width (*i.e.*, they have equivalent y-axis ranges) as illustrated by the gray shaded areas on the plots in A–D. Note that these regions are different from the q-ranges used to calculate the centers-of-mass which are indicated as dotted arrows near the horizontal axis in panels A–D.

thus provide the best approach for identifying a bilayer model with the correct structure.

Sensitivity of cryo-EM imaging and analysis

Cryo-EM enables the direct visualization of lipid bilayers at sub-nanometer resolution.[86,87] When spherical liposomes are imaged in projection, the defining characteristic is a pair of dark concentric circles roughly corresponding to the electron dense lipid headgroup layers (Fig. 5A). For laterally homogeneous bilayers, images of individual vesicles can be reduced to 1D intensity profiles normal to the bilayer by azimuthal integration and then ensemble averaged to produce data suitable for quantitative analysis.[88,89] As described in Methods, the cryo-EM image formation process for liposomes can also be accurately mimicked computationally using atomic number density profiles from MD simulations as input.[90] Fig. 5B shows cryo-EM intensity profiles calculated from the simulated asymmetric bilayers with varying phospholipid imbalance, as well as the symmetric scrambled bilayer with 30 mol% Chol.

Because the intensity profiles arise from electron density variation across the bilayer, their features are sensitive to details of the bilayer structure. For example, the distance between the two deep troughs has been shown to be strongly

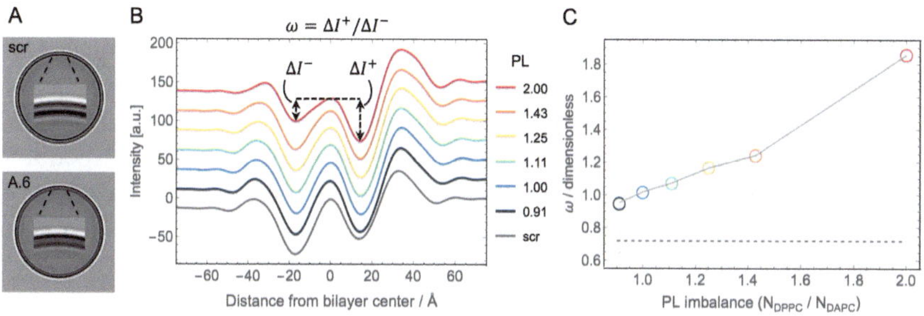

Fig. 5 Cryo-EM intensity profiles calculated from simulation trajectories. (A) Synthetic cryo-EM images of 100 nm-diameter vesicles corresponding to the scramble (top) and asymmetric A.6 (bottom, see Table 1) simulated bilayers. Shown in the middle of each vesicle is an enlargement of the intensity close to and across the bilayer, showing how the contrast between the outer and inner troughs in the two images is reversed, consistent with the change in PL imbalance in the bilayer leaflets. (B) Intensity profiles for the asymmetric bilayers (color-coded according to their PL imbalance), and the scrambled bilayer (gray) computationally derived from the simulated atomic number density profiles and membrane dipole potential as described in Methods. The outer (ΔI^+) and inner (ΔI^-) trough depths relative to the central peak which define the parameter ω, are illustrated for the curve corresponding to a PL imbalance of 2.0. Profiles have been shifted vertically for better visualization. (C) ω as a function of PL imbalance for all asymmetric bilayers (symbols). ω for the scramble membrane is also shown for comparison (dashed line).

correlated with bilayer thickness in both simulated and experimental data.[91] Similarly, the depth of the troughs can reveal differences in lipid packing density within the leaflets.[53] To explore this further, we defined a model-free parameter, $\omega = \Delta I^+/\Delta I^-$, calculated as the ratio of the outer (ΔI^+) and inner (ΔI^-) trough depths relative to the central peak as depicted in Fig. 5B. Fig. 5C shows that ω nearly doubled (from 0.95 to 1.86) when the phospholipid imbalance was varied from 0.9 to 2.0. It is important to note that ω for perfectly symmetric bilayers is not unity but instead ranges from ~0.65–0.8 (*i.e.*, the outer trough is shallower than the inner trough) due to the spherical geometry of the liposome and the nature of projection imaging, as shown by the scrambled bilayer in Fig. 5B (gray curve). These results demonstrate that a simple model-free analysis can be used to accurately estimate the phospholipid imbalance of liposomes, provided that a sufficient number of liposomes are analyzed to achieve high signal-to-noise in the intensity profile. Whether the trends in ω for asymmetric bilayers can be affected by variations in lipid composition, cholesterol concentration and electron contrast-enhancers (such as substitutions of hydrogen in the lipid molecules with halogen atoms) needs to be further explored.

Conclusion

While the compositional asymmetry of biological membranes has been known for over half a century, it has only recently been recognized that an asymmetric membrane cannot solely be defined as just having leaflets with different lipid compositions. Imbalances in the total lipid abundances of the two monolayers can accompany interleaflet compositional variations and have profound effects on the bilayer biophysical properties. This makes the detection and characterization of both compositional and number asymmetries critical for analysis of model and cell membranes. Existing biophysical techniques have different sensitivities to the precise interleaflet lipid organization in a bilayer, that can be maximized by optimizing experimental conditions and data analysis protocols. Coupling the *in vitro* approaches with simulations of corresponding atomistic models holds great promise for refining these methodologies, fully defining the state of the experimental systems and uncovering the full potential of membrane asymmetry for shaping and regulating bilayer properties.

Data availability

The data corresponding to this article have been stored at https://zenodo.org/records/14626488, https://zenodo.org/records/14655810, and https://zenodo.org/records/14627064.

Author contributions

MD: conceptualization, data curation, formal analysis, investigation, project administration, writing – original draft, writing – review & editing. FAH: formal analysis, methodology, writing – original draft, writing – review & editing.

Conflicts of interest

The authors declare no conflicts of interest.

Acknowledgements

This research was supported by NIH grant R01GM138887 (to FAH) and the SciLifeLab & Wallenberg Data Driven Life Science Program grant: KAW 2024.0159 (to MD). The table of contents entry graphic was created in https://BioRender.com.

References

1 G. van Meer, D. R. Voelker and G. W. Feigenson, Membrane lipids: where they are and how they behave, *Nat. Rev. Mol. Cell Biol.*, 2008, **9**(2), 112–124.
2 R. M. Venable, F. L. Brown and R. W. Pastor, Mechanical properties of lipid bilayers from molecular dynamics simulation, *Chem. Phys. Lipids*, 2015, **192**, 60–74.
3 S. D. Guler, D. D. Ghosh, J. Pan, J. C. Mathai, M. L. Zeidel, J. F. Nagle and S. Tristram-Nagle, Effects of ether vs. ester linkage on lipid bilayer structure and water permeability, *Chem. Phys. Lipids*, 2009, **160**(1), 33–44.
4 J. C. Mathai, S. Tristram-Nagle, J. F. Nagle and M. L. Zeidel, Structural determinants of water permeability through the lipid membrane, *J. Gen. Physiol.*, 2008, **131**(1), 69–76.
5 N. Kucerka, F. A. Heberle, J. Pan and J. Katsaras, Structural Significance of Lipid Diversity as Studied by Small Angle Neutron and X-ray Scattering, *Membranes*, 2015, **5**(3), 454–472.
6 D. Marsh *Handbook of Lipid Bilayers*; CRC Press, 2013.

7 I. Levental, K. R. Levental and F. A. Heberle, Lipid Rafts: Controversies Resolved, Mysteries Remain, *Trends Cell Biol.*, 2020, **30**(5), 341–353.

8 E. Sezgin, I. Levental, S. Mayor and C. Eggeling, The mystery of membrane organization: composition, regulation and roles of lipid rafts, *Nat. Rev. Mol. Cell Biol.*, 2017, **18**(6), 361–374.

9 I. Levental and E. Lyman, Regulation of membrane protein structure and function by their lipid nano-environment, *Nat. Rev. Mol. Cell Biol.*, 2023, **24**(2), 107–122.

10 H. A. Faizi, A. Tsui, R. Dimova and P. M. Vlahovska, Bending Rigidity, Capacitance, and Shear Viscosity of Giant Vesicle Membranes Prepared by Spontaneous Swelling, Electroformation, Gel-Assisted, and Phase Transfer Methods: A Comparative Study, *Langmuir*, 2022, **38**(34), 10548–10557.

11 G. Enkavi, M. Javanainen, W. Kulig, T. Rog and I. Vattulainen, Multiscale Simulations of Biological Membranes: The Challenge To Understand Biological Phenomena in a Living Substance, *Chem. Rev.*, 2019, **119**(9), 5607–5774.

12 R. Ashkar, H. Bilheux, H. Bordallo, R. Briber, D. J. Callaway, X. Cheng, X.-Q. Chu, J. E. Curtis, M. Dadmun and P. Fenimore, Neutron scattering in the biological sciences: Progress and prospects, *Acta Crystallogr., Sect. D:Struct. Biol.*, 2018, **74**(12), 1129–1168.

13 K. J. Mallikarjunaiah, J. J. Kinnun, H. I. Petrache and M. F. Brown, Flexible lipid nanomaterials studied by NMR spectroscopy, *Phys. Chem. Chem. Phys.*, 2019, **21**(34), 18422–18457.

14 A. J. Verkleij, R. F. Zwaal, B. Roelofsen, P. Comfurius, D. Kastelijn and L. L. van Deenen, The asymmetric distribution of phospholipids in the human red cell membrane. A combined study using phospholipases and freeze-etch electron microscopy, *Biochim. Biophys. Acta, Biomembr.*, 1973, **323**(2), 178–193.

15 P. F. Devaux, Static and dynamic lipid asymmetry in cell membranes, *Biochemistry*, 1991, **30**(5), 1163–1173.

16 J. H. Lorent, K. R. Levental, L. Ganesan, G. Rivera-Longsworth, E. Sezgin, M. Doktorova, E. Lyman and I. Levental, Plasma membranes are asymmetric in lipid unsaturation, packing and protein shape, *Nat. Chem. Biol.*, 2020, **16**(6), 644–652.

17 T. Yeung, G. E. Gilbert, J. Shi, J. Silvius, A. Kapus and S. Grinstein, Membrane phosphatidylserine regulates surface charge and protein localization, *Science*, 2008, **319**(5860), 210–213.

18 A. Gupta, T. Korte, A. Herrmann and T. Wohland, Plasma membrane asymmetry of lipid organization: fluorescence lifetime microscopy and correlation spectroscopy analysis, *J. Lipid Res.*, 2020, **61**(2), 252–266.

19 C. Patra, M. Samad, T. S. van Zanten, M. Singh, R. Vishwakarma, P. Pal Singh and S. Mayor, Solvatochromic reporter to image plasma membrane order leaflet by leaflet reveals a highly asymmetric bilayer locally modulated by transbilayer interactions, *bioRxiv*, 2024, preprint, DOI: 10.1101/2024.07.23.604763.

20 A. A. Gurtovenko and I. Vattulainen, Membrane potential and electrostatics of phospholipid bilayers with asymmetric transmembrane distribution of anionic lipids, *J. Phys. Chem. B*, 2008, **112**(15), 4629–4634.

21 D. W. Allender, A. J. Sodt and M. Schick, Cholesterol-Dependent Bending Energy Is Important in Cholesterol Distribution of the Plasma Membrane, *Biophys. J.*, 2019, **116**(12), 2356–2366.

22 A. Hossein and M. Deserno, Spontaneous Curvature, Differential Stress, and Bending Modulus of Asymmetric Lipid Membranes, *Biophys. J.*, 2020, **118**(3), 624–642.

23 B. Kollmitzer, P. Heftberger, M. Rappolt and G. Pabst, Monolayer spontaneous curvature of raft-forming membrane lipids, *Soft Matter*, 2013, **9**(45), 10877–10884.

24 M. Varma and M. Deserno, Distribution of cholesterol in asymmetric membranes driven by composition and differential stress, *Biophys. J.*, 2022, **121**(20), 4001–4018.

25 H. L. Scott, K. B. Kennison, T. A. Enoki, M. Doktorova, J. J. Kinnun, F. A. Heberle and J. Katsaras, Model Membrane Systems Used to Study Plasma Membrane Lipid Asymmetry, *Symmetry*, 2021, **13**(8), 1356.

26 J. Niu, Y. Ma, Y. Yang, H. Lv, J. Wang, T. Wang, F. Liu, S. Xu, Z. Jiang and W. Lin, Lighting up the changes of plasma membranes during apoptosis with fluorescent probes, *Coord. Chem. Rev.*, 2023, **476**, 214926.

27 J. R. St Clair, Q. Wang, G. Li and E. London, Preparation and Physical Properties of Asymmetric Model Membrane Vesicles, in *The Biophysics of Cell Membranes*, ed. Epand, R. and Ruysschaert, J. M., Springer, 2017, vol. 19, pp. 1–27.

28 M. Doktorova, J. L. Symons and I. Levental, Structural and functional consequences of reversible lipid asymmetry in living membranes, *Nat. Chem. Biol.*, 2020, **16**(12), 1321–1330.

29 M. Doktorova, F. A. Heberle, B. Eicher, R. F. Standaert, J. Katsaras, E. London, G. Pabst and D. Marquardt, Preparation of asymmetric phospholipid vesicles for use as cell membrane models, *Nat. Protoc.*, 2018, **13**(9), 2086–2101.

30 M. Doktorova and H. Weinstein, Accurate in silico modeling of asymmetric bilayers based on biophysical principles, *Biophys. J.*, 2018, **115**(9), 1638–1643.

31 S. Park, A. H. Beaven, J. B. Klauda and W. Im, How Tolerant are Membrane Simulations with Mismatch in Area per Lipid between Leaflets?, *J. Chem. Theory Comput.*, 2015, **11**(7), 3466–3477.

32 S. L. Foley, M. Varma, A. Hossein and M. Deserno, Elastic and thermodynamic consequences of lipid membrane asymmetry, *Emerging Top. Life Sci.*, 2023, 7(1), 95–110.

33 A. Sreekumari and R. Lipowsky, Large stress asymmetries of lipid bilayers and nanovesicles generate lipid flip-flops and bilayer instabilities, *Soft Matter*, 2022, **18**(32), 6066–6078.

34 L. Hua, M. Kaiser, I. Carabadjac, A. Meister, G. Hause and H. Heerklotz, Vesicle budding caused by lysolipid-induced asymmetry stress, *Biophys. J.*, 2023, **122**(20), 4011–4022.

35 N. Bettache, L. Baisamy, S. Baghdiguian, B. Payrastre, P. Mangeat and A. Bienvenue, Mechanical constraint imposed on plasma membrane through transverse phospholipid imbalance induces reversible actin polymerization via phosphoinositide 3-kinase activation, *J. Cell Sci.*, 2003, **116**(11), 2277–2284.

36 H. Heerklotz, Membrane stress and permeabilization induced by asymmetric incorporation of compounds, *Biophys. J.*, 2001, **81**(1), 184–195.

37 T. Fujii and A. Tamura, Shape change of human erythrocytes induced by phosphatidylcholine and lysophosphatidylcholine species with various acyl chain lengths, *Cell Biochem. Funct.*, 1984, **2**(3), 171–176.
38 Y. Elani, S. Purushothaman, P. J. Booth, J. M. Seddon, N. J. Brooks, R. V. Law and O. Ces, Measurements of the effect of membrane asymmetry on the mechanical properties of lipid bilayers, *Chem. Commun.*, 2015, **51**(32), 6976–6979.
39 B. W. Rickeard, M. H. L. Nguyen, M. DiPasquale, C. G. Yip, H. Baker, F. A. Heberle, X. Zuo, E. G. Kelley, M. Nagao and D. Marquardt, Transverse lipid organization dictates bending fluctuations in model plasma membranes, *Nanoscale*, 2020, **12**(3), 1438–1447.
40 M. P. K. Frewein, P. Piller, E. F. Semeraro, O. Czakkel, Y. Gerelli, L. Porcar and G. Pabst, Distributing aminophospholipids asymmetrically across leaflets causes anomalous membrane stiffening, *Biophys. J.*, 2023, **122**(12), 2445–2455.
41 S. Park, W. Im and R. W. Pastor, Developing initial conditions for simulations of asymmetric membranes: a practical recommendation, *Biophys. J.*, 2021, **120**(22), 5041–5059.
42 M. Doktorova, I. Levental and F. A. Heberle, Seeing the Membrane from Both Sides Now: Lipid Asymmetry and Its Strange Consequences, *Cold Spring Harbor Perspect. Biol.*, 2023, a041393.
43 S. Jo, T. Kim, V. G. Iyer and W. Im, CHARMM-GUI: a web-based graphical user interface for CHARMM, *J. Comput. Chem.*, 2008, **29**(11), 1859–1865.
44 J. Lee, X. Cheng, J. M. Swails, M. S. Yeom, P. K. Eastman, J. A. Lemkul, S. Wei, J. Buckner, J. C. Jeong, Y. Qi, *et al.*, CHARMM-GUI Input Generator for NAMD, GROMACS, AMBER, OpenMM, and CHARMM/OpenMM Simulations Using the CHARMM36 Additive Force Field, *J. Chem. Theory Comput.*, 2016, **12**(1), 405–413.
45 B. R. Brooks, C. L. Brooks, A. D. Mackerell, L. Nilsson, R. J. Petrella, B. Roux, Y. Won, G. Archontis, C. Bartels, S. Boresch, *et al.*, CHARMM: The Biomolecular Simulation Program, *J. Comput. Chem.*, 2009, **30**(10), 1545–1614.
46 M. Doktorova, J. L. Symons, X. Zhang, H.-Y. Wang, J. Schlegel, J. H. Lorent, F. A. Heberle, E. Sezgin, E. Lyman, K. R. Levental and I. Levental, Cell Membranes Sustain Phospholipid Imbalance Via Cholesterol Asymmetry, *Cell*, 2025, DOI: 10.1016/j.cell.2025.02.034.
47 J. C. Phillips, D. J. Hardy, J. D. C. Maia, J. E. Stone, J. V. Ribeiro, R. C. Bernardi, R. Buch, G. Fiorin, J. Henin, W. Jiang, *et al.*, Scalable molecular dynamics on CPU and GPU architectures with NAMD, *J. Chem. Phys.*, 2020, **153**(4), 044130.
48 J. B. Klauda, R. M. Venable, J. A. Freites, J. W. O'Connor, D. J. Tobias, C. Mondragon-Ramirez, I. Vorobyov, A. D. MacKerell Jr and R. W. Pastor, Update of the CHARMM all-atom additive force field for lipids: validation on six lipid types, *J. Phys. Chem. B*, 2010, **114**(23), 7830–7843.
49 M. Doktorova, G. Khelashvili, R. Ashkar and M. F. Brown, Molecular simulations and NMR reveal how lipid fluctuations affect membrane mechanics, *Biophys. J.*, 2023, **122**(6), 984–1002.
50 W. Humphrey, A. Dalke and K. Schulten, VMD: visual molecular dynamics, *J. Mol. Graphics*, 1996, **14**(1), 33–38.
51 J. N. Sachs, P. S. Crozier and T. B. Woolf, Atomistic simulations of biologically realistic transmembrane potential gradients, *J. Chem. Phys.*, 2004, **121**(22), 10847–10851.

52 N. Kucerka, J. Pencer, J. N. Sachs, J. F. Nagle and J. Katsaras, Curvature effect on the structure of phospholipid bilayers, *Langmuir*, 2007, **23**(3), 1292–1299.
53 K. D. Sharma, M. Doktorova, M. N. Waxham and F. A. Heberle, Cryo-EM images of phase-separated lipid bilayer vesicles analyzed with a machine-learning approach, *Biophys. J.*, 2024, **123**(17), 2877–2891.
54 E. J. Kirkland, *Advanced Computing in Electron Microscopy*, Springer, 1998.
55 F. Leeb and L. Maibaum, Spatially Resolving the Condensing Effect of Cholesterol in Lipid Bilayers, *Biophys. J.*, 2018, **115**(11), 2179–2188.
56 W. C. Hung, M. T. Lee, F. Y. Chen and H. W. Huang, The condensing effect of cholesterol in lipid bilayers, *Biophys. J.*, 2007, **92**(11), 3960–3967.
57 M. Alwarawrah, J. Dai and J. Huang, A molecular view of the cholesterol condensing effect in DOPC lipid bilayers, *J. Phys. Chem. B*, 2010, **114**(22), 7516–7523.
58 H. T. Cheng and E. London, Preparation and properties of asymmetric large unilamellar vesicles: interleaflet coupling in asymmetric vesicles is dependent on temperature but not curvature, *Biophys. J.*, 2011, **100**(11), 2671–2678.
59 B. Eicher, D. Marquardt, F. A. Heberle, I. Letofsky-Papst, G. N. Rechberger, M. S. Appavou, J. Katsaras and G. Pabst, Intrinsic Curvature-Mediated Transbilayer Coupling in Asymmetric Lipid Vesicles, *Biophys. J.*, 2018, **114**(1), 146–157.
60 M. A. Perez and P. A. Beales, Dynamics of asymmetric membranes and interleaflet coupling as intermediates in membrane fusion, *Biophys. J.*, 2023, **122**(11), 1985–1995.
61 B. L. Mui, H. G. Dobereiner, T. D. Madden and P. R. Cullis, Influence of transbilayer area asymmetry on the morphology of large unilamellar vesicles, *Biophys. J.*, 1995, **69**(3), 930–941.
62 R. Dimova, Giant vesicles and their use in assays for assessing membrane phase state, curvature, mechanics, and electrical properties, *Annu. Rev. Biophys.*, 2019, **48**(1), 93–119.
63 A. J. Sodt, R. M. Venable, E. Lyman and R. W. Pastor, Nonadditive Compositional Curvature Energetics of Lipid Bilayers, *Phys. Rev. Lett.*, 2016, **117**(13), 138104.
64 M. Deserno, Fluid lipid membranes: from differential geometry to curvature stresses, *Chem. Phys. Lipids*, 2015, **185**, 11–45.
65 F. Ragaller, E. Sjule, Y. B. Urem, J. Schlegel, R. El, D. Urbancic, I. Urbancic, H. Blom and E. Sezgin, Quantifying Fluorescence Lifetime Responsiveness of Environment-Sensitive Probes for Membrane Fluidity Measurements, *J. Phys. Chem. B*, 2024, **128**(9), 2154–2167.
66 D. M. Owen, P. M. Lanigan, C. Dunsby, I. Munro, D. Grant, M. A. Neil, P. M. French and A. I. Magee, Fluorescence lifetime imaging provides enhanced contrast when imaging the phase-sensitive dye di-4-ANEPPDHQ in model membranes and live cells, *Biophys. J.*, 2006, **90**(11), L80–L82.
67 K. Strakova, J. Lopez-Andarias, N. Jimenez-Rojo, J. E. Chambers, S. J. Marciniak, H. Riezman, N. Sakai and S. Matile, HaloFlippers: A General Tool for the Fluorescence Imaging of Precisely Localized Membrane Tension Changes in Living Cells, *ACS Cent. Sci.*, 2020, **6**(8), 1376–1385.

68 G. Gunther, L. Malacrida, D. M. Jameson, E. Gratton and S. A. Sánchez, LAURDAN since Weber: the quest for visualizing membrane heterogeneity, *Acc. Chem. Res.*, 2021, **54**(4), 976–987.

69 F. Ragaller, L. Andronico, J. Sykora, W. Kulig, T. Rog, Y. B. Urem, Abhinav, D. I. Danylchuk, M. Hof, A. Klymchenko, *et al.*, Dissecting the mechanisms of environment sensitivity of smart probes for quantitative assessment of membrane properties, *Open Biol.*, 2022, **12**(9), 220175.

70 M. Amaro, F. Reina, M. Hof, C. Eggeling and E. Sezgin, Laurdan and Di-4-ANEPPDHQ probe different properties of the membrane, *J. Phys. D: Appl. Phys.*, 2017, **50**(13), 134004.

71 M. N. Doktorova, *Biophysics of Asymmetric Membranes: Protocols and Revelations*, Weill Medical College of Cornell University, 2018.

72 R. Dasgupta, M. S. Miettinen, N. Fricke, R. Lipowsky and R. Dimova, The glycolipid GM1 reshapes asymmetric biomembranes and giant vesicles by curvature generation, *Proc. Natl. Acad. Sci. U.S.A.*, 2018, **115**(22), 5756–5761.

73 M. Doktorova, N. Kucerka, J. J. Kinnun, J. Pan, D. Marquardt, H. L. Scott, R. M. Venable, R. W. Pastor, S. R. Wassall, J. Katsaras, *et al.*, Molecular Structure of Sphingomyelin in Fluid Phase Bilayers Determined by the Joint Analysis of Small-Angle Neutron and X-ray Scattering Data, *J. Phys. Chem. B*, 2020, **124**(25), 5186–5200.

74 T. R. Molugu, X. Xu, A. Leftin, S. Lope-Piedrafita, G. V. Martinez, H. I. Petrache and M. F. Brown, Solid-state deuterium NMR spectroscopy of membranes, *Mod. Magn. Reson.*, 2017, 1–23.

75 A. A. Nevzorov, T. P. Trouard and M. F. Brown, Lipid bilayer dynamics from simultaneous analysis of orientation and frequency dependence of deuterium spin-lattice and quadrupolar order relaxation, *Phys. Rev. E:Stat. Phys., Plasmas, Fluids, Relat. Interdiscip. Top.*, 1998, **58**(2), 2259.

76 D. B. Fenske and P. R. Cullis, Acyl chain orientational order in large unilamellar vesicles: comparison with multilamellar liposomes: a 2H and 31P nuclear magnetic resonance study, *Biophys. J.*, 1993, **64**(5), 1482–1491.

77 O. Cruciani, L. Mannina, A. P. Sobolev, C. Cametti and A. Segre, An improved NMR study of liposomes using 1-palmitoyl-2-oleoyl-sn-glycero-3-phospatidylcholine as model, *Molecules*, 2006, **11**(5), 334–344.

78 W. K. Subczynski, J. Widomska and J. B. Feix, Physical properties of lipid bilayers from EPR spin labeling and their influence on chemical reactions in a membrane environment, *Free Radical Biol. Med.*, 2009, **46**(6), 707–718.

79 F. A. Heberle, J. Pan, R. F. Standaert, P. Drazba, N. Kucerka and J. Katsaras, Model-based approaches for the determination of lipid bilayer structure from small-angle neutron and X-ray scattering data, *Eur. Biophys. J.*, 2012, **41**(10), 875–890.

80 E. F. Semeraro, L. Marx, M. P. Frewein and G. Pabst, Increasing complexity in small-angle X-ray and neutron scattering experiments: from biological membrane mimics to live cells, *Soft Matter*, 2021, **17**(2), 222–232.

81 M. R. Brzustowicz and A. T. Brunger, X-ray scattering from unilamellar lipid vesicles, *J. Appl. Crystallogr.*, 2005, **38**(1), 126–131.

82 F. A. Heberle, D. Marquardt, M. Doktorova, B. Geier, R. F. Standaert, P. Heftberger, B. Kollmitzer, J. D. Nickels, R. A. Dick, G. W. Feigenson, *et al.*, Subnanometer Structure of an Asymmetric Model Membrane: Interleaflet CouplingInfluences Domain Properties, *Langmuir*, 2016, **32**(20), 5195–5200.

83 M. P. K. Frewein, P. Piller, E. F. Semeraro, K. C. Batchu, F. A. Heberle, H. L. Scott, Y. Gerelli, L. Porcar and G. Pabst, Interdigitation-Induced Order and Disorder in Asymmetric Membranes, *J. Membr. Biol.*, 2022, **255**(4–5), 407–421.

84 B. Eicher, F. A. Heberle, D. Marquardt, G. N. Rechberger, J. Katsaras and G. Pabst, Joint small-angle X-ray and neutron scattering data analysis of asymmetric lipid vesicles, *J. Appl. Crystallogr.*, 2017, **50**(2), 419–429.

85 E. F. Semeraro, M. P. Frewein and G. Pabst, Structure of symmetric and asymmetric lipid membranes from joint SAXS/SANS, *Methods Enzymol.*, 2024, **700**, 349–383.

86 Y. Tahara and Y. Fujiyoshi, A new method to measure bilayer thickness: cryo-electron microscopy of frozen hydrated liposomes and image simulation, *Micron*, 1994, **25**(2), 141–149.

87 K. D. Sharma, F. A. Heberle and M. N. Waxham, Visualizing lipid membrane structure with cryo-EM: past, present, and future, *Emerging Top. Life Sci.*, 2023, **7**(1), 55–65.

88 F. A. Heberle and M. N. Waxham, Phase separation in model lipid membranes investigated with cryogenic electron microscopy, *Methods Enzymol.*, 2024, **700**, 189–216.

89 E. Lafarge, C. Marques, M. Schmutz, P. Muller and A. Schroder, Thickness determination of hydroperoxidized lipid bilayers from medium-resolution cryo-TEM images, *Biophysical Approaches for the Study of Membrane Structure-Part A: Experimental*, 2024, vol. 700, pp. 329–348.

90 L. Wang, P. S. Bose and F. J. Sigworth, Using cryo-EM to measure the dipole potential of a lipid membrane, *Proc. Natl. Acad. Sci. U. S. A.*, 2006, **103**(49), 18528–18533.

91 F. A. Heberle, M. Doktorova, H. L. Scott, A. D. Skinkle, M. N. Waxham and I. Levental, Direct label-free imaging of nanodomains in biomimetic and biological membranes by cryogenic electron microscopy, *Proc. Natl. Acad. Sci. U.S.A.*, 2020, **117**(33), 19943–19952.

Faraday Discussions

ROYAL SOCIETY OF CHEMISTRY

PAPER

Measuring the mechanical properties of asymmetric membranes in computer simulations – new methods and insights

Oded Farago

Received 25th November 2024, Accepted 3rd January 2025
DOI: 10.1039/d4fd00182f

We present Monte Carlo simulations of an ultra coarse-grained lipid bilayer with different numbers of lipids on both leaflets. In the simulations, we employ a new method for measuring the elastic parameters of the membrane, including the area per lipid, area elasticity modulus, and bending rigidity. The method also allows measurement of the spontaneous curvature and non-local bending modulus, which are not accessible by standard computer simulations with periodic boundary conditions. For membranes with lipid densities much smaller than the liquid to gel transition density, ρ_g, we find a very good agreement between the simulation results and the theory expressing the bilayer elastic free energy as the sum of quadratic free energies in the strains associated with the area density and the local curvature of the monolayers. The theory fails when the lipid area density (in the symmetric reference case) is only slightly smaller than ρ_g. Increasing the degree of asymmetry and changing the density of the condensed leaflet to a value larger than ρ_g, causes the layer to phase separate between regions with distinct densities which, in turn, may also induce density variations in the dilated liquid layer. Moreover, the phase separation may also trigger local curvature variations along the membrane, which can be attributed to the disparity between the values of the elastic parameters of the coexisting bilayer segments that are mechanically coupled. This mechanism leading to density–curvature variations and instabilities may play a role in cellular processes occurring in liquid-ordered raft domains that are surrounded by the disordered liquid matrix of the cell.

1 Introduction

A key feature of biological membranes is their asymmetry.[1,2] The plasma membrane (PM) of eukaryotic cells, for example, exhibits an uneven distribution of most lipid species between the extracellular and cytoplasmic leaflets, with nearly all of their sphingomyelin and phosphatidylcholine in the former, and the phosphatidylethanolamine, phosphatidylserine, and phosphatidylinositol in the

Biomedical Engineering Department, Ben Gurion University of the Negev, 84105, Beer Sheva, Israel. E-mail: ofarago@bgu.ac.il

latter.[3,4] Membrane asymmetry is critical to many biological processes like cell signaling,[5] processes involving budding (endo- and exocytosis),[6] and apoptosis.[7] While this feature of the PM had been recognized already more than 50 years ago,[8] investigations of the biophysical consequences of the asymmetric nature of biological membranes have only recently gathered momentum. This is largely due to modern tools providing detailed information about the composition and properties (*e.g.*, packing and saturation) of lipids in the two monolayers of the PM.[9] Simultaneously, we witness the development of new protocols for the preparation of synthetic vesicles with controllable asymmetry,[10] which paves the way for a new set of research investigations addressing the relationship between composition, physical properties, and biological functions.[11–13]

A bilayer membrane with different lipid composition in the two monolayers is clearly out of thermal equilibrium. Maintaining the asymmetric state requires constant energy consumption. In the PM, this is done through ATP-driven action of membrane proteins like flippases, floppases, and scramblases,[14] which maintain and control the compositional asymmetry by translocating lipids between the layers, thus countering the passive thermodynamic forces driving the system toward equilibrium and equal compositions. The rate of passive lipid transbilayer movement (flip-flop) is slow (hours to days), and this is largely attributed to the free energy barrier involved in moving the hydrophilic headgroups of the lipids through the hydrophobic bilayer core.[15] The slow rates of spontaneous flip-flops imply that active translocation by proteins is essential not only for keeping the non-equilibrium asymmetry of the PM, but also to enable remodeling processes that require faster changes in the compositional asymmetry.[16] In addition to their direct impact on the lipids, proteins contribute to bilayer asymmetry through structural and functional adaptations. The asymmetric shapes and orientations of transmembrane proteins, established during their directional insertion in the endoplasmic reticulum, not only create uneven functional domains across the bilayer but also stabilize lipid asymmetry by interacting with specific lipid species.

In order to understand the mechanical behavior of biological membranes, one clearly must take their compositional asymmetry into account. Generally speaking, the fact that the outer monolayer is packed with saturated lipids makes it more rigid than the inner leaflet, which is largely populated by unsaturated lipids and is more fluid.[17,18] This mechanical disparity between the two monolayers modifies the flexibility and stability of asymmetric membranes compared to their symmetric counterparts,[19,20] and is critical to many of the above mentioned biological processes (*e.g.* endo- and exocytosis). One "positive" consequence of the slow rates of spontaneous lipid exchange between leaflets, is the fact there is a sufficiently long time frame for investigating the mechanical behavior of asymmetric biomimetic membranes. Explicitly, if a model membrane does not contain cholesterol (which is rapidly flip-flopping), the initial asymmetric distribution of the lipids is preserved long enough to enable measurements of quantities of interest characterizing their elastic response. This is the case, for example, in atomistic simulations or coarse-grained (CG) models like MARTINI,[21,22] but not in ultra CG simulations of models like the three-bead implicit-solvent Cooke–Deserno (CD) model.[23] Such ultra CG models provide a correct quantitative description of the elastic response of lipid bilayers (bending and stretching moduli), but have unphysical large flip-flop rates. This poses no problem in simulations of symmetric membranes, but prevents simulations of asymmetric ones. To overcome this problem, the CD model was amended recently by introducing an additional hydrophobic bead, *i.e.* making it a four-bead model, which reduces the flip-flopping rates exponentially.[24]

Here, we consider the original three-bead CD model and demonstrate how it can be used for simulations of asymmetric bilayers. The key difference between our work and other simulation studies of asymmetric membranes is the fact that we run Monte Carlo (MC) rather than Molecular Dynamics simulations, which allows us to introduce rejection rules that prohibit flip-flopping. The type of asymmetry considered herein is the simplest one: we simulate a membrane consisting of a single lipid type but with different numbers of lipids in the two monolayers.[25,26] With the MC simulations, we introduce a new method for measuring the bending modulus of the simulated bilayers, and validate it by simulating symmetric membranes and comparing our results to published data which is based on other approaches, *e.g.*, Fourier analysis of thermal fluctuations. A novel feature of the method is that although the bilayer is subjected to conventional periodic boundary conditions (PBCs) and is flat on average, the method also allows measurements of the so called non-local bending modulus which is associated with the area-difference elasticity of closed vesicles[27] (see next section), and which has hardly been measured computationally. In the following section, we briefly review the elastic theory of asymmetric bilayers. The theory is based on the assumption that the two monolayers are fluid and connected in parallel without modifying each other's individual elastic response. Our results are in agreement with this theoretical framework, but only if the model parameters are set such that the symmetric bilayer is far from the gel transition. Near the gel transition, we observe that the compressed monolayer may undergo a first order phase transition and show coexisting gel and liquid regions with distinct non-vanishing curvatures of opposite sign. This type of curvature instability is expected in raft-forming biological membranes,[28] and is likely to be relevant in several biological processes.

2 Elasticity of asymmetric membranes

The elastic theory of asymmetric membranes has been discussed in several recent publications[25,26,29] and is briefly reviewed here for completeness. We begin by writing the elastic free energy of the bilayer as the sum of the elastic free energies of its constituent monolayers ($i = 1, 2$):

$$F = \sum_{i=1}^{2}\left[\frac{K_i(A_i - A_{i,0})^2}{2A_{i,0}} + \int\frac{\kappa_i}{2}(C_i - C_{i,0})^2\mathrm{d}A_i\right]. \tag{1}$$

The first term in eqn (1) represents the area elasticity of the mono-layers with K_i, A_i and $A_{i,0}$ denoting the stretching/compression (two-dimensional) modulus, area, and relaxed area of the monolayers, respectively. The second term represents the bending energy, with κ_i, C_i and $C_{i,0}$ denoting, respectively, the monolayer's bending modulus, sum of principle curvatures (twice the mean curvature – to be henceforth called "the curvature"), and the spontaneous curvature. The Gaussian curvature term has been omitted by virtue of the Gauss–Bonnet theorem. Summing the monolayers contributions in eqn (1) and expressing the energy in terms of the so called neutral surface,[30] we can write

$$F = \frac{K(A - A_0)^2}{2A_0} + \frac{K_r(\Delta A - \Delta A_0)^2}{2A_0} + \int \frac{\kappa}{2}(C - C_0)^2 \mathrm{d}A. \tag{2}$$

The first and the last terms on the r.h.s of eqn (2) represent the area and bending elasticities of the bilayers, and have similar forms to the corresponding monolayers contributions in eqn (1) with[29]

$$A_0 = \frac{K_1 + K_2}{K_1/A_{1,0} + K_2/A_{2,0}} \tag{3}$$

$$K = K_1 + K_2 \tag{4}$$

$$C_0 = \frac{\kappa_1 C_{1,0} + \kappa_2 C_{2,0}}{\kappa_1 + \kappa_2} \tag{5}$$

$$\kappa = \kappa_1 + \kappa_2 \tag{6}$$

In the middle term, $\Delta A = A_1 - A_2$ is the area difference between the monolayers, and $\Delta A_0 = A_{1,0} - A_{2,0}$ is the optimal area difference. This contribution to the energy of closed membranes has been termed area difference elasticity.[31,32] Using the parallel surface theorem,[33] we can relate the area difference to the average curvature, $\bar{C}$, defined by

$$\Delta A = h\int C\mathrm{d}A \equiv hA\bar{C} \tag{7}$$

where h is the distance between the surfaces of the two leaflets. Thus, the area difference elasticity term can be rewritten in a form resembling the bending rigidity term, *i.e.*,

$$F = \frac{K(A - A_0)^2}{2A_0} + \frac{\kappa_{\mathrm{nl}}}{2}\left(\overline{C} - \overline{C}_0\right)^2 A + \int \frac{\kappa}{2}(C - C_0)^2 \mathrm{d}A. \tag{8}$$

The modulus κ_{nl} is the non-local bending rigidity of the vesicle.[34] In contrast to the local bending rigidity κ, it is not associated with the molecular shape of the lipids and the corresponding preferred curvature, but is linked to elastic frustration arising from the fact that it may not be possible to relax the areas of the two leaflets simultaneously. From eqn (2) we conclude that a closed tensionless bilayer ($A = A_0$) may be in a state of differential stress, *i.e.*, its two leaflets may experience non-vanishing stresses of equal size and opposite signs. This can be understood by considering a surface with constant curvature $C = \bar{C}$ (*e.g.*, a sphere or a cylinder). Differentiating E in eqn (8) w.r.t. $\bar{C}$, we find that the energy is minimized when

$$C^* = \frac{\kappa C_0 + \kappa_{\mathrm{nl}}\overline{C}_0}{\kappa + \kappa_{\mathrm{nl}}}, \tag{9}$$

i.e., for $C^* \neq \bar{C}$, which implies that while the bilayer is tensionless, the monolayers are not.

Since κ_{nl} is coupled to $\bar{C}$, it can be measured in experiments involving a change in the average curvature like micropipette aspiration,[35] but not in flicker experiments analyzing the thermal fluctuation of a quasi-spherical vesicle with a given radius $R \simeq 2/\bar{C}$. The same applies to computer simulations with PBCs, where the average curvature $\bar{C} = 0$ is fixed,[36] and the fluctuation spectrum is governed by κ but insensitive to $\kappa_{\rm nl}$. In the following section, we describe a computational method for measurements of both κ and $\kappa_{\rm nl}$ in ultra CG membrane simulations of asymmetric membranes.

3 Computational methodology

The method for determining the local and non-local bending rigidities of asymmetric bilayers is based on a useful feature of MC simulations, which is the ability to reject certain moves and thereby to sample a partial configurational phase space. Specifically, we consider membrane configurations similar to the one shown in Fig. 1, consisting of two cylindrical arcs having their axes lying along the y direction – one at $x = 0$ and the other in the middle of the simulation box at $x = L_x/2$ [Fig. 1(a)]. The two arcs, which are marked by green and yellow colors in Fig. 1(b), have opposite curvatures $\pm c$ and similar spanning angles θ and, therefore, they are joined smoothly to each other along the interfaces indicated by black lines in Fig. 1(b). PBCs are applied along both the x and y directions.

The green/yellow arcs in Fig. 1(b) show the mid-plane of the membrane. Above and below the mid-plane are the two leaflets with N_+ and N_- lipids, respectively. We use the CD model (see details in ref. 23), where each lipid is represented as a trimer consisting of one hydrophilic (red) and two hydrophobic (blue) beads of diameter σ. In the simulations, we set the CD model parameter $\omega_c = 1.35$ and tune the membrane fluidity by varying the other CD model parameter, ε. The position of each lipid (r_l, θ_l, y_l) is specified by the location of the middle bead, where r_l is the radial distance from the axis of the relevant cylinder, $-\theta/2 < \theta_l < +\theta/2$ is the

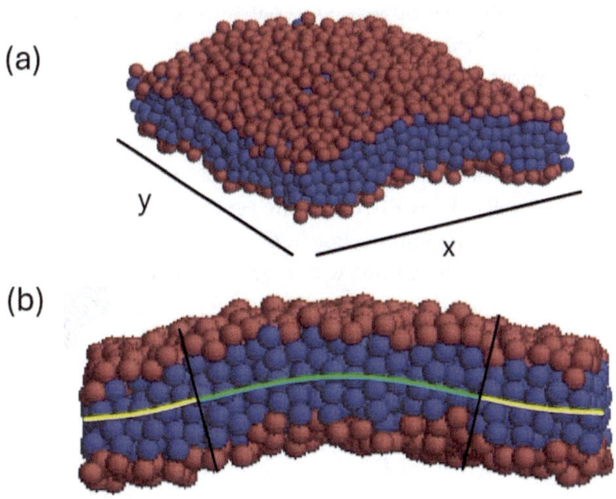

Fig. 1 Snapshot of an asymmetric bilayer membrane with different numbers of lipids in both leaflets. The membrane consists of two cylindrical arcs with opposite curvatures and the same spanning angle, such that they join smoothly to each other. (a) A view from the "corner" with y being the direction parallel to the axes of the cylinders, and x the direction along which the membrane has a "wiggly" shape. (b) A front view with the green and yellow arcs marking the mid-plane of the membrane, and the black lines indicating the interfaces along which the two arcs are joined together.

angular coordinate (with θ being the spanning angle of the arcs), and y_l is the coordinate along the cylinder axis. As noted above, the CD model "suffers" from a very high rate of flip-flops. To maintain the initial asymmetry of the bilayer, we use the following rejection rule to restrict the radial coordinate of the lipids: consider the snapshot in Fig. 1(b) with positive curvature c in the segment indicated by the green arc, and an opposite negative curvature in the other segment indicated by yellow. The radial coordinates of the lipids in the upper leaflet in the figure must satisfy $1/c < r_l < 1/c + 2.5\sigma$ in the segment with the positive curvature, and $1/c - 2.5\sigma < r_l < 1/c$ in the segment with negative curvature. This restriction prevents not only flip-flopping of lipids between the layers, but also their escape out of the membrane and, thus, guarantees the integrity of the membrane over long simulations.

We simulate tensionless membranes with $2N_0 = N_+ + N_- = 1200$ lipids in a simulation box whose length along the y axis is fixed to $L_y = 28.25\sigma$. The simulations involve displacements and rotations of randomly selected lipids, as well as occasional "collective" moves attempting changes in (i) the spanning angle θ, and (ii) the curvature of the two membrane segments $\pm c$ (in which case we change the angle so that $\theta|c|$ remains unchanged, or otherwise the area of the bilayer may change significantly and the acceptance will be nearly-vanishing). All moves are accepted according to the Metropolis criterion with the addition of the above-mentioned rejection rule for the locations of the lipids. In most of the simulations, the lipids are allowed to move between the two segments with the opposite curvatures. If a lipid is displaced from one segment into the other (and the move is accepted), then it is simply counted as part of the same layer but in the new segment. When both layers are liquid, the lipids diffuse between the two segments and establish a uniform density throughout each layer. In the asymmetric case, the densities will be different between the two layers, *i.e.*, the tensionless bilayer will have opposite stresses in the two monolayers. However, according to eqn (8), the differential stress is not expected to modify the mechanical behavior of the simulated membrane because all the sampled configurations have, by construction, $\bar{C} = 0$ and, therefore, the non-local bending term is a constant. The importance of the non-local bending term is expected to be negligible also for vesicles with radius $R = 1/\bar{C} \gg h$, where $2h$ [see definition of h in eqn (7)] is the molecular width of the bilayer ($h \sim 2$ nm), as long as they retain their quasi-spherical shape and do not bud. For the simulated phase space, the elastic free energy of configurations with a given area A and curvatures $\pm c$ in the two segments is

$$\begin{aligned} F(A,c) &= \frac{K(A-A_0)^2}{2A_0} + \frac{\kappa}{2}\left[(c-C_0)^2 + (-c-C_0)^2\right]\frac{A}{2} \\ &\simeq \frac{K(A-A_0)^2}{2A_0} + A_0\frac{\kappa}{2}c^2 + A_0\frac{\kappa}{2}{C_0}^2, \end{aligned} \tag{10}$$

and the last term is obviously an unimportant constant that can be discarded. This result implies that the area and curvature are independent quadratic degrees of freedom since there is no coupling between A and c in eqn (10), which reads $F(A, c) = F_1(A) + F_2(c)$ with

$$F_1(A) = \frac{K(A - A_0)^2}{2A_0} + E_1, \tag{11}$$

$$F_2(c) = A_0 \frac{\kappa}{2} c^2 + E_2. \tag{12}$$

This means that the elastic moduli of the bilayer can be measured by using the equipartition theorem: $\langle (A - A_0)^2 \rangle = A_0(k_BT/K)$, where $A_0 = \langle A \rangle$, and $\langle c^2 \rangle = (A_0)^{-1}(k_BT/\kappa)$. However, we can do better than that, and extract from the simulations the whole probability distributions of the area and curvature, $\Pi(A)$ and $\Psi(c)$, which if found to be Gaussian would prove the validity of eqn (11) and (12). The elastic moduli can then be determined by fitting these equations for the free energies to the logarithms of the probability distributions:

$$F_1(A) = -k_BT \ln[\Pi(A)] \tag{13}$$

$$F_2(c) = -k_BT \ln[\Psi(c)]. \tag{14}$$

To validate the method, we simulated the symmetric case $N_+ = N_- = N_0$ for the CD model parameter $\varepsilon = 1.05k_BT$. For this value of ε, the bilayer is fluid and very flexible, and in previous simulations of the same system we found $\kappa = 8 \pm 1k_BT$ based on Fourier analysis of the spectrum of thermal fluctuations.[37] This value of κ is characteristic of red blood cells.[38] Fig. (2) shows the MC simulations results for the logarithms of the probability densities of the (a) area – $\Pi(A)$ and (b) curvature – $\Psi(c)$ (black symbols), and their fits to free energies $F_1(A)$ and $F_2(c)$ (red dashed lines). The data fit very nicely to the quadratic forms of eqn (11) and (12), with some non-quadratic corrections visible away from the minima of the free energies. These corrections at large strains are expected because of the excluded volume interactions causing the response to compression to be stiffer than the response to stretching. From the fits at small strains, we derive the following values for the equilibrium area and the stretching/compression modulus: $A_0 = 804\sigma^2$ (area per lipid $a_0 = A_0/N_0 = 1.340\sigma^2$), $K = (17.2 \pm 0.2)k_BT/\sigma^2$. For the bending rigidity we find $\kappa = (7.0 \pm 0.5)k_BT$, consistent with the measurement in ref. 37.

We can also measure the non-local bending modulus of the membrane. This is done by introducing an additional rejection rule prohibiting displacement of lipids between the two membrane segments. By initially placing the same number of lipids N_+ in the upper layer of the segment with positive curvature c, and in the opposite lower layer of the segment with the opposite negative curvature [indicated, respectively, by green and yellow arcs in Fig. 1(b)], we create a bilayer composed of two replicas of cylindrical arcs having the same asymmetry. In this configuration phase space, the segments can relax individually into a state where the stresses in both monolayers vanish, *i.e.*, with no differential stress. In the symmetric case ($N_+ = N_0$), the bilayer will be on average flat, but in the asymmetric case ($N_+ > N_0$), the state of mechanical equilibrium will have a non-vanishing spontaneous curvature. Furthermore, curvature fluctuations around this value will be attenuated by both the local and non-local bending rigidities because they also involve changes in the area difference ΔA in each of the two segments. The free energy in this set of simulations is expected to have the form $F(A, c) = F_1(A) + F_2^{\text{nl}}$ (the superscript "nl" indicates that this free energy also involves a non-local

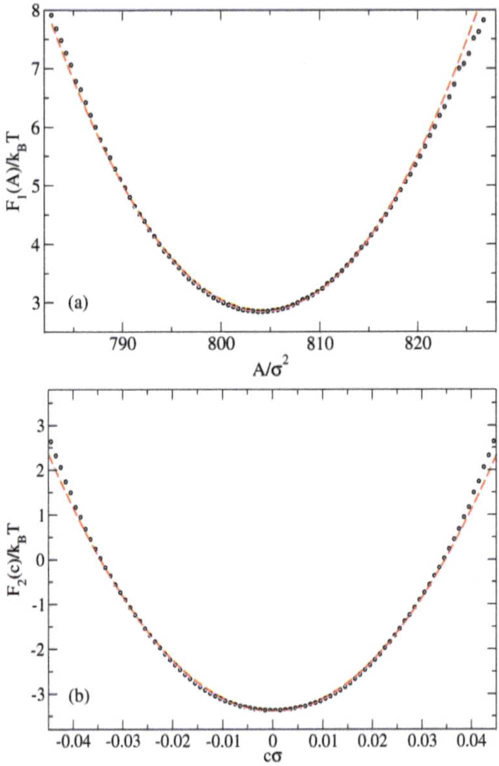

Fig. 2 The logarithms of the probability distribution functions of the (a) area A, and (b) curvature, of a symmetric membrane. Black circles – computational results; red dashed lines – fits of the results to the quadratic forms of the free energies eqn (11) and (12), respectively.

contribution) with $F_1(A)$ having the same form as in eqn (11), and replacing eqn (12) with

$$F_2^{\rm nl}(c) = A_0\frac{(\kappa + \kappa_{\rm nl})}{2}(c - C^*)^2 + E_2^{\rm nl}, \quad (15)$$

where c is the curvature of the segment with N_+ lipids in the upper layer, and C^* is the spontaneous curvature given by eqn (9). Fig. 3 displays the analysis of the results for this set of simulations, similar to Fig. 2 but with the curvature free energy fitted to eqn (15). We obtain exactly the same values $A_0 = 804\sigma^2$ and $K = (17.2 \pm 0.2)k_BT/\sigma^2$ for the area parameters, and a total bending rigidity of $\kappa_{\rm tot} = \kappa + \kappa_{\rm nl} = (32.4 \pm 0.3)k_BT$. Since κ is already known for this system, we conclude that $\kappa_{\rm nl} = \kappa_{\rm tot} - \kappa \simeq (25.4 \pm 0.4)k_BT$.

4 Asymmetric fluid membranes

Fig. 4 shows the variations in the mechanical properties of asymmetric membranes with N_+ lipids in the upper layer and $N_- \leq N_+$ in the lower layer, as a function of the asymmetry parameter defined as $\delta = (N_+ - N_-)/2N_0$. The simulation results shown here correspond to bilayers with CD parameter $\varepsilon =$

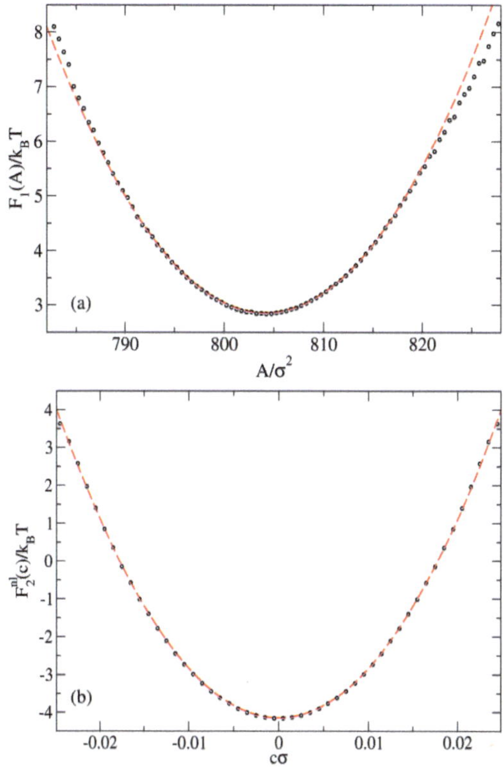

Fig. 3 Same as in Fig. 2, but for a symmetric membrane also having non-local bending rigidity. The simulation data (symbols) is fitted to the free energies eqn (11) for the area in (a), and (15) for the curvature in (b).

$1.05k_BT$, which are fluid and flexible, and which were simulated in the setting without non-local bending rigidity (see previous Section 3). Sub-figures (a), (b), and (c) present the results for the relaxed area per lipid a_0, bilayer area elasticity modulus K, and bilayer bending modulus κ, respectively. The agreement with the theoretical framework in Section 2 is very good. Explicitly, eqn (3), (4) and (6) suggest that if the monolayers properties vary linearly with δ in opposite manners to each other [*i.e.*, $\sim B_0 \pm B_1\delta + \mathcal{O}(\delta^2)$], then the corresponding bilayer properties should have only a weak δ^2 dependence on the asymmetry parameter. Indeed, Fig. 4(c) shows no significant variations in the bending modulus. The area elasticity modulus and the relaxed area exhibit a slight linear dependence on δ, which can be attributed to the weak non-linear elastic response, also observed at larger areal strains in Fig. 2(a).

Fig. 5 and 6 show the results for the same membrane (with $\varepsilon = 1.05k_BT$) but in the setting with non-local bending rigidity. The area elasticity parameters, A_0 and K, are plotted in Fig. 5(a) and (b), respectively. As expected, the relaxed area does not vary with δ, and is identical to the relaxed area derived in Fig. 4 for $\delta = 0$, *i.e.*, for the membrane with no differential stress. Moreover, according to the theory in Section 2, non-local rigidity (area-difference elasticity) represents an elastic deformation mode which is distinct from the area-elasticity and bending rigidity

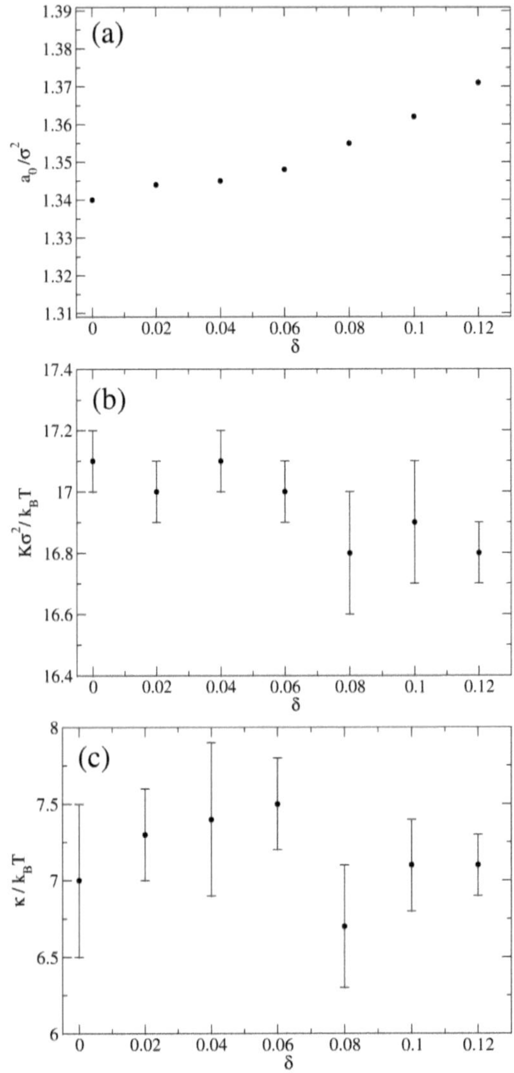

Fig. 4 (a) The relaxed area per lipid $a_0 = A_0/N_0$, (b) area elasticity modulus K, and (c) bending modulus κ, as a function of the asymmetry parameter $\delta = (N_+ - N_-)/2N_0$. The simulated bilayers have CD parameter $\varepsilon = 1.05k_BT$, and the measurements are conducted in the setting without non-local bending rigidity.

and, therefore, the results for the area elasticity modulus K in Fig. 5(b) should be similar to those displayed in Fig. 4(b) when the non-local rigidity is absent. Comparing the data in these two figures confirms that the measurements for K in both settings are in fairly good agreement with each other. Fig. 6(a) shows the total bending rigidity $\kappa_{tot} = \kappa + \kappa_{nl}$. The dependence of this elastic on δ is weak, suggesting that both κ and κ_{nl} vary only slightly with the degree of asymmetry. Fig. 6(b) shows the dependence of the spontaneous curvature C^* on δ. The displayed linear dependence is fully consistent with eqn (9), since C^* is proportional

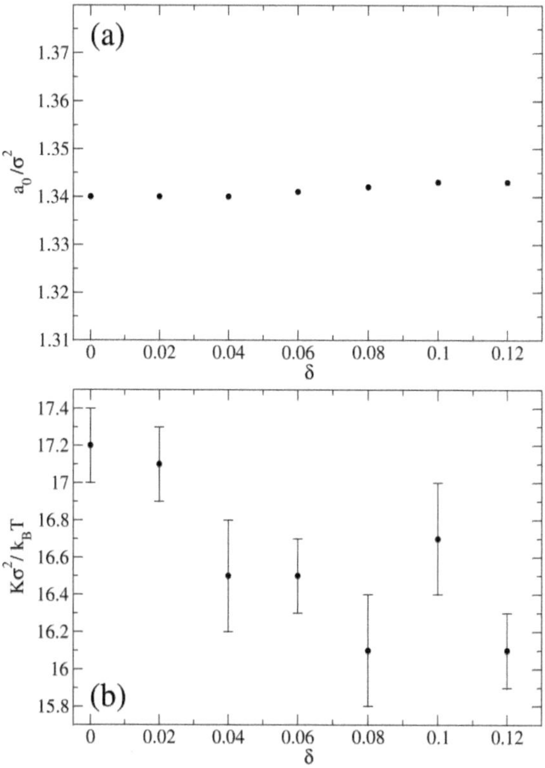

Fig. 5 (a and b) Same as in Fig. 4, but for the simulations setting with non-local bending rigidity.

to the optimal area difference between the monolayers: $C^* \propto \Delta A_0 \propto \delta$ (provided that $C_0 = 0$ for $\delta = 0$).

5 Membranes near a gel transition

Next, we consider the CD model with $\varepsilon = 1.3k_BT$, which in the symmetric case is located close to the liquid to gel transition of the membrane. The lipid diffusion coefficient for $\varepsilon = 1.3k_BT$ is almost an order of magnitude smaller than its corresponding value for the fluid bilayer with $\varepsilon = 1.05k_BT$, discussed in Section 4. We first consider the setting involving non-local rigidity, where we simulate two separate membrane segments that supposedly relax to a state without differential stress. The results from this set of simulations are summarized in Fig. 7. The elastic behavior displayed in this figure is quite similar to the one observed in Fig. 5 and 6 for the fluid membranes with the smaller value of $\varepsilon = 1.05k_BT$. The relaxed area, area elasticity modulus, and total bending rigidity [sub-figures Fig. 7(a–c), respectively] show weak dependence on δ, which should be largely related to non-quadratic corrections to the free energy eqn (8), while the spontaneous curvature [sub-Fig. 7(d)] exhibits a clear linear increase with δ.

To capture the behavior of a single bilayer, we turn to the set of MC simulations that allow transitions of lipids between the oppositely-curved segments, *i.e.*, the

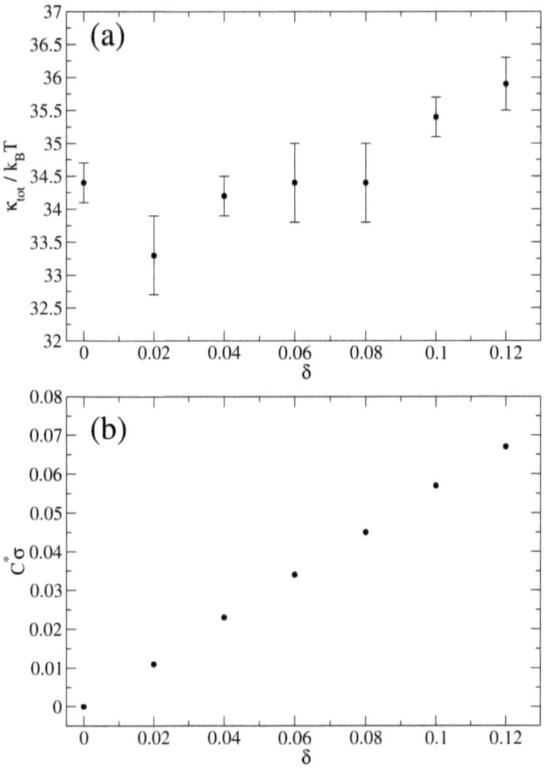

Fig. 6 (a) Total bending rigidity, and (b) spontaneous curvature as a function of the asymmetry parameter δ. The results are obtained in the same set of simulations as in Fig. 5 (*i.e.*, for the CD parameter $\varepsilon = 1.05k_BT$ and in a setting with non-local bending rigidity).

setting without non-local rigidity. These simulations reveal a very distinct elastic behavior. Specifically, significant deviations from the description outlined by the elasticity theory in Section 2 become noticeable for $\delta \geq 0.06$. This is demonstrated, for instance, in Fig. 8(a) showing the area free energy $F_1(A)$ obtained from the area distribution function $\Pi(A)$ *via* eqn (13). The black, red, and green symbols in Fig. 8(a) show $F_1(A)$ for $\delta = 0.06$, 0.08, and 0.10, respectively. The computed free energies exhibit strong deviations from the quadratic form of eqn (11). Two particularly interesting features emerge from the data: (i) the minimum of the free energy, corresponding to the relaxed area A_0, is reduced when the asymmetry parameter δ increases from 0.06 to 0.08, and (ii) there is a significant probability [associated with a relatively low free energy difference $F_1(A) - F_1(A_0)$] to find the membrane with an area $A < A_0$. These results suggest that the upper (compressed) monolayer may be undergoing a liquid to gel transition.[39,40] A strong indication for such a possible phase transition is the rapid decrease in the lipid diffusion coefficient measured in the upper layer, but not in the lower (dilated) layer, when δ is varied from 0.06 to 0.08 (*i.e.*, concurrently with the decrease in A_0). An additional striking observation that points to the phase change comes from the curvature probability distribution, $\Psi(c)$, and the associated free energy, $F_2(c)$, which is plotted in Fig. 8(b). The results reveal that the phase transition occurring

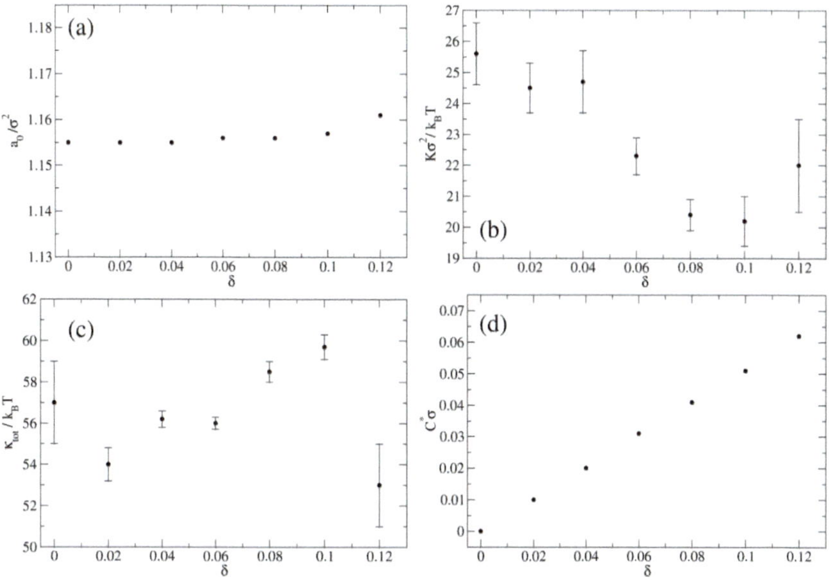

Fig. 7 (a) The relaxed area per lipid $a_0 = A_0/N_0$, (b) area elasticity modulus K, (c) total bending rigidity κ_{tot}, and (d) spontaneous curvature C^*, as a function of the asymmetry parameter δ. The simulated bilayers have CD parameter $\varepsilon = 1.3k_BT$, and the measurements are conducted in the setting with non-local bending rigidity.

at δ values between 0.06 and 0.08 involves not only a sudden change in the relaxed area A_0, but also a curvature instability. The bilayer state of mechanical equilibrium changes from flat to a state where the two segments have non-vanishing curvatures [which, by the construction of the configuration phase space in our simulations, are opposite to each other – see Fig. 1(b)]. From simple symmetry considerations it is clear that such a "buckled" state cannot be the equilibrium of a membrane with similar densities across each layer, and we readily conclude that the condensed layer does not gel uniformly, and the lipid densities at two oppositely-curved segments must be different.

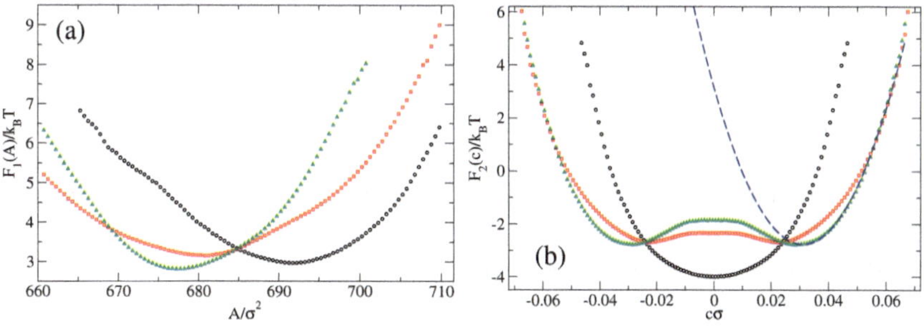

Fig. 8 The logarithms of the probability distribution functions of the (a) area A, and (b) curvature c, for membranes with CD parameter $\varepsilon = 1.3k_BT$ in the setting without non-local rigidity. Black circles, red squares, and green triangles show data for $\delta = 0.06$, 0.08, and 0.1, respectively. The blue dashed line in (b) is a fit of the green symbols to eqn (22).

To write the bilayer elastic free energy, we (i) express the elastic free energies of the phase-separated segments as the sum of quadratic area and curvature elasticities of their monolayers [compare with eqn (1)], then (ii) sum up the segments contributions:

$$F = \sum_{\alpha=1,2}\sum_{i=1,2}\left[\frac{K_i^\alpha\left(A_i^\alpha - A_{i,0}^\alpha\right)^2}{2A_{i,0}^\alpha} + A_i^\alpha\frac{\kappa_i^\alpha\left(c_i^\alpha - C_{i,0}^\alpha\right)^2}{2}\right], \tag{16}$$

where the superscript $\alpha = 1, 2$ refers to, respectively, the segments with positive and negative curvatures, and the subscript $i = 1, 2$ corresponds to the upper and lower monolayers respectively. To express the free energy (16) in terms of bilayer area A and the curvatures of the segments $\pm c$, we must use the following geometric relations:

$$A_1^1 = A_2^2 = \frac{A(1 + hc/2)}{2}, \tag{17}$$

$$A_2^1 = A_1^2 = \frac{A(1 - hc/2)}{2}, \tag{18}$$

where h is the distance between the monolayers [eqn (7)]. Further assuming that $h \ll 1/c$, permits to replace $c_1^1 \simeq \frac{1}{2} \simeq c$, and $c_1^2 \simeq c_2^2 \simeq -c$. To find the state of mechanical equilibrium of the tensionless bilayer, we must then minimize the free energy w.r.t. A and c [which are now coupled through eqn (17) and (18)], subject to the constraints that monolayer stresses are equal in both segments and opposite to each other, *i.e.*,

$$K_1^1(A_1^1 - A_{1,0}^1) = K_1^2(A_1^2 - A_{1,0}^2) = K_2^1(A_{2,0}^1 - A_2^1) = K_2^2(A_{2,0}^2 - A_2^2) \tag{19}$$

Also, we must take into account the fact that the lipids can move between the segments (despite their low diffusivity), which imposes the following relationship between the relaxed areas:

$$A_{1,0}^1/a_{1,0}^1 + A_{1,0}^2/a_{1,0}^2 = N_+ \tag{20}$$

$$A_{2,0}^1/a_{2,0}^1 + A_{2,0}^2/a_{2,0}^2 = N_- \tag{21}$$

where $a_{i,0}^\alpha$ is the equilibrium area per lipid of the phase in the monolayer i of segment α.

Due to the coupling between bilayer area A and curvatures $\pm c$, the elastic free energy (16) is *not* equal to the sum of the free energies displayed in Fig. 8: $F(A, c) \neq F_1(A) + F_2(c)$. Instead, we should regard $F_1(A)$ and $F_2(c)$ as operational free energies from which the effective bilayer moduli K and κ can be evaluated. Recent experiments found a considerable increase in the measured bending rigidity of substantially-asymmetric vesicles compared to their symmetric and weakly-asymmetric counterparts.[41,42] A sudden increase in κ was also measured in MARTINI simulations of membranes with $N_+ \neq N_-$.[26] It was speculated that this observation may be related to the gel transition of the compressed layer and the reasonable expectation that the bilayer then becomes stiffer than in the liquid state. To look for this phenomenon in our simulation results, we must first discuss how κ should be measured from the results in Fig. 8(b) for δ larger than the transition value. In this regime, the bending free energy $F_2(c)$ has two minima at $c = \pm C_0$, and κ is measured by fitting $F_2(c)$ to

$$F_2(c) = A_0 \frac{\kappa}{2}(c - C_0)^2 + E_2, \tag{22}$$

close to one of the minima. Furthermore, one must measure κ only along the stable $|c| > C_0$ branch of $F_2(c)$, and avoid the unstable branch $|c| < C_0$ representing flat transition states between the two equivalent sub-spaces of configurations obtained by reflection of the membrane w.r.t. the vertical z-direction. Such a fit to the results for $\delta = 0.1$ is shown by the blue dashed curve in Fig. 8(b). Fig. 9 summarizes our measurements of κ *vs.* δ. We find that $\kappa \sim 12.5k_BT$ for $\delta < 0.06$, and $\kappa \sim 18k_BT$ for $\delta > 0.08$. In the transition regime, the data does not fit well to eqn (12) (for $\delta = 0.06$), or eqn (22) (for $\delta = 0.08$), and therefore κ cannot be evaluated.

One must keep in mind that the results plotted in Fig. 9 do not show the bending rigidities of neither of the membrane segments, but the effective κ of the whole membrane which is, roughly, the average of the segments' rigidities. To take a closer look at the differences between the segments, we measured the average area densities, ρ_i^α, in the different monolayers of the two segments. Similarly to eqn (16), we use the superscript $\alpha = 1, 2$ to denote the segments with positive and negative curvatures respectively, and the subscript $i = 1, 2$ to refer to the upper and lower monolayers respectively. The results are summarized in Fig. 10. For $\delta < 0.06$, the densities in the condensed layer grow linearly with δ. The densities decrease, by the same amount, in the dilated layer. These exactly opposite variations in the densities imply that the monolayers are oppositely strained, and that the strain is linear in δ. Since the monolayers' elastic free energy is quadratic [see eqn (1)], the stress–strain relation of the monolayers in the small-δ regime is linear, which implies the monolayers are oppositely-stressed.

For $\delta > 0.06$, the data reveals a difference in the lipid densities at the upper monolayers of the positively (ρ_1^1) and negatively (ρ_1^2) curved segments. The latter continues with the linear dependence on δ, characteristic of liquid membranes,

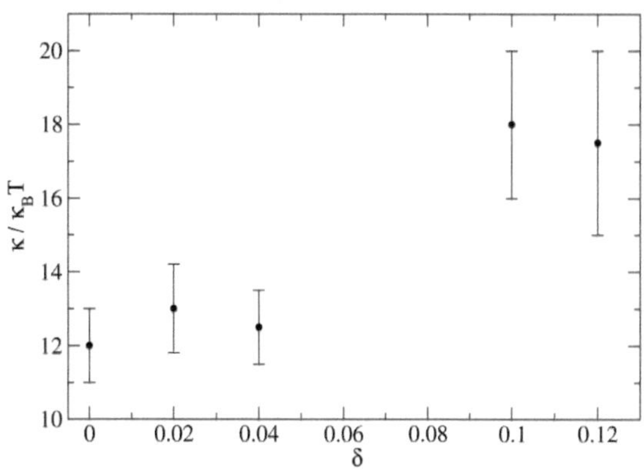

Fig. 9 The bending rigidity κ, and as a function of the asymmetry parameter δ. The simulated bilayers have CD parameter $\varepsilon = 1.3k_BT$, and the measurements are conducted in the setting without non-local bending rigidity.

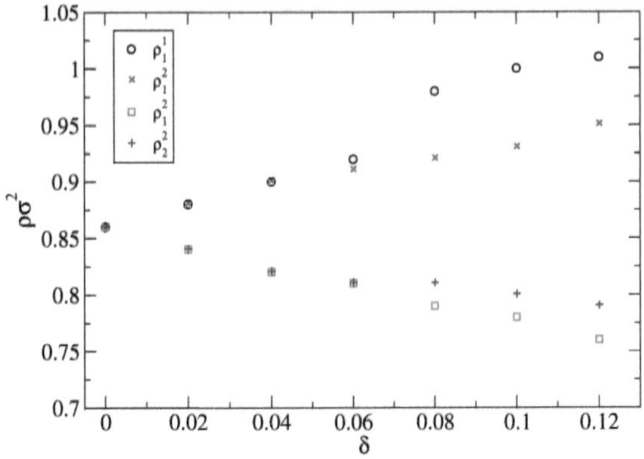

Fig. 10 The average area density, ρ_i^α, in the upper and lower layers (subscript $i = 1, 2$, respectively) of the segment with positive and negative curvature (superscript $\alpha = 1, 2$, respectively).

while the former exhibits a discontinuity which is a signature of a gel transition. Quite remarkably, we observe that the phase separation occurring at the condensed layer, induces a density difference between the oppositely-curved segments of the dilated layer as well. This is surprising because the dilated layer is in a liquid state, and the uniformity of the stress in this layer suggests that the density must also be uniform, *i.e.*, the same on both segments. A possible explanation for this seemingly odd observation is the coupling between monolayers, which has been completely ignored thus far, both in Sections 2 and 5. Explicitly, lipids in each layer also interact with lipids in the opposite layer and, therefore, a phase separation in one layer may indeed lead to density variations in the other. Possibly-related effects originating in cross-layer interactions may be encountered in raft-forming biological membranes,[17] but this discussion is beyond the scope of this paper.

6 Discussion and conclusions

The availability of simulation packages, like LAMMPS and GROMACS, enabling simple design and execution of Molecular Dynamics simulations, have contributed tremendously to the popularity of this computational method in many areas of molecular science and engineering. Regrettably, this has led to a decline in the usage of the alternative approach to molecular simulations, *i.e.*, MC sampling, which has some unique capabilities. In this paper, we presented a new method for measuring the mechanical properties of an ultra CG bilayer membrane model with different lipid area densities in the two monolayers. The method exploits a useful feature of MC dynamics, which is to make statistical sampling of a partial configuration phase space. This is achieved by supplementing the Metropolis criterion for importance sampling with certain rejection rules. In the present study, these rules ensure that the initial asymmetry of the bilayer is preserved during the course of simulations. Partial statistical sampling provides us with the route to measure the area, K, and bending, κ, elastic moduli, as well as the non-local bending modulus κ_{nl} that cannot be measured in full statistical sampling with PBCs.

Starting from simulations of symmetric membranes with the same number of lipids in the upper (N_+) and lower (N_-) leaflets, we investigated how their mechanical behavior changes with the asymmetry parameter $\delta = (N_+ - N_-)/(N_+ + N_-)$. We simulated membranes under two different conditions – one when the lipid area density in the symmetric case is much smaller than the density at the liquid to gel transition, ρ_g, and the other when it is only slightly smaller than ρ_g. In the former case, we find excellent agreement between the computational results and the elasticity theory for asymmetric liquid membranes, briefly reviewed in Section 2. The theory makes two simple assumptions: (i) the bilayer elastic free energy is the sum of the monolayers' contributions; (ii) for each monolayer, the elastic free energy is written as the sum of area and bending quadratic elastic terms. From (i) and (ii) follows that the bilayer elastic free energy can be also written as the sum of quadratic terms in the strains associated with area density and curvature, with an additional non-local bending rigidity (or, equivalently, area-difference elasticity) term associated with the elastic frustration arising from the fact that it may not be possible to relax the areas of the two leaflets simultaneously. Non-local bending rigidity, however, has null to negligible impact on the mechanics of weakly fluctuating vesicles. This is the reason why κ_{nl} cannot be extracted from the spectrum of thermal fluctuations, and is only important (and measurable) in processes like fusion and budding involving significant changes in the geometry of the vesicle. One of the outcomes of the theory is the well known fact that the monolayers of a tensionless bilayer may not be tensionless themselves, but experience stresses that are similar in size and opposite in sign. Small deviations of the computational measurements from the predictions of the linear elasticity theory have been observed at high values of δ, which can be attributed to corrections to the quadratic elastic free energies that are expected when the stresses in the monolayers grow.

When the density of a symmetric membrane is only slightly smaller than ρ_g, an increase in the degree of asymmetry, δ, may cause the density of the compressed layer to exceed ρ_g. In contrast to symmetric membranes where both layers undergo a first order gel transition, in the substantially-asymmetric case only the compressed monolayer gels, and only in one of the two segments of our simulation membranes (*i.e.*, at coexistence with a lower density in the other segment). Furthermore, the coexistence of regions with different densities in the compressed layer may also induce correlated density variations in the dilated layer. The two bilayer segments with different monolayer densities are mechanically coupled – the stresses in the upper layers of the segments must be identical, and (for a tensionless bilayer) be opposite to the stresses in their lower layers. The segments' elasticities are characterized by different values of the elastic parameters, a_0 (area per lipid), K (area elasticity modulus), κ (bending modulus), and C_0 (spontaneous curvature). The consequence of the mechanical coupling between the segments is a buckling instability into a new state of mechanical equilibrium, which is not flat as in the symmetric case. A theory for the elasticity of such inhomogeneous membranes is outlined in Section 5. Rather than attempting to minimize the elastic free energy of the asymmetric inhomogeneous membrane, we present an operational definition for the effective bending rigidity of the bilayer. Measurements of the effective κ as a function of δ reveal that its value increases abruptly at the gelling–buckling phase transition. A rapid increase in the value of κ has indeed been observed in experimental[41,42] and computational[26] studies, and our study confirms that it is related to the stiffening of the membrane when it becomes partially gel-like.

The bilayer membranes considered in this work are composed of a single type of lipid with different numbers of lipids in their leaflets. This type of asymmetry is far simpler than the compositional asymmetry of complex biological membranes that contain many different lipid species, and whose distribution within- and between-layers varies in time. Moreover, the bilayer in our simulations are also confined to have two segments of equal area with opposite curvatures. Yet, despite its simplicity, this model captures the mechanism of density-buckling instabilities originating from the mechanical coupling between the coexisting segments. We can now envision how this mechanism may be biologically important. Many cellular processes are initiated in liquid-ordered raft domains and, moreover, their progression involves variations in the compositions in the raft segments.[43] These density variations play a role not only in the biochemistry of cellular processes, but also in the biomechanics of correlated local curvature changes.[44,45] In the present model, the instability is driven by the liquid-to-gel first order transition and is, therefore, discontinuous in δ. However, the change in the density and the coupled curvature variations may be continuous. The mechanism of density–curvature instability depends on the dissimilarity of the elastic parameters of different membrane regions. It is not limited to coexisting liquid and gel phases, but is also expected to affect the curvature of the liquid-ordered raft domains that are surrounded by the disordered liquid matrix of the cell.[46] Experiments with large unilamellar vesicles composed of mixtures of saturated and unsaturated lipids with cholesterol exhibit a rich phase diagram,[47] including also second-order[48] and possibly other phase transitions involving macro- or micro-phase separations,[49] and even transitions from a macro- to micro-separation.[50,51] Several theoretical frameworks involving similar mechanisms of curvature–density coupling have been proposed to explain the puzzle of micro-phase separation in lipid mixtures,[52,53] which seems contradictory to the notion of thermodynamic phase separation. The molecular composition of biological membranes is far more complex, heterogeneous, and dynamic than of synthetic model mixtures. In such complex membranes, local density–curvature coupling can lead to many different curvature effects, some of which with relevance to specific biological processes. A major difference between large complex membranes and the simple CG model investigated herein, is the fact that the later considers a bilayer with an identically vanishing average curvature with only two segments of opposite curvatures. The average curvature of the plasma membrane is not fixed, but its temporal variations are small because the plasma membrane includes multiple domains, whose sizes are much smaller then the cell size. Thus, on the cell-scale, the non-local curvature elasticity is unimportant, and its influence is limited only to the local and transient curvature variations.

While this paper focuses on the connections between the lipidic heterogeneity and membrane asymmetry, it is important to remember that biological membranes are populated with embedded proteins and proteins that are associated with one of the leaflets. These proteins contribute to the membrane asymmetry in various ways. A prime example are ATP-dependent transporters, such as flippases and floppases, whose role in regulating lipid asymmetry has been already mentioned at the introduction of this manuscript.[14,15] In addition to their direct impact on lipids, proteins also contribute to asymmetry through structural and functional adaptations.[54] The asymmetric shapes and orientations of transmembrane proteins, established during their directional insertion in the endoplasmic reticulum, not only create uneven functional domains across the bilayer, but also stabilize lipid asymmetry by interacting with specific lipid species.[9,55] Peripheral proteins further amplify these effects by selectively binding to lipids, influencing membrane curvature and enhancing structural integrity.[1] Additionally, the attachment of the inner leaflet to the cytoskeleton and the outer leaflet's interactions with extracellular matrix proteins or ligands significantly influence the asymmetric distribution of membrane components and the membrane's spontaneous curvature.[4,56] These examples illustrate the complex interplay between lipids and proteins in achieving a stable yet adaptable asymmetric membrane architecture, underscoring the importance of also considering protein contributions when studying membrane asymmetry. We expect the growing research interest in asymmetric membranes to yield new exciting insights into the properties of such membranes, which might shed new light into the relationship between the asymmetric nature of biological membranes and their cellular functions.

Data availability

Atomic coordinates for Fig. 1 are available at https://tzin.bgu.ac.il/∼ofarago/files/config.pdb. Use "rasmol -noconnect config.pdb" to open file.

Conflicts of interest

There are no conflicts to declare.

Acknowledgements

This work was supported by the Israel Science Foundation (ISF), Grant No. 1258/22.

Notes and references

1 G. Pabst and S. Keller, *Trends Biochem. Sci.*, 2024, **49**, 333–345.
2 M. Bogdanov, *Emerging Top. Life Sci.*, 2023, **7**, 1–6.
3 A. Verkleij, R. Zwaal, B. Roelofsen, P. Comfurius, D. Kastelijn and L. van Deenen, *Biochim. Biophys. Acta, Biomembr.*, 1973, **323**, 178–193.
4 T. Fujimoto and I. Parmryd, *Front. Cell Dev. Biol.*, 2017, **4**(1–11), 155.
5 G. van Meer, *EMBO J.*, 2005, **24**, 3159–3165.
6 A. A. Mironov, A. Mironov, J. Derganc and G. V. Beznoussenko, *Int. J. Mol. Sci.*, 2020, **21**(20), 7594.
7 R. G. Hanshaw and B. D. Smith, *Bioorg. Med. Chem.*, 2005, **13**, 5035–5042.
8 M. S. Bretscher, *Nat. N. Biol.*, 1972, **236**, 11–12.
9 J. H. Lorent, K. R. Levental, L. Ganesan, G. Rivera-Longsworth, E. Sezgin, M. Doktorova, E. Lyman and I. Levental, *Nat. Chem. Biol.*, 2020, **16**, 644–652.
10 M. Krompers and H. Heerklotz, *Membranes*, 2023, **13**(3), 267.
11 E. London, *Acc. Chem. Res.*, 2019, **52**, 2382–2391.
12 M. Doktorova, J. Symons and I. Levental, *Nat. Chem. Biol.*, 2020, **16**, 1321–1330.

13 M. Deserno, *Curr. Opin. Struct. Biol.*, 2024, **87**, 102832.
14 F. X. Contreras, L. Sanchez-Magraner, A. Alonso and F. M. Goni, *FEBS Lett.*, 2010, **584**, 1779–1786.
15 T. G. Pomorski and A. K. Menon, *Prog. Lipid Res.*, 2016, **64**, 69–84.
16 S. Shukla and T. Baumgart, *Biochim. Biophys. Acta, Biomembr.*, 2021, **1863**, 183534.
17 P. Devaux and R. Morris, *Traffic*, 2004, **5**, 241–246.
18 J. C. M. Holthuis, G. van Meer and K. Huitema, *Mol. Membr. Biol.*, 2003, **20**, 231–241.
19 S. L. Foley, M. Varma, A. Hossein and M. Deserno, *Emerging Top. Life Sci.*, 2023, **7**, 95–110.
20 A. Sreekumari and R. Lipowsky, *Soft Matter*, 2022, **18**, 6066–6078.
21 M. Blumer, S. Harris, M. Li, L. Martinez, M. Untereiner, P. N. Saeta, T. S. Carpenter, H. I. Ingolfsson and W. F. D. Bennett, *Front. Cell Dev. Biol.*, 2020, **8**, 575.
22 E. H. Chaisson, F. A. Heberle and M. Doktorova, *Membranes*, 2023, **13**(7), 629.
23 I. R. Cooke and M. Deserno, *J. Chem. Phys.*, 2005, **123**, 224710.
24 S. Foley and M. Deserno, *J. Chem. Theory Comput.*, 2020, **16**, 7195–7206.
25 B. Rozycki and R. Lipowsky, *J. Chem. Phys.*, 2015, **142**, 054101.
26 A. Hossein and M. Deserno, *Biophys. J.*, 2020, **118**, 624–642.
27 T. Takiue, *Curr. Opin. Colloid Interface Sci.*, 2022, **62**, 101646.
28 M. Schick, *J. Phys. Chem. B*, 2018, **122**, 3251–3258.
29 S. Svetina and B. Žekš, *Handbook of Nonmedical Applicationsof Liposomes*, CRC Press, Boca Raton, USA, 1996, pp. 13–42.
30 W. Helfrich and M. Kozlov, *J. Phys. II*, 1993, **3**, 287–292.
31 U. Seifert, *Adv. Phys.*, 1997, **46**, 13–137.
32 X. Bian, S. Litvinov and P. Koumoutsakos, *Comput. Methods Appl. Mech. Eng.*, 2020, **359**, 112758.
33 M. Deserno, *Chem. Phys. Lipids*, 2015, **185**, 11–45.
34 S. Svetina and B. Žekš, *Adv. Colloid Interface Sci.*, 2014, **208**, 189–196.
35 R. E. Waugh, J. Song, S. Svetina and B. Žekš, *Biophys. J.*, 1992, **61**, 974–982.
36 H. Wang, D. Hu and P. Zhang, *Commun. Comput. Phys.*, 2013, **13**, 1093–1106.
37 O. Farago, *J. Chem. Phys.*, 2008, **128**, 184105.
38 S. Himbert, A. D'Alessandro, S. M. Qadri, M. J. Majcher, T. Hoare, W. P. Sheffield, M. Nagao, J. F. Nagle and M. C. Rheinstädter, *PLoS One*, 2022, **17**, 1–20.
39 R. F. Vazquez, E. Ovalle-Garcia, A. Antillon, I. Ortega-Blake, L. S. Bakas, C. Munoz-Garay and S. M. Mate, *Biochim. Biophys. Acta, Biomembr.*, 2021, **1863**, 183467.
40 S. L. Foley, A. Hossein and M. Deserno, *Biophys. J.*, 2022, **121**, 2997–3009.
41 Y. Elani, S. Purushothaman, P. J. Booth, J. M. Seddon, N. J. Brooks, R. V. Law and O. Ces, *Chem. Commun.*, 2015, **51**, 6976–6979.
42 K. Karamdad, R. V. Law, J. M. Seddon, N. J. Brooks and O. Ces, *Chem. Commun.*, 2016, **52**, 5277–5280.
43 D. Sviridov and Y. I. Miller, *J. Lipid Res.*, 2020, **61**, 598–600.
44 B. Kollmitzer, P. Heftberger, M. Rappolt and G. Pabst, *Soft Matter*, 2013, **9**, 10877–10884.

45 Y.-S. Ryu, L. R. Jordan, N. J. Wittenberg, S. M. Kim, D. Yoo, C. Jeong, A. E. Warrington, M. Rodriguez, S.-H. Oh and S.-D. Lee, *Adv. Mater. Interfaces*, 2018, **5**, 1801290.
46 K. Simons and E. Ikonen, *Nature*, 1997, **387**, 569–572.
47 G. W. Feigenson, *Biochim. Biophys. Acta, Biomembr.*, 2009, **1788**, 47–52.
48 S. L. Veatch, O. Soubias, S. L. Keller and K. Gawrisch, *Proc. Natl. Acad. Sci. U. S. A.*, 2007, **104**, 17650–17655.
49 L. S. Hirst, P. Uppamoochikkal and C. Lor, *Liq. Cryst.*, 2011, **38**, 1735–1747.
50 S. Goh, J. Amazon and G. Feigenson, *Biophys. J.*, 2013, **104**, 853–862.
51 T. Sarkar and O. Farago, *Soft Matter*, 2023, **19**, 2417–2429.
52 M. Schick, *Phys. Rev. E:Stat., Nonlinear, Soft Matter Phys.*, 2012, **85**, 031902.
53 S. Meinhardt, R. L. C. Vink and F. Schmid, *Proc. Natl. Acad. Sci. U. S. A.*, 2013, **110**, 4476–4481.
54 I. D. Pogozheva, S. Tristram-Nagle, H. I. Mosberg and A. L. Lomize, *Biochim. Biophys. Acta, Biomembr.*, 2013, **1828**, 2592–2608.
55 G. J. Schütz and G. Pabst, *BioEssays*, 2023, **45**, 2300116.
56 T. Lu, C.-H. Lee and B. Anvari, *ACS Appl. Mater. Interfaces*, 2022, **14**, 18219–18232.

Faraday Discussions

ROYAL SOCIETY OF CHEMISTRY

PAPER

Membrane potential fluctuations and water asymmetry on plasma cell and model lipid membranes: origins, implications and properties

Zhi Li,† Iwona Swiderska, † Lena Dalifoski, Seonwoo Lee, Nelson Alonso Correa-Rojas, David Roesel, Maksim Eremchev, Mischa Flor, Orly B. Tarun, Arianna Marchioro and Sylvie Roke *

Received 11th December 2024, Accepted 30th January 2025

DOI: 10.1039/d4fd00197d

Membrane potential fluctuations have previously been detected using second harmonic (SH) water imaging on neuronal cells and model lipid bilayer membranes. We report that such fluctuations are also visible when membrane potential-sensitive fluorophores are used as contrast agents, and fluctuations are imaged on both free-standing lipid membranes (FLMs) and on the plasma membranes of neuroblastoma cells. We show that upon K^+ depolarization, non-uniform recovery responses occur across cells and within single cells. We discuss the origins and implications of such fluctuations, and investigate the molecular-level details of membrane potential distributions on FLMs and compare it to those on giant unilamellar vesicles (GUVs). SH water imaging shows that the hydrated part of lipid membranes is most likely composed of regions having a diffuse double layer, and other regions having an additional condensed double layer, with a high concentration of ions/ionic groups. In terms of transmembrane potential distributions, FLMs and GUVs show similar signatures, as expected from electrostatics. Comparing passive ion transport, FLMs and GUVs of identical composition behave differently, with GUVs being more permeable for proton transport (~20×). This is likely caused by differences in the hydrophobic cores of the membranes, which create different energetic barriers for the proton transport.

1. Introduction

The inner and outer leaflets of cellular and organelle membranes are compositionally diverse, which is crucial for cell function. Leaflet composition is often non-random and occurs in eukaryotic membranes.[1–4] Transmembrane

Laboratory for Fundamental BioPhotonics (LBP), Institute of Bioengineering (IBI), and Institute of Materials Science (IMX), School of Engineering (STI), and Lausanne Centre for Ultrafast Science (LACUS), École Polytechnique Fédérale de Lausanne (EPFL), CH-1015, Lausanne, Switzerland. E-mail: sylvie.roke@epfl.ch

† These authors contributed equally to this work.

asymmetry can be achieved actively and passively.[5–7] Active pathways use regulating proteins and peptides to induce asymmetry, whereas passive pathways derive from a balance of intermolecular interactions.[8] Interactions that are strong enough to impart asymmetry are, in decreasing order of strength,[9] electrostatic, and hydrogen (H) bonding interactions, followed by dispersive and steric interactions. Although electrostatic and H-bonding interactions are the most relevant in terms of strength and responsible for stabilizing membranes in water, they are rarely studied in the context of membrane asymmetry.[10–12] These interactions are the topic of this work.

The part of lipid membranes that is immersed in water is the hydrated head group region. The charged head group moieties (*e.g.* $R_2PO_4^-$, COO^-, NH_3^+, and $N(CH_3)_3^+$) as well as their counterions (*e.g.* Na^+, K^+, Cl^-), and water and ionic species from the bulk aqueous phase, form the electric double layer (EDL, sketched in Fig. 1A).[13,14] The EDL comprises a concentration gradient of co- and counterions, as well as oriented and interacting water molecules, and is part of every aqueous interface and membrane.[15] In the vicinity of a charged interface,

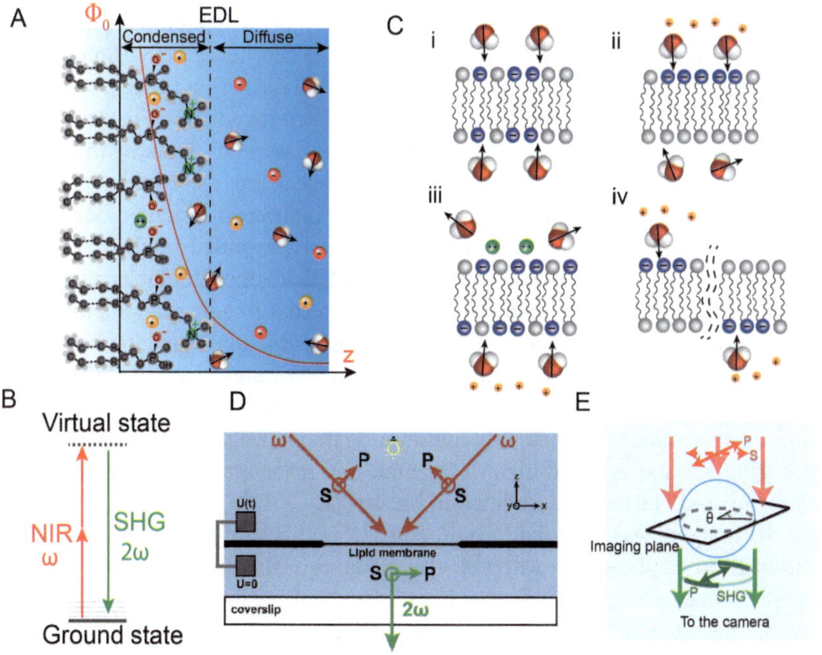

Fig. 1 Electric double layer, lipid hydration and SH imaging of hydration asymmetry. (A) Illustration of the molecular structure of the hydrated lipid headgroup region which forms an electric double layer in the aqueous phase. The electrostatic potential arising from the distribution of ions is modelled by the Gouy–Chapman–Stern model and a typical potential decay in the solution as a function of distance is depicted. (B) Energy level diagram of a non-resonant second harmonic process. (C) (i) Fully symmetric bilayer with symmetric hydration (no coherent SH emission), and three ways in which this symmetry can be broken (with coherent SH emission): (ii) on-average asymmetric lipid/protein distributions, (iii) specific ion binding or (iv) by spatiotemporal fluctuations. (D) SH imaging configuration for imaging free standing lipid membranes (FLMs) in aqueous solution.[36,40–44] (E) SH imaging configuration for SH imaging GUVs[45–47] and cells.[48]

dipolar molecules and ions distribute in a manner that lowers the free energy. A balance between enthalpic and entropic driving forces creates a layer of charge separation, as sketched in Fig. 1A. According to the Gouy–Chapman (GC) theory or Gouy–Chapman–Stern (GCS) theory, this gradient can be either gradual, giving rise to a diffuse double layer (DDL), or composed of a condensed part (Stern layer) in combination with a diffuse part. The GC and GCS models comprise linearized solutions to the Poisson–Boltzmann equation, which describe the distribution of point charges in the vicinity of a homogeneously charged interface in contact with a mean-field dielectric medium. In the GCS model, the surface potential (Φ_0) of a charged planar surface with surface charge density σ_0 in contact with an aqueous solution having a symmetric 1 : 1 electrolyte with concentration c is given by:[15]

$$\Phi_0 = \frac{\sigma_0 d_s}{\varepsilon_0 \varepsilon_r'} + \frac{2k_B T}{e}\sinh^{-1}\left(\frac{\sigma_0}{\sqrt{8000 k_B T N_A c \varepsilon_0 \varepsilon_r}}\right), \tag{1}$$

where d_s, ε_0, ε_r', ε_r, $k_B T$, e, N_A, are the Stern layer thickness, the permittivity of vacuum, the dielectric constant in the Stern layer, the dielectric constant of the bulk medium (water), the Boltzmann constant, the temperature, the elementary charge and Avogadro's number, respectively.

Although other idealized models of this charge separation phenomena exist,[16,17] many aspects remain unknown, such as when/how exactly a Stern layer is formed, how different ionic species influence this formation (known as specific ion effects), and what is the effect of interfacial heterogeneities. Historically, the understanding of the EDL on lipid membranes has been driven by theoretical/computational efforts,[14,18] aided primarily by electrical recordings.[17] This has led to the conceptual idea that electrostatic interfacial potentials are 'fixed' properties of membranes. This is perhaps best exemplified by the notion that every cell has a specific resting potential and that, upon (de)polarization, the whole cell changes its membrane potential.[19] Recent imaging developments are beginning to alter this picture. It has become possible to image interfacial water on sub-second time scales. This is important, as water is generally the main compound in any biological system, and the obtained data provides quantified information about the electric double layer on the molecular level.[20] The enabling imaging method is called high throughput non-resonant second harmonic (SH) imaging and is described in detail in ref. 20–22.

SH imaging was invented in the 1970s,[23–25] and became an important tool for biomedical applications, whereby either non-centrosymmetric bulk structures were probed, see *e.g.* ref. 26–31, or interfaces, combined with the use of resonant dyes, see *e.g.* ref. 32–35. These strategies were necessary to obtain sufficient imaging contrast for the generally weak nonlinear optical process. Non-resonant SH generation, is a process in which two photons with frequency ω are converted into a photon with frequency 2ω *via* a series of non-resonant interactions (illustrated in Fig. 1B). Using an approach with a medium-repetition rate near-infrared femtosecond illumination with gated detection, it became possible to perform non-resonant wide-field SH imaging of interfacial water on sub-second time scales.[20,36] SHG has a unique sensitivity to interfaces,[37] by virtue of a spatial symmetry selection rule that applies to all even-order nonlinear optical techniques within the electric dipole approximation.[37] Anisotropic molecular arrangements of non-isotropic molecules generate coherent SH photons, while isotropic arrangements do not. Therefore, coherent SH photons are uniquely generated in the phase matching direction from an interface if it is sandwiched between isotropic media. This offers a unique optical contrast mechanism for bilayer imaging. A fully symmetric bilayer (Fig. 1C(i)), with all components superimposable by reflection, generates no coherent SH photons.[38] An asymmetric bilayer that lacks such a symmetry does generate coherent SH photons.[36] The asymmetry can be static, by having 2 different leaflets with on-average asymmetric lipid/protein distributions (Fig. 1C(ii)), specific interactions in the EDL of one leaflet, *e.g.* with divalent ions or protons (Fig. 1C(iii)), or by spatio-temporal fluctuations (Fig. 1C(iv)). When a liquid interface is probed, coherent SH photons are emitted only from a 2–3 molecular diameter layer. It has been determined in ref. 20 that ~60% of all SH photons emerge from a 0.6 nm-thick region at high ionic strength. Note that the probed region is located in the aqueous phase and that the lipids do not contribute to the SHG signal intensity, which will be further explained in Section 2.4. The thickness of this layer is determined by the optical geometry and the ionic strength of the solution, not by the nature of the substance in contact with water.[39]

High-throughput SH imaging has shed new light on the electrostatics/EDL properties of interfaces. SH imaging of a variety of surfaces has generated surface potential videos of glass–water,[20] electrode–water,[49,50] membrane and cellular interfaces.[42,44,45,47,48,50] In terms of lipid membranes, SH imaging was applied to free-standing lipid membranes (FLMs)[40,41,44] and giant unilamellar vesicles (GUVs),[45,47] whose importance as a model membrane system is detailed in ref. 51 and 52. All interfaces displayed membrane potential fluctuations. The source of these fluctuations lies in the fact that interfacial surface chemistry is spatially varying and that transport of chemicals in the EDL is not a diffusion-limited motion but rather determined by jamming, field gradients and nano-scale fluctuations, which are also found in concentrated electrolyte solutions,[47] as further explained below in Section 3.4. The spatiotemporal variation of the charged state of lipids and ions along lipid leaflets induces transient electrostatic fields that aid in the formation of transient nanopores[46,47] and water wires/needles,[40] which have both been predicted previously by simulations.[40,47]

Here, we report membrane potential fluctuations as seen by water molecules as well as the intensity variations of a potential-sensitive fluorophore on both model and plasma membranes of cells. We also show that upon K^+ depolarization, non-uniform recovery responses occur across cells and within single cells. The origins and implications of such fluctuations are considered in detail, which warrant investigating the molecular level details of membrane potential distribution on FLMs and GUVs on the molecular level using water as a probe. The hydrated region of lipid membranes is composed of areas having a diffuse double layer, and other regions having an additional condensed double layer. The condensed regions have a high concentration of ionic groups. In terms of transmembrane potential distributions, FLMs and GUVs behave similarly, as expected from electrostatics. Comparing the passive ion transport properties, FLMs and GUVs of identical composition behave differently, with GUVs being much more permeable for ions, which is illustrated by examining proton transport. Protons transport on average ~20× faster through GUV membranes than through FLMs having identical composition. This transport difference is likely caused by differences in the hydrophobic cores of the membranes.

2. Experimental section

2.1. Chemicals

1,2-Diphytanoyl-*sn*-glaycero-3-phosphocholine (DPhPC), 1,2-diphytanoyl-*sn*-glycero-3-phospho L-serine (DPhPS), 1,2-dioleoyl-*sn*-glycero-3-phosphocholine (DOPC), 1,2-dioleoyl-*sn*-glycero-3-phospho-L-serine (DOPS), and 1,2-diphytanoyl-*sn*-glycero-3-phosphate were obtained in powder form (>99%, Avanti Polar Lipids, Alabama, USA). Hexadecane ($C_{16}H_{34}$, 99.8%, Sigma-Aldrich), hexane (C_6H_{14}, 99%, Sigma-Aldrich), chloroform ($CHCl_3$, 99.8%, Merck), hydrogen peroxide (H_2O_2, 30–32%, Reactolab SA), sulfuric acid (H_2SO_4, 95–97%, ISO, Merck), hydrochloric acid (HCl, 36–38%, Sigma-Aldrich), potassium chloride (KCl, 99.999%, Aros), monosodium phosphate ($NaH_2PO_4 \geq 99.0\%$, Sigma-Aldrich), disodium phosphate ($Na_2HPO_4 \geq 99.0\%$, Sigma-Aldrich), *N*-(3-triethylammoniumpropyl)-4-(6-(4-(diethylamino)phenyl)hexatrienyl)pyridinium dibromide (FM4-64 dye (Sigma-Aldrich)), calcium chloride ($CaCl_2$, 99.999%, Sigma-Aldrich), 4-(2-hydroxyethyl) piperazine-1-ethane-sulfonic acid (HEPES), poly-D-lysine (PDL), and agar ($(C_{12}H_{18}O_9)_n$, >99%, Sigma-Aldrich) were used as received. All aqueous solutions were made with 18.2 MΩ cm ultra-pure water (H_2O, Milli-Q UF plus, Millipore, Inc.), and filtered with 0.1 μm Millex filters. For each experiment, the coverslips were taken fresh from the packaging and were pre-cleaned by soaking them thoroughly in piranha solution (a 1 : 3 mixture of 30–32% H_2O_2 : 95–97% H_2SO_4) for 30 min after which they were thoroughly rinsed with ultrapure water. For FLM formation, 13 mM lipid in $CHCl_3$ solutions were used, which is sufficient to make a well-packed lipid monolayer on an air/water interface. The pH was adjusted (Fig. 6) by micro-pipetting a 3 M HCl solution to the bottom compartment of the membrane.

2.2. Sample preparation and characterization

2.2.1 FLMs. Freestanding horizontal planar lipid membranes (FLMs) were formed using the Montal–Müller procedure,[53,54] and subsequently characterized by wide-field imaging and electrical recordings as described in ref. 36. Briefly, two lipid monolayers were formed on two air/water interfaces that were separated by a 25 μm thick Teflon film having a ∼110 μm diameter central aperture. The aperture in the Teflon was treated with a 99.5 : 0.5 vol% mixture of *n*-hexane and *n*-hexadecane. The bilayer was created by apposition. Wide-field white light imaging, the electrical recording of resistance and SH imaging were performed after ∼15 min. Capacitance and resistance measurements were taken with a HEKA patch clamp amplifier. Only bilayers with specific capacitance, $C_m > 0.7$ μF cm^{-2} and specific resistance, $R_m \sim 10^8$ Ω cm^2 were used for SH imaging experiments.

2.2.2 GUVs. Giant unilamellar vesicles were formed by gel-assisted growth using polyvinyl alcohol (PVA) similar to the process described by Weinberger *et al.*,[55] and detailed in ref. 47. In brief, a 5% (w/w) solution of PVA in water was prepared and heated to 90 °C in a water bath. A rubber O-ring was bonded to a cleaned circular glass coverslip using a silicone elastomer (Kwik-Cast, World Precision Instruments) to form an open growth chamber. 50 μl of a heated 5% w/w PVA solution in water was spread on the coverslip and dried. Lipids dissolved in chloroform (5–10 μl, 1 mg ml^{-1}) were deposited, and the chloroform was evaporated, after which the growth chamber was filled with a solution composed of 45 mM sucrose in order to match the osmolarity of the observation solution in the SH imaging chamber. The formed vesicles were tracked by light microscopy and pipetted into an open observation chamber, assembled separately using a bovine serum albumin (BSA) coated coverslip, which was placed under the SH microscope. The solution in the chamber was adjusted to match the osmolarity of the inner solution and contained 30 mM glucose and 5 mM divalent salt solution.

2.2.3 Cells. Primary cultures of cortical neurons were prepared from E17 OF1 mice embryos. The average plated density of cells was 15 000 cells per cm^2 per coverslip. Cultured neurons were maintained at 37 °C in a humidified atmosphere of 95% air/5% CO_2 and were used for experiments after 17 days *in vitro* (Fig. 2D–G). Coverslips with plated cells were inserted in a 3 ml laminar flow (1 ml min^{-1}) Quick Change Imaging Chamber from Warner Instruments (Series 40 RC-41LP). The extracellular solution contained 140 mM NaCl, 5 mM KCl, 3 mM $CaCl_2·2H_2O$, 2 mM $MgCl_2·6H_2O$, 5 mM glucose and 10 mM Hepes. During the depolarization experiments the solution was switched for 2 min to one that had 93 mM NaCl, 5 mM KCl. For more details, see ref. 48. Neuroblastoma cells were cultured in the same extracellular solution and imaged following the same protocol. For imaging applications that require glass-bottomed dishes, the dish was coated with poly-D-lysine (PDL) before use. Cells were seeded evenly in

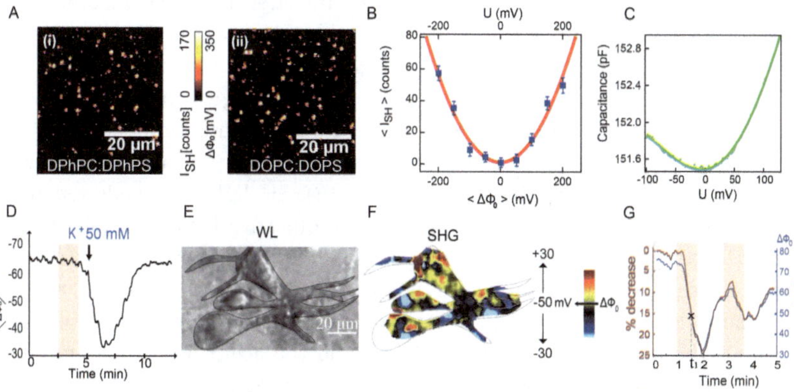

Fig. 2 Electrical recording and SH water imaging of FLMs and living mouse brain neurons. (A) SH intensity (I_{SH}) and corresponding transmembrane potential difference ($\Delta\Phi_0$) images (acquisition times 1 s) of a symmetric membrane composed of a 70 : 30 mol% (i) DPhPC : DPhPS and (ii) DOPC : DOPS lipids. (B) Image averaged SH intensity (blue squares) as a function of external bias (U) for the membrane (i) in A and image-averaged transmembrane potential $\Delta\Phi_0$. The red curve is a parabolic fit according to eqn (4). (C) Capacitance as a function of applied bias U for a symmetric membrane. (D) Whole cell current patch clamp potential recording. The orange area represents the time lapse application of the K^+ enriched solution and the arrow shows when the K^+ solution reaches the neurons. (E and F) SH imaging of K^+ depolarization: phase contrast image (E) of cortical neurons, 15 days *in vitro*. Three cell bodies are seen in the image. The black line indicates the area in which the membrane potential calculations were made. (F) Membrane potential map at a time point indicated by (*x*) in (G). The scale bar is 20 μm. (G) Spatially-averaged SH intensity changes (left axis) and extracted average membrane potential (right axis) as a function of time during two applications and recovery cycles of a 50 mM K^+ enriched extracellular solution (orange areas). The data and analysis procedure of (D–G) are from ref. 48.

a 35 mm glass-bottomed well dish, while avoiding air bubbles at the bottom of the dish. The cells were imaged 48–72 hours after seeding. For resonant SHG imaging 10 μM FM4-64 was added to the extracellular solution. For K^+ depolarization experiments, 50 mM aqueous KCl solution was added to the extracellular solution and flushed out after ~1 min.

2.3. SH imaging

SH images were obtained with custom-built wide-field second harmonic microscopes that have been previously characterized in detail, see ref. 21 and 36 for details on the microscope used to image FLMs, and ref. 45 and 47 for the microscopes used to image GUVs and cells. Briefly, in both cases the fundamental (ω) beam consisted of a pulsed 190 fs, 200 kHz, 1030 nm light source that is polarization controlled. The coherent SH light (2ω) is emitted in the phase-matched direction, spectrally filtered, polarization controlled and imaged with an electronically amplified intensified gated CCD camera. White-light imaging was performed in parallel.

2.3.1 FLMs. The FLMs were imaged by means of two weakly focused 100 mW counter-propagating polarization-controlled beams (having a 90° angle, see Fig. 1D), each incident at 45° with respect to the membrane. The chamber with the FLM is installed slightly out of the focus of both lenses to create a wide-field illumination of ~140 μm in diameter. The polarization state in which the images were recorded was PPP, where the first letter indicates the SH beam, and the last two the incoming photons. The signal was stable over hours and the images were recorded with 1 s integration time. A typical experiment lasts 2–3 hours.

2.3.2 GUVs and cells. The GUVs and cells were imaged in a different geometry, whereby a combination of a 25 cm lens and a 60× water immersion objective was used to excite an area of 90 μm on a sample plane at normal incidence angle using a 100 mW beam (illustrated in Fig. 1E). SH light was collected in the forward/phase matched direction with a 60× objective lens, spectrally filtered, polarization controlled and imaged. For all SH imaging experiments using water as a contrast agent, the error bar on the intensity is 1.5% of the average, while for resonant FM4-64 SH imaging the error bar is 0.9% of the mean. In the case of FM4-64, the power was reduced to 7 mW and the acquisition time and/or the intensifier amplification were reduced. Typical experimental times are up to several hours. For all types of SH imaging white light imaging is conducted in parallel.

2.4. Transmembrane potential obtained from non-resonant high throughput SH imaging

Wide-field high-throughput SH imaging increases the SH imaging throughput by a factor of >5000,[20] compared to scanning confocal approaches that are typically applied to study resonant SHG.[56] This approach allows for imaging interfacial water with ~400 nm spatial resolution and sub-second acquisition times. Although the non-resonant interaction scheme (Fig. 1B) is not chemically selective, it is used to specifically detect the orientation distribution of interfacial water, and the recorded intensity is proportional to the square of the transmembrane potential difference ($\sqrt{|\Delta\Phi_0|}$), and the transmembrane potential difference in mV can be determined from each pixel.

In a non-resonant nonlinear optical interaction, the molecular response is independent of the molecular species, and the number of non-zero nonlinear optical parameters is small, with each of these being frequency independent.[37] At most interfaces, water outnumbers every other molecule with typical ratios >10 : 1 (typically ∼10–50 for lipids[12,57]), resulting in $>10^2:1$ SH intensity ratio in favor of water. This ability to SH image water has been demonstrated and exploited for solid–liquid, liquid–liquid, and cellular membrane interfaces.[20,44,48] The surface potential information derives from the proportionality of the emitted SH field to the nonlinear optical polarization ($\boldsymbol{P}^{(\mathrm{NL})}(2\omega)$). $\boldsymbol{P}^{(\mathrm{NL})}(2\omega)$ depends on the incoming electromagnetic fields ($\boldsymbol{E}_1(\omega), \boldsymbol{E}_2(\omega)$) from the light source having frequency ω, and $\boldsymbol{E}_{\mathrm{DC}}(z)$, the electrostatic field in the aqueous phase that emerges from the surface (Fig. 1A). This field relates to the electrostatic potential, since $E_{\mathrm{DC}}(z) = -\dfrac{\mathrm{d}}{\mathrm{d}z}\Phi(z)$, and the expression for $\boldsymbol{P}^{(\mathrm{NL})}(2\omega)$ becomes:[39]

$$\boldsymbol{P}^{(\mathrm{NL})}(2\omega,\ x,\ y) = \varepsilon_0\left(\chi_{\mathrm{s}}{}^{(2)} : \boldsymbol{E}_1(\omega)\boldsymbol{E}_2(\omega) + \chi^{(3)'} \vdots \int_0^{+\infty} \boldsymbol{E}_1(\omega,\ z)\boldsymbol{E}_2(\omega,\ z)\boldsymbol{E}_{\mathrm{DC}}(z)\mathrm{d}z\right). \tag{2}$$

The parameters $\boldsymbol{\chi}_{\mathrm{s}}{}^{(2)}$, and $\boldsymbol{\chi}^{(3)'}$ are the second-order surface susceptibility and the third-order effective susceptibility, respectively. For non-resonant SH imaging of interfacial water, these tensors reduce to scalars ($\chi_{\mathrm{s}}{}^{(2)}$ and $\chi^{(3)'}$).[39,57] They report on the orientational distribution of interfacial water in the direction of the surface normal due to chemical interactions ($\chi_{\mathrm{s}}{}^{(2)}$), and due to electrostatic field–driven interactions ($\chi^{(3)'}$), respectively. The intensity is related to the polarizability by: $I(2\omega) \propto |\boldsymbol{P}^{(\mathrm{NL})}(2\omega)|^2$. For a lipid membrane with two opposing leaflets ($i = 1$ or 2) and $I(\omega) = |\boldsymbol{E}_1(\omega)|^2 = |\boldsymbol{E}_2(\omega)|^2$, the total emitted SH intensity $I(2\omega, x, y)$ originating from the spatial coordinates x, y is related to the membrane surface potential on the aqueous side of each leaflet ($\Phi_{0,i}$):[36]

$$I(2\omega,\ x,\ y) \propto I(\omega,\ x,\ y)^2 \left|\chi_{\mathrm{s},1}{}^{(2)}(x,\ y) - \chi_{\mathrm{s},2}{}^{(2)}(x,\ y) + \chi^{(3)'} f_3(\Phi_{0,1}(x,\ y) - \Phi_{0,2}(x,\ y))\right|^2 \tag{3}$$

where f_3 is an interference term determined by the geometry of the experiment, and $f_3 = 1$ for the cases discussed here.[39] For symmetric membranes that have, on average, an identical lipid composition on each leaflet, $\chi_{\mathrm{s},1}{}^{(2)} \cong \chi_{\mathrm{s},2}{}^{(2)}$. For an asymmetric leaflet, $|\chi_{\mathrm{s},1}{}^{(2)} - \chi_{\mathrm{s},2}{}^{(2)}|$ is around 10^{-24} m^2 V^{-1} and in the case of a symmetric bilayer this values becomes smaller. Therefore, the contribution of $|\chi_{\mathrm{s},1}{}^{(2)} - \chi_{\mathrm{s},2}{}^{(2)}|$ is much smaller than that of $\chi^{(3)'}$ (ref. 36 and 57)) and it does not change significantly upon the addition of ions. The SH intensity variations in the images are attributed to the transmembrane potential difference $\Delta\Phi_0(x, y) = \Phi_{0,1}(x, y) - \Phi_{0,2}(x, y)$. For a structurally symmetric bilayer, Eqn (3) becomes:

$$I(2\omega,\ x,\ y) \propto I(\omega,\ x,\ y)^2 \left|\chi^{(3)'}\Delta\Phi_0(x,\ y)\right|^2 \tag{4}$$

where eqn (4) shows that $\Delta\Phi_0(x, y)$ is proportional to $\sqrt{I(2\omega,\ x,\ y)}$. To obtain $\Delta\Phi_0(x, y)$ in voltage, the image-integrated SH intensity is recorded as a function of external electric bias across the membrane (U). Using this curve, the intensity is converted into a membrane potential difference in mV (see ref. 36 for details). An example of such a curve is shown in Fig. 2B. Because there is only one connecting parameter, which is the effective third-order susceptibility of water ($\chi^{(3)'}$), eqn (4) is the same for all on-average symmetric lipid membranes, no matter the membrane composition or type of solutes at physiological concentration in the aqueous solution, as they do not contribute to $\chi^{(3)'}$. For an asymmetric membrane, $|\chi_{s,1}^{(2)} - \chi_{s,2}^{(2)}|$ can be determined with a parabolic fit to a plot of the image averaged intensity recorded as a function of external bias.[36,40,42] With $\chi^{(3)'}$ and $|\chi_{s,1}^{(2)} - \chi_{s,2}^{(2)}|$ determined, the transmembrane potential $\Delta\Phi_0(x, y)$ of every pixel in a SH image can then be computed in mV using eqn (3) or (4). This procedure was previously tested in three ways: (1) by comparing the image averaged membrane potential difference to the one measured with capacitance minimization, a purely electrical recording.[36] This gave identical values. (2) By determining membrane-averaged binding constants of divalent ions (Ca^{2+}, Ba^{2+}) that complex with certain lipids; the image-averaged values compared well to reported values in the literature.[42] (3) By performing simulations.[36] In addition, inverse experiments in which a know DC field is interacting with bulk water were used to relate field strength to SH intensity.[20]

3. Results and discussion

3.1. Hydration/membrane potential asymmetry of model lipid and cellular membranes

We start by reviewing SH water imaging of model free-standing lipid membranes (FLMs), and then consider living mouse brain neurons. Fig. 2A–C shows several measurements of FLMs. Fig. 2A shows SH images of membranes having a symmetric composition, 70 : 30 mol% DPhPC : DPhPS (i) and DOPC : DOPS (ii), in contact with a pH 7.3, 50 mM KCl and 10 mM phosphate buffer solution, from a single frame with a 1 s acquisition time. The SH intensity is converted to membrane surface potential as described in Section 2.4. and published in ref. 36, 40–42. Fig. 2B shows image averaged SH intensity values (blue squares) plotted as a function of external bias (U) for membrane (i) in Fig. 2A. The computed image-averaged trans-membrane potential $\Delta\Phi_0$ values are also plotted on the bottom axis, and they are the same as the external bias.[36] The red curve is a parabolic fit to eqn (4). The symmetry of the membrane is experimentally confirmed by capacitance minimization measurements shown in Fig. 2C. The minimum of the capacitance for a symmetric membrane occurs when the applied external bias $U = 0$ mV. Transient membrane potential fluctuations are observed for both membrane compositions and reach absolute values of up to 350 mV. Such fluctuations agree with previous observations.[36,40,42,43,46,47]

Fig. 2D shows an electrophysiological recording in whole-cell current clamp mode, of a mouse brain neuron, 17 days *in vitro*, that is being subjected to a temporary increase in extracellular concentration of K^+ ions from 5 mM (normal concentration) to 50 mM (indicated by the shaded area). The membrane averaged potential of the neuron $\langle\Delta\Phi_0\rangle$ (which corresponds to the difference in surface potentials of the two membrane leaflets) changes from −65 to −34 mV, in response, after which it recovers back to its original value. The electrophysiological recording suggests that the membrane potential of the neuron is constant over the whole cell membrane. The neurons in Fig. 2E are SH imaged with 300 ms acquisition time (Fig. 2F), and the interfacial water response around the neuron is extracted and processed into membrane potential changes ($\Delta\Phi_0$). Fig. 2G shows the average membrane potential change over the entire image. This graph shows that the average SH response matches with the electrical recording in Fig. 2E, as it should. However, the spatiotemporal distribution of membrane potential values in the snapshot of Fig. 2F is not constant. This suggests that ion channels are opening and closing at different times, and with different efficiencies. It also implies that the charge distribution on the cell membrane is dynamically fluctuating. This data set was previously published in ref. 48.

Fig. 2 illustrates that both free-standing lipid membranes as well as the plasma membrane of neurons display drastic membrane potential fluctuations on sub-second time scales. These fluctuations have a spatial correlation of ~1 μm, a temporal correlation <300 ms, and membrane potential fluctuations reaching values of several hundreds of millivolts.

Next, the following aspects are examined:

• Do fluorescence imaging methods also display membrane potential fluctuations?

• Do the distributions depend on the alkyl chains and ionic composition of the adjacent solution?

• Are they different for FLMs and GUVs?

• Is passive proton transport, a process that depends on membrane potential fluctuations, different across FLMs and GUVs?

3.2. Membrane potential fluctuations imaged using a voltage sensitive fluorophore

To investigate if membrane potential fluctuations are also seen when imaged by other means, we use a membrane potential sensitive fluorophore, FM4-64 (Fig. 3A),[58] which resonantly enhances the SH intensity by a factor of $\sim 10^7$ compared to water.[45] FM4-64 is commonly used for neuro-imaging.[59–63] The spectrum shown in Fig. 3A shows that the SH emission (515 nm) is spectrally sharp and isolated from the 2-photon fluorescence of the dye (550–750 nm). Fig. 3B shows SH images of an asymmetric FLM composed of DPhPA/DPhPC (3 : 7 mol%) + 165 μM KCl on the top leaflet and DPhPC + 165 μM KCl on the bottom leaflet. Fig. 3B(i) was generated by SH imaging of membrane water (1 s snapshot), while Fig. 3B(ii) was SH imaged after 10 μM of FM4-64 in 165 μM KCl was added to the solution adjacent to the top leaflet. The square root of the SH intensity that is emitted by FM4-64 correlates linearly with an applied voltage bias[35] and is therefore used as an indicator of the transmembrane potential. Fig. 3C shows a superimposed bright field and SH image of neuroblastoma cells stained with 10 μM FM4-64. Several ROIs are indicated for which intensity histograms are shown in Fig. 3D. It can be seen that the intensity and thus also the membrane potential varies drastically across the membrane surface. Repeating the same K^+ depolarization experiment as in Fig. 2D–G, neuronal depolarization is SH imaged using FM4-64 in Fig. 3E, and analyzed in Fig. 3F. Different response curves are seen for different cells, and even different regions within the same cells. Interestingly, similar fluctuations in the intensity of FM4-64 have been observed previously with scanning confocal 2-photon imaging within the context of several neuroimaging studies[60,64] by the Yuste lab, but were not published (Yuste, personal communication, 2022). Scanning confocal measurements, in contrast to wide-field imaging, record one pixel at a time, with some time, on the order of micro-milliseconds between different pixel recordings. In such a measurement

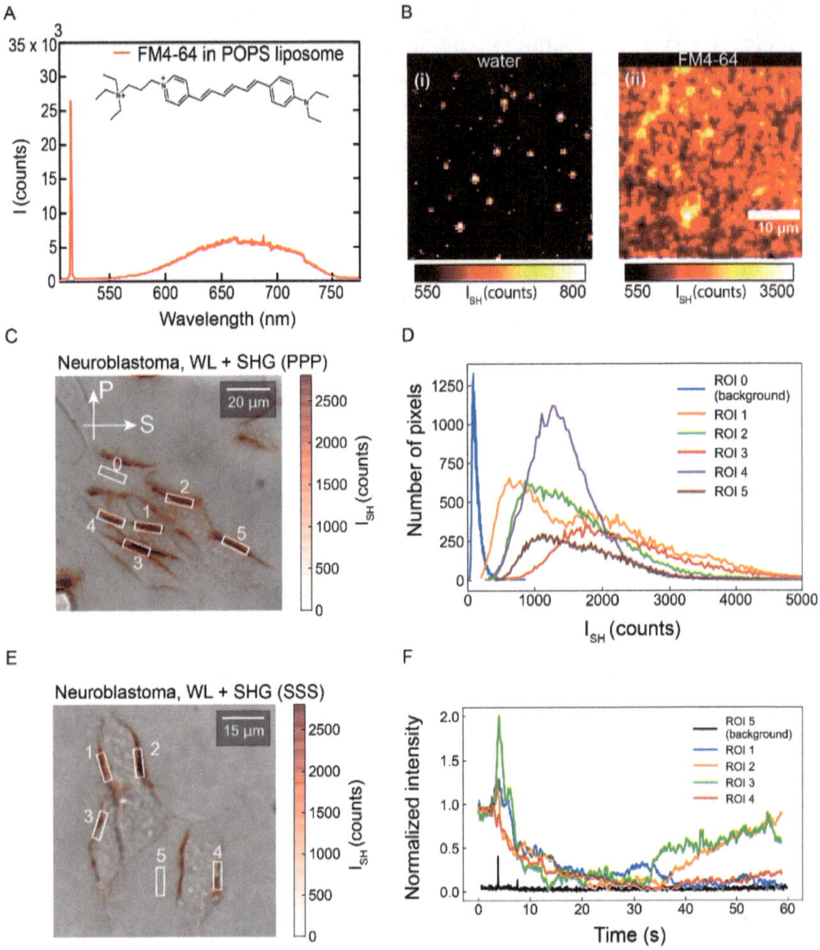

Fig. 3 Transmembrane potential distributions as seen by a membrane potential-sensitive dye: (A) structure of FM4-64 and emission spectrum of a DOPS liposome solution containing 10 μM FM4-64. (B) SH images of an asymmetric FLM in contact with 0.165 mM KCl aqueous solution, composed of a 7 : 3 mol% DPhPC : DPhPA lipid mixture on the top leaflet in contact with a DPhPC bottom leaflet. Image (i) is a non-resonant SH water image, and image (ii) is a resonant SH image with 10 μM FM4-64 added to the top leaflet. (C) Bright field image of neuroblastoma cells on a glass coverslip superimposed on a SH image after 10 μM FM4-64 is added to the solution. The SHG images are recorded with 50 ms integration time. (D) Histograms of several ROIs from the image, displaying large variations in the SH intensity, which is proportional to $\sqrt{|\Delta\Phi_0|}$. (E) Bright field and FM4-64 SH image superimposed and (F) time lapse depolarization response of the indicated ROIs after adding a 50 mM K^+ solution to the extracellular solution. Each trace is normalized by the first data point, the intensity counts of which are 1126, 1814, 1125, 1423, 621 for ROI 1, ROI2, ROI3, ROI4, and ROI5, respectively.

scheme transient effects are not easily spotted. Thus, both the membrane potential fluctuations observed at FLMs and neuronal cell membranes are seen with two different optical methods, and agree with three different types of electric recordings (patch clamp, free space electrical recording and capacitance minimization). The presence of membrane potential fluctuations in different membrane systems, model membranes, and cell membranes of living neurons indicates that the membrane potential and thus also the electrostatic free energy ($\Delta G = e\Delta\Phi_0$) is a dynamic feature of lipid membranes. Next, we consider the origin of these fluctuations in more detail, and then examine their dependence on several parameters such as clustering of ionic species and fluctuating gradient in concentration of ions in FLMs and GUVs.

3.3. The origin and implications of membrane potential fluctuations

3.3.1 Considerations involving the electric double layer. Having observed trans-membrane potential fluctuations in different model and living systems using different imaging modalities, we next discuss their origin and implication. The source of trans-membrane potential fluctuations is found in the structure of the hydrated lipid headgroup region (sketched in Fig. 1A), since this is what is imaged in the non-resonant SH experiments. This region consists of charged head group moieties ($R_2PO_4^-$, COO^-, NH_3^+, $N(CH_3)_3^+$, *etc.*) as well as their counterions (Na^+, K^+, Cl^- *etc.*); it also includes water and ionic species from the bulk aqueous phase. SH imaging probes this region by measuring the orientational distribution of the water, as explained in Section 2.4. Counting all the charges per lipid in a typical bilayer membrane, the local concentration of ions inside the hydrated head group region is easily >1 M. This also applies for zwitterionic lipids as they have charged groups on them. Such a concentration exceeds the Kirkwood transition[65,66] of ionic solutions (~0.1 M). Above this concentration the distribution of ions is no longer stochastic, leading to non-random distributions. X-ray diffraction experiments in the 1930's[67] and, more recently, also molecular dynamics simulations[68] show that at concentrations exceeding the Kirkwood limit, ions are partially and dynamically distributed in quasi-periodic lattices spaced by more dilute regions of ions.[69] Segregation over longer length scales in ionic solutions has also been observed with dynamic light scattering, where domains on the length scale of 0.1–0.5 microns are observed.[70] The autocorrelation function of scattered light decays on time scales ranging from 10 μs–0.1 s, depending on the type of salt and concentration used. This type of clustering behavior is a manifestation of the statistical distribution of matter.

Lipid membrane interfaces have hydrated headgroup regions that make up ~40% of a bilayer thickness.[12,71] Such regions can be considered as partially confined electrolyte solutions, that are open ended on the diffuse double layer side. Being a half-open highly concentrated electrolyte solution, such systems are likely displaying the same kind of behavior as observed in 3D concentrated ionic solutions. The relevant time scale for the observed transient membrane potential fluctuations thus falls in the same time scale as the one measured by DLS in concentrated salt solutions (10 μs–0.1 s). The clustering of ionic species and the dynamically fluctuating gradient in concentration leads to dynamic fluctuations in the distribution of ions, which manifests itself as membrane potential fluctuations.[47] To get to the bottom of this process, more research is needed.

Another aspect that contributes to charge segregation is that the fluctuation-induced electrostatic fields create strain in the membrane. This strain arises from a surface pressure gradient across the membrane in combination with steric pressure along the membrane.[42,72,73] Although the details have not been worked out completely, this interaction most likely leads to nano/micron scale curvature

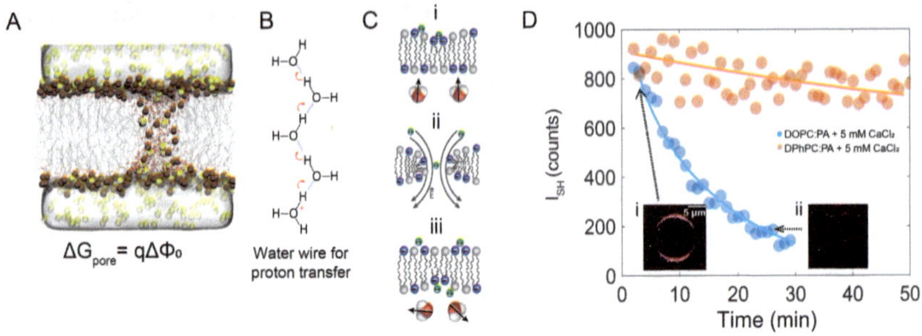

Fig. 4 Transient membrane potential fluctuations drive passive ion transport. (A) Molecular dynamics snapshot of a transient nanopore,[47] (B) illustration of a water wire or needle that facilitates H^+ transport, (C) illustration of the nanopore formation process proposed in ref. 47 and (D) single GUV averaged SH intensity over time (blue data points: 1 : 1 mole% DOPC : DOPA; orange data points: 1 : 1 mole% DPhPC : DPhPA) after a 5 mM $CaCl_2$ solution is added to the outer solution. The lines are exponential fits (dark blue). The SH images show two 1 s acquisitions taken of the GUV composed of unsaturated lipids at 2 min and 27 min after addition of $CaCl_2$.

changes and charge segregation.[42] The coupling between mechanical curvature changes and the electrostatic properties of the membrane in relation to its EDL structure is complex, and more research in this direction is needed as well.

3.3.2 Implications for membrane free energy and passive ion transport. Higher charge densities lead to on-average higher transmembrane potentials. Since the potential distribution is broad, fluctuations lead to significantly higher $\Delta\Phi_0$ values than the average. For chemical/physical transformations the high values at the tails of the distributions are likely more important than the average value. For a given transient transmembrane potential $\Delta\Phi_0$, there is also a transient electrostatic field across a membrane. For a membrane with thickness d, this has magnitude $E = \Delta\Phi_0/d$.

Besides curvature fluctuations,[42] another biophysical process that is strongly affected is passive ion transport.[46,47,74–76] This is illustrated in Fig. 4. A transient electrostatic field drastically lowers the free energy to form a transient nanopore (Fig. 4A) for the transport of cations, or a transient water wire/needle (Fig. 4B) for the transport of protons.[40,77,78] Water wire transport involves the transport of protons through water molecules that are connected by hydrogen bonds. The protons move across *via* the breaking and making of correlated hydrogen bonds along the water molecules whose connected hydrogen bond network forms the wire (the Grotthuss mechanism).[79] This passive ion transport – either ions through nanopores or protons through water wires – depends on the magnitude of the transmembrane potential[47] and on the hydrophobic barrier formed by the alkyl chains of the lipids and other molecules such as cholesterol.[46] The electrostatic free energy, $\Delta G = e\Delta\Phi_0$, contributes to the probability P_{pore} to form a transient pore *via*: $P_{pore} \sim e^{-\Delta G_{pore}/k_BT} \sim 1/\tau$, where ΔG_{pore} is the free energy of the pore relative to the unperturbed membrane, k_B is the Boltzmann constant, T the temperature, and τ is the translocation time of ions.[47] The transport of protons and divalent ions can be SH imaged. Ions that interact with the lipid headgroups on one leaflet, such as Ca^{2+}, associate/bind with charged (*e.g.* PS or PA) head groups and neutralize charges on one leaflet. This severely distorts the water orientation on this leaflet. The opposing leaflet, which is uncoupled to the first, has not been neutralized, and its water orientation has not been changed. This difference in the orientational distribution of water creates a SH contrast, which can be imaged and converted into a membrane potential value. This mechanism is illustrated in Fig. 4C(i), and a SH image of a GUV composed of a 1 : 1 mole% mixture of DOPC : DOPA immersed in an osmolarity-balanced solution with 5 mM $CaCl_2$ added to the outside, is shown in the inset of Fig. 4D(i). When a transient nanopore is formed, ions transport though the membrane (Fig. 4C(ii)), which will restore the imbalance of Ca^{2+} ions that are complexed on both leaflets. The difference in the orientational distributions of water molecules on both leaflets becomes smaller (Fig. 4C(iii)), which decreases the SH intensity. This process continues until the distribution of divalent ions on both leaflets is equal. Fig. 4D shows the GUV averaged reduction in SH intensity, finally resulting in a non-existent SH contrast. Adding cholesterol to the membrane slows down the transport process.[47] Branched lipids slow it down further, as can be seen in Fig. 4D, and no decrease in the SH intensity is seen on the time scale of the experiment (∼2–3 h).

Previously, symmetric and asymmetric FLMs of the following lipids were studied: DPhPC, DPhPA, DPhPS, DOPC, DOPA, DOPS, DPPC (1,2-dipalmitoyl-*sn*-glycero-3-phosphocholine), DPPS (1,2-dipalmitoyl-*sn*-glycero-3-phospho-L-serine), with ionic species Na^+, K^+, Mg^{2+}, Ca^{2+}, Ba^{2+}, Cu^{2+}, H^+.[40,41,46,47] While divalent cations change the membrane potential distribution of FLMs they do not lead to changes to the SHG intensity over time, which means they do not translocate through the membrane. Protons, however, do change the transmembrane potential distribution and also transfer through the FLM, albeit on a time scale of hours.[40] GUVs with compositions DOPC : PA interacting with Ca^{2+}, Ba^{2+}, Cu^{2+} ions have been investigated and simulated.[46,47] Here, as shown in Fig. 4D, divalent cations can transfer through the membrane on timescales of minutes, depending on the composition of the membrane and the type of ion, and the magnitude of the transmembrane potential. For both FLMs and GUVS the magnitude of the transmembrane potential changes that are induced by the divalent cations is determined by the strength of the divalent ion–headgroup complex, and follows the order $Cu^{2+} > Ca^{2+} > Mg^{2+} > Ba^{2+}$ and the density of the charged lipid headgroup. Membrane potential changes only occur when ions form complexes with the charged groups of the lipids, and this is higher for charged lipids that have smaller footprints. PA headgroups, with their small areas generate higher interfacial charge densities/transmembrane potential differences than PS headgroups. Adding monovalent K^+ ions does not change the membrane potential distribution up to and above physiological concentrations.

Having considered the origin of the membrane potential fluctuations, and discussed the implications for passive ion transport, we next investigate membrane potential distributions on FLMs, whether these depend on the alkyl chain length and the adjacent solution (with or without buffer), and what we can learn from the measured distributions. We then make a comparison with GUVs, considering the amount of charged lipids in the membrane. Finally, we will compare passive proton transport across FLM and GUV membranes having identical composition, and explain the observed differences.

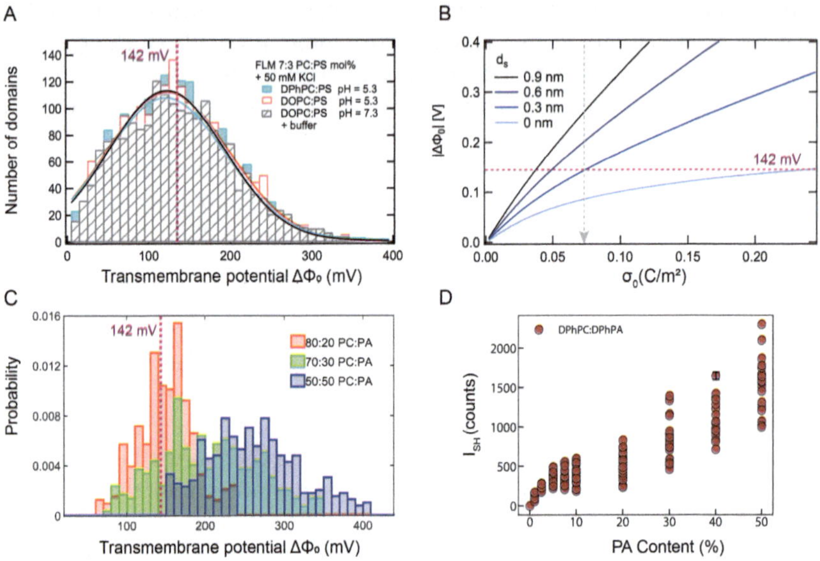

Fig. 5 Transmembrane potential distributions on FLMs and GUVs. FLMs: (A) transmembrane potential values per domain ($\langle \Delta\Phi_{\text{domain}} \rangle$, taken from 1 s SH images and averaged over 20 images) from 3 different symmetric FLMs composed of 70 : 30 mol% DPhPC : DPhPS (turquoise), 70 : 30 mol% DOPC : DOPS (red/white) immersed in a pH 5.3, 50 mM KCl solution, and the same DOPC : DOPS mixture but with 10 mM PBS buffer added to both sides (dashed grey). The lines are Gaussian fits to the data. (B) Computed curves (eqn (1)) of the surface potential difference ($|\Delta\Phi_0|$) as a function of the surface charge density (σ_0) plotted for different thicknesses of the Stern layer (d_s). The x axis spans a neutral surface up to a fully ionized DPhPS layer. (C) Transmembrane potential values per 20 μm^2 domains obtained for GUVs composed of DPhPC : DPhPA with 3 different mole% ratios of 1 : 1 (blue), 7 : 3 (green), and 8 : 2 (red). The solution inside the GUV contains 45 mM sucrose, and the solution outside the GUV contains 30 mM glucose and 5 mM $CaCl_2$. (D) Bar graph showing how the distribution of average SH intensity per GUV changes for different GUVs prepared according to the same protocol. Each marker represents one GUV. Between 9 and 29 GUVs were imaged per lipid composition.

3.4. The structure of the hydrated interface in FLMs and GUVS

3.4.1 FLMs. In Fig. 2A two 1 s snapshot SH images were presented of symmetric FLMs composed of 70 : 30 mol% DOPC : DOPS and DPhPC : DPhPS lipid mixtures surrounded by an aqueous solution containing 50 mM KCl. The SH contrast was shown to arise from an asymmetric distribution of charges per leaflet, which modifies the measured water structure. This is asymmetry caused by spatio-temporal fluctuations (Fig. 1C(iv)). The intensity scale as well as the converted transmembrane potentials ($\Delta\Phi_0$) were shown. Fig. 5A shows a histogram of the $\Delta\Phi_0$ values obtained from an analysis of the domain intensity of the image stack from which the images of Fig. 2A are taken. Fig. 5A also shows histograms obtained for a symmetric FLM composed of a 70 : 30 mol% DOPC : DOPS lipid mixture, surrounded by an aqueous solution containing 50 mM KCl or 10 mM PBS buffer. The lines are fits to a Gaussian distribution. All three bilayer membranes have very similar transmembrane potential distributions with an average domain transmembrane potential of $|\Delta\Phi_0| \sim 122$ mV, and a maximum (transient) value of $|\Delta\Phi_0| \sim 390$ mV. This confirms the notion that the transmembrane potential distribution is a phenomenon that is independent of the alkyl chain structure and relatively insensitive to pH (which is 5.3 without buffer and 7.3 with buffer).

Next, we analyze the structural implications of the $\Delta\Phi_0$ values using the GC and GCS models (eqn (1)). This determines the degree of ionization and whether there is a Stern layer present. In this analysis we assume that for a given domain, the charge is on one leaflet only. Fig. 5B shows the computed surface potential ($\Delta\Phi_0$) *vs.* surface charge density (σ_0) curves using the GC and GCS models in light and progressively darker blue curves, whereby the Stern layer is varied from 0 nm (absent, GC model) to one with a thickness of 0.3, 0.6, and 0.9 nm corresponding to a thickness of ~1, ~2 and ~3 hydration layers, respectively (GCS model). The graph shows that, in absence of a condensed charge or Stern layer, the maximum membrane potential value that can be reached is 142 mV. The potential measured above this value, can only be generated when a Stern layer is present. Considering a hypothetical 7 : 3 mol% PC : PS lipid leaflet with a uniform surface charge density and 100% ionization, the surface charge density is $\sigma_0 = 0.0735$ C m^{-2}, which corresponds to an area of 212 Å^2 per charge, indicated by the dashed grey arrow. Such a charge density leads to surface potentials of 85 mV, 142 mV, 199 mV and 259 mV for Stern layer thicknesses of 0, 0.3, 0.6 and 0.9 nm, respectively. A 100% PS membrane with 100% ionization, in which a PS headgroup has a cross-sectional area of ~65 Å^2 per charge ($\sigma_0 = 0.245$ C m^{-2}),[10] can have a surface potential of 142 mV ($d_s = 0$ nm, GC model), 338 mV ($d_s = 0.3$ nm), 530 mV ($d_s = 0.6$ nm), 730 mV ($d_s = 0.9$ nm).

Examining the measured domain values in Fig. 5A, ~73% of the $\Delta\Phi_0$ values are above 85 mV and ~40% of the $\Delta\Phi_0$ values are above 142 mV. This means that most of the charged domains will have charge condensation or Stern layers in them, as the measured potential exceeds the theoretical GC potential maximum. In addition, there needs to be a higher percentage of PS than present in the 'average' 7 : 3 mole% structure. Also, for the highest potentials, >300 mV, which comprises ~1.5% of the charged domains, the Stern layer is more than 1–2 hydration layers thick (>0.3 nm).

3.4.2 GUVs. Fig. 5C displays transmembrane potential ($\Delta\Phi_0$) distributions extracted from SH water imaging from GUVs composed of mixtures of DPhPC : DPhPA having 3 different mole ratios of 1 : 1, 7 : 3, and 8 : 2. Fig. 5D shows the GUV averaged intensity for different measurements on GUVs, in which between 9 and 29 GUVs were measured for each lipid composition indicated in the graph. In each case, the solution inside the GUV contains 45 mM sucrose, and the solution outside the GUV contains 30 mM glucose and 5 mM $CaCl_2$. In these experiments, the SH contrast observed in the GUV images is enhanced by the complexation of Ca^{2+} ions with the phosphate headgroups of the outer leaflet. The asymmetry here is caused by specific divalent ion–head group interactions on top of spatio-temporal fluctuations (Fig. 1C(iii)). Note that adding additional ionic strength to the aqueous solution does not change the observations in any way. Also note that, in using branched lipids, divalent ion transport is halted.

Increasing the amount of charged lipids increases the surface charge density, as well as the membrane potential, as expected (Fig. 5B). It is clear though that the spread per GUV (Fig. 5C) and between GUVs (Fig. 5D) is very large. The values for the 3 : 7 PA : PC membrane, $-117 < \Delta\Phi_0 < -235$ are consistent with those of the

PC : PS FLM membranes in Fig. 5A. The graphs also show that as with the FLMs, the degree of ionization varies across the domains as does the presence of a Stern layer. The potential limit above which a condensed charged layer is formed is indicated by the dashed line.

Thus, for both FLMs and GUVs the hydrated lipid membrane is most likely composed of some regions having a diffuse double layer, and regions having an additional condensed double layer, with a high concentration of ionic groups, and this is the same whether the asymmetry is due to having different amounts of charged lipids, or asymmetry induced by specific ion interactions. In terms of transmembrane potential distribution FLMs and GUVs behave similarly. This is expected as the determining factor is the amount of charge and the structure of the electric double layer, which are properties of the aqueous phase in contact with the lipid membrane.[17] As discussed in Section 3.3, transient transmembrane potentials of several 100 mVs are sufficient to induce passive ion transport, *via* transient nanopores (divalent cations) or transient water wires. The probability to create such nanopores or water nanowires/needles depends also on the nature of the hydrophobic core of the membrane. Divalent ion transport through GUVs is slowed down/halted when branched lipids are used or when cholesterol is added to the membrane,[46] and has not been observed for FLMs. Proton transport does occur through FLMs, and is faster in a symmetric FLM having unsaturated alkyl chains than in a symmetric FLM made with branched lipids that have the same head group composition.[40] Because of the difference between divalent ion transport through FLMs (not detected on the time scale of the measurement) and GUVs (depends on the strength of specific ion–head group interactions, and alkyl chain/membrane composition) is different, proton transport may be different through FLMs and GUVs as well. We investigate this next.

3.5. Passive proton translocation in FLMs *vs.* GUVs

To compare passive proton transport through FLMs and GUVs, we examine proton transport across both types of membranes using the same lipid membrane composition, and the same ionic strength. FLMs and GUVs were made having symmetric lipid membranes composed of 70 : 30 mol% DPhPC : DPhPS. Both leaflets are initially in contact with a 50 mM KCl solution having identical pH values for both leaflets (pH = 5.3/pH = 5.3). For the experiment on FLMs the pH of the bottom leaflet was changed by adding $(HCl)_{aq}$ to the solution, and for GUVs the pH was changed by changing the outer solution. In both cases the pH was brought down to 2.5.

After changing the pH of the solution in contact with one leaflet, the initially, on average, symmetric distribution of interfacial water is distorted by the interaction with $(H^+)_{aq}$, which changes the orientational distribution of water (Fig. 6A). Proton transport through the membrane diminishes the interfacial asymmetry, eventually resulting in symmetric interfacial orientational distributions of water. The SH contrast is very low before adding $(H^+)_{aq}$. It increases when protons reach the interface and then decays back to the initial value. The decay time is related to the proton translocation time across the membrane. Fig. 6B shows three SH images of an FLM (top row), and a GUV (bottom row) obtained before changing the pH (t_1), and at times t_2, t_3, and t_4 after changing the pH. Fig. 6C shows the image-averaged SH intensity over time after addition of $(HCl)_{aq}$, for the GUV (red

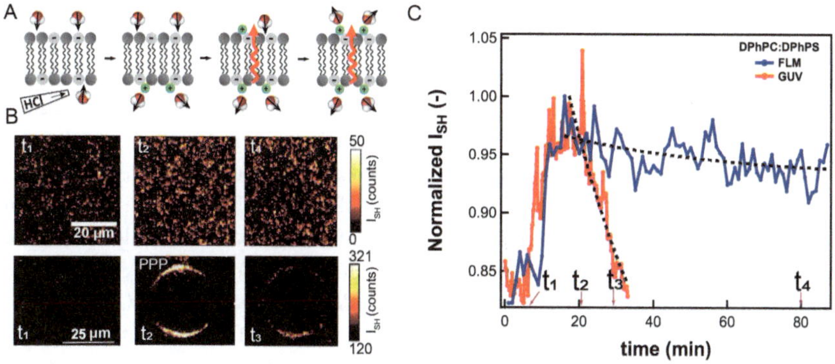

Fig. 6 Passive $(H^+)_{aq}$ translocation across FLMs and GUVs. (A) Illustration of proton translocation through a lipid membrane. The interfacial interactions change the orientational distribution of water, which provides the SH contrast. (B) SH intensity of an FLM (top row), and a GUV (bottom row) observed over time, before (t_1) and after (t_2, t_3, and t_4 indicated in (C)) adding $(HCl)_{aq}$. In both cases the membrane is composed of 70 : 30 mol% DPhPC : DPhPS. Both leaflets are initially in contact with a pH 5.3, 50 mM KCl solution. Adding $(HCl)_{aq}$, to the bottom compartment (outside solution) of the FLM (GUV) changes the pH from pH 5.3 to pH 2.5, while the pH value of the solution in contact with the top (inner) leaflet is pH 5.3. The SH images are obtained with all beams P-polarized. (C) The FLM (blue) and GUV (red) image average SH intensity after adding $(HCl)_{aq}$ to the solution (t_1). The SH intensity increases after the addition of protons, and then decays over time, which is a manifestation of the translocation of protons. Exponential fits provide average membrane translocation times (τ).

data), and an FLM (blue data). The times t_1–t_4 are indicated in the graph. Both bilayers display the above-described rise-and-fall behavior. The increase in intensity occurs on the same time scale, but the decrease is much faster for the GUV than for the FLM. Exponential fits provide average membrane translocation times (τ). The fitted decay time for DPhPC : DPhPS GUVs is $\tau_{GUV} = 14$ min, and for the FLM it is $\tau_{FLM} = 332$ min. Note that the distributions in translocation time for each system is ~50%, which is a manifestation of the spatial heterogeneity of the process. The value obtained for FLM is close to the value obtained previously for translocations of protons on membranes having the same composition but with the pH buffered to 7.3 and 4.1, respectively (442 min (ref. 40)).

This experiment shows that the translocation of protons is much faster across GUV membranes compared to FLMs. Even though both membranes are composed of the same lipids, there is a difference of a factor of ~20 in the translocation time. This difference is likely caused by the hydrophobic core being somewhat different, *e.g.* in thickness, for the FLM compared to the GUV. The FLMs in these experiments are produced using a tiny amount of hexadecane oil, which results in a sub-monolayer amount of oil that interdigitates with the membrane.[44] This likely results in a larger hydrophobic barrier for passive proton transport, as it increases the energetic barrier for creating a water wire needle.

4. Conclusions

We investigated membrane potential fluctuations as seen by water as well as a potential sensitive fluorophore on FLMs, GUVs and living cells. We also showed that upon K^+ depolarization, non-uniform recovery responses occur across cells and within single cells, which are detected both by SH water and SH fluorophore imaging. The origins and implications of such fluctuations are considered in detail, arising from molecular level processes in the EDL. Next, the molecular-level details of membrane potential distribution on FLMs and GUVs using water as a probe were considered. Hydrated lipid membranes are most likely heterogeneous in structure, and composed of regions having a diffuse double layer, and other regions having an additional condensed double layer. The condensed double layer/Stern regions have a high concentration of ionic groups. In terms of transmembrane potential distributions, FLMs and GUVs behave similarly, displaying the same magnitude and spread in fluctuations, as expected from electrostatics. However, comparing the passive ion transport properties, FLMs and GUVs of identical composition behave differently, with GUVs being much more permeable for proton transport ($\sim 20\times$). This is likely caused by differences in the hydrophobic cores of the membranes.

Data availability

The authors confirm that the raw data supporting the findings of this study are available from the corresponding author, upon reasonable request.

Conflicts of interest

There are no conflicts of interest to declare.

Acknowledgements

This work was supported by the Julia Jacobi Foundation, the European Union's Horizon 2020 research and innovation program under Marie Skłodowska-Curie grant agreement 860592 (H2020-MSCA-ITN, PROTON), and the European Research Council grant agreement No. 951324 (H2020, R2-tension).

References

1 P. F. Devaux, Static and dynamic lipid asymmetry in cell membranes, *Biochemistry*, 1991, **30**(5), 1163–1173.

2 J. H. Lorent, K. R. Levental, L. Ganesan, G. Rivera-Longsworth, E. Sezgin, M. Doktorova, *et al.*, Plasma membranes are asymmetric in lipid unsaturation, packing and protein shape, *Nat. Chem. Biol.*, 2020, **16**(6), 644–652.

3 G. van Meer, D. R. Voelker and G. W. Feigenson, Membrane lipids: where they are and how they behave, *Nat. Rev. Mol. Cell Biol.*, 2008, **9**(2), 112–124.

4 Z. Wang, J. M. Jumper, S. Wang, K. F. Freed and T. R. Sosnick, A Membrane Burial Potential with H-Bonds and Applications to Curved Membranes and Fast Simulations, *Biophys. J.*, 2018, **115**(10), 1872–1884.

5 D. Marquardt, B. Geier and G. Pabst, Asymmetric Lipid Membranes: Towards More Realistic Model Systems, *Membranes*, 2015, **5**(2), 180–196.

6 M. Markones, C. Drechsler, M. Kaiser, L. Kalie, H. Heerklotz and S. Fiedler, Engineering Asymmetric Lipid Vesicles: Accurate and Convenient Control of the Outer Leaflet Lipid Composition, *Langmuir*, 2018, **34**(5), 1999–2005.

7 M. Doktorova, F. A. Heberle, B. Eicher, R. F. Standaert, J. Katsaras, E. London, *et al.*, Preparation of asymmetric phospholipid vesicles for use as cell membrane models, *Nat. Protoc.*, 2018, **13**(9), 2086–2101.

8 G. van Meer and A. I. P. M. de Kroon, Lipid map of the mammalian cell, *J. Cell Sci.*, 2011, **124**(1), 5–8.

9 J. N. Israelachvili, *Intermolecular and Surface Forces*, Academic Press, 2010, p. 706.

10 H. I. Petrache, S. Tristram-Nagle, K. Gawrisch, D. Harries, V. A. Parsegian and J. F. Nagle, Structure and Fluctuations of Charged Phosphatidylserine Bilayers in the Absence of Salt, *Biophys. J.*, 2004, **86**(3), 1574–1586.

11 E. A. Disalvo, F. Lairion, F. Martini, E. Tymczyszyn, M. Frías, H. Almaleck, *et al.*, Structural and functional properties of hydration and confined water in membrane interfaces, *Biochim. Biophys. Acta, Biomembr.*, 2008, **1778**(12), 2655–2670.

12 E. A. Disalvo, *Membrane Hydration: the Role of Water in the Structure and Function of Biological Membranes*, Springer, 2015, p. 295.

13 S. Y. Bhide and M. L. Berkowitz, Structure and dynamics of water at the interface with phospholipid bilayers, *J. Chem. Phys.*, 2005, **123**(22), 224702.

14 R. Vácha, M. L. Berkowitz and P. Jungwirth, Molecular Model of a Cell Plasma Membrane With an Asymmetric Multicomponent Composition: Water Permeation and Ion Effects, *Biophys. J.*, 2009, **96**(11), 4493–4501.

15 R. J. Hunter, *Foundations of Colloid Science*, Oxford University Press, 2001, p. 824.

16 H. Ohshima, *Electrical Phenomena at Interfaces and Biointerfaces: Fundamentals and Applications in Nano-, Bio-, and Environmental Sciences*, John Wiley & Sons, 2012, p. 870.

17 S. McLaughlin, The Electrostatic Properties of Membranes, *Annu. Rev. Biophys.*, 1989, **18**, 113–136.

18 K. Bohinc, M. Špadina, J. Reščič, N. Shimokawa and S. Spada, Influence of Charge Lipid Head Group Structures on Electric Double Layer Properties, *J. Chem. Theory Comput.*, 2022, **18**(1), 448–460.

19 *Neuroscience*, ed. D. Purves, G. J Augustine, D. Fitzpatrick, L. C. Katz, A.-S. La Mantia, J. O. McNamara and S. Mark Williams, Sinauer Associates, Sunderland (MA), 2nd edn, 2001.

20 C. Macias-Romero, I. Nahalka, H. I. Okur and S. Roke, Optical imaging of surface chemistry and dynamics in confinement, *Science*, 2017, **357**(6353), 784–788.

21 C. Macias-Romero, M. E. P. Didier, P. Jourdain, P. Marquet, P. Magistretti, O. B. Tarun, *et al.*, High throughput second harmonic imaging for label-free biological applications, *Opt. Express*, 2014, **22**(25), 31102–31112.

22 C. Macias-Romero, M. E. P. Didier, V. Zubkovs, L. Delannoy, F. Dutto, A. Radenovic, *et al.*, Probing rotational and translational diffusion of nanodoublers in living cells on microsecond time scales, *Nano Lett.*, 2014, **14**(5), 2552–2557.

23 R. Hellwarth and P. Christensen, Nonlinear optical microscopic examination of structure in polycrystalline ZnSe, *Opt. Commun.*, 1974, **12**(3), 318–322.

24 C. Sheppard, J. Gannaway, R. Kompfner and D. Walsh, The scanning harmonic optical microscope, *IEEE J. Quantum Electron.*, 1977, **13**(9), 912.
25 B. R. Masters and P. So, *Handbook of Biomedical Nonlinear Optical Microscopy*, Oxford University Press, 2008, p. 895.
26 E. J. Gualtieri, F. Guo, D. J. Kissick, J. Jose, R. J. Kuhn, W. Jiang, *et al.*, Detection of Membrane Protein Two-Dimensional Crystals in Living Cells, *Biophys. J.*, 2011, **100**(1), 207–214.
27 E. L. DeWalt, V. J. Begue, J. A. Ronau, S. Z. Sullivan, C. Das and G. J. Simpson, Polarization-resolved second-harmonic generation microscopy as a method to visualize protein-crystal domains, *Acta Crystallogr., Sect. D: Biol. Crystallogr.*, 2013, **69**(1), 74–81.
28 P. J. Campagnola, A. C. Millard, M. Terasaki, P. E. Hoppe, C. J. Malone and W. A. Mohler, Three-dimensional high-resolution second-harmonic generation imaging of endogenous structural proteins in biological tissues, *Biophys. J.*, 2002, **82**(1), 493–508.
29 S. V. Plotnikov, A. C. Millard, P. J. Campagnola and W. A. Mohler, Characterization of the myosin-based source for second-harmonic generation from muscle sarcomeres, *Biophys. J.*, 2006, **90**(2), 693–703.
30 D. A. Dombeck, K. A. Kasischke, H. D. Vishwasrao, M. Ingelsson, B. T. Hyman and W. W. Webb, Uniform polarity microtubule assemblies imaged in native brain tissue by second-harmonic generation microscopy, *Proc. Natl. Acad. Sci.*, 2003, **100**(12), 7081–7086.
31 G. Latour, L. Kowalczuk, M. Savoldelli, J. L. Bourges, K. Plamann, F. Behar-Cohen, *et al.*, Hyperglycemia-induced abnormalities in rat and human corneas: the potential of second harmonic generation microscopy, *PLoS One*, 2012, 7(11), e48388.
32 B. A. Nemet, V. Nikolenko and R. Yuste, Second harmonic imaging of membrane potential of neurons with retinal, *J. Biomed. Opt.*, 2004, **9**(5), 873.
33 M. A. Kriech and J. C. Conboy, Imaging Chirality with Surface Second Harmonic Generation Microscopy, *J. Am. Chem. Soc.*, 2005, **127**(9), 2834–2835.
34 K. L. Sly, T. T. Nguyen and J. C. Conboy, Lens-less surface second harmonic imaging, *Opt. Express*, 2012, **20**(20), 21953–21967.
35 M. Nuriya, S. Fukushima, A. Momotake, T. Shinotsuka, M. Yasui and T. Arai, Multimodal two-photon imaging using a second harmonic generation-specific dye, *Nat. Commun.*, 2016, **7**(1), 11557.
36 O. B. Tarun, C. Hannesschläger, P. Pohl and S. Roke, Label-free and charge-sensitive dynamic imaging of lipid membrane hydration on millisecond time scales, *Proc. Natl. Acad. Sci*, 2018, **115**(16), 4081–4086.
37 R. W. Boyd, *Nonlinear Optics*, Academic Press, 2020, p. 636.
38 R. S. Ries, H. Choi, R. Blunck, F. Bezanilla and J. R. Heath, Black Lipid Membranes: Visualizing the Structure, Dynamics, and Substrate Dependence of Membranes, *J. Phys. Chem. B*, 2004, **108**(41), 16040–16049.
39 G. Gonella, C. Lütgebaucks, A. G. F. De Beer and S. Roke, Second Harmonic and Sum-Frequency Generation from Aqueous Interfaces Is Modulated by Interference, *J. Phys. Chem. C*, 2016, **120**(17), 9165–9173.
40 S. Lee, C. S. Poojari, A. Maznichenko, D. Roesel, I. Swiderska, P. Pohl, *et al.*, Dynamic Second Harmonic Imaging of Proton Translocation Through Water Needles in Lipid Membranes, *J. Am. Chem. Soc.*, 2024, **146**(29), 19818–19827.

41 S. Lee, D. Roesel and S. Roke, Imaging Cu2+ binding to charged phospholipid membranes by high-throughput second harmonic wide-field microscopy, *J. Chem. Phys.*, 2021, **155**(18), 184704.

42 O. B. Tarun, H. I. Okur, P. Rangamani and S. Roke, Transient domains of ordered water induced by divalent ions lead to lipid membrane curvature fluctuations, *Commun. Chem.*, 2020, **3**(1), 1–8.

43 O. B. Tarun, M. Y. Eremchev, A. Radenovic and S. Roke, Spatiotemporal Imaging of Water in Operating Voltage-Gated Ion Channels Reveals the Slow Motion of Interfacial Ions, *Nano Lett.*, 2019, **19**(11), 7608–7613.

44 O. B. Tarun, M. Y. Eremchev and S. Roke, Interaction of Oil and Lipids in Freestanding Lipid Bilayer Membranes Studied with Label-Free High-Throughput Wide-Field Second-Harmonic Microscopy, *Langmuir*, 2018, **34**(38), 11305–11310.

45 M. Eremchev, D. Roesel, P. M. Dansette, A. Michailovas and S. Roke, High throughput wide field second harmonic imaging of giant unilamellar vesicles, *Biointerphases*, 2023, **18**(3), 031202.

46 M. Eremchev, D. Roesel, C. S. Poojari, A. Roux, J. S. Hub and S. Roke, Passive transport of Ca2+ ions through lipid bilayers imaged by widefield second harmonic microscopy, *Biophys. J.*, 2023, **122**(4), 624–631.

47 D. Roesel, M. Eremchev, C. S. Poojari, J. S. Hub and S. Roke, Ion-Induced Transient Potential Fluctuations Facilitate Pore Formation and Cation Transport through Lipid Membranes, *J. Am. Chem. Soc.*, 2022, **144**(51), 23352–23357.

48 M. E. P. Didier, O. B. Tarun, P. Jourdain, P. Magistretti and S. Roke, Membrane water for probing neuronal membrane potentials and ionic fluxes at the single cell level, *Nat. Commun.*, 2018, **9**(1), 5287.

49 I. Nahalka, G. Zwaschka, R. K. Campen, A. Marchioro and S. Roke, Mapping Electrochemical Heterogeneity at Gold Surfaces: A Second Harmonic Imaging Study, *J. Phys. Chem. C*, 2020, **124**(37), 20021–20034.

50 H. I. Okur, O. B. Tarun and S. Roke, Chemistry of Lipid Membranes from Models to Living Systems: A Perspective of Hydration, Surface Potential, Curvature, Confinement and Heterogeneity, *J. Am. Chem. Soc.*, 2019, **141**(31), 12168–12181.

51 R. Dimova and C. Marques, *The Giant Vesicle Book*, CRC Press, 2019, p. 677.

52 R. Dimova, Giant Vesicles and Their Use in Assays for Assessing Membrane Phase State, Curvature, Mechanics, and Electrical Properties, *Annu. Rev. Biophys.*, 2019, **48**, 93–119.

53 M. Montal and P. Mueller, Formation of Bimolecular Membranes from Lipid Monolayers and a Study of Their Electrical Properties, *Proc. Natl. Acad. Sci*, 1972, **69**(12), 3561–3566.

54 A. Horner, S. A. Akimov and P. Pohl, Long and Short Lipid Molecules Experience the Same Interleaflet Drag in Lipid Bilayers, *Phys. Rev. Lett.*, 2013, **110**(26), 268101.

55 A. Weinberger, F. C. Tsai, G. H. Koenderink, T. F. Schmidt, R. Itri, W. Meier, *et al.*, Gel-Assisted Formation of Giant Unilamellar Vesicles, *Biophys. J.*, 2013, **105**(1), 154–164.

56 D. A. Dombeck, M. Blanchard-Desce and W. W. Webb, Optical Recording of Action Potentials with Second-Harmonic Generation Microscopy, *J. Neurosci.*, 2004, **24**(4), 999–1003.

57 C. Lütgebaucks, G. Gonella and S. Roke, Optical label-free and model-free probe of the surface potential of nanoscale and microscopic objects in aqueous solution, *Phys. Rev. B*, 2016, **94**(19), 195410.
58 E. Reeve J, L. Anderson H and K. Clays, Dyes for biological second harmonic generation imaging, *Phys. Chem. Chem. Phys.*, 2010, **12**(41), 13484–13498.
59 D. A. Dombeck, L. Sacconi, M. Blanchard-Desce and W. W. Webb, Optical Recording of Fast Neuronal Membrane Potential Transients in Acute Mammalian Brain Slices by Second-Harmonic Generation Microscopy, *J. Neurophysiol.*, 2005, **94**(5), 3628–3636.
60 M. Nuriya, J. Jiang, B. Nemet, K. B. Eisenthal and R. Yuste, Imaging membrane potential in dendritic spines, *Proc. Natl. Acad. Sci.*, 2006, **103**(3), 786–790.
61 M. Nuriya and M. Yasui, Membrane potential dynamics of axons in cultured hippocampal neurons probed by second-harmonic-generation imaging, *J. Biomed. Opt.*, 2010, **15**(2), 020503.
62 L. Sacconi, D. A. Dombeck and W. W. Webb, Overcoming photodamage in second-harmonic generation microscopy: Real-time optical recording of neuronal action potentials, *Proc. Natl. Acad. Sci*, 2006, **103**(9), 3124–3129.
63 S. Rama, L. Vetrivel and A. Semyanov, Second-harmonic generation voltage imaging at subcellular resolution in rat hippocampal slices, *J. Biophotonics*, 2010, **3**(12), 784–790.
64 J. Jiang and R. Yuste, Second-Harmonic Generation Imaging of Membrane Potential with Photon Counting, *Microsc. Microanal.*, 2008, **14**(6), 526–531.
65 J. G. Kirkwood, Statistical Mechanics of Liquid Solutions, *Chem. Rev.*, 1936, **19**(3), 275–307.
66 J. E. Enderby, W. S. Howells and R. A. Howe, The structure of aqueous solutions, *Chem. Phys. Lett.*, 1973, **21**(1), 109–112.
67 J. A. Prins, X-ray diffraction in ionic solutions, *Physica*, 1934, **1**(7), 1171–1173.
68 E. O. Fetisov, C. J. Mundy, G. K. Schenter, C. J. Benmore, J. L. Fulton and S. M. Kathmann, Nanometer-Scale Correlations in Aqueous Salt Solutions, *J. Phys. Chem. Lett.*, 2020, **11**(7), 2598–2604.
69 M. A. Marques, M. I. Cabaço, M. I. Marques, B. de and A. M. Gaspar, Intermediate-range order in aqueous solutions of salts constituted of divalent ions combined with monovalent counter-ions, *J. Phys.: Condens. Matter*, 2002, **14**(32), 7427.
70 Y. Georgalis, A. M. Kierzek and W. Saenger, Cluster Formation in Aqueous Electrolyte Solutions Observed by Dynamic Light Scattering, *J. Phys. Chem. B*, 2000, **104**(15), 3405–3406.
71 J. F. Nagle and S. Tristram-Nagle, Structure and Interactions of Lipid Bilayers: Role of Fluctuations, in *Lipid Bilayers: Structure and Interactions*, ed. J. Katsaras and T. Gutberlet, Springer, Berlin, Heidelberg, 2001, pp. 1–23, DOI: 10.1007/978-3-662-04496-4_1.
72 J. C. Stachowiak, F. M. Brodsky and E. A. Miller, A cost–benefit analysis of the physical mechanisms of membrane curvature, *Nat. Cell Biol.*, 2013, **15**(9), 1019–1027.
73 D. Marsh, Lateral Pressure Profile, Spontaneous Curvature Frustration, and the Incorporation and Conformation of Proteins in Membranes, *Biophys. J.*, 2007, **93**(11), 3884–3899.
74 M. Tarek, Membrane Electroporation: A Molecular Dynamics Simulation, *Biophys. J.*, 2005, **88**(6), 4045–4053.

75 E. Duboué-Dijon, P. E. Mason, H. E. Fischer and P. Jungwirth, Hydration and Ion Pairing in Aqueous Mg2+ and Zn2+ Solutions: Force-Field Description Aided by Neutron Scattering Experiments and *Ab Initio* Molecular Dynamics Simulations, *J. Phys. Chem. B*, 2018, **122**(13), 3296–3306.
76 M. Kohagen, P. E. Mason and P. Jungwirth, Accurate Description of Calcium Solvation in Concentrated Aqueous Solutions, *J. Phys. Chem. B*, 2014, **118**(28), 7902–7909.
77 H. L. Tepper and G. A. Voth, Protons May Leak through Pure Lipid Bilayers *via* a Concerted Mechanism, *Biophys. J.*, 2005, **88**(5), 3095–3108.
78 S. J. Marrink, F. Jähnig and H. J. Berendsen, Proton transport across transient single-file water pores in a lipid membrane studied by molecular dynamics simulations, *Biophys. J.*, 1996, **71**(2), 632–647.
79 N. Agmon, H. J. Bakker, R. K. Campen, R. H. Henchman, P. Pohl, S. Roke, *et al.*, Protons and Hydroxide Ions in Aqueous Systems, *Chem. Rev.*, 2016, **116**(13), 7642–7672.

Faraday Discussions

PAPER

Using phase-resolved vibrational sum-frequency imaging to probe the impact of head-group functionality on hierarchical domain structure in lipid membranes†

Ben John, Sarabjeet Kaur, Martin Wolf, Martin Thämer and Alexander P. Fellows *

Received 30th November 2024, Accepted 18th December 2024

DOI: 10.1039/d4fd00187g

The substantial diversity in phospholipids within a plasma membrane, varying in tail length, degree of saturation, and head-group functionality, generates widespread structural heterogeneity. This exists both laterally across the membrane through the spontaneous formation of condensed domains that differ from their surrounding expanded phase in density, composition, and molecular packing order, as well as between its two leaflets, which normally maintain significant compositional asymmetry. Of particular importance is the exposure of phosphatidylserine (PS) lipids which is a marker for important physiological processes *e.g.* apoptosis. Despite this, the molecular-level alterations to the phase-structure of the membrane that result from PS exposure remain generally unknown. In this work, we utilise recently developed phase-resolved azimuthal-scanned sum-frequency generation (SFG) microscopy to investigate structural changes that occur heterogeneously across model membranes as a result of PS-lipid exposure. Specifically, by probing mixed monolayers of 1,2-dipalmitoylphosphatidylcholine (DPPC) and deuterated 1-palmitoyl-2-oleoylphosphatidylcholine (dPOPC) in both the C–H and C–D stretching regions as well as equivalent films with DPPC exchanged with DPPS, we analyse the variations in the apparent phase distributions and domain morphologies, and quantitatively extract the density, composition, and relative out-of-plane packing order for both mixtures. We find that, in these mixtures, DPPS shows vast differences in the domain growth and coalescence behaviour compared to DPPC, as well as in the relative compositions and molecular ordering within each phase. This demonstrates the critical role the head-group plays in the heterogeneous phase structure of the membrane and may give insights into their impact on important physiological processes.

Fritz-Haber-Institut der Max-Planck-Gesellschaft, Faradayweg 4-6, Berlin, 14195, Germany. E-mail: fellows@fhi-berlin.mpg.de

† Electronic supplementary information (ESI) available. See DOI: https://doi.org/10.1039/d4fd00187g

Introduction

The phospholipids within cell membranes play a pivotal role in determining their physicochemical properties and physiological function. One of the most pertinent and characteristic features of the membranes is their widespread heterogeneity, which exists both laterally across the membrane as well as between its two leaflets.[1–3] The lateral heterogeneity is often expressed by spontaneous separation of the membrane constituents into distinct phases, which can differ in density, composition, and packing structure.[4–7] The condensed lipid phase domains (rafts) that form during this phase separation are at the heart of many important cellular processes *e.g.*, by providing specific binding sites for proteins and enzymes,[2,8] altering the adhesion properties of the membrane,[9] and displaying characteristic signalling markers.[8] Overall, the functional behaviour of these domains, and thus of the membrane as a whole, is highly reliant on the specific details of their molecular structure.

Beyond the lateral heterogeneity, membranes also possess substantial heterogeneity in the lipid composition between their two leaflets, with the proximal (internal) leaflet being enriched in lipids with phosphatidylethanolamine (PE) and phosphatidylserine (PS) head-groups, whilst the distal (external) leaflet is enriched in phosphatidylcholine (PC) head-group lipids.[3,10] This asymmetry is actively maintained by the balance between active 'flippase' and 'scramblase' transport proteins[11] but can easily be disturbed, leading to a mixing of lipids and thus substantial compositional changes between the two leaflets. These changes in the lipid constitution can hence have marked effects on the lateral structure of each leaflet, with potentially significant alterations to any formed domains and thus also to the macroscopic physiological membrane properties.

This loss in leaflet asymmetry is particularly critical for the PS lipids as their presence in the proximal leaflet is linked to specific docking sites for signalling proteins[12,13] as well as them being important enzyme cofactors.[14] Equally, the loss of internalised PS is concomitant with PS exposure in the distal leaflet, which has been implicated in many disease pathogeneses[15] as well as being an early marker for apoptosis.[16] Specifically, externalisation of PS lipids has been suggested to impact the susceptibility of the membrane for viral entry[13] and neuronal re-/degeneration,[17,18] as well as generally altering the adhesive properties of the membrane, leading to an increased likelihood of aggregation.[19,20] Nevertheless, despite the importance of this compositional heterogeneity in the membrane, relatively little is known about the molecular-level structural alterations induced by PS exposure, and thus about the mechanistic pathways leading to the wider changes in membrane behaviours that are crucial in these critical physiological phenomena.

One important route to understand these changes to the membrane structure due to the varying concentration of PS lipids is through fundamental studies of the inter-lipid interactions within the liquid-condensed (LC) and liquid-expanded (LE) phases in model systems comprising phospholipid mono- or bi-layers. By understanding the deterministic factors controlling the fluctuations in the hierarchical packing structure in these films, one gets a unique insight into the molecular-level origins of the aforementioned physiological processes. Techniques such as atomic force microscopy (AFM) and Brewster angle microscopy

(BAM) have been combined with this bottom-up approach to elucidate the structural characteristics in lipid membranes and have revealed many of the crucial details underlying the observed changes in phase behaviour *i.e.*, differences in the size and morphology of the formed LC domains, as well as their growth and coalescence mechanisms.[21–23] Beyond this, both Raman imaging[24,25] and fluorescence microscopy (FM)[26,27] have been used to study differences in their composition, showing a general enrichment of saturated lipids in the LC phase. Furthermore, through the use of polarised light, BAM[28] and FM[29–33] have demonstrated that certain lipid mixtures can lead to substantial crystallinity in the LC molecular packing arrangements. While these studies have indeed revealed important details about the molecular-level factors underlying PS lipid exposure, many crucial structural properties within these model membranes nevertheless remain elusive.[34]

In this context, one particularly important structural property of the lipid films is the relative packing order in the different phases. While the lipids are known to pack with their tail-groups generally oriented in an 'upright' fashion,[35] the extent to which they are conformationally ordered can be highly variable.[36,37] This out-of-plane packing order is a parameter that is intimately connected to other structural properties such as the local packing density and composition (which both modulate the local intermolecular environment), but itself can lead to substantial differences in the membrane behaviour *e.g.*, through changing its thickness or energy barrier to transport processes, as well as altering the exposed structural motifs that are linked to signalling interactions.[38–40] Therefore, without this important piece of information, our understanding of the changes in molecular-to-mesoscopic phase structure of the membrane due to PS exposure remains limited.

A technique that is sensitive to the orientational order is vibrational sum-frequency generation (SFG) spectroscopy. Like Raman, by probing the vibrational resonances of the different lipids, SFG can provide molecular recognition and is thus sensitive to the film composition and lipid density. As a second-order optical technique, however, the unique symmetry selection rules of SFG mean it is dependent on absolute orientations (instead of just molecular alignment), and thus the output signals are highly sensitive to the orientational distribution *i.e.*, the amount of orientational order.[41–47] These properties have made SFG spectroscopy a powerful tool in the structural elucidation of model lipid membranes,[48–57] but it was only recently that microscopic imaging of these systems has been demonstrated.[58–62] Through these latest technical advances, it was shown that SFG microscopy can reveal previously unattainable structural features in the lipid packing structure.[59,60]

In this work, we employ this state-of-the-art vibrational SFG microscopy to study phase-separated phospholipid monolayers with different compositions to gain an unprecedented insight into the impact of PS exposure on the membrane structure. Specifically, by using our recently developed phase-resolved azimuthal-scanned SFG microscope,[60] we elucidate the differences in domain size and morphology, lipid composition, and out-of-plane packing order between mixed PS/PC monolayers with ones containing only PC lipids. The obtained results give a new perspective on the structural changes that occur when saturated PC lipids are replaced with those possessing a negatively charged PS head-group, and thus also provide insight on the molecular-level mechanisms resulting in the wider membrane alterations associated with PS lipid exposure.

Results and discussion

To elucidate the structural impact of PS lipid exposure, phase-resolved hyperspectral SFG images were recorded in the PPP polarisation combination (SFG, visible, infrared) of mixed monolayers of either DPPC or DPPS with POPC. In order to separate the spectral signatures of the two constituent lipids within each film, deuterated POPC (dPOPC) was used, and the hyperspectral images recorded in both the C–H (~2800–3000 cm^{-1}) and C–D (~2050–2250 cm^{-1}) stretching regions. Furthermore, we isolate the out-of-plane structural information of the lipids from the overall PPP response that contains both in-plane (x-direction) and out-of-plane (z-direction) contributions by recording SFG images with azimuthal scanning of the sample. Full theoretical details of this procedure can be found elsewhere.[60] In short, the azimuthal scanning modulates the in-plane contributions whilst leaving the out-of-plane contributions unchanged. Therefore, taking the 0-frequency response of the rotational Fourier transform of this four-dimensional dataset yields hyperspectral images solely containing the out-of-plane signals. A brief schematic of this structural characterisation procedure is presented in Fig. 1a. The resulting out-of-plane hyperspectral information for

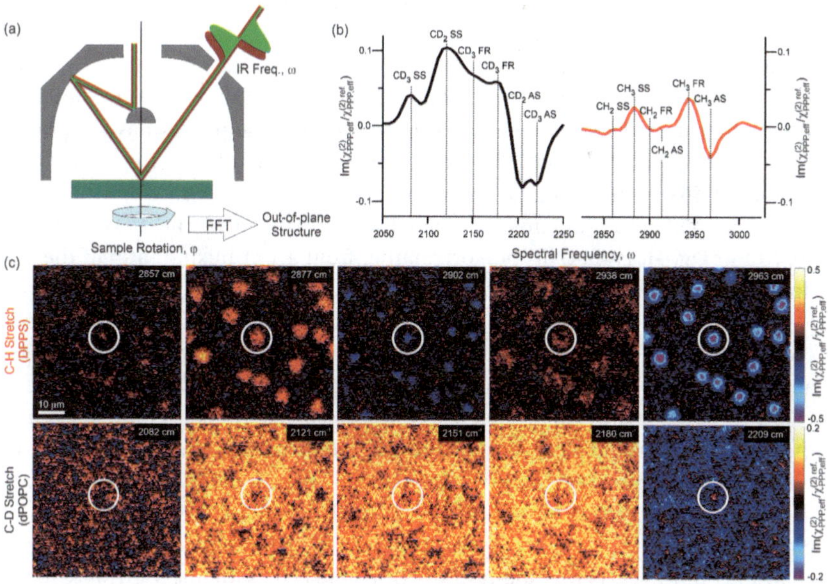

Fig. 1 Azimuthal-scanned phase-resolved SFG microscopy of a mixed lipid monolayer of 1 : 1 DPPS : dPOPC. (a) Schematic of the SFG microscope. (b) Imaginary parts of the out-of-plane spectra averaged over the entire image, in both the C–D (black) and C–H (red) frequency ranges, with labels showing the assignment of the resonant features. (c) Images at selected frequencies through the two frequency ranges. A single condensed lipid domain is encircled in each image for reference. All SFG spectra and images are from the PPP polarisation combination and are given relative to the response from a z-cut quartz reference.

a 1 : 1 mixture of DPPS with dPOPC is then shown in Fig. 1b and c. Specifically, Fig. 1b presents the imaginary parts (absorptive line-shapes) of the average spectra across the two frequency ranges, also showing their resonant assignments based on the literature[44,63] and a comparison between pure dPOPC and d_{31}-DPPC SFG spectra shown in the ESI.† Fig. 1c then shows images of the out-of-plane SFG response at selected frequencies through both spectral regions.

An initial comparison of the spatially averaged spectra shown in Fig. 1b highlights two important points about the structure of the membrane. Firstly, the C–H spectrum, corresponding to the DPPS lipid, contains predominantly CH_3 resonances, with only minimal contributions from the (more numerous) CH_2 groups. This strongly contrasts to the C–D spectrum of dPOPC which shows substantial CD_2 contributions amongst relatively modest CD_3 resonances. The presence of CH_2/CD_2 signals in such systems is typically indicative of decreased packing order given that an 'all-trans' tail conformation will present perfectly cancelling methylene signals (as there is an equal number of oppositely oriented groups, and implying an effective local centro-symmetry).[44] Only deviations from this well-ordered conformation *e.g.*, the introduction of gauche 'defects', can therefore give rise to a net CH_2/CD_2 response. Even considering the partial tail-group oxidation of dPOPC which can reduce the effective CD_3 density by up to a factor of 2 (both the saturated and unsaturated chains are still present after oxidation, but only the former maintains its terminal CD_3),[64] the observed methyl-to-methylene ratios strongly indicate that the DPPS lipid is more ordered than dPOPC. Such an observation aligns with previous SFG spectroscopy measurements of similar lipid systems[52,64–66] as well as with expectation as the fully saturated DPPS lipid can pack more closely and thus form highly restricted packing motifs that favour the all-trans 'pointing up' conformation with its smaller molecular footprint. On the other hand, the unsaturated dPOPC lipid (as well as its oxidised forms) cannot pack as efficiently[64] and thus has tail-groups with more spatial freedom, favouring the formation of the higher entropy conformations containing gauche defects.

The second important insight from the comparison of the C–H and C–D spectra in Fig. 1b is that, despite originating from a 1 : 1 mixing ratio of the two lipids and despite the effects of oxidation of dPOPC, the amplitude of the average C–D spectrum is notably larger than that of the average C–H spectrum. This is even more surprising given that aliphatic C–H resonances typically have larger amplitudes than their corresponding C–D resonances (by a factor of ~1.5, see ESI†). Nevertheless, as mentioned earlier, SFG signals are not simply related to the local molecular density, but also to the orientational order. Therefore, this observation could also originate from a reduced order in the DPPS (*cf.* dPOPC). In the discussion above, however, we determined that the DPPS showed signs of greater packing order given its relative lack of CH_2 signals. This apparent contradiction suggests that many of the DPPS molecules do not contribute to the observed spectrum. This could be due to them having largely in-plane contributions from their tail-groups (thus no out-of-plane signals), or equally that they are so disordered that their signals entirely cancel (including CH_2 signals from conformational defects). Both of these are characteristic of relatively low packing densities with greater spatial freedom. The combination of some DPPS lipids that contribute highly ordered SFG signals with others that appear to be highly disordered hence suggests a distinct two-phase structure of the monolayer.

Far greater insight into the phase structure of the model membrane can clearly be obtained by looking beyond the spatially averaged spectra and into the hyperspectral images in Fig. 1c. Unlike the average spectra, these showcase the significant and expected lateral heterogeneity present in the film. Through the C–H stretching frequencies, the images at selected resonant features show roughly circular surface regions with increased amplitude. This indicates that they possess increased molecular densities of DPPS and/or greater packing order than their surroundings. By contrast, the same surface regions appear to show a diminished amplitude at the C–D stretching frequencies, suggesting either a decrease in the local dPOPC density or a reduced packing order of this lipid. These observations are consistent with the expectations for condensed phase (LC) domains, which are typically enriched in saturated lipids like DPPS at the expense of unsaturated lipids like POPC, which predominantly occupy the surrounding expanded (LE) phase.[67,68] Overall, this confirms the presence of a clear two-phase, LC–LE equilibrium structure in the membrane.

Domain size and morphology

The hyperspectral images in Fig. 1c clearly contain a vast wealth of structural information that can be used to assess the impact of PS lipid exposure. In the next step, we directly compare the phase-separated structure of the membranes by contrasting images of the integrated C–H magnitudes taken from similar SFG microscopy measurements to those presented in Fig. 1. These comparisons are made between the DPPS:dPOPC and DPPC:dPOPC mixtures, thus essentially replacing the DPPS with DPPC, at both the 1 : 1 mixture that is shown for the DPPS mixture in Fig. 1, as well as the higher, 4 : 1 ratio of the saturated lipids. The resulting images are presented in Fig. 2, showcasing the different sizes and morphologies of the formed LC domains. For the two 1 : 1 mixtures, while both

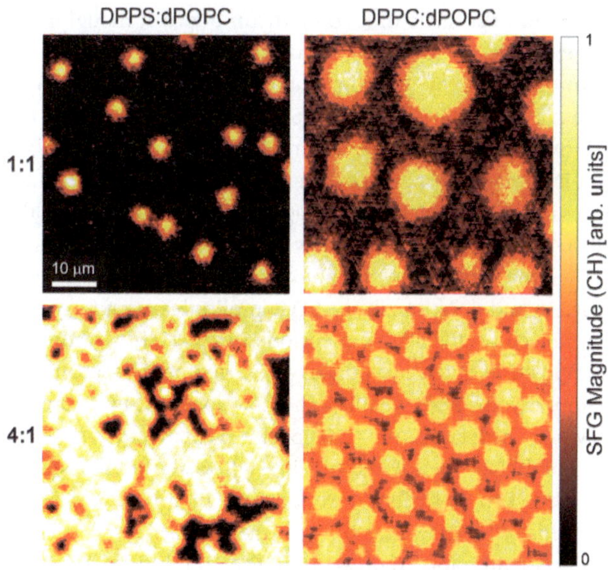

Fig. 2 SFG magnitude images for lipid mixtures of both DPPS and DPPC with dPOPC in 1 : 1 and 4 : 1 ratios, calculated by integrating across the C–H stretching region.

form roughly circular domains, those formed with DPPS are considerably smaller, with ∼5 μm diameters compared to ∼10–15 μm for those formed with DPPC. For the 4 : 1 mixtures, while the increased proportion of DPPC still presents roughly circular LC domains (albeit smaller than those in the corresponding 1 : 1 mixture), the DPPS sample essentially shows a continuous, interconnected condensed phase network.

These vastly contrasting phase structures within the two lipid mixtures (with and without PS) demonstrates that important differences exist in the intermolecular interactions between the lipids in each system. These different interactions are emphasised by the presence of the smaller isolated domains in the 1 : 1 mixture with DPPS (*cf.* with DPPC). This is because the domain size is linked to thermodynamic properties such as line tension which, in turn, relates to the lateral lipid interactions in the film.[69,70] Therefore, the exposure of PS lipids appears to vastly perturb the domain growth mechanism and thus also the resulting condensed structural motifs present in the membrane. Equally important are the differences between the two 4 : 1 mixtures, with the DPPS sample showing that the condensed domains have merged into a continuous LC network unlike those containing DPPC that maintain their separation. This suggests that the domains formed with PS lipids have complementary packing structures which present a much smaller energy barrier to coalescence.[69] In contrast, the apparent large coalescence energy barrier for the DPPC domains may well be an underlying factor determining the smaller domain sizes in the 4 : 1 mixture. This results from the combination of unfavourable coalescence and larger LC area coverage favouring smaller domains for a more optimal 2D packing. Overall, these observations accentuate the substantial structural impact that PS lipids can have on the phase-separated membrane structure.

Composition and packing order

As a further step in the structural comparison between model membranes with and without PS lipids, we look deeper into the structure of the LC and LE phases within the monolayers by probing their density, composition, and relative packing orders. Fig. 3 shows equivalent SFG magnitude images to those in Fig. 2, but for both the C–H and C–D spectral regions of the two 1 : 1 mixture samples. Additionally, the imaginary parts of the spectra averaged over the LC (red) and LE (black) phases are shown below each image. For each sample, both the magnitude images and spectra in Fig. 3a and c show that the LC phase (red trace) has greater C–H signal than the surrounding LE phase (black trace), and *vice versa* for the C–D signals in Fig. 3b and d. Although the SFG amplitudes are not solely a measure of the specific molecular density, as mentioned earlier, this is consistent with the expected composition of the two phases *i.e.* enrichment of saturated lipids in the LC phase. Similarly, for both samples, the C–H spectra (Fig. 3a and c) across both phases show more order than the C–D spectra (Fig. 3b and d, even accounting for the oxidation of dPOPC mentioned previously). As above, this is characterised by the ratio of methyl to methylene resonances, which is emphasised here by the green (methylene, CH_2/CD_2) and orange (methyl, CH_3/CD_3) bars shown within each of the spectral plots representing the amplitudes of their symmetric stretches in the LC phase spectra. Once again, this aligns with the expected packing structure of the saturated *vs.* unsaturated lipids.

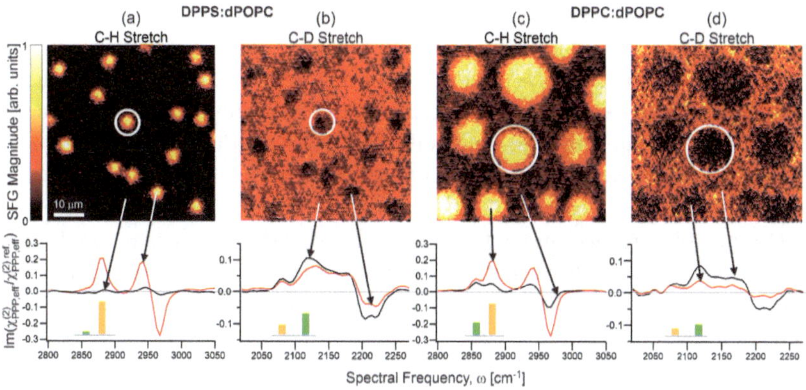

Fig. 3 SFG magnitude images for both DPPS:dPOPC ((a) and (b)) and DPPC:dPOPC ((c) and (d)) 1 : 1 lipid mixtures in the C–H ((a) and (c)) and C–D ((b) and (d)) stretching regions. A single LC domain in each system is encircled for reference. Also shown are the imaginary parts of the PPP spectra averaged over the LC (red) and LE (black) phases. The spectra are given relative to the response from a z-cut quartz reference. The amplitudes of the methyl (CH_3/CD_3, orange) and methylene (CH_2/CD_2, green) symmetric stretches within the LC spectra are indicated by bars at the bottom of each spectral plot.

Beyond these similarities, however, the two samples also show notable differences. Firstly, comparing the C–H spectra, the DPPS in the LC phase (Fig. 3a, red trace) appears to be generally more ordered than the LC phase DPPC (Fig. 3c, red trace), with the latter presenting larger relative CH_2 signals (see orange/green bars). For the LE phase, however, while the DPPC (Fig. 3c, black trace) shows a relatively weak spectrum with significant CH_2 signals (thus indicating more disorder than the corresponding LC phase DPPC), the DPPS sample (Fig. 3a, black trace) essentially shows no C–H amplitude. This could indicate a general lack of DPPS density in the LE phase, or equally that the DPPS is strongly disordered there. The above comparison of the average spectra for this sample (Fig. 1b) indicated that there must be a significant proportion of DPPS molecules not contributing to the SFG signals. This would hence suggest that the latter is true *i.e.*, the LE phase does contain DPPS, but it is largely disordered.

Comparing the C–D spectra shows generally larger amplitudes in the DPPS sample (Fig. 3b, *cf.* with DPPC in Fig. 3d). Again, this indicates either a higher density or greater packing order of dPOPC. Interestingly, the C–D spectra for the LC and LE phases shows a greater contrast in the DPPC sample (Fig. 3d) than for the mixture with DPPS (Fig. 3b). Specifically, the LC spectrum with DPPC (Fig. 3d, red trace) is much weaker than that for the LE phase (Fig. 3d, black trace), and has a notably different ratio of the CD_3 and CD_2 symmetric stretches, indicating greater order in the former. By contrast, while the LC spectrum for the DPPS sample (Fig. 3b, red trace) is also weaker and has a reduced CD_3 to CD_2 ratio (*cf.* LE), the differences are more modest and the spectra generally appear closer in overall amplitude and line-shape. This indicates that the structure of the dPOPC lipids within the two phases are likely more similar for the DPPS sample than with DPPC.

While the above analysis of the spectra from the LC and LE phases in the two samples does give some snippets of structural insight into the differences between the two membranes, it is nevertheless not straightforward to combine these pieces of information into a comprehensive overview of the different membrane structures. This is predominantly due to the qualitative nature of this comparison. For better insight, therefore, the measured signals should be quantitatively related to the structural details. To do this, the observed SFG signals need to be separated into their two main contributions: density and out-of-plane order.

Overall, SFG signals can be described as in eqn (1), where ρ is the molecular density, β is the molecular hyperpolarisability, and R represents a rotation matrix as part of the Euler transformation converting the molecular coordinates (a, b, c) to the lab frame (x, y, z).

$$S_{ijk}^{\mathrm{SFG}} \propto \rho \sum_{l,m,n} \langle R_{il} \cdot R_{jm} \cdot R_{kn} \rangle \beta_{lmn} \quad i, j, k \in \{x, y, z\};\ l, m, n \in \{a, b, c\} \tag{1}$$

In this expression, the $\langle\rangle$ brackets indicate averaging over the orientational distribution. As mentioned earlier, this orientational averaging will generally result in some signal cancellation. We therefore rewrite eqn (1) as in eqn (2) to include a general orientational order coefficient, O, that is hence related to the width of the orientational distribution and the amount of signal cancellation. Equally, we can include a general transformation parameter, τ, that represents the Euler transformation based on the average molecular orientation in the lab frame (indicated by their three average Euler angles, $\bar{\theta}$, $\bar{\phi}$, and $\bar{\psi}$).

$$S^{\mathrm{SFG}} \propto \rho \cdot O \cdot \tau(\bar{\theta}, \bar{\phi}, \bar{\psi}) \cdot \beta \tag{2}$$

For the comparison of the different lipid mixtures presented here, eqn (2) can be further simplified. Firstly, as we are solely isolating the out-of-plane contribution, the in-plane Euler angle, $\bar{\phi}$, can be neglected. Secondly, as the SFG signals are generally dominated by the lipid tail-groups,[52,65] they are modulated by the same hyperpolarisabilities, meaning β can be absorbed into the proportionality constant. Finally, as the lipids all present a general 'upright' orientation of their tail-groups, the variation in their two remaining average Euler angles is likely not particularly large. In fact, comparing the ratio of the symmetric and antisymmetric methyl stretches (which is highly dependent on both of these parameters[46,65]) across the spectra for both phases in both lipid mixtures shows very little variation (see ESI†). This means, to a reasonable approximation, the τ parameter can also be absorbed into the proportionality constant, leaving any variations in signals being dictated only by changes in density or the out-of-plane orientational order coefficient. This is shown in eqn (3), which forms the basis of the analysis of the SFG signals in this work.

$$S_{\mathrm{OP}}^{\mathrm{SFG}} \propto \rho \cdot O_{\mathrm{OP}} \tag{3}$$

With this description, and assuming these structural parameters are constant within the LC and LE phases (which appears reasonable given the relative homogeneity of their observed SFG amplitudes), there are still 10 unknown structural parameters that describe each membrane. These are the densities and order coefficients of each lipid within each phase, as well as the area coverages of the two phases (A^{LC} and A^{LE}). The microscopy measurements shown in Fig. 3

provide 7 observables for each membrane, namely the area coverages of each phase (thus directly reducing the number of unknown structural parameters to 8), the SFG amplitudes in both spectral regions for the LC and LE phases (defined here as the amplitude separation between the symmetric and antisymmetric methyl stretches), and the average mixing ratio of the two lipids. This latter observable gives us eqn (4) (given the 1 : 1 ratios), relating the densities of each lipid in phase to each other through the area coverages.

$$\rho^{\mathrm{LC}}_{\mathrm{DPP(S,C)}}A^{\mathrm{LC}} + \rho^{\mathrm{LE}}_{\mathrm{DPP(S,C)}}A^{\mathrm{LE}} = \rho^{\mathrm{LC}}_{\mathrm{POPC}}A^{\mathrm{LC}} + \rho^{\mathrm{LE}}_{\mathrm{POPC}}A^{\mathrm{LE}}\left(=\frac{\rho_{\mathrm{tot}}}{2}A_{\mathrm{tot}}\right) \tag{4}$$

As the SFG amplitudes are only given by proportionalities (see eqn (3)), however, the number of unknowns is then increased to 9 (the remaining 8 structural parameters plus the proportionality constant). Clearly, the aforementioned observables obtained by SFG microscopy alone are insufficient for the full

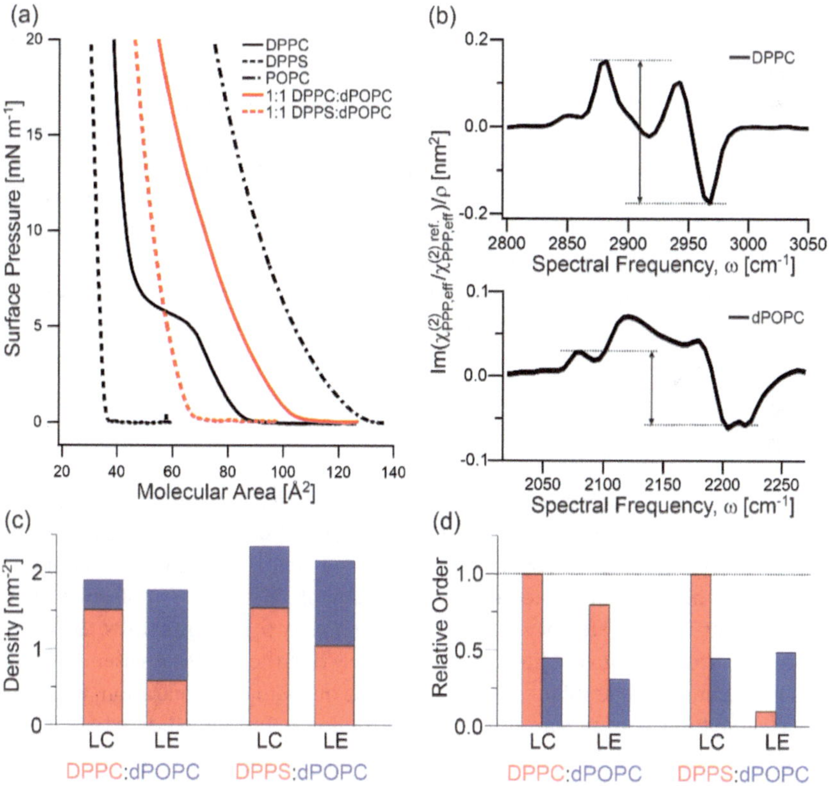

Fig. 4 Analysis of the molecular structure in the 1 : 1 DPPS:dPOPC and DPPC:dPOPC membranes. (a) Pressure–area isotherms for the three pure lipids as well as the two 1 : 1 mixtures. (b) SFG spectra in the PPP polarisation combination of a pure DPPC and dPOPC monolayer, referenced to quartz and divided by their average molecular density to provide compositional and structural reference spectra. (c) Calculated compositions of the LC and LE phases for the two lipid mixtures, given in absolute molecular densities. (d) Relative out-of-plane order coefficients in the LC and LE phases for the two lipids within each 1 : 1 mixture. The saturated lipid (DPPC or DPPS) within the LC phase is taken to have a coefficient of unity.

quantitative elucidation of the different structural parameters. Nevertheless, this can be overcome by combining these measurements with information from the pressure–area (Π–A) isotherms of the different monolayers shown in Fig. 4a, as well as reference spectra from pure DPPC and dPOPC monolayers which are shown in Fig. 4b. The isotherms can be used to determine the total (average) molecular density, ρ_{tot} (see eqn (4)), providing one additional observable. Equally, the average SFG signals from the two pure monolayers, when divided by their measured average densities (also obtained from their isotherm data), yield spectra that only depend on their orientational order coefficient, as shown in eqn (5) and (6) (based on eqn (3)).

$$\frac{S_{\mathrm{DPPC}}^{\mathrm{ref}}}{\rho_{\mathrm{DPPC}}^{\mathrm{ref}}} \propto O_{\mathrm{DPPC}}^{\mathrm{ref}} \tag{5}$$

$$\frac{S_{\mathrm{POPC}}^{\mathrm{ref}}}{\rho_{\mathrm{POPC}}^{\mathrm{ref}}} \propto O_{\mathrm{POPC}}^{\mathrm{ref}} \tag{6}$$

Therefore, by assuming that these condensed pure monolayers (reference measurements) have equal orientational coefficients to the LC domains in the mixed lipid samples, as in eqn (7) and (8), we obtain two further observables from the sizes of their SFG signals, without any further unknowns. This assumption is justified by the similarity between the spectral line-shapes of the pure monolayer references and the LC phases across both samples (see ESI†). Furthermore, the saturated lipids in the LC phase in the mixed systems and the lipids in the pure condensed monolayers both generally have well-packed structures. As they should therefore present narrow orientational distributions, it is reasonable to take their orientational order coefficients to be unity, as in eqn (7).

$$O_{\mathrm{DPPC}}^{\mathrm{ref}} = O_{\mathrm{DPP(S,C)}}^{\mathrm{LC}} = 1 \tag{7}$$

$$O_{\mathrm{POPC}}^{\mathrm{ref}} = O_{\mathrm{POPC}}^{\mathrm{LC}} \tag{8}$$

With these reasonable assumptions, the two density-normalised reference spectra shown in Fig. 4b can hence act as absolute density calibrations for the observed C–H and C–D signals. Of course, the validity of this C–D calibration reference also requires the relative oxidation of the dPOPC molecules to be equivalent in the reference sample and for the two lipid mixtures. Based on previous studies from the literature,[64] it is likely that the dPOPC is essentially fully oxidised under the ambient conditions during our sample preparation, which are the same for across all samples. This assumption is further validated by the fact that they also present highly overlapping average spectra (see ESI†), indicating relatively similar surface structures.

With this combined dataset, we can hence fully deconvolute the SFG signals and obtain the compositional densities and out-of-plane orientational order coefficients. Full details of this deconvolution are given in the ESI,† with the obtained results presented in Fig. 4c, which reports on the density, and Fig. 4d, which shows the out-of-plane orientational order coefficients. Firstly, the absolute molecular densities of each lipid in the two phases presented in Fig. 4c confirm the previous suggestion that the LC phase is enriched in DPPC/DPPS (shown in red) and depleted in dPOPC (shown in blue) for both sample systems. Secondly, for both samples, the LE phases show a lesser total (combined) density than the LC phases, and a greater relative proportion of dPOPC, also aligning with expectation. An overall comparison between the densities of both samples then shows that the DPPS mixture generally leads to less enrichment of the saturated lipids into the LC phase *i.e.*, there is less de-mixing of the two lipids. This hence indicates that the ratio of homo- to hetero-molecular interaction energies is also diminished, leading to more hetero-molecular mixing. This is not entirely surprising given that the PS head-group is doubly negatively charged and thus provides a strong electrostatic repulsion, unlike the zwitterionic PC head-group.

Another important comparison is the total (combined) density of each phase between the samples with the two different head-groups. Both phases of the DPPS sample have a higher total density than those with DPPC. This observation is likely the origin of the higher CH_3 to CH_2 ratio observed for DPPS *cf.* DPPC discussed above. The observed increase in density for the DPPS sample is also reflected in the isotherm data in Fig. 4a. This shows that a greater density is required to achieve the same surface pressure (which was the defining parameter used to form the samples) and means that at the same molecular density, the PS-containing membrane has a higher surface energy density, which is generally an indicator for a less ordered structure. This clearly demonstrates that PS exposure can have a pronounced effect on compositional heterogeneity within the proximal membrane leaflet as well as its overall thermodynamic properties.

The results on orientational ordering are highlighted in Fig. 4d, showing the difference in out-of-plane order coefficients in the two films obtained from these measurements. As mentioned above, the values for DPPC/DPPS in the respective LC domains are taken to be equal to that in the condensed DPPC reference measurement, and set to unity based on them having well-packed structures (eqn (7)). Equally, the values for dPOPC in the LC phase are also taken to be equal for both samples as well as the reference (eqn (8)), but are not set to unity. This value is instead determined by comparing the two reference measurements (DPPC and dPOPC), as discussed in ESI,† yielding a reduced value of ~0.45. This value assumes no loss in effective CD_3 density due to oxidation and thus represents the lower limit of its relative out-of-plane order. That means for the likely case of almost complete oxidation, we obtain a value of ~0.9, which suggests a surprisingly high order for dPOPC (although still being smaller than for the saturated lipids, as expected). However, as the extent to which the dPOPC has oxidised cannot be fully quantified from our measurements, we simply consider the minimum order parameter and present the corresponding values in the mixed lipid samples (which are determined relative to this) having been scaled by this reduced factor.

With these fixed order parameters for the LC phase, we can now compare the deviations observed for the LE phase. For the DPPC sample, the orientational order of both lipids is found to be somewhat lower in the LE phase compared to the LC phase, again fitting with expectation. This, however, strongly contrasts with the observations for the DPPS sample. Here, the DPPS order in the LE phase is shown to be substantially reduced to ~10% of the relative order in the LC phase. This observation aligns with our previous discussions relating to the seemingly low amplitudes of the C–H average spectra and the general lack of

DPPS signals from the LE phase. On the other hand, the dPOPC in the LE phase shows a very similar order coefficient to the LC phase. This also lines up with the above observation of similar C–D spectra for the two phases in this sample. Overall, these substantial differences demonstrate that the simple exchange of PC lipids with their PS equivalents can also lead to pronounced changes in the molecular packing order, particularly for DPPS within the LE phase.

Conclusions and outlook

Based on the individual findings presented and discussed above, a broader picture of the properties of the PS lipid in model membranes and its substantial impact on their structure and behaviour can be derived. Just as with solely PC lipids, the LC domains in the PS/PC mixture show the expected enrichment of saturated lipids, with a higher density and greater orientational order than the surrounding LE phase. However, the PS/PC mixture generally forms a higher energy monolayer and shows distinctly less compositional de-mixing between the two phases, pointing towards a lesser distinction between homo- and hetero-molecular interactions in these films. This shows that important thermodynamic properties in these model membranes are altered due to the presence of the PS lipids. Furthermore, the PS lipids clearly tend to show large disorder unless they are conformationally restricted in the LC phase, in direct contrast to PC lipids. On the other hand, it seems that there is a high barrier for forming these LC domains containing the PS lipids. These two observations may well be linked, since significant energy must be input to overcome both the entropic cost of ordering the largely disordered PS lipids in the LE phase, as well as the significant electrostatic repulsion of their negatively charged head-groups. Interestingly, the LC domains with DPPS, once they are formed, also show no significant barrier to coalescence at higher relative concentrations, in stark contrast to those with DPPC. For DPPC, it even seems that the high barrier to coalescence leads to the formation of smaller domains so as to optimise their 2D packing and avoid contact. This indicates that the boundary structures of the LC domains with DPPS are easily coupled without a significant disruption to either structure unlike those with DPPC.

While these observations lead to important insights into the molecular-to-mesoscopic structural changes that accompany the externalisation of PS lipids, many questions remain unanswered. It would thus be interesting to extend these studies by varying more sample parameters such as, mixing ratios, surface pressures, and other relevant head-groups, to shed more light on the thermodynamic and mechanistic origins of these structural effects. As demonstrated in this work, with our microscope, we nevertheless have a tool which is capable of obtaining advanced structural characterisation of these systems. Importantly, the accuracy of the quantitative structural parameters obtained here, namely densities and orientational order coefficients, are also highly supported by the complementary analysis of methyl-to-methylene ratios in the different mean spectra that provide independent confirmation. Therefore, this unprecedented insight into the molecular-level details within these systems can be extended to directly address many of these open questions. Furthermore, the elucidative potential of this technique opens possibilities for the structural investigation of a broader range of sample systems.

Experimental methods

Sample preparation

The lipids 1,2-dipalmitoyl-*sn*-glycero-3-phosphocholine (DPPC, >99%), 1,2-dipalmitoyl-*sn*-glycero-3-phosphoserine (DPPS, >99%), and d_{82}-1-palmitoyl-2-oleoyl-*glycero*-3-phosphocholine (dPOPC, >99%) were procured from Avanti Polar Lipids (Alabaster, AL, USA). Stock solutions of DPPC and dPOPC were prepared in chloroform (99.0–99.4% purity, VWR International GmbH, Darmstadt, Germany) at a concentration of 1 mg mL^{-1}. The DPPS stock solution was prepared in a 2 : 1 (v/v) mixture of chloroform and methanol (≥99.8% purity, VWR International GmbH, Darmstadt, Germany) also at a concentration of 1 mg mL^{-1}, as the addition of methanol aids in solubilising the polar and charged serine head-group of DPPS.

Mixed lipid monolayers were prepared by combining lipids in mass ratios of 4 : 1 and 1 : 1 (DPPC : dPOPC and DPPS : dPOPC, respectively) and depositing them dropwise onto the water surface of a PTFE Langmuir–Blodgett trough (MicroTrough G1, Kibron, Helsinki, Finland) containing ultrapure water (Milli-Q, 18.2 MΩ cm, <3 ppb total organic carbon). Prior to lipid deposition, the Langmuir–Blodgett trough was cleaned using chloroform and ultrapure water, followed by repeated aspiration of the water surface until the surface pressure variation was within 0.1 mN m^{-1} under full compression. The platinum Wilhelmy plate pressure sensor was flamed and rinsed with chloroform and ultrapure water to remove residual contamination. After deposition, the films were left undisturbed for approximately 5 minutes to allow for the evaporation of chloroform (or the chloroform–methanol mixture for DPPS). The monolayers were then compressed to a surface pressure of 20 mN m^{-1} at a barrier speed of 10 mm min^{-1} and equilibrated for 2 hours to ensure stabilisation and oxidation of the POPC lipid by ambient ozone exposure.[64,71]

To immobilise the monolayers for microscopy measurements, a Langmuir–Blodgett transfer method was employed to suppress both intrinsic dynamic motion and convection-driven flows, such as Bénard–Marangoni convection caused by localised subphase heating.[72] Monolayers were transferred at a rate of 2 mm min^{-1} using a LayerX dipper (Kibron, Helsinki, Finland) onto ultra-flat fused silica substrates (Korth Kristalle, Altenholz, Germany, 5 mm thickness, 25.4 mm diameter, <2 nm surface roughness). The substrates were cleaned with chloroform and ultrapure water, followed by UV-ozone treatment (UV/Ozone ProCleaner Plus, BioForce Nanosciences, Virginia Beach, VA, USA) for at least 30 minutes prior to use.

The surface pressure–area isotherms of pure DPPC, DPPS, and POPC, as well as the 1 : 1 mass mixtures of DPPC : dPOPC and DPPS : dPOPC, were recorded using the same Langmuir–Blodgett trough. The trough was cleaned following the procedure described above and the lipid stock solutions were deposited dropwise onto the water surface using an Eppendorf pipette, ensuring sufficient coverage. After deposition, the films were left undisturbed for a few minutes to ensure complete solvent evaporation. Isotherms were acquired using the compression isotherm module integrated into the Langmuir–Blodgett trough system, with the barriers compressed at a constant speed of 10 mm min^{-1}.

SFG microscope

The heterodyned phase-resolved widefield vibrational sum-frequency generation (vSFG) microscope used in this work is based on our previously developed system, details of which can be found in our prior publications.[59,60,73] In brief, the microscope operates fully in the time-domain, utilising a home-built interferometer driven by the broadband IR and visible outputs from a Ti:sapphire laser system.[74] These outputs, combined with a local oscillator (LO) generated from z-cut quartz, are aligned in a collinear geometry and focused onto the sample through a custom-designed reflective objective (0.78 NA, Pike Technologies, Madison, WI, USA). The incident beams (IR, visible, and LO) are frequency-filtered, temporally aligned, and focused onto the sample surface at an incidence angle of 36° using a 38 cm focal length off-axis parabolic mirror. The reflected vSFG signal is isolated and recorded on an electrically cooled CCD camera (ProEM-HS:1024BX3, Teledyne Princeton Instruments, Trenton, NJ, USA) using paired-pixel balanced imaging to reduce noise.[73] A PPP polarisation combination is employed throughout, probing both in-plane and out-of-plane molecular vibrations.

The hyperspectral SFG images were acquired as interferometric images recorded over time delays between the IR and other input beams (visible and LO) from −300 fs to 3000 fs in 2 fs steps. Fast Fourier transforms (FFT) of the time-domain data provided spectral images, which were normalised in phase and amplitude using a reference measurement of z-cut quartz (from −300 to 300 fs). Post-processing included removing dark counts, subtracting a linear non-resonant baseline, and averaging over 4–12 interferometric scans depending on the dataset.

For rotational analysis, vSFG images were obtained at rotational increments of 60° across the full azimuthal range. The processed images were back-rotated to align corresponding pixels, generating a 4D dataset correlating surface location, spectral frequency, and azimuthal angle. A complex Fourier transform was then applied directly to the azimuthal angle data to extract azimuthal frequencies. Two sets of measurements were performed to probe specific vibrational regions: one with the IR frequency centred at 2900 cm^{-1} to target CH stretching modes and another centred at 2100 cm^{-1} to target CD stretching modes.

Background correction

To accurately isolate the resonant vibrational signals, the SFG magnitude images were corrected for background contributions. Background correction was performed by subtracting the mean signal intensity from spectral regions outside the frequency ranges of interest, specifically the CD stretching region (~2100 cm^{-1}) and the CH stretching region (~2900 cm^{-1}). These out-of-interest regions primarily contain non-resonant contributions and noise, which do not correspond to the targeted vibrational modes. This correction step ensured that the processed images reflected only the resonant signals associated with the molecular vibrations under investigation.

Data availability

Due to the size, relative complexity, and multi-level processing of the raw data presented in this work, it has not been made available on a data repository, but any and all parts will be made available upon reasonable request to the corresponding author.

Conflicts of interest

There are no conflicts of interest.

Acknowledgements

We thank M. Krenz along with the technicians group at the Fritz Haber Institute for providing invaluable technical support along with the Bluhm group for use of their Langmuir–Blodgett trough in these measurements. Open Access funding provided by the Max Planck Society.

References

1 T. Harayama and H. Riezman, *Nat. Rev. Mol. Cell Biol.*, 2018, **19**, 281–296.
2 E. Sezgin, I. Levental, S. Mayor and C. Eggeling, *Nat. Rev. Mol. Cell Biol.*, 2017, **18**, 361–374.
3 W. Stillwell, *An Introduction to Biological Membranes: from Bilayers to Rafts*, Academic Press, London, UK, 2013.
4 D. Lingwood and K. Simons, *Science*, 2010, **327**, 46–50.
5 P. J. Quinn and D. Chapman, *CRC Crit. Rev. Biochem.*, 1980, **8**, 1–117.
6 M. Carquin, L. D'Auria, H. Pollet, E. R. Bongarzone and D. Tyteca, *Prog. Lipid Res.*, 2016, **62**, 1–24.
7 J. J. Kinnun, D. Bolmatov, M. O. Lavrentovich and J. Katsaras, *Chem. Phys. Lipids*, 2020, **232**, 104976.
8 D. A. Brown and E. London, *Annu. Rev. Cell Dev. Biol.*, 1998, **14**, 111–136.
9 S. Munro, *Cell*, 2003, **115**, 377–388.
10 Y. Ma, K. Poole, J. Goyette and K. Gaus, *Front. Immunol.*, 2017, **8**, 01513.
11 M. Doktorova, J. L. Symons and I. Levental, *Nat. Chem. Biol.*, 2020, **16**, 1321–1330.
12 T. Yeung, G. E. Gilbert, J. Shi, J. Silvius, A. Kapus and S. Grinstein, *Science*, 2008, **319**, 210–213.
13 E. M. Bevers and P. L. Williamson, *Physiol. Rev.*, 2016, **96**, 605–645.
14 J. E. Vance and R. Steenbergen, *Prog. Lipid Res.*, 2005, **44**, 207–234.
15 D. C. Calianese and R. B. Birge, *Cell Commun. Signal.*, 2020, **18**, 41.
16 B. Fadeel and S. Orrenius, *J. Intern. Med.*, 2005, **258**, 479–517.
17 H.-W. Shin and H. Takatsu, *Crit. Rev. Biochem. Mol. Biol.*, 2020, **55**, 166–178.
18 H. Faris, M. Almasieh and L. A. Levin, *Cell Death Discovery*, 2021, **7**, 247.
19 R. F. A. Zwaal, P. Comfurius and E. M. Bevers, *Cell. Mol. Life Sci.*, 2005, **62**, 971–988.
20 A. P. Fellows, M. T. L. Casford, P. B. Davies, J. S. Gibson, J. N. Brewin and D. C. Rees, *Colloids Surf., B*, 2021, **197**, 111383.
21 D. Vollhardt, *Curr. Opin. Colloid Interface Sci.*, 2014, **19**, 183–197.
22 R. I. S. Romão, Q. Ferreira, J. Morgado, J. M. G. Martinho and A. M. P. S. Gonçalves da Silva, *Langmuir*, 2010, **26**, 17165–17177.
23 J. Sanchez and A. Badia, *Chem. Phys. Lipids*, 2008, **152**, 24–37.
24 S. H. J. Donaldson and H. B. de Aguiar, *J. Phys. Chem. Lett.*, 2018, **9**, 1528–1533.

25 J. Ando, M. Kinoshita, J. Cui, H. Yamakoshi, K. Dodo, K. Fujita, M. Murata and M. Sodeoka, *Proc. Natl. Acad. Sci. U. S. A.*, 2015, **112**, 4558–4563.
26 R. F. M. de Almeida, L. M. S. Loura and M. Prieto, *Chem. Phys. Lipids*, 2009, **157**, 61–77.
27 I. C. Shieh and J. A. Zasadzinski, *Proc. Natl. Acad. Sci. U. S. A.*, 2015, **112**, E826–E835.
28 P. Krüger and M. Lösche, *Phys. Rev. E:Stat. Phys., Plasmas, Fluids, Relat. Interdiscip. Top.*, 2000, **62**, 7031–7043.
29 V. T. Moy, D. J. Keller, H. E. Gaub and H. H. McConnell, *J. Phys. Chem.*, 1986, **90**, 3198–3202.
30 V. T. Moy, D. J. Keller and H. M. McConnell, *J. Phys. Chem.*, 1988, **92**, 5233–5238.
31 U. Bernchou, J. Brewer, H. S. Midtiby, J. H. Ipsen, L. A. Bagatolli and A. C. Simonsen, *J. Am. Chem. Soc.*, 2009, **131**, 14130–14131.
32 J. Dreier, J. R. Brewer and A. C. Simonsen, *J. Phys. Chem. Lett.*, 2013, **4**, 2789–2793.
33 J. Dreier, J. Brewer and A. C. Simonsen, *Langmuir*, 2014, **30**, 10678–10685.
34 A. J. Sodt, M. L. Sandar, K. Gawrisch, R. W. Pastor and E. Lyman, *J. Am. Chem. Soc.*, 2014, **136**, 725–732.
35 B. Alberts, A. Johnson, J. Lewis, M. Raff, K. Roberts and P. Walter, *Garland Science*, New York, 2002.
36 J. L. Ranck, L. Mateu, D. M. Sadler, A. Tardieu, T. Gulik-Krzywicki and V. Luzzati, *J. Mol. Biol.*, 1974, **85**, 249–277.
37 K. Kjaer, J. Als-Nielsen, C. A. Helm, L. A. Laxhuber and H. Möhwald, *Phys. Rev. Lett.*, 1987, **58**, 2224–2227.
38 R. Lipowsky, *Nature*, 1991, **349**, 475–481.
39 J. F. Nagle and S. Tristram-Nagle, *Biochim. Biophys. Acta*, 2000, **1469**, 159–195.
40 J. Frallicciardi, J. Melcr, P. Siginou, S. J. Marrink and B. Poolman, *Nat. Commun.*, 2022, **13**, 1605.
41 Y. R. Shen, *Fundamentals of Sum-Frequency Spectroscopy*, Cambridge University Press, Cambridge, 2016.
42 A. Morita, *Theory of Sum Frequency Generation Spectroscopy*, Springer, Singapore, 2018.
43 R. W. Boyd, *Nonlinear Optics*, Elsevier Inc., London, UK, 4th edn, 2020.
44 A. G. Lambert, P. B. Davies and D. J. Neivandt, *Appl. Spectrosc. Rev.*, 2005, **40**, 103–145.
45 A. J. Moad and G. J. Simpson, *J. Phys. Chem. B*, 2004, **108**, 3548–3562.
46 H. F. Wang, W. Gan, R. Lu, Y. Rao and B. H. Wu, *Int. Rev. Phys. Chem.*, 2005, **24**, 191–256.
47 H.-F. Wang, L. Velarde, W. Gan and L. Fu, *Annu. Rev. Phys. Chem.*, 2015, **66**, 189–216.
48 G. L. Richmond, *Chem. Rev.*, 2002, **102**, 2693–2724.
49 S. Roke, *ChemPhysChem*, 2009, **10**, 1380–1388.
50 J. Zhang, W. Yang, J. Tan and S. Ye, *Phys. Chem. Chem. Phys.*, 2018, **20**, 5657–5665.
51 J. Liu and J. C. Conboy, *J. Am. Chem. Soc.*, 2004, **126**, 8376–8377.
52 G. Ma and H. C. Allen, *Langmuir*, 2006, **22**, 5341–5349.
53 J. F. D. Liljeblad, V. Bulone, E. Tyrode, M. W. Rutland and C. M. Johnson, *Biophys. J.*, 2010, **98**, L50–L52.

54 A. P. Fellows, M. T. L. Casford and P. B. Davies, *AIP Adv.*, 2021, **11**, 045119.
55 S. A. Goussous, A. P. Fellows, M. T. L. Casford and P. B. Davies, *Biochim. Biophys. Acta, Biomembr.*, 2019, **1861**, 1568–1578.
56 W. Liu, Z. Wang, L. Fu, R. M. Leblanc and E. C. Y. Yan, *Langmuir*, 2013, **29**, 15022–15031.
57 X. Chen, J. Wang, C. B. Kristalyn and Z. Chen, *Biophys. J.*, 2007, **93**, 866–875.
58 F. Yang and S. Baldelli, *J. Phys. Chem. B*, 2022, **126**, 5637–5645.
59 A. P. Fellows, B. John, M. Wolf and M. Thämer, *Nat. Commun.*, 2024, **15**, 3161.
60 A. P. Fellows, B. John, M. Wolf and M. Thämer, *J. Phys. Chem. Lett.*, 2024, **15**, 10849–10857.
61 S. M. Ortiz and S. Baldelli, *Surf. Sci.*, 2025, **751**, 122627.
62 H. Wang and W. Xiong, *Annu. Rev. Phys. Chem.*, 2021, **72**, 279–306.
63 E. Tyrode and J. Hedberg, *J. Phys. Chem. C*, 2012, **116**, 1080–1091.
64 L. Qiao, A. Ge, Y. Lang and S. Ye, *J. Phys. Chem. B*, 2015, **119**, 14188–14199.
65 R. J. Feng, X. Li, Z. Zhang, Z. Lu and Y. Guo, *J. Chem. Phys.*, 2016, **145**, 244707.
66 K. I. Inoue, C. Takada, L. Wang, A. Morita and S. Ye, *J. Phys. Chem. B*, 2020, **124**, 5246–5250.
67 P. J. Quinn and C. Wolf, *Biochim. Biophys. Acta, Biomembr.*, 2009, **1788**, 33–46.
68 G. Ma and H. C. Allen, *Langmuir*, 2007, **23**, 589–597.
69 M. Flörsheimer and H. Möhwald, *Chem. Phys. Lipids*, 1989, **49**, 231–241.
70 C. W. McConlogue and T. K. Vanderlick, *Langmuir*, 1997, **13**, 7158–7164.
71 A. P. Fellows, M. T. L. Casford and P. B. Davies, *Biophys. J.*, 2023, **122**, 2007–2022.
72 A. P. Fellows, M. T. L. Casford and P. B. Davies, *J. Chem. Phys.*, 2022, **156**, 164701.
73 T. Khan, B. John, R. Niemann, A. Paarmann, M. Wolf and M. Thämer, *Opt. Express*, 2023, **31**, 28792–28804.
74 M. Thämer, R. K. Campen and M. Wolf, *Phys. Chem. Chem. Phys.*, 2018, **20**, 25875–25882.

Faraday Discussions

ROYAL SOCIETY OF CHEMISTRY

PAPER

Phosphatidylserine affinity for and flip-flop dependence on Ca^{2+} and Mg^{2+} ions

Preston P. Hymas and John C. Conboy *

Received 21st December 2024, Accepted 4th February 2025
DOI: 10.1039/d4fd00206g

Ca^{2+} ions are believed to play a crucial role in regulating lipid membrane asymmetry by modulating the activity of flippases, floppases, and scramblases. Dysregulation of Ca^{2+} homeostasis, and subsequent loss of phosphatidylserine (PS) lipid asymmetry, is associated with physiological conditions such as blood clotting, neurodegeneration, and apoptosis. Yet, despite the prominence of Ca^{2+} with regards to PS flip-flop, the specific actions of Ca^{2+} are not fully understood and detailed mechanisms remain elusive. Much focus has been placed on enzymatic interactions, while the endogenous interactions of Ca^{2+} ions with PS and the direct role Ca^{2+} ions play on maintaining PS asymmetry have not been characterized in detail, a potentially crucial gap in understanding. In the current study the binding affinities of Ca^{2+} ions to planar supported lipid membranes containing PS were measured *via* sum-frequency vibrational spectroscopy (SFVS). Evaluation of binding affinity obtained from SFVS peak area analysis yielded an affinity of 1.3×10^5 M^{-1}. The rate of PS flip-flop was also measured in the presence and absence of Ca^{2+} *via* SFVS, with a nearly five-fold decrease in the rate of translocation when Ca^{2+} ions are present. Controls which tested Mg^{2+} with PS or phosphatidylcholine (PC) with Ca^{2+} did not show similar slowing effects, highlighting the specificity of the PS–Ca^{2+} interaction. For the binary lipid mixture tested, the disparity in the PS flip-flop rate would be sufficient to produce an 82% PS asymmetry if Ca^{2+} ions are localized to one side of the membrane. These studies have important implications for the non-enzymatic role Ca^{2+} ions may play in the maintenance of PS asymmetry.

Introduction

Lipid asymmetry in living cells has long been recognized,[1–5] though the details of how phospholipid asymmetry is established, maintained, and destroyed is the subject of continued debate. What is known is that lipids are asymmetrically distributed between the cytosolic and extracellular leaflets of the plasma membrane based on both headgroup structure and alkyl chain saturation.[6,7] Lipids facing the extracellular environment are predominantly phosphatidylcholine (PC) and sphingomyelin (SM) lipids; conversely, the aminophospholipids

Department of Chemistry, University of Utah, 315 S. 1400 E. RM 2020, Salt Lake City, UT 84112, USA. E-mail: John.Conboy@utah.edu

phosphatidyl-ethanolamine (PE) and phosphatidylserine (PS) are found enriched on the cytoplasmic leaflet. Of these lipid species, PC, SM, and PE are net neutral zwitterions, while PS headgroups are net negative. Notably, while detectable levels of externalized PE are known to be present, PS sequestration to the inner leaflet is essentially absolute. An outer neutral leaflet with an underlying negative inner leaflet contributes to a physiologically vital membrane potential.[8] The charge differential maintained by this arrangement has been shown to be important to intracellular fusion, budding, and ion diffusion processes which help maintain internal homeostasis.[4,9–15] Meanwhile, loss of asymmetry and the subsequent externalization of PS is essential to blood coagulation and the inability to express PS is a key characteristic of the clotting disorder Scott's syndrome.[9,12,15–18] Externalization of PS to the extracellular leaflet also serves as a signal for phagocytes to engulf and clear apoptotic cells.[19]

The origin of the asymmetric lipid distribution found in cells has been ascribed to the action of ATP-dependent enzymes known as flippases and floppases that flip or flop lipids inward or outward, respectively.[2,4,12,20–24] Several PS-active flippases have been identified, such as ATP11A, ATP11C, and aminophospholipid translocase (APLT).[22,24–28] The ability of these enzymes to deliver PS lipids to the inner leaflet has been established through various means, including monitoring the uptake of fluorescent PS lipid analogues such as 7-nitro-2-1,3-benzoxadiazol-4-yl–PS (NBD–PS) with or without active flippases or floppases present.[26–30] Yet, while deactivation of these enzymes will slow the uptake and transfer of newly delivered PS, loss of flippase activity does not itself lead to loss of existing PS asymmetry. PS externalization in dead or dying cells has therefore been attributed to a third class of proteins, scramblases, which are believed to actively destroy PS asymmetry as part of apoptosis.[2,9,12,24,27,30–37]

Activation of the scrambling process is routinely linked to Ca^{2+} ions. Elevation of cytosolic Ca^{2+} through either the incorporation of ionophores such as A23187, which increases inward flux, or deactivation of Ca-ATPase *via* agonists like thapsigargan, which decreases outward flux, reliably result in PS externalization.[1,16,33,35,38–41] Williamson *et al.* have, through a series of papers involving PS staining with FTIC-labeled Annexin V, quenching of NBD- or doxylpentanoyl-labeled lipids, and examination of phospholipase degradation products, established scrambling activity to affect a wide variety of lipid species.[4,28,31,35,36,42–44] Williamson's group has found that not only will elevated cytosolic Ca^{2+} lead to mixing of the native PC, SM, PE, and PS lipids, but also NBD or doxylpentanoyl-labeled variants. Further, Ca-induced mixing of even non-natural lipid analogues such as fatty-acid carnitine esters and trimethylammonium–diphenylhexatriene has also been shown.[29,40] These data establish the scrambling process to be quite indiscriminate toward lipid structure, yet the mechanism for how exactly Ca^{2+} enhanced scramblases lead to lipid translocation is not well understood.

Despite the demonstration of Ca-induced lipid mixing over a wide range of cell strains and lipid types, the identities of the specific enzymes responsible are currently unclear. Several potential scramblases have been identified, such as phospholipid scramblase (PLSCR) and ABCA1, yet studies on their activity are decidedly mixed. $PLSCR^{-/-}$ knockout in mice platelets did not inhibit PS exposure.[45] In addition, studies on lymphocytes, macrophages, and fibroblasts of

Tangier disease patients who do not express ABCA1 found Ca-induced scrambling of PS to remain active.[46,47]

The lack of a defined, well-established mechanism for how Ca^{2+} may lead to externalized PS is notable, a deficiency compounded by the fact that the direct effect Ca^{2+} ions have on the rates of protein-free PS flip-flop are unknown. There is a wide body of literature, beyond the discussion of scramblase activation, which suggests that PS–Ca pairing is quite important. When observing mixed monolayers and bilayers of PS and PC in the presence of Ca^{2+} ions, phase segregation of PS and PC was observed.[48–50] Janshoff *et al.* found the process can be reversed *via* removal of the Ca^{2+} with ethylene glycol tetraacetic acid (EGTA), a chelating agent with a strong affinity for Ca^{2+}. Similar results were found by Papahadjopoulos, who noted both Ca^{2+} and Mg^{2+} ions affect PS-containing membranes.[49,51,52] Interestingly, although PS was responsive to both ions, only Ca^{2+} leads to phase segregation, and while Mg^{2+} elevated the lipid melting point, Ca^{2+} completely abolished the transition within the tested temperature range. The ability for Ca^{2+} to induce membrane disruptions over Mg^{2+} has also been observed in lipid vesicles. PS-containing vesicles have enhanced fusion and aggregation in the presence of Ca^{2+}, behaviors which were either absent or reduced in the presence of Mg^{2+}.[10,14,50–53]

The divergence in Ca^{2+} and Mg^{2+} behavior toward PS is notable considering that, like PS, the concentrations of these species differ drastically over the cell membrane boundary. In the extracellular media, Ca^{2+} is in the range of 1–2 mM, yet exists in the 100–200 nM range within the cytoplasm.[54,55] Conversely, Mg^{2+} ions within the cytoplasm are in the range of 0.6–1 mM.[54] While Ca^{2+} influx is commonly stated as the cause of scramblase activation, there is data to suggest the existing concentration gradient of Ca^{2+} is an equally important factor. Hampton *et al.* found complexation of internal Ca^{2+} using the chelating agent 1,2-bis(2-aminophenoxy)ethane-*N*,*N*,*N*′,*N*′-tetraacetic acid tetrakis(acetoxymethyl ester) (BAPTA-AM) did not inhibit PS exposure, whereas chelation of external Ca^{2+} with EGTA lead to a reduction of exposed PS by 50% to 65%.[56] Further, Bratton *et al.* found PS expression could be achieved *via* UV irradiation of human leukaemia (HL-60) cells, but that the degree of expression was directly related to the external concentration of Ca^{2+}.[25] To test if irradiation had caused the cells to become more permeable to Ca^{2+} resulting in Ca^{2+} influx, the authors used fluorescence from fura-2-acetoxymethylester-loaded cells and found intracellular Ca^{2+} had only risen by 100 nM in an external media containing 1 mM Ca. Their results indicate the presence of Ca^{2+} was required for externalization even though Ca^{2+} flux was minimal.

The results of Hampton and Bratton, combined with the previously detailed evidence of strong a Ca–PS interaction, suggest the presence of Ca^{2+} may directly influence the rate of PS flip-flop and, by extension, the maintenance of PS lipid asymmetry in cellular membranes irrespective of postulated scramblase activity. The interaction of solvated ions and membrane lipids have previously been shown to have a pronounced effect on rates of translocation. Cheng and Conboy found that exposure to Yb^{3+} ions slowed 1,2-dipalmitoyl-*sn*-glycero-3-phosphocholine (DPPC) flip-flop by an order of magnitude, demonstrating the potency of ion–lipid interactions to modulate lipid flip-flop dynamics.[57] It is hypothesized here that PS externalization may, at least in part, be attributed to an equilibration of PS toward the Ca^{2+} exposed leaflet of the bilayer by way of unequal rates of native, nonenzymatic flip and flop. Such a hypothesis does not preclude enzymatic activity, but rather aims to assist in the interpretation of enzymatic data by clarifying what behaviors can be explained through direct Ca–PS and Mg–PS competitive effects in the extracellular and intercellular environments, respectively.

The importance of the nonenzymatic flip-flop of PS is less characterized than the enzymatic behaviors, as lipid mixing dynamics are often approached with the starting assumption that the non-enzymatic or "native" lipid flip-flop is too slow to be relevant. Leaflet mixing is therefore viewed largely through the lens of active transport, hence an overall focus on scramblase-driven PS movement. While there are several studies which find native flip-flop to be quite slow, the literature contains a wide variance in the reported rates of translocation. Methods such as electron paramagnetic resonance (EPR),[3] nuclear magnetic resonance (H^1 NMR),[58] small angle neutron scattering (SANS),[59–61] fluorescence,[27,44] and sum-frequency vibrational spectroscopy (SFVS)[62–67] have been applied to the problem, yet there is little consensus on specific rates.

Divergent results are generally ascribed to two experimental factors. Firstly, determined rates of flip-flop can be highly dependent on the presence of chemical labels like the 2,2,6,6-tetramethylpiperidine 1-oxyl (TEMPO) label used by Kornberg and McConnell's EPR study.[3,64] Analyzing the influence of the TEMPO label employed in Kornberg and McConnell's original work, Liu and Conboy showed that native DPPC lipids flip-flopped nearly 50-fold faster than the TEMPO-DPPC analogue. Liu and Conboy's study was performed using SFVS of planar supported lipid bilayers (PSLBs), a methodology advantaged by being capable of measuring the flip-flop rates of unlabelled lipids. The PSLBs methodology is, however, counter to Kornberg and McConnell who had utilized vesicle membranes. The membrane model, planar or vesicle, is another factor which can influence flip-flop measurements.

An example of a label-free method which has been employed for vesicle models is ^{1}HNMR analysis of asymmetric large unilamellar vesicles (aLUVs).[58,68–72] aLUV analysis provides DPPC flip-flop rates several orders of magnitude slower than those measured in PSLBs, hypothesized by Marquardt *et al.* to be due to defect-enhanced translocation in PSLBs near sub-micron holes in surface coverage.[69] The effect of surface defects can be minimized through the incorporation of a highly hydrated polymer cushion between the bilayer and surface support, but cannot be removed entirely.[73–77] The influence of surface defects on PSLB flip-flop kinetics may not, therefore, be entirely discounted. However, inconsistencies in reported flip-flop rates are not entirely explainable purely by planar *versus* vesicle membrane dynamics. The TEMPO-PC flip-flop observed by Kornberg and McConnell in DPPC is several orders of magnitude faster than DPPC flip-flop rates observed in aLUVs,[3,69] despite the presence of the bulky TEMPO label. As both studies utilized vesicle models, the discrepancy would not be explainable by surface coverage defects.

While it is clear that methodological differences can influence the flip-flop rates obtained, the study herein is interested primarily in relative kinetic shifts which may affect the equilibrium of PS expression. In order to validate the Ca-modulated PS flip-flop hypothesis, the present study used SFVS of PS-containing PSLBs in the presence of physiologic 1 mM concentrations of Ca^{2+} or Mg^{2+} ions to compare their effects on PS flip-flop dynamics. Solutions were ionic strength matched and use a baseline comparison to 100 mM KCl to mimic the cytosolic environment where K^+ ions predominate.[78] As the utilized methodology was consistent among all samples examined, the Ca^{2+} or Mg^{2+} induced changes in lipid flip-flop kinetics are used to make inferences on *in vivo* PS lipid behaviour. Gel phase mixtures of DPPC and 1,2-dipalmitoyl-*sn*-glycero-3-phosphoserine (DPPS) were utilized due to the inability to measure the rapid flip-flop mixing of liquid crystalline (LC) phase lipids by SFVS using PSLB model membranes. While binary mixtures of DPPC and DPPS are not physiologic, the simplified mixture allowed the variable of PS-selective Ca^{2+} or Mg^{2+} interactions to be isolated. Aside from headgroup identity these two lipids are structurally identical. Thus, any divergence in response to Ca^{2+} or Mg^{2+} exposure is attributed to differences in the binding of these two ions to PC and PS headgroups. Additionally, compression isotherms of lipid monolayers on subphases with or without Ca^{2+} and Mg^{2+} were used to probe the effects these ions have on the packing and compressibility of DPPS and DPPC lipids. Finally, SFVS of hybrid bilayers, composed of lipid monolayers deposited to a methylated silica surface, was utilized to determine the affinity of DPPS for Ca^{2+} and Mg^{2+} and characterize structural changes occurring within the headgroups, water solvation layer, and lipid alkyl chains.

Sum-frequency vibrational spectroscopy

Sum-frequency vibrational spectroscopy is a coherent nonlinear optical method which involves simultaneous infrared (IR) excitation and anti-stokes Raman scattering to generate an outgoing photon at the sum of the two input energies.[79] In the present case, the system employed uses a tunable IR (ω_{IR}) and a fixed 532 nm visible (ω_{532}) laser to generate the sum-frequency beam ($\omega_{S.F.}$):

$$\omega_{S.F.} = \omega_{532} + \omega_{IR} \tag{1}$$

Generation of the sum-frequency is dependent on the second order nonlinear susceptibility of the material ($\chi^{(2)}$) and scaled by the linear transmission Fresnel coefficients of the IR (f_{IR}) and 532 nm (f_{532}) beams, as well as the nonlinear Fresnel coefficient of the outgoing sum-frequency ($\tilde{f}_{S.F.}$):

$$I_{S.F.} = |\tilde{f}_{S.F}f_{532}f_{IR}\chi^{(2)}|^2 \tag{2}$$

The second order susceptibility is a third order, 27 element tensor which describes the coupling of the three electric field vectors (IR, visible, and sum-frequency) in three-dimensional space (x, y, and z). $\chi^{(2)}$ contains both resonant (${\chi_R}^{(2)}$) and non-resonant (${\chi_{N.R.}}^{(2)}$) contributions, however the non-resonant component is small compared to the resonant component when dealing with dielectric materials and may usually be ignored:

$$\chi^{(2)} = {\chi_R}^{(2)} + {\chi_{N.R.}}^{(2)} \approx {\chi_R}^{(2)} \tag{3}$$

The resonant component (${\chi_R}^{(2)}$) is proportional to the number of molecules (N) at the interface, scaled by the IR (A_i) and Raman (M_{jk}) transition probabilities (eqn (4)). Resonance will reach a maximum when the tunable IR (ω_{IR}) is matched to a vibrational mode (ω_v), though will be limited by the linewidth of the transition (Γ_v). Expressed as a sum over all possible vibrational modes, the sum-frequency dependence on the total resonant contribution can be expressed as follows:

$$I_{SF} \propto \left|\chi_R{}^{(2)}\right|^2 = \left|\sum_v \frac{NA_iM_{jk}}{\omega_v - \omega_{IR} - i\Gamma_v}\right|^2 \tag{4}$$

The orientation of the molecules is implicit *via* the bra (⟨) and ket (⟩) operators, which denote an average of all molecular orientations. In isotropic media where the orientational averaging results in no net directionality, the expected sum-frequency is zero. Only in anisotropic media, or at an interface, should sum-frequency be observed.[80] Due to the symmetry dependence of the sum-frequency, SFVS can be utilized to great advantage when observing interleaflet mixing. Bilayer systems contain a natural plane of inversion between the leaflets, and thus any sum-frequency generated is proportional to the number difference of species in the proximal ($N_{proximal}$) and distal (N_{distal}) leaflets:[63,81]

$$I_{SF} \propto (N_{proximal} - N_{distal})^2 \tag{5}$$

One technique to achieve lipid leaflet asymmetry is through the use of PSLBs prepared *via* the Langmuir–Blodgett/Langmuir–Schaffer (LB/LS) transfer methods. Selective deuteration of the lipid tails in one leaflet shifts the resonances out of the C–H region into the C–D region, removing any interference of sum-frequency from the antiparallel orientation of the tails. This bilayer asymmetry allows the alkyl chain C–H vibrations to become sum-frequency active. Fig. 1A illustrates this principal for the C–H vibrational region recorded for an asymmetric DPPC/DPPCd62 bilayer and a symmetric DPPC/DPPC bilayer. Five key vibrational modes are observed: the CH_2 symmetric stretch at 2850 cm^{-1}, CH_3 symmetric stretch at 2875 cm^{-1}, CH_2 Fermi resonance at 2905 cm^{-1}, CH_3 Fermi resonance at 2935 cm^{-1}, and CH_3 asymmetric stretch at 2960 cm^{-1}.[64,65] The spectra in Fig. 1 were collected with s-polarized sum-frequency, s-polarized 532 nm and p-polarized IR. This optical configuration probes the component of

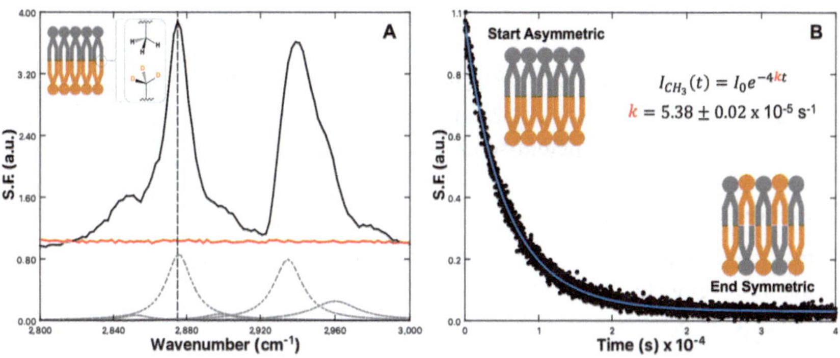

Fig. 1 (A) SFVS spectrum scan of an asymmetric DPPC/DPPC-d62 bilayer (black) and symmetric DPPC/DPPC bilayer (red). Inset image shows an illustration of an isotopically asymmetric lipid bilayer highlighting the terminal CH_3 symmetric stretch. The CH_3 symmetric stretch is located at 2875 cm^{-1}, denoted with a vertical grey dashed line. The grey dashed curves are the fits to the various C–H resonances obtained by fitting the SFVS spectrum to eqn (4), with the data offset for clarity. (B) SFVS intensity decay of the CH_3 symmetric stretch at 2875 cm^{-1} for an asymmetric DPPC/DPPCd62 bilayer recorded at 28 °C. The solid blue line is the fit of the decay data using eqn (7).

the vibrational transitions parallel to the surface normal. For this reason, the terminal CH_3 stretch appears most prominently in *ssp* spectra.

As the sum-frequency is proportional to the lipid number difference between the leaflets, changes in the sum-frequency (SF) intensity can therefore be attributed directly to mixing between the two leaflets:

$$N_{\text{distal}} \underset{k_{\text{flop}}}{\overset{k_{\text{flip}}}{\rightleftharpoons}} N_{\text{proximal}} \tag{6}$$

where k_{flip} and k_{flop} are the rates of lipid translocation. By measuring the decay in intensity emanating from the CH_3 symmetric stretch (ν_s) at 2875 cm^{-1}, lipid translocation can be observed without the use of fluorescent or spin-labelled lipids. A plot of the CH_3 ν_s sum-frequency decay (Fig. 1B) with time can be fit to eqn (7), derived by Liu and Conboy, which allows the rate constant of lipid flip-flop to be determined:[65]

$$I_{CH_3,t} = I_0 e^{-4kt} \tag{7}$$

where $I_{CH_3,t}$ is the sum-frequency at any time, t; I_0 is the starting intensity; and k is the rate of lipid flip-flop. Importantly, eqn (7) is derived under the assumption of equivalent environments within each leaflet such that $k_{\text{flip}} = k_{\text{flop}}$.

Experimental

Materials

All materials were used as received without further purification. 1,2-dipalmitoyl-*sn*-glycero-3-phosphocholine (DPPC), 1,2-dipalmitoyl-d62-*sn*-glycero-3-phosphocholine (DPPC-d62), 1,2-dipalmitoyl-*sn*-glycero-3-phosphoserine (DPPS), and 1,2-dipalmitoyl-d62-*sn*-glycero-3-phosphoserine (DPPS-d62) (Fig. 2) were purchased from Avanti Polar Lipids (Alabaster, AL). 99.995% pure KCl (metals basis), ultra dry, 99.9% pure $MgCl_2$ (metals basis), Tris Base, HPLC-grade chloroform, and deuterium oxide were purchased from Thermo Fisher Scientific (Waltham, MA). Nanopure, 18.2 MΩ cm water from a Barnstead Thermolyne (Dubuque, IA) system was used for all aqueous solutions.

Lipid stock solutions of DPPC, DPPC-d62, DPPS, and DPPS-d62 were made at 1 mg mL^{-1} in HPLC-grade chloroform. Mixtures of various PS : PC mol fractions were made using these parent solutions.

Compression isotherms

Pressure–area isotherms were collected on a Langmuir trough (KSV, Helsinki, Finland). Prior to lipid additions, each subphase was checked for surface

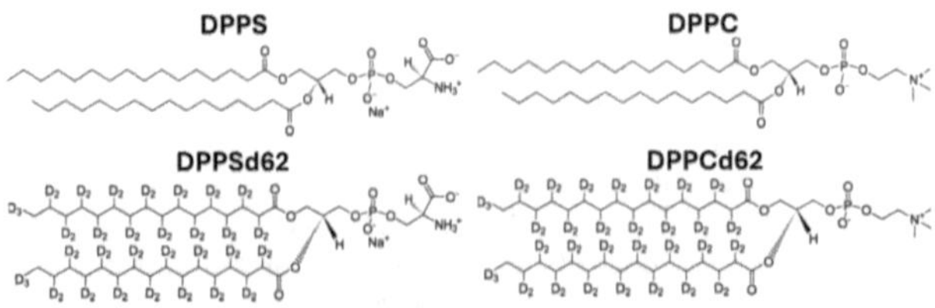

Fig. 2 Chemical structures of protonated and deuterated DPPS and DPPC lipids.

contaminants with a barrier sweep over the surface, with barrier compressions resulting in less than a 0.2 mN m^{-1} lateral surface pressure increase considered sufficient for lipid deposition. Lipid solutions in chloroform were deposited dropwise at the air–water interface and allowed to evaporate for 10 minutes. All monolayers were compressed at a rate of 4 mm min^{-1} at room temperature (21 °C).

Silica supports

Custom IR-grade fused silica prisms were purchased from Almaz Optics (Marlton, NJ). Before each use, prisms were cleaned for at least 10 minutes under ozone (Jetlight Co., Irvine, CA), followed by submersion in a 3 : 1 mixture of concentrated sulfuric acid and 30% H_2O_2 for at least one hour. CAUTION: Piranha solutions contain strong oxidants which react violently, potentially explosively, with organics. Submerged prisms should be absolutely free of any residual organic solvents, and extreme care with proper personal protective equipment should be used when handling these corrosive mixtures. Finally, each prism was rinsed with 18.2 MΩ cm water, dried under N_2, and treated with argon plasma (Harrick Scientific, Ithaca, NY) for at least 10 minutes.

Planar supported lipid bilayers

PSLBs were deposited to silica prisms at a surface pressure of 30 mN m^{-1} to mimic physiological conditions.[82] Compressions were performed following the same procedure listed in the 'Compression isotherms' section. To prevent interactions between the lipid headgroups and the silica surface, particularly salt bridging with PS, a 3% C16-PEG(5000)PE lipid fraction was incorporated for LB depositions (proximal leaflet). The highly water-soluble PEG groups attached to the lipid heads form a PEG polymer cushion to act as a spacer between the support and lipids.[74–76,83] For LB depositions the prism was pulled up through the interface at a rate of 4 mm min^{-1}. For the LS deposition the prism was rotated 90° and aligned to be as perfectly level with the trough as possible. The trough was reset with a fresh subphase, new lipids of the desired composition, and compressed to 30 mN m^{-1} as previously. The prism was then brought back through the interface to deposit the distal leaflet.

Hybrid bilayers

Silica prisms were surface modified by submersion in a solution of 2% trimethoxymethylsilane in dry toluene (v/v) with an incubation time of at least one hour. Incubation should be performed in a covered container to prevent evaporation of the toluene. Following incubation, the surface was rinsed with methanol and placed into an oven at >80 °C for at least one hour. Finally, a lipid monolayer was deposited *via* a single LS transfer at 30 mN m^{-1} on the methylated surface to create a hybrid bilayer.

Results and discussion

Compression isotherms

The effect of Ca^{2+} and Mg^{2+} ions on lipid packing and compression were examined by way of pressure–area isotherms for three lipid systems: control monolayers of 100% DPPC (Fig. 3A), monolayers of 10% DPPS in DPPC as

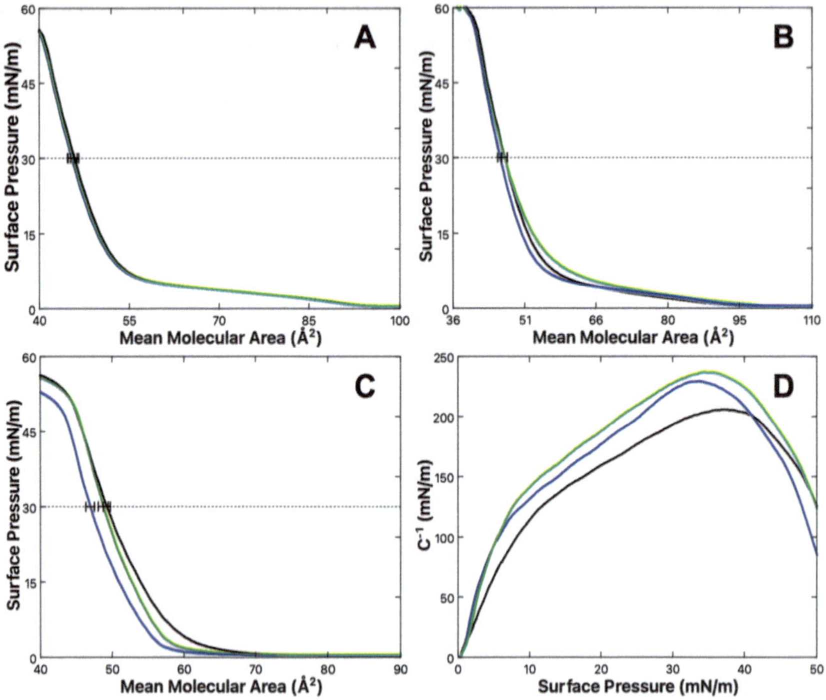

Fig. 3 Pressure–area isotherms of (A) DPPC, (B) 10% DPPS in DPPC, and (C) DPPS obtained with a subphases of 100 mM KCl (black), 97 mM KCl + 1 mM $CaCl_2$ (blue) or 1 mM $MgCl_2$ (green). Each line represents the average of at least three replicate measurements. All subphases are buffered with 50 mM Tris at pH 7.4. (D) Compression moduli of DPPS as a function of surface pressure, calculated from the isotherms displayed in (C) using eqn (7) for 100 mM KCl (black), 97 mM KCl + 1 mM $CaCl_2$ (blue) or 1 mM $MgCl_2$ (green). Compression moduli were obtained by employing a 50-point rolling average on the corresponding π–A isotherm.

a physiologically appropriate PS fraction (Fig. 3B), and monolayers of 100% DPPS (Fig. 3C). The subphases for all experiments were ionic strength matched, with a base solution of 100 mM KCl and 50 mM Tris buffer for a total ionic strength of 133.4 mM at pH 7.4. The amount of KCl present was reduced to 97 mM to compensate for the addition of 1 mM $CaCl_2$ or $MgCl_2$ when those species were present. Each experiment was performed at least three times and averaged.

In the absence of either Ca^{2+} or Mg^{2+}, the mean molecular areas (MMAs) at a physiological surface pressure of 30 mN m^{-1} were determined to be 49.3 ± 0.5, 45.9 ± 0.7, and 46.7 ± 0.7 Å^2 per molecule for the 100% PS, 100% PC, and 10% PS systems, respectively. The values determined for pure PC and PS agree favorably with those reported elsewhere.[84,85] Upon the addition of 1 mM Ca^{2+} or Mg^{2+}, no net change in the MMA of DPPC was observed. A previous report from Adams *et al.* found DPPC monolayers expand in the presence of Ca^{2+};[86] however, those studies were carried out at much higher concentrations of Ca^{2+}, starting at 300 mM $CaCl_2$ compared to 1 mM here. Additionally, their baseline of comparison was a pure water subphase rather than ionic strength matched buffer. The divergent results can therefore likely be attributed to experimental differences between their study and the results found here.

More notable results are found in the DPPS mixtures where, when in the presence of Ca^{2+}, but not Mg^{2+}, there is a reduction in the MMA of -0.2 ± 0.7 and -2.4 ± 0.6 Å^2 per molecule for 10% and 100% DPPS, respectively, at 30 mN m^{-1} surface pressure. In the case of the 10% DPPS system the differential is within the margin of error; however, the differential for the 100% DPPS system is significant and suggests a net condensation of the DPPS monolayer in the presence of Ca^{2+} ions. Given the low mol fraction in the 10% mixture, one might expect only one-tenth the response of that in the neat DPPS monolayer, corresponding to a shift of around -0.2 Å^2 per molecule, which is within the standard deviation of the experiment. The 10% DPPS data therefore agrees within error. The condensing behavior observed for Ca^{2+} with pure PS monolayers is additionally consistent with behavior observed by Narayanan who found a contraction of 1,2-dilauroyl-*sn*-glycero-3-phospho-L-serine (DLPS) in the presence of 10 mM Ca^{2+}.[87] Narayanan observed Mg^{2+} to have a weaker condensing effect than Ca^{2+}, with the condensation disappearing at higher surface pressures. A similar trend was observed here, with the Mg^{2+}-induced condensation effect disappearing above 20 mN m^{-1}.

There is additionally a shift in the compression moduli of the DPPS monolayer in the presence of both the Ca^{2+} and Mg^{2+}. The compression modulus was computed from the first derivative of the isotherm in Fig. 3C, using the following expression:[88]

$$C^{-1} = -A\frac{d\Pi}{dA} = -A_i\left(\frac{\Pi_{i+1} - \Pi_{i-1}}{A_{i+1} - A_{i-1}}\right) \tag{8}$$

where C^{-1} is the compression modulus, or inverse compressibility; A is mean molecular area; Π is surface pressure, and subscript i represents a particular point along the compression isotherm. In the presence of both the Ca^{2+} and Mg^{2+}, there is a shift toward higher compression moduli, or lower compressibility, for DPPS (Fig. 3D). While the effects on packing appear to diverge between the two ions, similar rigidification of the DPPS monolayer implies some similarity between Ca–PS and Mg–PS interactions. Analysis of the Π–A isotherms and the compression modulus show a combined condensation and rigidification of the monolayer which is an indication of a binding interaction between Ca^{2+} and Mg^{2+} with the PS lipids. The mode of the Mg–PS and Ca–PS interactions do not, however, appear identical as the effect of lipid monolayer condensation with Ca^{2+} ions is present at all surface pressures, as well as being more

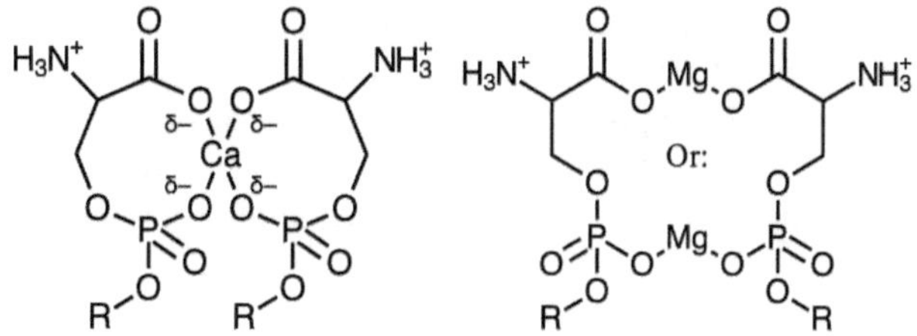

Fig. 4 Schematic diagrams comparing the simultaneous bridging of carboxylate and phosphate oxygens by Ca^{2+} (left) *versus* Mg^{2+} bridged carboxylate or phosphate regions (right). These represent the most favourable configurations determined by MD simulations by Martin-Molina *et al.*[89]

pronounced even in the 0 to 20 mN m^{-1} region where Mg-induced condensation is seen. Molecular dynamics simulations from Martin-Molina *et al.* predict a headgroup-bridging effect whereby Ca^{2+} and Mg^{2+} ions both bind two PS lipids simultaneously.[89] This bridging behavior can explain how both Ca^{2+} and Mg^{2+} result in higher compressibility, as individual lipids are held more tightly together, which will reduce the ease of deformation. Yet, while both ions were found to bind lipids in a 1 : 2 ion : lipid ratio, the binding modality was found to be quite different. The most stable conformation of Ca–PS complexation is a motif of four oxygens per Ca^{2+} ion, with each lipid coordinating one phosphate oxygen and one carboxylate oxygen to form a cage around the central ion (Fig. 4). Conversely, the two most stable forms of the Mg–PS complex were found to involve only two total oxygens, with Mg^{2+} bridging either the lipid phosphate or carboxylate regions, but not both simultaneously as calculated for Ca^{2+}. These conformations represented the most likely states, though Martin-Molina observed other binding modes at lower probabilities. The average of all possible states found Mg^{2+} to bind 2.7 oxygens, while Ca^{2+} bound 4.2. Overall, Mg^{2+} was found to occupy more superficial positions than Ca^{2+}, which penetrated more deeply into the membrane and led to a loss of more associated waters.

The MD simulations by Martin-Molina are consistent with X-ray diffraction studies by Papahadjopoulos which support the conclusions of a larger dehydration effect of Ca^{2+} ions on PS.[51] Tighter bridging of the headgroups by Ca^{2+} would additionally explain the Ca-induced abolishment of the PS lipid phase transition and enhanced fusogenic activity also observed by Papahadjopoulos.[10,52] These data make clear that while both ions impact the behavior of PS, an overall stronger binding interaction exists between PS and Ca^{2+} than with Mg^{2+}. SFVS was used to determine whether the binding of either species results in changes within the C–H stretch or H-bonding regions, which can elucidate how differences in binding modality influence the lipid monolayer structure, both in terms of tail packing and headgroup solvation.

Sum-frequency analysis

The vibrational spectra of DPPS and DPPC hybrid bilayers were examined to determine whether significant changes to the alkyl chain or hydrogen bonding structure were occurring as a result of Ca^{2+} and Mg^{2+} binding the DPPS headgroups. The use of hybrid bilayers (lipid monolayers) rather than lipid bilayers is important, as equivalent hydration environments across the bilayer interface would result in interference of sum-frequency generation between the two leaflets. The use of hybrid bilayers removes the antiposed hydration layer associated with headgroups on the opposite leaflet of a symmetric bilayer. Fig. 5A–C illustrate how the SFVS spectra of DPPC and DPPS hybrid bilayers change as a function of Ca^{2+} or Mg^{2+} concentration. The sum-frequency spectra of these hybrid bilayers can be broken into three general regions of interest: the C–H region from 2800 to 3000 cm^{-1}, the water region from 3000 to 3500 cm^{-1}, and an amine peak centered at 3300 cm^{-1} originating from the hydrogen bonded $-NH_3$ group of the PS headgroup. Each of these vibrational features will be analyzed in turn to determine how the structure of the headgroups, associated water layer, and lipid tails respond to bound Ca^{2+} and Mg^{2+}.

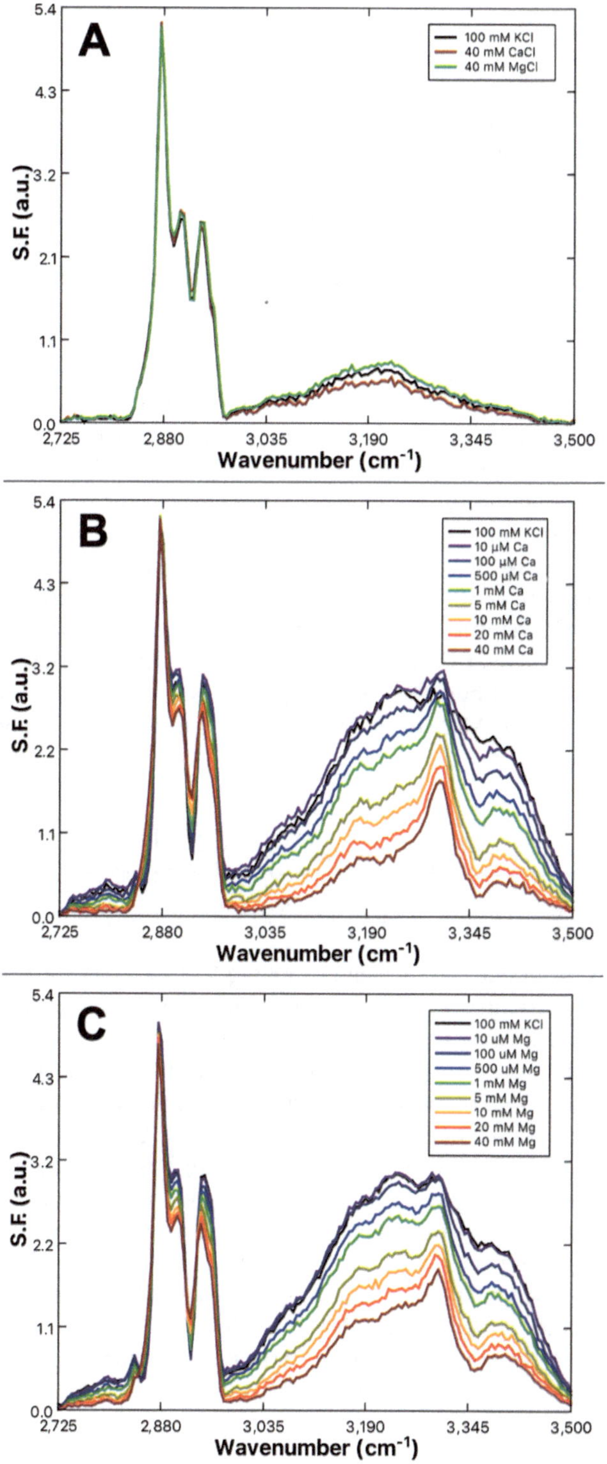

Fig. 5 SFVS spectra of hybrid bilayers composed of (A) 100% DPPC with 100 mM KCl (black), 40 mM $CaCl_2$ (red), and 40 mM $MgCl_2$ (green); (B) 100% DPPS with increasing

Amine peak analysis

The resonance observed at ∼3300 cm^{-1} is attributed to the symmetric stretch of the primary amine moiety present on the serine headgroup. The resonance observed here is similar to the sum-frequency spectra obtained from mixtures of ammonia and water, where a 3300 cm^{-1} band appears due to the ordering of ammonia near the air/water interface.[90] A very prominent 3300 cm^{-1} also appears within the sum-frequency spectra of fibrinogen-coated surfaces as reported by both the Chen and Cremer groups.[91,92] Chen and Cremer both attributed the peak to N–H stretches, though disagreed on whether the peak indicated an ordering of the backbone amides ($R{-}NH_2{-}R$) or side-chain amines ($R{-}NH_3$). Further analysis by Weidner and Castner using a lysine-containing protein analogue LKα14 supports the hypothesis of the peak being related to the amine groups, as replacement of the lysine nitrogens with ^{15}N resulted in a shift of the peak.[93] The amine peak, for the purposes of this study, was considered a reporter of the reordering of serine headgroup amines resulting from Ca^{2+} or Mg^{2+} complexation, in line with the lipid complexing behavior predicted by Martin-Molina *et al.*[89] as well as phospholipid reorientations observed by Seelig *et al.*[94]

The amine peak, present only in the PS system, gains in prominence as the bulk Ca^{2+} (Fig. 5B) or Mg^{2+} (Fig. 5C) content is increased. The prominence of the amine peak reports directly on structural changes resulting from Ca^{2+} or Mg^{2+} complexation. Therefore, correlation of the net area change in the amine peak to the concentration of bulk Ca^{2+} or Mg^{2+} present was used to derive the binding affinity of Ca^{2+} and Mg^{2+} to PS by way of a fit to a Frumkin model binding isotherm, eqn (9):[95,96]

$$A_X \propto \left(\frac{\sqrt{A_{\max}} K_a [X] e^{\left(\frac{g' \sqrt{A_X}}{\sqrt{A_{\max}}}\right)}}{1 + K_a [X] e^{\left(\frac{g' \sqrt{A_X}}{\sqrt{A_{\max}}}\right)}} \right)^2 \tag{9}$$

where A_X is the amine peak area at a given Mg^{2+} or Ca^{2+} concentration, $[X]$; K_a is the binding constant; A_{max} is the maximum area at surface saturation, and g' is a constant representing attractive or repulsive forces between binding species. A negative g' indicates a repulsive interaction, while a positive g' represents attraction. A g' of zero results in the standard Langmuir model, which assumes no net interactions of the binding species.

Here it was found that while both Ca^{2+} and Mg^{2+} bind to the PS monolayer, Ca^{2+} will do so with both a higher affinity and a greater impact on the headgroup structure. Fig. 6 shows the amine peak area as a function of Ca^{2+} and Mg^{2+} concentration, with the lines representing the best Frumkin model fit of each dataset to eqn (9). The increase in amine peak area for Ca^{2+} is roughly double that of Mg^{2+}. A more pronounced ordering of the amines in the case of Ca^{2+} is consistent with both the tighter caging of the ion as predicted in Martin-Molina's MD simulations as well as the condensation effect of Ca^{2+} on DPPS observed in

concentrations of $CaCl_2$; (C) 100% DPPS with increasing concentrations of $MgCl_2$. All solutions are buffered with Tris at pH 7.4 and maintain a constant ionic strength of 133.4 mM.

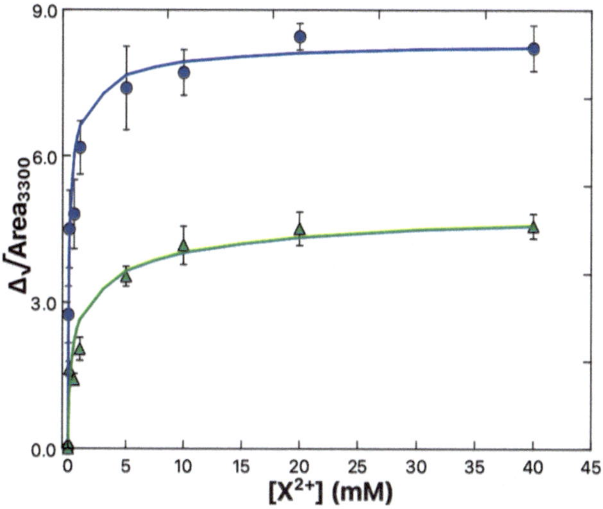

Fig. 6 Binding isotherm data for Ca^{2+} (blue) and Mg^{2+} (green) obtained from the areas of the 3300 cm^{-1} amine peak in the spectra shown in Fig. 5B and C respectively. The solid lines are the fits to the data using a Frumkin model (eqn (9)).

compression isotherms. The K_a values of $1.3 \pm 0.3 \times 10^5$ and $7.3 \pm 2.5 \times 10^3$ M^{-1} for Ca–PS and Mg–PS, respectively, additionally indicate Ca^{2+} ions have around a 20-fold higher affinity toward PS. Newton and Papahadjopoulos reported a 10-fold difference between Ca^{2+} and Mg^{2+}, roughly matching the order of magnitude relation found here, though their K_a values of 0.35 and 0.04 M^{-1} are much smaller.[51] These values were based on dialysis equilibrium experiments and assumptions were made regarding the permeability and "probable amount of lipid exposed to the bulk solution". It is possible there was less exposure of the lipids to the Ca^{2+} and Mg^{2+} solutions than anticipated, which could explain the variance.

An alternate comparison with the results of Hauser *et al.*, who examined Ca–PS affinity *via* measurement of the surface-enhanced radioactivity of solutions containing ^{45}Ca in the presence and absence of PS monolayers was also made.[97] Based on increases in surface radioactivity in the presence of PS, representing surface enrichment of $^{45}Ca^{2+}$ relative to the bulk, Hauser determined the Ca–PS affinity to be anywhere from 10^4 to 10^6 M^{-1}. Hauser could not go above 1 mM Ca^{2+} due to the cost and health concerns related to the use of radioactive isotopes. K_a values were therefore reported as single point measurements not fit to an isotherm model, however the results of Hauser are in relative agreement with the SFVS results. Mg^{2+} was not tested by Hauser, so a comparison of relative Mg^{2+}/Ca^{2+} affinities cannot be made.

Finally, Sinn *et al.* characterized the binding affinity of PS and Ca^{2+} by way of ion selective electrode (ISE) measurements.[98] Known amounts of Ca^{2+} were used to titrate a solution containing vesicles composed of 20% DOPS in 80% DOPC. The free concentration of Ca^{2+} measured by the probe was compared to the injected amount of Ca^{2+} to deduce the amount of Ca^{2+} ions bound to the vesicle membranes. Results indicated a binding constant of 650 M^{-1}, roughly 200-fold lower than that determined by SFVS. The authors do note however that repeat calibrations of the ISE probe showed drift indicative of lipids depositing to the hydrophobic surfaces of the probe and that the presence of charged lipid on the probe surface could shift the measured Ca^{2+} concentration. Sinn *et al.* hypothesized the lipid coating could either decrease the measured [Ca^{2+}] *via* reduced ion transport across the electrode membrane or increase the measured [Ca^{2+}] through electrostatic enrichment in the local Ca^{2+} concentration due to the presence of negatively charged PS lipids. The former would increase the determined K_a by reporting an overly large differential in the injected and detected amounts of bulk Ca^{2+} whereas the latter would lead to a reduction in the measured K_a. Given Hauser's finding of a great local enrichment of Ca^{2+} ions in the presence of PS lipids, it is possible the deviation in results between ISE and SFVS are due to the second scenario put forward by Sinn.

Water peak analysis

The water region from 3000 to 3500 cm^{-1} reports on the hydrogen bonding structure within the membrane associated water layer and contains two characteristic water peaks centered at 3200 cm^{-1} and 3400 cm^{-1}, representing the ice-like O–H symmetric and liquid-like O–H symmetric stretches.[80,99–101] While both result from the symmetric O–H stretch of water, the ice-like resonance is indicative of a strong, tetrahedral H-bonding structure like that seen in ice. Conversely, the liquid-like peak results from a weaker H-bonding structure, namely bifurcated H-bonding whereby hydrogens are shared unevenly between neighboring oxygens. It can be seen that the intensity of the water sum-frequency is far higher with DPPS (Fig. 5B and C) as compared to DPPC (Fig. 5A), results which are in line with previous observations by Chen *et al.*[102] Chen attributed sum-frequency enhancement in the presence of negatively charged lipids to the ordering of water at the lipid interface, with waters orienting their hydrogens toward the charged lipid headgroups. The observed reduction in water sum-frequency for a DPPS monolayer exposed to either Ca^{2+} or Mg^{2+} is therefore hypothesized to result from neutralization of the membrane surface charge and relaxation of the electrostatically imposed ordering of water. Conversely, the lack of an increase in sum-frequency from the water region of the DPPC hybrid bilayer when exposed to Ca^{2+} or Mg^{2+} ions implies the neutral DPPC lipids have not increased in charge. It is therefore hypothesized that either no binding of Ca^{2+} or Mg^{2+} has occurred, or that any bound Ca^{2+} and Mg^{2+} simply displace previously bound Na^{+} ions in a surface charge conserving manner. Both the hypothesized rationales for the DPPS and DPPC sum-frequency water band responses to Ca^{2+} or Mg^{2+} rely upon the interpretation that sum-frequency in the water region is sensitive to the charge state of the lipids.

The influence of interfacial charge on the observed sum-frequency intensity has been previously considered by others as a third order, $\chi^{(3)}$ effect.[103–110] Just as the $\chi^{(2)}$ tensor describes a material's response to the coupling of two incident electric fields, the $\chi^{(3)}$ tensor describes a material's response to three incident fields. In the present case, two are the E_{532} and E_{IR} from the incoming 532 nm and IR beams and the third is a static field, E_{DC}, which is produced by the charged DPPS/water interface. The total sum-frequency generated is expressed as a summation of the respective second and third order contributions:[105–108,110–113]

$$I_{SF} \propto |(\chi^{(2)} + \chi^{(3)}E_{DC})E_{IR}E_{532}|^2 \tag{10}$$

Depending on the material properties, E_{DC} may influence $\chi^{(3)}$ in various ways. As noted by Allen, Eisenthal, Levine and others, dipolar molecules, like water, will realign within a static electric field.[100,102,105–107,110,114] As $\chi^{(2)}$ depends on the molecular axis being probed, realignment of interfacial molecules can yield changes in sum-frequency generation without necessarily resulting directly from third-order coupling. Therefore, the static field (E_{DC}) contributions can be expressed as a summation of two components, one dependent on third-order molecular hyperpolarizability (γ) and the other dependent on molecular rearrangement of the second-order molecular hyperpolarizability (β). $\chi^{(3)}$ can then be expressed as:[105–107]

$$\chi^{(3)} = N^{(3)}\gamma + \frac{\mu N^{(2)}\beta}{5k_BT} \tag{11}$$

where $N^{(3)}$ is volumetric number density in molecules per cm^3, $N^{(2)}$ is the surface density in molecules per cm^2, μ is the dipole moment of the molecules, k_B is Boltzmann constant, and T is temperature in Kelvin. In cases where μ is large, molecular alignment within an applied E_{DC} field will magnify $\chi^{(3)}$ through the $\frac{\mu N^{(2)}\beta}{5k_BT}$ term.[107,110] Substitution of eqn (11) into eqn (10) yields the relationship between the generated sum-frequency and E_{DC} field strength:

$$I_{SF} \propto \left|\left(\chi^{(2)} + \left[N^{(3)}\gamma + \frac{\mu N^{(2)}\beta}{5k_BT}\right]E_{DC}\right)E_{IR}E_{532}\right|^2 \tag{12}$$

The magnitude of the static E_{DC} field is itself ultimately determined by the surface charge density and the ionic strength of the solution. The surface charge will generate a surface potential in accordance with eqn (13):[8,13,115–117]

$$\phi_0 = \frac{2k_BT}{e}\sinh^{-1}\left(\frac{\sigma}{\sqrt{8000k_BTN_AI\varepsilon_0\varepsilon_r}}\right) \tag{13}$$

where ϕ_0 is the surface potential; σ is the surface charge density in C m^2; I is the ionic strength of solution in molarity; N_A is Avogadro's number; ε_0 is the vacuum permittivity; and ε_r is the relative permittivity of the medium.[8,13,115–117] Gouy–Chapman theory predicts that the surface potential decays with distance away from the interface, with the effective potential at any given distance described by:

$$\phi_x = \phi_0 e^{-\kappa x} \tag{14}$$

where x is distance from the charged surface and κ is the inverse Debye screening length.[118] The E_{DC} field strength is taken to be the rate of change, or derivative, of the potential at one Debye length from the surface:

$$E_{DC} = \frac{d\phi_x}{dx}\text{ at } \kappa^{-1} = -\kappa\phi_0 e^{-\kappa\kappa^{-1}} = -\kappa\phi_0 e^{-1} \tag{15}$$

The charge density of the DPPS membrane was computed using the MMA of the DPPS monolayer obtained from the pressure–area isotherms. Given a mean

Table 1 Calculation of electric field strengths using the predicted surface charge densities from eqn (14) and (15). Ionic strength was held constant at 133.4 mM, with a Debye length of 0.828 nm. The percent surface saturation of Ca^{2+} is taken from the amine peak binding isotherm (Fig. 6)

[Ca] (mM)	$\Gamma_{Ca^{2+}}$	$\frac{\sigma_i}{\sigma_{max}}$	σ_i (C m^{-2})	ϕ_0 (V)	E_{DC} (V m^{-1}) $\times 10^{-7}$
0	0	100%	−0.33	−0.142	6.30
0.01	27%	73%	−0.25	−0.126	5.59
0.1	54%	46%	−0.16	−0.103	4.57
0.5	72%	28%	−0.094	−0.078	3.48
1	79%	21%	−0.070	−0.065	2.89
5	92%	8%	−0.028	−0.031	1.40
10	95%	5%	−0.017	−0.020	8.67
20	97%	3%	−0.009	−0.010	4.95
40	100%	0%	0	0	0

molecular area of 49.3 ± 0.5 Å^2 for DPPS at a deposition pressure of 30 mN m^{-1}, the theoretical maximum charge density for a pure DPPS monolayer is −0.33 C m^{-2}. Conversely, at a saturating concentration of Ca^{2+}, one would expect the PS lipids to have a charge state of zero, assuming a binding ratio of 2 : 1 PS : Ca. Using these surface charge limits and the isotherm data in Fig. 6, Table 1 lists the surface saturation, charge density (σ), surface potential (ϕ_0), and electric field strength one Debye length away from the surface ($E_{\kappa^{-1}}$) for each bulk concentration of Ca^{2+}. The Debye length for all calculations was taken to be 0.828 nm as the ionic strength was held constant at 133.4 mM.

The validity of the calculated surface saturation and surface charge density values were evaluated as follows. The spectra of Fig. 5B were fit using eqn (4) to obtain the transition probability amplitudes [PA = (NA_iM_{jk})] for the ice-like 3200 cm^{-1} and water-like 3400 cm^{-1} resonances. Equating eqn (4) and (12) at a specific resonance yields:

$$\mathrm{PA} \propto \left| \left(\chi^{(2)} + \left[N^{(3)}\gamma + \frac{\mu N^{(2)}\beta}{5k_B T} \right] E_{DC} \right) \right|. \tag{16}$$

From eqn (16), PA is expected to scale linearly with E_{DC}. Fig. 7 shows the relationship between the calculated E_{DC} and observed transition probabilities of the ice-like (3200 cm^{-1}) and water-like (3400 cm^{-1}) peaks, respectively. A linear relationship between E_{DC} for both the ice-like and water-like peak amplitudes is observed which, firstly, confirms E_{DC} scaling of $\chi^{(3)}$ for the DPPS associated water layer. Secondly, these data also validate the use of the 3300 cm^{-1} amine peak as a gauge of Ca^{2+} adsorption and the associated membrane charge density. The starting assumption stipulated that the amine peak resulted from reorientation of the amines on the serine headgroup due to a 2 : 1 PS : Ca complexation. Under such a motif, the E_{DC} should be reduced due to charge neutralization of PS upon binding Ca^{2+}. A linear drop in the ice- and water-like transition probabilities with respect to E_{DC} supports these previous assumptions. There does, however, seem to be a deviation from linearity beyond an E_{DC} of 5.59 × 10^7 V m^{-1}. The 10 μM Ca^{2+} and zero Ca^{2+} data have roughly equal water intensity despite the binding isotherm predicting around 27% neutralization of the surface charge should have

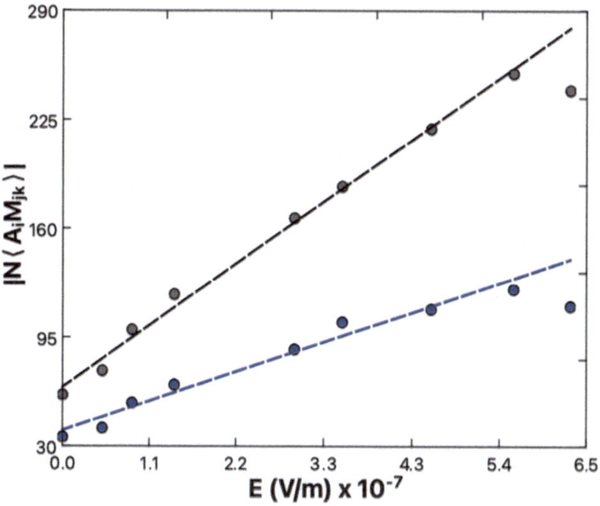

Fig. 7 3200 cm^{-1} (black) and 3400 cm^{-1} (blue) peak transition probabilities from spectral fits of Ca-exposed DPPS (Fig. 5B) plotted as a function of the electric field strength one Debye length away from the lipid surface (Table 1). The determined slopes, representing the relative $\chi^{(3)}$, are $3.4 \pm 0.1 \times 10^6$ m V^{-1} for the ice-like 3200 cm^{-1} and $1.6 \pm 0.1 \times 10^6$ for the water-like 3400 cm^{-1}.

occurred. The deviation from linearity seen in Fig. 7 is explained by considering the two types of contributions to $\chi^{(3)}$ detailed in eqn (11). As observed by both Richmond and Kielich, E_{DC} scaling reaches a saturation point beyond which there is no further change in intensity.[99,110] Saturation occurs in instances where $\chi^{(3)}$ has significant contributions from molecular reorientations, $\left(\frac{\mu N^{(2)}\beta}{5k_BT}\right)$, as opposed to being purely third-order coupling, $(N^{(3)}\gamma)$. Both Richmond and Kielich attributed saturation to a limit on the alignment of the interfacial water within the E_{DC} field.[99,110] Once the E_{DC} strength is sufficient to orient the water dipoles along the E_{DC} field vector, further increases in field strength will not yield further alignment. The realignment term, $\frac{\mu N^{(2)}\beta}{5k_BT}$, is thus maximized and higher E_{DC} fields further magnify only the $N^{(3)}\gamma$ component of $\chi^{(3)}$. The deviation seen here is therefore ascribed to maximal water alignment beyond $E_{DC} = 5.59 \times 10^7$ V m^{-1}. Accordingly, the line fits of Fig. 7 were limited to the linear region.

The ice-like peak results from tightly ordered water structures, whereas the water-like peak results from more transient and disordered H-bonding. As previously detailed by Gragson and Richmond, application of an electric field will lead to sequential conversion of disordered bulk water to the semi-ordered population resonant at 3400 cm^{-1} and from the semi-ordered population to the tetrahedrally ordered population resonant at 3200 cm^{-1}.[99] Gragson and Richmond altered the E_{DC} field strength through modulation of either the density of charged sodium dodecyl sulfate (SDS) molecules at the air/water interface or the bulk ionic strength. In both cases it was found the 3200 cm^{-1} peak responded more strongly to the electric field due to population migration of water molecules toward the more ordered state.

The rate of change of the ice-like and water-like amplitudes $|NA_iM_{jk}|$, with respect to E_{DC} were used to examine the relative $\chi^{(3)}$ of each peak. In Fig. 7, the $\chi^{(3)}$ for the ice-like and water-like peaks is represented by the slope of each line (eqn (10)). The data yield $\chi^{(3)}$ values of $3.4 \pm 0.1 \times 10^6$ m V^{-1} for the ice-like 3200 cm^{-1} and $1.6 \pm 0.1 \times 10^6$ m V^{-1} for the water-like 3400 cm^{-1}. The relative ice-like to water-like $\chi^{(3)}$ ratio is thus found to be 2.1 ± 0.1. A larger $\chi^{(3)}$ for the 3200 cm^{-1} peak is in agreement with Gragson and Richmond's findings of net population conversion of water molecules toward the more ordered ice-like state at higher E_{DC}. As the water population conversion requires reordering of the water molecules, it is concluded E_{DC} dependent molecular rearrangements are occurring within the solvation layer. These data are limited, however, in that while the presence of molecular rearrangement is suggested, the magnitude of $\frac{\mu N^{(2)}\beta}{5k_BT}$ relative to the static component, $N^{(3)}\gamma$, of $\chi^{(3)}$ remains unclear.

The relative magnitude of the molecular rearrangement and third-order contributions was assessed as follows. From eqn (12), only the molecular rearrangement term, $\frac{\mu N^{(2)}\beta}{5k_BT}$, of $\chi^{(3)}$ has a $1/T$ dependence. Determination of the relative contributions of the dynamic and static components of $\chi^{(3)}$, $\frac{\mu N^{(2)}\beta}{5k_BT}$ and $N^{(3)}\gamma$, respectively, was therefore performed by measurement of the sum-frequency intensity as a function of $1/T$ for the 3200 cm^{-1} ice-like peak, Fig. 8.[105,106] The ice-like resonance was chosen due to its larger $\chi^{(3)}$, as found in Fig. 7. A linear relationship between the square root of the sum-frequency intensity and $1/T$ should be observed if E_{DC} induced reorganization of water is occurring,[105] with the slope and intercept terms equal to:

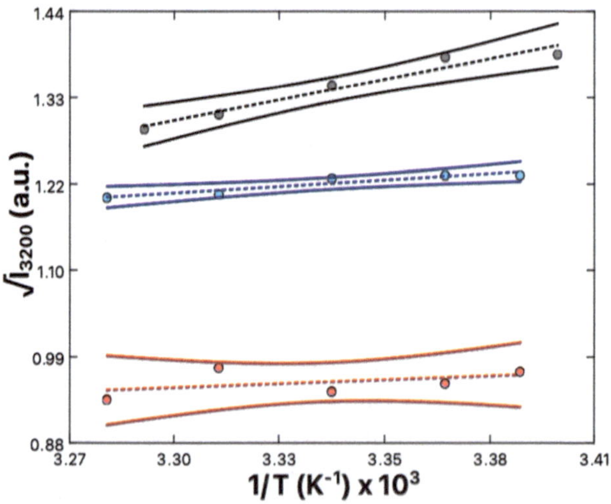

Fig. 8 Square root of SFVS intensity measured at 3200 cm^{-1} for a hybrid DPPS bilayer as a function of inverse temperature. The data were collected at concentrations of: 100 mM KCl + zero $CaCl_2$ (black), 97 mM KCl + 1 mM $CaCl_2$ (blue), and 40 mM Ca (red), The dashed lines are the linear fits to the data. The solid lines indicate the 95% confidence intervals for each linear regression. All solutions were buffered with Tris at pH 7.4 and maintain a constant ionic strength of 133.4 mM.

$$\text{Slope} = \frac{\mu N \beta}{5k_B} E_{DC} E_{532} E_{IR} \tag{17}$$

$$\text{Intercept} = (\chi^{(2)} + N^{(3)}\gamma E_{DC}) E_{532} E_{IR} \tag{18}$$

The relative influences of the dynamic realignment, $\frac{\mu N^{(2)}\beta}{5k_B T}$, and static third-order, $N^{(3)}\gamma$, contributions to $\chi^{(3)}$ were calculated as follows. Intensity data was collected in the absence of Ca^{2+}, as well as in the presence of 1 mM Ca^{2+} or 40 mM Ca^{2+}. These data contain three important trends. Firstly, a linear trend is observed in the zero and 1 mM Ca^{2+} systems. As the realignment contribution, $\frac{\mu N^{(2)}\beta}{5k_B T}$, is the only temperature dependent term in eqn (12), the results validate the hypothesis of electrostatically induced water reorganization. Secondly, it is shown the $1/T$ dependence decreases as the concentration of Ca^{2+} increases. As per eqn (17), the slope, representing the dynamic term $\frac{\mu N^{(2)}\beta}{5k_B T}$, is scaled by E_{DC}. The observation of reduced E_{DC} as Ca^{2+} content is increased is further evidence of charge neutralization of PS lipids as Ca^{2+} binds. Lastly, at 40 mM Ca^{2+} the slope of the line becomes indistinct from zero (180 ± 200). The conclusion is that complete neutralization of PS headgroup charges has occurred resulting in $E_{DC} = 0$. Complete neutralization supports the previous assumption of a 2 : 1 PS : Ca^{2+} binding ratio. Further, the neutralization point agrees with the amine peak-derived binding isotherm (Fig. 6), which indicates surface saturation at 40 mM Ca^{2+}.

From the previous analysis, it becomes possible to determine the relative contributions (eqn (12)) to $\chi^{(3)}$. The contribution from molecular rearrangements, $\frac{\mu N^{(2)}\beta}{5k_B T}$, at any temperature T is simply the product of the measured slope and $1/T$ (eqn (17)). As all sum-frequency spectra shown in Fig. 5A–C were collected at 21 °C, $T = 294.15$ K is used here. The determined slopes of 960 ± 130 K and 300 ± 60 K yield $\frac{\mu N^{(2)}\beta}{5k_B T}$ values of 3.26 ± 0.44 and 1.02 ± 0.35 for the zero Ca^{2+} and 1 mM Ca^{2+} data, respectively. The static third-order contributions, $N^{(3)}\gamma$, can be determined from rearrangement of eqn (18). To accomplish this, however, the value of $\chi^{(2)}$ must be known. $\chi^{(2)}$ is obtained from the 40 mM Ca^{2+} data set, as complete neutralization of surface charge results in an E_{DC} of zero. The average of all the measured 3200 cm^{-1} sum-frequency intensities at 40 mM Ca^{2+} yields a relative $\chi^{(2)}$ of 0.958 ± 0.008. Using this $\chi^{(2)}$ value, the static third-order contributions, $N^{(3)}\gamma$, were calculated for the zero Ca^{2+} and 1 mM Ca^{2+} systems. From the determined intercepts of −1.85 ± 0.42 and 0.21 ± 0.21, $N^{(3)}\gamma$ values of −2.81 ± 0.43 and −0.75 ± 0.22 are obtained for the zero Ca^{2+} and 1 mM Ca^{2+} data, respectively. The relative contribution of dynamic realignment $\left(\frac{\mu N^{(2)}\beta}{5k_B T}\right)$ to $\chi^{(3)}$ is therefore 54 ± 9% and 57 ± 8% for the zero Ca^{2+} and 1 mM Ca^{2+} data, respectively. The results suggest an equal contribution of the dynamic and static components to $\chi^{(3)}$, within error. At saturation, where maximum dipole alignment and maximum $\frac{\mu N^{(2)}\beta}{5k_BT}$ is reached, further increases in E_{DC} are predicted to increase the contribution from the negative $N^{(3)}\gamma$ term while contributions from $\frac{\mu N^{(2)}\beta}{5k_BT}$ would remain constant. Increases in E_{DC} beyond saturation are therefore expected to reduce the total $\chi^{(3)}$ and result in a drop in generated sum-frequency. From Fig. 7 it can be seen that maximum 3200 cm^{-1} amplitude is reached at 5.59×10^7 V m^{-1}, which then decreases at 6.30×10^7 V m^{-1}. These data are consistent with maximal water alignment at 5.59×10^7 V m^{-1}.

Alkyl chain analysis

The alkyl chain C–H stretches, found between 2800 and 3000 cm^{-1}, were characterized to determine whether the lipid tails undergo any structural changes as PS binds Ca^{2+} or Mg^{2+}. All five peaks previously detailed for the bilayer system of Fig. 1 are present, with an additional peak at 2920 cm^{-1} due to the CH_3 symmetric stretch of the methylated surface. The C–H region of the DPPC system in Fig. 5A displays no change when comparing the 100 mM KCl, 40 mM $CaCl_2$, and 40 mM $MgCl_2$ solutions. Conversely, changes are quite apparent within the C–H bands of DPPS as Ca^{2+} (Fig. 5B) or Mg^{2+} (Fig. 5C) content is increased. At first glance this may imply an alteration in the packing structure of the lipid tails; however, the C–H region of the DPPS sits atop the trailing edge of the water stretches, the magnitudes of which are highly influenced by both Ca^{2+} and Mg^{2+} association due to charge screening.

The direct effect of Ca^{2+} ions on DPPS tail structure was determined by isolation of the C–H region from the water region. SFVS spectra were recorded in buffered D_2O to remove the water features from the spectra. Fig. 9 illustrates the

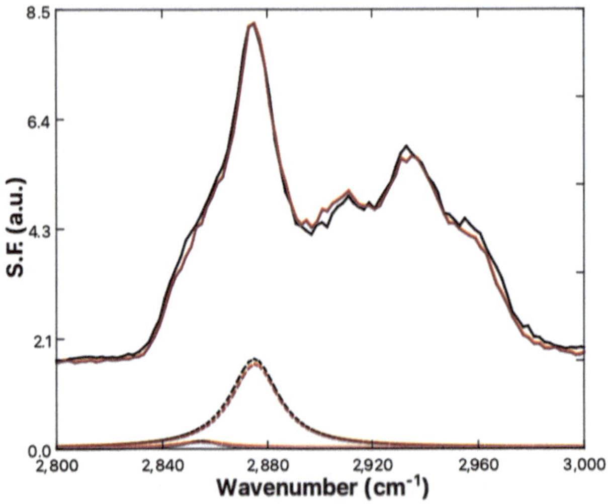

Fig. 9 SFVS spectra of DPPS in D_2O with 100 mM KCl (black) or 40 mM $CaCl_2$ (red). The offset curves (dashed lines) are the spectral fits of the CH_2 ν_s and CH_3 ν_s peaks at 2850 cm^{-1} and 2875 cm^{-1} respectively, obtained using eqn (4). All solutions were buffered with Tris at pH = 7.4 with a total ionic strength of 133.4 mM. Spectra were recorded at 21 °C.

influence Ca^{2+} has on the C–H region, absent any O–H bands. Comparing the 100 mM KCl (zero Ca^{2+}) and 40 mM Ca^{2+} solutions, it is clear the changes observed in Fig. 5B are not due to changes in the C–H bands intensities, but rather result from Ca^{2+} induced disruption of the underlying water resonance.

The ratio of the CH_2 symmetric stretch (ν_s) at 2850 cm^{-1} and the CH_3 ν_s at 2875 cm^{-1} has previously been used to evaluate net changes in *gauche* content within the chain structure.[81,119–122] Increased ordering of the alkyl chains is expected to result in reduced *gauche* content, which is observed in sum-frequency spectra as a reduced CH_2 ν_s character relative to the CH_3 ν_s due to cancellation of sum-frequency within the locally symmetric *trans* configuration.[79,119,120] Ordering of the alkyl chains from Ca^{2+} induced condensation of DPPS would be expected to result in a decrease in the CH_2 ν_s to CH_3 ν_s peak ratio. The sum-frequency spectra of Fig. 9 were fit using eqn (4) to calculate the $CH_2:CH_3$ ratio for the 100 mM KCl and 40 mM $CaCl_2$ data, which remains unchanged within error at 0.20 ± 0.05 and 0.17 ± 0.06, respectively. The observed invariance of the $CH_2:CH_3$ ratio with respect to Ca^{2+} induced condensation of DPPS is taken to simply be a result of minimal initial *gauche* content within the alkyl chains. The membranes under measurement here were deposited at 30 mN m^{-1}, within the gel phase where *gauche* content of saturated lipids is expected to be minimal, for example H^1NMR and FT-IR analysis place the *gauche vs. trans* conformation of DPPC at less than 3%.[123,124] Given DPPS has the same 16-carbon alkyl chain lengths as DPPC, a similarly low *gauche* content would be expected.

Flip-flop kinetics

The central hypothesis of the current work is that PS expression can be influenced by Ca^{2+}. At equilibrium, a net population difference in the number density of PS lipids is possible if the rate constants for PS flip and flop are not equal:

$$\frac{k_{\text{flip}}}{k_{\text{flop}}} = \frac{N_{\text{proximal}}}{N_{\text{distal}}}. \tag{19}$$

To achieve this, Ca^{2+} ions need to alter the flip-flop mixing behavior of PS lipids such that PS lipids will flip and flop at different rates in the presence and absence of Ca^{2+}. To test the hypothesis, DPPS flip-flop kinetics of asymmetric PSLBs containing physiologic levels of 10% PS were performed to determine if the binding interactions observed between PS and Ca^{2+} or Mg^{2+} affect the rate of DPPS flip-flop. The concentration of Ca^{2+} and Mg^{2+} were 1 mM to mimic the physiologic concentrations of the extracellular fluid and cytosol, respectively.[54,55] KCl controls without Ca^{2+} or Mg^{2+} were also collected as a baseline for comparison. Fig. 10A–C show representative DPPS flip-flop decays for both DPPC and DPPS with Ca^{2+} or Mg^{2+}, while Fig. 10D and E show linearized Eyring plots of all determined flip-flop rates.

Eyring plots were generated from the Eyring equation, which relates the rate of flip-flop, k, to the transition state free energy, $\Delta G^{\ddagger}$:[63,66,125]

$$\frac{hk}{k_B T} = e^{\frac{\Delta G^{\ddagger}}{RT}} \tag{20}$$

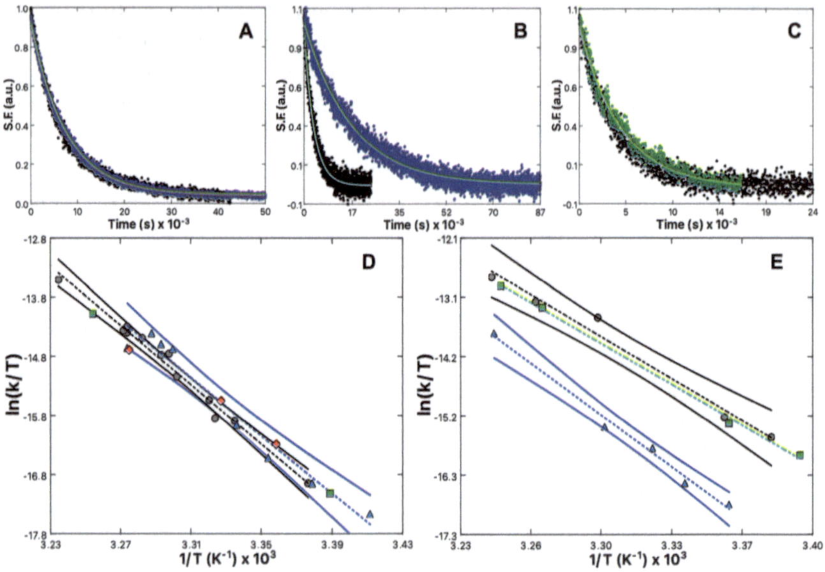

Fig. 10 Representative SFVS intensity decays of the CH_3 symmetric stretch at 2875 cm^{-1} for (A) DPPC/DPPCd62 bilayers in the presence of 100 mM KCl (black) and 97 mM KCl + 1 mM $CaCl_2$ (blue) recorded at 26.9 and 26.6 °C, respectively; (B) 10% DPPS in DPPCd62 bilayers in the presence of 100 mM KCl (black) and 97 mM KCl + 1 mM $CaCl_2$ (blue) recorded at 24.7 and 24.5 °C, respectively; (C) 10% DPPS in DPPCd62 bilayers in the presence of 100 mM KCl (black) and 97 mM KCl + 1 mM $MgCl_2$ (green) recorded at 24.7 and 24.5 °C, respectively. The solid lines in (A–C) are the fits to the data using eqn (7). (D and E) show the linearized Eyring plots of the measured flip-flop rates for DPPC and 10% DPPS in DPPCd62, respectively obtained in 100 mM KCl (gray circles), 97 mM KCl + 1 mM $CaCl_2$ (blue triangles) and 97 mM KCl + 1 mM $MgCl_2$ (green squares). The dashed lines are the fits to the data using eqn (21) with the corresponding confidence intervals (solid lines). (D) also contains measured flip-flop rates for DPPC in the presence of 10% DPPSd62 (red diamonds) with 97 mM KCl + 1 mM $CaCl_2$.

where h is Planck's constant, k_B is Boltzmann's constant, R is the molar gas constant, and T is the temperature in Kelvin. In the linearized form, the Eyring relation is expressed as:

$$\ln\left(\frac{k}{T}\right) = \ln\left(\frac{k_B}{h}\right) - \frac{\Delta G^{\ddagger}}{RT} \tag{21}$$

which allows for linear presentation of all kinetic data as well as determination of the change in the transition state free energy of the flip-flop due to DPPS or DPPC lipids binding Ca^{2+} or Mg^{2+} ions.

Considering the DPPC controls first, no effect of Mg^{2+} or Ca^{2+} on the rate of DPPC flip-flop was observed. A representative comparison of DPPC flip-flop behavior is shown in Fig. 10A, where the rate of DPPC flip-flop is found to be $3.35 \pm 0.03 \times 10^{-5}$ s^{-1} at 26.9 °C or $3.45 \pm 0.04 \times 10^{-5}$ s^{-1} at 26.6 °C in the presence and absence of 1 mM Ca^{2+}, respectively. The slight deviation in rate is considered to be within the experimental variance as, from Fig. 10D, the 95% confidence intervals of the DPPC flip-flop rates in the presence and absence of 1 mM Ca^{2+} overlap. Likewise, measured flip-flop rates for DPPC in the presence of

1 mM Mg^{2+} fall within the 95% confidence intervals for DPPC flip-flop rates both in the presence and absence of 1 mM Ca^{2+}. The lack of any influence of Ca^{2+} or Mg^{2+} on DPPC flip-flop rates is in agreement with the compression isotherms which found no change in the MMA in the presence of Ca^{2+} or Mg^{2}, results which indicate neither ion had bound DPPC. Likewise, bound Ca^{2+} or Mg^{2+} would be expected to increase the surface charge density of the neutral DPPC lipids and, through $\chi^{(3)}$ (eqn (13)), scale the sum-frequency generated by the 3200 cm^{-1} and 3400 cm^{-1} water peaks. No such effect was observed in sum-frequency scans of DPPC hybrid bilayers in the presence and absence of Ca^{2+} or Mg^{2+} (Fig. 5A). The sum-frequency spectra, compression isotherms, and flip-flop data are all consistent with no observable binding interaction between the PC headgroup and either Ca^{2+} or Mg^{2+}.

Determination of the flip-flop kinetics for 10% DPPS + 90% DPPC membranes was performed with respect to both the DPPS and DPPC fractions. The effect of 10% DPPS on the parent DPPC matrix was examined by measurement of the flip-flop rate of 90% DPPC in the presence of 10% DPPS with 1 mM Ca^{2+} (87.3% DPPC + 9.7% DPPSd62 + 3% PEG5000-PE/90% DPPCd62 + 10% DPPSd62, proximal/distal). The flip-flop rates fall within the 95% confidence intervals of the flip-flop rates obtained for a pure DPPC membrane shown in Fig. 10D. This result indicates that the presence of 10 mol% DPPS has no influence on DPPC flip-flop. The flip-flop behavior of the PS fraction, however, yields more interesting results. The rate of DPPS flip-flop in DPPC (87.3% DPPCd62 + 9.7% DPPS + 3% PEG5000-PE/90% DPPCd62 + 10% DPPSd62, proximal/distal) show a pronounced effect of Ca^{2+} on the rate of DPPS flip-flop. The rate of DPPS flip-flop decreased from $7.11 \pm 0.08 \times 10^{-5}$ s^{-1} (24.7 °C) in the absence of Ca^{2+} to $1.46 \pm 0.07 \times 10^{-5}$ s^{-1} (24.5 °C) when in the presence of 1 mM Ca^{2+}, roughly a 5-fold decrease. Alternatively, Fig. 10C shows a comparison of DPPS flip-flop measured in the presence of 1 mM Mg^{2+} at 24.5 °C compared to DPPS flip-flop in the absence of Ca^{2+} or Mg^{2+} at 24.7 °C (same as Fig. 10B). The measured flip-flop rate of $6.8 \pm 0.1 \times 10^{-5}$ s^{-1} for the 1 mM Mg^{2+} exposed DPPS is essentially identical to the $7.11 \pm 0.08 \times 10^{-5}$ s^{-1} flip-flop rate measured in the absence of Mg^{2+}. All flip-flop rates measured for DPPS in the presence of 1 mM Mg^{2+} fall within the 95% confidence intervals of DPPS flip-flop in the absence of Ca^{2+} or Mg^{2+}, whereas all flip-flop rates measured for DPPS in the presence of 1 mM Ca^{2+} fall outside (Fig. 10E). Ca^{2+} exposure has a pronounced and specific slowing effect on DPPS translocation. From eqn (19), the slowed flip-flop kinetics of DPPS in the presence of Ca^{2+} (Fig. 10B) results from an increase in $\Delta G^{\ddagger}$ from 96.6 ± 0.1 kJ mol^{-1} in the absence of Ca^{2+} to 100.5 ± 0.2 kJ mol^{-1} in the presence of 1 mM Ca^{2+}, an increase of 3.9 ± 0.2 kJ mol^{-1}. In the presence of 1 mM Mg^{2+}, the $\Delta G^{\ddagger}$ of 96.7 ± 0.2 kJ mol^{-1} remains within error of the $\Delta G^{\ddagger}$ in the absence of Ca^{2+} or Mg^{2+}.

The divergence in DPPS response to the presence of Ca^{2+} and Mg^{2+} can be attributed to two important factors. Firstly, Mg^{2+} was found to have a PS binding affinity roughly 20-fold lower than Ca^{2+}. Yet further, the binding data from Fig. 6 shows bound Mg^{2+} does not produce as great of an impact on the ordering of the serine headgroup amines. A 1 mM concentration of Mg^{2+} resulted in an amine peak area increase roughly equivalent to just 10 μM Ca^{2+}. These results track well with Papahadjopoulos, who observed a 10-fold differential in Ca^{2+} and Mg^{2+} binding affinity, and noted only Ca^{2+} lead to fusion of PS-containing vesicles or phase segregation of PS.[10,50–53] Mg^{2+} ions have lower PS affinity, more muted impact on headgroup structure, and lower denticity binding to the membrane as predicted by MD simulations,[89] and it is therefore not surprising to find that Mg's binding modality is simply too weak to influence translocation.

Ca^{2+} has a unique ability to slow the rate of DPPS translocation relative to Mg^{2+}, which has no measurable effect. The finding has potentially important implications for PS expression *in vivo*. The extracellular concentration of Ca^{2+} is far higher than within the cytosol, at 1 mM and 100 nM, respectively.[55] Some degree of Ca^{2+} induced PS externalization could therefore be explained in terms of the difference in the rate of PS flip and flop (eqn (19)).[63,65,126] In the case where $k_{\text{flip}} = k_{\text{flop}}$, one would expect the system to mix to homogeneity, where $N_{\text{proximal}} = N_{\text{distal}}$. Such is observed in the flip-flop measurements in Fig. 10A–C, where sum-frequency signal trends toward zero due to the increasing symmetry of the system (eqn (5)). However, in the case of unequal rates of k_{flip} and k_{flop}, the equilibrium expression (eqn (19)) dictates accumulation toward the side of lower k, as the rate of inward movement is greater than the rate of outward transfer. Using the flip-flop rates determined in Fig. 10C as an example, a Ca^{2+} concentration differential whereby the distal leaflet is exposed to 1 mM Ca^{2+} while the proximal leaflet is unexposed to Ca^{2+} would result in a DPPS k_{flip} of $1.46 \pm 0.07 \times 10^{-5}\ \text{s}^{-1}$ and k_{flop} of $7.11 \pm 0.08 \times 10^{-5}\ \text{s}^{-1}$ at 24.5 °C. Applied to eqn (19), DPPS would be expected to enrich in the Ca^{2+} exposed leaflet by a factor of 4.87 : 1, or approximately 80% of the total DPPS population (Fig. 11). Ongoing research aims to further validate the Ca-induced PS expression model by way of SFVS of surface-tethered vesicles where the concentration of Ca^{2+} of the external and internally encapsulated solution can be independently controlled. From the data shown here, an initially symmetric PS population should produce asymmetric vesicles upon exposure to a calcium concentration gradient and thus yield an increase in sum frequency. Such experiments would additionally allow a comparison of lipid flip-flop rates determined from PSLB and vesicle systems and aid in the examination of inconsistencies between methods.

While the flip-flop rates used in the prior calculation are particular to the non-physiologic binary mixture tested, Ca–PS interactions *in vivo* can reasonably be inferred to slow PS lipid flip-flop in an analogous fashion. The extent of the effect is very likely to depend on the membrane matrix and experimental parameters, which precludes any universal statement on how the effect may manifest. However, the theorized enrichment of PS toward the Ca-exposed leaflet is in agreement with the observations of both the Hampton and Bratton groups where the degree of PS expression was dependent on a Ca^{2+} concentration differential over the bilayer as opposed to Ca^{2+} influx.[25,56]

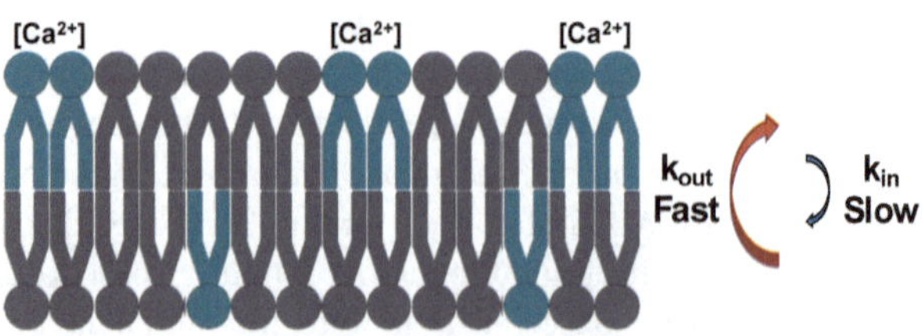

Fig. 11 Illustration of the enrichment of PS lipids (blue) on the Ca^{2+}-exposed distal leaflet, driven by unequal rates of slow inward flip (k_{in}) and fast outward flop (k_{out}).

On the surface, migration of PS lipids toward Ca^{2+} would seem incompatible with the usual observations of internally held PS, where Ca^{2+} content is minimal.[55] However, PS externalization is normally induced *in vivo* by some intentional perturbation of the cells. For example, common methods of achieving PS externalization include UV irradiation, incorporation of membrane-permeating ionophores, ATP-depletion, deoxygenation, and other treatments which may destabilize cell functioning in some unforeseen ways.[16,29,33,35,36,39,40,43,47,127] As the mechanisms by which PS asymmetry is maintained in healthy cells are not fully understood, the effects of such treatment on the cell's ability to maintain PS asymmetry is not fully detailed. Destabilized cell functionality may simply lead to PS lipids becoming free to equilibrate across the membrane boundary. Many studies on PS externalization utilize high concentrations of external Ca^{2+} to mimic physiologic conditions, make use of Ca-dependent Annexin V staining, or test the dependence of scramblases[2,4,11,17,19,24,25,28,31–36] on Ca^{2+} content.[39,43,44,47,127–138] Due to PS experiencing differing flip and flop rates, some amount of PS expression may simply be attributed to the high external concentrations of Ca^{2+}.

The hypothesized effect of Ca-induced PS enrichment does not preclude the effects of scramblases or signalling processes which may also serve to externalize PS. Multiple pathways of externalization may coexist. The potential presence of a direct PS–Ca^{2+} mechanism of externalization does however suggest that determinations of cooperative actions between Ca^{2+} and scramblases may first require disentanglement from effects driven purely by PS–Ca^{2+} interactions. The data reported here demonstrates the importance of considering the influence of Ca^{2+} concentration differentials when interpreting PS expression data.

Conclusions

Compression isotherms of DPPC and DPPS monolayers in the presence and absence of 1 mM Ca^{2+} or 1 mM Mg^{2+} revealed a specific interaction between DPPS and Ca^{2+} ions. When in the presence of Ca^{2+}, but not Mg^{2+}, a -2.4 ± 0.6 Å^2 per molecule reduction in the MMA of DPPS was observed at 30 mN m^{-1}. No change in MMA was observed for DPPC. A net condensation of the DPPS monolayer in the presence of Ca^{2+} is consistent with a headgroup-bridging binding interaction whereby neighboring headgroups are brought into closer contact. A 2 : 1 PS : Ca^{2+} binding ratio was hypothesized.

SFVS spectra of DPPS and DPPC hybrid bilayers were examined at concentrations between 0 and 40 mM Ca^{2+} or Mg^{2+} to determine whether the binding of either Ca^{2+} or Mg^{2+} to PS or PC headgroups impacted the membrane structure. No effect on DPPC was observed for either Ca^{2+} or Mg^{2+}, whereas DPPS responded to both Ca^{2+} and Mg^{2+}. As the concentration of Ca^{2+} or Mg^{2+} was increased, DPPS hybrid bilayers showed an increase in the amine peak sum-frequency at 3300 cm^{-1}, as well as decreases in sum-frequency of the ice- and water-like O–H peaks at 3200 cm^{-1} and 3400 cm^{-1}, respectively. The growth in the amine peak is attributed to complexation of the PS headgroup and ordering of the amine moiety. Ca^{2+} ions were shown to have a more pronounced influence on ordering of the amine groups, resulting in double the 3300 cm^{-1} peak area at saturation when compared to Mg^{2+}. Frumkin model binding isotherms were calculated through relation of the change in amine peak area with the bulk concentration of either Ca^{2+} or Mg^{2+} present. The binding isotherms showed both Ca^{2+} and Mg^{2+}

demonstrate binding behaviors toward PS lipids, with affinity values of $1.3 \pm 0.3 \times 10^5\ M^{-1}$ and $7.3 \pm 2.5 \times 10^3\ M^{-1}$, respectively.

The amine peak binding isotherm for Ca^{2+} was used in conjunction with the assumed 2 : 1 PS : Ca^{2+} binding ratio to calculate the membrane charge state and E_{DC} field strength. The peak amplitudes of both the ice- and water-like O–H peaks of the DPPS associated water layer were found to track linearly with E_{DC} values projected from the Ca–PS binding isotherm. These results are in line with E_{DC} scaling of $\chi^{(3)}$ and the reorientation of water within the surface charge dependent E_{DC} field. At saturating concentrations of Ca^{2+}, loss of temperature dependence of sum-frequency at the 3200 cm^{-1} ice-like water resonance was observed. These data indicate an E_{DC} of zero and complete charge neutralization at Ca^{2+} saturation, consistent with the assumed 2 : 1 PS : Ca^{2+} binding ratio.

SFVS measurements of DPPS and DPPC flip-flop rates in the presence and absence of physiologic 1 mM Ca^{2+} and 1 mM Mg^{2+} were utilized to determine the impact of bound Ca^{2+} or Mg^{2+} on DPPS and DPPC lipid translocation. The measured rate of translocation of 10% DPPS in DPPC decreased from $7.11 \pm 0.08 \times 10^{-5}\ s^{-1}$ (24.7 °C) in the absence of Ca^{2+} to $1.46 \pm 0.07 \times 10^{-5}\ s^{-1}$ (24.5 °C) when in the presence of 1 mM Ca^{2+}, roughly a 5-fold decrease. No effect of Mg^{2+} on DPPS flip-flop rate or Ca^{2+} or Mg^{2+} on DPPC flip-flop rate was observed. From these data it is inferred that a Ca^{2+} concentration differential over the bilayer would produce differing rates of PS lipid flip and flop between the leaflets. For the tested binary mixture of 10% DPPS in DPPC, the differential in the zero Ca^{2+} and 1 mM Ca^{2+} flip-flop rates at ~24.6 °C would be expected to result in an approximate 80% sequestration of DPPS toward the Ca^{2+}-exposed leaflet. Given the large differential in Ca^{2+} concentration over the cell membrane *in vivo*, the endogenous effects of Ca^{2+} on PS lipids may influence PS lipid externalization even absent enzymatic action. These results illustrate the importance of considering direct Ca–PS interactions when interpreting PS locality and expression.

Data availability

The data that supports these findings are available from the corresponding author upon reasonable request.

Author contributions

J. C. C and P. P. H designed the research. P. P. H. performed the research. P. P. H. and J. C. C. analysed the data and wrote the article. P. P. H. and J. C. C. edited the manuscript.

Conflicts of interest

There are no conflicts to declare.

Acknowledgements

The following work was supported by the National Science Foundation (NSF) (No. 2304682).

Notes and references

1 R. Chandra, P. C. Joshi, V. K. Bajpai and C. M. Gupta, *Biochim. Biophys. Acta*, 1987, **902**, 253–262.

2 D. L. Daleke, *Curr. Opin. Hematol.*, 2008, **15**, 191–195.

3 R. D. Kornberg and H. M. Mcconnell, *Biochemistry*, 1971, **10**, 1111–1120.

4 G. Lenoir, P. Williamson and J. C. Holthuis, *Curr. Opin. Chem. Biol.*, 2007, **11**, 654–661.

5 J. A. Op den Kamp, *Annu. Rev. Biochem.*, 1979, **48**, 47–71.

6 A. J. Verkleij, R. F. Zwaal, B. Roelofsen, P. Comfurius, D. Kastelijn and L. L. van Deenen, *Biochim. Biophys. Acta*, 1973, **323**, 178–193.

7 J. H. Lorent, K. R. Levental, L. Ganesan, G. Rivera-Longsworth, E. Sezgin, M. Doktorova, E. Lyman and I. Levental, *Nat. Chem. Biol.*, 2020, **16**, 644–652.

8 G. Cevc, *Biochim. Biophys. Acta*, 1990, **1031**, 311–382.

9 B. Fadeel and D. Xue, *Crit. Rev. Biochem. Mol. Biol.*, 2009, **44**, 264–277.

10 N. Duzgunes, J. Wilschut, R. Fraley and D. Papahadjopoulos, *Biochim. Biophys. Acta*, 1981, **642**, 182–195.

11 J. W. Gauer, K. J. Knutson, S. R. Jaworski, A. M. Rice, A. M. Rannikko, B. R. Lentz and A. Hinderliter, *Biophys. J.*, 2013, **104**, 2437–2447.

12 J. G. Kay and G. D. Fairn, *Cell Commun. Signal.*, 2019, **17**, 1–8.

13 S. Ohki and R. Sauve, *Biochim. Biophys. Acta*, 1978, **511**, 377–387.

14 A. Portis, C. Newton, W. Pangborn and D. Papahadjopoulos, *Biochemistry*, 1979, **18**, 780–790.

15 J. Wang, C. Yu, J. Zhuang, W. Qi, J. Jiang, X. Liu, W. Zhao, Y. Cao, H. Wu, J. Qi and R. C. Zhao, *Biomark. Res.*, 2022, **10**, 4.

16 E. M. Bevers, P. Comfurius and R. F. Zwaal, *Biochim. Biophys. Acta*, 1983, **736**, 57–66.

17 A. D. Ide, E. M. Wight and C. K. Damer, *PLoS One*, 2021, **16**, e0250710.

18 A. K. Kimura and T. Kimura, *FASEB J.*, 2021, **35**, e21177.

19 I. Vermes, C. Haanen, H. Steffens-Nakken and C. Reutelingsperger, *J. Immunol. Methods*, 1995, **184**, 39–51.

20 P. F. Devaux, A. Herrmann, N. Ohlwein and M. M. Kozlov, *Biochim. Biophys. Acta, Biomembr.*, 2008, **1778**, 1591–1600.

21 A. Papadopulos, S. Vehring, I. López-Montero, L. Kutschenko, M. Stöckl, P. F. Devaux, M. Kozlov, T. Pomorski and A. Herrmann, *J. Biol. Chem.*, 2007, **282**, 15559–15568.

22 K. Segawa, S. Kurata and S. Nagata, *J. Biol. Chem.*, 2016, **291**, 762–772.

23 K. Segawa, S. Kurata, Y. Yanagihashi, T. R. Brummelkamp, F. Matsuda and S. Nagata, *Science*, 2014, **344**, 1164–1168.

24 H. Takatsu, M. Takayama, T. Naito, N. Takada, K. Tsumagari, Y. Ishihama, K. Nakayama and H. W. Shin, *Nat. Commun.*, 2017, **8**, 1423.

25 D. L. Bratton, V. A. Fadok, D. A. Richter, J. M. Kailey, L. A. Guthrie and P. M. Henson, *J. Biol. Chem.*, 1997, **272**, 26159–26165.

26 T. Pomorski, A. Herrmann, P. Müller, G. Van Meer and K. Burger, *Biochemistry*, 1999, **38**, 142–150.

27 T. Pomorski, P. Müller, B. Zimmermann, K. Burger, P. F. Devaux and A. Herrmann, *J. Cell Sci.*, 1996, **109**, 687–698.

28 X. Tang, M. S. Halleck, R. A. Schlegel and P. Williamson, *Science*, 1996, **272**, 1495–1497.

29 E. M. Bevers, P. F. Verhallen, A. J. Visser, P. Comfurius and R. F. Zwaal, *Biochemistry*, 1990, **29**, 5132–5137.

30 E. F. Smeets, P. Comfurius, E. M. Bevers and R. F. Zwaal, *Biochim. Biophys. Acta*, 1994, **1195**, 281–286.

31 E. M. Bevers and P. L. Williamson, *FEBS Lett.*, 2010, **584**, 2724–2730.

32 M. Ikeda, A. Kihara and Y. Igarashi, *Biol. Pharm. Bull.*, 2006, **29**, 1542–1546.

33 A. Momchilova, L. Ivanova, T. Markovska and R. Pankov, *Arch. Biochem. Biophys.*, 2000, **381**, 295–301.

34 J. Suzuki, M. Umeda, P. J. Sims and S. Nagata, *Nature*, 2010, **468**, 834–838.

35 P. Williamson, E. M. Bevers, E. F. Smeets, P. Comfurius, R. A. Schlegel and R. F. Zwaal, *Biochemistry*, 1995, **34**, 10448–10455.

36 P. Williamson, A. Christie, T. Kohlin, R. A. Schlegel, P. Comfurius, M. Harmsma, R. F. Zwaal and E. M. Bevers, *Biochemistry*, 2001, **40**, 8065–8072.

37 M. Seigneuret and P. F. Devaux, *Proc. Natl. Acad. Sci. U. S. A.*, 1984, **81**, 3751–3755.

38 B. J. Abbott, D. S. Fukuda, D. E. Dorman, J. L. Occolowitz, M. Debono and L. Farhner, *Antimicrob. Agents Chemother.*, 1979, **16**, 808–812.

39 J. Dachary-Prigent, J. M. Pasquet, J. M. Freyssinet and A. T. Nurden, *Biochemistry*, 1995, **34**, 11625–11634.

40 U. Henseleit, G. Plasa and C. Haest, *Biochim. Biophys. Acta*, 1990, **1029**, 127–135.

41 A. Jaskulska, A. E. Janecka and K. Gach-Janczak, *Int. J. Mol. Sci.*, 2020, **19**, 131–135.

42 J. K. Paterson, K. Renkema, L. Burden, M. S. Halleck, R. A. Schlegel, P. Williamson and D. L. Daleke, *Biochemistry*, 2006, **45**, 5367–5376.

43 P. Williamson, A. Kulick, A. Zachowski, R. A. Schlegel and P. F. Devaux, *Biochemistry*, 1992, **31**, 6355–6360.

44 D. W. Dekkers, P. Comfurius, E. M. Bevers and R. F. Zwaal, *Biochem. J.*, 2002, **362**, 741–747.

45 Q. Zhou, J. Zhao, T. Wiedmer and P. J. Sims, *Blood*, 2002, **99**, 4030–4038.

46 J. R. Nofer, G. Herminghaus, M. Brodde, E. Morgenstern, S. Rust, T. Engel, U. Seedorf, G. Assmann, H. Bluethmann and B. E. Kehrel, *J. Biol. Chem.*, 2004, **279**, 34032–34037.

47 P. Williamson, M. S. Halleck, J. Malowitz, S. Ng, X. Fan, S. Krahling, A. T. Remaley and R. A. Schlegel, *PLoS One*, 2007, **2**, e729.

48 M. Ross, C. Steinem, H.-J. Galla and A. Janshoff, *Langmuir*, 2001, **17**, 2437–2445.

49 K. Jacobson and D. Papahadjopoulos, *Biochemistry*, 1975, **14**, 152–161.

50 D. Papahadjopoulos, G. Poste, B. E. Schaeffer and W. J. Vail, *Biochim. Biophys. Acta*, 1974, **352**, 10–28.

51 C. Newton, W. Pangborn, S. Nir and D. Papahadjopoulos, *Biochim. Biophys. Acta*, 1978, **506**, 281–287.

52 D. Papahadjopoulos and A. Portis, *Ann. N. Y. Acad. Sci.*, 1978, **05467**, 50–66.

53 D. Papahadjopoulos, W. J. Vail, C. Newton, S. Nir, K. Jacobson, G. Poste and R. Lazo, *Biochim. Biophys. Acta*, 1977, **465**, 579–598.

54 R. Vink and M. Nechifor, *Magnesium in the Central Nervous System*, University of Adelaide Press, South Australia, 2011.

55 D. K. Atchison and W. H. Beierwaltes, *Pfluegers Arch.*, 2013, **465**, 59–69.
56 M. B. Hampton, D. M. Vanags, M. I. Porn-Ares and S. Orrenius, *FEBS Lett.*, 1996, **399**, 277–282.
57 V. Cheng and J. C. Conboy, *J. Phys. Chem. B*, 2022, **126**, 7651–7663.
58 F. A. Heberle, D. Marquardt, M. Doktorova, B. Geier, R. F. Standaert, P. Heftberger, B. Kollmitzer, J. D. Nickels, R. A. Dick, G. W. Feigenson, J. Katsaras, E. London and G. Pabst, *Langmuir*, 2016, **32**, 5195–5200.
59 M. Nakano, M. Fukuda, T. Kudo, H. Endo and T. Handa, *Phys. Rev. Lett.*, 2007, **98**, 238101–238104.
60 M. Nakano, M. Fukuda, T. Kudo, N. Matsuzaki, T. Azuma, K. Sekine, H. Endo and T. Handa, *J. Phys. Chem. B*, 2009, **113**, 6745–6748.
61 B. Wah, J. M. Breidigan, J. Adams, P. Horbal, S. Garg, L. Porcar and U. Perez-Salas, *Langmuir*, 2017, **33**, 3384–3394.
62 J. S. Allhusen and J. C. Conboy, *Acc. Chem. Res.*, 2017, **50**, 58–65.
63 T. C. Anglin and J. C. Conboy, *Biochemistry*, 2009, **48**, 10220–10234.
64 J. Liu and J. C. Conboy, *Biophys. J.*, 2005, **89**, 2522–2532.
65 J. Liu and J. C. Conboy, *J. Am. Chem. Soc.*, 2004, **126**, 8376–8377.
66 A. Timothy, C. Michael, L. Hao, C. Katherine and C. John, *J. Phys. Chem. B*, 2010, **114**, 1903–1914.
67 K. L. Brown and J. C. Conboy, *J. Phys. Chem. B*, 2013, **117**, 15041–15050.
68 D. Marquardt, B. Geier and G. Pabst, *Membranes*, 2015, **5**, 180–196.
69 D. Marquardt, F. A. Heberle, T. Miti, B. Eicher, E. London, J. Katsaras and G. Pabst, *Langmuir*, 2017, **33**, 3731–3741.
70 M. Doktorova, F. A. Heberle, B. Eicher, R. F. Standaert, J. Katsaras, E. London, G. Pabst and D. Marquardt, *Nat. Protoc.*, 2018, **13**, 2086–2101.
71 L. Marx, M. P. K. Frewein, E. F. Semeraro, G. N. Rechberger, K. Lohner, L. Porcar and G. Pabst, *Faraday Discuss.*, 2021, **232**, 435–447.
72 M. P. K. Frewein, P. Piller, E. F. Semeraro, K. C. Batchu, F. A. Heberle, H. L. Scott, Y. Gerelli, L. Porcar and G. Pabst, *J. Membr. Biol.*, 2022, **255**, 407–421.
73 H. M. McConnell, L. K. Tamm and R. M. Weis, *Proc. Natl. Acad. Sci. U. S. A.*, 1984, **81**, 3249–3253.
74 M. L. Wagner and L. K. Tamm, *Biophys. J.*, 2000, **79**, 1400–1414.
75 V. Kiessling and L. K. Tamm, *Biophys. J.*, 2003, **84**, 408–418.
76 V. Kiessling, C. Wan and L. K. Tamm, *Biochim. Biophys. Acta*, 2009, **1788**, 64–71.
77 V. Kiessling, J. M. Crane and L. K. Tamm, *Biophys. J.*, 2006, **91**, 3313–3326.
78 M. Zacchia, M. L. Abategiovanni, S. Stratigis and G. Capasso, *Kidney Dis.*, 2016, **2**, 72–79.
79 P. Guyot-Sionnest, J. H. Hunt and Y. R. Shen, *Phys. Rev. Lett.*, 1987, **59**, 1597–1600.
80 V. Ostroverkhov, G. A. Waychunas and Y. R. Shen, *Chem. Phys. Lett.*, 2004, **386**, 144–148.
81 J. Liu and J. C. Conboy, *Langmuir*, 2005, **21**, 9091–9097.
82 D. Marsh, *Biochim. Biophys. Acta*, 1996, **1286**, 183–223.
83 A. J. Diaz, F. Albertorio, S. Daniel and P. S. Cremer, *Langmuir*, 2008, **24**, 6820–6826.
84 O. Bouffioux, A. Berquand, M. Eeman, M. Paquot, Y. F. Dufrene, R. Brasseur and M. Deleu, *Biochim. Biophys. Acta*, 2007, **1768**, 1758–1768.

85 J. Miñones, J. M. Rodríguez Patino, O. Conde, C. Carrera and R. Seoane, *Colloids Surf., A*, 2002, **203**, 273–286.
86 E. M. Adams, C. B. Casper and H. C. Allen, *J. Colloid Interface Sci.*, 2016, **478**, 353–364.
87 S. Narayanan, PhD thesis, The Pennsylvania State University, 2019.
88 J. T. Davies and E. K. Rideal, *Interfacial Phenomena*, Academic Press, New York, 2nd edn, 1963.
89 A. Martin-Molina, C. Rodriguez-Beas and J. Faraudo, *Biophys. J.*, 2012, **102**, 2095–2103.
90 D. Simonelli, S. Baldelli and M. J. Shultz, *Chem. Phys. Lett.*, 1998, **298**, 400–404.
91 M. L. Clarke, J. Wang and Z. Chen, *J. Phys. Chem. B*, 2005, **109**, 22027–22035.
92 S. Y. Jung, S. M. Lim, F. Albertorio, G. Kim, M. C. Gurau, R. D. Yang, M. A. Holden and P. S. Cremer, *J. Am. Chem. Soc.*, 2003, **125**, 12782–12786.
93 T. Weidner, N. F. Breen, G. P. Drobny and D. G. Castner, *J. Phys. Chem. B*, 2009, **113**, 15423–15426.
94 J. Seelig, P. M. Macdonald and P. G. Scherer, *Biochemistry*, 1987, **26**, 7535–7541.
95 T. T. Nguyen and J. C. Conboy, *Anal. Chem.*, 2011, **83**, 5979–5988.
96 T. T. Nguyen, K. Rembert and J. C. Conboy, *J. Am. Chem. Soc.*, 2009, **131**, 1401–1403.
97 H. Hauser, A. Darke and M. C. Phillips, *Eur. J. Biochem.*, 1976, **62**, 335–344.
98 C. G. Sinn, M. Antonietti and R. Dimova, *Colloids Surf., A*, 2006, **282–283**, 410–419.
99 D. E. Gragson and G. L. Richmond, *J. Am. Chem. Soc.*, 1998, **120**, 366–375.
100 G. L. Richmond, *Chem. Rev.*, 2002, **102**, 2693–2724.
101 A. G. de Beer and S. Roke, *J. Chem. Phys.*, 2016, **145**, 044705.
102 X. Chen, W. Hua, Z. Huang and H. C. Allen, *J. Am. Chem. Soc.*, 2010, **132**, 11336–11342.
103 P. E. Ohno, H. F. Wang and F. M. Geiger, *Nat. Commun.*, 2017, **8**, 1032.
104 P. E. Ohno, H. F. Wang, F. Paesani, J. L. Skinner and F. M. Geiger, *J. Phys. Chem. A*, 2018, **122**, 4457–4464.
105 S. Ong, X. Zhao and K. B. Eisenthal, *Chem. Phys. Lett.*, 1992, **191**, 327–335.
106 X. Zhao, S. Ong and K. B. Eisenthal, *Chem. Phys. Lett.*, 1993, **202**, 513–520.
107 B. F. Levine and C. G. Bethea, *J. Chem. Phys.*, 1976, **65**, 2429–2438.
108 P. D. Maker and R. W. Terhune, *Phys. Rev.*, 1965, **137**, A801–A818.
109 R. W. Terhune, P. D. Maker and C. M. Savage, *Phys. Rev. Lett.*, 1962, **8**, 404–406.
110 S. Kielich, *IEEE J. Quantum Electron.*, 1969, **5**, 562–568.
111 G. Gonella, C. Lütgebaucks, A. G. F. de Beer and S. Roke, *J. Phys. Chem. C*, 2016, **120**, 9165–9173.
112 J. A. Armstrong, N. Bloembergen, J. Ducuing and P. S. Pershan, *Phys. Rev.*, 1962, **127**, 1918–1939.
113 P. A. Franken and J. F. Ward, *Rev. Mod. Phys.*, 1963, **35**, 23–39.
114 M. A. Habib and J. O. Bockris, *Langmuir*, 2002, **2**, 388–392.
115 A. Chachaj-Brekiesz, J. Kobierski, A. Wnetrzak and P. Dynarowicz-Latka, *Membranes*, 2021, **11**, 53.
116 M. Eisenberg, T. Gresalfi, T. Riccio and S. McLaughlin, *Biochemistry*, 1979, **18**, 5213–5223.

117 Y. A. Ermakov, *Membranes*, 2023, **13**, 883–905.
118 P. Debye and E. Hückel, *Phys. Z.*, 1923, **24**, 185–206.
119 J. C. Conboy, M. C. Messmer and G. L. Richmond, *Langmuir*, 1998, **14**, 6722–6727.
120 J. C. Conboy, M. C. Messmer and G. L. Richmond, *J. Phys. Chem.*, 1996, **100**, 7617–7622.
121 R. N. Ward, D. C. Duffy, P. B. Davies and C. D. Bain, *J. Phys. Chem.*, 2002, **98**, 8536–8542.
122 X. Zhuang, P. B. Miranda, D. Kim and Y. R. Shen, *Phys. Rev. B*, 1999, **59**, 12632–12640.
123 R. Mendelsohn, M. A. Davies, J. W. Brauner, H. F. Schuster and R. A. Dluhy, *Biochemistry*, 1989, **28**, 8934–8939.
124 J. r. Tuchtenhagen, W. Ziegler and A. Blume, *Eur. Biophys. J.*, 1994, **23**, 323–335.
125 H. Eyring, *J. Chem. Phys.*, 1935, **3**, 107–115.
126 C. M. Guldberg and P. Waage, *Forh. Vidensk.-Selsk.*, **1864**, 35–45.
127 E. Weiss, D. C. Rees and J. S. Gibson, *Anemia*, 2011, **2011**, 379894.
128 P. Meers, D. Daleke, K. Hong and D. Papahadjopoulos, *Biochemistry*, 1991, **30**, 2903–2908.
129 P. Raynal and H. B. Pollard, *Biochim. Biophys. Acta*, 1994, **1197**, 63–93.
130 M. A. Swairjo, N. O. Concha, M. A. Kaetzel, J. R. Dedman and B. A. Seaton, *Nat. Struct. Biol.*, 1995, **2**, 968–974.
131 G. Kohler, U. Hering, O. Zschornig and K. Arnold, *Biochemistry*, 1997, **36**, 8189–8194.
132 K. Balasubramanian, E. M. Bevers, G. M. Willems and A. J. Schroit, *Biochemistry*, 2001, **40**, 8672–8676.
133 K. G. Maffey, L. B. Keil and V. A. DeBari, *Ann. Clin. Lab. Sci.*, 2001, **31**, 85–90.
134 J. F. Tait, D. F. Gibson and C. Smith, *Anal. Biochem.*, 2004, **329**, 112–119.
135 M. A. Lizarbe, J. I. Barrasa, N. Olmo, F. Gavilanes and J. Turnay, *Int. J. Mol. Sci.*, 2013, **14**, 2652–2683.
136 H. Yang, A. Kim, T. David, D. Palmer, T. Jin, J. Tien, F. Huang, T. Cheng, S. R. Coughlin, Y. N. Jan and L. Y. Jan, *Cell*, 2012, **151**, 111–122.
137 Y. Furuta, O. Pena-Ramos, Z. Li, L. Chiao and Z. Zhou, *PLoS Genet.*, 2021, **17**, e1009066.
138 T. Sakuragi and S. Nagata, *Nat. Rev. Mol. Cell Biol.*, 2023, **24**, 576–596.

Faraday Discussions

ROYAL SOCIETY OF CHEMISTRY

PAPER

Fusion of asymmetric membranes: the emergence of a preferred direction†

Petr Shendrik, [a] Raya Sorkin *[a] and Gonen Golani *[b]

Received 8th December 2024, Accepted 3rd February 2025

DOI: 10.1039/d4fd00189c

The fusion of lipid membranes progresses through a series of intermediate steps with two significant energy barriers: hemifusion-stalk formation and fusion-pore expansion. The cell's ability to tune these energy barriers is crucial as they determine the rate of many biological processes involving membrane fusion. However, a mechanism that allows the cell to manipulate both barriers in the same direction remains elusive, since membrane properties that the cell could dynamically tune during its life cycle, such as the lipids' spontaneous curvatures and membrane tension, have an opposite effect on the two barriers: tension inhibits stalk formation while promoting fusion-pore expansion. In contrast, increasing the total membrane concentration of lipids with negative intrinsic curvatures, such as cholesterol, promotes hemifusion-stalk formation while inhibiting pore expansion, and *vice versa* for lipids with positive intrinsic curvatures. Therefore, changes in these membrane properties increase one energy barrier at the expense of the other, resulting in a mixed effect on the fusion reaction. A possible mechanism to change both barriers in the same direction is by inducing lipid composition asymmetry, which results in tension and spontaneous curvature differences between the monolayers. To test the feasibility of this mechanism, a continuum elastic model was used to simulate the fusion intermediates and calculate the changes in the energy barriers. The calculations showed that a reasonable lipid composition asymmetry could lead to a 10–20k_BT difference in both energy barriers, depending on the direction from which fusion occurs. We further provide experimental support to the model predictions, demonstrating changes in the time to hemifusion upon asymmetry introduction. These results indicate that biological membranes, which are asymmetric, have a preferred direction for fusion.

Introduction

Membrane fusion is crucial for the function of cells and organisms. It plays a pivotal role in the cellular internal trafficking system, neurotransmission,

[a]*School of Chemistry, Faculty of Exact Sciences, Tel Aviv University, Tel Aviv, Israel. E-mail: rsorkin@tauex.tau.ac.il*

[b]*Department of Physics, University of Haifa, Haifa, 3498838, Israel. E-mail: ggolani@univ.haifa.ac.il*

† Electronic supplementary information (ESI) available. See DOI: https://doi.org/10.1039/d4fd00189c

exocytosis, and the homeostasis of organelles.[1,2] Cells have also developed mechanisms to inhibit fusion, for example, to prevent the entry of enveloped viruses.[3,4] The malfunctioning of membrane fusion is associated with multiple pathologies, notably neurodegenerative diseases such as Alzheimer's and Parkinson's.[5,6] Therefore, controlling the location and efficiency of fusion is crucial.

The fusion of lipid membranes progresses through a series of non-bilayer intermediate steps with two significant barriers: hemifusion-stalk ('stalk' for brevity) formation and fusion-pore expansion.[2,7–12] The energy barriers, E_{stalk} and E_{pore}, respectively, have typical magnitudes of several dozens of k_BT.[13–16] As a result of these high energy barriers, membranes under physiological conditions require specialized proteins to initiate and promote the fusion reaction.[17] The rate and efficiency of fusion can be optimized by maximizing the performance of these fusion proteins, as shown in SNARE,[18,19] and hemagglutinin[20] mediated fusion. Alternatively, the magnitude of the energy barriers can be reduced by manipulating the membrane's mechanical properties, making it more fusogenic.

In practice, cells can change their membranes' properties, such as rigidity and tension, by controlling lipid composition and their abundance through the rate of lipid synthesis, metabolism, and transport between organelles.[21] However, changing the membrane properties has two significant drawbacks. First, the change in mechanical properties has the same effect on the fusion reaction from both sides of the membrane. For example, increasing membrane rigidity inhibits the fusion of pathogens from the extracellular side but will also impede exocytosis from the intracellular side, along with other favorable membrane remodeling events.

Second, some properties have a conflicting effect on the energy barriers, increasing one but decreasing the other. For example, membrane tension increases the stalk energy[22] but reduces the pore energy.[23,24] Altering the membrane lipid composition also results in a conflicting effect: a higher abundance of lipids with positive intrinsic curvature,[25] such as lysophosphatidylcholine (LPC),[26] in both monolayers increases E_{stalk} and decreases E_{pore}, while lipids with more negative intrinsic curvatures, such as phosphatidylethanolamine (PE) and cholesterol,[27] have the opposite effect.[14,28,29]

A potential resolution to these conflicts, and possibly one of their underlying causes, could be the asymmetry in lipid composition between the two monolayers of the membrane.[30,31] The lipid asymmetry is actively maintained by flippases that flip certain lipids much faster than the passive back-flip-flop rate,[32] creating an asymmetric steady state. This asymmetry and its regulation, predominantly of charged lipids such as phosphatidylserine (PS) and phosphatidylinositol 4,5-bisphosphate (PiP_2), was shown to play a crucial role in apoptosis[33] and interaction with soluble proteins.[34–36]

Interestingly, not only are charged lipids asymmetrically dispersed but also lipids with larger negative intrinsic curvatures, such as PE, which is more abundant in the intracellular monolayer of red blood cells.[30] Moreover, there is a significant difference in the unsaturation level of the lipids in the two monolayers, with the intracellular monolayer being much richer in unsaturated lipids.[30] These two observations indicate that the spontaneous curvature of the intracellular monolayer is more negative than that of the extracellular monolayer. The fact that the cell has developed mechanisms to maintain such asymmetry suggests that these have a biological function.

It is tempting to speculate that this asymmetry has a role in controlling membrane fusion as it is strongly affected by the intrinsic curvatures of the fusing membranes' lipids.[28] However, little is known about the effect of asymmetry on this process. To bridge this gap, we numerically calculate the impact of monolayer lipid composition asymmetry on the fusion pathway, focusing on its impact on the fusion energy barriers from the two sides of the membrane. We further provide experimental support to the model predictions.

Model

The deformations of membranes and monolayers during the fusion pathway are modeled in the framework of continuum elastic theory[37–40] as described in previous publications for the stalk,[14,41] the hemifusion diaphragm,[42] and the opening of the fusion pore.[29,43] The full details can be found in the references. The following offers a brief overview of the key aspects of this theory, along with the modifications arising from lipid composition asymmetry.

Lipid-monolayer elastic theory

The basic component in the model is the lipid monolayer, which deforms during the fusion reaction. The model accounts for elastic deformations up to quadratic order in lipid splay, saddle-splay, and tilt. The tilt vector, $\vec{t} = \dfrac{\vec{n}}{\vec{n}\cdot\vec{N}} - \vec{N}$, quantifies the magnitude of shear in the hydrocarbon tails, with $\vec{N}$ the normal to the monolayer neutral plane and $\vec{n}$ the lipid director – a vector pointing from the base to the head of the lipid.[38,39] Lipid splay and saddle-splay are generalizations of total and Gaussian curvatures, respectively, in the presence of spatially changing lipid tilt,[38] which occurs in the vicinity of the fusion site. These are calculated based on each of the monolayers' splay tensors, given by $S_\alpha^\beta = \nabla_\alpha n^\beta$, with ∇_α being a covariant derivative. Lipid splay is the trace of the splay tensor, $\hat{J} = S_\alpha^\alpha$, while saddle splay is its determinant, $\hat{K} = \det S_\alpha^\beta$.

The energy density associated with deforming a lipid monolayer is[39]

$$u_\mathrm{m} = \gamma_\mathrm{m} + \frac{1}{2}\kappa_\mathrm{m}\tilde{J}^2 - \kappa_\mathrm{m}\tilde{J}\zeta_\mathrm{sm} + \kappa_\mathrm{G,m}\tilde{K} + \frac{1}{2}\kappa_\mathrm{t}\vec{t}^{\,2}. \tag{1}$$

κ_m, $\kappa_\mathrm{G,m}$ and κ_t are the monolayer splay, saddle-splay, and tilt rigidities, respectively. The parameters used in the following simulation are $\kappa_\mathrm{m} = 10k_\mathrm{B}T$,[44] $\kappa_\mathrm{G,m} = -5k_\mathrm{B}T$,[45,46] and $\kappa_\mathrm{t} = 40\ \mathrm{mN\ m^{-1}}$,[47,48] and are assumed to be identical in both monolayers. ζ_sm is the spontaneous monolayer curvature[49–53] and γ_m is the monolayer tension.[40] The fusion site is much smaller than the surrounding membranes, which act as a reservoir, setting the tension in each monolayer, individually.

The coupling of the monolayers and the membrane elastic energy

The membrane consists of two lipid monolayers with a joint mid-plane whose position in space is given by $\vec{r}$. The monolayers are free to slide on one another along the joint mid-plane. The orientation of lipids within each monolayer is independent of the orientation of lipids in the opposing monolayer. These two orientations are given by the local lipid director, denoted as $\hat{n}_p$ and $\hat{n}_d$ for the proximal and distal monolayers, respectively. The positions of the monolayers' neutral planes are given by $\vec{r}_p = \vec{r} + \delta \cdot \hat{n}_p$ and $\vec{r}_d = \vec{r} + \delta \cdot \hat{n}_d$ for the proximal and distal monolayers, respectively. δ is the distance from the membrane mid-plane to the monolayer neutral plane, located roughly between the lipids' head group region and the hydrophobic tails.[54] It is set here to 1.5 nm in both monolayers.

The elastic energy of a membrane is given by the sum of the elastic energy-density integrals over the monolayer areas,

$$U = \int u_p \mathrm{d}A_p + \int u_d \mathrm{d}A_d. \tag{2}$$

$\mathrm{d}A_p$ and $\mathrm{d}A_d$ are the area elements on the proximal and distal neutral planes and u_p and u_d are the respective energy densities given by eqn (1).

Tension and spontaneous curvature of the monolayers

This work focuses on the effect of lipid composition asymmetry between the monolayers, rather than the impact of the total membrane lipid composition, which has been addressed in numerous previous publications.[12,14,28,29,41,42,55–59] Therefore, the following calculations assume that the asymmetry is induced by swapping lipids of different species between the monolayers while keeping the total membrane lipid composition fixed (Fig. 1). The asymmetry is not in equilibrium, but is maintained in a steady state by the balance between the passive entropy-driven lipid flip-flop and the active work of proteins (Fig. 1B).

The swapping of lipid species with different intrinsic curvatures (Fig. 1A) gives rise to a difference in the monolayer's spontaneous curvatures,

$$\Delta\zeta = \frac{1}{2}\left(\bar{\zeta}_p - \bar{\zeta}_d\right). \tag{3}$$

$\bar{\zeta}_p$ and $\bar{\zeta}_d$ are the proximal and distal monolayer spontaneous curvatures, respectively. In fact, $\Delta\zeta$ can be induced not only by the swapping of lipids between the monolayers but also by any process that creates asymmetry between the monolayers, such as the transport of lipids to one monolayer but not the other, or by lipases acting on one side of the membrane. However, these processes will change the membrane lipid compositions as well, and as indicated above, this is not the focus of this work.

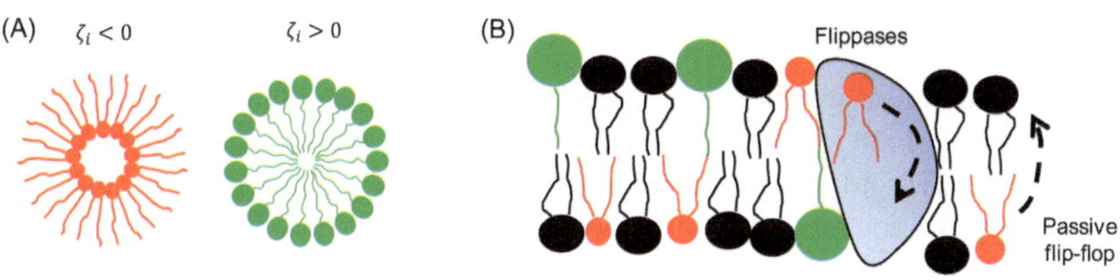

Fig. 1 Membrane asymmetry and lipid intrinsic curvatures. (A) Illustration of lipids with negative (red) and positive (green) intrinsic curvatures. (B) Membrane asymmetry is maintained by the balance between the active work of flippases and the passive entropy-driven back-flip-flop.

The average of the monolayers' spontaneous curvatures is independent of $\Delta\zeta$, and is given by

$$\bar{\zeta} = \frac{1}{2}\left(\bar{\zeta}_p + \bar{\zeta}_d\right). \tag{4}$$

With eqn (3) and (4), the distal and proximal monolayer spontaneous curvature are related to $\Delta\zeta$ and $\bar{\zeta}$ with

$$\bar{\zeta}_p = \bar{\zeta} + \Delta\zeta \text{ and } \bar{\zeta}_d = \bar{\zeta} - \Delta\zeta. \tag{5}$$

It is worth noting that using this definition, $\Delta\zeta$ changes sign if fusion occurs from the other side of the membrane.

The tension in the two monolayers, given by γ_p and γ_d for the proximal and distal monolayers, respectively, may differ because these are in contact with different area reservoirs. The membrane tension is given by the sum of the monolayers' tensions,[60–62]

$$\gamma = \gamma_p + \gamma_d. \tag{6}$$

The tension difference between the monolayers is

$$\Delta\gamma = \gamma_p - \gamma_d. \tag{7}$$

The proximal and distal monolayer tensions are respectively given by

$$\gamma_p = \frac{1}{2}(\gamma + \Delta\gamma) \text{ and } \gamma_d = \frac{1}{2}(\gamma - \Delta\gamma), \tag{8}$$

The asymmetry in monolayer spontaneous curvature and tension induces a torque on the membrane, which manifests in a non-vanishing membrane spontaneous curvature,[37,63] given by[60–62]

$$J_0 = \Delta\zeta - \frac{\delta}{2\kappa_m}\Delta\gamma \tag{9}$$

The effect of membrane tension on fusion has been discussed in previous works.[22,23] Therefore, to simplify the analysis, the membrane tension is set to zero in all simulations so that $\gamma_p = -\gamma_d$.

Membrane state prior to fusion

The membranes are in a mechanical equilibrium state with respect to their shapes before the initiation of fusion. That is, all torques and forces applied are balanced. Two possible membrane pre-fusion configurations that satisfy this condition are considered in this work:

(A) Two flat membranes with monolayer tension, such that $J_0 = 0$ (eqn (9)):

$$\Delta\gamma = \frac{2\kappa_m}{\delta}\Delta\zeta. \tag{10}$$

(B) Spherical membranes with a radius that matches the membrane's spontaneous curvature and no monolayer tension, $\Delta\gamma = 0$:

$$R = 2\Delta\zeta^{-1}. \tag{11}$$

An alternative configuration where the fusing membrane has a cylindrical pre-fusion shape is also possible but is out of the scope of this work. A calculation of the effect of curvature in cylindrical and spherical configurations was presented previously.[29,64]

The stalk

The stalk is an intermediate structure in which the proximal monolayers are fused, but the distal monolayers are still separated (Fig. 2A). The formation of the stalk is not spontaneous under physiological conditions. It is directly driven by the work of fusion proteins that bring a small patch of the two fusing compartments into proximity against strong repulsive dehydration forces.[65] This excess energy is relaxed by the merger of the proximal monolayers. Its formation energy is comparable to the dehydration repulsive force and is the first major energy barrier in the reaction.[13,14,16]

The numerical simulations of the stalk are based on previous works describing the structure using a continuum elasticity model.[14] The angle between the mid-planes of the merging membranes is fixed at 90° for symmetry reasons. To prevent voids in the hydrocarbon moiety, the lipids are sheared at the stalk center so that the tilt magnitude is $|\vec{t}| = 1$. This tilt decays at a distance of a few nanometers from the stalk at a cost of lipid splay. The stalk energy, E_{stalk}, is the minimum work needed to create this structure from non-interacting membranes in the pre-fusion state,

$$E_{\text{stalk}} = \min[U(\rho_D = 0)]. \tag{12}$$

U is the sum of the fusing membranes' energies presented in eqn (2) and ρ_D the diaphragm radius, which is set to zero. The stalk has an axially symmetric configuration.

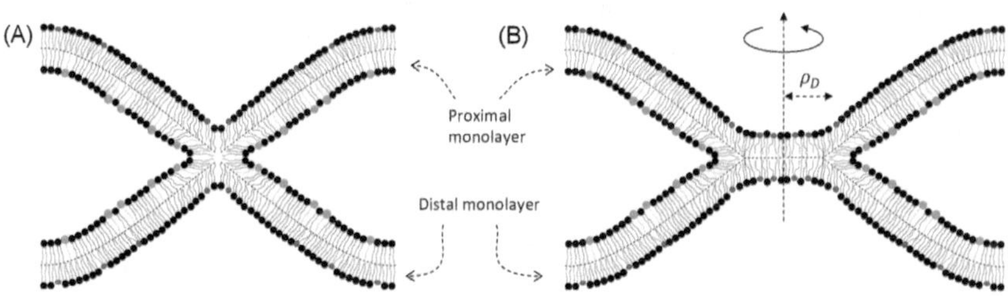

Fig. 2 Illustration of hemifusion intermediates. (A) The hemifusion stalk and (B) the hemifusion diaphragm. The black lipids have a cylindrical shape with close-to-zero intrinsic curvature, the red lipids have a strong negative intrinsic curvature, and the green lipids have a positive curvature. The lipid asymmetry is positive in the cartoon ($\Delta\zeta > 0$). In the opposite case, $\Delta\zeta < 0$, the proximal monolayers would be enriched with negatively curved lipids (red) and the distal with positively curved lipids (green).

The hemifusion diaphragm

The stalk is unstable and expands rapidly[66] by bringing the distal monolayers together to form a bilayer[15,42] (Fig. 2B). The newly formed membrane has a typical radius of a few nanometers,[4,67,68] depending on $\bar{\zeta}$ (eqn (4)).[42] This intermediate is called the hemifusion diaphragm (Fig. 2B). Numerically, the hemifusion diaphragm shape is found by minimizing the elastic energy with the constraint of fused proximal monolayers and no constraint on the radius of the diaphragm, ρ_D, or the angle between the mid-planes at the diaphragm rim. For simplicity of the calculations, the diaphragm is assumed to be axially symmetric. Cases where this constraint is not enforced are discussed in previous works.[29,42,57]

Importantly, the diaphragm is formed by the distal monolayers and retains their lipid composition, spontaneous curvature, and monolayer tension.

Fusion-pore formation

The diaphragm is subjected to high mechanical stress, which relaxes by forming small membrane pores, usually close to the diaphragm edge where stress is maximal.[69–71] The expansion of these pores is resisted by the cost of the highly curved pore rim. Since the energy related to stress relaxation is proportional to the pore area, while the cost of the pore rim is linear with its dimension, small pores tend to shrink below a critical size and expand above it. Once the pore defect forms, it stochastically flickers until it reaches the critical size, beyond which its expansion is favored, completing the fusion reaction. The elastic energy of the pore at this critical size is the fusion-pore expansion energy, E_{pore}.

The balance between the membrane stress and the pore rim line tension determines E_{pore}. The membrane stress is the energy per unit area gained by removing the lipids from the diaphragm and placing them in the surrounding reservoir. To first order, asymmetry changes this stress (u_m, given by eqn (1)) as follows:

$$\Delta\sigma = 2(u_d(\Delta\zeta) - u_d^0) = -\Delta\gamma + 2\kappa_m \tilde{J}\Delta\zeta. \tag{13}$$

The factor of 2 comes from the sum of the two monolayers of the diaphragm, which are symmetric in the cases considered here. u_d^0 is the spatially dependent stress in one of the diaphragm's monolayers under the condition of $\Delta\zeta = 0$.[72]

The pore rim is modeled as having a circular half-cross-section[28] (Fig. 3). Using the monolayer's elastic continuum description presented in eqn (1) with the distal monolayer's spontaneous curvature (eqn (5)), it is straightforward to see that the contribution of lipid asymmetry to energy per unit length, or line tension, is given by

$$\Delta\lambda = \pi\kappa_m\Delta\zeta. \tag{14}$$

Here, terms proportional to δ/ρ are neglected, with ρ being the pore radius and δ the monolayer thickness.

The energy of opening a pore with a radius ρ is given by

$$U_{\text{pore}}(\rho) = 2\pi\left[\rho(\lambda_0 + \pi\kappa_m\Delta\zeta) - \int_{\rho'=0}^{\rho'=\rho}\left(2u_d^0\left(\rho'\right) - \Delta\gamma + 2\kappa_m\tilde{J}\left(\rho'\right)\Delta\zeta\right)\rho' \mathrm{d}\rho'\right]. \tag{15}$$

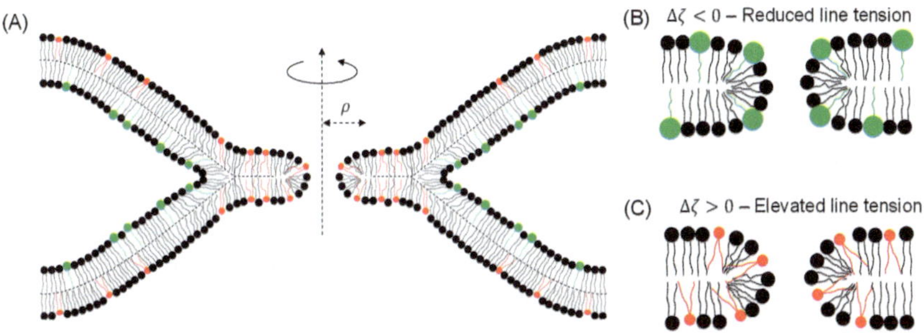

Fig. 3 Pore formation. (A) Illustration of a hemifusion diaphragm with a membrane pore at its center of radius ρ. The distal monolayer is enriched with negatively curved lipids, while the proximal monolayer is enriched with positively curved lipids, resulting in $\Delta\zeta > 0$ as defined in eqn (3). (B and C) Illustration of the pore rim. (B) The diaphragm is enriched with positively curved lipids that reduce the pore rim line tension, $\Delta\zeta < 0$. (C) The negatively curved lipids in the diaphragm increase the pore line tension, $\Delta\zeta > 0$.

λ_0 is pore line tension in a symmetric membrane. Its value depends on the membrane lipid composition, typically 10–30 pN.[73,74] Here, it is set to 20 pN.

The fusion-pore formation energy barrier is the maximum of $U_{\text{pore}}(\rho)$,

$$E_{\text{pore}} = \max[U_{\text{pore}}(\rho)]. \tag{16}$$

To simplify the computation, the initial pore defect is assumed to form at the center of the diaphragm, and the flickering in pore size does not change the hemifusion diaphragm shape.

Numerical calculations

The shape and formation energies of the stalk and hemifusion diaphragm are calculated using a self-written, open-source software called the 'Fusion Solver'. The code is available at https://github.com/GonenGolani/Fusion_Solver, and a detailed explanation of the algorithm can be found in previous publications.[29,75]

Results

The following analysis aims to determine the effect of lipid composition asymmetry on the fusion energy barriers. Therefore, the *change* in the energy barriers, $\Delta E_{\text{stalk}} = E_{\text{stalk}}(\Delta\zeta) - E_{\text{stalk}}(\Delta\zeta = 0)$ and $\Delta E_{\text{pore}} = E_{\text{pore}}(\Delta\zeta) - E_{\text{pore}}(\Delta\zeta = 0)$ as a function of $\Delta\zeta$, are calculated with respect to a symmetric membrane $(\Delta\zeta = 0)$. Discussion on the absolute magnitude of these energy barriers and how the variation of membrane lipid composition and the membrane mechanical properties affect them have been discussed in previous publications.[12,14,28,29,41,42,55–59]

Stalk formation between initially flat membranes

The above numerical and theoretical framework is first used to calculate the shape and formation energy of the hemifusion stalk between two initially flat, identical membranes with $\bar{\zeta}$ (eqn (4)) values between $-0.12\ \text{nm}^{-1}$ to $-0.30\ \text{nm}^{-1}$, corresponding to typical phosphatidylcholine (PC)- and cholesterol-rich membranes, respectively. The stalk energy in the absence of asymmetry, $E_{\text{stalk}}(\Delta\zeta = 0)$, as a function of $\bar{\zeta}$ is presented in a previous publication[29] and is in the 65–130k_BT range, with the lower energy for $\bar{\zeta} = -0.3$ nm^{-1} and the higher for $\bar{\zeta} = -0.12$ nm^{-1}. The magnitude of $\Delta\zeta$ was varied from -0.05 nm^{-1} to 0.05 nm^{-1}, and the shape and change in energy of the stalk compared to a symmetrical membrane were calculated for each $\bar{\zeta}$ value. An example of a stalk configuration is presented in the inset of Fig. 4A.

To understand the relationship between $\Delta\zeta$ and ΔE_{stalk}, the energetic contributions are separately plotted (Fig. 4A): splay, tension, tilt, and saddle-splay. The changes in these four energy terms (with respect to $\Delta\zeta = 0$) were calculated as a function of $\Delta\zeta$. The dominant change in E_{stalk} is due to the splay energy with an increase (or decrease, depending on the sign of $\Delta\zeta$) of up to 18k_BT in the relevant $\Delta\zeta$ range. This result agrees with a previous study showing that elevating ζ_P, while ζ_d is fixed, increases the stalk formation energy by a similar magnitude, while changing ζ_d has a much smaller effect.[14] This is due to a higher negative splay in the proximal monolayer compared to the distal one.

The contribution of tension asymmetry to ΔE_{stalk} is an order of magnitude smaller than that of splay (Fig. 4A). It is due to the different areas of the proximal and distal monolayers in the stalk. In general, the formation of a stalk requires additional area to be taken from the surrounding membrane in both monolayers compared to the pre-fusion configuration. Hence, membrane tension increases the stalk formation energy.[22] However, here, the tension has an opposite sign in the distal and proximal monolayers (eqn (8) and (10)), which partially compensates for this additional energy, resulting in a weak effect on E_{stalk} compared to that of splay.

Lastly, saddle-splay and tilt energies are only indirectly affected by $\Delta\zeta$ due to changes in the stalk shape. As a result, the changes in these two energetic

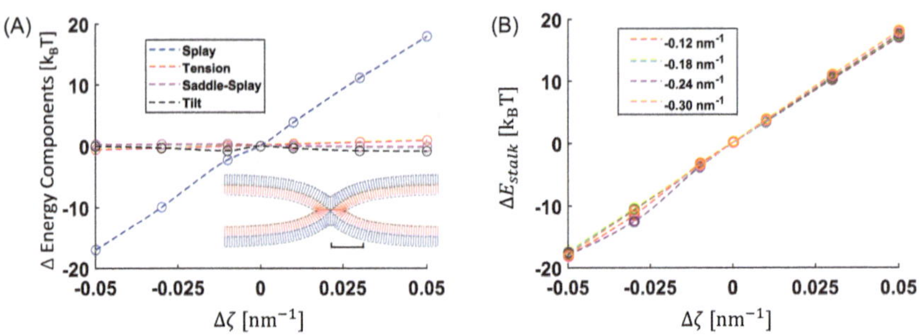

Fig. 4 Stalk formation. (A) The change in the contribution of the different monolayer deformations to the stalk energy as a function of $\Delta\zeta$. The integration is performed on the respective monolayers' neutral plane, and the plotted result is the sum of the deformation energies of both monolayers: splay $-\frac{1}{2}\int \kappa_m[\tilde{J}^2 - 2\zeta\tilde{J}]\cdot dA$, tension $-\gamma_p\Delta A_p + \gamma_d\Delta A_d$, tilt $-\frac{1}{2}\kappa_t\int t^2 dA$ and saddle splay $-\int \bar{\kappa}_m\tilde{K}dA$. The simulation is done at $\bar{\zeta} = -0.24$ nm^{-1}. The inset shows a simulation result for $\Delta\zeta = 0.05$ nm^{-1}. The scale bar is 5 nm. The red lines represent the averaged orientation of the lipids in the proximal monolayer, and the blue are the distal monolayer lipids. (B) Change in the stalk formation energy, $\Delta E_{\text{stalk}} = E_{\text{stalk}}(\Delta\zeta) - E_{\text{stalk}}(\Delta\zeta = 0)$ as a function of $\Delta\zeta$ for different $\bar{\zeta}$ (indicated in the legend). Parameters in all simulations: $\kappa_m = 10k_BT$, $\kappa_{G,m} = -5k_BT$, $\kappa_t = 40$ mN m^{-1} and $\delta = 1.5$ nm.

contributions are negligible. The sum of the change in these four energy terms is presented in Fig. 4B for different $\bar{\zeta}$ values, all of which collapse to a single close-to-linear master curve with a slope of $\sim 360 k_B T$ nm.

To conclude, the stalk energy increases if positively curved lipids are flipped from the distal to the proximal monolayer and are replaced by negatively curved lipids. This is mostly due to the mismatch between the negative splay in the proximal monolayer and the lipids' intrinsic curvatures. The mismatch increases for positive values of $\Delta\zeta$ and decreases for negative values.

Lipid mixing assay with asymmetric GUVs supports theoretical predictions

To validate the results of the theoretical model on the effect of lipid asymmetry on stalk formation energy, we used a lipid mixing assay between asymmetric giant unilamellar vesicles (GUVs) and membrane-coated polystyrene microparticles (beads) based on previously described work.[22] Asymmetric GUVs were prepared using the inverted-emulsion method,[76,77] and further adjusted to increase the vesicle yield for our lipid composition (see ESI Experimental materials and methods section†). Asymmetry was confirmed using the NBD (nitrobenzoxadiazol) labeled lipid reduction method (Fig. S1A†), resulting in ~28% interleaflet mixing of the NBD-DSPE lipid during vesicle formation, which is standard for the inverted-emulsion method.[78,79]

The $\Delta\zeta$ in our setup was modified by adjusting the positively curved 14:0 LPC concentration in the distal and proximal monolayers while maintaining the overall membrane lipid concentrations identical between conditions (90% DOPC, 8% DOPS, 1% LPC, and 1% Rh-PE). The lipid composition on the bead was symmetrical and contained positively charged DOTAP (1,2-dioleoyl-3-trimethyl-ammonium-propane) lipids (98% DOPC and 2% DOTAP), which allowed lipid mixing initiation following membrane contact.[80] To minimize the effect of membrane tension on lipid mixing, the GUVs were aspirated using a micropipette and held under a fixed membrane tension of 0.041 ± 0.014 mN m^{-1}. We monitored the time delay to lipid mixing initiation to probe the effect of lipid asymmetry on the reaction energy barrier (Fig. 5A).

Our findings demonstrate that lipid asymmetry significantly influences the kinetics of lipid mixing initiation during membrane fusion (Fig. 5B). Specifically, the presence of LPC in the proximal monolayer ($\Delta\zeta > 0$) increases the time delay for lipid mixing, indicating a higher energy barrier (Fig. 4B), consistent with prior observations of delayed fusion following the addition of LPC micelles to the extracellular environment of fusing membranes.[28] In contrast, reversed asymmetry ($\Delta\zeta < 0$) reduces the time delay, lowering the energy barrier (Fig. 4B) compared to the case where the LPC was uniformly distributed ($\Delta\zeta = 0$). These results, combined with the established correlation between lipid mixing energy barriers and stalk formation energy,[22,28] provide strong experimental support for the theoretical prediction that lipid asymmetry modulates the energy landscape of stalk formation.

Hemifusion diaphragm between flat membranes

Next, the stalk was allowed to expand to a hemifusion diaphragm and assume a local elastic energy minimum configuration. An example of a simulation result is presented in the inset of Fig. 6A. A detailed discussion of the shapes of these

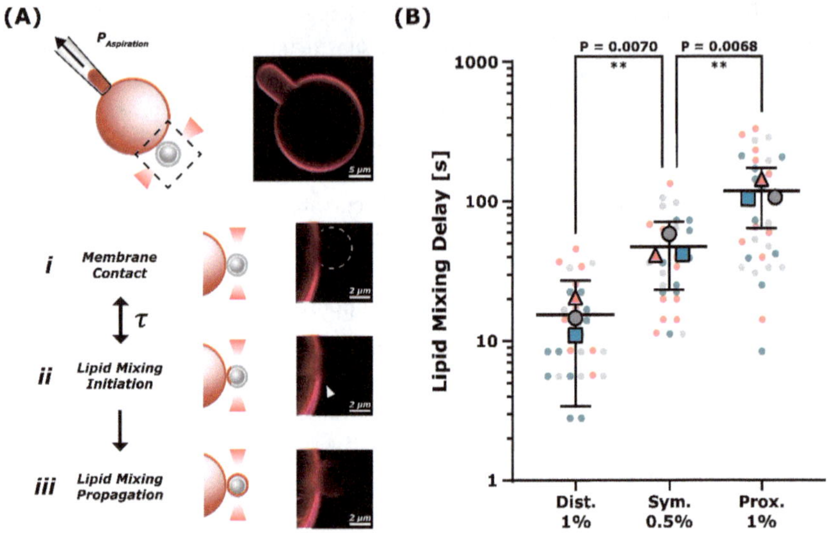

Fig. 5 Lipid mixing assay. (A) Experimental setup – a GUV was aspirated with a micropipette to maintain constant tension in all experiments. (i) An optically-trapped bead was brought into contact with the aspirated GUV (bead position is marked with a dashed line); (ii) the lipid mixing time delay, τ, was evaluated from the fluorescence increase on the bead edge (white arrow), (iii) followed by verification of lipid mixing propagation. (B) LPC localization on the distal monolayer decreased the time delay to lipid mixing, 15.3 ± 11.9 s (P value $= 0.0070$), while localization in the proximal monolayer increased it, 117.7 ± 54.0 s (P value $= 0.0068$), in comparison to the symmetric condition 47.0 ± 23.9 s. Three independent experiments were performed for each condition, with small color-coded dots representing a single measurement and large symbols of the same color showing the average of the experiment; the data is displayed on a logarithmic scale. The error bars correspond to 5–95% confidence intervals ($n = 3$), and the statistical significance was assessed using a two-tailed t-test.

diaphragms and how they depend on the various elastic moduli can be found in previous publications.[29,42]

The different contributions to the diaphragm formation energy are presented in Fig. 6A. Similarly to the stalk, the change in the splay energy is dominant due to the larger negative splay in the proximal monolayer. In contrast to the stalk (Fig. 4A), the tension contribution is also significant, up to $\pm 12k_BT$, depending on the sign of $\Delta\zeta$. This is due to a difference in the monolayer area taken from the distal and proximal monolayers to form the hemifusion diaphragm, with more area required in the proximal monolayer. Specifically, for $\bar{\zeta} = -0.24\ \text{nm}^{-1}$ this difference is 36 nm^2. To compare, this area difference was 2 nm^2 in the stalk configuration.

The change in the diaphragm's radius, which strongly affects the mechanical stress that promotes pore opening,[72] is presented in Fig. 6B. The change in radius is small, up to 0.2 nm for $\bar{\zeta} = -0.3\ \text{nm}^{-1}$, with even smaller changes for more positive $\bar{\zeta}$ values. This shrinkage represents a maximum of 5% of the diaphragm radius. As a result of this shortening, the length of the circular three-way junction between the diaphragm and the surrounding membranes is reduced, diminishing the tilt-related deformations that originate from this junction (Fig. 6A). Saddle-splay energy is almost unaffected by $\Delta\zeta$.

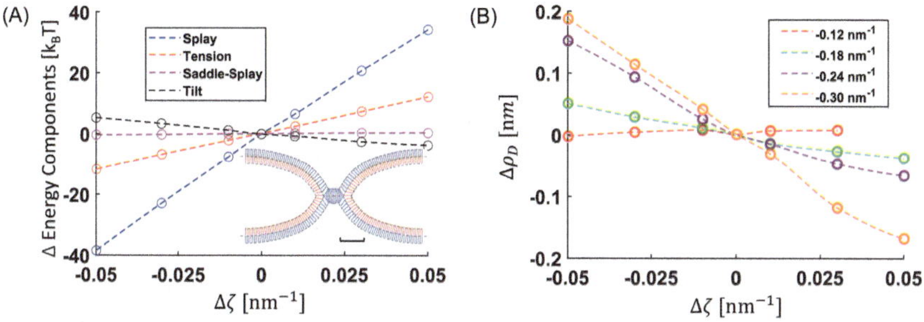

Fig. 6 Expansion to hemifusion diaphragm. (A) The change in the contribution of the different monolayer deformation energies to the hemifusion diaphragm energy as a function of $\Delta\zeta$. The integration is performed on the respective monolayers' neutral plane, and the plotted result is the sum of the deformation energies of both monolayers: splay – $\frac{1}{2}\int \kappa_m[\tilde{J}^2 - 2\zeta\tilde{J}]\cdot dA$, tension – $\gamma_p\Delta A_p + \gamma_d\Delta A_d$, tilt – $\frac{1}{2}\kappa_t\int t^2 dA$ and saddle splay – $\int \bar{\kappa}_m\tilde{K}dA$. The simulation is done at $\bar{\zeta} = -0.24$ nm^{-1}. The inset shows a simulation result for $\Delta\zeta = 0.05$ nm^{-1}. The scale bar is 5 nm. The red lines represent the averaged orientation of the lipids in the proximal monolayer, and the blue are the distal monolayer lipids. (B) Change in the hemifusion diaphragm radius, $\Delta\rho_D = \rho_D(\Delta\zeta) - \rho_D(\Delta\zeta = 0)$ as a function of $\Delta\zeta$ for different $\bar{\zeta}$ (indicated in the legend). Parameters in all simulations: $\kappa_m = 10k_BT$, $\kappa_{G,m} = -5k_BT$, $\kappa_t = 40$ mN m^{-1} and $\delta = 1.5$ nm.

Expansion of the fusion pore

The expansion energy of a fusion pore in the diaphragm was calculated using eqn (13)–(16). Increasing $\Delta\zeta$ has a conflicting effect on the stress in the diaphragm (Fig. 7A): first, the diaphragm is characterized by negative splay;[72] positive values of $\Delta\zeta$ reduce the mismatch between the lipids' intrinsic curvatures and geometric splay, thereby reducing the stress (eqn (13)). In addition, the tension in the diaphragm, $\gamma_d = -\frac{\kappa_m}{\delta}\Delta\zeta$ (eqn (8) and (10)), is negative (for $\Delta\zeta > 0$) and further

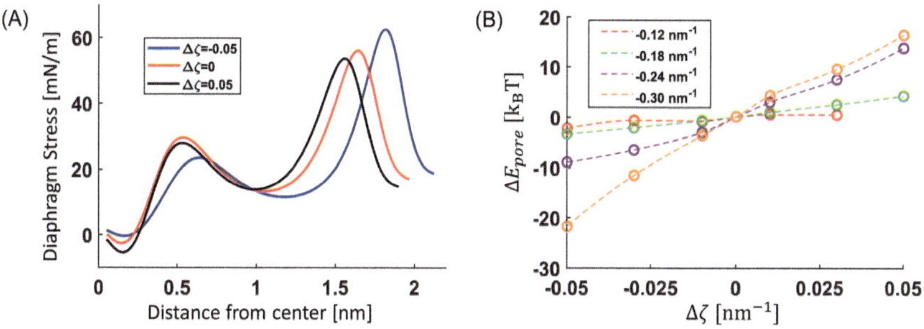

Fig. 7 The pore formation energy barrier. (A) Radial distribution of stress in the diaphragm for different $\Delta\zeta$, as indicated in the legend. $\bar{\zeta} = -0.24$ nm^{-1}. The local maxima are the points in the diaphragm where stress is at its highest, making these locations more likely for pore initiation. (B) Change in pore formation energy, $\Delta E_{pore} = E_{pore}(\Delta\zeta) - E^0_{pore}$, as a function of $\Delta\zeta$ for different mean lipid intrinsic curvatures, $\bar{\zeta}$ (indicated in the legend). Parameters in all simulations: $\kappa_m = 10k_BT$, $\kappa_{G,m} = -5k_BT$, $\kappa_t = 40$ mN m^{-1} and $\delta = 1.5$ nm.

inhibits the pore formation. This effect is counteracted by the above-mentioned reduction in diaphragm radius (Fig. 6B), which increases the stress in the diaphragm.[72] Due to these conflicting effects of $\Delta\zeta$ on the diaphragm stress, its effect on the pore formation energy is minor.

In contrast, the increase of the line tension, λ, with $\Delta\zeta$ (eqn (14)) is a significant factor leading to the almost linear correlation between ΔE_{pore} and $\Delta\zeta$ (Fig. 6B). These numerical results are in agreement with a simple analytical argument: for uniform monolayer stress, the pore formation energy is $E^0_{\text{pore}} \sim \frac{{\lambda_0}^2}{2\sigma_0}$.[81] Assuming that the stress, σ_0, is unchanged with $\Delta\zeta$ and that $\Delta\lambda \ll \lambda_0$, it is straightforward to show, to first order, that the change in pore formation energy is

$$\Delta E_{\text{pore}} \sim E^0_{\text{pore}} \cdot \frac{2\pi\kappa_{\text{m}}}{\lambda_0} \Delta\zeta. \tag{17}$$

The slope of ΔE_{pore} depends on the reference pore formation energy E^0_{pore}, which is larger for more negative values of $\bar{\zeta}$,[29] like PE- and cholesterol-rich membranes. Specifically, E^0_{pore} increases from $5k_{\text{B}}T$ at $\bar{\zeta} = -0.12\ \text{nm}^{-1}$ to $79k_{\text{B}}T$ at $\bar{\zeta} = -0.3\ \text{nm}^{-1}$, resulting in a 16-fold larger impact of lipid asymmetry at more negative values of $\bar{\zeta}$, as also seen in the numerical calculation (Fig. 7B).

Fusion of curved membranes

In this section, the fusion of two spherical membranes with no monolayer tension is simulated. The curvature radius, R, is given by eqn (11). In the absence of membrane tension, the pre-fusion configuration is stable if $\zeta_{\text{p}} > \zeta_{\text{d}}$, so only

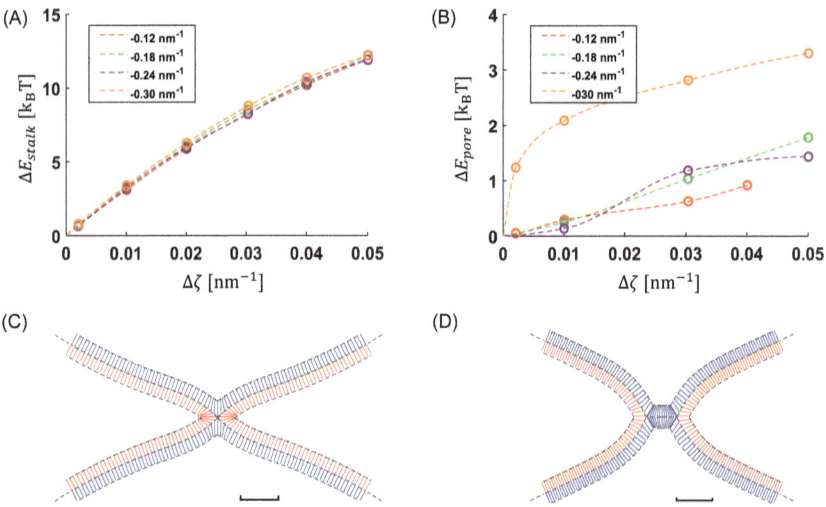

Fig. 8 Energy barriers for fusion between spherical membranes. The curvature radius $R = 2\Delta\zeta^{-1}$ (eqn (11)) for different $\bar{\zeta}$ as indicated in the legends. (A) Change in stalk formation energy, ΔE_{stalk}. (B) Change in pore formation energy, ΔE_{pore}. (C and D) Simulation results at $\bar{\zeta} = -0.24\ \text{nm}^{-1}$ and $\Delta\zeta = -0.05\ \text{nm}^{-1}$ for (C) the stalk and (D) the hemifusion diaphragm. The red lines represent the averaged orientation of the lipids in the proximal monolayer, and the blue are the distal monolayer lipids. The scale bars are 5 nm. Parameters in all simulations: $\kappa_{\text{m}} = 10k_{\text{B}}T$, $\kappa_{\text{G,m}} = -5k_{\text{B}}T$, $\kappa_{\text{t}} = 40\ \text{mN m}^{-1}$ and $\delta = 1.5$ nm.

positive values of $\Delta\zeta$ are calculated. The maximal curvature radius considered is 40 nm, corresponding to $\Delta\zeta = 0.05\ \text{nm}^{-1}$. The change in energy barriers, ΔE_{stalk} and ΔE_{pore}, is with respect to a curved membrane with an identical curvature radius, but with a symmetric lipid composition and no tension.

Similar to fusion between initially flat membranes, ΔE_{stalk} (Fig. 8A) and ΔE_{pore} (Fig, 8B) increase with $\Delta\zeta$, but more modestly. This difference in the stalk is due to the elastic energy accumulated in curved membranes prior to fusion, which is partially relaxed when the stalk is formed.[82] A typical simulation result is presented in Fig. 8C. The weak effect of $\Delta\zeta$ on ΔE_{pore} is due to lower E^0_{pore} values in curved membranes compared to flat membranes (eqn (17)).[29] As in the case of flat membrane fusion, E^0_{pore} is larger for more negative values of $\bar{\zeta}$ and so is ΔE_{pore}. A typical simulation result is presented in Fig. 8D.

It is worthwhile noting that the tensions of both monolayers vanish as they are in contact with reservoirs that buffer the change in the monolayer area as the fusion reaction progresses. However, if the monolayers are separated from the reservoir, such as in the case of closed vesicles, tension would accumulate in the distal monolayers as the diaphragm expands, increasing E^0_{pore} and ΔE_{pore}.[29]

Discussion

This work investigates the effect of lipid composition asymmetry on the energy barriers for membrane fusion. The primary finding is that asymmetry changes both energy barriers in the same way (Fig. 9A): stalk formation and pore expansion energy barriers increase (Fig. 9B) if negatively curved lipids from the proximal monolayer are exchanged with less negatively or positively curved lipids from the distal monolayer ($\Delta\zeta > 0$). Both energy barriers decrease if the exchange is in the opposite direction ($\Delta\zeta < 0$). The difference can reach dozens of k_BT, a significant fraction of the absolute magnitude of the energy barriers.

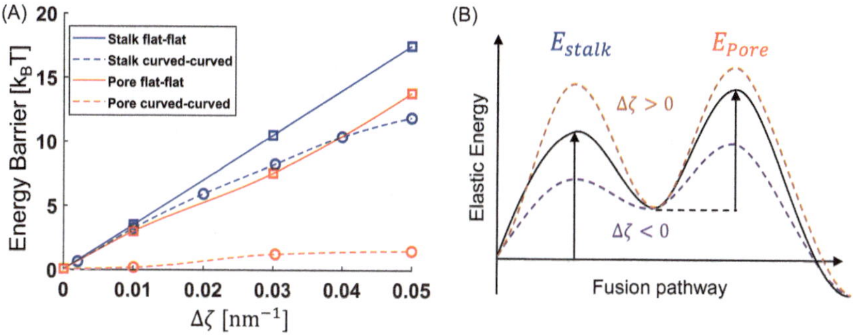

Fig. 9 The effect of lipid composition asymmetry on fusion energy barriers. (A) Change in stalk formation (blue) and pore expansion (red) energies in the two scenarios considered in this work: solid lines with square data points are for fusion between initially flat membranes, and the dashed lines with circular data points are for initially curved membranes (whose curvature is given by eqn (11)). Parameters: $\kappa_m = 10k_BT$, $\kappa_{G,m} = -5k_BT$, $\kappa_t = 40\ \text{mN m}^{-1}$, $\delta = 1.5$ nm and $\bar{\zeta} = -0.24\ \text{nm}^{-1}$. (B) A cartoon illustrating the effect of asymmetry on the fusion energy barriers from both sides of the membrane. In the direction where $\Delta\zeta > 0$, both energy barriers increase, while if fusion occurs from the other side, $\Delta\zeta < 0$ and both energy barriers decrease.

Importantly, these findings show, for the first time, that the direction from which fusion occurs has a significant effect on the energy barriers and, as a result, also on the rate of the fusion reaction. The biological significance of these results is that the cell has a mechanism to promote fusion from one side of a membrane and inhibit it from the other. This is likely to be most important in the plasma membrane. There, fusion should be promoted from the intracellular side, for example, to assist the fusion of vesicles associated with the secretory pathway, and inhibited from the extracellular side to prevent unwanted fusion events, for example, with enveloped viruses. In addition, some cell–cell fusion pathways involve deliberate lipid scrambling, which may not only expose negatively charged lipids essential for fusion but also alter lipid asymmetry to lower the energy barrier for fusion from the extracellular side.

Lipid mixing assays performed here (Fig. 5B) support the theoretical predictions – a constant amount of LPC was added to the membrane: upon addition to the proximal monolayer, the lipid mixing delay time was significantly longer (117.7 ± 54.0 s) compared to that for the symmetric distribution (47.0 ± 23.9 s) and to the case where the LPC was added to the distal monolayer (15.3 ± 11.9 s). The increase in lipid mixing time delay corresponds to an increase in the lipid mixing energy barrier, which previous works have shown to correlate with stalk formation energy.[22,28] These results qualitatively agree with the theoretical predictions. Due to the limitations of the inverted-emulsion formation method, the precise monolayer composition and exact mechanical properties of the formed GUVs are challenging to predict, mainly due to potential oil contamination in the bilayer along with interleaflet mixing during vesicle formation, which varies between different lipids.[76] Therefore, further experimental work is needed, including testing the effect of negatively curved lipids on lipid mixing and the effect of lipid composition asymmetry on content mixing.

In addition, it was recently shown that the uptake efficiency of PC-PS liposomes containing mRNA strongly depends on the lipid asymmetry between the monolayers:[79] the uptake was significantly faster when the majority (90%) of the outer monolayer was composed of PS and the inner by PC, compared to either pure PC or PS liposomes, showing that the surface charge on the outer monolayer of the liposome is not the significant factor determining the uptake efficiency. Interestingly, the uptake efficiency was significantly reduced (5-fold) when the monolayer compositions were flipped, and most of the outer monolayer was composed of PC and the inner of PS. These findings demonstrated that the direction of the asymmetry plays a significant role.

Finally, the numerical results from this work can be used to estimate the difference in fusion energy barriers from the two sides of the plasma membrane based on existing monolayer lipidomic data. Unfortunately, such data is currently available for only a few cell types, one of which is a highly detailed lipidomic profile of red blood cells.[30] Based on typical values available in the literature for the different lipids' intrinsic curvatures and the lipidomic data (Table 1 in ESI†), the monolayers' spontaneous curvatures are estimated as $-0.22\ \mathrm{nm}^{-1}$ for the inner monolayer and $-0.26\ \mathrm{nm}^{-1}$ for the outer monolayer of the plasma membrane of the red blood cell. This estimate should be considered a rough approximation, as the membrane comprises many lipid species with poorly characterized intrinsic curvatures and does not account for the asymmetric distribution of cholesterol and membrane proteins. However, since the lipids' tail unsaturation level is not fully accounted for here, the intracellular monolayer spontaneous curvature is likely to be even more negative.

Based on these estimations, $\bar{\zeta}$ and $\Delta\zeta$ are -0.24 nm^{-1} and -0.02 nm^{-1}, respectively, with $\Delta\zeta$ defined as half the difference between the extracellular and intracellular monolayer spontaneous curvatures. According to the numerical calculations and these estimations, the difference in stalk formation energy between the extracellular and intracellular sides of a red blood cell is $15k_BT$ (Fig. 4B), and the difference in fusion-pore expansion energy is $10k_BT$ (Fig. 7B).

This finding supports the hypothesis that the lipid composition asymmetry favors fusion from the intracellular side and inhibits it from the extracellular side. It is tempting to speculate that the large difference in the monolayers' spontaneous curvature is maintained to facilitate this large difference in fusion energy barriers. However, further experiments and high-resolution lipidomic data are required to establish this as a widely used mechanism to control fusion in different cell types and organisms.

Data availability

The code used in this study is available as open-source software. It can be accessed at the following GitHub repository: https://github.com/GonenGolani/Fusion_Solver. This repository contains all the necessary scripts and documentation required to reproduce the results presented in the paper.

Conflicts of interest

There are no conflicts to declare.

Acknowledgements

R. S. acknowledges support by the Israel Science Foundation (Grant No. 1289/20), the NSF-BSF (Grant No. 2021793), and by the European Union (ERC ReMembrane 101077502). Views and opinions expressed are however those of the authors only and do not necessarily reflect those of the European Union or the European Research Council Executive Agency. Neither the European Union nor the granting authority can be held responsible for them.

References

1 P. A. Beales, B. Ciani and A. J. Cleasby, Nature's lessons in design: nanomachines to scaffold, remodel and shape membrane compartments, *Phys. Chem. Chem. Phys.*, 2015, **17**(24), 15489–15507.

2 L. V. Chernomordik, J. Zimmerberg and M. M. Kozlov, Membranes of the world unite!, *J. Cell Biol.*, 2006, **175**(2), 201–207.

3 X. Guo, J. Steinkühler, M. Marin, X. Li, W. Lu, R. Dimova and G. B. Melikyan, Interferon-Induced Transmembrane Protein 3 Blocks Fusion of Diverse Enveloped Viruses by Altering Mechanical Properties of Cell Membranes, *ACS Nano*, 2021, **15**(5), 8155–8170.

4 S. Klein, G. Golani, F. Lolicato, C. Lahr, D. Beyer, A. Herrmann, M. Wachsmuth-Melm, N. Reddmann, R. Brecht and M. Hosseinzadeh,

IFITM3 blocks influenza virus entry by sorting lipids and stabilizing hemifusion, *Cell Host Microbe*, 2023, **31**, 616–633.

5 H. Chen and D. C. Chan, Mitochondrial dynamics-fusion, fission, movement, and mitophagy-in neurodegenerative diseases, *Hum. Mol. Genet.*, 2009, **18**(R2), R169–R76.

6 S. H. Tan, V. Karri, N. W. R. Tay, K. H. Chang, H. Y. Ah, P. Q. Ng, H. San Ho, H. W. Keh and M. Candasamy, Emerging pathways to neurodegeneration: dissecting the critical molecular mechanisms in Alzheimer's disease, Parkinson's disease, *Biomed. Pharmacother.*, 2019, **111**, 765–777.

7 R. Jahn and H. Grubmüller, Membrane fusion, *Curr. Opin. Cell Biol.*, 2002, **14**(4), 488–495.

8 S. C. Harrison, Mechanism of membrane fusion by viral envelope proteins, *Adv. Virus Res.*, 2005, **64**, 231–261.

9 L. V. Chernomordik and M. M. Kozlov, Mechanics of membrane fusion, *Nat. Struct. Mol. Biol.*, 2008, **15**(7), 675.

10 M. M. Kozlov, H. T. McMahon and L. V. Chernomordik, Protein-driven membrane stresses in fusion and fission, *Trends Biochem. Sci.*, 2010, **35**(12), 699–706.

11 M. Fuhrmans, G. Marelli, Y. G. Smirnova and M. Müller, Mechanics of membrane fusion/pore formation, *Chem. Phys. Lipids*, 2015, **185**, 109–128.

12 S. A. Akimov, R. J. Molotkovsky, P. I. Kuzmin, T. R. Galimzyanov and O. V. Batishchev, Continuum models of membrane fusion: evolution of the theory, *Int. J. Mol. Sci.*, 2020, **21**(11), 3875.

13 D. P. Siegel, Energetics of intermediates in membrane fusion: comparison of stalk and inverted micellar intermediate mechanisms, *Biophys. J.*, 1993, **65**(5), 2124–2140.

14 Y. Kozlovsky and M. M. Kozlov, Stalk model of membrane fusion: solution of energy crisis, *Biophys. J.*, 2002, **82**(2), 882–895.

15 K. Katsov, M. Mueller and M. Schick, Field theoretic study of bilayer membrane fusion. I. Hemifusion mechanism, *Biophys. J.*, 2004, **87**(5), 3277–3290.

16 C. S. Poojari, K. C. Scherer and J. S. Hub, Free energies of membrane stalk formation from a lipidomics perspective, *Nat. Commun.*, 2021, **12**(1), 6594.

17 J. M. White, S. E. Delos, M. Brecher and K. Schornberg, Structures and mechanisms of viral membrane fusion proteins: multiple variations on a common theme, *Crit. Rev. Biochem. Mol. Biol.*, 2008, **43**(3), 189–219.

18 T. C. Südhof and J. E. Rothman, Membrane fusion: grappling with SNARE and SM proteins, *Science*, 2009, **323**(5913), 474–477.

19 Y. Zhang, L. Ma and H. Bao, Energetics, kinetics, and pathways of SNARE assembly in membrane fusion, *Crit. Rev. Biochem. Mol. Biol.*, 2022, **57**(4), 443–460.

20 N. Sriwilaijaroen and Y. Suzuki, Molecular basis of the structure and function of H1 hemagglutinin of influenza virus, *Proc. Jpn. Acad. B Phys. Biol. Sci.*, 2012, **88**(6), 226–249.

21 G. Van Meer, D. R. Voelker and G. W. Feigenson, Membrane lipids: where they are and how they behave, *Nat. Rev. Mol. Cell Biol.*, 2008, **9**(2), 112–124.

22 P. Shendrik, G. Golani, R. Dharan, U. S. Schwarz and R. Sorkin, Membrane Tension Inhibits Lipid Mixing by Increasing the Hemifusion Stalk Energy, *ACS Nano*, 2023, **17**(19), 18942–18951.

23 M. M. Kozlov and L. V. Chernomordik, Membrane tension and membrane fusion, *Curr. Opin. Struct. Biol.*, 2015, **33**, 61–67.

24 L. Gao, R. Lipowsky and J. Shillcock, Tension-induced vesicle fusion: pathways and pore dynamics, *Soft Matter*, 2008, **4**(6), 1208–1214.

25 W. Kunz, F. Testard and T. Zemb, Correspondence between curvature, packing parameter, and hydrophilic – lipophilic deviation scales around the phase-inversion temperature, *Langmuir*, 2009, **25**(1), 112–115.

26 N. Fuller and R. P. Rand, The influence of lysolipids on the spontaneous curvature and bending elasticity of phospholipid membranes, *Biophys. J.*, 2001, **81**(1), 243–254.

27 B. Kollmitzer, P. Heftberger, M. Rappolt and G. Pabst, Monolayer spontaneous curvature of raft-forming membrane lipids, *Soft Matter*, 2013, **9**(45), 10877–10884.

28 L. Chernomordik, M. Kozlov, G. Melikyan, I. Abidor, V. Markin and Y. A. Chizmadzhev, The shape of lipid molecules and monolayer membrane fusion, *Biochim. Biophys. Acta Biomembr.*, 1985, **812**(3), 643–655.

29 G. Golani and U. S. Schwarz, High curvature promotes fusion of lipid membranes: predictions from continuum elastic theory, *Biophys. J.*, 2023, **122**, 1868–1882.

30 J. Lorent, K. Levental, L. Ganesan, G. Rivera-Longsworth, E. Sezgin, M. Doktorova, E. Lyman and I. Levental, Plasma membranes are asymmetric in lipid unsaturation, packing and protein shape, *Nat. Chem. Biol.*, 2020, **16**(6), 644–652.

31 J. A. Op den Kamp, Lipid asymmetry in membranes, *Annu. Rev. Biochem.*, 1979, **48**(1), 47–71.

32 M. M. Sperotto and A. Ferrarini, Spontaneous lipid flip-flop in membranes: a still unsettled picture from experiments and simulations, *The Biophysics of Cell Membranes: Biological Consequences*, 2017, pp. 29–60.

33 S. Nagata, J. Suzuki, K. Segawa and T. Fujii, Exposure of phosphatidylserine on the cell surface, *Cell Death Differ.*, 2016, **23**(6), 952–961.

34 J. G. Kay and G. D. Fairn, Distribution, dynamics and functional roles of phosphatidylserine within the cell, *Cell Commun. Signal.*, 2019, **17**(1), 126.

35 K. Mandal, Review of PIP2 in cellular signaling, functions and diseases, *Int. J. Mol. Sci.*, 2020, **21**(21), 8342.

36 S. L. Winter, G. Golani, F. Lolicato, M. Vallbracht, K. Thiyagarajah, S. S. Ahmed, C. Lüchtenborg, O. T. Fackler, B. Brügger and T. Hoenen, The Ebola virus VP40 matrix layer undergoes endosomal disassembly essential for membrane fusion, *EMBO J.*, 2023, **42**, e113578.

37 W. Helfrich, Elastic properties of lipid bilayers: theory and possible experiments, *Z. Naturforsch. C Biosci.*, 1973, **28**(11–12), 693–703.

38 M. Hamm and M. Kozlov, Tilt model of inverted amphiphilic mesophases, *Eur. Phys. J. B*, 1998, **6**(4), 519–528.

39 M. Hamm and M. M. Kozlov, Elastic energy of tilt and bending of fluid membranes, *Eur. Phys. J. E*, 2000, **3**(4), 323–335.

40 M. Deserno, Fluid lipid membranes: from differential geometry to curvature stresses, *Chem. Phys. Lipids*, 2015, **185**, 11–45.

41 Y. Kozlovsky, A. Efrat, D. A. Siegel and M. M. Kozlov, Stalk phase formation: effects of dehydration and saddle splay modulus, *Biophys. J.*, 2004, **87**(4), 2508–2521.

42 Y. Kozlovsky, L. V. Chernomordik and M. M. Kozlov, Lipid intermediates in membrane fusion: formation, structure, and decay of hemifusion diaphragm, *Biophys. J.*, 2002, **83**(5), 2634–2651.

43 R. J. Ryham, T. S. Klotz, L. Yao and F. S. Cohen, Calculating transition energy barriers and characterizing activation states for steps of fusion, *Biophys. J.*, 2016, **110**(5), 1110–1124.

44 R. Dimova, Recent developments in the field of bending rigidity measurements on membranes, *Adv. Colloid Interface Sci.*, 2014, **208**, 225–234.

45 R. H. Templer, B. J. Khoo and J. M. Seddon, Gaussian curvature modulus of an amphiphilic monolayer, *Langmuir*, 1998, **14**(26), 7427–7434.

46 M. M. Terzi, M. F. Ergüder and M. Deserno, A consistent quadratic curvature-tilt theory for fluid lipid membranes, *J. Chem. Phys.*, 2019, **151**(16), 164108.

47 M. Kozlov and M. Winterhalter, Elastic moduli for strongly curved monolayers. Position of the neutral surface, *J. Phys. II*, 1991, **1**(9), 1077–1084.

48 J. F. Nagle, Experimentally determined tilt and bending moduli of single-component lipid bilayers, *Chem. Phys. Lipids*, 2017, **205**, 18–24.

49 M. M. Kozlov and W. Helfrich, Effect of a Cosurfactant on the Stretching and Bending Elasticities – a Surfactant Monolayer, *Langmuir*, 1992, **8**(11), 2792–2797, DOI: 10.1021/la00047a035.

50 S. L. Keller, S. M. Bezrukov, S. Gruner, M. W. Tate, I. Vodyanoy and V. A. Parsegian, Probability of alamethicin conductance states varies with nonlamellar tendency of bilayer phospholipids, *Biophys. J.*, 1993, **65**(1), 23–27.

51 G. Khelashvili, D. Harries and H. Weinstein, Modeling membrane deformations and lipid demixing upon protein-membrane interaction: the BAR dimer adsorption, *Biophys. J.*, 2009, **97**(6), 1626–1635.

52 E. E. Kooijman, V. Chupin, N. L. Fuller, M. M. Kozlov, B. de Kruijff, K. N. Burger and P. R. Rand, Spontaneous curvature of phosphatidic acid and lysophosphatidic acid, *Biochemistry*, 2005, **44**(6), 2097–2102, DOI: 10.1021/bi0478502.

53 M. Kaltenegger, J. Kremser, M. P. Frewein, P. Ziherl, D. J. Bonthuis and G. Pabst, Intrinsic lipid curvatures of mammalian plasma membrane outer leaflet lipids and ceramides, *Biochim. Biophys. Acta Biomembr.*, 2021, **1863**(11), 183709.

54 M. Kozlov, S. Leikin and R. Rand, Bending, hydration and interstitial energies quantitatively account for the hexagonal-lamellar-hexagonal reentrant phase transition in dioleoylphosphatidylethanolamine, *Biophys. J.*, 1994, **67**(4), 1603–1611.

55 A. Grafmüller, J. Shillcock and R. Lipowsky, The fusion of membranes and vesicles: pathway and energy barriers from dissipative particle dynamics, *Biophys. J.*, 2009, **96**(7), 2658–2675.

56 S. Kawamoto, M. L. Klein and W. Shinoda, Coarse-grained molecular dynamics study of membrane fusion: curvature effects on free energy barriers along the stalk mechanism, *J. Chem. Phys.*, 2015, **143**(24), 243112.

57 Y. G. Smirnova, H. J. Risselada and M. Müller, Thermodynamically reversible paths of the first fusion intermediate reveal an important role for membrane anchors of fusion proteins, *Proc. Natl. Acad. Sci. U. S. A.*, 2019, **116**(7), 2571–2576.

58 R. B. Lira and R. Dimova, Fusion assays for model membranes: a critical review, *Advances in Biomembranes and Lipid Self-Assembly*, 2019, vol. 30, pp. 229–270.

59 A. Joardar, G. P. Pattnaik and H. Chakraborty, Effect of phosphatidylethanolamine and oleic acid on membrane fusion: phosphatidylethanolamine circumvents the classical stalk model, *J. Phys. Chem. B*, 2021, **125**(48), 13192–13202.

60 M. Deserno, Biomembranes balance many types of leaflet asymmetries, *Curr. Opin. Struct. Biol.*, 2024, **87**, 102832.

61 S. L. Foley, M. Varma, A. Hossein and M. Deserno, Elastic and thermodynamic consequences of lipid membrane asymmetry, *Emerg. Top, Life Sci.*, 2023, **7**(1), 95–110.

62 A. Hossein and M. Deserno, Spontaneous curvature, differential stress, and bending modulus of asymmetric lipid membranes, *Biophys. J.*, 2020, **118**(3), 624–642.

63 O. Y. Zhongcan and W. Helfrich, Bending Energy of Vesicle Membranes - General Expressions for The 1st, 2nd, and 3rd Variation of the Shape Energy and Applications to Spheres and Cylinder, *Phys. Rev. A:At., Mol., Opt. Phys.*, 1989, **39**(10), 5280–5288, DOI: 10.1103/PhysRevA.39.5280.

64 Y. Smirnova and M. Müller, How does curvature affect the free-energy barrier of stalk formation? Small vesicles vs apposing, planar membranes, *Eur. Biophys. J.*, 2021, **50**(2), 253–264.

65 R. Rand and V. Parsegian, Hydration forces between phospholipid bilayers, *Biochim. Biophys. Acta*, 1989, **988**(3), 351–376.

66 P. M. Kasson, N. W. Kelley, N. Singhal, M. Vrljic, A. T. Brunger and V. S. Pande, Ensemble molecular dynamics yields submillisecond kinetics and intermediates of membrane fusion, *Proc. Natl. Acad. Sci. U. S. A.*, 2006, **103**(32), 11916–11921.

67 V. Mangala Prasad, J. S. Blijleven, J. M. Smit and K. K. Lee, Visualization of conformational changes and membrane remodeling leading to genome delivery by viral class-II fusion machinery, *Nat. Commun.*, 2022, **13**(1), 4772.

68 L. Milojević, Z. Si, X. Xia, L. Chen, Y. He, S. Tang, M. Luo and Z. H. Zhou, Capturing intermediates and membrane remodeling in class III viral fusion, *Sci. Adv.*, 2024, **10**(49), eadn8579.

69 L. V. Chernomordik and M. M. Kozlov, Protein-lipid interplay in fusion and fission of biological membranes, *Annu. Rev. Biochem.*, 2003, **72**(1), 175–207.

70 A. Pabis, R. J. Rawle and P. M. Kasson, Influenza hemagglutinin drives viral entry via two sequential intramembrane mechanisms, *Proc. Natl. Acad. Sci. U. S. A.*, 2020, **117**(13), 7200–7207.

71 H. J. Risselada, Y. Smirnova and H. Grubmüller, Free energy landscape of rim-pore expansion in membrane fusion, *Biophys. J.*, 2014, **107**(10), 2287–2295.

72 G. Golani, E. Leikina, K. Melikov, J. M. Whitlock, D. G. Gamage, G. Luoma-Overstreet, D. P. Millay, M. M. Kozlov and L. V. Chernomordik, Myomerger promotes fusion pore by elastic coupling between proximal membrane leaflets and hemifusion diaphragm, *Nat. Commun.*, 2021, **12**(1), 495.

73 T. Portet and R. Dimova, A new method for measuring edge tensions and stability of lipid bilayers: effect of membrane composition, *Biophys. J.*, 2010, **99**(10), 3264–3273, DOI: 10.1016/j.bpj.2010.09.032.

74 S. A. Akimov, P. E. Volynsky, T. R. Galimzyanov, P. I. Kuzmin, K. V. Pavlov and O. V. Batishchev, Pore formation in lipid membrane I: continuous reversible trajectory from intact bilayer through hydrophobic defect to transversal pore, *Sci. Rep.*, 2017, 7(1), 1.
75 B. Zucker, G. Golani and M. M. Kozlov, Model for ring closure in ER tubular network dynamics, *Biophys. J.*, 2022, **122**, 1974–1984.
76 D. Drabik, P. Hinc, M. Stephan, R. R. Cavalcanti, A. Czogalla and R. Dimova, Effect of leaflet asymmetry on the stretching elasticity of lipid bilayers with phosphatidic acid, *Biophys. J.*, 2024, **123**, 2406–2421.
77 A. Moga, N. Yandrapalli, R. Dimova and T. Robinson, Optimization of the inverted emulsion method for high-yield production of biomimetic giant unilamellar vesicles, *ChemBioChem*, 2019, **20**(20), 2674–2682.
78 F. S. Leomil, M. Stephan, S. Pramanik, K. A. Riske and R. Dimova, Bilayer charge asymmetry and oil residues destabilize membranes upon poration, *Langmuir*, 2024, **40**(9), 4719–4731.
79 C. Yang, J. Menge, N. Zhvania, D. Chen, D. A. Weitz and K. Jahnke, Engineering asymmetric nanoscale lipid vesicles for drug delivery, *bioRxiv*, 2024, preprint, DOI: 10.1101/2024.08.30.610290.
80 R. B. Lira, T. Robinson, R. Dimova and K. A. Riske, Highly efficient protein-free membrane fusion: a giant vesicle study, *Biophys. J.*, 2019, **116**(1), 79–91.
81 C. Taupin, M. Dvolaitzky and C. Sauterey, Osmotic pressure-induced pores in phospholipid vesicles, *Biochemistry*, 1975, **14**(21), 4771–4775.
82 R. M. Markosyan, G. B. Melikyan and F. S. Cohen, Tension of membranes expressing the hemagglutinin of influenza virus inhibits fusion, *Biophys. J.*, 1999, 77(2), 943–952.

Faraday Discussions

PAPER

Elucidating the mechanical properties of asymmetric membranes by direct derivation of their energetics†

Giacomo Fiorin *[ab] and Lucy R. Forrest [a]

Received 14th January 2025, Accepted 13th February 2025
DOI: 10.1039/d5fd00006h

The asymmetry between the two leaflets of a plasma membrane (PM) is widely accepted as an essential condition for most PM-associated biochemical processes. However, recent work has also shown that asymmetric bilayers can be significantly stiffer upon bending than symmetric ones, suggesting that the same asymmetry may hinder the ability of the PM to remodel itself. Here, we address this issue by combining all-atom molecular dynamics (MD) simulations with an enhanced sampling scheme that explicitly induces membrane deformations to quantify their free-energy cost. Examining small asymmetric POPC/DOPC bilayers, we find that a small density imbalance between the leaflets increases their bending rigidity compared to bilayers with minimal imbalance, or to symmetric bilayers of the same two lipids. This result is consistent with recently proposed theoretical models that identify differential stress as the main source of stiffening in asymmetric membranes. The first-principles approach used in this study is broadly applicable to other types of membrane, enabling further exploration of the interplay between asymmetry and curvature, or the simulation of specific biological conditions of the PM.

Introduction

One of the most notable characteristics of the plasma membranes in animal cells is the difference in chemical composition and physical properties between its leaflets.[1,2] Considerable effort has been invested in identifying how the leaflets differ in detail, and which biochemical processes are responsible for creating leaflet differences[3,4] or are controlled by them.[5] Nonetheless, the precise mechanism by which plasma membrane asymmetry supports the cell's lifecycle is largely unknown. It is likely that this mechanism is physical in nature,[2] which

[a]*Computational Structural Biology Section, National Institute for Neurological Disorders and Stroke, National Institutes of Health, Bethesda, MD 20894, USA. E-mail: giacomo.fiorin@nih.gov*
[b]*Theoretical Molecular Biophysics Section, National Heart, Lung and Blood Institute, National Institutes of Health, Bethesda, MD 20894, USA*

† Electronic supplementary information (ESI) available. See DOI: https://doi.org/10.1039/d5fd00006h

would be consistent with the fact that membrane asymmetry can be found in increasingly distant species.[6]

Investigation of the biophysical properties of membranes is most informative when experiments are performed on synthetic vesicles or other lipid bilayer preparations with controlled composition. For asymmetric bilayers, such constructs have traditionally been challenging to produce, until a variety of novel techniques were developed in recent years.[7–9] In one such experiment, it was shown that asymmetric vesicles made of 1,2-dioleoyl-*sn-glycero*-3-phosphocholine (DOPC) in one leaflet and 1-palmitoyl-2-oleoyl-*sn-glycero*-3-phosphocholine (POPC) in the other leaflet, have much higher bending stiffness than symmetric counterparts made of a mixture of the same two lipids.[10,11] Therefore, this result and others[12–14] suggest that the plasma membrane may be much harder to remodel than was previously inferred from symmetric bilayers that mimic each of its leaflets.

To address this and other questions regarding the mechanics of asymmetric membranes, theoretical and computational approaches can prove essential.[15–20] In one of these studies, Hossein and Deserno[21] showed how an imbalance in curvature stress can cause one leaflet to compress and the other to expand. At first glance, these two changes seem likely to perfectly complement each other: however, van der Waals forces increase in magnitude much more quickly at short distances than they do at large ones. Therefore, the increased stiffness of the compressed leaflet dominates the bending mechanics of the whole bilayer.[21] Even more significantly, molecular dynamics (MD) simulations with coarse-grained molecular representations have suggested that the structure of the compressed leaflet deviates from a homogeneous fluid, requiring the adoption of a theoretical framework compatible with the lipid gel phase to model its bending mechanics.[22]

In this study, atomically detailed MD simulations are used to investigate symmetric and asymmetric bilayers composed of DOPC and POPC, both at resting conditions and under bending strain. The latter condition is achieved by applying an enhanced-sampling approach based on the Multi-Map collective variable,[23] which was shown to accurately describe membrane remodeling with heterogeneous lipid compositions[24] or under applied tension.[25] Through direct quantification of the bending free-energy of each bilayer, we find that bilayers with asymmetry between the densities of their leaflets exhibit a higher bending rigidity than bilayers where those two densities are equivalent. These results shed further light on the mechanics of asymmetric bilayers, which in turn provides a foundation for understanding plasma membrane remodeling.

Methods

Molecular dynamics simulations of atomistic hydrated lipid bilayers

Initial atomic coordinates for lipid bilayers were prepared for each composition using the CHARMM-GUI membrane builder,[26] containing approximately 100 lipid molecules per leaflet. The simulation cell dimensions for these bilayers are $80 \times 80 \times 80$ Å^3. Two groups of these bilayers were prepared, all with an upper leaflet composed entirely of POPC, and with lower leaflets made of either POPC or DOPC. In each asymmetric bilayer the number of POPC molecules in the upper leaflet was kept constant, whereas the number of lipid molecules in the lower leaflet was varied in increments of 5 molecules between systems.

To allow comparison of derived bending moduli with those obtained in previous work,[10,11] two larger bilayers, containing about 400 lipid molecules per leaflet, were also prepared by replicating pre-equilibrated smaller bilayers twice along each longitudinal dimension. An additional layer of water molecules was added to these bilayers, resulting in simulation cells of approximately 160 × 160 × 90 Å^3. The first bilayer was prepared with 400 POPC molecules and 380 DOPC molecules in the upper and lower leaflets, respectively. The second bilayer was prepared as a symmetric homogeneous mixture of 200 POPC and 200 DOPC molecules in each leaflet.

During all simulations, the center of mass of each bilayer was restrained to the midplane of the simulation cell (*i.e.*, around the Cartesian plane $z = 0$) using a harmonic potential with force constant equal to 5 kcal mol^{-1} Å^{-2}. The position of the center of mass was defined as the geometric center of the phosphorus atoms of all lipid molecules in both leaflets. Besides facilitating analysis, this measure also allowed the definition of Multi-Map collective variables[23] (see following section), which are expressed in absolute atomic coordinates. Interatomic forces were represented *via* the CHARMM36 and TIP3P force fields.[27,28] Short-ranged Lennard-Jones forces were gradually switched to zero for distances between 10 and 12 Å, and all covalent bonds involving hydrogen atoms were constrained to their experimental length.

All simulations were performed with NAMD v3.0 [29] using a 2 fs integration step, with all short-ranged forces (bonded and non-bonded) computed at every step and long-range electrostatics computed every 3 steps. Although some of the simulations were carried out on GPUs, the GPU-resident scheme of NAMD was not used, owing to a software bug in NAMD version 3.0, later corrected in version 3.0.1. A target temperature of 300 K was maintained by applying a Langevin thermostat[29] onto all heavy atoms with damping coefficient of 1 ps^{-1}. A target pressure of 1 atm was maintained using a Langevin piston[30,31] with period 0.2 ps and decay time 0.1 ps. Atomic coordinates were stored every 20 ps of simulated time.

Analysis of lipid bilayer structure

Simulated trajectories were analyzed using MDAnalysis[32] for the purpose of computing leaflet-wise properties such as the mean area per lipid molecule (APL), electron density profiles (EDPs), and carbon–hydrogen order parameters (S_{CH}). Additionally, the configurational entropy of the acyl chains was computed (in units of free-energy) by integrating the function $-k_B T p(\theta) \log(p(\theta))$ where $p(\theta)$ is the histogram, averaged over all identical molecules in the same leaflet, of the torsional angle θ of four consecutive carbon atoms. In the Results section, $-T\Delta S_{mol}$ indicates the sum of all configurational entropies over both acyl chains of a given lipid molecule.

Spatially resolved properties of each leaflet, such as the center of mass of its lipid headgroups and the mean orientation vector of lipid molecules **n**, were accumulated onto two-dimensional grids with a spatial resolution of approximately 3 Å. The orientation vector **n** was defined as the vector connecting the midpoint of the phosphorus and glycerol "C2" atoms to the midpoint of the terminal carbons of the two acyl chains, as done in earlier work.[24,33,34]

Aggregate statistics between the two leaflets (for example, the position of the membrane's midplane) were calculated as the average of the two corresponding maps from the leaflets, weighted by their respective numbers of lipid molecules. For deformed bilayers, each 2D map was linearly regressed against a sinusoidal function to estimate the magnitude of deformation and, from the latter value, the root-mean-squared (RMS) curvature, c. Similarly, the longitudinal components of the lipid orientation vector (n_x, n_y) evaluated over the membrane plane were linearly regressed against the longitudinal components of the normal vector of the midplane (N_x, N_y).

Molecular graphics were rendered with VMD.[35]

Enhanced-sampling simulations of deformed bilayers

To compute the free energy landscape of bilayers of varying composition and asymmetry, we used the Multi-Map method[23] to enhance the sampling of the molecular dynamics simulations. This approach requires definition of collective variables that reflect the magnitude of curvature of a bilayer, represented using three-dimensional volumetric maps of individual leaflets. Such maps are then used, collectively, as targets for Umbrella Sampling simulation windows with increasing magnitudes of curvature.

To create volumetric maps with physically meaningful shapes, simulated trajectories of each bilayer composition at rest were used to record transmembrane electron density profiles $\rho(z)$ of the phosphorous atoms in each leaflet (Fig. S1†). These profiles were then multiplied by sinusoidal functions to create three-dimensional volumetric phosphorous densities $\varphi(\mathbf{x})$ for each leaflet.

For each bilayer composition, an ensemble of curved membranes was simulated, guided by such maps, where the collective variable ξ_{leaflet} describes the ensemble properties of those bilayers. In practice, we used sinusoidal functions whose period was equal to the simulation cell length along the first Cartesian axis and with increasing amplitudes, I_k ranging from −18 Å to +18 Å in steps of 4 Å ($k = 1, \ldots, 10$). Thus, the Multi-Map collective variable is:

$$\xi_{\text{leaflet}} = \sum_{k=1}^{10} I_k \sum_{i=1}^{N} \varphi_k(\mathbf{x}_i) \tag{1}$$

where N is the number of phosphorus atoms in the leaflet and $\mathbf{x}_i$ are their Cartesian coordinates. In other words, the value assumed by the collective variable ξ_{leaflet} during a simulation is a weighted average of I_k with weights defined by the volumetric maps $\varphi_k(\mathbf{x})$ at atomic positions. For the entire bilayer, the collective variable becomes $\xi = \xi_{\text{upper}} + \xi_{\text{lower}}$.

Target values of ξ were computed over the simulated trajectories of each bilayer at rest, using VMD[35] and the integrated Colvars library.[36,37] The distributions of ξ were used to identify trajectory snapshots with increasingly high magnitudes of deformation. These snapshots were then used to initialize umbrella-sampling simulation windows[38] with uniformly spaced restraints on ξ, totaling about 14 windows for each 80 Å^2 bilayer; each window was simulated independently for approximately 0.5 μs using NAMD[29] and Colvars.[36,37] The two larger POPC/DOPC bilayers were simulated with the same protocol, but with 16 windows of 0.7 μs each. In all simulation windows, the harmonic force constant applied to ξ was 0.6 kcal mol^{-1}. Trajectories were visually inspected with the Colvars Dashboard[39] to ensure that deformations of the bilayer midplane are consistent with the sinusoidal shape, within thermal fluctuations. For each window, trajectory frames after the first 0.1 μs of simulated time were used to compute histograms of ξ, which were then unbiased using WHAM[40] to derive the corresponding free-energy landscapes. Having confirmed that ξ and the bilayer curvature c are directly proportional to each other (Fig. S2†), free-energy landscapes were plotted as a function of the latter.

Derivation of elastic moduli from free-energy landscapes and atomic structure

Membrane bending deformations (and the corresponding elastic moduli) were evaluated using the combined theory of curvature and lipid orientation,[41,42] which uses the following expression for the bending energy density:

$$f(\mathbf{n},\mathbf{N}) = \frac{1}{2}k_\mathrm{c}(\nabla\cdot\mathbf{n} - c_0)^2 + \frac{1}{2}k_t(\mathbf{n} - \mathbf{N})^2 \quad (2)$$

where $\mathbf{n}$ is the lipid orientation vector, its divergence $\nabla\cdot\mathbf{n}$ defines the local curvature of the bilayer, c_0 is the bilayer's curvature preference, $\mathbf{N}$ is the vector normal to the bilayer's local midplane surface, k_c is the bending modulus and k_t the tilt modulus. The scalar field $\nabla\cdot\mathbf{n}$ and the vector fields $\mathbf{n}$ and $\mathbf{N}$ are all defined for two-dimensional Cartesian coordinates (x, y) over the bilayer.

Note that eqn (2) treats the bilayer as a single medium with elastic moduli resulting from the sum of the moduli of the two leaflets, *i.e.* $k_\mathrm{c} = k_\mathrm{c}^{(\mathrm{upper})} + k_\mathrm{c}^{(\mathrm{lower})}$. Furthermore, although $c_0 \neq 0$ in asymmetric membranes, c_0 only contributes an additive constant to the integral of eqn (2) over the bilayer's area owing to the periodic boundary conditions of a MD simulation. Similarly, other terms whose contribution would amount to a constant are not discussed for clarity.

Because the bilayers used here were simulated in a curved state, the vector fields $\mathbf{n}$ and $\mathbf{N}$ can both be quantified directly from atomic coordinates of the lipid molecules in simulation, as done previously.[24] This quantification is further simplified by leveraging the fact that the imposed membrane deformation is sinusoidal, and therefore the longitudinal components of the two vectors, $n_{\parallel}$ and $N_{\parallel}$, are also sinusoidal functions whose ratio $n_{\parallel}/N_{\parallel}$ may be quantified by linear regression over all grid points. By using the latter ratio when fitting eqn (2) against the energy landscapes computed from the simulations, k_c and k_t are derived for each system.

Results

Quantifying asymmetry using measured area per lipid

Ten bilayers with $\approx 80 \times 80$ Å^2 lateral dimensions and two bilayers with $\approx 160 \times 160$ Å^2 dimensions as listed in Table 1 were prepared and simulated at rest (*i.e.*, without any externally applied forces) for at least 3 μs each. In all asymmetric bilayers, the upper leaflet contains POPC molecules whereas the composition of the lower leaflet differs in every bilayer. Note that, in all simulations reported here, the dimensions of the simulation cell parallel and orthogonal to the membrane plane were each controlled by targeting a pressure of 1 atm. Therefore, by construction, the total tension imposed on the bilayer, which is the sum of the two leaflet tensions, $\Sigma^{(\mathrm{upper})} + \Sigma^{(\mathrm{lower})}$, is zero.

For a bilayer with identical lipid types in both leaflets, the differential tension $\Delta\Sigma = \Sigma^{(\mathrm{upper})} - \Sigma^{(\mathrm{lower})}$ is automatically zero when equal numbers of lipid

Table 1 Lipid bilayers simulated in this study. Systems vary in the composition of the upper and lower leaflets and the bilayer's asymmetry as estimated by the relative change in the area per lipid (APL) measured in that simulation

Lipid type		Num. molecules		
Upper	Lower	Upper	Lower	Area asymmetry
POPC	POPC	100	100	0
		100	95	0.029
		100	90	0.051
		100	85	0.067
POPC	DOPC	100	105	−0.057
		100	100	−0.027
		100	95	0.003
		100	90	0.027
		100	85	0.048
		400	380	0.003
POPC : DOPC 1 : 1	POPC : DOPC 1 : 1	100	100	0
		400	400	0

molecules are used in the two leaflets. However, no equivalent criterion for quantifying differential tension exists for bilayers with different lipid types between their leaflets. To estimate the value of $\Delta\Sigma$ in that case, one could in principle compute the lateral pressure profile along the bilayer normal and integrate it over each leaflet to calculate the tension that they are subject to. While this approach is straightforward in continuum mechanics, when applied to MD simulations it becomes sensitive to the choice of integration contour,[43,44] the choice of geometric boundary between the leaflets, and the implementation details of the MD simulation software used.

As a reasonable alternative to estimating $\Delta\Sigma$, one may also quantify the asymmetry in terms of the area that each lipid molecule (APL) occupies given the dimensions of the simulation box after equilibration, which can vary when maintaining the system using a constant pressure regime. Here we define such a value, termed "area asymmetry" using the cell dimensions extracted from the simulation divided by the number of lipids in each leaflet, relative to the value that such a leaflet would assume in a symmetric bilayer, APL_{ref}. Thus area asymmetry is computed using the expression $(APL_{ref} - APL)/APL_{ref}$. Among the five 80×80 Å^2 POPC/DOPC asymmetric bilayers simulated here, the bilayer whose measured APL best approximates APL_{ref} in both its leaflets contains 100 POPC molecules in the upper leaflet and 95 DOPC molecules in the lower leaflet (Table 1). This composition is thus used as the best approximation of the $\Delta\Sigma \approx 0$ condition for asymmetric POPC/DOPC bilayers.

We note that using a molecular system with periodic boundary conditions precludes analysis of potentially competing forces, in particular curvature stress, which could significantly violate the $\Delta\Sigma \approx 0$ condition.[21] Although there are alternative protocols available that aim to enforce that condition and related conditions,[45–48] we here focus on characterizing the membrane mechanics of bilayers in a broad interval around the $\Delta\Sigma \approx 0$ condition to minimize the imposed bias.

Compositional asymmetry alone does not significantly affect bending mechanics

To establish a baseline, we first test the hypothesis that asymmetric POPC/DOPC bilayers with very small estimated differential tension ($\Delta\Sigma \approx 0$) are mechanically similar to symmetric bilayers. To this end, an asymmetric 160 × 160 $Å^2$ POPC/DOPC bilayer (see Table 1) was simulated at multiple curvature states, from approximately flat (root-mean-squared curvature $c = 0$) to a sinusoidal shape with $c \approx 0.004$ $Å^{-1}$. This was achieved by running multiple independent simulations with harmonic biasing potentials applied on the Multi-Map collective variable,[23] as described in the Methods. Briefly, this methodology supports the simulation of an atomically-detailed lipid membrane under conditions of increasingly high curvature, as well as the direct determination (from atomic forces) of the corresponding free-energy landscape. The same protocol was applied to a symmetric bilayer composed of a 1 : 1 POPC : DOPC mixture in both leaflets.

The free-energy landscapes (Fig. 1A) show that the cost of inducing curvature is almost identical between the bilayers independent of whether the DOPC and POPC lipids were mixed, or separated in the leaflets, supporting the hypothesis that the bending energetics are equivalent. Additionally, the two energy landscapes were analyzed using the unified theory of curvature and lipid orientation:[41,42] the derived bending moduli k_c are shown in Fig. 1B, alongside those of

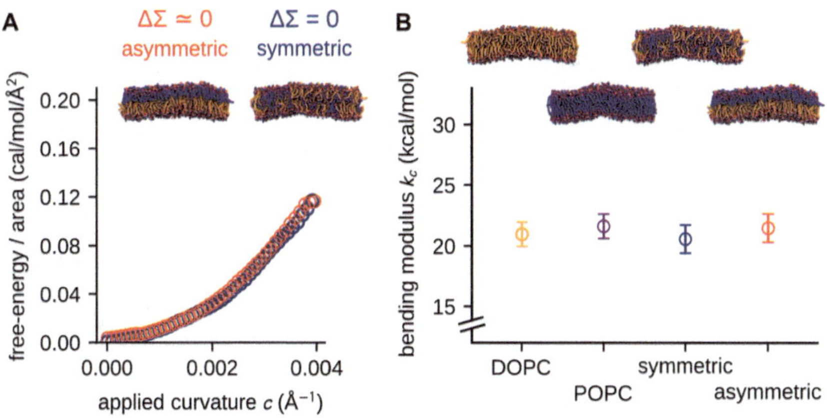

Fig. 1 Bending free-energy landscapes of an asymmetric POPC/DOPC bilayer with approximately equal leaflet tensions and of a fully symmetric bilayer comprising the same two lipids; POPC and DOPC molecules are shown as purple and orange sticks, respectively. (A) Free-energy landscapes as a function of the root-mean-squared magnitude of the curvature c of sinusoidal deformations applied to the asymmetric bilayer (red circles) and to the symmetric bilayer (blue circles). The highest magnitude ($c \approx 0.004$ $Å^{-1}$) describes a sinusoid with peak curvature ≈ 0.008 $Å^{-1}$, which is equivalent to a radius $\approx$ 125 Å. Error bars are within the size of the symbols. (B) Values of the bending modulus k_c derived from the data in panel (A) are shown alongside those derived for pure DOPC (orange circle) and pure POPC (purple circle) bilayers using the same approach.[24] Error bars indicate standard sampling errors estimated by block analysis.

symmetric POPC or DOPC bilayers of equivalent sizes computed previously using the same method.[24] All these values of k_c are within the estimated statistical error, estimated by block analysis, again indicating that compositional asymmetry alone does not significantly affect bending mechanics.

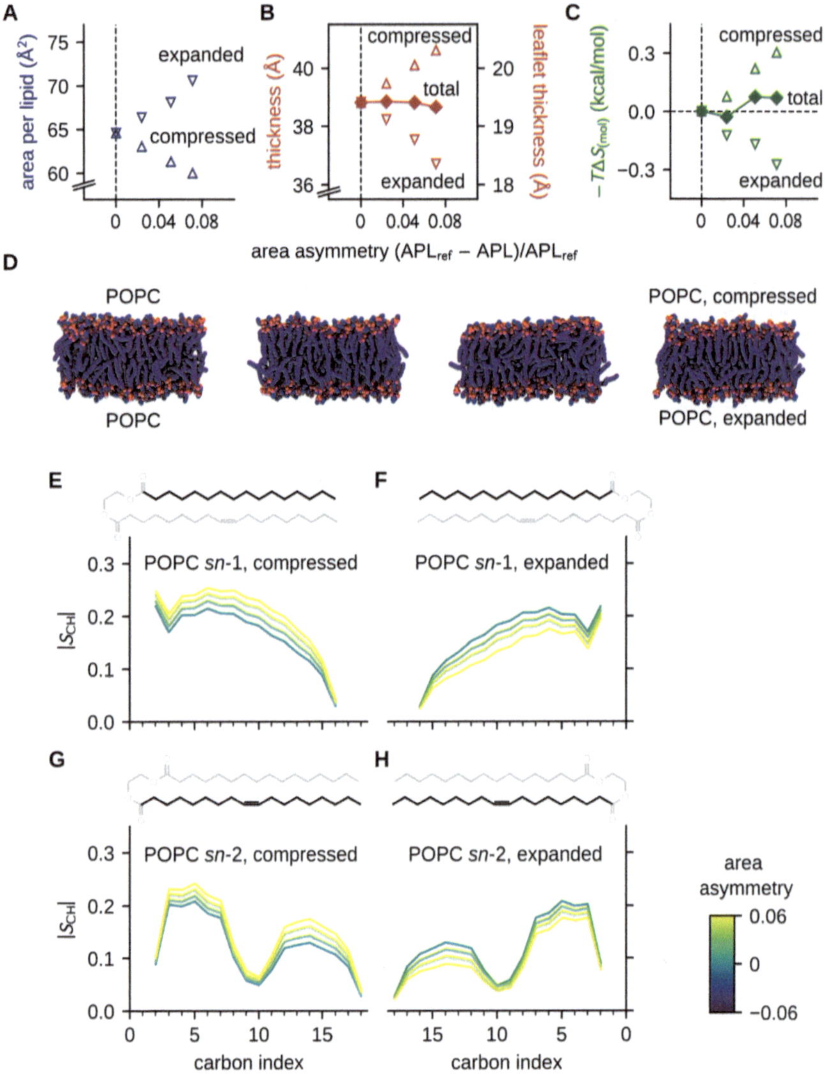

Fig. 2 Biophysical properties of four POPC bilayers, one symmetric and three with asymmetry in the numbers of lipid molecules in each leaflet. (A) Areas per lipid molecule (APL) in the compressed leaflet (upper triangles) and expanded leaflet (lower triangles); the dashed vertical line indicates the value for a symmetric bilayer. (B) Hydrophobic thickness of the whole bilayer (diamonds) and of each leaflet (triangles). (C) Net change in entropy per molecule in each leaflet (triangles) and the whole bilayer (diamonds). (D) Simulation snapshots of the four bilayers, with increasing asymmetry from left to right. (E–H) Segmental order parameters, S_{CH}, as a function of the carbon atom's position along each chain, with data from multiple bilayers overlaid and color-coded based on the level of area asymmetry in that bilayer. In panels (A–C) and (E–H), sampling errors are comparable to or smaller than the size of symbols.

Besides providing the values of k_c themselves, the above analysis suggests that a single theoretical framework can apply equally well to the bending mechanics of bilayers with either symmetric or asymmetric composition. Consequently, we proceeded to examine the consequences of asymmetry by computing the bending energy landscape of bilayers with smaller sizes (80 × 80 Å^2), thus reducing computational cost and allowing the sampling of multiple configurations with varying asymmetry. Specifically, we generated bilayers that were asymmetric with respect to either the number of lipids per leaflet in pure-POPC bilayers; or in the ratio of DOPC to POPC lipids (Table 1).

Complementarity of effects on compressed and expanded leaflets

Asymmetric lipid compositions are expected to result in unique structural and mechanical properties, even at rest, which must be characterized prior to understanding the response of that membrane to bending strain. In particular, one leaflet is expected to be more compressed while the other is expanded, as reflected in the computed APL in each leaflet (Fig. 2A and 3A). To assess the impact of the asymmetry on membrane structure, we computed the hydrophobic thickness of each bilayer using the distance between the phosphate groups of the two leaflets, based on electron density profiles (Fig. S1†), while the thickness of each leaflet was defined as the distance between the respective phosphate peaks and the peak of the terminal methyl carbon atoms. As expected, as the asymmetry increases, the compressed leaflet becomes thicker, whereas the expanded (in this case, lower) leaflet becomes thinner, independent of the origins of the asymmetry (Fig. 2B and 3B). However, the two changes do not appear to compensate each other fully, resulting in a slight decrease in the total bilayer thickness for the most extreme asymmetries.

To quantify the local impact of the asymmetry on the orientation of the acyl chains, segmental order parameters, S_{CH}, between acyl carbon atoms and their bound hydrogens were calculated over the same trajectories (Fig. 2E–H and 3E–H). Their changes as a function of asymmetry largely confirm the trends already seen in the APL and thickness: the C–H bond vectors in the acyl chains of the compressed leaflet become increasingly more aligned with the membrane plane, while the reverse is true for the expanded leaflet.

To analyze the internal dynamics of each lipid molecule separately from their orientational order, the configurational entropy of each acyl chain torsional angle was calculated: the results are shown in Fig. S3† as a function of the position of the carbon atoms along the chains. The total change in entropy per lipid molecule, in units of free energy, $-T\Delta S_{mol}$, is shown in Fig. 2C and 3C. Here, a significant difference appears depending on lipid composition: in bilayers made only of POPC, the changes in entropy occurring in the two leaflets upon increasing asymmetry appear to either compensate each other or create an unfavorable imbalance (Fig. 2C). However, in the POPC/DOPC bilayers the change in the entropy of the DOPC molecules is consistently smaller in magnitude than that of the POPC molecules (Fig. 3C). This leads to a favorable imbalance (*i.e.* $-T\Delta S_{mol} < 0$) when the DOPC leaflet is increasingly compressed. This result is consistent with the hypothesis that asymmetric POPC/DOPC bilayers may exhibit differential tension under the effect of two competing driving forces, here being the planar boundary conditions and the difference in preferred curvature.[21]

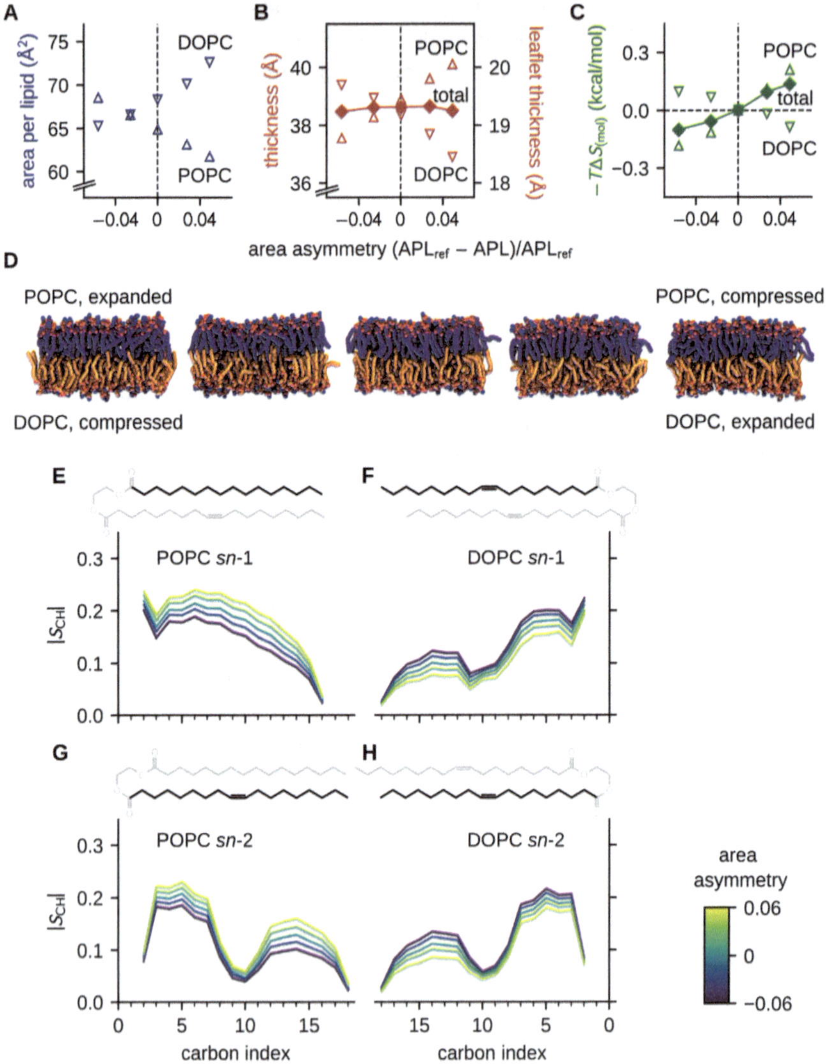

Fig. 3 Biophysical properties of asymmetric bilayers with POPC molecules in the upper leaflet and varying numbers of DOPC molecules in the lower leaflet. See Fig. 2 for details. (A–C) APLs, hydrophobic thickness, and entropy changes of each bilayer; symbols are as in Fig. 2, with data from each leaflet labeled based on its constituent lipid. (D) Simulation snapshots, with POPC and DOPC molecules as orange and purple sticks, respectively. (E–H) Segmental order parameters, S_{CH}, for each acyl chain in the corresponding bilayers.

Compressed lipid leaflets remain predominantly fluid

It has previously been suggested that asymmetric lipid bilayers remain in a fluid phase even when significant compressive strain is applied to one of their leaflets, *i.e.* that the "stiffening transition" precedes the fluid-to-gel transition of the compressed leaflet and that the two transitions are well separated.[21] However, that observation was based on coarse-grained MD simulations, and should be verified for simulations with atomistic detail. This is a key test specifically for atomistic

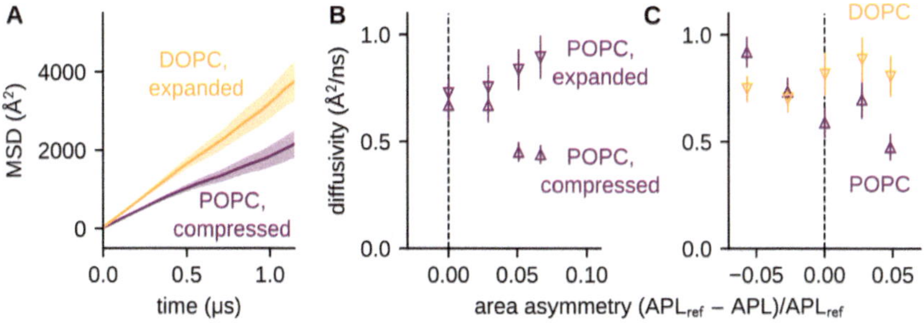

Fig. 4 Diffusivity of lipid molecules in asymmetric bilayers. (A) Mean squared displacement of lipid molecules in the POPC/DOPC bilayer with the largest area asymmetry; shaded areas indicate the standard error of the mean, computed over all lipid molecules. (B) Diffusion coefficients D extracted from the MSD plots of all POPC/POPC bilayers, with upper and lower triangles showing values for lipids in the compressed and expanded leaflet, respectively. Error bars indicate standard errors computed from block averaging. (C) Values of D extracted from POPC/DOPC bilayers, averaged over either POPC (purple) or DOPC (orange) lipid molecules.

MD simulations, also because the long relaxation time scales required to achieve the gel phase could severely compromise the effectiveness of a MD simulation-based approach. To this end, two characteristic descriptors of fluid membranes, namely the acyl chain order (as shown above) and lateral diffusion, were inspected for features of the transition to gel phase.

The segmental order parameters, S_{CH}, shown in Fig. 2E–H and 3E–H, exhibit no significant discontinuities as a function of asymmetry. The largest change in S_{CH} observed between the most extreme asymmetries was about ~0.03 in magnitude. For comparison, this change is about 3-fold smaller than the change observed when comparing atomistic simulations of each lipid type in its pure phase with a 30 mol% cholesterol mixture,[24] which is also a fluid-phase system. Thus, the order parameters show no indication of gel phase behavior in these simulations.

As further confirmation that the bilayers remain fluid, mean squared displacements (MSDs) of the center of mass of lipid molecules were computed over continuous trajectory segments of simulations of the asymmetric bilayers at rest (Fig. 4A). By averaging over all molecules in each leaflet, diffusion coefficients D were extracted using the Einstein relation (MSD $\sim 2Dt$) and plotted as a function of area asymmetry (Fig. 4B and C). Relative to the fluid-phase symmetric bilayers (area asymmetry $\approx$ 0), only small (at most 30%) variations in diffusivity were found among the asymmetric systems, again consistent with the absence of fluid to gel transitions. Note that the values of D reported here are not extrapolated to infinite systems,[49] because only relative variations in D are examined.

Bilayer bending rigidity increases when asymmetry is introduced

The structural and dynamical properties analyzed above provide a molecular description of the asymmetric bilayers at rest, and a starting point for understanding their mechanical response to bending strain. To quantify the latter, we applied the Multi-Map procedure,[23] using the collective variable reflecting bilayers of increasing curvature, as described in the Methods. Specifically, snapshots were

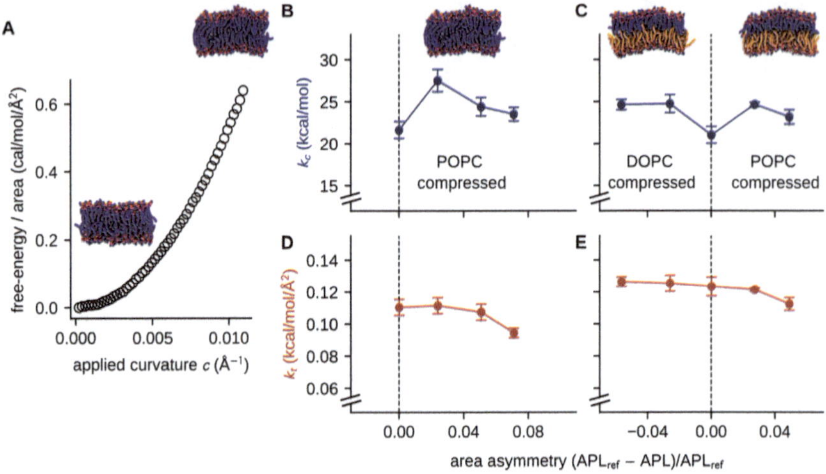

Fig. 5 Elastic moduli of asymmetric bilayers with varying asymmetry. (A) Bending free-energy landscape as a function of the curvature c computed by numerical sampling of atomic forces for the POPC bilayer with area asymmetry $\approx$ 0.03. Error bars are within the magnitude of the symbols. Also shown are two simulation snapshots at $c \approx 0$ and $c \approx 0.12$ Å^{-1}. (B and C) Bending moduli k_c of POPC/POPC and POPC/DOPC bilayers as estimated from the free-energy landscape of each bilayer; error bars are standard errors computed from block analysis. Also shown in each panel are snapshots of three bilayers exhibiting large changes in k_c. (D and E) Tilt moduli k_t of the same bilayers, as derived concurrently with k_c.

extracted from the simulated trajectory of each bilayer under resting conditions, based on increasing magnitude of sinusoidal curvature. These snapshots were used to initialize biased simulations that included harmonic restraints on the variable aimed at maintaining the existing curvature in a steady state, while allowing all other atomic fluctuations of the system. The resulting histograms from each set of simulations were used to derive free-energy landscapes as a function of the curvature c. Fig. 5A shows one such free-energy landscape. Bending moduli k_c and tilt moduli k_t were derived from such free-energy landscapes for the POPC/POPC bilayers (Fig. 5B and D) and for the POPC/DOPC bilayers (Fig. 5C and E). As in the previous figures, values are reported as a function of the area asymmetry.

The most significant observation regarding the dependence on asymmetry is a sharp $\approx$25% rise in k_c from the bilayers that approximate the $\Delta\Sigma \approx 0$ condition, to the bilayers immediately adjacent. This stiffening is observed for all three such situations: when POPC is compressed in the POPC/POPC bilayer (Fig. 5B), when POPC is compressed in the POPC/DOPC bilayer (right side of Fig. 5C) and finally when DOPC is compressed in the same bilayer (left side of Fig. 5C).

Remarkably, the change in k_c is not accompanied by a corresponding change in k_t for the same configurations (Fig. 5D and E), which remains very close to the value of the reference bilayer ($\approx$0.11 kcal mol^{-1} Å^{-2} for POPC/POPC and $\approx$0.12 kcal mol^{-1} Å^{-2} for POPC/DOPC). It is important to underline that with the protocol used here, each free-energy curve measures the combined magnitude of the two terms of eqn (2), which are linear in k_c and k_t, respectively. However, the *ratio* between the two parameters, k_t/k_c, is independently determined from the analysis of the lipid orientation vector (see Methods). Therefore, the observation that one parameter changes significantly while the other remains constant is unlikely to be attributable, say, to limited statistical sampling, and rather appears to stem directly from membrane dynamics.

Discussion

In this study, atomistic simulations were used to quantify the bending stiffness of lipid bilayers with asymmetry that varied in either lipid density or chemical composition between their leaflets. To quantify the energy cost of curvature, forces were applied onto each bilayer to increase the probability of observing membrane shapes with high curvature.[23]

The above approach yields bending free-energy landscapes directly from atomic forces in a "model-free" approach, without need to, *e.g.*, extrapolate from the magnitude of thermally-excited fluctuations. However, extending these findings to length scales other than those explicitly simulated requires the use of continuum mechanics and the derivation of material constants, such as the bending modulus k_c. The theory used to that end (eqn (2))[41,42] has been shown to consistently describe the magnitude of curvature fluctuations at small length scales[50,51] as well as the orientational dynamics of the lipid molecules.[24,33] Nonetheless, it also assumes, as many continuum theories do, that the membrane is a homogeneous elastic medium.

As was shown previously for asymmetric bilayers simulated at coarse-grained resolution,[21] compressed lipid leaflets may experience non-homogeneous variations in density along the bilayer, even manifesting in a few cases as visible gel-phase domains.[52] The impact of these variations on the bending energy was empirically modeled through the "softening length" ℓ,[22] a length scale at or below which the deviations from a homogeneous fluid are most relevant. It is thus worth considering whether the asymmetry-dependent $\approx 25\%$ increase in k_c derived from small bilayers (Fig. 5), whose dimensions are comparable to ℓ, might correspond to the $\approx 50\%$ or $\approx 100\%$ increases measured from large vesicles,[10,11] whose bending fluctuations extend over far greater lengths.

In the context of this work, determining the softening length ℓ from simulation data would entail replacing the term $\frac{1}{2}k_c(\nabla\cdot\mathbf{n})^2$ in eqn (2) with the expression $k_c\ell^{-2}(\sqrt{(1+(\nabla\cdot\mathbf{n})^2\ell^2)}-1)$,[22] which is straightforward in principle. However, when adopting this approach, the level of statistical sampling of each free-energy landscape proved insufficient to reliably determine both k_c and ℓ, as shown by large estimated errors for both parameters (Table S2†). This outcome is not at all surprising, considering that the simulations reported here entail atomistic bilayers of modest size ($L \sim 80$ Å), and thus the level of statistical sampling is limited both in time and length scale compared to coarse-grained simulations.

Similarly to the computed energy landscapes, the simulated trajectories in this study fail to provide evidence in favor or against the notion of non-homogeneous softening.[22] For example, no gel-phase domains, nor significant variations in lipid density are observed in the simulated bilayers (Fig. S4 and S5†). Again, this is expected to some degree because the bending strain applied on a small area is unlikely to outweigh the line tension associated with the boundary between gel-phase and fluid-phase domains. On the other hand, the fact that k_c rapidly increases at small asymmetry, but decreases again at larger asymmetry (Fig. 5) is not immediately explained by the increased tension, suggesting that additional theoretical modeling may be required to compare the results of this study with macroscopic measurements.[10,11] Pending further work, we opted to simply report the results obtained using the fluid-phase theory (eqn (2)), while also alerting the reader to the fact that the inclusion of non-homogeneous softening[22] when analyzing data from this work would likely yield increased bending moduli for asymmetric bilayers.

Notwithstanding the value of thoroughly characterizing the mechanics of synthetic phospholipid bilayers, it is important to remember that the plasma membrane contains a much broader variety of lipid types, as well as a large fraction of cholesterol. It has already been observed that due to their high rate of flip–flop, cholesterol molecules have the potential to significantly affect the differential tension of the membrane.[53,54] Furthermore, cholesterol itself could amplify the membrane's inherent mechanical and compositional asymmetry, by associating itself preferentially with one leaflet over the other.[55–57] In this context, we also note that the general effect of cholesterol is to increase lipid packing while also preserving the membrane's fluid phase,[58] even at conditions where fluid/fluid phase separation occurs.[59] Therefore, it remains unclear whether the bending mechanics of the plasma membrane would be best described by mechanical theories suitable for the fluid phase,[41,42] for the gel-phase,[22] or by different assumptions altogether. As is often the case in biomembrane dynamics, this question is most likely to be answered in future work by complementing experiments with simulations. To this end, the ability to compute free-energy landscapes from simulations at atomic detail provides an important technological advance.

Conclusions

Characterizing the mechanics of asymmetric bilayers is key to understanding the fundamental functions of the plasma membrane. Using all-atom MD simulations, we show that the biophysical properties of compositionally asymmetric POPC/DOPC bilayers are strongly correlated with the density imbalance between the leaflets. Notwithstanding the fixed boundary conditions of the simulations, analysis of the lipid dynamics among these bilayers suggests that the imbalance is entropically favorable, consistent with the presence of curvature stress. Furthermore, quantification of the bilayers' energy landscapes reveals that density asymmetry leads to increased stiffness. These results demonstrate how the direct simulation of membrane bending in increasingly realistic models of the plasma membrane can provide valuable insights into the mechanisms underlying its remodeling.

Data availability

Input files and representative simulation snapshots are available in Zenodo (https://doi.org/10.5281/zenodo.14861695).

Conflicts of interest

There are no conflicts to declare.

Acknowledgements

This research was supported by the Division of Intramural Research of the National Institutes of Health (NIH), National Institute of Neurological Disorders and Stroke (ZIA NS003143). Computational resources were provided by the NIH Biowulf cluster and the "Blue Waters" supercomputer (National Science Foundation awards OCI-0725070 and ACI-1238993, the State of Illinois, and the National Geospatial-Intelligence Agency). The content is solely the responsibility of the authors and does not necessarily represent the official views of the National Institutes of Health.

References

1 A. J. Verkleij, R. F. A. Zwaal, B. Roelofsen, P. Comfurius, D. Kastelijn and L. L. M. van Deenen, *Biochim. Biophys. Acta, Biomembr.*, 1973, **323**, 178–193.
2 P. F. Devaux, *Biochemistry*, 1991, **30**, 1163–1173.
3 T. Kobayashi and A. K. Menon, *Curr. Biol.*, 2018, **28**, R386–R391.
4 J. H. Lorent, K. R. Levental, L. Ganesan, G. Rivera-Longsworth, E. Sezgin, M. Doktorova, E. Lyman and I. Levental, *Nat. Chem. Biol.*, 2020, **16**, 644–652.
5 P. Piller, E. F. Semeraro, G. N. Rechberger, S. Keller and G. Pabst, *PNAS Nexus*, 2023, **2**, pgad126.
6 M. Bogdanov, K. Pyrshev, S. Yesylevskyy, S. Ryabichko, V. Boiko, P. Ivanchenko, R. Kiyamova, Z. Guan, C. Ramseyer and W. Dowhan, *Sci. Adv.*, 2020, **6**, eaaz6333.
7 T. A. Enoki and G. W. Feigenson, *Biophys. J.*, 2019, **117**, 1037–1050.
8 F. S. C. Leomil, M. Stephan, S. Pramanik, K. A. Riske and R. Dimova, *Langmuir*, 2024, **40**, 4719–4731.
9 S. Chiantia and E. London, *Biophys. J.*, 2012, **103**, 2311–2319.
10 Y. Elani, S. Purushothaman, P. J. Booth, J. M. Seddon, N. J. Brooks, R. V. Law and O. Ces, *Chem. Commun.*, 2015, **51**, 6976–6979.
11 K. Karamdad, R. V. Law, J. M. Seddon, N. J. Brooks and O. Ces, *Chem. Commun.*, 2016, **52**, 5277–5280.
12 L. Lu, W. J. Doak, J. W. Schertzer and P. R. Chiarot, *Soft Matter*, 2016, **12**, 7521–7528.
13 M. P. Frewein, P. Piller, E. F. Semeraro, O. Czakkel, Y. Gerelli, L. Porcar and G. Pabst, *Biophys. J.*, 2023, **122**, 2445–2455.
14 D. Drabik, P. Hinc, M. Stephan, R. R. M. Cavalcanti, A. Czogalla and R. Dimova, *Biophys. J.*, 2024, **123**, 2406–2421.
15 A. A. Gurtovenko and I. Vattulainen, *J. Am. Chem. Soc.*, 2007, **129**, 5358–5359.
16 A. Polley, S. Vemparala and M. Rao, *J. Phys. Chem. B*, 2012, **116**, 13403–13410.
17 J. Agudo-Canalejo and R. Lipowsky, *ACS Nano*, 2015, **9**, 3704–3720.
18 J. Tian, J. Nickels, J. Katsaras and X. Cheng, *J. Phys. Chem. B*, 2016, **120**, 8438–8448.
19 M. Siggel, R. M. Bhaskara and G. Hummer, *J. Phys. Chem. Lett.*, 2019, **10**, 6351–6354.
20 M. L. Karlsen, D. S. Bruhn, W. Pezeshkian and H. Khandelia, *RSC Adv.*, 2021, **11**, 22677–22682.
21 A. Hossein and M. Deserno, *Biophys. J.*, 2020, **118**, 624–642.

22 P. Diggins, Z. A. McDargh and M. Deserno, *J. Am. Chem. Soc.*, 2015, **137**, 12752–12755.
23 G. Fiorin, F. Marinelli and J. D. Faraldo-Gomez, *J. Comput. Chem.*, 2020, **41**, 449–459.
24 G. Fiorin, L. R. Forrest and J. D. Faraldo-Gómez, *PNAS Nexus*, 2023, **2**, pgad269.
25 W. Zhou, G. Fiorin, C. Anselmi, H. A. Karimi-Varzaneh, H. Poblete, L. R. Forrest and J. D. Faraldo-Gómez, *eLife*, 2019, **8**, e50576.
26 S. Jo, J. B. Lim, J. B. Klauda and W. Im, *Biophys. J.*, 2009, **97**, 50–58.
27 W. L. Jorgensen, J. Chandrasekhar, J. D. Madura, R. W. Impey and M. L. Klein, *J. Chem. Phys.*, 1983, **79**, 926–935.
28 J. B. Klauda, R. M. Venable, J. A. Freites, J. W. O'Connor, D. J. Tobias, C. Mondragon-Ramirez, I. Vorobyov, A. D. MacKerell and R. W. Pastor, *J. Phys. Chem. B*, 2010, **114**, 7830–7843.
29 J. C. Phillips, D. J. Hardy, J. D. C. Maia, J. E. Stone, J. V. Ribeiro, R. C. Bernardi, R. Buch, G. Fiorin, J. Henin, W. Jiang, R. McGreevy, M. C. R. Melo, B. K. Radak, R. D. Skeel, A. Singharoy, Y. Wang, B. Roux, A. Aksimentiev, Z. Luthey-Schulten, L. V. Kale, K. Schulten, C. Chipot and E. Tajkhorshid, *J. Chem. Phys.*, 2020, **153**, 044130.
30 G. J. Martyna, D. J. Tobias and M. L. Klein, *J. Chem. Phys.*, 1994, **101**, 4177–4189.
31 S. E. Feller, Y. Zhang, R. W. Pastor and B. R. Brooks, *J. Chem. Phys.*, 1995, **103**, 4613–4621.
32 N. Michaud-Agrawal, E. J. Denning, T. B. Woolf and O. Beckstein, *J. Comput. Chem.*, 2011, **32**, 2319–2327.
33 M. C. Watson, E. G. Brandt, P. M. Welch and F. L. H. Brown, *Phys. Rev. Lett.*, 2012, **109**, 028102.
34 M. F. Erguder and M. Deserno, *J. Chem. Phys.*, 2021, **154**, 214103.
35 W. Humphrey, A. Dalke and K. Schulten, *J. Mol. Graphics*, 1996, **14**, 33–38.
36 G. Fiorin, M. L. Klein and J. Henin, *Mol. Phys.*, 2013, **111**, 3345–3362.
37 G. Fiorin, F. Marinelli, L. R. Forrest, H. Chen, C. Chipot, A. Kohlmeyer, H. Santuz and J. Henin, *J. Phys. Chem. B*, 2024, **128**, 11108–11123.
38 G. M. Torrie and J. P. Valleau, *J. Comput. Phys.*, 1977, **23**, 187–199.
39 J. Henin, L. J. S. Lopes and G. Fiorin, *J. Chem. Theory Comput.*, 2022, **18**, 1945–1956.
40 S. Kumar, D. Bouzida, R. H. Swendsen, P. A. Kollman and J. M. Rosenberg, *J. Comput. Chem.*, 1992, **13**, 1011–1021.
41 M. Hamm and M. M. Kozlov, *Eur. Phys. J. E:Soft Matter Biol. Phys.*, 2000, **3**, 323–335.
42 E. R. May, A. Narang and D. I. Kopelevich, *Phys. Rev. E:Stat., Nonlinear, Soft Matter Phys.*, 2007, **76**, 021913.
43 A. Harasima, in *Advances in Chemical Physics*, 1957, pp. 203–237, DOI: 10.1002/9780470143476.ch7.
44 J. H. Irving and J. G. Kirkwood, *J. Chem. Phys.*, 1950, **18**, 817–829.
45 H. I. Ingolfsson, M. N. Melo, F. J. van Eerden, C. Arnarez, C. A. Lopez, T. A. Wassenaar, X. Periole, A. H. de Vries, D. P. Tieleman and S. J. Marrink, *J. Am. Chem. Soc.*, 2014, **136**, 14554–14559.
46 R. Vacha, M. L. Berkowitz and P. Jungwirth, *Biophys. J.*, 2009, **96**, 4493–4501.
47 M. Doktorova and H. Weinstein, *Biophys. J.*, 2018, **115**, 1638–1643.
48 S. Park, W. Im and R. W. Pastor, *Biophys. J.*, 2021, **120**, 5041–5059.

49 B. A. Camley, M. G. Lerner, R. W. Pastor and F. L. Brown, *J. Chem. Phys.*, 2015, **143**, 243113.

50 G. Brannigan and F. L. Brown, *Biophys. J.*, 2006, **90**, 1501–1520.

51 M. S. Jablin, K. Akabori and J. F. Nagle, *Phys. Rev. Lett.*, 2014, **113**, 248102.

52 A. Hossein and M. Deserno, *J. Chem. Phys.*, 2021, **154**, 014704.

53 M. S. Miettinen and R. Lipowsky, *Nano Lett.*, 2019, **19**, 5011–5016.

54 M. Varma and M. Deserno, *Biophys. J.*, 2022, **121**, 4001–4018.

55 T. L. Steck and Y. Lange, *Traffic*, 2018, **19**, 750–760.

56 D. W. Allender, A. J. Sodt and M. Schick, *Biophys. J.*, 2019, **116**, 2356–2366.

57 M. Doktorova, J. L. Symons, X. Zhang, H. Y. Wang, J. Schlegel, J. H. Lorent, F. A. Heberle, E. Sezgin, E. Lyman, K. R. Levental and I. Levental, *bioRxiv*, 2024, preprint, DOI: 10.1101/2023.07.30.551157.

58 H. M. McConnell and A. Radhakrishnan, *Biochim. Biophys. Acta, Biomembr.*, 2003, **1610**, 159–173.

59 T. Baumgart, A. T. Hammond, P. Sengupta, S. T. Hess, D. A. Holowka, B. A. Baird and W. W. Webb, *Proc. Natl. Acad. Sci. U. S. A.*, 2007, **104**, 3165–3170.

Faraday Discussions

ROYAL SOCIETY OF CHEMISTRY

PAPER

The influence of higher order geometric terms on the asymmetry and dynamics of membranes†

Jan Magnus Sischka,[a] Ingo Nitschke[a] and Axel Voigt*[abc]

Received 18th December 2024, Accepted 31st January 2025
DOI: 10.1039/d4fd00202d

We consider membranes as fluid deformable surfaces and allow for higher order geometric terms in the bending energy related to the Gaussian curvature squared and the mean curvature minus the spontaneous curvature to the fourth power. The evolution equations are derived and numerically solved using surface finite elements. The two higher order geometric terms have different effects. While the Gaussian curvature squared term has a tendency to stabilize tubes and enhance the evolution towards equilibrium shapes, thereby facilitating rapid shape changes, the mean curvature minus the spontaneous curvature to the fourth power destabilizes tubes and leads to qualitatively different equilibrium shapes but also enhances the evolution. This is demonstrated in axisymmetric settings and fully three-dimensional simulations. We therefore postulate that not only surface viscosity but also higher order geometric terms in the bending energy contribute to rapid shape changes which are relevant for morphological changes of cells.

1 Introduction

Membranes are ubiquitous and essential in biology; they compartmentalize biomaterials, separate the cell from its exterior and organelles from the cytoplasm, dynamically remodel and change conformation. Geometric properties of the membrane have been identified as key players for such processes.[1–3] As the typical thickness of a membrane is orders of magnitude smaller than its lateral extension, treating the membrane as a two-dimensional surface embedded in a three-dimensional space is a reasonable approximation. This separation of length scales allows for a mesoscopic modeling where details related to membrane molecular structure are considered in effective material parameters and geometric quantities and led to the success of the classical

[a]Faculty Mathematics, TU Dresden, 01062 Dresden, Germany. E-mail: axel.voigt@tu-dresden.de
[b]Center for Systems Biology Dresden (CSBD), Pfotenhauerstr. 108, 01307 Dresden, Germany
[c]Cluster of Excellence, Physics of Life, TU Dresden, Arnoldstr. 18, 01307 Dresden, Germany

† Electronic supplementary information (ESI) available. See DOI: https://doi.org/10.1039/d4fd00202d

Canham–Helfrich model,[4,5] which builds on a bending energy $\mathcal{F}_{\mathrm{BE}}(\mathcal{H},\mathcal{K}) = \int_{\mathcal{S}} k_{2,0}(\mathcal{H} - \mathcal{H}_0)^2 + k_{2,1}\mathcal{K}\, d\mathcal{S}$ with mean curvature $\mathcal{H}$, Gaussian curvature $\mathcal{K}$, bending rigidity parameters $k_{2,0}$ and $k_{2,1}$, a spontaneous curvature $\mathcal{H}_0$ and additional (local or global) area and volume constraints. For definitions see Section 2.1. Assuming constant values of $k_{2,1}$, the second term reduces to a topological measure and thus a constant as long as the topology does not change. We will therefore neglect this term.

Equilibrium shapes, resulting from minimizing the bending energy $\mathcal{F}_{\mathrm{BE}}$, have been extensively studied, see ref. 6 for a review. For $\mathcal{H}_0 = 0$ and wide ranges of the reduced volume, which is the ratio of the volume and the volume of an equivalent sphere with the same area, they are dominated by prolate and oblate shapes.[7] The spontaneous curvature $\mathcal{H}_0$, which accounts for the asymmetry of the membrane, is able to modify these shapes.[7] A wealth of studies also aims to produce tubes as minimizing shapes. Tubes are ubiquitous in membranes and play crucial roles in trafficking, ion transport, and cellular motility. For idealized situations this is rather simple as $\mathcal{F}_{\mathrm{BE}} = 0$ if $\mathcal{H} = \mathcal{H}_0$, which is achieved for an infinite tube of radius r and $\mathcal{H}_0 = -1/r$. However, this solution is not unique, a sphere of radius $2r$ also leads to $\mathcal{F}_{\mathrm{BE}} = 0$. More realistic situations with finite volume and area require further considerations, *e.g.*, introducing a spontaneous curvature deviator.[8–10] Also, various ideas have been proposed to consider higher order geometric terms in the bending energy $\mathcal{F}_{\mathrm{BE}}$ to enforce the stability of tubes.[2,11–14] For example, fourth order terms proportional to $\mathcal{K}^2$ seem plausible as $\mathcal{K} = 0$ for tubes.

Most of these studies only focus on equilibrium shapes, comparing the bending energy $\mathcal{F}_{\mathrm{BE}}$ of different configurations and addressing their stability. But the dynamic evolution of the membrane and associated shape changes are also of interest, *e.g.* in the context of the formation of bulges with pinch-offs, which can be associated with endocytosis and exocytosis[15–17] or shape changes associated with cell motility.[18–20] All these processes require additional phenomena, which are not considered in the Canham–Helfrich model. But there is growing interest in the role of membranes in these processes. One striking example where membranes at least participate are frequently forming and retracting filopodia.[21] This process requires rapid shape changes in cells and according to ref. 22 the membrane and the underlying cortex act as an integrated system to globally coordinate such changes in cell shape. To facilitate these rapid morphological changes, cells maintain an excess of membrane that is organized in membrane reservoirs and is available to the cell on the order of seconds.[23] To understand such processes thus not only requires unveiling the secretes of equilibrium shapes but also to consider the flow of membranes which facilitates the rapid shape changes. To model such processes is a different story and even if various models for the cellular cortex exist[24–27] and also first attempts to couple them with membrane models to form the mentioned integrated system have been proposed,[22,28] we here refrain from such couplings and only aim first for a minimal model of the membrane alone, which facilitates rapid shape changes.

Surface viscosity has been identified as a key player[29] and considering membranes as fluid deformable surfaces[30–33] opened new perspectives on the description of the dynamics of membranes. Fluid deformable surfaces can be viewed as two-dimensional viscous fluids with bending elasticity. Due to this solid–fluid duality, any shape change contributes to tangential flow, and *vice versa*, any tangential flow on a curved surface induces shape deformations. This tight coupling between shape and flow makes curvature a natural element of the governing equations. As demonstrated by numerical studies of the equations for fluid deformable surfaces, surface hydrodynamics can significantly speed up the evolution[34,35] and can enhance bulging and furrow formation in membranes.[17]

Combined with higher order geometric terms in the bending energy, models for fluid deformable surfaces has the potential to enable the rapid shape changes in the release and formation of membrane reservoirs and the formation and retraction of filopodia. We computationally explore the effect of these terms on the equilibrium shapes and the dynamics to reach them and analyze their impact to facilitate rapid shape changes. Any coupling with the cortex,[22] interaction with proteins that induce curvature[36] or to forces exerted on the membrane[37,38] are not considered. We also neglect any interaction with the surrounding bulk phases. This results from the theoretical interest in exploring the membrane properties without any additional influence and the limit of a large Saffman–Delbrück number. This number describes the relation between the viscosities of the membrane and the typically less viscous bulk fluid and if large allows the decoupling of surface and bulk flows.[39]

The rest of the paper is structured as follows. In Section 2 we introduce the full model and briefly mention the considered numerical approach. More details on the derivation of the model and on the numerical approach including convergence studies are provided in Appendices (Model derivation, Numerical method and Validation). Computational results are described in Section 3. They contain axisymmetric and full three-dimensional computations addressing the dynamic evolution and the equilibrium shapes. In Section 4 we draw conclusions and mention possible directions to extend the described model to enable simulations of rapid morphological changes of cells.

2 Model

2.1 Notation

The considered mesoscale model requires basic notation from differential geometry and geometric partial differential equations. Besides classical mathematical text books in these fields we refer to ref. 2 for an introduction in the context of membranes. We follow the same notation as in ref. 17 which is here repeated for convenience. We consider a time dependent smooth and oriented surface $\mathcal{S} = \mathcal{S}(t)$ without boundary, embedded in $\mathbb{R}^3$. The enclosed volume is denoted by $\Omega = \Omega(t)$. We denote by $\boldsymbol{\nu}$ the outward pointing surface normal, the surface projection is $\boldsymbol{P} = \boldsymbol{I} - \boldsymbol{\nu} \otimes \boldsymbol{\nu}$, with $\boldsymbol{I}$ the identity matrix, the shape operator is $\mathcal{B} = -\nabla_{\boldsymbol{P}} \boldsymbol{\nu}$, the mean curvature $\mathcal{H} = \text{tr}\mathcal{B}$, and the Gaussian curvature $\mathcal{K} = \frac{1}{2}(\mathcal{H}^2 - \|\mathcal{B}\|^2)$. We consider time-dependent Euclidean-based n-tensor fields in $T^n\mathbb{R}^3|_{\mathcal{S}}$. We call $T^0\mathbb{R}^3|_{\mathcal{S}} = T^0\mathcal{S}$ the space of scalar fields, $T^1\mathbb{R}^3|_{\mathcal{S}} = T\mathbb{R}^3|_{\mathcal{S}}$ the space of vector fields, and $T^2\mathbb{R}^3|_{\mathcal{S}}$ the space of 2-tensor fields. Important subtensor fields are tangential n-tensor fields in $T^n\mathcal{S} \leq T^n\mathbb{R}^3|_{\mathcal{S}}$. Let $p \in T^0\mathcal{S}$ be a continuously differentiable scalar field, $\boldsymbol{u} \in T\mathbb{R}^3|_{\mathcal{S}}$ a continuously differentiable $\mathbb{R}^3$-vector field, and $\boldsymbol{\sigma} \in T^2\mathbb{R}^3|_{\mathcal{S}}$ a continuously differentiable $\mathbb{R}^{3\times3}$-tensor field defined on $\mathcal{S}$. We define the different surface gradients by $\nabla_{\boldsymbol{P}} p = \boldsymbol{P}\nabla p^e$, $\nabla_{\boldsymbol{P}} \boldsymbol{u} = \boldsymbol{P}\nabla \boldsymbol{u}^e \boldsymbol{P}$ and $\nabla_C \boldsymbol{\sigma} = \nabla \boldsymbol{\sigma}^e \boldsymbol{P}$, where p^e, $\boldsymbol{u}^e$ and $\boldsymbol{\sigma}^e$ are arbitrary smooth extensions of p, $\boldsymbol{u}$ and $\boldsymbol{\sigma}$ in the normal direction and ∇ is the gradient of the embedding space $\mathbb{R}^3$. The corresponding divergence operators for a vector field $\boldsymbol{u}$ and a tensor field $\boldsymbol{\sigma}$ are $\mathrm{div}_{\boldsymbol{P}}\, \boldsymbol{u} = \mathrm{tr}(\nabla_{\boldsymbol{P}}\, \boldsymbol{u})$ and $\mathrm{div}_C\, (\boldsymbol{\sigma P}) = \mathrm{tr}\nabla_C\, (\boldsymbol{\sigma P})$, where tr is the trace operator. The relations to the covariant derivatives $\nabla_{\mathcal{S}}$ and the covariant divergence $\mathrm{div}_{\mathcal{S}}$ on $\mathcal{S}$, with $\Delta_{\mathcal{S}} = \mathrm{div}_{\mathcal{S}} \cdot \nabla_{\mathcal{S}}$ the Laplace–Beltrami operator, read $\nabla_{\boldsymbol{P}}\, p = \nabla_{\mathcal{S}}\, p$ and $\mathrm{div}_{\boldsymbol{P}}\, \boldsymbol{u} = \mathrm{div}_{\mathcal{S}}(\boldsymbol{Pu}) - (\boldsymbol{u}\cdot\boldsymbol{\nu})\mathcal{H}$, respectively.

2.2 Governing equations

The material velocity $\boldsymbol{u} \in T\mathbb{R}^3|_{\mathcal{S}}$ can be decomposed into $\boldsymbol{u} = u_N\boldsymbol{\nu} + \boldsymbol{u}_T$, with $u_N = \boldsymbol{u}\cdot\boldsymbol{\nu}$ and $\boldsymbol{u}_T = \boldsymbol{Pu}$, the normal and the tangential part, respectively. The pressure $p \in T^0\mathcal{S}$ serves as Lagrange multiplier for the inextensibility constraint. The governing equations for these unknowns read

$$\partial_t \boldsymbol{u} + \nabla_w \boldsymbol{u} = -\nabla_{\mathcal{S}} p - p\mathcal{H}\,\boldsymbol{\nu} + \frac{2}{\mathrm{Re}}\mathrm{div}_C \boldsymbol{\sigma} - \gamma \boldsymbol{u} + \boldsymbol{b} - \lambda \boldsymbol{\nu} \tag{1}$$

$$\mathrm{div}_{\boldsymbol{P}} \boldsymbol{u} = 0 \tag{2}$$

$$\int_{\mathcal{S}} \boldsymbol{u}\cdot\boldsymbol{\nu}\, d\mathcal{S} = 0, \tag{3}$$

where $[\nabla_{\mathbf{w}}\boldsymbol{u}]_i = (\nabla_{\mathcal{S}}\boldsymbol{u}_i, \mathbf{w})$, $i = 1, 2, 3$, with $\mathbf{w} = \boldsymbol{u} - \partial_t \mathbf{X}$ is the relative velocity and $\mathbf{X}$ a parameterisation of $\mathcal{S}$, $\boldsymbol{\sigma}(\boldsymbol{u}) = \frac{1}{2}(\nabla_{\boldsymbol{P}}\boldsymbol{u} + (\nabla_{\boldsymbol{P}}\boldsymbol{u})^{\mathrm{T}}) \in T^2\mathbb{R}^3\big|_{\mathcal{S}}$ is the rate of deformation tensor, $\mathrm{Re} > 0$ is the Reynolds number, $\gamma \geq 0$ is a friction coefficient, $\lambda \in \mathbb{R}$ is a Lagrange multiplier to ensure a constant enclosed volume, and $\mathbf{b}$ denotes a bending force, defined by

$$\begin{aligned}\boldsymbol{b} = &-2k_{2,0}\left(\Delta_{\mathcal{S}}\mathcal{H} + (\mathcal{H} - \mathcal{H}_0)\left(\|\mathcal{B}\|^2 - \frac{1}{2}\mathcal{H}(\mathcal{H} - \mathcal{H}_0)\right)\right)\boldsymbol{\nu}\\ &-12k_{4,0}\mathrm{div}_{\mathcal{S}}\left((\mathcal{H} - \mathcal{H}_0)^2\nabla_{\mathcal{S}}\mathcal{H}\right)\boldsymbol{\nu}\\ &-k_{4,0}\left((\mathcal{H} - \mathcal{H}_0)^3((3\mathcal{H} + \mathcal{H}_0)\mathcal{H} - 8\mathcal{K})\right)\boldsymbol{\nu}\\ &-k_{4,2}\left(2\mathrm{div}_{\mathcal{S}}((\mathcal{H}\boldsymbol{P} - \mathcal{B})\nabla_{\mathcal{S}}\mathcal{K}) + \mathcal{H}\mathcal{K}^2\right)\boldsymbol{\nu},\end{aligned} \tag{4}$$

where $\mathcal{H}_0$ is a spontaneous curvature and $k_{2,0}, k_{4,0}, k_{4,2} \in \mathbb{R}$ are bending rigidity parameters. The system of equations considers a model for fluid deformable surfaces. Such models consist of incompressible surface Navier–Stokes equations with bending forces and a constraint on the enclosed volume. The highly nonlinear model accounts for the tight interplay between surface evolution, surface curvature and surface hydrodynamics and allows for the modelling of membranes with surface viscosity. For derivations of the model (with $k_{4,0} = k_{4,2} = 0$) we refer to ref. 17, 31, 34, 40 and 41. They consider different principles and build on a nonlinear Onsager formalism,[31] thin film limits from three-dimensional models[34,40] and a Lagrange–D'Alembert approach.[17,41] For a comparison of derivations for $\boldsymbol{b} = 0$ and without the constraint on the enclosed volume we refer to ref. 42–44.

The mentioned solid–fluid duality of these models, which leads to a tight coupling between shape and flow, and lets any shape change contribute to tangential flow, and *vice versa*, can best be seen by rewriting eqn (1)–(3) as coupled equations for $u_N = \boldsymbol{u}\cdot\boldsymbol{\nu}$ and $\boldsymbol{u}_T = \boldsymbol{P}\boldsymbol{u}$.[34,42,43] One out of several of the resulting coupling terms results from eqn (2), which leads to $\mathrm{div}_{\boldsymbol{P}}\boldsymbol{u} = u_N\mathscr{H} + \mathrm{div}_{\boldsymbol{P}}\boldsymbol{u}_T = 0$ and explicitly demonstrates the relation between the mean curvature and the tangential velocity.

Considering the overdamped limit, formally letting $\gamma \to \infty$, leads to the classical dynamic equations

$$\boldsymbol{u} = -\nabla_{\mathscr{S}} p - p\mathscr{H}\boldsymbol{\nu} + \boldsymbol{b} - \lambda\boldsymbol{\nu} \tag{5}$$

$$\mathrm{div}_{\boldsymbol{P}}\boldsymbol{u} = 0 \tag{6}$$

$$\int_{\mathscr{S}} \boldsymbol{u}\cdot\boldsymbol{\nu} d\mathscr{S} = 0, \tag{7}$$

for an inextensible membrane with constant volume. Using (5) in (6) provides the equation for the Lagrange multiplier for the inextensibility constraint $-\Delta_{\mathscr{S}} p + p\mathscr{H}^2 + \lambda = \boldsymbol{b}\cdot\boldsymbol{\nu}\mathscr{H}$ and corresponds to previous models, if $\boldsymbol{b}$ only contains second order geometric terms.[45–49] Further relaxing the constraint on inextensibility leads to the classical Canham–Helfrich models with area and volume constraints.[48]

In contrast to previous approaches for fluid deformable surfaces[17,31,34,35,50] the bending force $\boldsymbol{b}$ also contains higher order geometric terms.

2.3 Bending forces

The bending force eqn (4) results from the bending energy

$$\mathscr{F}_{\mathrm{BE}}(\mathscr{H},\mathscr{K}) = \int_{\mathscr{S}} f_{\mathrm{BE}}(\mathscr{H},\mathscr{K}) d\mathscr{S} \tag{8}$$

with bending energy density f_{BE}. Formally we can derive f_{BE} *via* Taylor expansion at the spontaneous curvature $\mathscr{H}_0$ leading to

$$\begin{aligned} f_{\mathrm{BE}}(\mathscr{H},\mathscr{K}) &= \sum_{n=0}^{N}\sum_{\alpha=0}^{\lfloor \frac{n}{2} \rfloor} k_{n,\alpha}(\mathscr{H}-\mathscr{H}_0)^{n-2\alpha}\mathscr{K}^{\alpha} \\ &= \underbrace{k_{0,0}}_{n=0} \underbrace{+k_{1,0}(\mathscr{H}-\mathscr{H}_0)}_{n=1} \underbrace{+k_{2,0}(\mathscr{H}-\mathscr{H}_0)^2 + k_{2,1}\mathscr{K}}_{n=2} \\ &\quad \underbrace{+k_{3,0}(\mathscr{H}-\mathscr{H}_0)^3 + k_{3,1}(\mathscr{H}-\mathscr{H}_0)\mathscr{K}}_{n=3} \\ &\quad \underbrace{+k_{4,0}(\mathscr{H}-\mathscr{H}_0)^4 + k_{4,1}(\mathscr{H}-\mathscr{H}_0)^2\mathscr{K} + k_{4,2}\mathscr{K}^2}_{n=4} + \ldots \end{aligned}$$

in terms of geometric orders $n \leq N \in \mathbb{N}$ for different bending rigidity parameters $k_{n,\alpha} \in \mathbb{R}$. This expression corresponds to the generalized form of the classical Canham–Helfrich energy ($n = 2$) introduced in ref. 11 and also considered in ref. 2, 12 and 13 if gradient terms are neglected. It can be simplified assuming certain properties of the bending energy: as we are interested in variations of the energy, we can omit the constant contribution $k_{0,0}$. We will further not allow for topological changes, thus, we can omit the $k_{2,1}\mathcal{K}$ term, applying Gauss–Bonnet's theorem. Furthermore, we only allow for terms which guarantee boundedness from below. This excludes all odd geometric orders as well as $k_{4,1}$. Considering these points and only contributions up to geometric order $N = 4$ leads to the following bending energy density

$$f_{\mathrm{BE}}(\mathcal{H},\mathcal{K}) = k_{2,0}(\mathcal{H} - \mathcal{H}_0)^2 + k_{4,0}(\mathcal{H} - \mathcal{H}_0)^4 + k_{4,2}\mathcal{K}^2,$$

which has been considered in ref. 14 to study the stability of tubular shapes considering $k_{2,0} > 0$, $k_{4,0} = 0$ and $k_{4,2} \gtrless 0$ as parameters to stabilize cylindrical shapes ($k_{4,2} > 0$) or to enforce pearling ($k_{4,2} < 0$). The same form is also considered in ref. 12 arguing that a cylinder with radius R is stable if $k_{2,0} < 0$ and $k_{4,0} = -\frac{R^2}{2}k_{2,0}$ with $k_{4,2}$ not determined. We here restrict the parameter space to $k_{2,0} > 0$, $k_{4,0} \geq 0$ and $k_{4,2} \geq 0$. We would like to remark that more general parameter combinations are possible, still leading to well-posed bending energies and stable solutions. However, $\mathcal{K} = 0$ is a property of a tube and thus $k_{4,2}\mathcal{K}^2$ seems to be the most plausible higher order extension in the bending energy to stabilize tubular structures. $k_{4,0}(\mathcal{H} - \mathcal{H}_0)^4$ is considered for completeness. In any case the bending force $\boldsymbol{b}$ is derived as the negative of the variational derivative of the bending energy (8)

$$\boldsymbol{b} = -\frac{\delta \mathcal{F}_{\mathrm{BE}}}{\delta \boldsymbol{X}}.$$

A detailed derivation of the bending force is done in the Appendix: Model derivation.

2.4 Numerical approach

The numerical approach extends the approach used in ref. [17] and [35] which is based on surface finite elements[51,52] and builds on a Taylor–Hood element for the surface Navier–Stokes equations, higher order surface parametrizations, appropriate approximations of the geometric quantities, mesh redistribution, a semi-implicit discretization in time and an iterative approach to deal with the non-local constraint on the enclosed volume. Additional challenges emerge from the higher order geometric terms. In the Appendix: Numerical method we provide a detailed description and in Appendix: Validation convergence studies for these terms. The implementation is done in DUNE/AMDiS.[53,54] Throughout the paper we consider Re = 1.0 and $\gamma = 0$ and only vary $\mathcal{H}_0$, $k_{2,0}$, $k_{4,0}$ and $k_{4,2}$. Numerical parameters, such as mesh size h and time step τ are chosen to resolve the highest curvature values and to meet stability constraints, following ref. 35.

3 Results

3.1 Axisymmetric simulation without surface viscosity

We begin by describing the membrane using cylindrical coordinates (r, θ, z) and consider a rotational symmetric tube with periodic boundary conditions. With

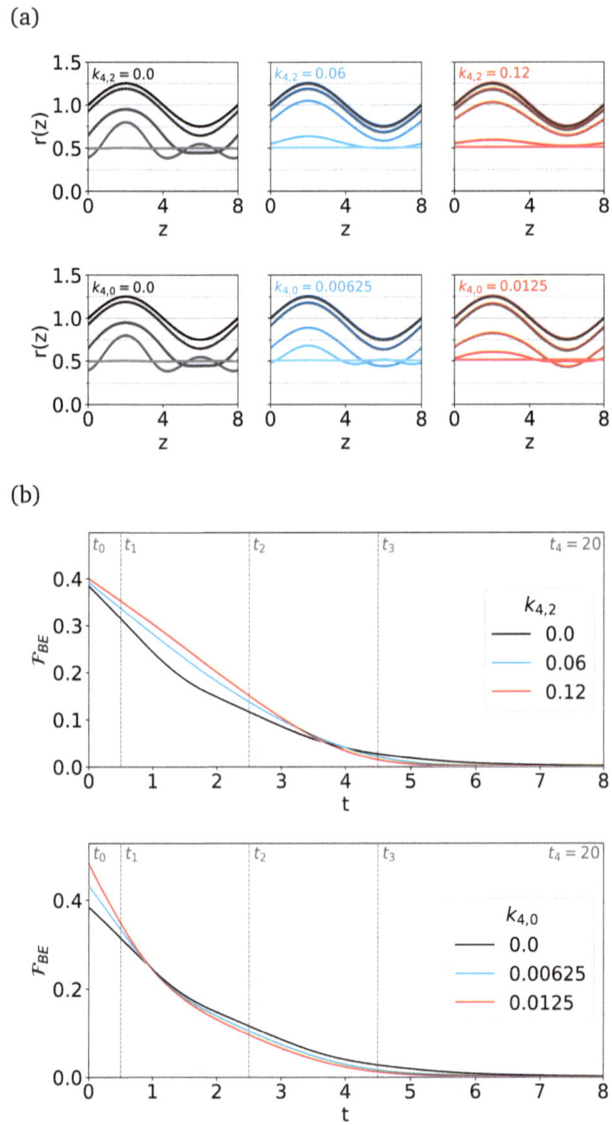

Fig. 1 (a) Time evolution of axisymmetric simulations for different parameters for $k_{4,2}$ and $k_{4,0}$ starting from the same periodic solution $r_0(z) = \frac{1}{4}\sin\left(\frac{\pi}{4}z\right) + 1$ at t_0. The evolution goes from dark to light converging to a tube with $r(z) = 0.5$. The depicted time instances are marked in (b). (b) Corresponding time evolution of the bending energy for different values of $k_{4,2}$ and $k_{4,0}$. Parameters are $k_{2,0} = 0.05$ and $\mathscr{H}_0 = -2$.

$$\mathcal{H} = \frac{1}{\sqrt{1+(\partial_z r)^2}}\left(\frac{\partial_{zz} r}{1+(\partial_z r)^2} - \frac{1}{r}\right), \quad \mathcal{K} = -\frac{\partial_{zz} r}{r\left(1+(\partial_z r)^2\right)^2} \tag{9}$$

the bending energy density $f_{\mathrm{BE}}(\mathcal{H},\mathcal{K})$ can be reformulated and considered as a function of $r(z, t)$, the radial distance of the axisymmetric membrane from the cylindrical symmetry axis, where z measures the coordinate along that axis and t is time. For the evolution we consider the corresponding equations to eqn (5)–(7) but drop the constraints on inextensibility and volume. The resulting equation to solve reads

$$\frac{1}{\sqrt{1+(\partial_z r)^2}}\partial_t r = u_N = -\frac{\delta \mathcal{F}_{\mathrm{BE}}}{\delta \boldsymbol{X}}\cdot \boldsymbol{\nu} = b, \tag{10}$$

with $b = \boldsymbol{b}\cdot\boldsymbol{\nu}$. While lengthy if fully written down, the resulting model can be solved using standard approaches. We again use finite elements in space, a semi-implicit discretization in time, smoothing of the geometric properties and consider DUNE/AMDiS[53,54] for the realization. Fig. 1 shows the time evolution of a periodically perturbed tube for different values of $k_{4,2}$ and $k_{4,0}$ together with the evolution of the bending energy $\mathcal{F}_{\mathrm{BE}}$ over time. For $k_{4,2} = 0$ and $k_{4,0} = 0$ this is a well studied problem of the stability of a tube.[55–57] The parameters are chosen to remain within the stability region. The results clearly indicate a stronger damping of the perturbations for increasing values of $k_{4,2}$ and $k_{4,0}$. However, the behaviour slightly differs with respect to the damping of the perturbations and adjustment of $r(z)$. This difference is also seen in the decrease of the bending energy.

Results of a linear stability analysis[58] for the case of $k_{4,2} = 0$ and $k_{4,0} = 0$ but considering surface viscosity using a Stokes approximation of eqn (1) indicate a similar stability region as in the overdamped limit and we assume this holds also for $k_{4,2} > 0$ and $k_{4,0} > 0$ and the full problem.

3.2 Equilibrium shape of tubular cell

The shape of a tubular cell can be approximated by a cylinder with hemispherical caps, as shown in Fig. 2. For this simple shape, we can compute the bending energy as

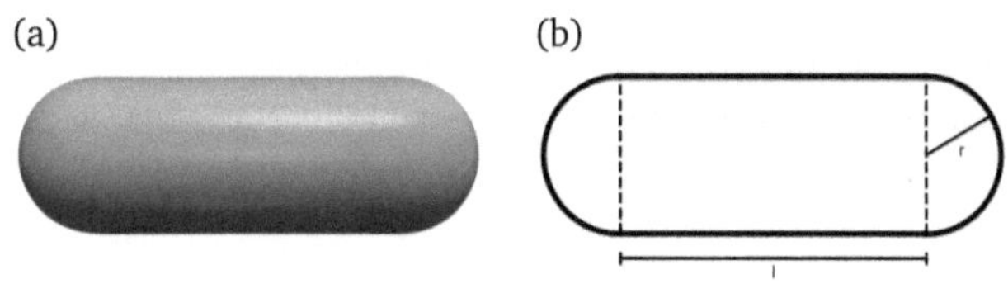

Fig. 2 (a) As initial surface we consider a cylinder with hemispherical caps. (b) The geometry is determined by the length of the cylinder l and the radius r of the cylinder and hemispherical caps.

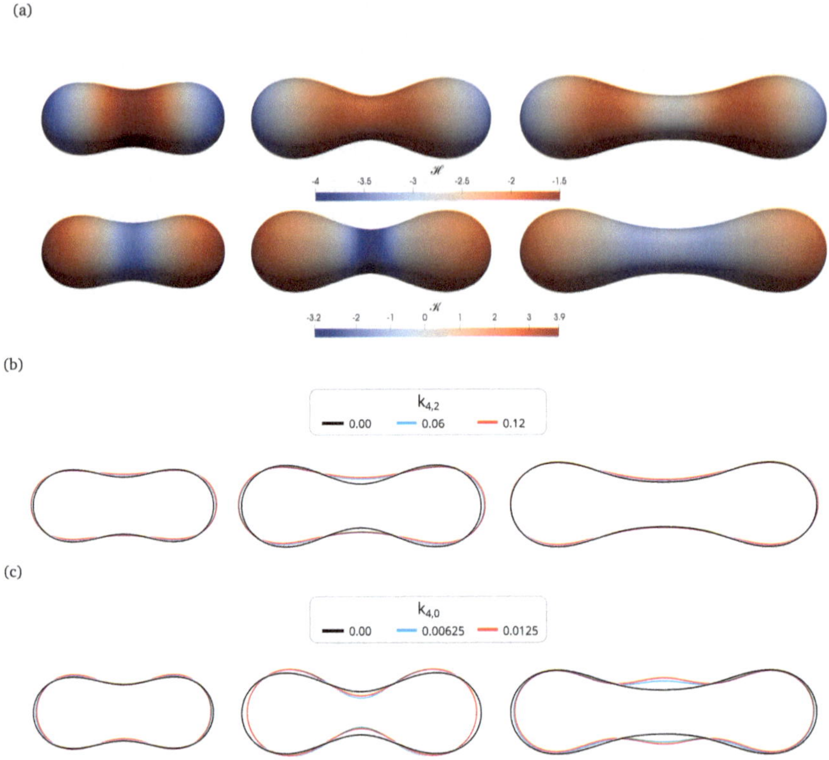

Fig. 3 (a) Equilibrium shapes for $k_{2,0} = 0.125$, $\mathcal{H}_0 = -2$ and $k_{4,2} = 0$. As initial condition we consider a cylinder of radius $r = 0.5$ with hemispherical caps and vary the length l of the cylindrical part. From left to right the lengths are $l = 2$, 3 and 4. The surfaces are colored by mean curvature $\mathcal{H}$ (upper row) and Gaussian curvature $\mathcal{K}$ (lower row). (b) Equatorial cuts through the equilibrium shapes for different values of $k_{4,2}$ and l, keeping $k_{2,0} = 0.125$ and $\mathcal{H}_0 = -2$. (c) Equatorial cuts through the equilibrium shapes for different values of $k_{4,0}$ and l, keeping $k_{2,0} = 0.125$ and $\mathcal{H}_0 = -2$.

$$\mathcal{F}_{\mathrm{BE}} = k_{2,0}\left(2\pi r l\left(-\frac{1}{r} - \mathcal{H}_0\right)^2 + 4\pi r^2\left(-\frac{2}{r} - \mathcal{H}_0\right)^2\right) + k_{4,0}\left(2\pi r l\left(-\frac{1}{r} - \mathcal{H}_0\right)^4 + 4\pi r^2\left(-\frac{2}{r} - \mathcal{H}_0\right)^4\right) + k_{4,2}\frac{4\pi}{r^2}$$

where r is the radius and l the length of the cylinder. For $\mathcal{H}_0 = -\frac{1}{r}$, the energy of the cylindrical part vanishes and the energy simplifies to

$$\mathcal{F}_{\mathrm{BE}} = 4\pi\left(k_{2,0} + \frac{k_{4,0}}{r^2} + \frac{k_{4,2}}{r^2}\right).$$

Note that this energy is independent of the length of the cylinder. We consider this shape as the initial configuration and the energy as a quantity for comparison.

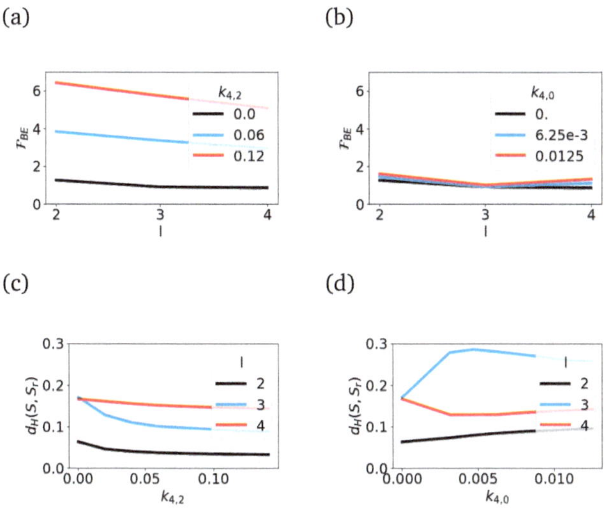

Fig. 4 (a) and (b) Bending energy of the final shape for fixed values of $k_{4,2}$ (a) and $k_{4,0}$ (b) for different lengths l. (c) and (d) Hausdorff distance between the equilibrium shape and the initial shape for different lengths $l = 2$, 3 and 4 as a function of $k_{4,2}$ (c) and $k_{4,0}$ (d).

Fig. 3 shows the final configurations achieved by solving the full problem, eqn (1)–(3), for three different values l. In all cases, the shape deviates from the initial configuration. For $k_{4,2} = 0$ and $k_{4,0} = 0$ the resulting shapes correspond to the equilibrium prolate shapes in ref. 7. However, for $k_{4,2} > 0$ and $k_{4,0} > 0$, these shapes deviate. For $k_{4,2} > 0$ the bending energy is reduced by increasing l and the Hausdorff distance $d_H(\mathcal{S},\mathcal{S}_0)$ to the idealized initial shape of a cylinder with hemispherical caps $\mathcal{S}_0$ is reduced by increasing $k_{4,2}$, see Fig. 4. The Hausdorff distance is a measure of the distance between sets of points, in our case the discretization points of the equilibrium shape and the initial shape. For a mathematical definition and implementation issues see Appendix: Hausdorff distance. The observed trends are far from being general. This can also not be expected due to the highly nonlinear coupling of the geometric terms and the considered constraints on volume and inextensibility. However, they confirm the intuition of the potential impact of $k_{4,2}\mathcal{K}^2$ in the bending energy on the emerging equilibrium shapes. The effect of $k_{4,0} > 0$ differs. The increased tendency to enforce $\mathcal{H} = \mathcal{H}_0$ leads to undulations with a wavelength related to $-2/\mathcal{H}_0$. How well this can be achieved strongly depends on the length l. Therefore also the Hausdorff distance $d_H(\mathcal{S},\mathcal{S}_0)$ to the idealized initial shape of a cylinder with hemispherical caps $\mathcal{S}_0$ does not decrease but leads to a non-monotonic behaviour if considered as a function of l. The dependency of l on the bending energy $\mathcal{F}_{BE}$ is negligible for the considered parameters. To summarize, within the stability region of the classical Canham–Helfrich model, higher order geometric terms related to $k_{4,2}\mathcal{K}^2$ seem to stabilize tubular shapes and terms related to $k_{4,0}(\mathcal{H} - \mathcal{H}_0)^4$ seem to destabilize these shapes.

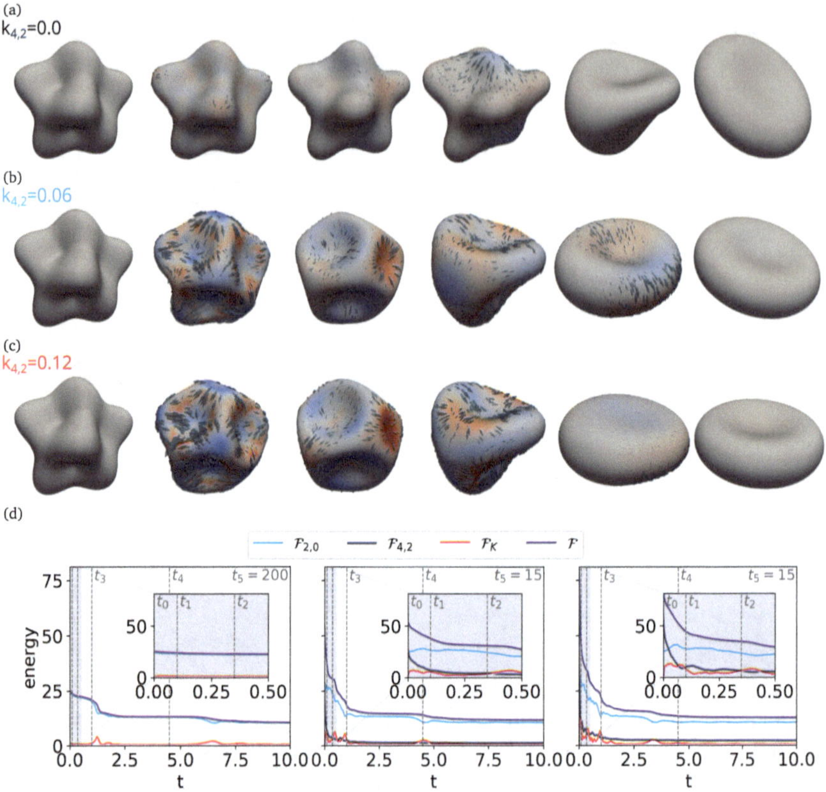

Fig. 5 Evolution of a perturbed sphere considering parameters $k_{2,0} = 0.125$, $\mathscr{H}^0 = 0$ and different values of $k_{4,2}$. (a)–(c) Time instances for the evolution for $k_{4,2} = 0$, 0.06 and 0.12, respectively. The color coding indicates movement in the normal direction (red – movement outwards, blue – movement inwards) and the arrows indicate the tangential velocity. The time evolves from left to right; the time instances except for the last one are equal and are depicted in (d). (d) Evolution of the different energy contributions: $\mathscr{F}^{2,0}$ and $\mathscr{F}^{4,2}$ are the energies linked to the corresponding bending terms, $\mathscr{F}^{\mathrm{K}}$ is the kinetic energy. The bending energy is $\mathscr{F}^{\mathrm{BE}} = \mathscr{F}^{2,0} + \mathscr{F}^{4,2}$ and the total energy $\mathscr{F} = \mathscr{F}^{\mathrm{BE}} + \mathscr{F}^{\mathrm{K}}$. Shown is the evolution for $k_{4,2} = 0$, 0.06 and 0.12 (from left to right). Corresponding videos to the evolution in (a)–(c) are provided in the SI† using a LIC filter for visualization of the tangential flow.

3.3 Dynamic evolution

In order to further explore the influence of the higher order geometric terms related to $k_{4,2}\mathscr{K}^2$ and $k_{4,0}(\mathscr{H} - \mathscr{H}_0)^4$ on the dynamics, we consider an initial surface $\mathscr{S}_0$ as a perturbed unit sphere

$$\mathscr{S}_0 = \left\{1 + r_0 Y_l^m(\phi,\vartheta):\ \phi\in[0,\pi],\ \vartheta\in[-\pi,\pi]\right\},\quad r_0 > 0, \tag{11}$$

$$Y_l^m(\phi,\vartheta) = \sqrt{\frac{2l+1(l-m)!}{4\pi(l+m)!}}P_l^m(\cos\vartheta)e^{im\phi} \tag{12}$$

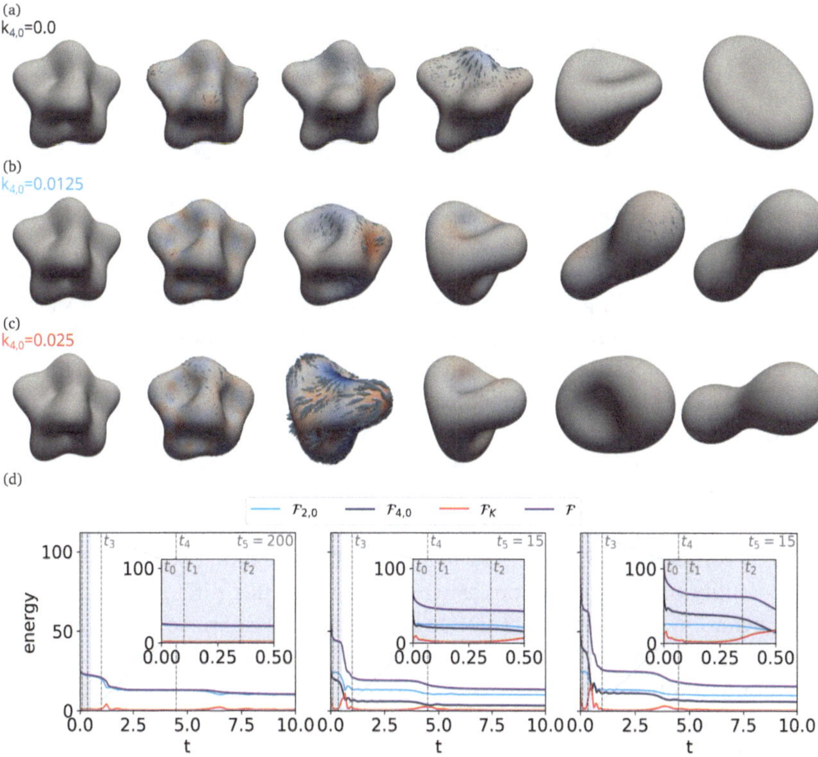

Fig. 6 Evolution of a perturbed sphere considering parameters $k_{2,0} = 0.125$, $\mathscr{H}^0 = 0$ and different values of $k_{4,0}$. (a)–(c) Time instances for the evolution for $k_{4,0} = 0$, 0.00625 and 0.0125, respectively. The color coding indicates movement in the normal direction (red – movement outwards, blue – movement inwards) and the arrows indicate the tangential velocity. The time evolves from left to right; the time instances except for the last one are equal and are depicted in (d). (d) Evolution of the different energy contributions: $\mathscr{F}^{2,0}$ and $\mathscr{F}^{4,0}$ are the energies linked to the corresponding bending terms, $\mathscr{F}^{\mathrm{K}}$ is the kinetic energy. The bending energy is $\mathscr{F}^{\mathrm{BE}} = \mathscr{F}^{2,0} + \mathscr{F}^{4,0}$ and the total energy $\mathscr{F} = \mathscr{F}^{\mathrm{BE}} + \mathscr{F}^{\mathrm{K}}$. Shown is the evolution for $k_{4,0} = 0$, 0.00625 and 0.0125 (from left to right). Corresponding videos to the evolution in (a)–(c) are provided in the SI† using a LIC filter for visualization of the tangential flow.

with spherical harmonics Y_l^m and Legendre polynomials P_l^m and the case $l = 5$, $m = 3$ and $r_0 = 0.5$. The velocity field is initialized with $\boldsymbol{u}_0 = 0$. The initial surface has non-zero mean and Gaussian curvature and is out of equilibrium. It serves as a prototypical example to study the rapid shape changes featuring moderate geometric properties of a cell with membrane reservoirs. The resulting bending force induces shape deformations in the normal direction. However, the curvature terms also induce tangential flows, which also contribute to shape deformations. This coupling between tangential flow and shape deformations is well understood and shown to enhance the evolution towards equilibrium shapes.[35] Here, we explore the evolution for different values of $k_{4,2}$ and $k_{4,0}$, which are $k_{4,2} = 0$, 0.06 and 0.12 and $k_{4,0} = 0$, 0.00625 and 0.0125, respectively. Fig. 5 and 6 show snapshots of the evolutions. The color coding corresponds to shape deformations (red – movement outwards, blue – movement inwards) and the arrows indicate the tangential velocity, where the length scales with the magnitude. Furthermore, the energy contributions are shown over time. Here, the bending energy is split into the different contributions $\mathscr{F}_{\mathrm{BE}} = \mathscr{F}_{2,0} + \mathscr{F}_{4,0} + \mathscr{F}_{4,2}$ and the total energy $\mathscr{F} = \mathscr{F}_{\mathrm{BE}} + \mathscr{F}_{\mathrm{K}}$ is the sum of the bending energy and the kinetic energy $\mathscr{F}_{\mathrm{K}} = \int_{\mathscr{S}} \frac{1}{2} \boldsymbol{u}^2 d\mathscr{S}$. All evolutions converge to an equilibrium shape, which for the cases $k_{4,2} = 0$ and $k_{4,0} = 0$ correspond to the associated Seifert shape,[7] which here corresponds to an oblate. For $k_{4,2} > 0$ the equilibrium shapes only slightly differ. The difference in orientation might result from the different dynamics. However, due to the decoupling from the surrounding bulk phases and the considered parameter $\gamma = 0$, force-free rigid body rotations are also possible.[59] While the equilibrium shapes are similar, significant changes can be observed in the dynamics. The close coupling between the bending energy $\mathscr{F}_{\mathrm{BE}}$ and the kinetic energy $\mathscr{F}_{\mathrm{K}}$ can be observed and related to significant shape changes. But their appearance differs. The plateau in $\mathscr{F}_{\mathrm{BE}} = \mathscr{F}_{2,0}$ for $k_{4,2} = 0$ between $t \approx 2$ and $t \approx 6$ is reduced for $k_{4,2} = 0.06$ and 0.12 and already ends at $t \approx 4.5$ and $t \approx 3$, respectively. It should be noted that the absolute values of the energies cannot be directly compared as $k_{4,2}$ varies by definition. The situation changes for $k_{4,0} > 0$. The equilibrium shapes are prolate shapes. Besides this difference in the long time behavior, the short time evolution also changes. We observe similar behavior as for $k_{4,2} > 0$ with an enhanced influence of the kinetic energy and a faster evolution towards intermediate shapes. The final convergence towards the equilibrium shape at late times probably results form similar values for the local minima on the oblate and prolate branch as also known for the classical Canham–Helfrich model.[6] Again, the absolute values of the energies cannot be directly compared as $k_{4,0}$ varies in the definition.

Qualitatively the higher order geometric terms further enhance the evolution and lead to alternative pathways to dissipate energy. This is most pronounced in the inlets highlighting the initial evolution; the drastic decrease of $\mathscr{F}^{4,2}$ and $\mathscr{F}^{4,0}$ is associated with large fluctuation of $\mathscr{F}^{\mathrm{K}}$. This behavior increases with increasing values of $k_{4,2}$ and $k_{4,0}$. Furthermore, while $\mathscr{F}^{4,2}$ is roughly one order of magnitude smaller than $\mathscr{F}^{2,0}$, the influence on the dynamics is dramatic. This is less pronounced for $\mathscr{F}^{4,0}$, which is at the same order as $\mathscr{F}^{2,0}$. We expect these fast shape changes to be even enhanced for less regular real geometries of membrane reservoirs, due to the appearance of larger curvature gradients.

4 Conclusions

Motivated by rapid shape changes of cells, where an excess of membrane that is organized in membrane reservoirs is made available to the cell on the order of seconds,[23] we formulated a minimal mesoscopic membrane model which helps to facilitate this behavior. This, on the one side, includes an extension of the classical Canham–Helfrich model towards higher order geometric terms, and on the other side, includes the explicit treatment of the fluid properties of the membrane by considering membranes as fluid deformable surfaces. The first aspect adds to the classical bending energy $\mathscr{F}_{2,0} = \int_{\mathscr{S}} k_{2,0}(\mathscr{H} - \mathscr{H}_0)^2 d\mathscr{S}$, terms proportional to the Gaussian curvature squared, $\mathscr{F}_{4,2} = \int_{\mathscr{S}} k_{4,2}\mathscr{K}^2 d\mathscr{S}$, and the mean curvature minus the spontaneous curvature to the fourth power,

$\mathscr{F}_{4,0} = \int_{\mathscr{S}} k_{4,0}(\mathscr{H} - \mathscr{H}_0)^4 d\mathscr{S}$. The effects of these higher order terms are different. While the one related to the Gaussian curvature squared not only helps to damp perturbations of tubes, it also has a tendency to stabilize them, the term related to the mean curvature minus the spontaneous curvature to the fourth power has a tendency to destabilize tubes. Both effects have been considered in idealized rotational symmetric and full three-dimensional situations by analyzing the evolution and the emerging equilibrium shapes. If combined with the second aspect, which takes the surface viscosity of the membrane into account and combines the bending in the normal direction with the properties of an inextensible surface fluid, as a fluid deformable surface, the dynamic drastically changes. Already for the classical bending energy $\mathscr{F}_{2,0} = \int_{\mathscr{S}} k_{2,0}(\mathscr{H} - \mathscr{H}_0)^2 d\mathscr{S}$ an enhanced evolution towards the equilibrium shape has been observed if the effects of surface viscosity are taken into account.[35] With the higher order geometric terms this is further enhanced. The considered numerical experiments for the relaxation of a perturbed sphere showed alternative pathways to dissipate energy and strong tangential flows inducing fast shape changes. However, besides enabling rapid shape changes, the evolution and also the emerging equilibrium shapes qualitatively differ for the considered higher order terms. While the model with the Gaussian curvature squared term also converges to similar oblate-like shapes as the corresponding Seifert shapes[6] for the classical Canham–Helfrich model, the model with the mean curvature minus the spontaneous curvature to the fourth power term converges to prolate-like shapes.

However, the focus of this paper is not to classify the increased phase space of equilibrium shapes resulting from the higher order terms, but to address additional mechanisms which facilitate rapid shape changes. While certainly more research is needed to fully explore the potential of the higher order geometric terms, *e.g.*, with respect to the stability of tubes extending the analysis in ref. 58 the numerical studies already clearly indicate the potential for rapid shape changes. Even if only passive contributions of a homogeneous membrane are considered, and a full model for morphological changes of a cell requires the taking of inhomogeneities, active processes of the underlying cortex, adhesion between the membrane and the cortex and probably even more phenomena into account, the study contributes to identifying underlying general mechanical principles which might help to predict and control the dynamics of cells.[20] Surface viscosity and higher order geometric terms in the bending energy provide mechanical cues and probably support active processes to enable rapid shape changes of cells.

A full model able to address the mentioned example of frequently forming and retracting filopodia and the associated fast shape changes of membrane reservoirs will require the coupling of models for the cellular cortex[24–27] with membrane models of the considered type, which requires additional modeling and numerical efforts. One intermediate step to link this research closer to biology could be efforts to resolve the dynamics of membrane reservoirs and compare them with simulations of the proposed model.

Data availability

Data are available from Zenodo at https://doi.org/10.5281/zenodo.14503545.

Conflicts of interest

There are no conflicts to declare.

Appendices

Model derivation

We here compute the variational derivatives of the higher order geometric terms in the bending energy. We therefore write the bending energy as

$$\mathscr{F}_{\text{BE}} = \int_{\mathscr{S}} k_{2,0} f_{\text{BE}}^{2,0} + k_{4,0} f_{\text{BE}}^{4,0} + k_{4,2} f_{\text{BE}}^{4,2} d\mathscr{S}$$

with bending density components $f_{\text{BE}}^{n,\alpha} = (\mathscr{H} - \mathscr{H}_0)^{n-2\alpha}\mathscr{K}^{\alpha}$. The corresponding forces $\boldsymbol{b}^{n,\alpha} \in T\mathbb{R}^3|_{\mathscr{S}}$ are given by

$$\langle \boldsymbol{b}^{n,\alpha}, \boldsymbol{W} \rangle_{L^2(T\mathbb{R}^3|_{\mathscr{S}})} := -\left\langle \frac{\delta}{\delta \boldsymbol{X}} \int_{\mathscr{S}} f_{\text{BE}}^{n,\alpha} d\mathscr{S}, \boldsymbol{W} \right\rangle_{L^2(T\mathbb{R}^3|_{\mathscr{S}})}$$

for all $\boldsymbol{W} \in T\mathbb{R}^3|_{\mathscr{S}}$. The result for $f_{\text{BE}}^{2,0}$ is well known

$$\boldsymbol{b}^{2,0} = -2\left(\Delta_{\mathscr{S}}\mathscr{H} + (\mathscr{H} - \mathscr{H}_0)\left(\|\mathscr{B}\|^2 - \frac{1}{2}\mathscr{H}(\mathscr{H} - \mathscr{H}_0)\right)\right)\boldsymbol{\nu}.$$

For the remaining terms we use the deformation derivative $\eth_{\boldsymbol{W}}$[60] and obtain the deformation formula

$$\langle \boldsymbol{b}^{n,\alpha}, \boldsymbol{W} \rangle_{L^2(T\mathbb{R}^3|_{\mathscr{S}})} = -\int_{\mathscr{S}} \eth_{\mathbf{W}} f_{\text{BE}}^{n,\alpha} + f_{\text{BE}}^{n,\alpha} \text{div}_{\text{C}} \boldsymbol{W} d\mathscr{S}. \tag{13}$$

Moreover,

$$\boldsymbol{P}(\eth_{\boldsymbol{W}}\mathscr{B})\boldsymbol{P} = \nabla_{\mathscr{S}}(\boldsymbol{\nu}\cdot\nabla_{\text{C}}\boldsymbol{W}) - \mathscr{B}\nabla_{\text{C}}\boldsymbol{W}, \tag{14}$$

$$\eth_{\boldsymbol{W}}\mathscr{H} = \text{tr}(\boldsymbol{P}(\eth_{\boldsymbol{W}}\mathscr{B})\boldsymbol{P}) = \text{div}_{\mathscr{S}}(\boldsymbol{\nu}\cdot\nabla_{\text{C}}\boldsymbol{W}) - \mathscr{B} : \nabla_{\text{C}}\boldsymbol{W} \tag{15}$$

are valid.[60] As a consequence, the deformation formula (13) and integrations by parts result in

$$\begin{aligned}\langle \boldsymbol{b}^{4,0}, \boldsymbol{W} \rangle_{L^2(T\mathbb{R}^3|_{\mathscr{S}})} &= -\int_{\mathscr{S}} 4(\mathscr{H} - \mathscr{H}_0)^3 \eth_{\mathbf{W}}\mathscr{H} + (\mathscr{H} - \mathscr{H}_0)^4 \text{div}_{\text{C}} \boldsymbol{W} d\mathscr{S} \\ &= \int_{\mathscr{S}} 12(\mathscr{H} - \mathscr{H}_0)^2 (\nabla_{S}\mathscr{H})\cdot(\boldsymbol{\nu}\cdot\nabla_{\text{C}}\boldsymbol{W}) \\ &\quad + (\mathscr{H} - \mathscr{H}_0)^3 (4\mathscr{B} - (\mathscr{H} - \mathscr{H}_0)\boldsymbol{P}) : \nabla_{\text{C}}\boldsymbol{W} d\mathscr{S}\end{aligned}$$

Another application of integrations by parts w.r.t. ∇_{C}, yields

$$\boldsymbol{b}^{4,0} = -\text{div}_{\text{C}}(12(\mathscr{H} - \mathscr{H}_0)^2 \boldsymbol{\nu} \otimes \nabla_{\mathscr{S}}\mathscr{H} + (\mathscr{H} - \mathscr{H}_0)^3 (4\mathscr{B} - (\mathscr{H} - \mathscr{H}_0)\mathbf{P}),\Big)$$

where

$$\mathrm{div}_C(\boldsymbol{\sigma P}) = \mathrm{div}_{\mathcal{S}}(\boldsymbol{P\sigma P}) - \boldsymbol{\nu}\cdot\boldsymbol{\sigma}\mathcal{B} + (\mathrm{div}_{\mathcal{S}}(\boldsymbol{\nu}\cdot\boldsymbol{\sigma P}) + \mathcal{B} : \boldsymbol{\sigma})\boldsymbol{\nu} \tag{16}$$

holds for all $\boldsymbol{\sigma} \in T\mathbb{R}^3|_{\mathcal{S}}$.[41,60] Therefore, the tangential part of $\boldsymbol{b}^{4,0}$ cancels out and we obtain

$$\boldsymbol{b}^{4,0} = -\left(12\mathrm{div}_{\mathcal{S}}\left((\mathcal{H}-\mathcal{H}_0)^2\nabla_{\mathcal{S}}\mathcal{H}\right) + (\mathcal{H}-\mathcal{H}_0)^3((3\mathcal{H}+\mathcal{H}_0)\mathcal{H} - 8\mathcal{K})\right)\boldsymbol{\nu}.$$

With (14) and (15), the deformation derivative of $\mathcal{K} = 2(\mathcal{H}^2 - \|\mathcal{B}\|^2)$ reveals

$$\eth_W\mathcal{K} = \mathcal{H}\eth_W\mathcal{H} - \mathcal{B} : \eth_W\mathcal{B} = (\mathcal{H}\boldsymbol{P} - \mathcal{B}) : (\nabla_{\mathcal{S}}(\boldsymbol{\nu}\cdot\nabla_C\boldsymbol{W}) - \mathcal{B}\nabla_C\boldsymbol{W}).$$

As a consequence, the deformation formula (13), integrations by parts, and $\mathcal{B}^2 = \mathcal{H}\mathcal{B} - \mathcal{K}\boldsymbol{P}$, result in

$$\left\langle\boldsymbol{b}^{4,2}, \boldsymbol{W}\right\rangle_{L^2\left(T\mathbb{R}^3|_{\mathcal{S}}\right)} = -\int_{\mathcal{S}} 2\mathcal{K}\eth_{\mathrm{W}}\mathcal{K} + \mathcal{K}^2\mathrm{div}_C\boldsymbol{W}d\mathcal{S}$$

$$= \int_{\mathcal{S}} 2\mathrm{div}_{\mathcal{S}}(\mathcal{K}(\mathcal{H}\mathbf{P} - \mathcal{B}))\cdot(\boldsymbol{\nu}\cdot\nabla_C\boldsymbol{W}) + \mathcal{K}^2\boldsymbol{P} : \nabla_C\boldsymbol{W}d\mathcal{S}.$$

Since $\mathrm{div}_{\mathcal{S}}(\mathcal{H}\boldsymbol{P} - \mathcal{B}) = 0$ holds, integrations by parts yields

$$\boldsymbol{b}^{4,2} = -\mathrm{div}_C\left(2\boldsymbol{\nu}\otimes(\mathcal{H}\boldsymbol{P} - \mathcal{B})\nabla_{\mathcal{S}}\mathcal{K} + \mathcal{K}^2\boldsymbol{P}\right).$$

Using (16), the tangential part of $\boldsymbol{b}^{4,2}$ cancels out and we obtain

$$\boldsymbol{b}^{4,2} = -(2\mathrm{div}_{\mathcal{S}}((\mathcal{H}\boldsymbol{P} - \mathcal{B})\nabla_{\mathcal{S}}\mathcal{K}) + \mathcal{H}\mathcal{K}^2)\boldsymbol{\nu}.$$

Putting everything together yields eqn (4). For the derivation of the other parts of the model we refer to.[17,41]

Numerical method

We consider a surface finite element method (SFEM)[51,52] to solve the highly nonlinear set of geometric and surface partial differential equations (1)–(3), using the approaches in ref. 17 and 35.

We combine the system (1)–(3) with a mesh redistribution approach.[61] These are equations for the parametrization

$$\partial_t\boldsymbol{X}\cdot\boldsymbol{\nu} = \boldsymbol{u}\cdot\boldsymbol{\nu} \tag{17}$$

$$\mathcal{H}\boldsymbol{\nu} = \Delta_C\boldsymbol{X}, \tag{18}$$

which generate a tangential mesh movement to maintain the shape regularity and additionally provide an implicit representation of the mean curvature $\mathcal{H}$. We consider a discrete k-th order approximation $\mathcal{S}_h^k$ of $\mathcal{S}$, with h the size of the mesh elements, *i.e.* the longest edge of the mesh. We use the DUNECurvedGrid library[62] and consider each geometrical quantity like the normal vector $\boldsymbol{\nu}_h$, the shape operator $\mathcal{B}_h$, the Gaussian curvature $\mathcal{K}_h$, and the inner products $(\cdot,\cdot)_h$ with respect to the $\mathcal{S}_h^k$. In the following, we will drop the index k. We define the discrete function spaces for scalar functions by $V_k(\mathcal{S}_h) = \{\psi \in C^0(\mathcal{S}_h)|\psi|_T \in \mathcal{P}_k(T)\}$ and for vector fields by $\mathbf{V}_k(\mathcal{S}[h]) = [V_k(\mathcal{S}[h])]^3$. Within these definitions T is the mesh element and $\mathcal{P}_k$ are the polynomials of order k. We consider $\boldsymbol{u}_h, \boldsymbol{X} \in \mathbf{V}_3(\mathcal{S}_h)$, $\mathcal{H}_h \in V_3(\mathcal{S}_h)$, and $p_h \in V_2(\mathcal{S}_h)$, which leads to an isogeometric setting for the velocity and a $\mathcal{P}_3 - \mathcal{P}_2$ Taylor-Hood element for the surface Navier–Stokes-like equations. We discretize in time using constant time stepping with step size τ. In each time step we solve the surface Navier–Stokes-like equations and the mesh redistribution together. We define a discrete surface update variable $\boldsymbol{Y}^n = \boldsymbol{X}^n - \boldsymbol{X}^{n-1}$, which is considered as unknown instead of the surface parametrization $\boldsymbol{X}^n$. The system to solve reads:

Find $(\boldsymbol{u}_h^n, p_h^n, \mathscr{H}_h^n, \boldsymbol{Y}^n) \in [\boldsymbol{V}_3 \times V_2 \times V_3 \times \boldsymbol{V}_3](\mathscr{S}_h^{n-1})$ such that:

$$\frac{1}{\tau}\left(\boldsymbol{u}_h^n - \boldsymbol{u}_h^{n-1}, \boldsymbol{v}_h\right)_h + \left(\nabla_{\boldsymbol{w}_h^{n-1}} \boldsymbol{u}_h^n, \boldsymbol{v}_h\right)_h = \left(p_h^n, \mathrm{div}_{\boldsymbol{P}} \boldsymbol{v}_h\right)_h - \frac{2}{\mathrm{Re}}\left(\boldsymbol{\sigma}\left(\boldsymbol{u}_h^n\right), \nabla_{\boldsymbol{P}} \boldsymbol{v}_h\right)_h - \gamma\left(\boldsymbol{u}_h^n, \boldsymbol{v}_h\right)_h$$

$$+\left(\left(2k_{2,0} \nabla_{\mathscr{S}} \mathscr{H}_h^n\right) \nabla_{\mathscr{S}}\left(\boldsymbol{v}_h \cdot \boldsymbol{v}_h^{n-1}\right)\right)_h$$

$$+\left(2k_{2,0}\left(\mathscr{H}_h^n - \mathscr{H}_0^{n-1}\right) B^{n-1} \boldsymbol{v}_h^{n-1}, \boldsymbol{v}_h\right)_h$$

$$+\left(k_{4,0}\left(\mathscr{H}_h^n - \mathscr{H}_0^{n-1}\right)\left(\mathscr{H}_h^{n-1} - \mathscr{H}_0^{n-1}\right)^2\right.$$

$$\left.\left(4\mathscr{B}^{n-1} - \left(\mathscr{H}_h^{n-1} - \mathscr{H}_0^{n-1}\right)\boldsymbol{P}\right), \nabla_{\mathrm{C}} \boldsymbol{v}_h\right)_h$$

$$+\left(12k_{4,0}\left(\mathscr{H}_h^{n-1} - \mathscr{H}_0^{n-1}\right)^2 \boldsymbol{v}_h^{n-1} \otimes \nabla_{\mathrm{S}} \mathscr{H}, \nabla_{\mathrm{C}} \boldsymbol{v}_h\right)_h$$

$$+\left(2k_{4,2} \mathscr{H}_h^n \boldsymbol{v}_h^{n-1} \otimes \left(\boldsymbol{P} \nabla_{\mathscr{S}} \mathscr{K}_h^{n-1}\right), \nabla_{\mathrm{C}} \boldsymbol{v}_h\right)_h$$

$$-\left(2k_{4,2} \boldsymbol{v}_h^{n-1} \otimes \left(\mathscr{B}^{n-1} \nabla_{\mathscr{S}} \mathscr{K}_h^{n-1}\right), \nabla_{\mathrm{C}} \boldsymbol{v}_h\right)_h$$

$$+\left(k_{4,2}\left(\mathscr{K}_h^{n-1}\right)^2 \boldsymbol{P}, \nabla_{\mathrm{C}} \boldsymbol{v}_h\right)_h$$

$$+\lambda\left(\boldsymbol{v}_h^{n-1}, \boldsymbol{v}_h\right)_h$$

$$\left(\mathrm{div} \boldsymbol{P} \boldsymbol{u}_h^n q_h\right)_h = 0$$

$$\frac{1}{\tau}\left(\boldsymbol{Y}^n \cdot \boldsymbol{v}_h^{n-1}, h_h\right)_h = \left(\boldsymbol{u}_h^n \cdot \boldsymbol{v}_h^{n-1}, h_h\right)_h$$

$$\left(\mathscr{H}_h^n \boldsymbol{v}_h^{n-1}, \boldsymbol{Z}_h\right)_h + \left(\nabla_{\mathrm{C}} \boldsymbol{Y}^n, \nabla_{\mathrm{C}} \boldsymbol{Z}_h\right)_h = -\left(\nabla_{\mathrm{C}} \boldsymbol{X}^{n-1}, \nabla_{\mathrm{C}} \boldsymbol{Z}_h\right)_h$$

for all $(\boldsymbol{v}_h, q_h, h_h, \boldsymbol{Z}_h) \in [\boldsymbol{V}_3 \times V_2 \times V_3 \times \boldsymbol{V}_3](\mathscr{S}_h^{n-1})$, where $B^{n-1} = \left(\|\mathscr{B}_h^{n-1}\|^2 - \frac{1}{2}\mathrm{tr}\mathscr{B}_h^{n-1}(\mathrm{tr}\mathscr{B}_h^{n-1} - \mathscr{H}_0(\phi_h^n))\right)$.

In the above formulation we used the identity $(-\nabla_{\mathscr{S}} p_h^n - p_h^n \mathscr{H}_h^n \boldsymbol{v}_h, \boldsymbol{v}_h)_h = (p_h^n, \mathrm{div}_{\boldsymbol{P}} \boldsymbol{v}_h)_h$. Note that the Lagrange multiplier λ is unknown, which leads to an underdetermined system. To resolve that problem, we follow the approach introduced in ref. 35. In order to fulfill the volume constraint, λ has to be chosen such that

$$\Phi(\lambda) := \int_{\mathscr{S}_h^{n-1}} \boldsymbol{u}_h^n(\lambda) \cdot \boldsymbol{v}_h d\mathscr{S} = 0. \tag{19}$$

We consider $\Phi(\lambda) = 0$ as an equation in λ and apply a Newton iteration $\lambda^{j+1} = \lambda^j - \Phi(\lambda^j)/\Phi'(\lambda^j)$. After convergence is achieved the new surface $\mathscr{S}_h^n$ needs to be computed by updating the parametrization $\boldsymbol{X}^n = \boldsymbol{X}^{n-1} + \boldsymbol{Y}^n$, lifting the solutions $\boldsymbol{u}_h^n$, p_h^n and $\mathscr{H}_h^n$ to the new surface and computing the remaining geometric quantities $\boldsymbol{\nu}_h^n$, $\mathscr{B}_h^n$, $\nabla_{\mathscr{S}}\mathscr{H}_h^n$, $\mathscr{K}_h^n$ and $\nabla_{\mathscr{S}}\mathscr{K}_h^n$ for the new surface. While this approach showed the (expected) optimal order of convergence,[35] with respect to computational and numerical analysis results for the underlying surface Stokes equations on stationary surfaces[63,64] for bending terms up to second order, the approach is not sufficient if higher order geometric terms are included.

We therefore introduce a smoothing step of the surface quantities $\mathscr{H}_h^n$, $\nabla_{\mathscr{S}}\mathscr{H}_h^n$, $\mathscr{K}_h^n$ and $\nabla_{\mathscr{S}}\mathscr{K}_h^n$. For each surface quantity or its components $a_h = \mathscr{H}_h^n$, $[\nabla_{\mathscr{S}}\mathscr{H}_h^n]_i$, $\mathscr{K}_h^n$ and $[\nabla_{\mathscr{S}}\mathscr{K}_h^n]_i$, $i = 1, 2$ we solve one time step of the diffusion equation

$$a_s - \varepsilon\Delta_{\mathscr{S}}a_s = a_h,$$

where $\varepsilon > 0$ is a smoothing parameter and a_s the smoothed surface quantity.

Validation

Instead of a full convergence study of the numerical approach we only test the smoothing of surface quantities. We consider a surface for which the surface quantities can be computed analytically. The surface is parametrized by $\boldsymbol{X}: [0,2\pi) \times [0,\pi) \rightarrow \mathbb{R}^2$,

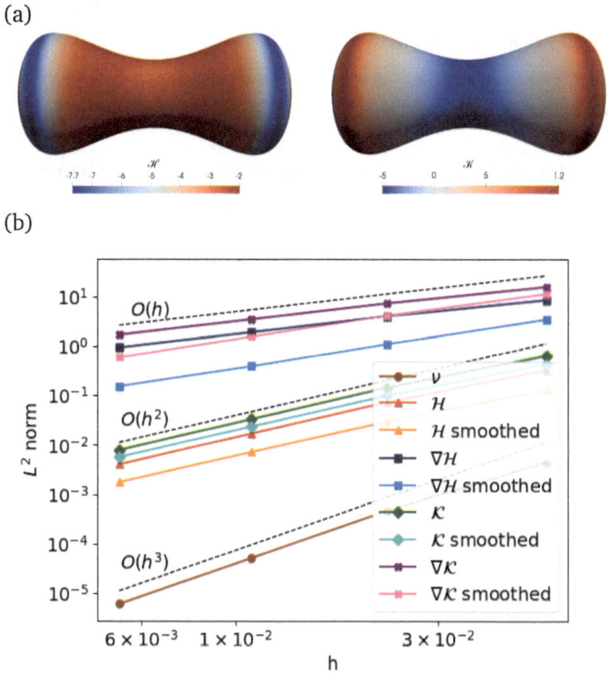

Fig. 7 (a) Reference surface for the tests. Depicted are the mean curvature $\mathscr{H}$ and Gaussian curvature $\mathscr{K}$. (b) Errors of geometric quantities with and without additional smoothing step for different grid widths h. The orders of convergence are indicated by the dashed lines and are optimal orders for the numerical implementation.

$$\mathbf{X}(\phi,\theta) = \begin{pmatrix} \frac{1}{4}+\frac{3}{4}\cos^2\theta\sin\phi\sin\theta \\ \cos\phi\left(\frac{1}{4}+\frac{3}{4}\cos^2\theta\sin\theta\right) \\ \cos\theta \end{pmatrix}.$$

We compute the L^2-error of the surface quantities $\mathscr{H}$, $\mathscr{K}$, $\nabla_{\mathscr{S}}\mathscr{H}$ and $\nabla_{\mathscr{S}}\mathscr{K}$ for different grid widths h. The smoothing parameter ε is chosen experimentally such that it shows optimal results. Fig. 7 shows that for this test case the surface quantities converge with the optimal orders and that the additional smoothing step for all quantities improves the approximation. Together with the convergence studies in ref. 17 and 35 these results provide enough confidence in the numerical approach for the full problem including the higher order geometric terms. They require an appropriate resolution of $\mathscr{K}$ and $\nabla_{\mathscr{S}}\mathscr{K}$, which is achieved in $O(h^2)$ and $O(h)$, respectively. This motivated the considered discrete function spaces.

Hausdorff distance

For two sets X, $Y \subset \mathbb{R}^3$, we consider the Hausdorff distance given by

$$d_H(X,Y) = \max\left\{\sup_{x\in X} d(x,Y), \sup_{y\in Y} d(X,y)\right\},$$

where $d(x, M) = \sup_{m\in M}\|x - m\|$ and $\|\cdot\|$ is the Euclidean norm. In the implementation we use the *VTK Hausdorff distance point set filter*[65] with the target distance method point to cell.

Acknowledgements

We acknowledge fruitful discussions with Patrick Zager (UCSF) and Veit Krause (TUD). This work was funded by DFG within FOR3013 "Vector- and tensor-valued surface PDEs". We further acknowledge computing resources at FZ Jülich under Grant No. MORPHO and at ZIH under Grant No. WIR.

References

1 H. T. McMahon and J. L. Gallop, *Nature*, 2005, **438**, 590–596.
2 M. Deserno, *Chem. Phys. Lipids*, 2015, **185**, 11–45.
3 R. Lipowsky and R. Dimova, *Soft Matter*, 2021, **17**, 214–221.
4 W. Helfrich, *Z. Naturforsch. C.*, 1973, **28**, 693.
5 P. Canham, *J. Theor. Biol.*, 1970, **26**, 61.
6 U. Seifert, *Adv. Phys.*, 1997, **46**, 13–137.
7 U. Seifert, K. Berndl and R. Lipowsky, *Phys. Rev. A:At., Mol., Opt. Phys.*, 1991, **44**, 1182–1202.
8 N. Bobrovska, W. Góźdź, V. Kralj-Iglič and A. Iglič, *PLoS One*, 2013, **8**, e73941.
9 N. Walani, J. Torres and A. Agrawal, *Phys. Rev. E:Stat., Nonlinear, Soft Matter Phys.*, 2014, **89**, 062715.
10 A. Mahapatra and P. Rangamani, *Soft Matter*, 2023, **19**, 4345–4359.
11 M. D. Mitov, *C. R. Acad. Bulg. Sci.*, 1978, **31**, 513–515.

12 J. B. Fournier and P. Galatola, *Europhys. Lett.*, 1997, **39**, 225.
13 R. Capovilla, J. Guven and J. A. Santiago, *J. Phys. A: Math. Gen.*, 2003, **36**, 6281.
14 T. Shemesh, A. Luini, V. Malhotra, K. N. Burger and M. M. Kozlov, *Biophys. J.*, 2003, **85**, 3813–3827.
15 M. Kaksonen and A. Roux, *Nat. Rev. Mol. Cell Biol.*, 2018, **19**, 313–326.
16 S. C. Al-Izzi, P. Sens and M. S. Turner, *Phys. Rev. Lett.*, 2020, **125**, 018101.
17 E. Bachini, V. Krause, I. Nitschke and A. Voigt, *J. Fluid Mech.*, 2023, **977**, A41.
18 M. Arroyo, L. Heltai, D. Millán and A. DeSimone, *Proc. Natl. Acad. Sci. U. S. A.*, 2012, **109**, 17874–17879.
19 B. Winkler, I. S. Aranson and F. Ziebert, *Commun. Phys.*, 2019, **2**, 82.
20 R. K. Sadhu, A. Iglič and N. S. Gov, *J. Cell Sci.*, 2023, **136**, jcs260744.
21 V. T. Ruhoff, N. Leijnse, A. Doostmohammadi and P. M. Bendix, *Trends Cell Biol.*, 2024, **35**, 129–140.
22 H. De Belly and O. D. Weiner, *Curr. Opin. Cell Biol.*, 2024, **89**, 102392.
23 H. De Belly, S. Yan, H. B. da Rocha, S. Ichbiah, J. P. Town, P. J. Zager, D. C. Estrada, K. Meyer, H. Turlier, C. Bustamante, *et al.*, *Cell*, 2023, **186**, 3049–3061.
24 S. R. Naganathan, S. Fürthauer, M. Nishikawa, F. Jülicher and S. W. Grill, *eLife*, 2014, **3**, e04165.
25 A.-C. Reymann, F. Staniscia, A. Erzberger, G. Salbreux and S. W. Grill, *eLife*, 2016, **5**, e17807.
26 H. B. da Rocha, J. Bleyer and H. Turlier, *J. Mech. Phys. Solids*, 2022, **164**, 104876.
27 A. Bhatnagar, M. Nestler, P. Gross, M. Kramar, M. Leaver, A. Voigt and S. W. Grill, *Curr. Biol.*, 2023, **33**, 5096–5108.
28 T. Itoh and K. Tsujita, *Curr. Opin. Cell Biol.*, 2023, **81**, 102173.
29 H. A. Faizi, R. Dimova and P. M. Vlahovska, *Biophys. J.*, 2022, **121**, 910–918.
30 M. Arroyo and A. DeSimone, *Phys. Rev. E:Stat., Nonlinear, Soft Matter Phys.*, 2009, **79**, 031915.
31 A. Torres-Sánchez, D. Millán and M. Arroyo, *J. Fluid Mech.*, 2019, **872**, 218–271.
32 A. Voigt, *J. Fluid Mech.*, 2019, **878**, 1–4.
33 S. C. Al-Izzi and R. G. Morris, *Semin. Cell Dev. Biol.*, 2021, **120**, 44–52.
34 S. Reuther, I. Nitschke and A. Voigt, *J. Fluid Mech.*, 2020, **900**, R8.
35 V. Krause and A. Voigt, *J. Comput. Phys.*, 2023, **486**, 112097.
36 M. Simunovic, G. A. Voth, A. Callan-Jones and P. Bassereau, *Trends Cell Biol.*, 2015, **25**, 780–792.
37 I. Derényi, F. Jülicher and J. Prost, *Phys. Rev. Lett.*, 2002, **88**, 238101.
38 A. Roux, G. Cappello, J. Cartaud, J. Prost, B. Goud and P. Bassereau, *Proc. Natl. Acad. Sci. U. S. A.*, 2002, **99**, 5394–5399.
39 P. G. Saffman and M. Delbrück, *Proc. Natl. Acad. Sci. U. S. A.*, 1975, **72**, 3111–3113.
40 I. Nitschke, S. Reuther and A. Voigt, *Phys. Rev. Fluids*, 2019, **4**, 044002.
41 I. Nitschke and A. Voigt, *Adv. Differ. Equ.*, 2025, **30**, 335–420.
42 S. Reuther and A. Voigt, *Multiscale Model. Simul.*, 2015, **13**, 632–643.
43 S. Reuther and A. Voigt, *Multiscale Model. Simul.*, 2018, **16**, 1448–1453.
44 P. Brandner, A. Reusken and P. Schwering, *Interface. Free Bound.*, 2022, **24**, 533–563.
45 T. Biben and C. Misbah, *Phys. Rev. E:Stat., Nonlinear, Soft Matter Phys.*, 2003, **67**, 031908.
46 Y. Kim and M.-C. Lai, *J. Comput. Phys.*, 2010, **229**, 4840–4853.

47 D. Salac and M. J. Miksis, *J. Fluid Mech.*, 2012, **711**, 122–146.
48 F. Haußer, W. Marth, S. Li, J. S. Lowengrub, A. Rätz and A. Voigt, *Int. J. Biomath. Biostat.*, 2013, **2**, 19–48.
49 S. Aland, S. Egerer, J. Lowengrub and A. Voigt, *J. Comput. Phys.*, 2014, **277**, 32–47.
50 V. Krause and A. Voigt, *J. R. Soc. Interface*, 2024, **21**, 20240056.
51 G. Dziuk and C. M. Elliott, *Acta Numer.*, 2013, **22**, 289–396.
52 M. Nestler, I. Nitschke and A. Voigt, *J. Comput. Phys.*, 2019, **389**, 48–61.
53 S. Vey and A. Voigt, *Comput. Visualization Sci.*, 2007, **10**, 57–67.
54 T. Witkowski, S. Ling, S. Praetorius and A. Voigt, *Adv. Comput. Math.*, 2015, **41**, 1145–1177.
55 T. R. Powers, *Rev. Mod. Phys.*, 2010, **82**, 1607–1631.
56 R. E. Goldstein, P. Nelson, T. Powers and U. Seifert, *J. Phys. II*, 1996, **6**, 767–796.
57 S. C. Al-Izzi, G. Rowlands, P. Sens and M. S. Turner, *Phys. Rev. Lett.*, 2018, **120**, 138102.
58 M. Janssen, S. Liese, S. C. Al-Izzi and A. Carlson, *Phys. Rev. E:Stat., Nonlinear, Soft Matter Phys.*, 2024, **109**, 044403.
59 M. Nestler and A. Voigt, *Proc. Appl. Math. Mech.*, 2023, **23**, e202300044.
60 I. Nitschke, S. Sadik and A. Voigt, *IMA Journal of Applied Mathematics*, 2023, **88**, 917–958.
61 J. W. Barrett, H. Garcke and R. Nurnberg, *SIAM J. Sci. Comput.*, 2008, **31**, 225–253.
62 S. Praetorius and F. Stenger, *Archive of Numerical Software*, 2022, 6, 1–27.
63 P. Brandner, T. Jankuhn, S. Praetorius, A. Reusken and A. Voigt, *SIAM J. Sci. Comput.*, 2022, **44**, A1807–A1832.
64 H. Hardering and S. Praetorius, *IMA Journal of Numerical Analysis*, 2023, **43**, 1543–1585.
65 F. Commandeur, J. Velut and O. Acosta, *VTK Journal*, 2011.

Faraday Discussions

ROYAL SOCIETY OF CHEMISTRY

DISCUSSIONS

Structure and dynamics of asymmetric membranes: general discussion

Damiano Andreghetti, Francisco N. Barrera, John C. Conboy, Simon Connell, Aileen Cooney, Zhibo Deng, Markus Deserno, Rumiana Dimova, Milka Doktorova, Thais A. Enoki, Oded Farago, Gerald W. Feigenson, Alexander P. Fellows, Giacomo Fiorin, Alberto Giacomello, Gonen Golani, Felix M. Goni, Heiko Heerklotz, Andreas Heuer, Daniel Huster, Di Jin, Ben John, Michael Kaltenegger, Manpreet Kaur, Sarabjeet Kaur, Syma Khalid, Michael M. Kozlov, Robert V. Law, Ilya Levental, Zhi Li, Reinhard Lipowsky, Fabio Lolicato, Erwin London, Patricia Losada Pérez, Tiemei Lu, Edward Lyman, Nasim Mirzajani, Walter Nickel, Georg Pabst, Madan Rao, Tyler Reagle, Sylvie Roke, Christian Schwieger, John M. Seddon, Iwona Swiderska, Robert Vacha, Malavika Varma, Martín E. Villanueva and Axel Voigt

DOI: 10.1039/d5fd90020d

Milka Doktorova opened discussion of the paper by Markus Deserno: It has been shown that under some conditions asymmetric vesicles can exhibit more than two types of 'phases' or 'environments' as evidenced by the differential partitioning of fluorescent lipid probes in experiments (*e.g.* Fig. 2 in ref. 1). This is not captured in conventional phase diagrams robustly characterized with symmetric model membranes and used here as a basis for examining the effects of differential stress on asymmetric 3-component membranes. How can the potential asymmetry-driven formation of a third 'bilayer phase' be incorporated into this analysis and is this something that the coarse-grained simulations can capture and provide the means to investigate further?

1 T. A. Enoki and F. A. Heberle, *Proc. Natl. Acad. Sci. U. S. A.*, 2023, **120**, e2308723120.

Markus Deserno responded: In the paper you reference,[1] Enoki and Heberle examine asymmetric binary membranes comprising two types of phospholipids (DPPC/DOPC at 295 K in their case). Disregarding differential stress for a moment, this situation is characterized by a single variable in each leaflet – say, the DPPC mole fraction. This nevertheless results in an already fairly subtle two-dimensional compositional phase diagram with 3 pure phases (shown in Fig. 4 of their paper): (i) both sides DPPC rich and ordered, (ii) both sides DOPC rich and disordered, and (iii) one side DPPC rich and ordered while the other side is DOPC rich and disordered. Between these phases lie three coexistence regions with generally tilted tie-lines (*e.g.*, two coexisting disordered phases on one side meeting an ordered phase coexisting with a disordered one on the other, each in mutual registration) – meaning, we observe four different leaflet mixing proportions. Moreover, under certain conditions we even get a three-phase coexistence region, bordering the three two-phase regions, which then amounts to six different mixing proportions. Once we add differential stress to this scenario, we have a new variable and hence a three-dimensional phase diagram. I am not aware that this has ever been studied, either in experiment or in theory. Much of this clearly goes beyond the physics we have learned in symmetric systems, and so we obviously do not expect trivial modifications of the "old" phase diagrams. But we do expect that the physical phenomenology we have learned to be important must be part of any explanatory models we wish to develop.

In the asymmetric ternary case we discuss in our paper (https://doi.org/10.1039/d4fd00196f) things are evidently even more complicated. However, our goal is not simply to extrapolate from ternary symmetric systems to ternary asymmetric ones. Instead, we use very generic thermodynamic reasoning (equilibrium conditions, Gibbs phase rule, mechanical stability, *etc.*) to suss out how many degrees of freedom we even have. We argue that there are four (in addition to temperature and net tension), and only if we fix two of them (say, the saturation ratios in both leaflets) do we get something we can easily plot. But you will notice that in most of the examples I have discussed in some detail I have not yet included the possibility of phase separation in each leaflet, by picking the saturation ratios to be sufficiently large or small so that they fall outside the coexistence region (at least the coexistence region as we know it from the symmetrical case). Once they go inside, one (or both) leaflets can laterally phase separate. This will not change the abundance asymmetry α nor the cholesterol mole fraction φ, but the compositions of the coexisting leaflet phases will split – first tentative guess: along the tie lines known from symmetric phase diagrams. But once this happens, inter-leaflet domain coupling needs to be included, as well as curvature effects (since the two laterally coexisting phases need not have the same torque). All this will quantitatively (and likely also qualitatively) change the observed behavior compared to what we know from symmetric systems. I must admit that at this moment I have no idea how this will all play out, or what a good set of thermodynamic variables is, or what observables we could measure that would be particularly informative about the underlying state. We need to develop better theories to address these situations and hope that the structure of these theories gives us some idea of the right variables or observables.

Of course, computational models would in principle include all these effects, or at the very least we typically know what molecular parameters we would have to adjust in order to account for them. But seeing things happen in simulation is not yet the same as understanding *why* they happen. We still require models, and given that the present situation is indeed quite complex – both theoretically and experimentally – I suspect we will have a lot on our plate for the foreseeable future.

1 T. A. Enoki and F. A. Heberle, *Proc. Natl. Acad. Sci. U. S. A.*, 2023, **120**, e2308723120.

John M. Seddon asked: If you move the lipid composition closer to the critical point in the ternary diagram, will you observe large curvature fluctuations in the bilayers, and will there be any difference between symmetric and asymmetric bilayers?

Markus Deserno replied: By the submission deadline of our contribution to this *Faraday Discussions* meeting we had not yet looked for this effect, but since then we have started to examine this question. I still have no data, but my firm expectation is that we will see enhanced curvature fluctuations – by the well-known composition–curvature coupling argument first made by Stan Leibler.[1] One of its predictions is that non-ideal mixing enhances this effect, and in a naïve quadratic mean-field theory the effective shape fluctuations actually diverge as we approach the critical point (of course, they do not in reality – we just need a more careful theory).

While we have not yet seen this in simulations, it appears that at least two other groups have: Pöhnl *et al.*[2] have examined binary mixtures of phospholipid and cholesterol, at the MARTINI level, while Hossein *et al.*[3] have looked at phospholipid mixtures featuring phosphatidylcholine (PC) *versus* phosphatidyl-ethanolamine (PE) headgroups, using both MARTINI 2 and 3 as well as an atomistic force field (CHARMM C36).

I am unsure whether asymmetry would entail any obvious qualitative changes. Clearly, the effect relies on leaflet-specific composition–curvature coupling, but it does not have to be "symmetric" in magnitude to "work". There are quite a number of interesting effects to probe with some detail, and we are starting to look at this. The biggest challenges I expect to face are sampling issues: getting good statistics on shape fluctuations in the precious long wavelength regime is already difficult. And compositional fluctuations come with an even slower relaxation time that you need to exceed in order to sample correctly. To measure the effect precisely enough that we do not merely see its correct "direction" but can also quantitatively compare it to its predicted magnitude will be difficult.

1 S. Leibler, *J. Phys.*, 1986, **47**, 507.
2 M. Pöhnl, *et al.*, *Nat. Commun.*, 2023, **14**, 8038.
3 A. Hossein, *et al.*, *J. Phys. Chem. B*, 2024, **128**, 6317.

Madan Rao enquired: What ensemble have you chosen to represent the phase diagrams you presented? Do you fix the tension in the two leaflets, or do you fix the overall bilayer tension?

Markus Deserno responded: The Gibbs phase diagrams shown in our paper (https://doi.org/10.1039/d4fd00196f) are meant to refer to the zero bilayer tension situation, but I have not yet thought carefully about how they would change under an applied tension. Similarly, the torque- and differential stress diagrams in Fig. 7 and 10 of our paper, respectively, were based on actual simulations that were run at zero bilayer tension.

We have argued previously that it is convenient to look at bilayer tension and differential stress as the obvious symmetrized and antisymmetrized linear combinations of the individual leaflet tensions. This is useful because the bilayer tension is often zero (or in some well-defined sense "small") so that only the differential stress is an observable we need to explicitly account for. But, of course, the value of its complementary variable, the total bilayer tension, still must be stated, and in virtually all cases we have looked at so far that value was zero.

Madan Rao queried: Do you explicitly include a transbilayer interaction between components across the bilayer, either at a microscopic scale or at a coarse grained scale (for instance, as an "interfacial" contribution)? This clearly has both entropic and enthalpic parts and is strictly an emergent statistical quantity.

Markus Deserno replied: We do not yet have a theoretical description of asymmetric ternary systems in terms of, say, a phenomenological free energy. But our coarse-grained simulations must of necessity "commit" to some form of transbilayer interaction. Specifically, since we use a solvent-free model, lipids from the two different leaflets must attract one another, for otherwise the leaflets would simply unbind. The dominant contribution of this attraction comes from the last tail beads (recall that our coarse-grained (CG) lipids consist of a single hydrophobic head bead and three consecutive tail beads, of which the last ones directly "touch" one another across the leaflets).

What your question is probably driving at is whether we choose this enthalpic contribution to the transbilayer interaction to be dependent on lipid type, and the answer is that we do not. After all, it is just methyl groups touching! Whether a lipid is of the "ordered" or "disordered" type (as determined by intra-lipid interactions that affect how much it can "wiggle") does not influence the strength of the enthalpic interaction.

I wish to emphasize that we could have easily enforced domain registration if we had chosen that interaction to be stronger between like-type lipids, but we felt that this would mostly just be putting in by hand the known experimental outcome, instead of exploring whether it would naturally arise for different reasons. And these different reasons exist, because effects such as thickness mismatches, or the specific form of line tensions, depend on the type of registration. Possibly even more interesting, there is an *entropic* contribution to registration, because a rigid phase leaflet loses undulation entropy, and if it lies opposite a soft phase leaflet, it forces that leaflet to suffer a similar loss because they must ultimately share the same overall bilayer shape. If, instead, rigid phases align with rigid phases and soft phases with soft ones, then the overall entropy is larger. Years ago I had offered this simple argument as an explanation why adhesion between two vesicles at the brink of phase separation might actually demix their membranes in the adhesion zone, as experimentally observed, even if all lipids have the same head groups.[1]

1 V. Gordon, *et al.*, *Europhys. Lett.*, 2008, **84**, 48003.

Madan Rao said: This is related to the question on transbilayer coupling. How would you discuss issues of bilayer registry, partial registry or anti-registry? What would be the tuning parameter and how would it appear in your phase diagram?

Markus Deserno responded: Assuming you now want me to comment on a suitable theoretical approach, I would consider, as a first step, a phenomenological Landau free energy that includes a coupling term between the compositions of the two leaflets.

For the binary (or effective binary) case, this has been done before.[1–4] These papers really set the stage for how to think about coupled phase separation in opposing leaflets. In fact, recently Enoki and Heberle[5] have used a simple version of such theories to explain the experimental phase diagram of asymmetric vesicles in which each side is a binary DPPC/DOPC mixture. The (single!) phenomenological coupling parameter between the local order parameter in the two leaflets can be tuned so that the morphology of the theoretically predicted phase diagram is compatible with what is experimentally observed.

In all these prior theories, there is effectively only one important order parameter in each leaflet, which results in just a single coupling constant. Of course, for ternary systems each side has three different concentrations, which upon symmetric quadratic coupling gives 6 possible coupling constants (*i.e.*, the number of independent entries of a symmetric 3×3 matrix). This number is too big, of course, because *principally* the concentrations are not independent, and because *practically* not all of these couplings might be equally important. We have not yet looked carefully into this, though, and – to the best of my knowledge – neither have others. So I cannot really answer your question more definitively.

1 A. J. Wagner, *et al.*, *Biophys. J.*, 2007, **93**, 4268, DOI: 10.1529/biophysj.107.115675.
2 G. Garbès Putzel and M. Schick, *Biophys. J.*, 2008, **94**, 869, DOI: 10.1529/biophysj.107.116251.
3 S. May, *Soft Matter*, 2009, **5**, 3148, DOI: 10.1039/b901647c.
4 J. J. Williamson and P. D. Olmsted, *Biophys. J.*, 2018, **115**, 1956, DOI: 10.1016/j.bpj.2018.10.003.
5 T. A. Enoki and F. A. Heberle, *Proc. Natl. Acad. Sci. U. S. A.*, 2023, **120**, e2308723120, DOI: 10.1073/pnas.2308723120.

Heiko Heerklotz requested: Can you please comment on the order of magnitude of the torque caused by cholesterol redistribution in a hypothetic plasma membrane containing the same number of phospholipids in the outer and inner leaflet? Would such a torque be likely to interfere with membrane structure and function?

Markus Deserno answered: Let's start by getting a feel for the orders of magnitude. Consider as an example the asymmetric membranes discussed in Elani *et al.*,[1] which were made from DOPC and POPC lipids. Computational results from Venable *et al.*[2] give spontaneous leaflet curvatures of $J_{0\text{m}}(\text{DOPC}) = -0.071\ \text{nm}^{-1}$ and $J_{0\text{m}}(\text{POPC}) = -0.032\ \text{nm}^{-1}$. If for a start we ignore any additional leaflet stresses, then the spontaneous bilayer curvature of an asymmetric DOPC/POPC membrane would be $J_0 \approx -0.02\ \text{nm}^{-1}$, and the associated spontaneous curvature torque would be $|\kappa J_0| \approx 30k_\text{B}T \times 0.02\ \text{nm}^{-1} \approx 2.5$ pN. How "big" is this?

Spherical vesicles made from such asymmetric DOPC/POPC leaflets would have a vanishing bending energy at a radius $R = 2/|J_0| = 100$ nm, with DOPC on the inside. But the actual giant unilamellar vesicles (GUVs) in the experiments of Elani *et al.*[1] had radii between 10 μm and 20 μm, at least 2 orders of magnitude larger. As a consequence, their bending energy would have to be around 4 orders of magnitude larger than that of a relaxed vesicle, and so they should have an extremely strong tendency to vesiculate or tubulate into shapes with a curvature closer to their actual preferred one. We have argued that this might not happen because these vesicles are actually under a differential stress that creates a counter-torque. In the present case this requires a differential stress of around $\Delta\Sigma = \kappa J_{0,\mathrm{b}}/z_0 \approx 2$ mN m^{-1}, and that order of magnitude ends up being typical. This shows that an unbalanced torque of just a few pN would give rise to dramatic shape changes.

Now, how much torque is created when cholesterol redistributes? In a previous paper[3] Dr Varma and I had examined this situation. As a simple example, consider an asymmetric membrane containing two species of phospholipids on its two sides, with vanishing differential stress but a difference $\delta\chi$ between the non-ideal mixing parameters of cholesterol and the respective phospholipids, such that one of the sides is preferred (by essentially $\delta\chi \times k_\mathrm{B}T$). Combining some of the results in the paper I just cited, one can show that in the simplified case where all lipids (including cholesterol) have the same area per lipid, a, the differential stress $\Delta\Sigma$ as a function of cholesterol concentration ϕ is given by

$$\Delta\Sigma \approx \frac{1}{2} K_\mathrm{A} \delta\chi \frac{\phi(1-\phi)}{1 + \frac{1}{2}\frac{\phi}{1-\phi} K_\mathrm{A} a / k_\mathrm{B} T}$$

Fig. 1 here plots this as the blue curve for the special case where $\delta\chi = 1$, area modulus $K_\mathrm{A} = 240$ mN m^{-1}, and a specific lipid area $a = 0.6$ nm^2. In our previous paper,[3] we also discuss the more difficult case where cholesterol is smaller than the lipids, and then we get the slightly reduced red curve in Fig. 1 here. Qualitatively, the outcome is the same though: the asymmetric partitioning of added cholesterol ramps up a differential stress that reaches a maximum value at around 20 mol% cholesterol and a magnitude that depends on the partitioning preference, but is again roughly a few mN m^{-1}. This is of the same magnitude as the "Elani case"[1] I discussed above. Hence, the associated torque would also be comparable, so we should again expect a strong tendency towards tubulation.

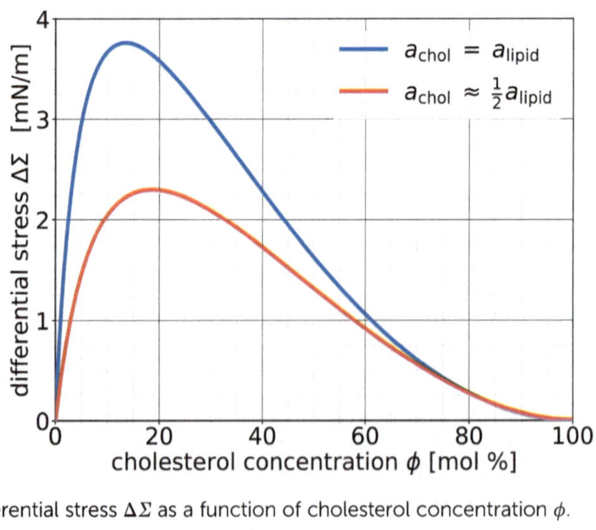

Fig. 1 Differential stress $\Delta\Sigma$ as a function of cholesterol concentration ϕ.

With all this in mind – yes, I believe that the torques arising from cholesterol redistribution can be large enough to drive very noticeable shape deformations, even though the details will always depend on exactly how much cholesterol we add or re-distribute. As a rule of thumb, which we justify in our paper, we expect cholesterol to nevertheless be a less potent driver of shape deformations than phospholipids, because it divides its impact between the two leaflets.

1 Y. Elani, *et al.*, *Chem. Commun.*, 2015, **51**, 6976–6979, DOI: 10.1039/c5cc00712g.
2 R. M. Venable, *et al.*, *Chem. Phys. Lipids*, 2015, **192**, 60–74, DOI: 10.1016/j.chemphyslip.2015.07.014.
3 M. Varma and M. Deserno, *Biophys. J.*, 2022, **121**, 4001–4018, DOI: 10.1016/j.bpj.2022.07.032.

Oded Farago asked: You consider changes in the thermodynamic variables that bring one of the leaflets into the two-phase region. This, in turn, should lead to changes in the mechanical properties of the bilayer. How are these changes handled in the calculations, and how do they show in the torque foliation curves?

Markus Deserno replied: I do not know yet, because we have not pushed our calculations that far. At this point our fairly generic thermodynamic reasoning avoids referring to a specific equation of state. Once phase coexistence occurs, some additional degrees-of-freedom counting arguments (basically the Gibbs phase rule) should enforce further constraints. All this might lead to important qualitative changes even before we consider mechanical aspects such as changes in rigidity, which will affect the torque, or that such asymmetric systems would also have a spontaneous curvature. For instance, if the spontaneous bilayer curvature is not the same in different regions of the bilayer, then what do we mean by a "zero torque condition" in the first place? Zero *average* torque, maybe?

How does all of this affect the torque foliations? Again, I cannot say for sure, because we have no theory yet for the specific details. I have thought a little bit about what would happen if the "cyan point" in the +-leaflet (Fig. 4 of our paper; https://doi.org/10.1039/d4fd00196f), while moving along its torque foliation curve, enters a two-phase region. My initial guess was that the torque foliation curve might become a straight line between the "entry" and "exit" points of the phase coexistence region. But since the system now is mechanically asymmetric, with the complications I just mentioned above, it is not clear to me whether this naïve linear expectation will hold, considering all the other effects that now matter and which have not been part of the currently proposed thermodynamic framework.

Oded Farago remarked: The nano-scale liquid ordered domains that you observe in the one-phase region of the phase space are local density fluctuations that always appear, especially near the binodal of the two-phase region (see, *e.g.*, ref. 1).

1 T. Sarkar and O. Farago, *Soft Matter*, 2023, **19**, 2417, DOI: 10.1039/d2sm01025a.

Markus Deserno responded: I agree that these systems should always show density fluctuations, or maybe more relevant here, *compositional* fluctuations, but I would expect them to be essentially molecular in size, just a few times as big as a single lipid's footprint. (Unfortunately I do not know how to make the vague term "a few" more precise!) Of course, near critical points these fluctuations can grow truly macroscopic, but that is because the respective susceptibilities diverge. This is not true as we merely approach a binodal sufficiently far away from the critical point. In fact, even slightly entering the coexistence region (sampling metastable states before nucleation sets in), the local free energy will give you no reason to expect enhanced fluctuations – not until you actually reach the spinodal. So I would disagree with the expectation that in the vicinity of the binodal compositional fluctuations are noticeably more pronounced than deeper into the pure phase.

That said, it is also possible that our unusually large "flakes" are small, ordered domains *inside* the two-phase region. That is possible because we do not know the location of the binodal very precisely. Measuring the composition of a phase requires partitioning all available molecules between the coexisting phases, which in our case is accomplished by a hidden Markov model. However, its output depends – among other things – on the training point, the chosen features, and the trajectory one actually trained on. The underlying systematic uncertainties are challenging to quantify, and we have not yet tried to do so.

Rumiana Dimova asked: How would an external mechanical stress (tension) couple with this thermodynamic framework of asymmetric membranes and do you expect a difference in the response of symmetric *vs.* asymmetric membranes?

Markus Deserno answered: I have not yet thought carefully about what would happen if we permit a nonzero net tension, but I expect several new phenomena to arise. At the very least, it is clear that tension can affect the bilayer *via* (i) any type of "area physics" and (ii) net curvature. I generally expect these to be coupled.

Let me illustrate this with a simple thought experiment: if the two leaflets differ in their (effective) area moduli, imposing an area strain will induce different leaflet stresses, which will then create a torque. Indeed, since the two leaflets need to stay together, any net tension can only ever change the areas in the same way, and so I expect a net tension to always create a torque proportional to the difference in leaflet area moduli. Incidentally, since this all makes sense within linear response, we should conversely expect torques to affect the net tension.

It gets more interesting if the two leaflets contain ordered domains that can melt under tension, especially if they cover different amounts of the area. This is closely related to the *Faraday Discussions* paper by Krompers *et al.* that was presented yesterday by Dr Heerklotz (https://doi.org/10.1039/d5fd00003c). He focused on temperature-driven melting or gelling, but tension would also be able to melt gel domains. Of course, in that case it is less likely that we can observe a curvature response; but if instead of fluid and gel phases we have L_o and L_d domains, then there is a different way in which this could manifest: melting an L_o domain in one leaflet would free up its cholesterol, some of which might be pushed into the other leaflet. Hence, a net tension could affect the trans-leaflet cholesterol distribution. We have briefly checked in our simulations whether tension could change the relative area fraction of coexisting L_o and L_d domains, but our domain identification algorithm was not precise enough to ascertain an effect.

Finally, should we be able to realize a scenario where one leaflet could phase separate but the other does not (because its composition is so far from the coexistence region that even the "imprinting" from the opposite side is not strong enough to trigger demixing), then we would have a system in which we have laterally coexisting asymmetric domains that almost certainly differ in their spontaneous curvature. As demixing progresses, *i.e.*, as domain coarsening sets in, this triggers ever larger deviations from a membrane's initial (say, flat) state. Hence, a net applied tension should at some point arrest coarsening and the system should end up with a compositionally and geometrically modulated phase, with a characteristic length scale set by the tension.

Andreas Heuer enquired: Does it help to include the information about the average area of the lipids for both leaflets in the general analysis, *e.g.*, in relation to the differential stress? Possibly as a reasonable approximation which is easy to access?

Markus Deserno responded: Yes, this information is both useful and important, because it is a basic factor in determining differential stress. In fact, in one of our recent papers[1] Dr Varma and I have accounted for specific lipid areas when it came to mixing between different lipid species. We examined a membrane that had two different types of phospholipids in its two leaflets (say, POPC and DOPC), plus additional cholesterol, rendering each side a binary mixture. Since the relative leaflet areas affect the stress, and cholesterol is significantly smaller than a typical phospholipid, we assigned different areas per lipid to each species. The equilibrium area of each leaflet was then postulated to be additive: $A_{eq} = N_p a_p + N_c a_c$, where N_p and N_c are the number of phospholipids and cholesterol molecules in the leaflet, respectively, and a_p and a_c are their specific areas. We also included an empirical area-correction in the Flory–Huggins type mixing free energy, and the results we get for many observables reflect obvious expectations.

As plausible as this looks, it remains questionable how accurate this approach is when applied to real systems, because "area additivity" is a fickle assumption to make, especially when applied to cholesterol. The notion that a lipid even *has* an area – *i.e.*, a molecular property independent of the specific bilayer it is part of – is a major simplification. This is of course generally true when thinking about molecules (recall the well-known example that mixing 50 ml of water with 50 ml of ethanol does not yield 100 ml of vodka), but it is even more subtle in the case of lipids: unlike water or ethanol, lipids have a rich ensemble of molecular conformations, *and this ensemble depends on a lipid's neighbors*. If a disordered lipid has relatively rigid cholesterol molecules nearby, its tail fluctuations get suppressed and it takes up a smaller area than it would if it were surrounded by other lipids of its own type. Hence, if we were to replace 30% of all phospholipids in such a membrane by cholesterol, the area would not merely go down because cholesterol is intrinsically smaller than the phospholipid. It would go down *further* than that because cholesterol orders neighboring phospholipids, *i.e.*, it changes the molecular ensemble of a different neighboring lipid species. Obviously, such effects are not included when we rely on area additivity. A few years back this condensing effect has been discussed very nicely in a simulation study by Leeb and Maibaum.[2]

To make a quantitative theoretical improvement to the simple additivity ansatz, the only practical solution I see at this point is to replace it by an empirical equation of state, which presumably would have to be determined from experiment or, likely easier, from simulation.

1 M. Varma and M. Deserno, *Biophys. J.*, 2022, **121**, 4001.
2 F. Leeb and L. Maibaum, *Biophys. J.*, 2018, **115**, 2179.

Andreas Heuer asked: Is it possible to use the properties of the fluctuations of the membrane system during the MD simulations, to learn about the parameters, entering the analysis of the general model with the four degrees of freedom?

Markus Deserno replied: Yes, this is definitely possible. Of course, the challenge is to develop quantitative theories that permit us to interpret the experimentally measured fluctuations.

Let me give you an example, though, that suggests we can obtain a useful but difficult to measure parameter from the shape undulations of a mixed membrane. Picture a symmetric binary membrane in which the two phospholipids form a well-mixed phase at all mixing ratios. What is the bending rigidity of that membrane? One might expect some linear interpolation between the rigidities of the two pure phases, but one typically finds that it dips below that. A possible explanation is softening due to composition curvature coupling, as first discussed by Stan Leibler.[1] In that case, the extent of softening depends on the spontaneous curvature difference between the two lipid species. This parameter is very difficult to reliably measure in lamellar phases, so one typically resorts to experiments performed on inverted hexagonal phases, which curvature-wise are orders of magnitude removed from the local conditions in fluctuating lamellar phases. I have seen experimental data on this from my colleague Stephanie Tristram-Nagle, which beautifully align with theoretical predictions, and which will appear later this year in a special issue of the *Biophysical Journal* honoring the late Adrian Parsegian. This, in my view, is a vivid example where fluctuations in mixed systems allow you to measure an observable that is otherwise quite difficult to obtain.

Of course, in reality everything is more subtle. "Lipid spontaneous curvature" is yet another one of those mythical lipid properties, just like "area", which only make sense in the context of an entire ensemble, not as an immutable molecular property that Avanti or Sigma could print on their sample vials. This has been well documented by Sodt *et al.*,[2] and yet the theoretical analysis I have just outlined straight up assumes that a lipid bilayer's local spontaneous curvature can be obtained as an average of individual molecular properties weighted by the local lipid concentrations. It is very hard to make analytical progress without such approximations, and so many scientists (me included) make them as a first step; but one should remain alert to complications arising from such simple assumptions.

1 S. Leibler, *J. Phys.*, 1986, **47**, 507.
2 A. J. Sodt, *et al.*, *Phys. Rev. Lett.*, 2016, **117**, 138104.

Michael M. Kozlov enquired: Does the effective spontaneous (intrinsic) curvature of cholesterol in your computations depends on the surrounding lipid? What is its characteristic value?

Markus Deserno responded: When we developed the ternary model system in ref. 1, we focused on lipid interactions that affect mixing but left all lipid shapes as essentially "straight" (*i.e.*, we did not "by hand" impose any spontaneous curvatures). However, as you correctly remind us, cholesterol affects the conformational ensemble of its surrounding lipids, which is really how we should think of spontaneous curvature arising – see ref. 2. We have not yet measured what its value is, but we briefly mention some indirect findings in our paper. Eqn (20) in the paper (https://doi.org/10.1039/d4fd00196f) recalls that the difference between the torque coming from differential stress and the total torque must be the torque due to spontaneous curvature, $\kappa J_{0,\mathrm{b}}$. Since we have measured both the differential stress and the torque, we can deduce what $\kappa J_{0,\mathrm{b}}$ is, and we found it to vary relatively little even as we change cholesterol from 0 to 40 mol%. We take this to mean that "our" coarse-grained cholesterol affects the conformational ensemble of lipids only weakly – not surprising considering how simple our lipids are. That said, to get the spontaneous curvature from its associated torque, we need to divide out the bending rigidity κ, and we have not yet measured its value as a function of cholesterol content – something which all by itself is a surprisingly subtle question that especially for unsaturated lipids has given rise to a very lively debate over the past few years.

1 M. Varma, *et al.*, *J. Chem. Phys.*, 2024, **161**, 114103.
2 A. J. Sodt, *et al.*, *Phys. Rev. Lett.*, 2016, **117**, 138104.

Di Jin asked: How do you relate the simulation results to experimental observations? In particular, would there be a way to apply the insights from nanometric simulations to interpret experimental observations of micron-sized phase-separated domains by techniques such as AFM?

Markus Deserno answered: This is an excellent question, and one which I am frequently asked by new students: How do we relate our highly coarse-grained simulations to experiment? Our lipids do not look like actual lipids, our values for area, length, order parameters, moduli, relaxation rates, *etc.*, do not match an actual system, and sometimes (especially with dynamics) they might even be far off. How would we distinguish a saturated from an unsaturated lipid if all we have is four beads? These and other issues may make one legitimately wonder: what is the point? I suppose your question was not meant to be quite that far reaching, but please grant me a soap box moment.

You likely know the aphorism "All models are wrong. But some models are useful." It reminds us that the goal of modeling is not to capture reality in a 1-to-1 fashion. The famous short story "On Exactitude in Science" by Jorge Luis Borges really nails this sentiment. Models leave out detail, and some very good models leave out almost everything. To describe the motion of Earth around the Sun, we can model both celestial objects as point particles. To pull yet another authority: Einstein supposedly said, "Make things as simple as possible, but no simpler." Alas – who decides what is "as simple as possible", and when do we know we made it "too simple"?

These decisions are made by the modeler, and later again by the person who wants to learn something from the model. A good rule is: to make it into the model, a certain property must matter for what we wish to explain. Nothing gets in gratuitously. But there is a subtlety: say I want to describe the flow of water through a pipe. Do I need to model hydrogen bonds? Probably not. Do I need to account for density and viscosity? Almost surely yes. But wait – these are determined to no small part by water's hydrogen bonds! True, but we only need the macroscopic emergent properties of "density" and "viscosity" to describe flow, we do not need to understand how they emerge from a more fine-grained atomistic reality. Modeling hence very often means radically short-circuiting pathways of emergence.

Finally – back to membranes. Our coarse-grained model does not contain hydrogen bonds. It does not even contain atoms! (Or water, for that matter.) But it captures a few key (emergent!) properties of lipids that matter for its physics: a lipid's approximate shape, its amphipathic nature, the fact that tails have some conformational entropy. These, it turns out, suffice to make these lipids aggregate into two-dimensional fluid bilayer membranes that exhibit stretching- and curvature elastic properties with values of the moduli that are surprisingly meaningful. If we distinguish different lipid species by their size and tail "wiggliness", we can represent saturated and unsaturated lipids, and even some bravely simplified version of cholesterol. If we adjust their pair interactions, we can affect their mixing behavior and, for instance, make cholesterol melt gel phases or let ternary systems exhibit L_o/L_d phase separation. All these "successes" are examples where emergent membrane properties that are largely determined by generic lipid properties are reproduced, and that means I can reason about them using such simplified models. I can now go on to explore other aspects of the L_o/L_d physics, such as what happens when both sides of the bilayer are ternary mixtures but one of the components, cholesterol, can rapidly flip-flop. I can ask whether phase segregation or domain melting can be driven by differential stress, or membrane tension, or a change in a membrane's curvature state, or adsorption to a substrate.

Since you asked specifically about AFM imaging – what aspects of such experiments could we mimic/explain/predict? For starters, micron-sized domains will be difficult, simply because this would be a very large system. A 10 μm × 10 μm field of view would contain about 300 million lipids, which even in our highly coarse-grained model would mean more than a billion particles. But maybe we do not need micron domains? Maybe a domain with a size of 30 nm would be OK for what you want to understand? We could for instance ask whether adsorption to a substrate that flattens the membrane and reduces its thermal undulations would increase the temperature at which it phase-separates. We could ask whether preferential absorption of one phospholipid species *versus* another would change the leaflet compositions enough to see domains on one side, but maybe not on the other. We could try to pin some lipids on the substrate and ask whether somehow "shearing" the upper leaflet would get any possibly matching domains out of registration, and if so, by how much – thereby mimicking the beautiful experiments done some time ago by Blosser *et al.*[1] But wait, you say, in our model we usually see domain anti-registration! And here we come to a point where we reach a problem: maybe we reached the limits of the model? Or we have not yet simulated big enough systems? Or have not yet done some necessary tuning? At any rate, we need to step back and think before we blindly apply our model to this problem.

1 M. C. Blosser, *et al.*, *Biophys. J.*, 2015, **109**, 2317–2327.

Simon Connell said: This is more a comment than a question, Markus. In Fig. 11 and 12 in your paper (https://doi.org/10.1039/d4fd00196f), you look at a composition in the co-existence region near the L_d binodal, and the effect of changing the composition of one leaflet thus driving the opposing leaflet into (or further away from) the coexistence region (Section 4.3.2). You observed a stable L_o domain in the $\alpha = -10\%$ (16.7% Chol) as expected in the co-existence region, but on the boundary and in the single phase ($\alpha = 0\%$, 25.5% Chol and $\alpha = +10\%$, 32.5% Chol) you saw remnant transient "flakes" dynamically fluctuate in and out of existence. We work on the dynamics of lipid phases using high resolution and high speed AFM, along with fluorescence microscopy, and in a paper currently under review[1] we have observed these transient "flakes" near phase boundaries. This is with symmetric and not with asymmetric bilayers, so I understand the system is a little different. However, we observe these "transient flakes" a long way from the critical point in terms of composition, near phase boundaries, and if starting from the critical point and increasing temperature, they are seen up to 10 °C higher. This opens up the possibility of testing some of your ideas in real experiments, with observation of transient flakes down to 5 nm in size.

1 G. R. Heath, S. D. Evans and S. D. Connell, Critical fluctuations of lipid domains display raft-like behavior, under review.

Markus Deserno responded: Thank you for this very interesting comment. As Dr Farago has pointed out in an earlier question, we should always see transient compositional fluctuations, except I do not know what size scale I should expect (other than "molecular"). My personal sense is that the flakes we observe, and whose size and dynamics we really have not yet characterized more quantitatively, are bigger than what could be justified based on compositional fluctuations far away from a critical point.

That said, there have long been claims in the literature that complex lipid mixtures may have nanodomains, whose origin and thermodynamic justification I have not yet spent enough time to understand. To my knowledge, the first experimental evidence was reported by Heberle *et al.*,[1] who observed what they describe as (nano-) "domains" with an effective size in the 5–15 nm range in small unilamellar vesicles of ~60 nm diameter (albeit for quaternary mixtures). The domain sizes were determined by fitting a model of randomly dispersed circular patches to small angle neutron scattering data and were found to correlate with bilayer thickness mismatch of the coexisting L_o and L_d phases. At the risk of displaying too much of my own ignorance, but here is why I find this confusing: if these are true equilibrium domains, then what stops them from growing? And if they are transient fluctuations, why are they so big?

1 F. A. Heberle, *et al.*, *J. Am. Chem. Soc.*, 2013, **135**, 6853.

Thais Enoki communicated: I would like to include a minor comment to highlight the importance of ternary phase diagrams. The use of ternary phase diagrams in *in vitro* experiments does not aim to ignore more than 99% of the lipid species in the plasma membrane; instead, all lipids are represented by categories. The extensive repertoire of lipids is classified into three categories, each representing key characteristics that are of considerable significance. The two phospholipids in the phase diagram represent a group of lipids with high and low melting temperatures, and the third category is cholesterol. Indeed, as mentioned in the paper (https://doi.org/10.1039/d4fd00196f), the model can capture some emergent phase behavior, not an intricate lipidomic fingerprint.

Markus Deserno communicated in reply: I believe we agree on the physics, so this is almost surely just a language issue, regarding the precise meaning of "ignore". Of course we do not restrict to ternary systems out of ignorance about reality, or because we are too lazy to do "the real thing". We model ternary systems because we have the strong expectation that the classification into ordered lipids, disordered lipids, and cholesterol captures the essence of the effects we are presently curious about. My point is that we do not necessarily have an irrefutable prior reason to believe that this will actually work. The fact that it appears to do is a deep observation about the physics of the situation, about exactly what type of emergent variables matter, and how the process of emergence ends up identifying lipid order as the dominant variable.

But even if it does, we know that all the different lipid types in a real biomembrane present us with a whole spectrum of such melting temperatures, not just two. And it is not obvious to me that such a spectrum can always be described as if we only had two main categories of phospholipids, ordered and disordered. Clearly, as we tune the temperature, and as we cross several such main transition temperatures for some of our lipids in the mix, one would think we ought to notice that. Maybe if two representatives of the "saturated lipid class" in our mix have similar but not quite identical main transition temperatures, we would observe something strange if only we could examine the critical point very carefully.

I do not know what the answer is, and I suspect there will be more subtle observables that would be sensitive to such details, observables where an effective ternary model would fail. When we wrote that in our paper (https://doi.org/10.1039/d4fd00196f), we merely wanted to draw attention to the fact that ternary mixtures are well-motivated models of nature, but as with all models, it is always healthy to remember what we ignored.

Zhi Li opened discussion of the paper by Reinhard Lipowsky: Have you done some research about how ions affect the membrane tension? To what extent do you think the ions could change the membrane tension compared with mechanical bending tension?

Reinhard Lipowsky answered: Most ions are expelled from the hydrophobic core of the bilayer but can bind to or intercalate into the two headgroup layers. If the ions bind to one of the headgroup layers, they will increase or decrease the area per lipid in this leaflet. If the ions repel each other and increase the area per lipid, they increase the leaflet tension as well. On the other hand, if the lipids condense around the ions and attain a smaller area per lipid, the leaflet becomes compressed corresponding to a decreased or even negative leaflet tension.

The effect of ions and other small solutes on the bilayer asymmetry and the associated spontaneous curvature has been studied by coarse-grained molecular dynamics simulations of planar bilayers[1] and nanovesicles[2] as well as by experimental studies of giant unilamellar vesicles (GUVs).[3] In the simulations, the ions were modelled as small solutes and it was demonstrated that they generate a large transbilayer asymmetry and spontaneous curvature in planar bilayers.[1] The binding of small solutes to nanovesicle membranes leads to the formation of closed membrane necks, to recurrent shape fluctuations between open and closed necks, and to the fission of the closed membrane necks.[2] The experimental studies in ref. 3 used different concentrations of LiCl, NaCl, and KCl in the interior and exterior aqueous solution and measured the spontaneous curvature by pulling tubes from aspirated GUVs.

1 B. Różycki and R. Lipowsky, *J. Chem. Phys.*, 2015, **142**, 054101, DOI: 10.1063/1.4906149.
2 R. Ghosh, *et al.*, *ACS Nano*, 2021, **15**, 7237, DOI: 10.1021/acsnano.1c00525.
3 M. Karimi, *et al.*, *Nano Lett.*, 2018, **18**, 7816, DOI: 10.1021/acs.nanolett.8b03584.

Georg Pabst added: Several experimental studies have reported that monovalent and divalent ions can modulate the intrinsic curvature of lipids. These ion-induced changes influence the spontaneous curvature of membrane leaflets, thereby altering the associated mean torque. In our recent work,[1] we demonstrated how such curvature modifications can affect the activity of an integral membrane enzyme in an asymmetric bilayer. This study also provides references to additional research exploring the impact of ions on intrinsic lipid curvature.

1 P. Piller, P. Reiterer, E. F. Semeraro and G. Pabst, *RSC Appl. Interfaces*, 2025, **2**, 69, DOI: 10.1039/d4lf00309h.

Aileen Cooney asked: You discussed membrane–condensate interactions, have you looked how at other systems interact with membranes? Specifically hydrogel or colloidal systems?

Reinhard Lipowsky replied: For systems consisting of biomembranes interacting with condensate droplets, we use the theoretical framework of what we call "fluid elasticity". This framework uses two basic assumptions to characterize the elastic and morphological responses of the vesicle–droplet systems.[1] The first assumption is that the condensate droplets are enclosed by a liquid–liquid interface and that the shape of this interface is governed by its interfacial tension. The second assumption is that the biomembranes are in a fluid state, which implies that their morphology is determined by a few curvature-elastic parameters such as bending rigidity and spontaneous (or preferred) curvature, in addition to the mechanical tensions within the membranes (https://doi.org/10.1039/d4fd00184b).

Hydrogels consist of networks of polymers (or peptides) in aqueous solutions. The macromolecules can be cross-linked by chemical bonds or by physical entanglement. Our theoretical framework can also be applied to hydrogel droplets interacting with biomembranes provided the shape of these droplets is governed, at least on long time scales, by a liquid–liquid interface with a well-defined interfacial tension. From an experimental point of view, this tension may be difficult to measure.

As far as rigid colloids are concerned, we also looked at the interactions of rigid nanoparticles with biomembranes. In this case, the interfacial tension of the condensate droplets is replaced by the adhesion free energy between membrane and particle surface. The membranes can again be described within the framework of curvature elasticity.[2]

1 R. Lipowsky, *Membranes*, 2023, **13**, 223, DOI: 10.3390/membranes13020223.
2 J. Agudo-Canalejo and R. Lipowsky, *ACS Nano*, 2015, **9**, 3704, DOI: 10.1021/acsnano.5b01285.

John M. Seddon said: The Gaussian curvature modulus κ_G is more difficult to measure experimentally than the bending modulus κ. In previous work, you showed that the profile of a vesicle phase-separated into two domains (for example, L_o and L_d) is determined by the difference in κ_G between the two bilayer domains. Could this provide a useful method for determining κ_G of a bilayer?

Reinhard Lipowsky responded: Consider a vesicle enclosed by a lipid bilayer that contains several molecular components and can undergo phase separation into two intramembrane domains, a and b. Let us denote the corresponding Gaussian curvature moduli by $\kappa_{G,a}$ and $\kappa_{G,b}$. Applying the Gauss–Bonnet theorem to such a vesicle, the domain boundary between the a and b domains generates a term that is proportional to the difference $\kappa_{G,a} - \kappa_{G,b}$ of the two Gaussian curvature moduli.[1] If the two Gaussian curvature moduli happen to be equal to each other, the domain boundary remains located within the membrane neck during the closure of this neck. On the other hand, if $\kappa_{G,a}$ is different from $\kappa_{G,b}$, the domain boundary is displaced from the closing neck. This displacement was originally obtained by numerical integration of the shape equations for axisymmetric vesicles,[2] has been observed experimentally,[3] and can be obtained in a simple and transparent manner from the matching condition for the mean curvature along the domain boundary.[4]

It seems possible to extend these theoretical considerations in such a way that one could determine the individual values of the Gaussian curvature moduli $\kappa_{G,a}$ and $\kappa_{G,b}$ rather than their difference but, at present, I am not aware of a feasible method to do so. It is interesting to note that different Gaussian curvature moduli will also displace the contact line between a vesicle membrane and a condensate droplet away from the closing membrane neck adjacent to the droplet.[5]

1 F. Jülicher and R. Lipowsky, *Phys. Rev. Lett.*, 1993, **70**, 2964, DOI: 10.1103/PhysRevLett.70.2964.
2 F. Jülicher and R. Lipowsky, *Phys. Rev. E: Stat. Phys., Plasmas, Fluids, Relat. Interdiscip. Top.*, 1996, **53**, 2670, DOI: 10.1103/PhysRevE.53.2670.
3 T. Baumgart, *et al.*, *Biophys. J.*, 2005, **89**, 1067, DOI: 10.1529/biophysj.104.049692.
4 R. Lipowsky, *Eur. Phys. J. E: Soft Matter Biol. Phys.*, 2024, **47**, 4, DOI: 10.1140/epje/s10189-023-00399-z.
5 R. Lipowsky, *Soft Matter*, under review.

Manpreet Kaur asked: How does the shape of the lipid (especially conical and inverted-conical shape lipids) affect the membrane tension? And what about the regions, where such non-bilayer lipids are clustered by strong lipid–lipid interactions?

Reinhard Lipowsky replied: To simplify the discussion, let us focus on the bilayer tension of a symmetric lipid bilayer, which exhibits the same lipid composition in both leaflets. This bilayer tension is a membrane-elastic parameter that describes the elastic response of the bilayer membrane to external constraints or forces that lead to the stretching (or compression) of the bilayer. Imagine, for instance, that we pull at the bilayer by external tangential forces in order to stretch it. As long as the bilayer does not rupture, these tangential forces must be balanced by the molecular forces between the different molecules within the membrane. In general, the membrane molecules experience a variety of molecular interaction forces such as van der Waals and electrostatic forces or forces arising from ions that adsorb onto the head groups of the lipids.

In order to determine the bilayer tension for a specific lipid bilayer with a certain lipid composition, one has to compute the local stress profile across the bilayer. By definition, the local stress within a lipid bilayer is equal to the negative local pressure. Therefore, stress profiles are equivalent to pressure profiles. From an intuitive point of view, the local stress is positive close to the interfaces between the lipid headgroups and the adjacent aqueous solutions, because these interfaces can reduce their interfacial free energy by reducing their areas, whereas the local stress is negative in the hydrophobic core, corresponding to a positive pressure arising from the confined hydrocarbon chains. More precisely, the stress profile is obtained from the anisotropic pressure tensor, which has a normal component P_N and a tangential component P_T, as recently reviewed in Appendix B of ref. 1. The bilayer tension is then obtained by integrating the stress profile over the coordinate perpendicular to the bilayer.

The local stress profile for a planar lipid bilayer in aqueous solution has been computed using three molecular models: binary Lennard-Jones fluids;[2] bead models for dissipative particle dynamics;[3–5] and the Martini force field for lipids.[6] Furthermore, in all of these molecular models, the hydrophobic chains of the lipids were treated as semiflexible polymers consisting of molecular groups that experience both a harmonic bond potential and a three-body bending potential, which favors a certain tilt angle between neighboring molecular groups along the hydrocarbon chains of the lipids. It is important to note that these interaction potentials between the hydrocarbon groups also affect the local stress profile across the bilayer.

Therefore, the bilayer tension of a planar lipid bilayer depends on all interaction potentials between the molecular groups within the membrane. The dependence of this tension on the shape of the lipids can be studied by using different architectures of the lipid molecules.[1–6] In ref. 5, we studied a binary mixture of two lipids with a small and a large headgroup. The lipid with the large headgroup may be regarded as a conical lipid. We focussed on the stress profiles for tensionless bilayers with zero bilayer tension, which are required for the computation of the bilayer's spontaneous curvature. Even for tensionless bilayers, the stress profiles vary significantly with the mole fraction of the large-headgroup lipids, see Fig. 2 in ref. 5.

1 R. Lipowsky, *et al.*, *Biomolecules*, 2023, **13**, 926, DOI: 10.3390/biom13060926.
2 R. Goetz and R. Lipowsky, *J. Chem. Phys.*, 1998, **108**, 7397, DOI: 10.1063/1.476160.
3 J. C. Shillcock and R. Lipowsky, *J. Chem. Phys.*, 2002, **117**, 5048, DOI: 10.1063/1.1498463.
4 B. Różycki and R. Lipowsky, *J. Chem. Phys.*, 2015, **142**, 054101, DOI: 10.1063/1.4906149.
5 A. Sreekumari and R. Lipowsky, *J. Chem. Phys.*, 2018, **149**, 084901, DOI: 10.1063/1.5038427.
6 M. S. Miettinen and R. Lipowsky, *Nano Lett.*, 2019, **19**, 5011, DOI: 10.1021/acs.nanolett.9b01239.

Sylvie Roke enquired: Most lipids have charges, when one does measurements of bending and tension, electrostatic interactions cannot be avoided, and these are strong, especially when compared to other interactions. How are such interactions incorporated/connected to membrane tension, bending computations that you presented?

Reinhard Lipowsky responded: The mechanical tension within a lipid bilayer, for which I use the short term "bilayer tension", is a membrane-elastic parameter that describes the elastic response of the bilayer membrane to external constraints or forces. Imagine, for instance, that we stretch a planar bilayer by tangential forces. As long as the bilayer does not rupture, these tangential forces must be balanced by the molecular forces between the different molecules within the membrane. In general, the membrane molecules experience a variety of molecular interaction forces such as van der Waals and electrostatic forces as well as forces arising from ions that adsorb onto the head groups of the lipids.

In order to compute the bilayer tension for a specific lipid bilayer with a certain lipid composition, one has to compute the local stress profile across the bilayer. By definition, the local stress is equal to the negative local pressure within the bilayer. From an intuitive point of view, the local stress is positive close to the two interfaces between the two lipid headgroup layers and the adjacent aqueous solutions, because this interface can reduce its interfacial free energy by reducing its area, whereas the local stress is negative in the hydrophobic core, corresponding to a positive pressure arising from the confinement of the hydrocarbon chains. More precisely, the stress profile is obtained from the anisotropic pressure tensor, which has a normal component P_N and a tangential component P_T, as reviewed in Appendix B of ref. 1. The bilayer tension is then obtained by integrating the stress profile over the coordinate perpendicular to the bilayer.

The local stress profile for planar lipid bilayers in aqueous solution has been computed using three different molecular models: binary Lennard-Jones fluids;[2] bead models for dissipative particle dynamics;[3–5] and the Martini force field.[6] Furthermore, in all of these molecular models, the hydrophobic chains of the lipids were treated as semiflexible polymers consisting of molecular groups that experience both a harmonic bond potential and a three-body bending potential, which favors a certain tilt angle between neighboring molecular groups along the hydrocarbon chains of the lipids. All of these interactions contribute to the stress profile as shown in Fig. 11 of ref. 2, which depicts the individual contributions from the different interactions. The total stress profile is obtained from the sum of the individual contributions as shown in Fig. 12a of ref. 2.

Stress profiles as obtained from dissipative particle dynamics of planar lipid bilayers are displayed in Fig. 4 and 10 of ref. 3, in Fig. 7 of ref. 1 based on data from ref. 4, and in Fig. 2 and 3 of ref. 5. Finally, a stress profile as obtained from the Martini force field is shown in Fig. 14b of ref. 1 based on data from ref. 6. The Martini force field includes Coulomb potentials for electrostatic interactions. Compared to the stress profiles as obtained from dissipative particle dynamics without Coulomb potentials, the stress profile as obtained from the Martini force field with Coulomb potentials exhibits additional fine structure, corresponding to small additional minima and maxima of the profile. However, when we compute the bilayer tension by integrating over the coordinate perpendicular to the bilayer, these additional minima and maxima will average out to a large extent.

1 R. Lipowsky, *et al.*, *Biomolecules*, 2023, **13**, 926, DOI: 10.3390/biom13060926.

2 R. Goetz and R. Lipowsky, *J. Chem. Phys.*, 1998, **108**, 7397, DOI: 10.1063/1.476160.
3 J. C. Shillcock and R. Lipowsky, *J. Chem. Phys.*, 2002, **117**, 5048, DOI: 10.1063/1.1498463.
4 B. Różycki and R. Lipowsky, *J. Chem. Phys.*, 2015, **142**, 054101, DOI: 10.1063/1.4906149.
5 A. Sreekumari and R. Lipowsky, *J. Chem. Phys.*, 2018, **149**, 084901, DOI: 10.1063/1.5038427.
6 M. S. Miettinen and R. Lipowsky, *Nano Lett.*, 2019, **19**, 5011, DOI: 10.1021/acs.nanolett.9b01239.

Sylvie Roke asked: Has anybody tried to look at tension parameters and relate them to actual interactions?

Reinhard Lipowsky answered: As mentioned in my reply to the previous question, the contributions of the different molecular interactions to the stress profile have been analyzed in ref. 1 for a Lennard-Jones fluid, as displayed in Fig. 11 of ref. 1. In general, it is always possible to decompose the total stress profile into the individual contributions arising from the different molecular interactions. Thus, it is also possible to identify the individual contribution arising from the Coulomb potentials in the Martini force field.[2] However, as emphasized in my reply to the previous question, this Coulombic contribution is expected to make a relatively small contribution to the bilayer tension as obtained by integrating the total stress profile $s = s(z)$ over the coordinate z perpendicular to the planar bilayer.

1 R. Goetz and R. Lipowsky, *J. Chem. Phys.*, 1998, **108**, 7397, DOI: 10.1063/1.476160.
2 M. S. Miettinen and R. Lipowsky, *Nano Lett.*, 2019, **19**, 5011, DOI: 10.1021/acs.nanolett.9b01239.

Milka Doktorova requested: Can you elaborate in a little more detail about the origins of the fluctuation tension? In the abstract it is stated that it governs the excess membrane area stored in the vesicle shape fluctuations. In that sense, does fluctuation tension vary with lipid composition?

Reinhard Lipowsky responded: Membranes in aqueous solutions undergo thermally-excited shape fluctuations in the form of bending undulations in order to increase their configurational entropy. Thus, when we look at the shape of a membrane, we will, in general, observe shape fluctuations that represent normal displacements of the membrane from its average shape. The analysis of these shape fluctuations is particularly simple for a planar bilayer. In this case, the average shape is provided by a plane and the shape fluctuations lead to local and transient displacements from this planar shape, see the simulation snapshot in Fig. 2 of the paper (https://doi.org/10.1039/d4fd00184b). This snapshot was obtained for a planar bilayer with vanishing bilayer tension. The bilayer tension can be increased by changing the projected area of the planar bilayer, corresponding to the base area of the simulation box, without changing the number of lipid molecules, see Fig. S5 in the supplementary information of the paper (https://doi.org/10.1039/d4fd00184b). The bilayer tension is increased by increasing this base area as in Fig. S5c, which acts to flatten the bending undulations and to decrease the excess area stored in these shape fluctuations.

For planar bilayers with periodic boundary conditions, the shape fluctuations can be decomposed into a discrete set of Fourier modes. More precisely, such a decomposition is possible provided the lateral size of the bilayer exceeds a certain minimal size. The Fourier analysis in ref. 1 revealed that this minimal size is about 6 nm for a lipid bilayer with a thickness of about 4 nm. On smaller scales, the shape fluctuations represent molecular protrusions corresponding to the normal displacements of single lipids or small groups of neighboring lipids. Thus, one has to consider a sufficiently large segment of a planar bilayer with vanishing bilayer tension in order to observe the bending undulations of this bilayer. Furthermore, the number of discrete Fourier modes increases as one increases the lateral size of this bilayer.

As described in Section 3.5 of the paper (https://doi.org/10.1039/d4fd00184b), the fluctuation tension of planar bilayers is defined *via* the spectrum of these discrete Fourier modes, see eqn (16) of the paper. The fluctuation tension vanishes for vanishing bilayer tension as demonstrated for one-component planar bilayers in ref. 1 and further corroborated by the fluctuation spectra of planar bilayers with two lipid components.[2] One interesting and open question is whether or not these two tensions are also equal for non-zero tension values, see the references cited in Section 3.5 of the paper.

1 R. Goetz, *et al.*, *Phys. Rev. Lett.*, 1999, **82**, 221, DOI: 10.1103/PhysRevLett.82.221.
2 A. Sreekumari and R. Lipowsky, *J. Chem. Phys.*, 2018, **149**, 084901, DOI: 10.1063/1.5038427.

Gerald W. Feigenson opened discussion of the paper by Michael M. Kozlov: I am interested in the calculation and thoughts on mid plane energy where two phases are touching each other. This is simple chemically, with the main difference being density, so I wonder if you have any comments?

Michael M. Kozlov replied: Should be both. Computational models reveal such differential energy (see *e.g.* ref. 1 and references therein and in our paper (https://doi.org/10.1039/d4fd00186a)).

1 T. A. Enoki and F. A. Heberle, *Proc. Natl. Acad. Sci. U. S. A.*, 2023, **120**, e2308723120, DOI: 10.1073/pnas.2308723120.

Oded Farago asked: Can you point to the differences between the membrane-mediated interactions between domains discussed in your work, and membrane-mediated interactions between cone-shaped proteins – a problem that has been investigated in prior studies.

Michael M. Kozlov answered: The difference is in the effects of the lipid tilt/splay deformations. They come on top of and interfere with the curvature deformations and modify the resulting interaction forces.

Zhibo Deng enquired: Would you consider incorporating species such as cholesterol into one of the monolayers to induce the formation of a lipid domain? If so, would you expect the resulting differential energy to initiate membrane curvature?

Michael M. Kozlov responded: We analyzed the effect of cholesterol and other lipid molecules with relatively large intrinsic curvatures. It boils down to a decrease in the effective splay (bending) modulus.

Ilya Levental asked: Would this set of differential transbilayer interactions be more relevant for bilayer domains or protein complexes?

Michael M. Kozlov answered: The interaction we analyzed might be relevant to the formation of bilayer domains out of the monolayer ones. Bilayer domains might interact as well if, *e.g.*, they are intrinsically curved. Yet I believe, based on the previous work, that similar bilayer domains would repel each other (I am talking about the membrane-mediated interaction only).

Concerning the protein complexes, the logic is the same. Similar protein complexes are expected to exercise an attractive membrane-mediated interaction (at least for large separations) if they are located in opposing leaflets, and a repulsive interaction in case they sit in the same leaflet.

Ilya Levental requested: About the attractive *vs.* repulsive energies, could you please expand on this?

Michael M. Kozlov responded: The model predicts that the interaction between domains located in opposing leaflets is attractive for large inter-domain distances. For small separations, the interaction can be either attractive or repulsive, depending on the ratio between the domain's rigidities and those of the surrounding lipid monolayers. The relevant rigidities are those for the monolayer bending (splay) and tilt (the average tilting of the lipid hydrocarbon chains with respect to the normal to the monolayer surface). For technical simplicity, we took this ratio to be the same for the bending rigidities and the tilt ones. If this rigidity ratio is larger than a certain "critical" value, *i.e.* the domains are sufficiently rigid, the interaction at small separations is predicted to be repulsive. Otherwise (for relatively soft domains), the interaction for small separations is attractive, which means that attraction persists for all separations. The critical rigidity ratio depends on the membrane tension and the domain's size. For example, for large tensions (flat membranes) and relevant domain sizes (say larger than a hundred nm), the predicted critical rigidity ratio is 3, as presented in the paper (https://doi.org/10.1039/d4fd00186a). We have the results for the critical ratio and the separation after which the interaction turns repulsive (if it does) for the whole range of tensions (not shown in this paper).

Gonen Golani enquired: What is the tension required to fulfill the requirement of a flat membrane compared to the rupture tension of the membrane?

Michael M. Kozlov replied: The required tension must be much larger than the tilt modulus, the latter constituting a few tens of mN m^{-1}. Therefore, the case of a completely flat membrane is interesting theoretically to start understanding the system, while the analysis of the experimentally feasible situations requires the accounting of the effects of small tensions. We have got these results but I did not have time to include them in the manuscript.

Damiano Andreghetti asked: The model you propose seems to share some features with the O(2) model (*XY* model) or theoretical frameworks used for liquid crystals. An interesting aspect of these theories is the presence of Goldstone modes (or gapless excitations) arising from continuous symmetry breaking. Do you expect your model to exhibit this behavior as well?

Michael M. Kozlov answered: I never thought about it.

Thais Enoki said: In your paper (https://doi.org/10.1039/d4fd00186a), you mentioned that you are not aware of any experimental data enabling the evaluation of the differential contact energy, $\Delta\varepsilon$. Although I do not have a model-free estimation of differential contact energy, we estimate this parameter by comparing an experimentally determined phase diagram of asymmetric bilayers of DPPC/DOPC[1] to the phase diagrams proposed by Wagner *et al.* (2007)[2] and later May (2009).[3] I want to share with you this paper for your evaluation and consideration to be cited along with the paper from May that you are already citing. It is important to note that our estimation agrees with the values in the range you mention in your paper.

If I understand it correctly, and as exploited in the 1D model, in the presence of one domain, there is a differential contact energy that is less favorable than the contact energy of the blue lipids (disordered lipids). Thus, this configuration tends to reduce the number of lipid molecules opposing the domains or make them tilt to decrease the contact area. These changes create deformations that propagate in the membrane (x-axis).

In the model's extension, there are two domains in opposite leaflets, and in this case, the model mimics a symmetric bilayer (I believe). The deformations mentioned above raise repulsive or attractive interactions that drive anti-registration or registration of domains in a symmetric-like bilayer. Can you add a line tension to this model, and how would that affect the results?

Could you add domains into the same leaflets, mimicking an asymmetric bilayer in lipid composition and phase behavior? In other words, one leaflet would have one phase, and the other would have two phases. In this case, would domains attract or repel each other? If domains repel each other, could this suggest that the interactions drive small domains, as they will never coalesce into larger domains? By adding a line tension to this model, how do you expect to change the deformations and interactions between domains? Would you expect the predictions from these models to be different in a spherical geometry compared to a plane geometry?

1 T. A. Enoki and F. A. Heberle, Experimentally determined leaflet–leaflet phase diagram of an asymmetric lipid bilayer, *Proc. Natl. Acad. Sci. U. S. A.*, 2023, **120**, e2308723120, DOI: 10.1073/pnas.2308723120.
2 A. J. Wagner, S. Loew and S. May, Influence of monolayer–monolayer coupling on the phase behavior of a fluid lipid bilayer, *Biophys. J.*, 2007, **93**, 4268–4277, DOI: 10.1529/biophysj.107.115675.
3 S. May, Trans-monolayer coupling of fluid domains in lipid bilayers, *Soft Matter*, 2009, **5**, 3148–3156, DOI: 10.1039/b901647c.

Michael M. Kozlov responded: Many thanks for sending me your paper.[1] It is great to have an estimation deduced from experiments rather than MD simulations. I will cite your paper in my upcoming paper on domain interaction in membranes subject to finite tensions and undergoing therefore bending deformations in addition to the tilt/splay deformations.

"If I understand it correctly, and as exploited in the 1D model, in the presence of one domain, there is a differential contact energy that is less favorable than the contact energy of the blue lipids (disordered lipids). Thus, this configuration tends to reduce the number of lipid molecules opposing the domains or make them tilt to decrease the contact area. These changes create deformations that propagate in the membrane (x-axis)." Correct. In the manuscript submitted to the meeting (https://doi.org/10.1039/d4fd00186a), I considered only a flat membrane, implying that the membrane is subject to a large enough tension. In this case the deformations are tilt/splay of hydrocarbon chains. I already have results for a more general case of arbitrary tensions, including the experimentally relevant ones. In addition to tilt/splay there are also bending deformations, which change the picture but on a qualitative level the predictions are similar. I am going to write it down and include a reference to your work as one of the central ones.

"In the model's extension, there are two domains in opposite leaflets, and in this case, the model mimics a symmetric bilayer (I believe). The deformations mentioned above raise repulsive or attractive interactions that drive anti-registration or registration of domains in a symmetric-like bilayer. Can you add a line tension to this model, and how would that affect the results?" The interaction is predicted to be always attractive at large inter-domain distances. But at vanishing and small distances it can be either attractive or repulsive depending on the domain rigidity. The former case corresponds to a spontaneous registration. The latter case corresponds to some equilibrium non-vanishing distance between the domains. In order to register the domains have to overcome this distance by "jumping" over an energy barrier.

"Could you add domains into the same leaflets, mimicking an asymmetric bilayer in lipid composition and phase behavior? In other words, one leaflet would have one phase, and the other would have two phases. In this case, would domains attract or repel each other?" The domains in the same leaflet would repel each other. I did not explicitly calculate that since the answer is obvious based on previous works (including ours) on the bending-mediated interactions between cap-like membrane inclusions.

"If domains repel each other, could this suggest that the interactions drive small domains, as they will never coalesce into larger domains? This means that separately nucleated domains would have to overcome large energy barriers to coalesce. So coalescence would be kinetically hindered. By adding a line tension to this model, how do you expect to change the deformations and interactions between domains?" The long-range interaction considered by the model would be independent of the line tension. Yet, the "microscopic" origin of the line tension itself would need to be understood and modeled.

"Would you expect the predictions from these models to be different in a spherical geometry compared to a plane geometry?" Definitely, there will be additional features of the interaction such as a different distance dependence. But I hope that the qualitative essence will remain the same. My student is trying to compute the 3D model. It is technically more challenging.

1 T. A. Enoki and F. A. Heberle, Experimentally determined leaflet–leaflet phase diagram of an asymmetric lipid bilayer, *Proc. Natl. Acad. Sci. U. S. A.*, 2023, **120**, e2308723120, DOI: 10.1073/pnas.2308723120.

Malavika Varma opened discussion of the paper by Daniel Huster: From an experimental technique perspective, how substantial is the difference in measurements between symmetric and asymmetric systems in Fig. 2 of the paper (https://doi.org/10.1039/d4fd00192c)?

Daniel Huster responded: The error bars for the determination of ^{2}H NMR order parameters are below symbol size in the plots. So the differences are significant.

Robert V. Law asked: In Fig. 1A & B, D & E of your paper (https://doi.org/10.1039/d4fd00192c), there is a comparison between the asymmetric and symmetrical bilayers. What is the origin of the isotropic line in the middle of the asymmetric (Fig. 1B and 1E) in the ^{2}H NMR spectra?

If these isotropic signals are from microvesicles or micelles, did you also confirm the presence of these microvesicles or micelles by static ^{31}P NMR spectroscopy by looking for any isotropic lines at 0 ppm?

Were there any issues *e.g.* Fourier artefacts in signal digitisation when you had a sharp isotropic signal combined with the broader ^{2}H NMR signals?

Daniel Huster replied: The isotropic line in the ^{2}H NMR spectra comes from highly curved lipid structures, possibly very small (*i.e.* budded) vesicles. Overall, this contribution amounts to 7% (for POPC in the outer layer) and 9% (for POPS in the outer layer) of the ^{2}H NMR spectra of deuterated lipids. We have also carried out ^{31}P NMR studies of the multilamellar vesicle (MLV), which detect *all* lipids (not just the outer leaflet). There was only a very small isotropic contribution in these spectra.

Robert V. Law questioned: If these signals are from microvesicles or micelles, can they possibly recombine with the membranes during ^{2}H NMR T_2 of the experiment? Would this affect the values of the order parameters and relaxation times? If these signals are from microvesicles or micelles, why do they occur only in the asymmetric system?

Daniel Huster answered: I guess budded vesicles cannot recombine with the membranes as this would increase the number asymmetry which is unfavorable. The order parameters and relaxation measurement come ONLY from the anisotropic part of the ^{2}H NMR spectrum, so the isotropic signal is not considered in either of these measurements. The symmetric system is prepared under no stress, lipid powder is hydrated in buffer and the membranes are in full equilibrium. Due to the exchange with cyclodextrin (CD), some (number) asymmetry could be introduced, which may lead to budding.

Edward Lyman asked: The differences in order parameter plots between the symmetric and asymmetric vesicles are intriguing. It would be extremely informative to be able to assign the S_{CD}'s to particular carbons. Is there some way to do this, short of carbon-by-carbon deuteration?

Daniel Huster replied: We are doing order parameter measurements with better site-specificity. We need ^{13}C labeled lipids to measure order parameters under magic-angle spinning conditions. We have done this for some symmetric systems and will soon provide these order parameters for asymmetric membranes. The key idea is to measure the motional averaging of the ^{1}H–^{13}C dipolar coupling, which probes the same C–D bond vector as measured in ^{2}H

NMR. Right now, we only have ^{13}C labeled PE but will get ^{13}C labeled PC, so we will do experiments on PC_{out}/PE_{in} large unilamellar vesicles (LUVs).

Edward Lyman enquired: Since the samples are multilamellar, only the outermost leaflet gets exchanged. Does that give enough of a signal (it apparently does), and is it feasible to get enough signal out of LUVs?

Daniel Huster answered: Yes, we can get sufficient signal for ^{2}H NMR and also for ^{13}C NMR. The latter will be done in LUVs. Due to the lateral diffusion of lipids in LUVs, the order parameters will also be lower as we have to consider averaging by diffusion over the curved surface. But comparing the lipids in the outer and in the inner leaflet will be straightforward in these preparations. We are working on it…

Manpreet Kaur said: A 2021 study,[1] with CHO cells, shows symmetric membranes have reduced tension relative to asymmetric membranes. Your work also shows that artificial asymmetric membranes are stiffer than symmetric ones. I would expect the artificial membrane with lipids you chose to have more mis-packing. Why is an artificial asymmetric membrane more stiff, is it because of multivesicle bodies affecting membrane properties?

1 A. Shiomi, *et al.*, Extreme deformability of insect cell membranes is governed by phospholipid scrambling, *Cell Rep.*, 2021, **35**, 109219, DOI: 10.1016/j.celrep.2021.109219.

Daniel Huster responded: Well, a cell is of course a much more complicated system with many more lipids and other species in the membrane. In model membranes, it has been shown experimentally[1,2] and theoretically[3] that asymmetric membranes are stiffer. Also, the experimental systems so far did not contain cholesterol, which plays an important role in membrane stiffness.

1 Y. Elani, S. Purushothaman, P. J. Booth, J. M. Seddon, N. J. Brooks, R. V. Law and O. Ces, *Chem. Commun.*, 2015, **51**, 6976, DOI: 10.1039/c5cc00712g.
2 K. Karamdad, R. V. Law, J. M. Seddon, N. J. Brooks and O. Ces, *Chem. Commun.*, 2016, **52**, 5277, DOI: 10.1039/c5cc10307j.
3 A. Hossein and M. Deserno, *Biophys. J.*, 2020, **118**, 624, DOI: 10.1016/j.bpj.2019.11.3398.

Heiko Heerklotz asked: Could a linewidth analysis of NMR (^{31}P?) be used as an alternative approach to AF4 to quantify the fraction of lipid in budding-induced daughter vesicles (<30 nm) as compared to mother vesicles (>80 nm?).

Daniel Huster replied: Yes, this could be possible, however, better than ^{31}P NMR where the anisotropies are almost completely averaged by lateral diffusion, ^{2}H NMR of DPPG-d_{62} should work better. Fig. 2 here is a NMR lineshape simulation considering averaging in 30 nm and 70 nm vesicles.

Erwin London enquired: Would it be worthwhile to repeat these experiments at a lower temperature at which the POPE enriched leaflet would be in the ordered gel state? It might be expected to result in a stronger effect on the lipids in the opposite leaflet than at higher temperatures.

Daniel Huster answered: Absolutely right, and we have already started these experiments. Will report more on this as soon as possible.

Markus Deserno asked: As has by now been pointed out numerous times at this meeting, the two leaflets of an asymmetric membrane do not have to be under the same tension. Looking at the order parameters in your NMR experiments, shown in Fig. 2 of your paper (https://doi.org/10.1039/d4fd00192c), I was wondering whether the curious *z*-dependent shifts might be caused by this differential stress? I know you are not a simulator, but have you discussed this with your simulation colleagues? If that were the reason for these shifts, it would basically amount to a (proxi-) measurement of differential stress *via* NMR, which would be amazingly useful.

Daniel Huster responded: All agreed. We collaborate with Robert Vacha on this issue. I am not sure how to measure leaflet-specific stress, though. I know Koenig and Gawrisch have measured the lateral compressibility by NMR/X-ray.[1]

1 B. W. Koenig, H. H. Strey and K. Gawrisch, Membrane lateral compressibility determined by NMR and X-ray diffraction: effect of acyl chain polyunsaturation, *Biophys. J.*, 1997, 73, 1954–1966, DOI: 10.1016/S0006-3495(97)78226-2.

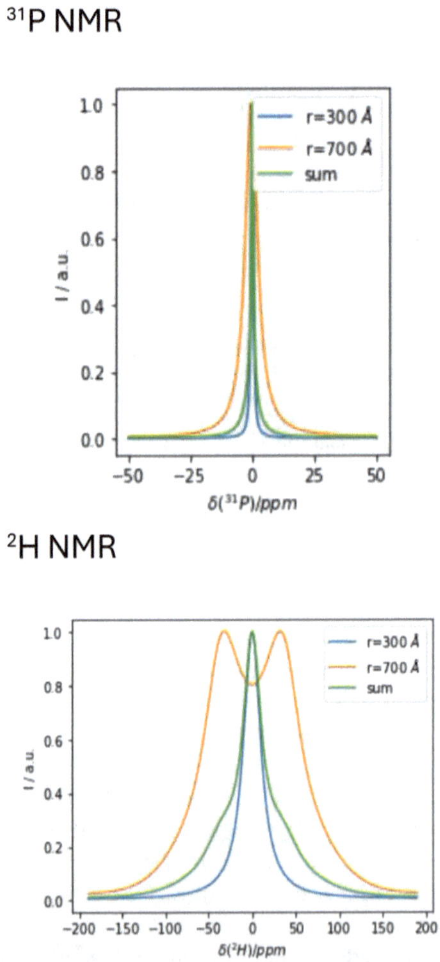

Fig. 2 NMR lineshape simulation considering averaging in 30 nm and 70 nm vesicles.

Robert Vacha remarked: We have calculated lipid order parameters from simulations with membranes at different tensions and we have not observed changes in order parameters similar to the ones obtained in NMR experiments so far.

Daniel Huster added: Yes, the interesting question to answer is: are the changes in order parameters we see in ^{2}H NMR related to cyclodextrin binding, a problem in the MD force field or periodic boundary conditions, or a new feature of asymmetry between the two leaflets? We will do experiments in POPC MLVs with the outer leaflet exchanged with POPC-d_{31}.

Milka Doktorova said: Obtaining structural data in the form of order parameters and relaxation rates for CH bond dynamics in asymmetric membranes with NMR provides a great opportunity to validate theoretical predictions and computational observations for asymmetric bilayers. In that respect, it is important to ensure that the experimentally measured parameters are not influenced by factors other than the asymmetric distribution of the bilayer lipids. Since the exchange in the NMR samples was performed with methyl-β-cyclodextrin (CD) and there is evidence that some of the CD is present in the sample after the exchange and washes, it is important to verify that the remaining CD is not affecting the results by *e.g.* binding to the bilayer and contributing to the apparent differential effect of asymmetry on the top and bottom parts of the chains. Have you compared the order parameter of POPC exchanged into POPC MLVs following the same protocol? This would serve as a direct control in this experimental setup and should reveal any potential artifactual effects from the protocol on the measured order parameters and relaxation rates.

Daniel Huster responded: This is indeed a very good suggestion. We have done this experiment and compared the ^{2}H NMR order parameters of symmetric POPC-d_{31} MLVs (in the absence of CD) with asymmetric MLVs POPC$_{in}$/POPC-$d_{31}{}^{out}$. Within experimental error, the order parameters are identical. This suggests that (i) CD does not influence POPC order parameters and (ii) the new features we see in the asymmetric membranes could indeed be caused by different tension in the individual leaflets of the asymmetric membranes.

Milka Doktorova asked: As mentioned in the text of the paper (https://doi.org/10.1039/d4fd00192c), one limitation of the current protocol is the limited accessibility only to the outer leaflet properties of the asymmetric outermost bilayer in the multilamellar vesicles. Can this type of NMR measurement be done on large or giant unilamellar vesicles instead where potentially the inner leaflet can also be labeled and examined? What are the minimal requirements for obtaining such structural data with regards to vesicle diameter and lipid concentration?

Daniel Huster answered: Yes, in principle, GUVs are possible. However, for good sensitivity, the lipid concentration should be really high, so GUVs should be not so diluted and concentrated. If they do not break while gently centrifuging them, I think this could work. Lipid concentration around 1 mg of ^{2}H-labeled lipid in the NMR volume (~0.5 ml) should be fine. A diameter of 200 nm or bigger should also work, but extruded LUVs of a diameter larger than 100 nm already contain multilayers…

Edward Lyman commented: Since the papers of Hossein, Varma, and Deserno[1,2] have come out the simulation community has been grappling with the problem of initializing asymmetric sims from a "correct" (or at least reasonable) membrane stress state. It seems to me like these S_{CD} measurements may be an incredibly informative window into this problem. One could run a series of simulations that replicate (*e.g.*) the compositional asymmetry of your sample, but over a range of number imbalance conditions. Because the force fields do a quite good job at capturing S_{CD}'s at a quantitative level, one could use the NMR data to select the simulation model that best matches the data. If you can't talk Markus Miettinen (or someone else) into this let me know and we will do it for you!

1 A. Hossein and M. Deserno, *Biophys. J.*, 2020, **118**, 624, DOI: 10.1016/j.bpj.2019.11.3398.
2 M. Varma and M. Deserno, *Biophys. J.*, 2022, **121**, 4001, DOI: 10.1016/j.bpj.2022.07.032.

Daniel Huster responded: Yes, absolutely. Robert Vacha has done preliminary simulations on our system but could not recover the order parameters so far. We will keep working on this issue for sure. We will also do ^{1}H–^{13}C order parameters where the assignment is clear, which may also help.

Michael M. Kozlov enquired: To what extent does the order parameter depend on the tension? I am asking since the in-plane area of lipid molecules should be rather independent of the tension because of the high value of the stretching rigidity.

Daniel Huster replied: I think this has not been worked out theoretically. I would guess that order parameter profile and the lateral stress profile ($\sigma(z)$) and the lateral pressure profile ($p(z)$) are related but I am not aware that this has been studied.

Gonen Golani opened discussion of the paper by Milka Doktorova: Can cryo-EM *in vivo* be sensitive enough to detect density differences between the monolayers in a realistic biological setting? In *in vitro* systems, will we still observe a difference in the electron density if the membrane is rich with cholesterol that can flip-flop to 'fill the gaps' in the monolayer with a lower phospholipid density?

Milka Doktorova responded: While cryo-EM can reveal structural details of biological membranes (as done for example, in a recent study[1]), it is not presently clear whether that can be done at the resolution needed to analyze the intensity variation across the bilayer as done in liposomes, and measure parameters like ω that inform on the lipid abundance imbalance between the two leaflets. One of the main difficulties in this endeavor comes from the requirement for relatively thin samples, *i.e.* a few hundred nanometers in thickness, which allows for imaging the membranes without 'squishing' them and potentially distorting the projected bilayer profiles. Ongoing technical advances could make this easier in the future but for now, this remains a challenge.

Regarding the *in vitro* systems, since lipid headgroups are more electron dense than any part of cholesterol, we do not expect Chol to interfere with the phospholipid (PL) imbalance that can be detected in the intensity profiles of the cryo-EM bilayer images. In fact, our computational analysis directly shows how even when Chol interleaflet distribution changes dramatically, ω varies systematically with PL imbalance. However, while we do not expect trends with PL imbalance to be affected, it is possible that Chol can change the absolute value of ω in bilayers with different lipid compositions *via* its effects on bilayer properties (and consequently, changes in the shape of the intensity profiles).

1 D. Glushkova, S. Böhm and M. Beck, *bioRxiv*, 2025, preprint, DOI: 10.1101/2025.03.13.642912.

Fabio Lolicato remarked: It is worth noting a methodological point regarding the calculation of area per lipid. Specifically, in a symmetric bilayer, if the area is normalized by the total number of molecules (phospholipids + cholesterol), the average area per molecule decreases as cholesterol is added. This is because cholesterol is smaller than a typical phospholipid, and mixing molecules of different sizes reduces the average area. In this context, one might infer that the volume per molecule also decreases. However, if the area is instead normalized by the number of phospholipids only, the interpretation becomes more complex. M. Javanainen[1] has shown that when the area is divided by the number of phospholipids alone, the average area per phospholipid remains nearly constant up to 20 mol% cholesterol. This indicates that cholesterol contributes no excess area up to this concentration. Beyond 20 mol%, however, the area per phospholipid begins to increase.

1 M. Javanainen, P. Heftberger, J. J. Madsen, M. S. Miettinen, G. Pabst and O. H. S. Ollila, *J. Chem. Theory Comput.*, 2023 **19**, 6342, DOI: 10.1021/acs.jctc.3c00648.

Milka Doktorova responded: Analysis of leaflet packing can be done by either counting or disregarding Chol. One case underestimates, while the other overestimates, the actual area occupied by the phospholipids. Our analysis shows that, considering Chol, the trend in leaflet packing does not follow trends observed in symmetric bilayers. If we calculate the area per phospholipid, A_P, instead (without counting Chol), the area of DAPC for example increases with Chol concentration in the leaflets of both symmetric and asymmetric membranes as shown in Fig. 3 here. However, the smaller effect on A_P up to ~20 mol% Chol observed in symmetric bilayers is not seen in the asymmetric ones. Instead, A_P increases steeply even at low Chol concentrations, demonstrating that trends in lipid packing in the leaflets of asymmetric membranes do not correspond to those of symmetric stress-free bilayers regardless of how the lipid packing is quantified.

Fabio Lolicato commented: Simulating highly asymmetric bilayers is still a developing area, and I don't yet know what the best practice is in this context. In particular, I'm concerned about the physical accuracy of such simulations. In a bilayer with lipid imbalance between the two leaflets, the (xy) dimensions of the simulation box are typically dictated by the leaflet with the higher area per lipid. This sets the overall box size and can lead to an artificially increased area per lipid in the opposing leaflet. Since lipid force fields are generally parameterized to

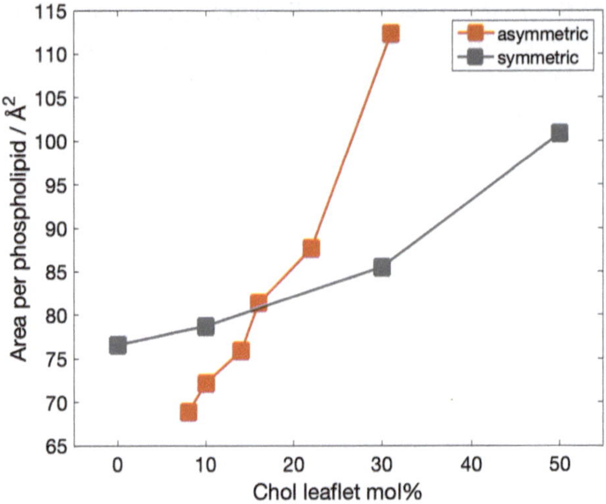

Fig. 3 Area per phospholipid, A_p, *versus* Chol concentration in a DAPC leaflet opposite of either DAPC leaflet (in symmetric bilayers, gray) or DPPC leaflet (in asymmetric bilayers, red).

reproduce a specific area per lipid, forcing lipids into a non-ideal packing environment may alter their physical properties and introduce artifacts.

Milka Doktorova responded: The bilayer area in typical *NPT* simulations of asymmetric bilayers is free to fluctuate and is governed not by the area per lipid of individual leaflets but by the differential stress in the membrane which arises from the mutual adaptation of the leaflets to the same shared total lateral area. Therefore, the simulation conditions represent a membrane state in which the two leaflets are constrained to have the same area. This is indeed the case for almost any existing model and biological membrane which are closed membrane surfaces.

Replacing one bilayer leaflet with a different one in simulations is analogous to replacing one protein binding partner with another one. If simulations can reproduce the behavior of the individual components, they present an ideal platform for studying the behavior of the assembly. It is true that lipid force fields have been calibrated against the bulk properties of symmetric bilayers for which there is extensive structural data from experiments (in contrast to asymmetric membranes). However, even though the parameterization relies on tension-free bilayers for example, simulations can reproduce the effects of net tension on bilayer area and dynamics, indicating that they are capable of capturing the properties of leaflets experiencing tension or compression such as those arising from differential stress.

It is important to note that in simulations of asymmetric bilayers there are important considerations that have to be taken into account. For instance, due to the presence of periodic boundary conditions a small bilayer patch is not able to relieve differential stress by *e.g.* curving or allowing lipids to flip between leaflets; there are no extra membrane pools and the number of lipids is fixed. This can occur for instance, in local regions of the plasma membrane, segregated from the bulk membrane *via* diffusion barriers. Analysis of the lateral pressure distribution in the simulated bilayer can reveal the amount of differential stress and propensity for curvature, and these can be related to the observed biophysical properties of the membrane. Until we are able to measure lipid number imbalance and fully characterize the asymmetry of membranes *in vitro*, we won't be able to directly validate the computational results with experiments. However, by specifically accounting for the constraints of the simulation conditions, we can use simulation results to improve our understanding of the effects of differential stress on bilayer structure and dynamics, and formulate testable hypotheses.

Fabio Lolicato queried: Do you think that a difference in area per lipid actually exists between the two leaflets in living cells? Or could it be that this imbalance is somehow compensated, either by the asymmetric shape of membrane proteins, differential protein crowding, or varying numbers of membrane-associated proteins on each side of the bilayer?

Milka Doktorova answered: Based on existing data, we believe there is a difference in the lipid packing (average area per lipid) in the two leaflets of living plasma membranes.[1] Some of the most direct evidence comes from experimentally measured differences in the readout of environment-sensitive probes confined to the exoplasmic or cytoplasmic leaflet (*e.g.* ref. 2–4). There are many advantages to a potential imbalance in the packing densities of the leaflets. Having a relatively compressed and impermeable exoplasmic leaflet, for instance can protect the cell while the ability to dynamically and locally change the relative packing of the two leaflets provides much greater functionality in terms of both the internal properties of the bilayer (*e.g.* lateral stress profile) and its ability to modulate differential interactions with peripheral and inserting proteins from the cytoplasm *versus* the extracellular medium. Therefore, considering the distinct lipid compositions of the two leaflets, equal areas per lipid would seem to be a very special limiting case with no clear advantages and thus unlikely to be actively maintained.

1 M. Doktorova, J. L. Symons, X. Zhang, H.-Y. Wang, J. Schlegel, J. H. Lorent, F. A. Heberle, E. Sezgin, E. Lyman, K. R. Levental and I. Levental, *Cell*, 2025, **188**, 2586, DOI: 10.1016/j.cell.2025.02.034.
2 C. Patra, M. Samad, T. van Zanten, M. Singh, R. A. Vishwakarma, P. P. Singh and S. Mayor, *bioRxiv*, 2024, preprint, DOI: 10.1101/2024.07.23.604763.
3 J. H. Lorent, K. R. Levental, L. Ganesan, G. Rivera-Longsworth, E. Sezgin, M. Doktorova, E. Lyman and I. Levental, *Nat. Chem. Biol.*, 2020, **16**, 644, DOI: 10.1038/s41589-020-0529-6.
4 A. Gupta, T. Korte, A. Herrmann and T. Wohland, *J. Lipid Res.*, 2020, **61**, 252, DOI: 10.1194/jlr.D119000364.

Giacomo Fiorin asked: Previously, when examining symmetric bilayers containing cholesterol you had collaborated with scientists who performed neutron spin echo (NSE), which is a powerful method to determine dynamic properties of a membrane, such as viscosity and interleaflet friction. For asymmetric bilayers like the plasma membrane, which are made of one ordered leaflet and a disordered leaflet, measuring such properties would be of great interest. Besides informing NMR experiments on asymmetric membranes, what are your thoughts about informing NSE experiments as well?

Milka Doktorova replied: Calculating representative NSE data from simulations is certainly feasible. This has been demonstrated for example, with coarse-grained simulations of flat bilayer patches and vesicles in ref. 1. Calculating the intermediate scattering functions from all-atom simulations should technically be doable as well, although the bilayer would have to be larger (*e.g.*, on the order of a few thousand lipids per leaflet). This is necessary to be able to obtain the decay of the dynamics at smaller wavevectors q (corresponding to larger real distances). For instance, a typical q range for analysis of bilayer dynamics with NSE is 0.037–0.12 $Å^{-1}$ (corresponding to length scales of ~50–170 Å) and a bilayer with dimensions of 35 × 35 nm would have ~2000 lipids per leaflet, assuming area per lipid of 60 $Å^2$. Since the dynamic range extends to ~100 ns, the simulations should also be at least 1–2 μs long. The intermediate scattering functions would yield the decay rates which can then be related to various bilayer parameters such as viscosity and bending rigidity. Doing this analysis for asymmetric bilayers with all-atom simulations would allow for selective deuteration of various lipids/lipid components (as done in the generation of representative SANS data) and should reveal the sensitivity of the measurement to the dynamics of bilayers with various degrees and types of asymmetry. It may also be interesting just from the point of view of analyzing the dynamics (thickness fluctuations) of the simulated bilayers *via* the NSE-based framework of parsing the intermediate scattering functions.

1 J.-M. Y. Carrillo, J. Katsaras, B. G. Sumpter and R. Ashkar, *J. Chem. Theory Comput.*, 2017, 13, 916, DOI: 10.1021/acs.jctc.6b00968.

Alberto Giacomello enquired: It is interesting to see that the area per lipid goes up as the cholesterol content in the leaflet is increased in asymmetric membranes. I was wondering whether you expect this behaviour to have consequences on the phase behaviour of asymmetric membranes as opposed to that of symmetric membranes?

Milka Doktorova answered: What our results suggest is that the interleaflet distribution of Chol, driven by differential stress, can produce leaflet properties that do not follow trends observed in symmetric bilayers with these compositions. It is therefore difficult to make predictions for the phase behavior of the asymmetric leaflets based on our understanding of phase separation in symmetric bilayers. If we were to draw any parallels to symmetric systems, the closest analog would probably be a bilayer with the lipid composition of one of the asymmetric leaflets, which is subjected to different amounts of net bilayer tension. This analysis would ignore any possible effects of interleaflet coupling in the asymmetric environment, but would show the effects of tension on the lateral mixing of the lipids.

Alberto Giacomello asked: Have you tried simulating cholesterol contents above 30%?

Milka Doktorova responded: We have not tried simulating these particular membranes with higher cholesterol concentration. Generally, adding more Chol would be easy, *i.e.* the bilayers should be able to easily tolerate it, and we would not expect big differences in Chol interleaflet distribution. We did a similar study in a different set of coarse-grained simulations, presented in Fig. 2A in ref. 1.

There, the outer leaflet had again saturated lipids while the inner leaflet was enriched in PUFA lipids and the phospholipid (PL) imbalance was fixed with the inner leaflet having 35% more PLs than the outer leaflet. We varied the overall Chol concentration in the bilayers and saw similar Chol enrichment in the underpopulated saturated leaflet across all simulations (Fig. S2A in ref. 1); however the bilayer remained in a stable flat morphology only when Chol concentration was ≥30 mol% of all bilayer lipids.

1 M. Doktorova, J. L. Symons, X. Zhang, H.-Y. Wang, J. Schlegel, J. H. Lorent, F. A. Heberle, E. Sezgin, E. Lyman, K. R. Levental and I. Levental, *Cell*, 2025, **188**, 2586, DOI: 10.1016/j.cell.2025.02.034.

Georg Pabst said: (i) Achieving a phospholipid (PL) imbalance of up to 2 experimentally seems quite challenging. How do you envision realizing this in a controlled and reproducible manner?

(ii) I am confused by the approach used to calculate the scattering form factor in eqn (1) in the paper (https://doi.org/10.1039/d4fd00200h). Why not simply compute the scattering amplitude directly from the scattering length density of a flat bilayer – that is, skip the squaring step, followed by taking the square root? A more rigorous calculation would also yield the form factor (and hence the scattered intensity) on an absolute scale. Note: this also makes modelling more reliable (see ref. 1 and 2).

(iii) I do not believe that the center of mass (COM) for SAXS/SANS is a straightforward or reliable measure of PL imbalance, at least not as suggested. Several issues need to be considered:

(a) Membrane thickness fluctuations contribute to the scattering signal in the relevant q-range, as discussed in ref. 1 and 2.

(b) Experimental noise, including incoherent scattering in SANS at high H_2O content, needs to be properly accounted for and can introduce significant uncertainties.

(c) The choice of q-range for the analysis appears somewhat arbitrary and could easily bias or mislead the conclusions.

1 M. P. K. Frewein, M. Doktorova, F. A. Heberle, H. L. Scott, E. F. Semeraro, L. Porcar, and G. Pabst, *Symmetry*, 2021, **13**, 1441, DOI: 10.3390/sym13081441.
2 M. P. K. Frewein, P. Piller, E. F. Semeraro, K. C. Batchu, F. A. Heberle, H. L. Scott, Y. Gerelli, L. Porcar and G. Pabst, *J. Membr. Biol.*, 2022, **255**, 407, DOI: 10.1007/s00232-022-00234-0.

Milka Doktorova responded: The bilayer with imbalance of 2 is very stressed and while this is tolerated in the simulations, it is unlikely that such state can be realized in a synthetic model system *in vitro* without causing budding, rupture or other morphological responses, as we point out in the text (https://doi.org/10.1039/d4fd00200h). Still, in our opinion, it is informative to examine it as a limiting case in this simulation study to gain better understanding of the trends and extents to which differential stress can affect bilayer properties.

Regarding question (ii), we apologize if the explanation we provide in the text is confusing. In retrospect, it would have been more straightforward to give the equation for the form factor as a function of the scattering length density (SLD) profile and the intensity as a function of the form factor (rather than the other way around, as we did it), but the result is the same.

Regarding absolute scale, we first note that the analytical scattering models used in the mentioned references are not derived on an absolute scale, nor does either paper evaluate or discuss whether or how analyzing data on an absolute scale makes the modelling more reliable. In our opinion, the larger issue is that extrusion (the typical method for preparing LUVs for scattering) can change the sample concentration either by dilution from water/buffer in the dead volume of the extruder or by lipid adsorption to the filter, so even if scattering curves are put on an absolute scale, there is still sample-to-sample variability in concentration. In our experience, fitting always requires an arbitrary scale factor for each dataset to account for this (generally unknown) variability in lipid concentration. Therefore, we do not see advantages in analyzing the scattering data (both from experiments and simulations) on an absolute scale.

Regarding question (iii), we agree that many factors can complicate the interpretation of the scattering data. This is true whether one uses an analysis based on fitting to models (as in the mentioned references) or a model-free approach like we propose. These confounding factors include not only thickness fluctuations, but also lateral lipid clustering (including phase separation), especially in mixtures of deuterated and protiated lipids (see ref. 1). While neither of these factors is accounted for in the simulated form factors presented here, in principle it is possible to include them with a hybrid MD/Monte Carlo approach similar to that in ref. 1.

We fully agree that the proposed analysis would not be advisable for SANS data of vesicles in a high-H_2O medium due to the large increase in incoherent scattering.

As for the *q*-range, it is not arbitrary but was instead chosen to encompass the region of the scattering curves that is most reliable (*i.e.*, least affected by background scattering); of course, other ranges could be explored using the provided methodology. We note that this analysis is much cruder than modeling-based approaches which can provide detailed structural parameters of the bilayers. Yet, this simple representation is able to quantify differences between asymmetric bilayers directly from the scattering form factors, allowing for potential calibration of the underlying PL number imbalances with simulations.

1 F. A. Heberle, V. N. P. Anghel and J. Katsaras, *J. Appl. Crystallogr.*, 2015, **48**, 1391, DOI: 10.1107/S160057671501362X.

Daniel Huster queried: You observed an increase in area of lipid as a function of cholesterol; normally, highly unsaturated lipid does not mix well with cholesterol. So I could see that the area stays the same as adding more cholesterol due to de-mixing. But how can it increase? In the upper leaflet with saturated chains only, one would expect a decrease in area due to the condensation effect of cholesterol.

Milka Doktorova replied: Cholesterol has a clear condensing effect in symmetric bilayers: it not only has smaller area than a phospholipid which naturally decreases bilayer area when it replaces lipids, but in the case of saturated lipids it also effectively orders their chains further compressing the area. However, in asymmetric membranes, the Chol concentration in a leaflet is driven by differential stress (among other factors), and is not the same as simply adding or removing Chol in the leaflet, as in symmetric bilayers. Instead, in asymmetric membranes the area of the leaflets (which essentially determines the area per lipid) is determined by the area of the bilayer which is a direct consequence of the stored differential stress; it is not determined by Chol concentration alone. In other words, the Chol concentration in the leaflets is an after-effect of PL imbalance and differential stress, and that is why the resulting area per lipid that we observe does not have to (and does not, in our simulations) follow the condensing effect that Chol has in stress-free symmetric bilayers.

Gerald W. Feigenson enquired: I enjoyed the new MD simulations. Have you tried, or could you try, instead of DAPC, cytoplasmic-type lipids with 16 or 18 *sn*1 chain? Such lipids are very different in interactions with cholesterol, *e.g.* as measured by maximum cholesterol solubility. Such cytoplasmic lipids could very well change your MD results.

Milka Doktorova answered: We can certainly try using different lipids in the simulations. For example, in ref. 1 we ran a series of coarse-grained bilayer simulations in which the two leaflets had lipid compositions representative of those of red blood cell plasma membranes, *i.e.* with sphingomyelin and PC in the top leaflet and mixed chain lipids in the bottom leaflet. There, we kept the phospholipid imbalance between the leaflets fixed (with the cytoplasmic leaflet having 35% more lipids than the exoplasmic leaflet) and varied the total cholesterol concentration in the membrane. In the bilayer with 30 mol% Chol, we saw Chol enrichment in the top leaflet but it was noticeably smaller compared to bilayer A.1 (the closest analog in the present study). This illustrates that, as suggested, the presence of mixed chain lipids can change the interleaflet distribution of Chol. However, the relative apparent effects of Chol on lipid packing (*e.g.* the trends in Fig. 1B in the paper (https://doi.org/10.1039/d4fd00200h)) are driven mainly by differential stress and we do not expect that to change with variations in lipid composition, but we would have to run the simulations to confirm.

1 M. Doktorova, J. L. Symons, X. Zhang, H.-Y. Wang, J. Schlegel, J. H. Lorent, F. A. Heberle, E. Sezgin, E. Lyman, K. R. Levental and I. Levental, *Cell*, 2025, **188**, 2586, DOI: 10.1016/j.cell.2025.02.034.

Robert V. Law asked: Do you always assume cholesterol is colinear with lipids (with its hydroxyl group pointing towards the water layer)? Some work with PUFA lipids indicates that cholesterol can be perpendicular to bilayer, between the two bilayer leaflets.[1]

1 N. Kučerka, *et al.*, *J. Am. Chem. Soc.*, 2009, **131**(45), 16358–16359, DOI: 10.1021/ja907659u.

Milka Doktorova responded: Generally, Chol is rarely seen to partition to the bilayer midplane in all-atom simulations. To verify that, we calculated the distribution of Chol tilt angles in our trajectories (defined by the angle between the vector connecting atoms C3 and C17 in CHARMM36 notation, and the bilayer normal, *i.e.* the positive direction of the *z* axis of the simulation box). The majority of Chol was either in a tight upright orientation in the DPPC leaflet (tilt angles around 10°) or adopting a wider range of tilt angles in the opposite DAPC leaflet (tilt angles around 150°). This is illustrated in Fig. 4 here for the two bilayers with most extreme PL imbalances, A.1 and A.6. There is no density at 90° confirming the absence of molecules lying flat at the midplane. In coarse-grained simulations of these bilayers, we do see some density of tilt angles around 90°, however analysis of the time evolution of the tilts of individual Chol molecules shows that these are states that Chol adopts only transiently as it flips between the leaflets. Therefore, we do not have reasons to believe that Chol at the midplane is affecting our results.

Markus Deserno said: In Fig. 1B of your paper (https://doi.org/10.1039/d4fd00200h) you argue that the effect of cholesterol concentration in a single leaflet can be quite different in symmetric *versus* asymmetric bilayers – namely, the area per lipid of both the saturated DPPC and the unsaturated DAPC leaflet increases with increasing leaflet cholesterol concentration, as opposed to the observations in a symmetric system, where the trend quite naturally goes the other way. I am not sure I entirely follow the explanation given in the text, but I would like to suggest what I think might be a simpler rationalization of this observation, and I am curious whether you would agree with it. For ease of reference, I have included as Fig. 5 here a version of your Fig. 1B with a few extra annotations in it, and with the symmetric trends removed for clarity.

My argument relies on the fact that the driving force in your sequence of simulations is not cholesterol but differential stress (as created by abundance asymmetry; cholesterol then just goes along for the ride). The bilayer in which both cholesterol leaflet fractions are close to 30 mol% features a very large abundance asymmetry: the DPPC leaflet has about twice as many phospholipids as the DAPC leaflet, as we can see in your Fig. 1A of the same panel. Much cholesterol is hence squeezed out of its more "natural" DPPC environment, leading to its unusual 50 : 50 partitioning between the DPPC and the DAPC side.

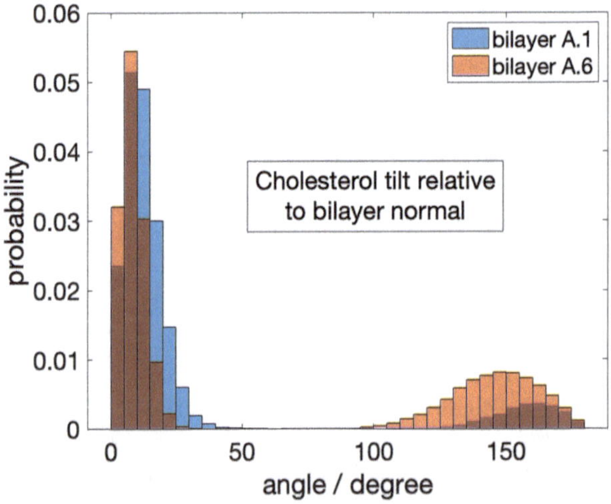

Fig. 4 Cholesterol tilt relative to bilayer normal (positive direction of the *z* axis of the simulation box) for the two simulator bilayers with most extreme PL imbalances, A.1 and A.6. Shown are the probability distributions of Chol tilt angles in the two bilayers.

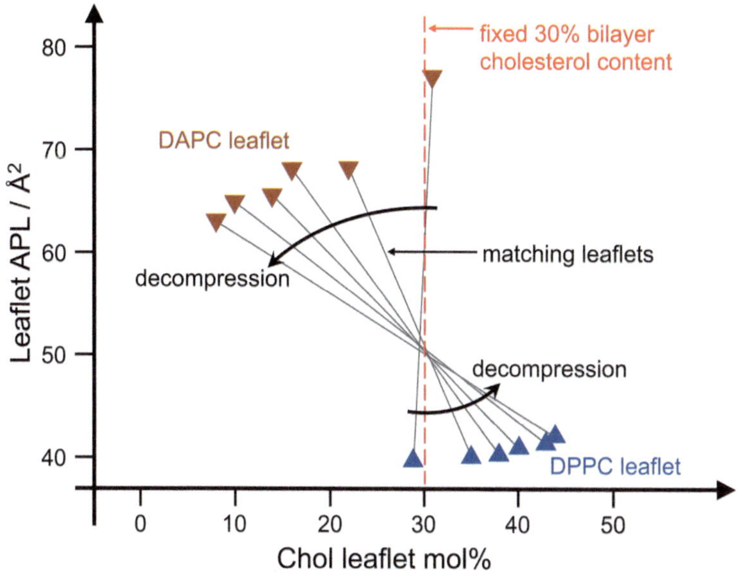

Fig. 5 Annotated version of Fig. 1B from https://doi.org/10.1039/d4fd00200h.

For the ease of discussion, an equivalent but easier to interpret realization of your abundance scan is to say that we run simulations at a fixed number of cholesterol lipids, which then also means a fixed number of phospholipids (say, 200 lipids overall, of which 60 (*i.e.*, 30%) are cholesterol and 140 are phospholipids, whose relative DPPC/DAPC abundance we now tune). The "decompression" that happens upon lowering the abundance asymmetry (indicated by the two black curved arrows in my Fig. 5 here) then effectively amounts to removing DPPC lipids from the DPPC leaflet, turning them into DAPC lipids, and inserting those into the DAPC leaflet. The number of DAPC lipids in the DAPC leaflet therefore *increases* upon decompression, while the area of the bilayer also shrinks (because the huge compressive stress in the DPPC leaflet is relieved). Both effects imply a reduction in the area per lipid in the DAPC leaflet, and the only confounding effect that could counter that is that you also lose cholesterol lipids from there into the DPPC leaflet. However, this is a smaller correction that cannot undo the bigger effect of differential stress relaxation. This means that plotting the effect against the emergent cholesterol leaflet fractions might lead one to the possibly misleading view that cholesterol "acts funny" when added to a DAPC leaflet in an asymmetric membrane. Of course, there is nothing wrong with your observed correlation, but it strikes me as a somewhat more indirect consequence of a bigger driver that is hiding in the background. Would you agree with this attempt at an alternative rationalization?

Milka Doktorova replied: We completely agree that differential stress is the main driving force here. Within the representation of fixed numbers of lipids and cholesterol, it is not entirely clear why decompression of the DPPC leaflet by turning DPPC into DAPC lipids and moving them to the DAPC leaflet would *shrink* the area of the bilayer since that would be associated with a much larger compressive force on the DAPC leaflet. We think this is more likely to slightly *increase* the bilayer area. However, the result is the same in that cholesterol movement to the DPPC leaflet is not able to counter the effect of differential stress relaxation and instead the area per lipid would increase in the DPPC leaflet and decrease in the DAPC leaflet in a way opposite of what we would expect from the corresponding changes in the leaflets' cholesterol concentrations. Plotting the area per lipid as a function of Chol leaflet mol% may be misleading if perceived as 'adding' Chol to a leaflet and seeing its response (as opposed to varying the relative numbers of lipids and analyzing the Chol distribution resulting from that). However, we believe the trends are still informative as they demonstrate that if we were able to measure the Chol concentration in one of the leaflets of an asymmetric membrane, for example, and see that it changes under different conditions or between membranes, we cannot make inferences about the relative packing of the leaflet based solely on Chol concentration. In the presence of differential stress, Chol does not behave as Chol in stress-free symmetric bilayers, and this is one of the main points that this analysis illustrates.

Francisco N. Barrera asked: In different simulations you can observe that the lower membrane leaflet gets thinner as the DAPC lipid chains bend more appreciably. Do you expect that this thinning affects the partition of cholesterol to this leaflet due to the changing local hydrophobic mismatch? This question is assuming that cholesterol will remain in a "vertical" state in all cases.

Milka Doktorova answered: The DAPC leaflet does indeed become thinner as the tension in the leaflet increases in the simulation series, and this is accompanied by an increase in the cholesterol concentration in that leaflet. This preferential partitioning of cholesterol to the DAPC leaflet is likely influenced by a number of different factors which affect its chemical potential. The changes in thickness are driven by the differential stress in the bilayer (compressing the top leaflet and expanding the bottom one). While cholesterol concentration is correlated with thickness in the DAPC leaflet, this is not the case in the DPPC leaflet where for bilayers A.4, A.5 and A.6 the leaflet thickness remains constant at 24.1 Å but Chol decreases from 38 to 29 mol%. Therefore, it would be hard to attribute the observed changes in Chol distribution to thickness in general, however it is possible that when it comes to the unsaturated DAPC leaflet, its thinning (driven by differential stress) contributes to making the environment more favorable for Chol relative to the stress-free state of the leaflet.

Michael M. Kozlov opened discussion of the paper by Oded Farago: By calculating the energies of the membrane leaflets, you are using the expression for the energy proportional to the square of the difference between the mean and spontaneous curvatures. This means that the reference state for the energy computations is always the state in which the leaflet's curvature is equal to the spontaneous curvature. This can be done if the spontaneous curvature of each leaflet is a constant. Yet, if there is a lipid exchange between the leaflets, the spontaneous curvatures of the leaflets are changing. In that case a direct comparison of energies computed by using the quadratic expression before and after the exchange appears problematic since these energies are related to different reference states. What would be your comment on that?

Oded Farago replied: Indeed, we assume that there is no exchange of lipids between the leaflets, which is applicable on times much smaller than the flip-flop times of lipids. The theory definitely needs to be extended when cholesterol, which is rapidly flip-flopping, is present. With that said, the key finding is that local phase separation may be accompanied by curvature instability because of the *lateral* inhomogeneities in the spontaneous curvature and bending rigidity. It is true that the variations in these parameters due to lipid monolayer exchange must be considered, but these will not necessarily change the outcome (*i.e.*, curvature instability) as long as the membrane is laterally inhomogeneous. In other words, the theory assumes quadratic curvatures energetics, and the instability arises from changes in the free energy parameters that are coupled to local phase transitions (and may involve also lipid exchanges).

Markus Deserno said: We had also considered the nature of the fluid–gel transition in a number-asymmetric membrane[1] and found a phase diagram that contains both continuous and discontinuous transitions. I cannot quite tell from the specific plots you display in your paper (https://doi.org/10.1039/d4fd00182f) whether the scenario you discover is similar or not, so I wanted to inquire. For ease of reference, I append a cleaned-up phase diagram we later published in ref. 2 (Fig. 6 here). As a function of relative lipid number asymmetry δn and (essentially) temperature T, we find that both the upper and the lower leaflet of the membrane can be in three different states: completely fluid ("f"), completely gel ("g"), or in a state where both fluid and gel phases coexist ("c"). We only observe a first order transition when, for not too large asymmetry, we heat or cool the system across the temperature where the "usual" (symmetric) gel transition used to be. In that case we see a "coexistence swap": a bilayer where, for instance, the upper leaflet is in a coexistence region and the lower leaflet is fluid would upon cooling go into a bilayer in which the upper leaflet fully turns into a gel and the lower leaflet enters fluid/gel coexistence. All other transitions – which

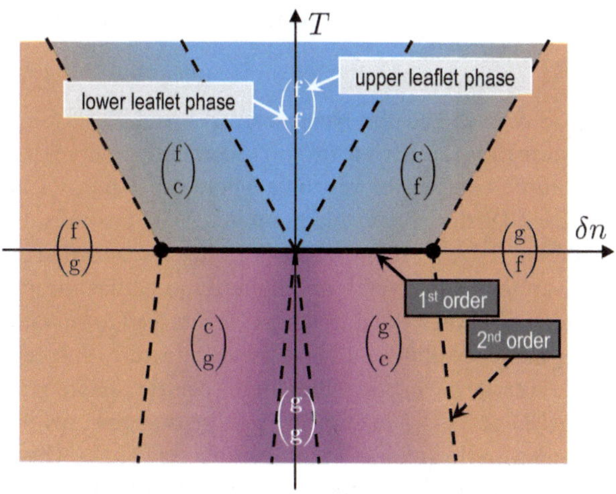

Fig. 6 Phase diagram as a function of temperature T and relative lipid number asymmetry δn.

basically amount to the emergence or dissolution of leaflets with coexistence – are continuous.

We have a theoretical model that is essentially equivalent to two coupled van der Waals gases, but this model assumes that the membrane always stays flat – there are no curvature contributions in our theory or any of the accompanying simulations. I understand that you have a slightly different situation where the bilayer has two oppositely curved regions, and this seems to matter in your case. Could you clarify how this pans out?

1 S. L. Foley, A. Hossein and M. Deserno, *Biophys. J.*, 2022, **121**(16), 2997–3009.
2 S. L. Foley, M. Varma, A. Hossein and M. Deserno, *Emerging Top. Life Sci.*, 2023, 7(1), 95–110.

Oded Farago responded: Fig. 10 in the paper (https://doi.org/10.1039/d4fd00182f) suggests that both monolayers transform simultaneously into a state of coexistence: the upper one – coexistence between gel and dense fluid; and the lower one – between two dilute fluid phases with slightly distinct densities. Thus, the short answer to your question is that this kind of a phase transition deviates from the ones proposed in your work because your phase diagram does not show a state where both leaflets are at coexistence. A close look at our Fig. 10 reveals another interesting thing: the monolayer densities in one of the segments seem to follow a linear dependence on δn with respect to the monolayer densities prior to the transition, while the densities in the monolayer of the other segment exhibit discontinuities. These observations suggests that coupling between the lateral (density) and normal (curvature) degrees of freedom must be considered in order to understand the phase behavior of asymmetric membranes. Eqn (16)–(21) in the paper highlight some of the considerations that one must consider when attempting to minimize the bilayer free energy with coupling between area and curvature.

Rumiana Dimova asked: You find that near the gel transition, asymmetry can induce phase separation and curvature instabilities. What are the potential biological implications of this finding, particularly in the context of lipid rafts? Are there particular systems close to gel transitions to which the findings will apply directly?

Oded Farago answered: The observed curvature instability is not limited to the first order gel transition occurring in single-component lipid bilayers. It can happen also in multi-component mixtures where other phase transitions take place, *e.g.*, the second order phase separation in ternary mixtures. In general, the transition is driven by the inter-molecular interactions. When it happens, different regions in the membrane have different mechanical properties and the mechanical coupling between the different segments lead to curvature instability (*i.e.*, different segments exhibiting distinct mean curvatures subject to the global constraint on the average mean curvature). In the context of lipid rafts, it is known that many biological processes take place in these domains, involve changes in local molecular distributions as well as curvature extrusions. The work demonstrates that the local changes in composition, lead not only to domain formation and local phase separation, but also to curvature instability arising from the membrane asymmetry and the variations in the mechanical parameters characterizing the different phases. It may be relevant to several biological processes like endo- and exo-cytosis.

Giacomo Fiorin enquired: One of the main advantages of Monte Carlo simulations, as you have highlighted in your paper (https://doi.org/10.1039/d4fd00182f), is the ability to fine-tune the moves of individual particles to implement complex physical forces on them.

Do you think that your approach could be extended to engineer different curvature preferences in the lipid molecules of each leaflet, possibly without changing their respective areas per lipid molecule (APLs)?

This would be greatly advantageous compared to deterministic MD simulations, both coarse-grained or atomistic, which currently do not provide a method to describe an asymmetric membrane where the preferred curvature is the only parameter that changes between the leaflets.

Oded Farago replied: A model of cone-shaped lipids with preferred curvature can be easily generated by changing the sizes of the beads forming the coarse-grained lipids. This has, in fact, already been done in previous works. What Monte Carlo allows you to do is to: (i) preserve the lipid asymmetry in simulations runs longer than the flip-flop times of the lipids (by rejecting moves that cause flip-flops), and (ii) perform simulations in a restricted configuration phase space in order to measure the free energies of specific membrane structures. With Molecular Dynamics you must follow the dynamics of the system over the whole phase space, subject to periodic boundary conditions.

Sarabjeet Kaur opened discussion of the paper by Sylvie Roke: How does second harmonic imaging accurately capture signals at the water surface, given that water molecules are inherently unstable, especially considering that water is transparent and its molecules are free-floating, making signal detection seemingly challenging?

Sylvie Roke responded: The second harmonic (SH) photons are created by tiny charge oscillations within molecules that occur at the second harmonic frequency. These oscillations are very weak but present in all water molecules, and if one detects very carefully, indeed SH scattering can be recorded from every substance, including water. This process was discovered in 1965,[1] and was termed hyper Rayleigh scattering. The emission can be enhanced when molecules are correlated in their orientational distribution, which creates a so-called coherent enhancement. On a surface, there is such a correlation, since the plane of the surface together with the separation between molecular species leads to a non-zero orientational distribution of molecules. This coherent emission is what is detected in SH imaging of water, and it is $\sim 10^7$ times weaker than resonant two-photon emission, under comparable imaging conditions.[2]

1 R. W. Terhune, P. D. Maker and C. M. Savage, Measurements of nonlinear light scattering, *Phys. Rev. Lett.*, 1965, **14**, 681–684, DOI: 10.1103/PhysRevLett.14.681.
2 M. Eremchev, D. Roesel, P.-M. Dansette, A. Michailovas and S. Roke, High throughput wide field SH imaging of giant unilamellar vesicles, *Biointerphases*, 2023, **18**, 031202, DOI: 10.1116/6.0002640.

Sarabjeet Kaur asked: Referring to the plot Fig. 6c in your paper (https://doi.org/10.1039/d4fd00197d) that shows second harmonic (SH) intensity *versus* time, is the observed SH signal decay a reliable and direct proxy for proton translocation across the membrane, or could it be influenced by other interfacial dynamics that affect the measured timescale?

Sylvie Roke replied: It is unlikely that other interactions/processes are changing the SH response, besides proton translocation. SH imaging probes changes in the orientational distribution of water in the narrow volume adjacent to the interface that is ~1 nm thick. Protons will change this water structure and neutralize/interact with PS head groups. These interactions initially reduce the amount of charges on one leaflet, which change the orientational distribution of water molecules on that leaflet, creating a bigger contrast (at t_2). Once the protons reach the other leaflet the orientational distribution of water will be impacted in the same way as the opposing leaflet, which creates a more symmetric orientational distribution of water molecules on the bilayer as a whole.

Felix M. Goni questioned: You mentioned that the ionic concentration near the membrane surface is 1 M. In my view, units of concentration are only meaningful when applied to bulk volumes. Can you really apply concentration units to a two-dimensional nanovolume?

Sylvie Roke responded: Indeed, the unit of concentration is in content per volume. The hydrated region of the bilayer has a thickness of several water layers and so we can consider this region as a volume.

Rumiana Dimova asked: Have you considered whether some of the second harmonics hotspots observed in GUVs or cells might originate from highly curved membrane regions – such as nanotubes or invaginations or even submicroscopic membrane undulations – rather than from charge or hydration asymmetry? Note that membrane nanotubes are not detected with transmission-light microscopy such as in phase contrast observations. It would be great if you are able to compare a tense *vs.* flaccid GUV held by a micropipette (tubulation and fluctuations are suppressed in tense GUVs) rather than under osmotic shocks, which can lead to membrane poration and subsequent tension relaxation.

Sylvie Roke answered: Yes, we have observed tubules growing out of the free-standing lipid membranes (FLMs) and they do look rather differently than the surface potential fluctuations. We have performed both osmotic shock experiments as well as aspiration experiments on GUVs, but the results are not published yet.

Rumiana Dimova said: In one of your slides, you showed a movie of the signal from the giant vesicle membrane, but there also appeared to be non-negligible signal inside and outside the vesicle, even though the surrounding medium shouldn't contribute. This may be due to signal enhancement. Could you comment on the potential risks of over-enhancement leading to artifacts or over-interpretation of the data?

Sylvie Roke replied: There are a few aspects to consider here, namely (i) the origin of the SH emission and (ii) how the images/videos were created. (i) SH photons are created by tiny charge oscillations within molecules that occur at the second harmonic frequency. These oscillations are very weak but present in all water molecules. SH scattering can be recorded from every molecularly anisotropic substance, including neat water. The emission is enhanced when (water) molecules are correlated in their orientational distribution, which creates a so-called 'coherent enhancement'. On a surface, there is such a correlation, since the interface creates a non-zero orientational distribution of molecules. These two effects, 'incoherent' and 'coherent' emission, both take place (see *e.g.* ref. 1 for an explanation) with the latter being bigger than the former. When GUVs are SH imaged, both are detected, albeit with different intensities. It is therefore possible that the high intensity tail of the bulk water SH emission distribution gets to be on the same order of magnitude of the low intensity tail of the interfacial SH emission. (ii) The imaging of GUVs is done with linearly polarized light, and only the component that is normal to the surface contributes to the emission. This is why the raw data shows a 'ring' of vanishing intensity. Dividing the intensity image by the normal component of the electrostatic field squared provides the true surface contribution. This is shown in detail in Fig. S4 of ref. 2.

1 M. Flór, D. M. Wilkins, M. de la Puente, D. Laage, G. Cassone, A. Hassanali and S. Roke, Dissecting the hydrogen bond network of water: charge transfer and nuclear quantum effects, *Science*, 2024, **386**(110), eads4369, DOI: 10.1126/science.ads4369.
2 D. Roesel, *et al.*, Ion-induced transient potential fluctuations facilitate pore formation and cation transport through lipid membranes, *J. Am. Chem. Soc.*, 2022, **144**(51), 23352–23357.

Nasim Mirzajani asked: In relating the SH intensity to the transmembrane potential difference, $\chi^{(2)}$ contributions in eqn (3) of your paper (https://doi.org/10.1039/d4fd00197d) have been neglected on the basis that the value of the difference in $\chi^{(2)}$ across the membrane is smaller than the value of $\chi^{(3)}$. However, to make a direct comparison, $\chi^{(3)}$ would have to be multiplied with the potential difference, as it appears in the equation. The potential difference values (<200 mV) would seem to bring the values of $\chi^{(2)}$ and $\chi^{(3)}$ contributions closer to each other. How would taking $\chi^{(2)}$ contributions into account change the subsequent interpretation of the data? Are there further studies that can be done to evaluate the role of surface chemical interactions in this system, without the field-driven interactions?

Sylvie Roke responded: The unit of the surface $\chi^{(2)}$ tensor is $m^2\ V^{-1}$ while that of the $\chi^{(3)}$ tensor is $m^2\ V^{-2}$. For an asymmetric lipid membrane, the first one has been determined to be on the order of $\sim 10^{-24}\ m^2\ V^{-1}$, while the second one is $\sim 10^{-21}\ m^2\ V^{-2}$.[1] Multiplying this by 0.2 V (we need to consider the unit V, not mV), gives $0.2 \times 10^{-21}\ m^2\ V^{-1}$. For a symmetric lipid bilayer the $\chi^{(2)}$ contribution is even smaller than the value for an asymmetric bilayer. So, for water imaging in all cases the third-order contribution is bigger than the second-order one, which results in a parabolic curve when the image-integrated SH intensity is measured as a function of applied bias (see Fig. 2B and 3D in the paper (https://doi.org/10.1039/d4fd00197d)). For a resonant process *e.g.* when applying SH resonant fluorophores, the resonant enhancement results in a situation where the $\chi^{(2)}$ contribution is larger than the $\chi^{(3)}$ contribution. This should result in different (linear) dependence when the integrated intensity is plotted as a function of applied bias.[2]

1 O. Tarun, C. Hannesschläger, P. Pohl and S. Roke, Label-free and charge-sensitive dynamic imaging of lipid membrane hydration on millisecond time scales, *Proc. Natl. Acad. Sci. U. S. A.*, 2018, **115**, 4081–4086.
2 B. A. Jemet, V. Nikolenko and R. Yuste, Second harmonic imaging of membrane potential of neurons with retinal, *J. Biomed. Opt.*, 2004, **9**(5), 873–881.

Edward Lyman enquired: I am really interested in new measurements of lipid flip-flop rates catalyzed by scramblases, especially with better time resolution. So, is it possible to reconstitute a scramblase (like TMEM16F) into an asymmetric black lipid membrane prep? Then activate it with Ca^{2+}, and observe how lipid asymmetry is lost?

Sylvie Roke answered: If the scramblase can be embedded in the free-standing lipid bilayer then this experiment should be possible. From the optical side there are no reasons why this could not be done.

Di Jin asked: When you have symmetric ion concentrations across a membrane containing PA lipids, you have shown you observed localised transmembrane potential. In the absence of PA lipids, would you still expect to observe the signal?

Sylvie Roke responded: The intensity is there because the PA head groups are charged, and these charges create an orientational distribution of water molecules, which has a non-zero component along the surface normal, by means of charge–water interactions. If the PA groups were not present, and the membrane had only PC groups, it would be on average charge neutral on each leaflet. There might still be some other interactions which provide an interfacial non-isotropic distribution, but since these interactions are much weaker, they are currently not easily detectable with SH imaging. As an example, see Fig. 1B and 2D in ref. 1.

1 O. Tarun, C. Hannesschläger, P. Pohl and S. Roke, Label-free and charge-sensitive dynamic imaging of lipid membrane hydration on millisecond time scales, *Proc. Natl. Acad. Sci. U. S. A.*, 2018, **115**, 4081–4086.

Di Jin enquired: Can you explain how the localised transmembrane potential is related to fusion?

Sylvie Roke replied: Membrane potential fluctuations are accompanied by transient electrostatic field fluctuations that might aid in the creation of membrane nanopores, or defects that can aid the fusion process. There is no investigation at this time available but I can imagine this as a potential scenario.

Michael Kaltenegger questioned: I was wondering how comparable FLMs and GUVs are as systems for inferring proton translocation rates from SH-intensity measurements. Specifically, I'm considering the impact of the larger volume of solvent on the other side of a FLM, compared to the smaller volume inside a GUV, on the time required for the pH difference to diminish. Wouldn't it inherently take more time for the pH contrast to disappear in the FLM system compared to GUVs due to the differences in solvent volumes?

Sylvie Roke replied: The contrast that appears in the SH images is determined by the water molecules adjacent to the membrane, in a layer that is typically 2–3 water molecules thick. When adding H^+ to the outside (GUV)/upper leaflet (FLM), the SH intensity increases on the same time scale (Fig. 6C in the paper (https://doi.org/10.1039/d4fd00197d)) in both systems. This means that protons have started interacting with the interfacial water/lipid head groups of the leaflet nearest to where they were injected. The SH intensity decreases again because the opposing leaflet begins to undergo a similar structural transformation, which restores the proton induced asymmetry that was maximum at t_2. Protons leaving/not in the ~1 nm thick interfacial region do not impact the SH intensity. This means that the diffusion into the bulk is unimportant for the experiment.

Iwona Swiderska opened discussion of the paper by Alexander P. Fellows: (i) You were comparing PC to PS lipids in terms of their domain formation. Do you expect this domain behavior difference to be purely electrostatic, *i.e.* related to the charge of the headgroup, and can it be therefore generalized to other lipid types with the same charge?

(ii) Do you expect the domain behavior to be different between a monolayer and a bilayer system and if so, how?

Alexander P. Fellows replied: I certainly think that electrostatics plays an important role in the domain formation with charged lipids. Clearly bringing together charged head-groups in such close proximity creates substantial electrostatic interactions that will impact the packing structure. However, the fact that these charged lipids overcome these repulsive interactions to form condensed domains shows that other factors are at play, particularly the lateral van der Waals interactions between the tail-groups. In this sense, I would generally expect other lipids with the same head-group charge and tail functionality to behave in a similar fashion (*i.e.* not depending that strongly on the specific chemical structure of the head-group), but just having the same charge on the head-group with tails of different chain length or degree of saturation would likely result in a different phase structure in the membrane. These are, however, definitely important questions that need to be addressed going forward.

In regards to the differences between mono- and bi-layer systems, it is definitely the case that the trans-leaflet interactions will play a role in the domain behaviour. We have even heard quite a few talks emphasising the importance of these interactions, with them acting to slightly perturb the domain structures and, importantly, lead to correlated packing structures in the two leaflets. Nevertheless, despite these differences, monolayer studies give significant insight into the fundamental lateral interactions unpinning domain formation, which can then be extended to bilayer studies to independently assess the impact of the inter-leaflet interactions. While the studies we present here are not currently feasible in bilayer systems, this is an area we are actively pursuing.

Edward Lyman addressed Alexander P. Fellows, Sylvie Roke and John C. Conboy: Is it possible to detect any of these spectroscopic observables in the near field in the liquid state? Maybe with a tip enhanced detection, or with some kind of patterned substrate? The goal is to obtain better spatial resolution. I know that Markus Raschke at University of Colorado Boulder was saying that he could do this in liquid state at some point, but I didn't follow this so closely.

Alexander P. Fellows responded: I completely agree that gaining better spatial resolution is an important factor for the future characterisation of these sorts of systems. It is definitely the case the tip-enhanced measurements have started to become a research focus,[1] and we have even applied such a method to model membrane systems in the past.[2] The complication with these methods, however, is primarily associated with the impossibly small signals being generated from the nanoscopic region, not to mention the uncertainty in polarisation generated from the tip. I believe that other super-resolution methods, such as those implemented in other nonlinear optical techniques,[3,4] might be more promising for improving resolution, but will still fall far away from the resolution that is possible in near-field approaches.

1 A. Sakurai, S. Takahashi, T. Mochizuki and T. Sugimoto, Tip-enhanced sum frequency generation for molecular vibrational nanospectroscopy, *Nano Lett.*, 2025, **25**(16), 6390–6398.
2 A. P. Fellows, Using Vibrational Laser Spectroscopy to Study Biologically Relevant Interfaces, PhD thesis, University of Cambridge, 2022.
3 J. H. Park, S.-W. Lee, E. S. Lee and J. Y. Lee, A method for super-resolved CARS microscopy with structured illumination in two dimensions, *Opt. Express*, 2014, **22**, 9854–9870.
4 A. Zhitnitsky, E. Benjamin and O. Bitton, *et al.*, Super-resolved coherent anti-Stokes Raman scattering microscopy by coherent image scanning, *Nat. Commun.*, 2024, **15**, 10073.

Sylvie Roke replied: Near-field and tip-enhanced approaches are possible under very well-defined situations, and the need for the tip or the patterned surface is not easily compatible with also having a freely suspended membrane in the aqueous phase. Another complication is the separation of the response of the tip and the sample, and the way in which the tip modifies the sample.

John C. Conboy replied: Measurement of liquid state flip-flop could not be measured due to rapid mixing of the lipids, gel phase lipids were therefore examined. However, for sum-frequency vibrational spectroscopy (SFVS) of hybrid bilayer systems which contain only one monolayer of lipid, there is theoretically no impediment to characterization of Ca^{2+} and Mg^{2+} binding and interfacial water structure for LC phase lipids. Markus Raschke uses infrared scattering scanning nearfield optical microscopy (IR-sSNOM) for nanoscale characterization of surface structures, which may be applicable toward spectroscopy of hybrid bilayers.[1–3] Flip-flop measurement requires the ability to track the movement of lipids between leaflets, which in the case of sum-frequency is derived from changes in asymmetry with time. Unlike sum-frequency, IR would be unable to distinguish between the vibrational resonances of lipids contained within proximal or distal leaflets and would not be applicable to flip-flop measurement. However, there have been recent reported cases of tip enhanced sum-frequency which would be applicable.[4,5]

1 Y. Wang, S. C. Johnson, N. Nookala, J. F. Klem, S. R. Turner, R. L. Puro, M. Hu, I. Brener, E. A. Muller, A. Belyanin, M. A. Belkin and M. B. Raschke, *Laser Photonics Rev.*, 2024, **18**, 2301148.

2 R. L. Puro, T. P. Gray, T. A. Kapfunde, G. B. Richter-Addo and M. B. Raschke, *Nano Lett.*, 2024, **24**, 1909–1915.
3 R. Wilcken, B. L. Esses, R. S. Nithyananda Kumar, L. A. Hurley, S. E. Shaheen and M. B. Raschke, *Sci. Adv.*, 2025, **11**, eads3706.
4 A. Sakurai, S. Takahashi, T. Mochizuki and T. Sugimoto, *Nano Lett.*, 2025, **25**, 6390–6398.
5 P. Roelli, I. Pascual, I. Niehues, J. Aizpurua and R. Hillenbrand, *arXiv*, 2025, preprint, arXiv:2501.01784, DOI: 10.48550/arXiv.2501.01784.

Syma Khalid said: Regarding the distinct patches of lipids you observe, in a plasma membrane, by mass ∼50% is composed of proteins – these membrane proteins often have distinct lipid fingerprints. It is unlikely a patch of lipid of the size that you report from *in vitro* studies would be present within a plasma membrane as there would always be a native membrane protein nearby. How would this alter the image?

Alexander P. Fellows responded: You are certainly correct that the lipid domains formed in these pure lipid systems are substantially larger than what would be formed in physiological settings and that proteins play an important role in the macroscopic structure of the membrane. Nevertheless, the underlying aim of such work on model lipid systems is to understand the fundamental lipid–lipid interactions governing domain formation that will still be prevalent in the presence of other membrane components. In this sense, observing differences in the domain line tension, compositional enrichment, *etc.* gives crucial insights into the favourability of certain structural motifs that help understand the wider impact on membrane structure due to processes such as PS exposure.

Ilya Levental addressed Alexander P. Fellows and Sylvie Roke: What is the applicability to biological systems? Would you see the asymmetries this way or would the proteins get in the way?

Alexander P. Fellows replied: The question of applicability to biological systems really comes down to what can be learned. In relatively simple model systems such as the two-component monolayers that we have studied here (https://doi.org/10.1039/d4fd00187g), we can really isolate the signals from the individual components and learn a lot about their specific 3D orientational distributions and intermolecular environments. As you add more and more complexity to the system, you generally sacrifice some of the information content. For ‘real’ biological systems, these kinds of second-order techniques can certainly tell you a lot about the amount and distribution of anisotropic molecular structure, but it would be very difficult to get this kind of information on the individual component level.

Sylvie Roke answered: The SH imaging method as presented here enables the time- and spatially-resolved recording of membrane potentials of living cells as demonstrated in ref. 1, in which membrane potential videos were made of living mouse brain neurons that were undergoing de- and repolarization events. It can be thought of as a future all-optical electrode-free, non-invasive version of electrophysiological recordings. The spatial resolution is determined by the optical resolution, which offers a drastic improvement compared to electrodes, which typically only record one data point per cell.

1 M. E. P. Didier, O. Tarun, P. Jourdain, P. Magistretti and S. Roke, *Nat. Commun.*, 2018, **9**, 5287.

Martín E. Villanueva opened discussion of the paper by John C. Conboy: Have you calculated the compressibility modulus for the 10% DPPS/DPPC mixture based on the obtained Langmuir monolayer isotherms? Additionally, how do you correlate the interfacial behavior observed in the monolayer with the molecular translocation (flip-flop) events detected in the corresponding lipid bilayers? This would be particularly relevant in the context of establishing monolayer–bilayer correspondence.

John C. Conboy replied: The compressibility moduli for monolayers of DPPC, DPPS, and 10% DPPS in DPPC were all calculated (Fig. 3) in the manuscript (https://doi.org/10.1039/d4fd00206g). The DPPC monolayer was unaffected by Ca^{2+}. Conversely, the pure DPPS monolayer displayed a decrease in compressibility upon exposure to Ca^{2+}. The change in compressibility of the 10% DPPS mixture was too small to quantify with statistical certainty; however, from the behaviors of the single component monolayers, a 10% DPPS in 90% DPPC monolayer would be expected to have a slight decrease in compressibility, roughly 1/10th the response of the 100% DPPS system if a linear response is assumed. As flip-flop motions require a local deformation of the bilayer to accommodate headgroup transfer through the hydrophobic core, the flip-flop rate is expected to decrease as the rigidity of the monolayer increases.[1] The slowed flip-flop kinetics of the 10% DPPS in DPPC is therefore consistent with an overall rigidification of the bilayer in response to Ca^{2+} ions.

1 T. C. Anglin and J. C. Conboy, *Biochemistry*, 2009, **48**, 10220–10234.

Ben John asked: Given that you used a PEG spacer, does it interact or mix with the lipids upon casting onto the substrate? If so, could it contribute to the signal you're detecting? And have you examined the system without PEG to compare?

John C. Conboy answered: The PEG forms a highly hydrated "mushroom" phase spacer between the bilayer and underlying support.[1–4] The PEG–PE lipid is fully protonated and will contribute to the sum-frequency generated in an asymmetric bilayer but will not flip-flop due to the inability to move the large PEG polymer through the membrane. The sum-frequency generated by the PEG is therefore constant over the course of the experiment and simply factored into the baseline. The PEG is not expected to influence the behavior of surrounding lipids, as the polymer is within the aqueous phase adjacent the membrane rather than intercalated within the membrane itself. The non-perturbing nature of the spacer was verified with DPPC controls where the flip-flop rate of DPPC was measured in the presence and absence of added PEG–PE lipid. No statistically significant difference was noted.

The notable way in which the PEG spacer influenced lipid behavior was the abolishment of Ca^{2+} salt bridging between the PS headgroups and silica support. 10% DPPS in DPPC deposited without the PEG spacer resulted in a biexponential decay in the presence of Ca^{2+} (see Fig. 7 here). Comparison of the flip-flop behavior of DPPS deposited in the proximal *versus* distal leaflets confirms the biexponential decay is due to Ca^{2+} salt bridging of the PS headgroups with the support. DPPS starting in the proximal leaflet near the silica was found to translocate far more slowly than distal DPPS and would not mix to completion, as indicated by significant residual asymmetry. Conversely, DPPS starting in the distal leaflet would mix quickly to homogeneity, with the generated sum-frequency appearing to reach a minimum, but would then re-establish asymmetry as evident from a subsequent increase in sum-frequency intensity. The observed behaviors imply DPPS can flip freely from the distal leaflet, whereas flops out of the proximal leaflet are hampered. PS lipids which salt bridge to the surface can effectively be captured and accumulate within the proximal leaflet. Consistent with the hypothesis is the observation of both single exponential behavior as well as complete mixing of the DPPS lipids with identical rates of PS flip-flop for both proximal and distal leaflets upon incorporation of the PEG spacer.

1 A. J. Diaz, F. Albertorio, S. Daniel and P. S. Cremer, *Langmuir*, 2008, **24**, 6820–6826.
2 M. L. Wagner and L. K. Tamm, *Biophys. J.*, 2000, **79**, 1400–1414.
3 V. Kiessling and L. K. Tamm, *Biophys. J.*, 2003, **84**, 408–418.
4 V. Kiessling, C. Wan and L. K. Tamm, *Biochim. Biophys. Acta*, 2009, **1788**, 64–71.

Manpreet Kaur enquired: In the artificial membranes, how is Ca^{2+} causing the PS flip-flop? How do you propose the hydrophilic head group flip-flops across the membrane?

John C. Conboy replied: There are two possible mechanisms for how Ca-exposed PS lipid flip-flop could proceed. Firstly, complexed PS could simply flip

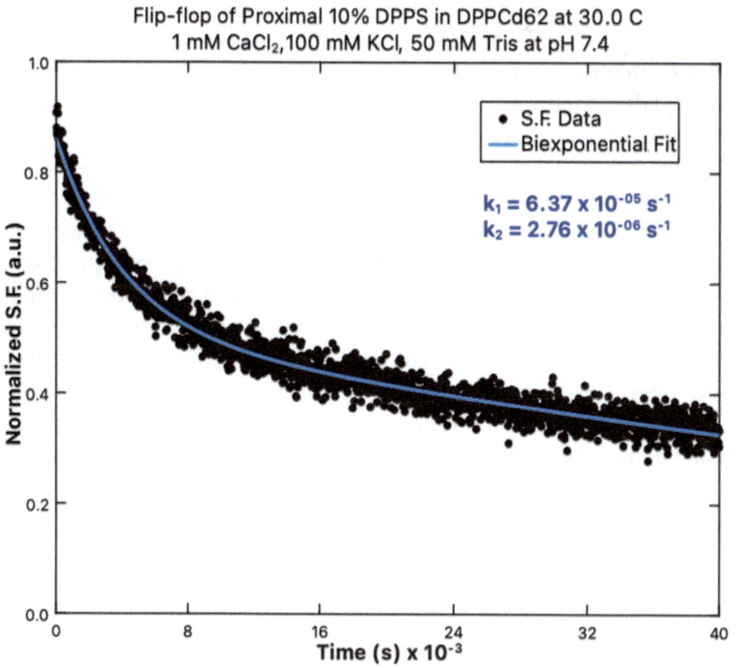

Fig. 7 Flip-flop behavior of proximal 10% DPPS in DPPC-d_{62} in the presence of Ca^{2+}.

as a $[PS]_2Ca$ unit and the slowed PS flip-flop rates observed in the presence of Ca^{2+} would therefore be attributed to the higher energetic cost of moving a larger complex through the bilayer. More likely however, is that flip-flop is preceded by a dissociation of the complex and the slowed rate of Ca-exposed PS flip-flop is influenced by the rate of the prerequisite dissociation step. Some preliminary data from induction of PS asymmetry in tethered vesicles may support this hypothesis. Asymmetry of PS in vesicles extruded without any internal Ca^{2+} can be induced by exposure to external 1 mM Ca^{2+}, presumably due to accumulation of PS toward the Ca-exposed leaflet. Preliminary data has shown that induced asymmetry can be cycled *via* removal of the external Ca^{2+} through complexation with EGTA. Such would theoretically not be the case if $[PS]_2Ca$ complexes could flip as a unit, as one would expect the internal and external Ca^{2+} concentrations to eventually equilibrate due to shuttling of Ca^{2+} ions over the bilayer. If Ca^{2+} were to transport into the vesicle, removal of the external Ca^{2+} would then result in a Ca^{2+} concentration differential and PS lipids would be expected to accumulate on the interior rather than return to baseline.

Robert V. Law said: Many of the phospholipids (*e.g.* DMPC) you use in your flip-flop measurements are in the gel phase as your measurements are done at 21 °C. From the literature, there is a well-established understanding that in the gel phase, lateral diffusion is very slow.[1,2]

Is it possible that due to the method that you use in the preparation of your supported bilayers that they are unusually defective (*e.g.* many pores *etc.*) or slightly dehydrated? Might the possibility of these contribute to the unusually fast flip-flop rates that you find?

1 D. Marquardt, *et al.*, *Langmuir*, 2017, **33**(15), 3731–3741.
2 J. Sot, *et al.*, *J. Mol. Liq.*, 2022, **363**, 119874.

John C. Conboy responded: Defect-mediated enhancement of lipid flip-flop rates in planar supported lipid bilayers (PSLBs), as postulated by Marquardt *et al.*,[1] cannot be ruled out entirely without high fidelity, sub-micron resolution AFM to characterize surface defects. However, flip-flop measurements of PSLBs from Langmuir–Blodgett/Langmuir–Schaffer (LB/LS) deposition are highly reproducible in a way which seems inconsistent with defect-mediated flip-flop. From the calculations of Marquardt, fluctuations in surface defects of less than a percent of surface coverage should have pronounced effects on the measured flip-flop rates. A reduction in surface coverage from 99.9% to 99.5% is predicted to speed flip-flop by an order of magnitude. Highly reproducible flip-flop measurements would imply highly reproducible defects, which, while not impossible, does seem unlikely. Further, flip-flop measurements performed by Kornberg and McConnell in vesicles are far faster than those reported for aLUVs, a discrepancy which cannot be attributed to surface defects.[2]

1 D. Marquardt, F. A. Heberle, T. Miti, B. Eicher, E. London, J. Katsaras and G. Pabst, *Langmuir*, 2017, **33**, 3731–3741.
2 R. D. Kornberg and H. M. McConnell, *Biochemistry*, 1971, **10**, 1111–1120.

Georg Pabst asked: (i) Based on your results, is it correct to conclude that, in the absence of flippases, phosphatidylserine (PS) would predominantly reside in the exoplasmic leaflet? If so, can you estimate the energetic barrier that the enzyme must overcome to facilitate its translocation to the cytoplasmic leaflet?

(ii) How would you expect the flip-flop rates you measured in the gel phase to extrapolate to the fluid phase? Would you anticipate any changes in the dynamics of Mg^{2+} and Ca^{2+} interactions with the lipids that might significantly affect the flip-flop behavior?

John C. Conboy replied: Yes, in the absence of flippases and other cellular control mechanisms the native lipid flip-flop dynamics measured here would predict an enrichment of PS toward the Ca-exposed exoplasmic leaflet. Without specific action by the cell to counteract the differential rates of flip and flop, a slow rate of inward flip combined with a fast rate of outward flop should result in an equilibration such that $N_{\text{external}}/N_{\text{internal}} = k_{\text{flop}}/k_{\text{flip}}$. The theorized enrichment of PS toward the Ca-exposed leaflet is in agreement with observations of both the Hampton and Bratton groups where the degree of PS externalization was dependent on the Ca^{2+} concentration differential over the bilayer.[1,2] The barrier to flip-flop can be estimated through extrapolation from the best line fit of the Ca-exposed 10% DPPS flip-flop rates in Fig. 10 in the paper (https://doi.org/10.1039/d4fd00206g) to physiologic temperature (37 °C) to then calculate $\Delta G^{\ddagger}$ using the Eyring relation (eqn (20) in the paper). The predicted $\Delta G^{\ddagger}$ in this case would be ~95 kJ mol^{-1}. However, this value is particular to the binary DPPS/DPPC mixture as tested, which is simplified and non-physiologic. A more holistic matrix would include a blend of saturated and unsaturated phospholipids which would be in the liquid crystalline state at physiologic temperature.[3,4] If the tested mixture is extrapolated out to the phase transition of DPPS, 54 °C, the LC phase flip-flop barrier would then be ~88 kJ mol^{-1}. However, there is difficulty in predicting the thermodynamic and kinetic properties of the LC phase using only data collected within the gel phase. Saturated, gel phase lipids had to be used here in order to slow the flip-flop process sufficiently to be measured. As a general assessment, higher temperatures and fluid lipids are expected to result in faster rates of flip-flop for both the Ca-exposed and Ca-free PS lipids. The specific rates of translocation within each environment are unknown, however, the relative kinetics of Ca-exposed and Ca-free PS lipid flip-flop [$k_{\text{PS}}/k_{(\text{Ca:PS})}$] are hypothesized to be similar even in the LC phase. Phase segregation of PS from PC in the presence of Ca^{2+} has been noted in both gel and LC phases,[5,6] which indicates PS–Ca interactions are not phase-specific.

1 M. B. Hampton, D. M. Vanags, M. I. Pörn-Ares and S. Orrenius, *FEBS Lett.*, 1996, **399**, 277–282.
2 D. L. Bratton, V. A. Fadok, D. A. Richter, J. M. Kailey, L. A. Guthrie and P. M. Henson, *J. Biol. Chem.*, 1997, **272**, 26159–26165.
3 A. J. Verkleij, R. F. Zwaal, B. Roelofsen, P. Comfurius, D. Kastelijn and L. L. van Deenen, *Biochim. Biophys. Acta*, 1973, **323**, 178–193.
4 J. H. Lorent, K. R. Levental, L. Ganesan, G. Rivera-Longsworth, E. Sezgin, M. Doktorova, E. Lyman and I. Levental, *Nat. Chem. Biol.*, 2020, **16**, 644–652.
5 C. P. Tilcock, P. R. Cullis and S. M. Gruner, *Biochemistry*, 1988, **27**, 1415–1420.
6 M. Ross, C. Steinem, H.-J. Galla and A. Janshoff, *Langmuir*, 2001, **17**, 2437–2445.

Rumiana Dimova addressed John C. Conboy and Sylvie Roke: Calcium and magnesium differ not only in their binding modes to PS, but also in their position within the Hofmeister series, with Ca^{2+} being more chaotropic than Mg^{2+}. To what extent do you think the differences you observe in lipid packing, headgroup ordering, and flip-flop dynamics might arise from broader Hofmeister effects on interfacial water structure or membrane hydration, beyond specific ion–lipid interactions? And do you think techniques like those developed by S. Roke could help disentangle these contributions?

John C. Conboy responded: The more chaotropic nature of Ca^{2+} and its ability to more effectively displace water from the lipid headgroups is very likely a large factor in the pronounced effect Ca^{2+} ions have on PS behavior. Papahadjopoulos noted that Mg-complexed PS lost significant weight when vacuum dried, whereas Ca-complexed PS did not, with this difference in weight being attributed to water content.[1] Papahadjopoulos therefore attributed the fusogenic activity of Ca^{2+} and the ability of Ca^{2+} ions to phase segregate PS from PC to the displacement of water and more effective binding compared to Mg^{2+}, which did not display the same effects.[1–7] The reduced ability of Mg^{2+} to effectively coordinate the PS headgroups, as calculated by Martin-Molina *et al.*[8] and inferred from our observed reduced amine peak ordering compared to Ca^{2+} in the SFVS data, could very well be due to Mg^{2+} being insufficiently chaotropic to disrupt and displace waters associated with the PS headgroup. The capacity of each ion to coordinate the lipid headgroup appears tightly linked with the relative chaotropic nature of each species and the two effects may not be entirely distinguishable from one another. Second harmonic generation could be utilized as a complementary technique for tracking net water structure at the lipid/water interface, however in our current study sum-frequency provides useful structural information (ice- and water-like bonding characteristics) beyond just a simple water ordering parameter achievable with SHG. A comparative analysis of the ice- and water-like peaks, combined with the amine peak derived Ca–PS binding isotherm data, also allowed for the separation of molecular reorientation and third-order coupling contributions to the $\chi^{(3)}$ effect. Knowledge of the net water structure alone would have been insufficient to the derivation of the Ca–PS and Mg–PS binding isotherms key to this assessment due to the tendency of water order to maximize at high field strengths near highly charged surfaces.[9,10] The ability to detect and track the reorientation of head-group amines during Ca^{2+} and Mg^{2+} complexation was crucial for characterizing the charge state of the PS lipids at low concentrations of Ca^{2+} and Mg^{2+} where the water ordering was beyond saturation and insensitive to changing surface charge.

1 C. Newton, W. Pangborn, S. Nir and D. Papahadjopoulos, *Biochim. Biophys. Acta*, 1978, **506**, 281–287.
2 D. Papahadjopoulos, G. Poste, B. E. Schaeffer and W. J. Vail, *Biochim. Biophys. Acta*, 1974, **352**, 10–28.
3 K. Jacobson and D. Papahadjopoulos, *Biochemistry*, 1975, **14**, 152–161.
4 D. Papahadjopoulos, W. J. Vail, C. Newton, S. Nir, K. Jacobson, G. Poste and R. Lazo, *Biochim. Biophys. Acta*, 1977, **465**, 579–598.
5 D. Papahadjopoulos and A. Portis, *Ann. N. Y. Acad. Sci.*, 1978, **05467**, 50–66.
6 A. Portis, C. Newton, W. Pangborn and D. Papahadjopoulos, *Biochemistry*, 1979, **18**, 780–790.
7 N. Duzgunes, J. Wilschut, R. Fraley and D. Papahadjopoulos, *Biochim. Biophys. Acta*, 1981, **642**, 182–195.
8 A. Martin-Molina, C. Rodriguez-Beas and J. Faraudo, *Biophys. J.*, 2012, **102**, 2095–2103.
9 D. E. Gragson and G. L. Richmond, *J. Am. Chem. Soc.*, 1998, **120**, 366–375.
10 S. Kielich, *IEEE J. Quantum Electron.*, 1969, **5**, 562–568.

Sylvie Roke replied: The observation that Ca^{2+}, Mg^{2+}, and Ba^{2+} when interacting with partially charged phospholipid membranes all create different membrane potential fluctuations[1,2] is a manifestation of ion specific interactions. For example, Mg^{2+} interacts with lipid head groups through an ion pair complex which still involves a hydration shell (solvent separated ion pair), whereas Ca^{2+} directly binds to the lipid head group (contact ion pair). The first results in weaker interactions that have a smaller impact on the membrane potential distribution/interfacial water orientation, while the latter leads to larger interactions that have a larger impact on the membrane potential distribution/interfacial water orientation.

1 O. B. Tarun, *et al.*, Transient domains of ordered water induced by divalent ions lead to lipid membrane curvature fluctuations, *Commun. Chem.*, 2020, **3**(1), 17.
2 D. Roesel, *et al.*, Ion-induced transient potential fluctuations facilitate pore formation and cation transport through lipid membranes, *J. Am. Chem. Soc.*, 2022, **144**(51), 23352–23357.

Markus Deserno said: You made a remark in your presentation that cells meticulously regulate the intracellular calcium concentration, but that the same would not be true for magnesium. I found that puzzling, because in view of the ~70 mV electrostatic potential difference across the plasma membrane, I would then expect the magnesium concentration – purely by the electrostatic contribution to their chemical potential – to be about 200 times as concentrated inside cells relative to the extracellular fluid. Is this really true, or am I missing something here?

John C. Conboy responded: The predominant contribution to the electrostatic potential of animal cells is the regulation of K^+ and Na^+ ions within the cytosol.[1] Na^+ ions are removed from the interior and replaced with K^+ ions in a $3Na^+ : 2K^+$ ratio, which results in a deficit of positive internal charges. Additionally, K^+ ions are free to diffuse through ion channels. As the internal concentration of K^+ is far higher than the external K^+ concentration due to active ion shuttling, diffusion favors outward efflux. The final equilibrium concentration is a balance between the negative potential which attracts K^+ toward the interior and the concentration gradient driven efflux. Theoretically, it is true that Mg^{2+} ions would accumulate toward the interior in response to the electrical driving force, however the cell membrane is not permeable to Mg^{2+} ions.[2] For eukaryotes, a large variety of transporters exist which serve to shuttle magnesium between the plasma and cytosol, as well as the cytosol and organelles (see page 17 of Vink and Nechifor).[3] However, the end result is simply that the cytosolic and exoplasmic concentrations of Mg^{2+} ions are roughly equal.

1 B. Alberts, D. Bray, J. Lewis, M. Raff, K. Roberts and J. Watson, *Molecular Biology of the Cell*, Garland Publishing, New York, 3rd edn, 1994.
2 E. Murphy, *Circ. Res.*, 2000, **86**, 245–248.
3 R. Vink and M. Nechifor, *Magnesium in the Central Nervous System*, University of Adelaide Press, South Australia, 2011.

Tyler Reagle asked: Have you investigated (i) whether the rate of DPPS flip-flop is linear with respect to the mole fraction of DPPS in the distal leaflet (denoted by X_x where the subscript "x" refers to either the "distal" or "proximal" leaflet) and (ii) whether the 10% DPPS used for the flip-flop measurements falls within such a linear-regime? Significant deviation from linearity might reflect non-ideal behavior arising from Coulombic-repulsion and/or accumulation of mechanical-tensions within the leaflets (which would be incurred if there is net transport of DPPS into the distal leaflet that is not balanced by a counter-flow of DPPC lipids);[1] the condensing effect of Ca^{2+} upon the molecular area of DPPS could incur an additional contribution to a mechanical-tension in the distal leaflet. Elementary rate-laws for the flip-flop of DPPS could be expressed in terms of the chemical activity of this lipid in a given leaflet (denoted by Λ_x, where the subscript "x" refers to the "distal" or "proximal" leaflet),[2] given by $R_{flip} = k_{flip}\Lambda_{distal}$ and $R_{flop} = k_{flop}\Lambda_{proximal}$, where R_{flip} and R_{flop} are the rates of DPPS flip and flop, respectively. Following the Debye–Hückel limiting law, Λ_x and X_x are related by $\Lambda_x = \gamma_x X_x$, where γ_x is the activity coefficient of DPPS in a given leaflet (which presumably depends on X_x, the density of adsorbed cation, and the mechanical conditions experienced by the lipid). Accounting for non-ideality in the rate-laws leads to $R_{flip} = k_{flip}\gamma_{distal}X_{distal}$ and $R_{flop} = k_{flop}\gamma_{proximal}X_{proximal}$, where γ_{distal} and $\gamma_{proximal}$ might not be equal. If DPPS does behave non-ideally under a given set of conditions but ideal behavior is assumed, then the products $k_{flip}\gamma_{distal}$ and $k_{flop}\gamma_{proximal}$ are the "measured" or "perceived" rate-constants for DPPS flip and flop, respectively, which would be obtained from fitting of kinetic data. If DPPS does not deviate significantly from ideal behavior (*i.e.* $\gamma_x \rightarrow 1$), then these products yield unbiased estimates of k_{flip} and k_{flop}. If (a) DPPS does deviate significantly from ideal behavior under certain conditions and (b) the fitting equation for the time-dependent spectroscopic data (eqn (7) in the paper (https://doi.org/10.1039/d4fd00206g)) could be reformulated to account for this in some way, it would be very interesting to know whether the Ca^{2+}-dependence of DPPS flipping could be explained by changes in the activity coefficient(s) of DPPS or if this effect comes from changes to the activation-energy barrier for flip-flop (conveyed by changes in k_{flip} after non-ideality has been accounted for).

1 T. Reagle, *et al.*, *Soft Matter*, 2024, **20**, 4291, DOI: 10.1039/d3sm01772a.
2 A. Radhakrishnan and H. M. McConnell, *Biochemistry*, 2000, **39**, 8119, DOI: 10.1021/bi0005097.

John C. Conboy responded: The present study did not examine different mole fractions of DPPS, however a previous publication by Brown and Conboy determined flip-flop rates for DSPC mixtures containing varying fractions of DSPG, which, like DPPS, is also net negative at neutral pH.[1] It was determined the transition state free energy, $\Delta G^{\ddagger}$, is linear with respect to the mole fraction of PG. The rate of flip-flop, however, responds exponentially [rate $= \exp(-\Delta G^{\ddagger}/RT \ln((k_BT)/h))$]. As altering the mole fraction of negative lipid within the membrane modulates the average distance (x) between the charged headgroups, and potential (ϕ) scales exponentially with distance [$\phi/\phi_0 = e^{-\kappa x}$], the exponential scaling in the rate of flip-flop with respect to PG mole fraction was attributed to coulombic interactions between the lipids. The influence of charge repulsion on rates of flip-flop was further confirmed through modulation of the inverse Debye length (κ) such that 25% PG in low ionic strength PBS would experience the same coulombic forces as 75% PG in high ionic strength PBS. When compared, these two systems had statistically identical rates of flip-flop. While chemical activities were not calculated, these data demonstrate charged lipid flip-flop is indeed dependent on the local ionic environment, in other words the charge screening (activity) of the PG lipids does modulate the rate of translocation. However, changes in PS activity (charge screening) does not explain the data obtained in the present study, as all 10% DPPS flip-flop measurements were performed using a fixed ionic strength in the presence and absence of Ca^{2+}. In addition, as all 10% DPPS bilayers contain one leaflet of 10% DPPS in DPPC-d_{62} and one leaflet of 10% DPPS-d_{62} in DPPC-d_{62}, all coulombic effects are symmetric over the bilayer and no disruptive mechanical tension should occur between the two leaflets. The slowing effect of Ca^{2+} on DPPS flip-flop must therefore be attributed to a direct interaction rather than effects due to altered PS activity. These effects would however need to be addressed in cases where the starting coulombic forces are unequal, in which case the rate equation could be rederived with non-ideality and unequal rates of apparent k_{flip} and k_{flop} (R_{flip} and R_{flop} in the above notation) in mind.

1 K. L. Brown and J. C. Conboy, *J. Phys. Chem. B*, 2015, **119**, 10252–10260.

Christian Schwieger questioned: Looking at the water spectra, why are the two water components, *i.e.* free and bound water, increasing in the same way?

John C. Conboy replied: The two water resonances centered at 3200 and 3400 cm^{-1} represent the ice- and liquid-like O–H symmetric stretches, respectively.[1–4] These represent all water which is ordered near the surface. While both result from the symmetric O–H stretch of water, the ice-like resonance is indicative of a strong, tetrahedral H-bonding structure like that seen in ice. Conversely, the liquid-like peak results from a weaker H-bonding structure, namely bifurcated H-bonding whereby hydrogens are shared unevenly between neighboring oxygens. The sum-frequency intensity of the water resonances are expected to scale with interfacial charge density due to a third-order, $\chi^{(3)}$, effect (eqn (10) in the paper (https://doi.org/10.1039/d4fd00206g)).[5–12] The $\chi^{(3)}$ effect represents two phenomena: the direct coupling of the three electric fields, dependent on the third-order hyperpolarizability of the material (γ), and rearrangement of dipolar molecules in response to changing surface charge, which is merely dependent on the second-order hyperpolarizability of the material (β) (eqn (11) in the paper). While both water peaks respond the "same" in the sense that both gain in intensity as surface charge density increases, the response of the ice-like peak is actually greater due to the aforementioned reorganization term. In short, an increase in surface charge will induce some number of disordered waters to migrate to the semi-ordered 3400 cm^{-1} population, while some number of the pre-existing semi-ordered waters will migrate to the tetrahedrally ordered 3200 cm^{-1} population. As the tetrahedrally ordered population represents a terminal state, where further population conversion is not possible, the number of waters contained within the tetrahedrally ordered state is expected to grow more rapidly than as with the semi-ordered state, which will gain waters from the disordered population, but also lose waters to the maximally ordered state. A fit of the transition probabilities of the 3200 and 3400 cm^{-1} resonances *versus* the interfacial electric field strength (Fig. 7 in the paper) found that the response of the ice-like peak is roughly double that of the liquid-like.

1 D. E. Gragson and G. L. Richmond, *J. Am. Chem. Soc.*, 1998, **120**, 366–375.
2 G. L. Richmond, *Chem. Rev.*, 2002, **102**, 2693–2724.

3 A. G. de Beer and S. Roke, *J. Chem. Phys.*, 2016, **145**, 044705.
4 V. Ostroverkhov, G. A. Waychunas and Y. R. Shen, *Chem. Phys. Lett.*, 2004, **386**, 144–148.
5 P. E. Ohno, H. F. Wang and F. M. Geiger, *Nat. Commun.*, 2017, **8**, 1032.
6 P. E. Ohno, H. F. Wang, F. Paesani, J. L. Skinner and F. M. Geiger, *J. Phys. Chem. A*, 2018, **122**, 4457–4464.
7 S. Ong, X. Zhao and K. B. Eisenthal, *Chem. Phys. Lett.*, 1992, **191**, 327–335.
8 X. Zhao, S. Ong and K. B. Eisenthal, *Chem. Phys. Lett.*, 1993, **202**, 513–520.
9 B. F. Levine and C. G. Bethea, *J. Chem. Phys.*, 1976, **65**, 2429–2438.
10 P. D. Maker and R. W. Terhune, *Phys. Rev.*, 1965, **137**, A801–A818.
11 R. W. Terhune, P. D. Maker and C. M. Savage, *Phys. Rev. Lett.*, 1962, **8**, 404–406.
12 S. Kielich, *IEEE J. Quantum Electron.*, 1969, **5**, 562–568.

Walter Nickel opened discussion of the paper by Gonen Golani: In calculations regarding energy barriers for membrane fusion, was this considering protein-dependent membrane fusion, *e.g.* catalysed by SNARE proteins? Or considered in context of membrane fusion *in vitro* that can also be triggered in a protein-independent manner? The principal question here would be as to whether your conclusions can be applied to cellular systems in which membrane fusion occurs by dedicated protein machineries such as SNARE proteins or viral fusion proteins.

Gonen Golani answered: The calculation is general and applies to both protein-dependent and protein-independent fusion. Specifically, we compute the energy barrier for fusion and how it is affected by membrane asymmetry. The driving force required to overcome this barrier is beyond the scope of this work, as it is not directly determined by lipid asymmetry. In other words, we focus on one side of the process, the energy barrier to fusion, without addressing the mechanism by which this barrier is overcome (*e.g.*, SNARE). The implicit assumption is that most biologically relevant fusion events proceed through a similar pathway, which is similarly influenced by induced asymmetry.

John M. Seddon remarked: I was surprised that the saddle-splay term doesn't vary with $\Delta\zeta$ (Fig. 4 in the paper; https://doi.org/10.1039/d4fd00189c). Is this to be expected?

Gonen Golani responded: The meaning of $\Delta\zeta$ is the difference between the spontaneous curvatures of the two monolayers. Since the elastic model is quadratic in tilt and splay, the first-order effect of $\Delta\zeta$ enters through the splay term in the energy density (which is the analog of mean curvature in the absence of tilt). This is why the splay energy is most strongly affected by $\Delta\zeta$. Saddle-splay energy is only indirectly influenced: changing $\Delta\zeta$ alters the minimal energy configuration of the hemifusion intermediates compared to a symmetric membrane, and as a result, the saddle-splay energy also changes. However, this effect is relatively small.

John M. Seddon asked: Will your work help to design lipid nanoparticles with more efficient endosomal escape, for delivery of nucleic acids or other therapeutics into cells?

Gonen Golani replied: Possibly. Our work showed that inducing a certain type of asymmetry can reduce the energy barriers for fusion. If fusion is indeed the rate-limiting step in delivery, this could offer a way to optimize the process.

Ilya Levental enquired: Two types of asymmetry, compositional and density. Which is most relevant? Splaying lipids promoting fusion, *e.g.* PE accumulating on cytosolic leaflet?

Gonen Golani answered: To my understanding, the most important form of asymmetry affecting the energy barriers for fusion is the level of lipid unsaturation. A higher concentration of unsaturated lipids in the cytoplasmic monolayer is expected to raise the energy barrier for fusion from the outside while lowering it from the inside. Other types of asymmetry play a lesser role. For example, we calculated the effect of a density difference, which manifests as differential stress. This had a minimal impact on the energy barrier for lipid mixing and a somewhat greater effect on content mixing, though still less significant than the influence of intrinsic lipid curvature.

Patricia Losada Pérez said: You mentioned that lipid splay is the dominant contribution. Excluding the action of lipids that induce non-zero spontaneous curvature, what is the driving mechanism of lipid splay? Can one predict or quantify lipid splay in some way?

Gonen Golani responded: The energetic contribution of splay arises from the difference between the actual geometrical splay of lipids near the fusion site and their intrinsic splay (curvature), which is an inherent property of the lipids. Lipids with bulkier head groups and narrower tails typically exhibit positive intrinsic splay (curvature), while lipids with smaller heads and broader tail regions have negative intrinsic splay. The intrinsic splay of lipids can be measured experimentally (*e.g.*, ref. 1), and the actual splay adopted by the lipids can be calculated numerically, as done in this work (https://doi.org/10.1039/d4fd00189c). However, to the best of my knowledge, there is currently no experimental method to directly measure the difference between the actual and intrinsic splay near the fusion site.

1 M. K. Dymond, *Chem. Phys. Lipids*, 2021, **239**, 105117.

Michael M. Kozlov asked: How does the overall tension affect the energy of the fusion intermediate?

Gonen Golani answered: Membrane tension has opposing effects on fusion intermediates. The energy required for stalk formation increases, as forming a stalk demands additional membrane area compared to the pre-fusion configuration. This was addressed both experimentally and theoretically in previous work,[1] where we found that the energy barrier increased by a few *kT*. In contrast, expansion into a hemifusion diaphragm is favored because it involves lipid flow into the surrounding membranes, releasing energy. Therefore, we expect the hemifusion diaphragm radius to increase with tension. Finally, the expansion of the fusion pore is also favored, as membrane tension lowers the energy barrier for pore formation.

1 P. Shendrik, *et al.*, *ACS Nano*, 2023, **17**, 18942.

Michael M. Kozlov questioned: Does the tension change the inter-membrane distance?

Gonen Golani replied: Yes, in general, tension reduces membrane undulations and, as a result, decreases the repulsive interactions between membranes. This effect was considered in recent work by the same authors.[1] However, in the current study (https://doi.org/10.1039/d4fd00189c), the membrane tension was not changed – only the monolayer tension was varied. Specifically, to avoid addressing this effect, we explicitly assumed that the sum of the monolayer tensions, which defines the membrane tension, vanishes. In other words, the two monolayers have tensions of equal magnitude but opposite sign. Under these conditions, the membrane's undulation spectrum remains unchanged, and the inter-membrane distance stays the same.

1 P. Shendrik, *et al.*, *ACS Nano*, 2023, **17**, 18942.

Rumiana Dimova enquired: In your experiments, membrane fusion involves two bilayers of different composition, whereas in the simulations, the membranes have identical composition. How do you think accounting for this compositional asymmetry in the simulation model would affect the predicted energy barriers and potentially influence the preferred directionality of fusion?

Gonen Golani responded: We have tested fusion between symmetric and asymmetric membranes in simulations, although these results were not included in the paper to maintain focus on the principal idea. In these simulations, the qualitative behavior remained the same, although the magnitude of the effect was reduced. We note that the goal of the simulations was not to achieve an exact fit to the experimental system – several factors are not known with sufficient accuracy – but rather to demonstrate qualitative agreement, which is unchanged if one membrane is symmetric.

Heiko Heerklotz asked: A membrane asymmetry that favors fusion with vesicles from the interior and opposes fusion from the outside seems advantageous in most but not all cases. How do you envisage the cell is dealing with the few cases where fusion events from the outside are favorable?

Gonen Golani answered: The hypothesis is that fusion from the extracellular side can be induced by lipid scrambling. Indeed, several reports indicate that plasma membrane scrambling – mediated, for example, by TMEM16 – promotes cell–cell fusion, a process involved in tissue development (*e.g.*, ref. 1).

1 Y. Zhang, *et al.*, *Sci. Adv.*, 2020, **6**, eaba0310, DOI: 10.1126/sciadv.aba0310.

Georg Pabst requested: Could you speculate on the role of long-range interactions – such as screened electrostatics, van der Waals forces, and membrane undulations – in promoting the preferential fusion of asymmetric vesicles? For instance, some asymmetric membranes have been found to be significantly more rigid than their symmetric counterparts.

Gonen Golani responded: Membrane interaction and rigidity play important roles in fusion. Membranes are usually repulsive at long range and one of the roles of the work of fusion proteins is to overcome this repulsive force and bring the membranes to close enough distance so they could fuse. This process, and the possible effect of asymmetry in this context, was out of the scope of the present work but could definitely be important. For example, scrambling of charged lipids could dramatically change the interaction profile between the membranes. Membrane rigidity is another important factor that was not considered here and plays an important role – a more rigid membrane is harder to fuse. However, I would speculate that if one monolayer becomes stiffer and the other softer there might be more complex effects. These should be quantified using numerical calculations.

Oded Farago opened discussion of the paper by Giacomo Fiorin: Technical questions: How do you measure the bending modulus in the simulations? How do you impose curvature on the simulated membranes (is it done using the buckling method for measuring the bending rigidity – see ref. 1)?

1 M. Hu, P. Diggins IV and M. Deserno, *J. Chem. Phys.*, 2013, **138**, 214110.

Giacomo Fiorin answered: Firstly, free-energy landscapes were computed by applying biasing potentials on the membrane curvature and unbiasing the resulting probability distributions using established MD simulation techniques. Secondly, each computed free-energy landscape was fitted against the mathematical expression of the same landscape as predicted from continuum theory. For very large bilayers, this theory would be the Helfrich–Canham model, and its only parameter would be the bending modulus, k_c. However, given the very small bilayer sizes it is also necessary to account for lipid orientation, as originally proposed by Hamm and Kozlov[1] and confirmed by MD simulations and X-ray diffraction measurements.[2,3] The additional parameter, k_t, was derived by fitting the resulting maps of lipid orientations against the theoretical prediction, as previously done in Fiorin *et al.*[4]

Rather than the buckling method used by Deserno, our computational approach[5] is more similar to the one by den Otter and Briels,[6] but with the following two differences:

(i) We used atomistic simulations and constant-pressure conditions, which allow the bilayer's projected area to adapt to increased local curvature without adding membrane tension.

(ii) Sinusoidal deformations were used in this work purely for mathematical convenience, but the same scheme supports arbitrary types of curvature,[5,7] as well as changes in other structural parameters such as thickness (work to be published).

1 M. Hamm and M. M. Kozlov, *Eur. Phys. J. E: Soft Matter Biol. Phys.*, 2000, **3**, 323–335.
2 M. C. Watson, *et al.*, *Phys. Rev. Lett.*, 2012, **109**, 028102.
3 M. S. Jablin, *et al.*, *Phys. Rev. Lett.*, 2014, **113**, 248102.
4 G. Fiorin, *et al.*, *PNAS Nexus*, 2023, **2**(8), pgad269.
5 G. Fiorin, *et al.*, *J. Comput. Chem.*, 2020, **41**, 449–459.
6 W. K. den Otter and W. J. Briels, *J. Chem. Phys.*, 2003, **118**(10), 4712–4720.
7 W. Zhou, *et al.*, *eLife*, 2019, **8**, e50576.

Malavika Varma asked: What are your thoughts on extending this work to mixtures with and without cholesterol?

Giacomo Fiorin replied: Quantification of the bending mechanics of asymmetric membranes containing cholesterol is critical to understand the dynamics of the plasma membrane, and will likely be the subject of future work. To that end, some challenges remain to be addressed: for example, our atomistic computations of simple asymmetric bilayers are so far consistent with proposed theoretical models, but not yet in quantitative agreement with experimental data. This apparent discrepancy needs to be addressed.

Furthermore, in multi-component bilayers the densities of each lipid species may also vary with the local curvature. This effect did not appear to be significant in symmetric POPC or DOPC bilayers with cholesterol,[1] but this behavior may not necessarily apply to other lipid compositions. For asymmetric bilayers, this is probably even more true due to differential stress, which would dramatically affect the lipid packing of each leaflet and thus the propensity of cholesterol to flip-flop. Although these effects would be accounted for in our atomistic simulations, theoretical modeling would also be required to correlate simulations with experiments.

1 G. Fiorin, *et al.*, *PNAS Nexus*, 2023, **2**(8), pgad269.

Walter Nickel queried: Are your conclusions valid in the context of native cellular membranes? There is quite a complexity observed for cellular membranes with a number of relevant parameters. Plasma membranes are among the stiffest bilayers, but they are not only characterised by lipid asymmetry but also high levels of cholesterol. By contrast, the endoplasmic reticulum is less asymmetric but also contains low levels of cholesterol. In addition, what are the relative contributions of membrane proteins to membrane stiffness? In particular, membranes from different organelles contain vastly different protein to lipid ratios. So what do you think is the contribution of the reported effects on membrane stiffness in native membranes with different lipid compositions and different protein to lipid ratios?

Giacomo Fiorin replied: The conclusions of this study are directly applicable to asymmetric lipid bilayers with homogeneous leaflets. These bilayers are, in turn, synthetic models of the plasma membrane used to infer its mechanical properties from *in vitro* experiments. Quantification of bending mechanics of symmetric bilayers with cholesterol using this method was done previously in ref. 1. Estimation of the effect of embedded membrane proteins would probably require computations on larger systems, and with varying protein to lipid ratios.

1 G. Fiorin, *et al.*, *PNAS Nexus*, 2023, **2**(8), pgad269.

Milka Doktorova said: The zero-tension state in the bilayers is estimated from the defined area asymmetry parameter. As stated in the text (https://doi.org/10.1039/d5fd00006h), the calculation of lateral pressure profiles from the simulation trajectories has caveats but standard protocols exist including within NAMD which was the software used for the simulations in this study. Have you tried to calculate the leaflet tension in the bilayers with some of these protocols to see how it corresponds to the area asymmetry, or do you have some specific observations indicating that it is unreliable for estimating the differential stress in the bilayers?

Giacomo Fiorin replied: We thank Dr Doktorova for bringing up calculations of lateral stress profiles as an approach to quantify the tension of each leaflet. As acknowledged in our paper (https://doi.org/10.1039/d5fd00006h), this would be an alternative to considering each leaflet individually and using the changes in area per lipid molecule to infer changes in tension. We have not used the former approach in our work, neither using the NAMD protocol, nor any others, because our focus on bilayers with homogeneous leaflet compositions allowed us to avoid the associated caveats.

To provide some context, the direct computation of local stress and tension from MD simulations may be done using a variety of schemes, most of which have not yet been implemented in mainstream software packages such as NAMD. It should be noted that these schemes can differ in how they generalize continuum-mechanics equations to the atomic scale. As a consequence, the numerical results of these methods may differ significantly from each other. A few examples can be found in ref. 1–4.

The method implemented in NAMD and raised in this question by Dr Doktorova is based on the theoretical framework used by Lindahl and Edholm,[5] but with modifications in the way that σ_{ij}, *i.e.* the virial stress contribution due to a force between two atoms i and j, is apportioned in space. Specifically, when recording σ_{ij} onto geometric "slabs" to compute a stress profile, NAMD makes an additional approximation: it assigns σ_{ij} only to the slabs containing atoms i and j, omitting any slabs between them. This discretization scheme may impact the calculation of lateral pressure in individual leaflets, particularly for atoms near the bilayer midplane. That is, if atoms i and j are in the same leaflet, but j extends into the opposing leaflet due to interdigitation, NAMD always attributes half of the value of σ_{ij} to the opposing leaflet, irrespective of how deeply atom j penetrates into it. (Another technical limitation to the NAMD implementation is that reciprocal-space contributions to the virial stress are estimated using a different algorithm than the one used to run the simulations, *i.e.* Ewald *vs.* Particle-Mesh-Ewald.)

On the other hand, direct computation of local stress can be applied to heterogeneous leaflets, unlike the strategy used in our work. We note that the approximations mentioned above do not necessarily render the calculated pressure profiles "unreliable" *per se*. For example, the NAMD discretization strategy mostly affects the region near the midplane, where the two leaflets interact, but the majority of the atoms can be neatly attributed to either leaflet and thus their contribution is dominant. This assumption is similar to the one adopted in this work, where the area asymmetry of individual leaflets was used. For future work, we are in fact considering the direct computation of local stress, but with methods designed to reduce or eliminate discretization errors.

1 O. H. S. Ollila, *et al.*, *Phys. Rev. Lett.*, 2009, **102**, 078101.
2 T. Nakamura, *et al.*, *Comput. Phys. Commun.*, 2015, **190**, 120–128.
3 J. M. Vanegas, *et al.*, *J. Chem. Theory Comput.*, 2014, **10**(2), 691–702.
4 R. M. Elder, *et al.*, *Chem. Phys. Lett.*, 2019, **731**, 136580.
5 E. Lindahl and O. Edholm, *J. Chem. Phys.*, 2000, **113**, 3882–3893.

Alberto Giacomello questioned: When fitting the free energy profile as a function of the curvature one obtains the thermodynamic variable conjugate to it. So what is the thermodynamic observable that you are obtaining by fitting the free energy profiles? To my understanding, it should be a combination of the bending modulus and the non-local bending modulus, which is in turn connected to the area compressibility of two leaflets.

Giacomo Fiorin responded: I assume that you are referring to the non-local curvature energy term as defined in recent work by the Deserno lab (eqn (16a–d) in ref. 1). This represents the combined penalties associated with the compression or expansion of the two leaflets. In that work, the non-local curvature energy term is expressed as a function of $\langle c \rangle$, the mean over the membrane's surface of the local curvature c. However, in our MD simulations the bilayers are subject to periodic boundary conditions, which constrain $\langle c \rangle$ to be zero; furthermore, the numbers of lipids in the two leaflets are fixed during all simulations of the same bilayer. As a result, all parameters influencing the non-local curvature energy are constant, and this term does not contribute to the computed free-energy profile in a manner that is accounted for by the theory.

It is important to note that the values of the non-local curvature energy may vary significantly between the bilayers considered in this work, but these bilayers were all simulated independently and free-energy differences between them were not computed. Performing such comparisons would likely require simulations without periodic boundary conditions, which in turn would necessitate significantly larger system sizes.

1 A. Hossein and M. Deserno, *J. Chem. Phys.*, 2021, **154**, 014704.

Oded Farago asked: What is the origin of the increasing bending rigidity at moderate asymmetry, and why does it decrease at larger asymmetry. Is the stiffening of the membrane related to a phase transition or partial gelation of compressed layers?

Giacomo Fiorin replied: Although the increase in bending modulus is correlated with asymmetry, we do not currently have direct evidence of a phase transition from a quantifiable order parameter. Note that this fact does not necessarily reflect the absence of a phase transition, especially considering that we have so far only examined variations in lipid density but not yet in other measures of lipid packing. Furthermore, attributing the observed stiffening to gelation would arguably also require a clearer relationship between the proposed cause and effect, *i.e.* an explanation to why compression of POPC leaflets appears to cause first a stiffening and then a softening.

Edward Lyman said: Fig. 1A and 5A in your paper (https://doi.org/10.1039/d5fd00006h) report the free energy when deforming the membrane. If I understand the method correctly, the deformation is applied by imposing a long wavelength undulation *via* a collective variable (a sinusoidal deflection of the membrane), and the free energy obtained by standard collective variable methods.

In these panels there are an awful lot of data points. Is each one a separate simulation? There also are no error bars on any of the points. You have indicated that this is because the errors are smaller than the points. Is this correct, and if so, how were the errors analyzed? If indeed the errors are this small I find it quite surprising, because these are very small amplitudes, and I would expect that a lot of sampling is needed to overcome the thermal noise.

Giacomo Fiorin responded: Thank you for pointing out that the data points in the free-energy plots could be interpreted as individual umbrella-sampling simulations: in fact, those points represent the discretized bins of the free-energy profile computed from the simulated histograms using the WHAM method.

There are specific reasons why the estimated sampling error is very small. For each bilayer, the umbrella-sampling windows were initialized by extracting random snapshots of a 3 µs conventional MD simulation, and the aggregated simulation time between all windows amounts to an additional 7 µs. At the wavelength corresponding to the bilayers' lateral sizes (*i.e.* below 10 nm) the variations in bilayer curvature are tightly coupled to the variations in lipid orientation. The correlation times of the latter are of the order of nanoseconds for homogeneous phospholipid bilayers (see *e.g.* ref. 1). Therefore, even within a single block of simulated data (five blocks for each bilayer were used to estimate errors), the total simulated time exceeds the relevant correlation times by at least two orders of magnitude. Lastly, due to the stratification strategy of umbrella sampling, each simulated window was restrained to fluctuate near a given curvature, resulting in near-uniform sampling across the free-energy landscape.

1 R. M. Venable, *et al.*, *Chem. Phys. Lipids*, 2015, **192**, 60–74.

Markus Deserno said: Let me preface this question with just a little bit of "historical context". When a few years back we found out that abundance asymmetry alone, and hence differential stress, could increase a membrane's bending modulus, we got the idea that this might be related to some type of ordering in the compressed leaflet. We then looked for instance at the hexatic order parameter and found this to be true. This led to a prediction: it should be easier to order the compressed leaflet if the membrane itself is already closer to the gel transition! We tested this in a follow-up publication[1] and indeed could confirm it – at least within the limited data we could obtain back then.

But here is the sad part of this story: when we looked at published experimental results from groups who observed asymmetric membranes to stiffen, this correlation is not very apparent. For instance, Lu *et al.*[2] see only about 40% stiffening in systems where the high melting lipid (DMPC) is essentially at its gel transition, while Elani *et al.*[3] see 150% stiffening at a temperature (295 K)[4] around 24 K above that of the high melting lipid (POPC). This is not at all what we expected, and this made us worry whether our idea even holds water.

So here is my question: with your more detailed membrane model (compared to our highly coarse-grained one), have you investigated this correlation between increased ordering and magnitude of the stiffening? Do you maybe plan to do this at different temperatures, *i.e.*, distances from the main phase transition? I am asking because I want to assure you that what we had claimed back then is by no means the final word on this question. I have a sense it might be more complicated than that.

1 A. Hossein and M. Deserno, *J. Chem. Phys.*, 2021, **154**, 014704, DOI: 10.1063/5.0028255.
2 L. Lu, *et al.*, *Soft Matter*, 2016, **12**, 7521–7528, DOI: 10.1039/c6sm01349j.
3 Y. Elani, *et al.*, *Chem. Commun.*, 2015, **51**, 6976–6979, DOI: 10.1039/c5cc00712g.
4 J. M. Seddon, personal communication.

Giacomo Fiorin answered: Thank you for this valuable historical note, which is both informative and helpful in framing future investigations. Coarse-grained (CG) models in general, and the MARTINI force field in particular, have known limitations in describing the transition between the liquid and gel phases. For example, the work by Salerno *et al.*[1] on long-chain hydrocarbons above their melting point has shown that MARTINI potentials differ from atomistic ones in ways that go beyond a simple reduction in the number of degrees of freedom. While developing a CG force field for lipids with a high melting point,[2] we also found that a single set of CG parameters could not simultaneously reproduce experimental properties of both the liquid and gel phases: eventually, we settled on simulating only the former. It appears that current CG models can represent the key characteristics of either phase reasonably well, but the predicted transition mechanism may still require further validation.

Regrettably, our answer to whether we correlated stiffening with curvature-induced ordering is no: due to time constraints we did not include a computation of the hexatic order parameter. We will likely use these computations in future work. However, we did examine local lipid density, another quantity influenced by molecular order, and found no correlation with curvature (see Fig. S4 and S5 in the supplementary information of the paper; https://doi.org/10.1039/d5fd00006h). That said, we cannot rule out the possibility that gel-phase domains may emerge over timescales or length scales beyond those explored in this study (several microseconds and ~8 nm). Given the computational cost of atomistic simulations, particularly when conducted across a range of temperatures, further work in this direction will require careful planning.

1 K. M. Salerno, *et al.*, *Phys. Rev. Lett.*, 2016, **116**, 058302.
2 C. M. MacDermaid, *et al.*, *Biophys. J.*, 2020, **118**(7), 1588–1601.

Tiemei Lu asked: Is it possible to simulate a full spherical (GUV) membrane instead of a flat membrane? (Not just patches of full vesicles.)

Giacomo Fiorin answered: In principle, MD simulations can tackle systems of any size. In practice, though, the longest simulation times attainable by typical MD simulations are limited (currently, 10–100 μs), and this limit also decreases slightly with increasing system sizes. Furthermore, the relevant correlation times of curvature deformations increase very rapidly with the bilayer size (see *e.g.* ref. 1 and related work). Therefore, attaining the necessary sampling would be very challenging for atomistic simulations of even relatively small vesicles (~100 nm), and barely approachable with coarse-grained simulations. For micrometer-sized giant vesicles, MD simulations of any kind are not possible for the foreseeable future. Instead, MD simulations can be used to study GUVs indirectly, *e.g.* by providing the parameters used in continuum-mechanics equations from first-principles computations.

1 A. G. Zilman and R. Granek, *Phys. Rev. Lett.*, 1996, 77(23), 4788–4791.

Rumiana Dimova addressed Giacomo Fiorin and Markus Deserno: Your studies suggest that compositional asymmetry alone does not significantly alter bending rigidity when the two leaflets experience equal stress. However, introducing stress or packing asymmetry appears to increase the bending rigidity. In contrast, Elani *et al.* (2015)[1] reported a 2–3-fold increase in bending rigidity for asymmetric DOPC/POPC membranes, which is considerably higher than the modest changes observed in your work. How do you interpret this discrepancy?

I would also like to offer a speculative comment: the emulsion transfer method used for GUV preparation in Elani's study[1] may result in a very different stress asymmetry between the leaflets, which is lipid specific. Since the two monolayers are formed separately (each at its own interface), differences in interfacial tension and lipid packing may lead to unequal stresses when the bilayer is finally assembled. In addition, the emulsion-transfer method may result in some lipids intended for one leaflet migrating to the "wrong" one, as we seem to observe in our own work.[2] Could such differences in (i) the final leaflet stress and (ii) composition account for the more pronounced effect seen in their measurements compared to your simulations?

1 Y. Elani, *et al.*, *Chem. Commun.*, 2015, **51**, 6976–6979, DOI: 10.1039/c5cc00712g.
2 D. Drabik, *et al.*, *Biophys. J.*, 2024, **123**, 2406–2421, DOI: 10.1016/j.bpj.2024.05.031.

Giacomo Fiorin replied: These are valuable considerations when comparing experimental and simulated data. In the case of asymmetric POPC/DOPC vesicles, it is noteworthy that the same laboratory has used fluctuation analysis to estimate different bending moduli for asymmetric bilayers: one study reported a value 2.2 times higher than that of the symmetric mixture,[1] while another found a factor of 1.5.[2] It is noteworthy that the latter result also has a higher experimental uncertainty, and even agrees with our own estimate of 1.25. However, as we acknowledge in our paper (https://doi.org/10.1039/d5fd00006h), our estimate is not consistent with the one of Elani *et al.*[1] Given the relatively small size of the bilayers in our simulations, our initial approach to reconcile this discrepancy has been to explore the potential influence of length scale.

Your first point suggests that, in the study by Elani *et al.*,[1] the densities of the two leaflets may be influenced not only by the intrinsic biophysical properties of the lipids but also by the specifics of the preparation method – implying that the vesicles could be subject to significant curvature strain. This appears to be a plausible interpretation, although we lack sufficient technical insight into the transfer method used to speculate further. It is important to note, however, that MD simulations can be biased in a similar way due to periodic boundary conditions, which impose a zero mean curvature. If the asymmetric bilayer configuration observed by Elani *et al.*[1] could be reproduced in a future simulation, it might require either a higher degree of leaflet asymmetry than we have considered or the introduction of non-zero membrane tension.

We appreciate your observation regarding potential spurious lipid transfer between leaflets and the insertion of oils, as suggested in your work.[3] While it is difficult to predict the precise effect of a partial loss of asymmetry, it could plausibly soften the bilayer, thus increasing the discrepancy between experimental measurements and simulations and reinforcing the importance of understanding its cause. Moreover, the increased bilayer thickness and reduced interdigitation due to oil insertion present additional questions that are worth investigating.

Finally, although we have not yet examined bilayers with charge asymmetry, your findings on POPC/POPA systems offer a potentially valuable benchmark – while recognizing the limited correlation between area modulus (K_A) and bending modulus (k_c).

1 Y. Elani, *et al.*, *Chem. Commun.*, 2015, **51**, 6976–6979, DOI: 10.1039/c5cc00712g.
2 K. Karamdad, *et al.*, *Chem. Commun.*, 2016, **52**, 5277–5280, DOI: 10.1039/c5cc10307j.
3 D. Drabik, *et al.*, *Biophys. J.*, 2024, **123**, 2406–2421, DOI: 10.1016/j.bpj.2024.05.031

Markus Deserno answered: Let me start with the Elani work.[1] It is true that we never saw a stiffening driven by compositional asymmetry alone, *i.e.*, in the absence of differential stress. But we have no proof either that Elani *et al.*[1] ever did. Recall that we simply do not know the differential stress in their experiments; but as we had explained in our original paper,[2] and as I have revisited in my earlier answer to a question of Dr Heerklotz, the very fact that Elani *et al.*[1] actually could observe stable fluctuating GUVs in their experiments suggests that their bilayers are under a differential stress of a few mN m^{-1} in order to cancel the spontaneous curvature torque.

I agree that the extent of stiffening observed by them is larger than what we see in our simulations, and there is an easy and a more complicated take on this. The easy answer is that I am not surprised that we see less stiffening, considering that we attribute this effect to a change in lipid order, and our highly coarse-grained lipids have much less entropy to lose than an actual lipid. It is not uncommon that order-affiliated phenomenology is underestimated in coarse-grained models, just as the differences in diffusion constants between liquid and gel phases, or liquid ordered and disordered phases, is much less pronounced at that level. The more complicated answer is that our original hypothesis ("ordering of the compressed leaflet drives stiffening") makes a clear prediction ("stiffening should be more visible if the high melting lipid is closer to its main phase transition") which does *not* appear to be borne out by the available experimental studies. It is therefore quite possible that we have not yet figured out the main reason why such membranes stiffen.

Regarding the comment that the specific protocol for making the asymmetric vesicles might create additional lipid-specific sources of differential stress: yes, this is definitely possible, and we had in fact thought about this for a while. In the end, my take is that the details of the creation process are likely very messy, and trying to theoretically understand at a quantitatively predictive level what kind of additional stresses might sneak in during the complex non-equilibrium emulsification process might be too difficult for us to explain with our current state of understanding. In contrast, it is much easier to summarily reason about the final outcome, and this leads to something I facetiously call the "anthropic principle" of asymmetric vesicle creation. Recall that in philosophy and cosmology the anthropic principle typically goes as follows: we might ask why our universe is "set up" in such a miraculously "fine-tuned" way that it supports the evolution of intelligent life. And one disarmingly disappointing answer to this is: "Well, if it weren't, nobody would be around to ask that question." So in our GUV creation example we could ask: how is it that the very messy production process [insert your favorite protocol] gives rise to stable gently fluctuating asymmetric vesicles which therefore have to be under close to vanishing torque? And a possible answer is: "If it didn't, we wouldn't be talking about this production process, because it wouldn't create objects of interest." In short, I believe whatever complicated processes are going on that preferentially partition lipids or affect tension, torque, and differential stress, if at the end of the day we get stable GUVs, then these GUVs must satisfy the conditions conducive to their own existence (such as, I submit, close to vanishing torque).

1 Y. Elani, *et al.*, *Chem. Commun.*, 2015, **51**, 6976–6979, DOI: 10.1039/c5cc00712g.
2 A. Hossein and M. Deserno, *Biophys. J.*, 2020, **118**, 624, DOI: 10.1016/j.bpj.2019.11.3398.

Michael M. Kozlov opened discussion of the paper by Axel Voigt: How can one estimate the values of the elastic coefficients of the higher order contributions to the energy?

Axel Voigt replied: To the best of my knowledge the bending rigidities of the higher order terms have not been measured. An estimate has been considered in ref. 1, where the term related to $\mathcal{K}^2$ is used to model periodic structures. This requires $k_{4,2}$ to be negative, which is stabilized by an even higher order term related to $\mathcal{K}^4$ with a positive coefficient. The same approach is used in ref. 2 to model neck formation. In our context we assume $k_{4,2}$ to be positive and to have a stabilizing effect. Instead of favouring the formation of saddles, we favour developable surfaces, surfaces with zero Gaussian curvature.

1 R. Goetz and W. Helfrich, *J. Phys. II*, 1996, **6**, 215–223.
2 T. Shemesh, *et al.*, *Biophys. J.*, 2003, **85**, 3813–3827.

Michael M. Kozlov said: If one compares your expression for that energy with that presented in Marin Mitov's 1978 paper,[1] one wonders why in the expression you are using the coefficients of the higher order terms depend on the spontaneous curvature.

1 M. D. Mitov, *C. R. Acad. Bulg. Sci.*, 1978, **31**, 513–515.

Axel Voigt responded: In ref. 1 an elastic energy as a function of director derivatives is considered and expanded leading to the higher order terms. A spontaneous curvature is not explicitly considered in this setting. We take a functional form which depends on the mean and the Gaussian curvature and consider a Taylor expansion at the spontaneous curvature. While setting the spontaneous curvature to zero leads to the same terms also considered in M. Mitov,[1] the methodology is different. To some extend M. Mitov considers a "physical correction" and we a "geometric correction" to the small curvature assumption in the classical Canham/Helfrich model.

1 M. D. Mitov, *C. R. Acad. Bulg. Sci.*, 1978, **31**, 513–515.

Markus Deserno said: Dr Kozlov expressed worries about using some spontaneous mean curvature as the expansion point also for higher order terms. I am not so terribly worried about that – of course you could decide to write these terms in this form, and it would amount to certain choices for some of the expansion coefficients, or how some cubic terms show up. As long as all terms contributing to any given order are present, this just amounts to "renaming" the various moduli. But in your expression all terms are *not* present, which then amounts to nontrivial statements about the underlying physics, and I was wondering what your thought process was for leaving out certain terms. For instance, you do not have a cubic term of the form $\mathcal{H}^3$, and you also do not have a term of the form $\mathcal{K}(\mathcal{H} - \mathcal{H}_0)^2$, which is quartic. I presume you made some conscious choices here – can you help me understand what considerations went into this?

Axel Voigt responded: We agree that there are additional terms with the same order. A complete list of third and fourth order terms can be found in ref. 1 or 2. We derive the higher order terms by a Taylor expansion of the bending energy density as a function of the mean and the Gaussian curvature at the spontaneous curvature. We only keep terms which are bounded from below. This excludes all odd geometric orders, *e.g.* $\mathcal{H}^3$ but also the quartic term $\mathcal{K}(\mathcal{H} - \mathcal{H}_0)^2$, as the mean and the Gaussian curvature also can be negative. While in different context such terms are considered to initiate instabilities, we are interested in stabilizing effects of the higher order terms. Therefore all higher order bending rigidities are nonnegative and only nonnegative terms in the expansion are considered. Including the missing terms would require to specify certain relations between the bending rigidities in order to guarantee boundedness from below.

1 M. D. Mitov, *C. R. Acad. Bulg. Sci.*, 1978, **31**, 513–515.
2 M. Deserno, *Chem. Phys. Lipids*, 2015, **185**, 11–45.

Gonen Golani enquired: Lower-order terms such as membrane tension, which can arise from pressure differences, appear to be absent from the energy considerations at equilibrium. These terms could stabilize cylindrical shapes, as observed in tether-pulling experiments. How would the introduction of membrane tension or a pressure difference affect the results?

Axel Voigt replied: As the membrane is considered to be inextensible (locally area preserving) an additional membrane tension does not have any effect. The pressure already serves this role, here it is the Lagrange multiplier to enforce the inextensibility constraint.

John M. Seddon asked: I assume that the modulus $k_{4,2}$ is positive in sign, and that this favours parabolic surfaces in general, rather than specifically cylindrical tubes?

Axel Voigt answered: Yes, in our investigation (https://doi.org/10.1039/d4fd00202d) we only consider terms which guarantee boundedness from below. Therefore $k_{4,2} \geq 0$. Without any constraints this would lead to smooth surfaces whose Gaussian curvature vanishes everywhere, so-called developable or ruled surfaces. Cylindrical tubes belong to this class. However, due to the constraints on inextensibility and enclosed volume such surfaces cannot be realised for any closed surface.

Markus Deserno questioned: You mentioned that you are not just interested in energy-minimized shapes (*i.e.*, solution to the shape equation) but also the time evolution of shape transformations. Your dynamical equations indeed contain an in-plane membrane viscosity, but there are other sources of dissipation as well. The viscosity of the embedding solvent of course comes to mind, but the friction between the two leaflets – which might have to slide past one another as the vesicle deforms – is another. Do you have any plans on incorporating them, or alternatively a good reason why you can ignore them?

Axel Voigt responded: Indeed there are other dissipation mechanisms. Our goal was to consider a minimal model, which puts the effect of surface viscosity and higher order curvature terms in the bending energy into the focus. Any coupling with the surrounding bulk phases (the embedding solvent) is neglected. This is justified in the limit of a large Saffman–Delbrück number. As the viscosity of the membrane is typically larger than the viscosity of the bulk phases, this is typically the case. To neglect interleaflet friction is probably less justified. An "effective" membrane viscosity, expressed by the Reynolds number Re, or the general friction term with the friction coefficient γ probably provide zeroth order approximations. A better approximation requires to model both leaflets, approaches in this direction can be found in ref. 1 building on ref. 2.

1 A. Torres-Sanchez, *et al.*, *J. Fluid Mech.*, 2018, **871**, 218–271.
2 U. Seifert, *et al.*, *Europhys. Lett.*, 1993, **23**, 71–76.

Markus Deserno commented: It would seem that inter-leaflet friction would require the explicit modeling of two leaflets, but I believe there is a way around that: the parallel surface theorem allows us to calculate the local area difference between the two leaflets as a function of local curvatures, and I presume a little bit of differential geometry will then show that the local inter-leaflet strain rate scales like the time evolution of curvature (or curvature gradients). I suspect this could be put into your model without too much extra effort. It is conceivable, though, that this does not greatly contribute at the large scales you appear to be interested in.

Axel Voigt replied: This sounds like a meaningful and numerically feasible approach worth exploring.

Di Jin asked: How would you interpret Reynolds number in terms of membrane properties and what is the typical range of Reynolds number, and what does it imply?

Axel Voigt answered: By interpreting membranes as fluid deformable surfaces, essentially two-dimensional incompressible fluids with bending properties, the Reynolds number follows the same definition as in classical fluid dynamics. It relates to the membrane viscosity which has been measured for various bilayer systems, see ref. 1, *e.g.* for DPPC : DOPC : Chol (1 : 1 : 1) the reported membrane viscosity is 17.7 $\pm$ 3.06 nPa s m. Together with low fluid velocities in the membrane and small length scales this leads to low Reynolds numbers. All numerical simulations are therefore considered in a regime which also would allow for a Stokes approximation.

1 H. A. Faizi, R. Dimova and P. M. Vlahovska, A vesicle microrheometer for high-throughput viscosity measurements of lipid and polymer membranes, *Biophys. J.*, 2022, **121**, 910–918.

Michael M. Kozlov opened a general discussion: A differential stress of a few mN m^{-1} should be able to drive formation of membrane pores and relaxation of the stress difference due to the trans-monolayer lipid exchange through these pores. Do you think this may happen?

Rumiana Dimova answered: In principle, I would expect that a differential stress of a few mN m^{-1} could indeed be sufficient to drive pore formation and thereby allow for trans-monolayer lipid exchange to relax the stress difference. However, I am not aware of direct experimental evidence confirming this mechanism.

I suppose that some of the most widely used experimental methods for creating asymmetric membranes, such as leaflet exchange *via* hemifusion or individual lipid exchange assisted by cyclodextrin, likely proceed without introducing large differential stresses. In contrast, the phase-transfer method, where two monolayers formed at oil–water interfaces are combined into a bilayer, could intuitively result in significant differential stress due to differences in the initial interfacial tensions of these monolayers.

Our own experience with this method, especially when using charged lipids, suggests that the resulting asymmetric GUVs are indeed less stable compared to their symmetric counterparts.[1,2] Unlike symmetric GUVs, these asymmetric vesicles tend to rupture spontaneously over time. Furthermore, we have observed that the pore edge tension of these asymmetric membranes is lower,[2] and they are prone to collapse when subjected to external stresses such as electric fields.

These observations are consistent with the idea that differential stress may indeed promote pore formation, but to my knowledge, a direct, real-time observation of stress relaxation *via* pores in this context is still lacking.

1 D. Drabik, *et al.*, *Biophys. J.*, 2024, **123**, 2406–2421, DOI: 10.1016/j.bpj.2024.05.031.
2 F. S. C. Leomil, *et al.*, *Langmuir*, 2024, **40**, 4719–4731, DOI: 10.1021/acs.langmuir.3c03370.

Conflicts of interest

There are no conflicts to declare.

Faraday Discussions

ROYAL SOCIETY OF CHEMISTRY

PAPER

Lipid–GPCR interactions in an asymmetric plasma membrane model†

Jingjing Ji[a] and Edward Lyman *[ab]

Received 26th December 2024, Accepted 23rd January 2025
DOI: 10.1039/d4fd00210e

We report simulations and analysis of the A_{2A} adenosine receptor in its fully active state, in two different membrane environments. The first is a model in which the lipids are distributed asymmetrically according to recent lipidomics, simulations, and biophysical measurements, which together establish the distribution of lipids and cholesterol between the two leaflets. The second is the symmetrized version, which captures the membrane state following loss of lipid asymmetry. By comparing lipid–protein interactions between these two cases we show that solvation by phosphatidyl serine (PS) is insensitive to the loss of asymmetry—an abundance of positively charged sidechains around the cytoplasmic side of the receptor enriches solvation by PS in both membrane states. Cholesterol interactions are sensitive to the loss of asymmetry, with the abundance of cholesterol in the exoplasmic leaflet driving long-lived cholesterol interactions in the asymmetric state. However, one cholesterol interaction site on helix 6 is observed in both cases, and was also observed in earlier work with different membrane models, supporting its identification as a *bona fide* cholesterol binding site.

Introduction

The first direct observations of lipid–protein interactions were reported in 1999 by Mitsuoka *et al.*, using electron crystallography of naturally occurring two-dimensional crystalline arrays of bacteriorhodopsin from purple membranes.[1] Using protein reconstituted into a single component phospholipid bilayer, Gonen *et al.* reported (in 2005) high resolution structures of lipids surrounding an aquaporin, reconstituted into a single component phospholipid membrane.[2] Around the same time, several publications reported lipids (including cholesterol) bound to membrane proteins observed by X-ray crystallography,[3–6] but these were either crystallized from concentrated detergent solutions or from the lipidic cubic phase—in either case, a not-very-native membrane environment. Thus, in these early years of lipid–protein interactions the work of Mitsuoka *et al.* stands apart as the most representative of the protein in its native environment.

[a]*Department of Physics and Astronomy, University of Delaware, Newark, DE, USA. E-mail: elyman@udel.edu*
[b]*Department of Chemistry and Biochemistry, University of Delaware, Newark, DE, USA*
† Electronic supplementary information (ESI) available. See DOI: https://doi.org/10.1039/d4fd00210e

Very recently, two groups have reported high resolution structures of membrane proteins in their native membranes, with stunning results. Tao *et al.* reported the structure of a K^+ channel in two different native membrane vesicle preparations, observing dozens of lipids including cholesterol, and finding that the native membrane environment orders previously unresolvable regions of the protein.[7] The cholesterols are observed around the receptor in the exoplasmic leaflet. Coupland *et al.* obtained the structure of a V-ATPase in native synaptic vesicles, finding regularly spaced cholesterols arrayed on the lumenal side of the transmembrane domain, which becomes the exoplasmic leaflet upon fusing with the plasma membrane.[8] Thus, in both cases the sterols are observed in the *exoplasmic* plasma membrane leaflet.

The lipid interactions observed in these two structures support an emerging model of plasma membrane asymmetry. By combining headgroup specific phospholipases with shotgun lipidomics, Lorent *et al.* reported the asymmetric distribution of glycero-phospholipids and sphingolipids in the plasma membrane of human red blood cells.[9] The results were consistent with prior reports obtained using lower chemical resolution methods.[10,11] The exoplasmic leaflet is more saturated in the hydrocarbon region, and contains essentially all of the sphingolipids, while the cytoplasmic leaflet contains all of the negatively charged headgroups (mostly phosphatidyl serine, abbreviated PS), nearly all of the ethanolamine headgroups (including plasmalogens with ether linked chains), and is significantly more unsaturated in the hydrocarbon region than the exoplasmic leaflet. Combining extensive simulations and leaflet-resolved biophysical measurements, Doktorova *et al.* reported recently that phospholipid asymmetry drives an asymmetric distribution of cholesterol, about two-thirds of which is located in the exoplasmic leaflet—more than half of all of the lipids in the exoplasmic leaflet are cholesterol.[12]

In this submission we use the recently reported asymmetric membrane model of Doktorova *et al.* for simulations of a G-protein coupled receptor (GPCR) (the A_{2A} adenosine receptor). Because this protein has been studied by many simulation groups to assess lipid–protein interactions,[13–26] it is a good test case to identify how such interactions change (or not) when the receptor is in a native-like membrane model. Visually, the receptor with its first shell lipids looks remarkably similar to the two recent native membrane structures mentioned above (Fig. 1), with substantially more cholesterol density observed around the receptor in the exoplasmic leaflet. Because the controlled loss of lipid asymmetry is associated with many different cellular functions,[27] we also consider how lipid–protein interactions change following loss of asymmetry by simulating the receptor in the symmetrized version of the asymmetric membrane model. Based on a pair of 10 μs all atom simulations (one in each membrane state) of the fully active receptor, we find that PS headgroups are enriched around the cytoplasmic half of the receptor regardless of membrane state, driven by a corresponding enrichment of positively charged sidechains (19 in all), following the "positive inside rule".[28,29] A recently described motif of conserved positive charge that allosterically favors receptor activation is occupied by PS headgroups in both membrane states.[15] We also find that cholesterol binds a previously described motif in the outer leaflet region of helix 6, again in both membrane states.[13]

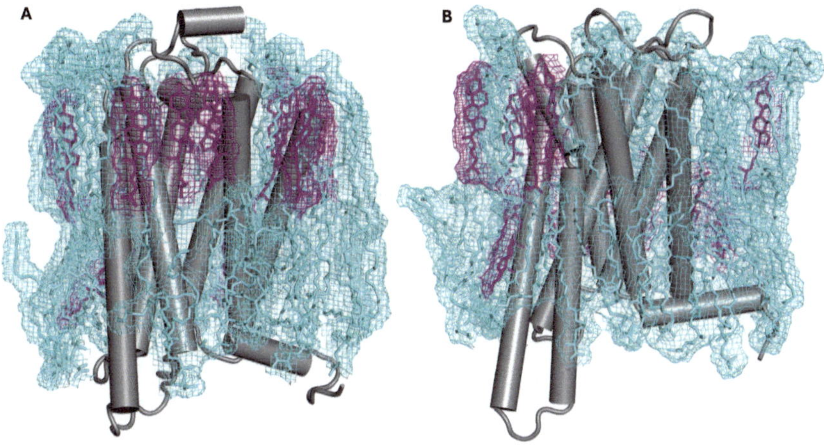

Fig. 1 Simulation snapshot showing the A_{2A} adenosine receptor in an asymmetric plasma membrane model (panel A) and its symmetrized counterpart (panel B). First shell cholesterols are rendered in pink and other first shell lipids in cyan. Wireframe mesh shows the outline of the lipid mass density. Each snapshot is the final configuration of a 10 μs all-atom simulation.

Methods

System setup for molecular dynamics simulations

Initial coordinates of the protein were based on a high-resolution structure of the adenosine A_{2A} receptor in complex with the full agonist NECA and an intracellular engineered G protein (PDB: 5G53).[30] The engineered G protein was deleted *in silico*, allowing lipids to access the G protein-binding interface. Modeller 10.5 was used to rebuild residues 147 to 158 and residues 212 to 223 that were not resolved in the receptor PDB structure.[31] The protein complex was embedded in two different membrane models. The first is a bilayer with an asymmetric composition containing 12 lipids,[12] containing a mixture of sphingolipids (SM), phosphatidylcholine (PC), phosphatidylethanolamine (PE), ether-linked (plasmalogen) PE (abbreviated PLAS), phosphatidylserine (PS), and cholesterol. The other is a symmetrized version of the same membrane, in which the lipid compositions of the two leaflets are identical. The symmetric membrane bilayer has the same overall lipid compositions as the asymmetric model and models the membrane environment following a loss of lipid asymmetry. The composition of both models is given in Table 1 and its caption.

Each system was prepared using the membrane builder in CHARMM-GUI.[32] The protein complex was embedded in the mixtures of lipids mentioned above, solvated with TIP3P[33] water molecules to provide a 30 Å-thick layer above and below the bilayer. Na^+ ions were added to neutralize the system, and additional Na^+ and Cl^- ions were added to maintain 0.15 M ionic concentration. Both systems contained approximately 464 000 atoms and measured 18.5 × 18.5 × 14.3 nm. Protein and lipids were modeled with the CHARMM36 force field,[34,35] and the ligand NECA was modeled with the CHARMM general force field.[36]

Table 1 Number of lipids in asymmetric and symmetric membrane systems. The value in the parentheses is the mole fraction % in each leaflet. Abbreviations are as follows: PC – phosphatidylcholine; PE – phosphatidyl ethanolamine; PS – phosphatidyl serine; PSM – 18:1,18:0 sphingomyelin; LSM – 18:1,24:0 sphingomyelin; NSM – 18:1,24:1 sphingomyelin; PAPC – 16:0,20:4 PC; SOPC – 18:0,18:1 PC; PLPC – 16:0,18:2 PC; POPC – 16:0,18:1 PC; OAPE – 18:1,20:4 PE; PDPE – 16:0,22:6 PE; PLQS – 18:0,22:4 plasmalogen PE; PAPS – 16:0,20:4 PS; CHOL – cholesterol

		Asymmetric		Symmetric	
Lipid class	Abbreviation	Exo	Cyto	Exo	Cyto
SM	LSM	48 (6.6)	0	24 (3.8)	24 (3.8)
	NSM	56 (7.7)	0	28 (4.5)	28 (4.5)
	PSM	72 (9.9)	0	36 (5.7)	36 (5.7)
PC	PAPC	24 (3.3)	0	12 (1.9)	12 (1.9)
	SOPC	40 (5.5)	0	20 (3.2)	20 (3.2)
	PLPC	80 (11.0)	80 (15.0)	80 (12.7)	80 (12.7)
	POPC	0	32 (6.0)	16 (2.5)	16 (2.5)
PE	OAPE	0	32 (6.0)	16 (2.5)	16 (2.5)
	PDPE	0	72 (13.5)	36 (5.7)	36 (5.7)
PLAS	PLQS	0	80 (15.0)	40 (6.4)	40 (6.4)
PS	PAPS	0	120 (22.6)	60 (9.6)	60 (9.6)
CHOL	CHOL	404 (55.8)	116 (21.8)	260 (41.4)	260 (41.4)

Simulation details

The equilibration of the two systems was run with NAMD version 3,[37] following CHARMM-GUI's default six-step protocol. The initial configuration was relaxed by 10 000 steps of steepest descent. The first three simulations had 1 fs timestep for a total of 375 ps (125 ps × 3) of simulation time, and the last three steps had a 2 fs timestep for a total of 1500 ps (500 ps × 3) simulation length. During the six-step equilibration protocol, velocities were reassigned every 500 steps. The heavy atoms of the receptor backbone and the ligand NECA were positionally restrained but with a decreasing force constant from 10.0 to 0.1 kcal mol^{-1} Å^{-2} following CHARMM-GUI's default. The force constant was gradually reduced from 5.0 to 0 kcal mol^{-1} Å^{-2} for the heavy atoms of the receptor sidechain. Similarly, the harmonic restraints on the heavy atoms of lipid headgroups were reduced from 5.0 to 0 kcal mol^{-1} Å^{-2} in a stepwise manner. The simulation box volume was allowed to change semi-isotropically *via* a Langevin piston with a barostat damping time scale of 25 fs and an oscillation period of 50 fs.[38] The target pressure and temperature were 1.01325 bar and 310 K, respectively. All covalently bonded hydrogens were constrained by SHAKE.[39] A cutoff of 1.2 nm was used for the electrostatic and van der Waals interactions, while the switching algorithm was on if the distance between two atoms was between 1.0 nm and 1.2 nm, ensuring the potential was smoothly reduced to 0 at the cutoff distance. The long-range electrostatic interactions were computed using the Particle Mesh Ewald (PME) method[40] on a 1 Å grid, with a tolerance of 10^{-6} and sixth order interpolation. The nonbonded interactions were not computed only if two atoms were connected within 3 covalent bonds. The neighbor list was updated every 10 steps, including all pairs of atoms whose distances were 1.2 nm or less. The sixth equilibration step was run for another 30 ns to ensure the receptor was well equilibrated. An additional 70 ns production run was performed in which all restraints were removed.

The equilibrated binary restart files from NAMD were converted into DMS format for further production simulation on Anton2. Viparr 4.7.49c7 was used to add force field information. Integration was performed under constant pressure (1 atm), temperature (310 K), and particle number with the multigrator[41] method with a 2.5 fs time step. Temperature was controlled by a Nosé–Hoover[42] thermostat coupled every 24 timesteps and pressure was controlled by the Martyna–Tobias–Klein barostat (semi-isotropic) coupled every 480 timesteps.[43] Nonbonded interactions were cut off at 1.2 nm. Long-range electrostatic interactions were computed using the *u*-series method[44] following Anton2's default. Hydrogens were constrained by M-SHAKE.[45] Production simulations for each system were performed for 10 μs, and configurations were stored every 480 ps.

Simulation analysis

Visual Molecular Dynamics (VMD)[46] was used for molecular visualization. Data analysis and plotting were performed using in-house python scripts based on publicly hosted python packages: NumPy,[47] MDAnalysis,[48] and matplotlib.[49]

We conducted Voronoi analysis with the Voronoi module from the python package SciPy.[50] For the entire 10 μs trajectory, the configurations were sampled at 0.48 ns intervals. In each of the 20 834 frames, for each leaflet, the center of mass of each residue from the seven transmembrane domains (above or below the midplane) and the center of mass of each lipid were projected onto the midplane ($z = 0$) to form a 2D point set. Following Voronoi analysis, lipids sharing an edge with receptor residues were identified as belonging to the first shell.

To characterize the interactions between cholesterol and the receptor residues, contacts between protein sidechains and cholesterol were identified for each residue, using a definition of a contact as ≤4.5 Å between the closest heavy atoms of any cholesterol and the protein sidechain. The interaction frequency of cholesterol with each residue was then classified into three tiers: "high" if the residue contacts a cholesterol for more than 80% of the duration of the 10 μs simulation, "medium" greater than 50%, and low if below 50%. Interactions between the receptor and PS were defined by a distance of ≤4.5 Å between the closest heavy atoms of any PS headgroups and the receptor residue sidechain, using the same classification of the interaction frequency.

Results

A PS interaction motif is occupied in both membrane states

Two simulations were performed of the receptor in its fully active state, but in the absence of the G-protein (see Methods for details on how the initial structure was prepared). This state of the receptor was chosen because in a recent publication[15] we showed that a motif of positive charge on the intracellular face of the receptor mediates interaction with a PS headgroup, favoring the fully active state of the receptor in the absence of the G-protein. Here, we simulated this receptor state in two different membrane environments—an asymmetric membrane model (in which all of the PS is located on the cytoplasmic leaflet), and its symmetrized counterpart, in which the concentration of PS is reduced by more than a factor of two. By designing the simulations in this way we were able to assess changes in PS interactions following loss of lipid asymmetry.

The lipid compositions of the first solvation shell in both leaflets were obtained by a Voronoi analysis, as described in Methods. To simplify the analysis the lipids were grouped into classes based on headgroup; the lipid classes are defined in Table 1. Most (though not all) lipids that begin the production simulation in the first shell have exchanged out of the first shell by the end of the 10 μs production simulations (Fig. S1†). The average number of first shell lipids in each class was then computed over 1 μs trajectory segments, these are reported as mole fractions in Fig. S2,† and the averages of each class (computed over the final 5 μs of each simulation) are reported as mole fractions in Fig. 2 and as numbers of each lipid class in Table 2.

Fig. 2 shows that the composition of the first shell does not differ significantly from the bulk for most lipid classes. This can be seen by comparing the average first shell composition (gray bars with errors) to the bulk composition (pink bars). Cases for which the bulk lipid composition is outside the 95% confidence interval of the first shell composition are indicated by an asterisk. A clear exception is PS, which is enriched in the cytoplasmic first shell in both the asymmetric model and its symmetrized counterpart. In the asymmetric membrane simulation it averages 43.8 ± 10.1 mol%, compared to the average value in the bulk of about 23 mol%. In the symmetrized membrane simulation it is still enriched compared to bulk, at 31.4 ± 8.1 mol% compared to about 10 mol% in the bulk. In the symmetrized model this comes at the expense of sphingolipids, which are depleted by a statistically significant amount, and to a lesser extent PC. In the asymmetric model the depletion is distributed over all of the other lipid classes, none rising to a 95% confidence level.

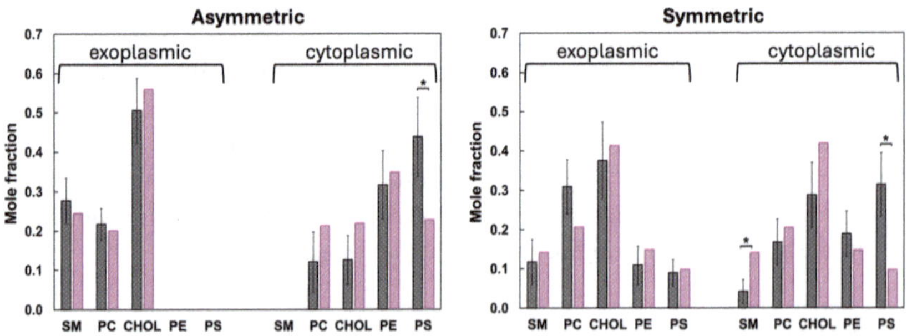

Fig. 2 The mole fraction of lipids in each class in the first solvation shell in the exoplasmic and cytoplasmic leaflets, and in the asymmetric membrane simulation (left panel) and in the symmetric membrane simulation (right panel), averaged over the last 5 μs of each trajectory simulations. Gray filled bars are the results for the first solvation shell of the receptor, and the magenta filled bars are the composition of that species averaged over the entire leaflet. An asterisk (*) indicates cases for which the lipid composition of the bulk is outside the 95% confidence interval of its composition in the first shell. The PE class includes plasmalogen.

Table 2 The average number of lipids in the first shell of the receptor in asymmetric and symmetric membrane systems over the last 5 μs simulations. The value in the parentheses is the standard deviation

	Asymmetric		Symmetric	
Lipid class	Exo	Cyto	Exo	Cyto
SM	6.9 (1.4)	0	2.7 (1.3)	0.9 (0.7)
PC	5.4 (1.0)	2.1 (1.4)	7.0 (1.6)	3.7 (1.3)
CHOL	12.7 (2.1)	2.2 (1.1)	8.5 (2.2)	6.3 (1.8)
PE + PLAS	0	5.6 (1.5)	2.5 (1.1)	4.1 (1.3)
PS	0	7.7 (1.8)	2.0 (0.8)	6.8 (1.8)

These interactions are driven by the "positive inside rule":[28] the receptor has 19 positively charged sidechains (Lys and Arg) located on its cytoplasmic half; it is these interactions that drive enrichment of PS headgroups around the receptor. Ten of the positive charges interact with a PS headgroup at least 80% of the time in both the asymmetric and symmetric membrane models (Fig. 3): R111$^{3.59}$, R120$^{4.41}$, R199$^{5.60}$, R205$^{5.66}$, R206$^{5.67}$, K233$^{6.35}$, R291$^{7.56}$, R293$^{7.58}$, R296$^{7.61}$, R300$^{7.65}$. Two additional residues interact with PS in the asymmetric simulation: R107$^{3.55}$ and R304$^{7.69}$. The fact that enrichment of PS is reproducible across two independent simulations with very different lipid compositions lends significance to the result.

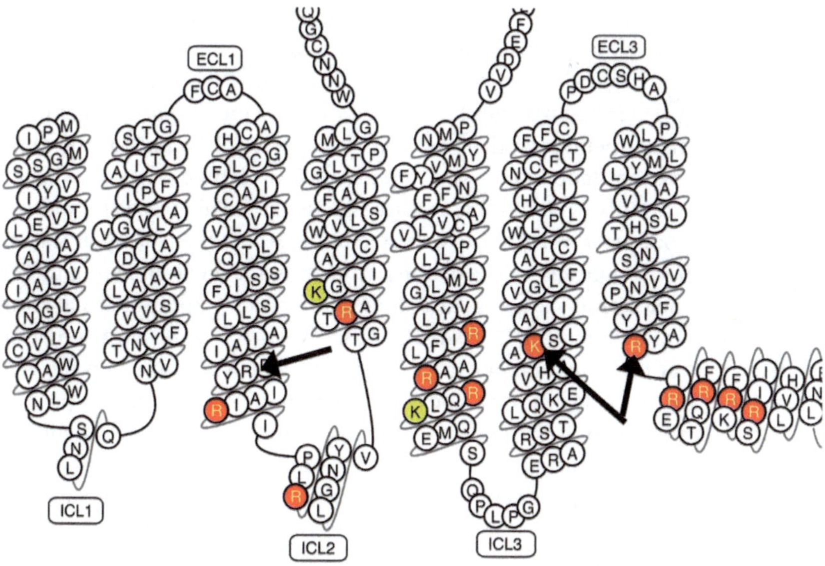

Fig. 3 Positively charged side chains that interact with a PS headgroup at least 80% of the time in the asymmetric membrane simulation (red) or at least 50% of the time (yellow). The arginine of the D/ERY motif is marked on helix 3 with a black arrow. K233$^{6.35}$ and R291$^{7.56}$ mediate a PS interaction motif that allosterically activates the receptor; these are indicated by black arrows on helices 6 and 7. For the corresponding plot from the symmetric membrane simulation, see ESI Fig. S3.†

A pair of side chains (K233$^{6.35}$ and R291$^{7.56}$, black arrows in Fig. 3) are of particular interest, as we recently reported that they mediate an interaction with negatively charged headgroups, preorganizing the intracellular face for G-protein interaction.[15] Both residues are interacting with PS headgroups nearly all of the time in both the asymmetric and symmetric simulations, suggesting that this interaction remains saturated even following loss of asymmetry.

Cholesterol interactions favor the exoplasmic leaflet

According to Fig. 2, the cholesterol composition of the first shell around the exoplasmic side of the receptor does not differ significantly from the bulk composition (defined again as a first shell composition that is outside the 95% CI of the bulk average). This is true of both the asymmetric model and its symmetrized counterpart. However, in the asymmetric model, because cholesterol is so abundant in the exoplasmic leaflet, the first shell of the protein in the exoplasmic leaflet contains many cholesterols—more than twelve in total, and more than 50% of all first shell lipids. This is apparent from the image in Fig. 1 (left panel) that shows the exoplasmic surface of the receptor decorated with individual sterols. In the cytoplasmic leaflet cholesterol is slightly depleted relative to its bulk concentration in both membrane models, with the PS enrichment described above displacing one to two cholesterols in the cytoplasmic leaflet. However, in neither case does the cholesterol depletion rise to a 95% confidence level.

Several long-lived cholesterol interactions are observed with specific locations on the receptor surface. In the asymmetric membrane model, three individual cholesterols remain bound to the receptor for the entire duration of the 10 μs simulation. The three locations are all in the exoplasmic leaflet on TM2, TM6, or TM7, and are indicated on the snake plot in Fig. 4 in purple. The binding disposition of all three cholesterols is shown also in Fig. 4 and a time series of the residues' interaction with cholesterol is shown in Fig. S4.† The configurations shown in Fig. 4 are the final ones from the asymmetric membrane simulation. In the symmetrized membrane there is a single cholesterol bound for the entire duration of the simulation to the *same location* on TM6; a time series showing the interaction with this cholesterol is reported in Fig. S5.† This same site was also identified in our prior work, where we named it "h6o" for "helix 6, outer leaflet". The prior observation was in a very different membrane environment (a ternary mixture of cholesterol, saturated, and di-unsaturated PC), and in both all atom and coarse-grained Martini 2.0 simulations.[13] We do not observe long-lived interactions at the "cholesterol consensus motif", which is located on the cytoplasmic leaflet in between TM2 and TM4. This motif has a lysine at position K122$^{4.43}$; in these simulations negatively charged headgroups are abundant, and this residue is interacting primarily with PS headgroups (see lysine colored in yellow on H4 in Fig. 3). However, in the symmetrized membrane simulation we find that this lysine and several residues form the consensus motif on helix 4 interacting with cholesterol instead of PS (ESI Fig. S6†).

Discussion

There is an extensive literature documenting interactions between lipids and integral membrane proteins, and how such interactions modulate function

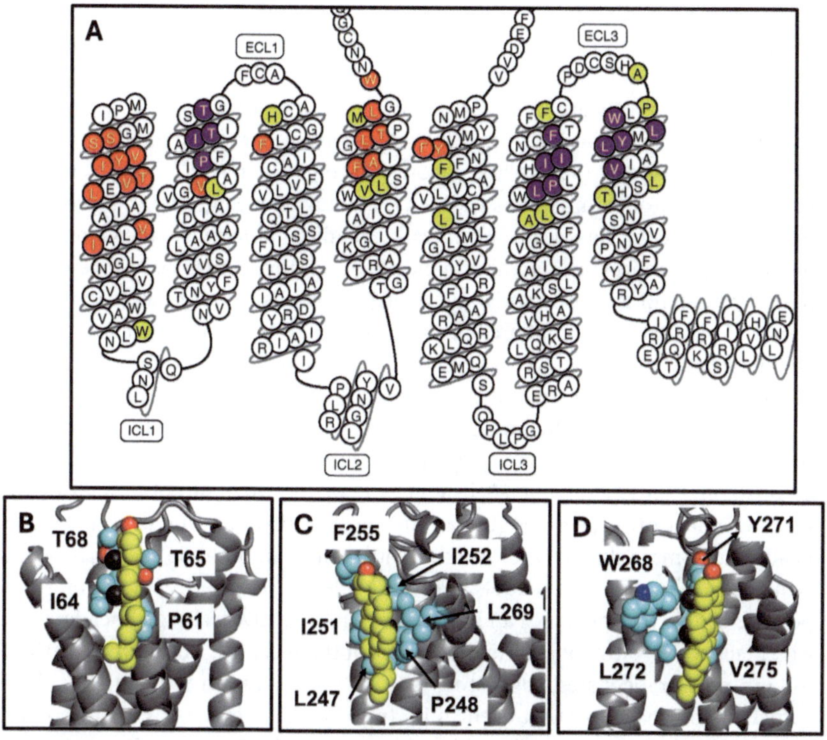

Fig. 4 Cholesterol interactions with the receptor in the asymmetric membrane simulation. Interactions are colored by residue in panel (A), with purple indicating interaction with a single cholesterol for the entire duration of the simulation, red indicating interaction with any cholesterol for at least 80% of the simulation, and yellow any cholesterol for at least 50% of the simulation. The binding disposition of the cholesterol on helix 2 is shown in panel (B), on helix 6 in panel (C), and on helix 7 in panel (D). The heavy atoms of the sidechains indicated in purple in panel A are represented with cyan spheres and are labeled in panels (B–D). Carbon atoms in cholesterol are yellow spheres, the red sphere shows the hydroxyl oxygen, and the black spheres represent the methyls of the beta face. The same analysis is shown for the symmetrized membrane simulation in ESI Fig. S6.†

across many classes of membrane proteins.[51] Since the earliest work on rhodopsin,[52] simulations of GPCRs have played an important role in establishing such interactions.[25,53,54] Most simulation reports (with the notable exception of the early rhodopsin work) have focused on interactions with negatively charged headgroups like PS and inositides,[18,55,56] or on cholesterol.[13,26,57–60] There is experimental evidence that both negative charge and cholesterol modulate GPCR activity. However, most of the experimental reports of cholesterol modulation are obtained in cell-based assays, using methyl-β-cyclodextrin (MβCD) to deplete cholesterol from the plasma membrane. As shown recently by Doktorova *et al.*, this reagent also triggers loss of membrane asymmetry,[12] which results in a drastic change in the membrane environment of the receptor, including redistribution of half of the PS from the cytoplasmic to the exoplasmic leaflet. Thus in order to develop a fuller picture of lipid-dependent mechanisms it is necessary to compare lipid–protein interactions across symmetric and asymmetric lipid distributions. Since transient loss of asymmetry is a common cellular response, results from such comparisons have relevance beyond interpreting assays employing MβCD.

As far as we are aware, very few GPCR simulations have used an asymmetric membrane model,[18] and all used the coarse-grained Martini model, with an asymmetric lipid distribution first reported by Ingólfsson *et al.*[61] In this work we performed all-atom simulations of the A_{2A} adenosine receptor in an asymmetric membrane model based on chemically detailed lipidomics measurements, including differences in lipid abundance between the leaflets as well as cholesterol. Cholesterol is significantly asymmetric in its distribution, with about two-thirds of the total membrane cholesterol in the exoplasmic leaflet, where it is about 50% of all lipids. We compared lipid interactions in the asymmetric membrane and its symmetrized counterpart.

The receptor follows the positive inside rule—there is an abundance of positively charged side chains on the cytoplasmic half of the receptor, which drives a substantial enrichment of PS headgroups in the first solvation shell of the receptor. Most of these are arginine, and most of them are interacting with PS headgroups (more than 80% of the simulation time) in both the asymmetric and symmetrized membrane models. This includes a recently reported motif of positive charge that mediates an interaction with negatively charged headgroups, which act as a positive allosteric modulator of the receptor.[15] It interacts with PS in both membrane states, suggesting that the concentration of PS might still be sufficient to drive activation even following loss of asymmetry. One important exception to the interaction of positive sidechains with PS is the arginine of the D/ERY motif—due to its positioning, we do not observe it to interact with PS headgroups in either membrane state.

In the asymmetric simulation many locations on the protein interact with cholesterol on the exoplasmic half of the receptor—this is hardly surprising, given the abundance of cholesterol in the exoplasmic leaflet. Visually, the cholesterol solvation of the receptor looks strikingly similar to recent cryo-EM structures of membrane proteins obtained in native membrane environments (Fig. 1). Three locations (shown in purple in Fig. 4) each bind a single cholesterol during the entire duration of the 10 μs asymmetric simulation. None of these sites has the characteristics of a CRAC/CARC motif. One location on helix 6 also binds a single cholesterol for the entirety of the symmetrized membrane simulation. This same location was also observed to bind a single cholesterol in simulations using both an all-atom and a Martini representation of a ternary lipid mixture. The concurrence of observations in very different membrane environments suggests that this is a *bona fide* cholesterol binding site.

Many cholesterols are observed to interact with A_{2A} in several different structures, all solved by X-ray diffraction on protein crystallized from the cubic phase.[18] Four of these are in the exoplasmic leaflet and five are in the cytoplasmic half of the receptor (counting only unique locations). The locations in the exoplasmic leaflet overlap partially with the residues that are observed to interact with cholesterol, but this observation needs to be balanced against the fact that much of the exoplasmic surface of the receptor interacts with cholesterol, which is to be expected based on its relative abundance in that leaflet. Many of these interactions are lost upon symmetrization.

The discrepancy between cholesterol interactions observed in X-ray structures and in our simulations raises an important question: to what extent are these interactions representative of the native membrane environment? We anticipate that the situation will become clearer as more native membrane structures are published, and as membrane protein simulations continue moving toward more native membrane models.

Data availability

The data supporting this article have been included as part of the supplementary information.†

Conflicts of interest

The authors declare no conflicts of interest.

Acknowledgements

EL and JJ were supported by the US National Institute of General Medical Sciences by award number R35-GM153273. Computational work utilized the Extreme Science and Engineering Discovery Environment supported by National Science Foundation Grant ACI-1548562. Anton2 computer time was provided by the Pittsburgh Supercomputing Center through grant R01GM116961 from the National Institutes of Health. The Anton2 machine was made available by D. E. Shaw Research.

References

1 K. Mitsuoka, T. Hirai, K. Murata, A. Miyazawa, A. Kidera, Y. Kimura and Y. Fujiyoshi, *J. Mol. Biol.*, 1999, **286**, 861–882.
2 T. Gonen, Y. Cheng, P. Sliz, Y. Hiroaki, Y. Fujiyoshi, S. C. Harrison and T. Walz, *Nature*, 2005, **438**, 633–638.
3 M. A. Hanson, V. Cherezov, M. T. Griffith, C. B. Roth, V.-P. Jaakola, E. Y. T. Chien, J. Velasquez, P. Kuhn and R. C. Stevens, *Structure*, 2008, **16**, 897–905.
4 V. Cherezov, D. M. Rosenbaum, M. A. Hanson, S. G. F. Rasmussen, F. S. Thian, T. S. Kobilka, H.-J. Choi, P. Kuhn, W. I. Weis, B. K. Kobilka and R. C. Stevens, *Science*, 2007, **318**, 1258–1265.
5 D. M. Rosenbaum, C. Zhang, J. A. Lyons, R. Holl, D. Aragao, D. H. Arlow, S. G. F. Rasmussen, H.-J. Choi, B. T. DeVree, R. K. Sunahara, P. S. Chae, S. H. Gellman, R. O. Dror, D. E. Shaw, W. I. Weis, M. Caffrey, P. Gmeiner and B. K. Kobilka, *Nature*, 2011, **469**, 236–240.
6 T. Warne, R. Moukhametzianov, J. G. Baker, R. Nehmé, P. C. Edwards, A. G. W. Leslie, G. F. X. Schertler and C. G. Tate, *Nature*, 2011, **469**, 241–244.
7 X. Tao, C. Zhao and R. MacKinnon, *Proc. Natl. Acad. Sci. U. S. A.*, 2023, **120**, e2302325120.
8 C. E. Coupland, R. Karimi, S. A. Bueler, Y. Liang, G. M. Courbon, J. M. Di Trani, C. J. Wong, R. Saghian, J.-Y. Youn, L.-Y. Wang and J. L. Rubinstein, *Science*, 2024, **385**, 168–174.

9 J. H. Lorent, K. R. Levental, L. Ganesan, G. Rivera-Longsworth, E. Sezgin, M. Doktorova, E. Lyman and I. Levental, *Nat. Chem. Biol.*, 2020, **16**, 644–652.

10 A. J. Verkleij, R. F. A. Zwaal, B. Roelofsen, P. Comfurius, D. Kastelijn and L. L. M. van Deenen, *Biochim. Biophys. Acta, Biomembr.*, 1973, **323**, 178–193.

11 A. Sandra and R. E. Pagano, *Biochemistry*, 1978, **17**, 332–338.

12 M. Doktorova, J. L. Symons, X. Zhang, H.-Y. Wang, J. Schlegel, J. H. Lorent, F. A. Heberle, E. Sezgin, E. Lyman, K. R. Levental and I. Levental, *Cell*, 2025, DOI: 10.1016/j.cell.2025.02.034.

13 E. Rouviere, C. Arnarez, L. Yang and E. Lyman, *Biophys. J.*, 2017, **113**, 2415–2424.

14 L. Yang and E. Lyman, *Biochemistry*, 2019, **58**, 4096–4105.

15 N. Thakur, A. P. Ray, L. Sharp, B. Jin, A. Duong, N. G. Pour, S. Obeng, A. V. Wijesekara, Z.-G. Gao, C. R. McCurdy, K. A. Jacobson, E. Lyman and M. T. Eddy, *Nat. Commun.*, 2023, **14**, 794.

16 A. N. Leonard and E. Lyman, *Biophys. J.*, 2021, **120**, 1777–1787.

17 C. McGraw, L. Yang, I. Levental, E. Lyman and A. S. Robinson, *Biochim. Biophys. Acta, Biomembr.*, 2019, **1861**, 760–767.

18 W. Song, H.-Y. Yen, C. V. Robinson and M. S. P. Sansom, *Structure*, 2019, **27**, 392–403.e3.

19 W. Song, A. L. Duncan and M. S. P. Sansom, *Structure*, 2021, **29**, 1312–1325.e3.

20 A. Bruzzese, J. A. R. Dalton and J. Giraldo, *PLoS Comput. Biol.*, 2020, **16**, e1007818.

21 H. W. Ng, C. A. Laughton and S. W. Doughty, *J. Chem. Inf. Model.*, 2014, **54**, 573–581.

22 R. A. Corey, O. N. Vickery, M. S. P. Sansom and P. J. Stansfeld, *J. Chem. Theory Comput.*, 2019, **15**, 5727–5736.

23 S. Genheden, J. W. Essex and A. G. Lee, *Biochim. Biophys. Acta, Biomembr.*, 2017, **1859**, 268–281.

24 D. Sengupta, X. Prasanna, M. Mohole and A. Chattopadhyay, *J. Phys. Chem. B*, 2018, **122**, 5727–5737.

25 E. Lyman, C. Higgs, B. Kim, D. Lupyan, J. C. Shelley, R. Farid and G. A. Voth, *Structure*, 2009, **17**, 1660–1668.

26 J. Y. Lee and E. Lyman, *J. Am. Chem. Soc.*, 2012, **134**, 16512–16515.

27 T. Sakuragi and S. Nagata, *Nat. Rev. Mol. Cell Biol.*, 2023, **24**, 576–596.

28 G. Von Heijne, *EMBO J.*, 1986, **5**, 3021–3027.

29 S. H. White, G. V. Heijne and D. M. Engelman, *Cell Boundaries: How Membranes and Their Proteins Work*, Garland Science, New York, 1st edn, 2021.

30 B. Carpenter, R. Nehmé, T. Warne, A. G. W. Leslie and C. G. Tate, *Nature*, 2016, **536**, 104–107.

31 B. Webb and A. Sali, *Curr. Protoc. Protein Sci.*, 2016, **86**, 2.9.1–2.9.37.

32 J. Lee, D. S. Patel, J. Ståhle, S.-J. Park, N. R. Kern, S. Kim, J. Lee, X. Cheng, M. A. Valvano, O. Holst, Y. A. Knirel, Y. Qi, S. Jo, J. B. Klauda, G. Widmalm and W. Im, *J. Chem. Theory Comput.*, 2019, **15**, 775–786.

33 W. L. Jorgensen, J. Chandrasekhar, J. D. Madura, R. W. Impey and M. L. Klein, *J. Chem. Phys.*, 1983, **79**, 926–935.

34 J. B. Klauda, R. M. Venable, J. A. Freites, J. W. O'Connor, D. J. Tobias, C. Mondragon-Ramirez, I. Vorobyov, A. D. MacKerell and R. W. Pastor, *J. Phys. Chem. B*, 2010, **114**, 7830–7843.

35 J. Huang, S. Rauscher, G. Nawrocki, T. Ran, M. Feig, B. L. De Groot, H. Grubmüller and A. D. MacKerell, *Nat. Methods*, 2017, **14**, 71–73.
36 K. Vanommeslaeghe, E. Hatcher, C. Acharya, S. Kundu, S. Zhong, J. Shim, E. Darian, O. Guvench, P. Lopes, I. Vorobyov and A. D. Mackerell, *J. Comput. Chem.*, 2010, **31**, 671–690.
37 J. C. Phillips, R. Braun, W. Wang, J. Gumbart, E. Tajkhorshid, E. Villa, C. Chipot, R. D. Skeel, L. Kalé and K. Schulten, *J. Comput. Chem.*, 2005, **26**, 1781–1802.
38 S. E. Feller, Y. Zhang, R. W. Pastor and B. R. Brooks, *J. Chem. Phys.*, 1995, **103**, 4613–4621.
39 J.-P. Ryckaert, G. Ciccotti and H. J. C. Berendsen, *J. Comput. Phys.*, 1977, **23**, 327–341.
40 T. Darden, D. York and L. Pedersen, *J. Chem. Phys.*, 1993, **98**, 10089–10092.
41 R. A. Lippert, C. Predescu, D. J. Ierardi, K. M. Mackenzie, M. P. Eastwood, R. O. Dror and D. E. Shaw, *J. Chem. Phys.*, 2013, **139**, 164106.
42 S. Nosé, *J. Chem. Phys.*, 1984, **81**, 511–519.
43 G. J. Martyna, D. J. Tobias and M. L. Klein, *J. Chem. Phys.*, 1994, **101**, 4177–4189.
44 C. Predescu, A. K. Lerer, R. A. Lippert, B. Towles, J. P. Grossman, R. M. Dirks and D. E. Shaw, *J. Chem. Phys.*, 2020, **152**, 084113.
45 V. Kräutler, W. F. Van Gunsteren and P. H. Hünenberger, *J. Comput. Chem.*, 2001, **22**, 501–508.
46 W. Humphrey, A. Dalke and K. Schulten, *J. Mol. Graphics*, 1996, **14**, 33–38.
47 C. R. Harris, K. J. Millman, S. J. Van Der Walt, R. Gommers, P. Virtanen, D. Cournapeau, E. Wieser, J. Taylor, S. Berg, N. J. Smith, R. Kern, M. Picus, S. Hoyer, M. H. Van Kerkwijk, M. Brett, A. Haldane, J. F. Del Río, M. Wiebe, P. Peterson, P. Gérard-Marchant, K. Sheppard, T. Reddy, W. Weckesser, H. Abbasi, C. Gohlke and T. E. Oliphant, *Nature*, 2020, **585**, 357–362.
48 N. Michaud-Agrawal, E. J. Denning, T. B. Woolf and O. Beckstein, *J. Comput. Chem.*, 2011, **32**, 2319–2327.
49 J. D. Hunter, *Comput. Sci. Eng.*, 2007, **9**, 90–95.
50 P. Virtanen, R. Gommers, T. E. Oliphant, M. Haberland, T. Reddy, D. Cournapeau, E. Burovski, P. Peterson, W. Weckesser, J. Bright, S. J. Van Der Walt, M. Brett, J. Wilson, K. J. Millman, N. Mayorov, A. R. J. Nelson, E. Jones, R. Kern, E. Larson, C. J. Carey, İ. Polat, Y. Feng, E. W. Moore, J. VanderPlas, D. Laxalde, J. Perktold, R. Cimrman, I. Henriksen, E. A. Quintero, C. R. Harris, A. M. Archibald, A. H. Ribeiro, F. Pedregosa, P. Van Mulbregt, SciPy 1.0 Contributors, A. Vijaykumar, A. P. Bardelli, A. Rothberg, A. Hilboll, A. Kloeckner, A. Scopatz, A. Lee, A. Rokem, C. N. Woods, C. Fulton, C. Masson, C. Häggström, C. Fitzgerald, D. A. Nicholson, D. R. Hagen, D. V. Pasechnik, E. Olivetti, E. Martin, E. Wieser, F. Silva, F. Lenders, F. Wilhelm, G. Young, G. A. Price, G.-L. Ingold, G. E. Allen, G. R. Lee, H. Audren, I. Probst, J. P. Dietrich, J. Silterra, J. T. Webber, J. Slavič, J. Nothman, J. Buchner, J. Kulick, J. L. Schönberger, J. V. De Miranda Cardoso, J. Reimer, J. Harrington, J. L. C. Rodríguez, J. Nunez-Iglesias, J. Kuczynski, K. Tritz, M. Thoma, M. Newville, M. Kümmerer, M. Bolingbroke, M. Tartre, M. Pak, N. J. Smith, N. Nowaczyk, N. Shebanov, O. Pavlyk, P. A. Brodtkorb, P. Lee, R. T. McGibbon, R. Feldbauer, S. Lewis, S. Tygier, S. Sievert, S. Vigna,

S. Peterson, S. More, T. Pudlik, T. Oshima, T. J. Pingel, T. P. Robitaille, T. Spura, T. R. Jones, T. Cera, T. Leslie, T. Zito, T. Krauss, U. Upadhyay, Y. O. Halchenko and Y. Vázquez-Baeza, *Nat. Methods*, 2020, **17**, 261–272.
51 I. Levental and E. Lyman, *Nat. Rev. Mol. Cell Biol.*, 2023, **24**, 107–122.
52 A. Grossfield, S. E. Feller and M. C. Pitman, *Proc. Natl. Acad. Sci. U. S. A.*, 2006, **103**, 4888–4893.
53 R. A. Corey, P. J. Stansfeld and M. S. P. Sansom, *Biochem. Soc. Trans.*, 2020, **48**, 25–37.
54 B. I. Sejdiu and D. P. Tieleman, *Biophys. J.*, 2020, **118**, 1887–1900.
55 H.-Y. Yen, K. K. Hoi, I. Liko, G. Hedger, M. R. Horrell, W. Song, D. Wu, P. Heine, T. Warne, Y. Lee, B. Carpenter, A. Plückthun, C. G. Tate, M. S. P. Sansom and C. V. Robinson, *Nature*, 2018, **559**, 423–427.
56 M. Damian, M. Louet, A. A. S. Gomes, C. M'Kadmi, S. Denoyelle, S. Cantel, S. Mary, P. M. Bisch, J.-A. Fehrentz, L. J. Catoire, N. Floquet and J.-L. Banères, *Nat. Commun.*, 2021, **12**, 3938.
57 R. Guixà-González, J. L. Albasanz, I. Rodriguez-Espigares, M. Pastor, F. Sanz, M. Martí-Solano, M. Manna, H. Martinez-Seara, P. W. Hildebrand, M. Martín and J. Selent, *Nat. Commun.*, 2017, **8**, 14505.
58 S. Genheden, J. W. Essex and A. G. Lee, *Biochim. Biophys. Acta, Biomembr.*, 2017, **1859**, 268–281.
59 A. G. Lee, *Biophys. J.*, 2019, **116**, 1586–1597.
60 D. Sengupta and A. Chattopadhyay, *Biochim. Biophys. Acta, Biomembr.*, 2015, **1848**, 1775–1782.
61 H. I. Ingólfsson, M. N. Melo, F. J. van Eerden, C. Arnarez, C. A. Lopez, T. A. Wassenaar, X. Periole, A. H. de Vries, D. P. Tieleman and S. J. Marrink, *J. Am. Chem. Soc.*, 2014, **136**, 14554–14559.

Faraday Discussions

ROYAL SOCIETY OF CHEMISTRY

PAPER

Does plasma membrane transbilayer asymmetry coupled to lipid nanodomains drive fast kinetics of FGF2 membrane translocation into the extracellular space?

Fabio Lolicato, *[ab] Manpreet Kaur, [a] Ana Marija Knez, [a] Roberto Saleppico [a] and Walter Nickel *[a]

Received 22nd December 2024, Accepted 17th February 2025
DOI: 10.1039/d4fd00208c

Fibroblast Growth Factor 2 (FGF2) is a potent mitogen secreted from mammalian cells through an unconventional secretory pathway. This process is mediated by direct translocation of FGF2 across the plasma membrane into the extracellular space. It requires several components that are asymmetrically distributed between the two leaflets of the plasma membrane. At the inner plasma membrane leaflet, FGF2 undergoes sequential interactions with the Na,K-ATPase, Tec kinase, and the phosphoinositide PI(4,5)P_2. While the Na,K-ATPase, and Tec kinase are auxiliary factors, interactions of FGF2 with PI(4,5)P_2 trigger the core mechanism of FGF2 membrane translocation, inducing FGF2-oligomerization-dependent formation of lipidic membrane pores. At the outer plasma membrane leaflet, membrane-inserted FGF2 oligomers are captured and disassembled by Glypican-1 (GPC1), resulting in translocation of FGF2 to the cell surface. In a cellular context, a single FGF2 membrane translocation event occurs within 200 milliseconds. In contrast, in an *in vitro* system, which uses a fully reconstituted liposomal inside-out system with FGF2 added from the outside and luminal encapsulation of high-affinity heparin molecules, FGF2 membrane translocation takes several minutes. Here, we hypothesize that the observed difference is, at least in part, due to the asymmetrical membrane lipid distribution and the spatial organization of the FGF2 translocation machinery in native plasma membranes. We suggest that the molecular machinery mediating FGF2 membrane translocation assembles in ordered nanodomains, characterized by sphingomyelin (SM), cholesterol and phosphoinositide PI(4,5)P_2 coupled together. The transbilayer asymmetry of these lipids likely plays a crucial role in regulating the thermodynamics and kinetics of FGF2-induced membrane pore formation. Therefore, succeeding in reconstituting the FGF2 translocation machinery in artificial membranes with an asymmetric transbilayer distribution of SM, PI(4,5)P_2 and other membrane lipids may reveal a direct impact on

[a]*Heidelberg University Biochemistry Center, Heidelberg, Germany. E-mail: fabio.lolicato@bzh.uni-heidelberg.de; walter.nickel@bzh.uni-heidelberg.de*
[b]*Department of Physics, University of Helsinki, Helsinki, Finland*

pore-opening kinetics. Similarly, disrupting lipid asymmetry in cells may significantly impact FGF2 secretion rates, a finding that would underscore the importance of the spatial organization of lipids in membrane dynamics. Testing this hypothesis may advance our understanding of how membrane asymmetry and ordered lipid nanodomains regulate critical biological processes, such as the unconventional secretion of FGF2.

Introduction

The molecular machinery and mechanism driving unconventional secretion of FGF2

The first pieces of evidence suggesting the existence of ER/Golgi-independent pathways of protein secretion from mammalian cells date back to the late 1980s.[1] Along with, for example, Interleukin 1β and members of the galectin family, FGF2 was recognized as one of the first candidates of proteins transported into the extracellular space in a signal peptide-independent manner, suggesting alternative pathways of protein secretion that bypass the classical ER/Golgi route.[2] This was experimentally verified by transport of FGF2 into the extracellular space and shown to be insensitive to brefeldin A, an inhibitor of ER/Golgi-dependent protein secretion.[3–5] Another early observation was a possible involvement of the plasma membrane resident Na,K-ATPase. This finding was based on the inhibition of unconventional secretion of FGF2 by ouabain, a cardenolide that was known to inhibit the membrane-potential-generating function of this ATP-dependent ion pump.[6] However, all of these observations were solely based upon pharmacological evidence and, therefore, not fully conclusive. Due to the lack of advanced techniques at the time, such as genome-wide screening systems based upon RNA interference or CRISPR gene knock-outs, as well as the unavailability of sophisticated *in vitro* reconstitution systems, mechanistic insights into the molecular machinery mediating unconventional secretion of FGF2 became available only much later starting in the mid-2000s.

A key observation in elucidating the FGF2 secretion pathway was the ability of FGF2 to physically traverse the membranes of affinity-purified plasma membrane vesicles with an inside-out topology.[7] This suggested unconventional secretion of FGF2 from cells to be mediated by direct translocation across plasma membranes. This conclusion could be confirmed by TIRF imaging of individual FGF2 membrane translocation events at the plasma membrane in real time with single-molecule resolution.[8,9] Further key insights that led to the unraveling of this secretory mechanism were its dependence on the phosphoinositide $PI(4,5)P_2$ at the inner leaflet[10] and heparan sulfate proteoglycans at the outer leaflet of the plasma membrane.[11] The interaction of FGF2 with $PI(4,5)P_2$ was then shown to trigger FGF2 oligomerization, which led to the formation of lipidic membrane pores implicated in FGF2 membrane translocation.[12] Using giant unilamellar vesicles with an inside-out topology containing $PI(4,5)P_2$ on their surfaces and luminal encapsulation of heparin molecules with high affinity towards FGF2, the minimal machinery required for directional translocation of FGF2 across membranes could be defined.[13] These studies revealed $PI(4,5)P_2$ and heparan sulfates from proteoglycans on opposing sides of the plasma membrane along with the ability of FGF2 to oligomerize and to form lipidic membrane pores to be the minimal machinery driving FGF2 membrane translocation. NMR binding studies of FGF2 towards $PI(4,5)P_2$ *versus* heparin established these interactions to be mutually exclusive, providing a compelling data set that explains the directionality of FGF2 membrane translocation from the cytoplasm into the extracellular space.[13] The functional oligomeric state of FGF2 was found to be highly dynamic, with about 4 to 8 FGF2 subunits required to trigger the formation of a membrane pore.[13–15] The building blocks of pore-forming FGF2 oligomers were found to be disulfide-bridged dimers that form at the inner plasma membrane leaflet, followed by assembly into higher oligomers with a membrane pore-forming activity.[16] It has been hypothesized that FGF2 oligomers cause the local and asymmetric accumulation of the non-bilayer lipid $PI(4,5)P_2$ at the inner plasma membrane leaflet, which triggers the opening of a lipidic membrane pore with a toroidal architecture. The discovery of $PI(4,5)P_2$-induced oligomerization triggering the formation of a membrane pore through which FGF2 reaches the extracellular space also explained earlier findings demonstrating that FGF2 membrane translocation does not involve unfolded intermediates of FGF2.[17,18] With all of these aspects in mind, the unusual secretory pathway of FGF2, schematized in Fig. 1, can be considered a protein self-translocation mechanism that, depending on $PI(4,5)P_2$ and heparan sulfates on opposing sides of the plasma

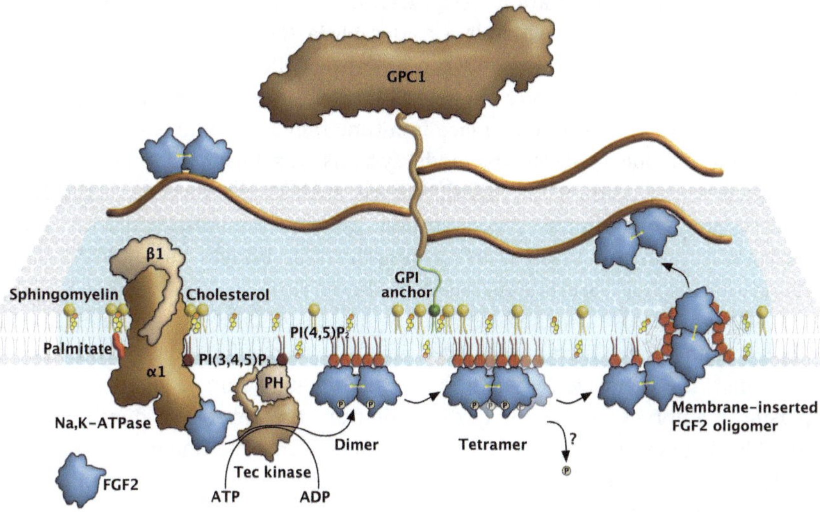

Fig. 1 FGF2 unconventional secretion mechanism. FGF2 secretion follows a unique mechanism involving direct translocation across the plasma membrane. Initially, soluble FGF2 monomers interact directly with the inner leaflet of the plasma membrane *via* binding to the Na,K-ATPase. Subsequently, FGF2 is transferred to $PI(4,5)P_2$ within cholesterol-enriched nanodomains, forming FGF2 oligomers. During this process, disulfide-linked FGF2 dimers, mediated by C95, act as intermediates and further self-assemble into tetramer ring structures. The $PI(4,5)P_2$-binding site in FGF2 also binds an additional four to five $PI(4,5)P_2$ molecules through weaker interactions, leading to localized clustering of $PI(4,5)P_2$ molecules under the FGF2 tetramer. Since $PI(4,5)P_2$ is a non-bilayer lipid, this clustering is thought to compromise the plasma membrane's integrity, facilitating the formation of a toroidal lipidic pore. Finally, the oligomers are captured and disassembled into dimers at the cell surface through interaction with heparan sulfate proteoglycan Glypican-1, resulting in the translocation of FGF2 to the cell exterior.

membrane, is based upon the ability of FGF2 to trigger a membrane pore through which it is capable of physically traversing the plasma membrane.[19]

While the early findings on a potential role of the Na,K-ATPase could not be further interpreted in terms of a mechanistic model for quite some time, it gained new attention when the Na,K-ATPase, and Tec kinase were identified as factors whose RNAi-mediated down-regulation limits the efficiency of FGF2 secretion.[20] In follow-up studies, a direct interaction of FGF2 with the cytoplasmic domain of the α1 subunit of the Na,K-ATPase could be demonstrated that was found to be sensitive to ouabain.[21] It was then shown that the Na,K-ATPase acts upstream of $PI(4,5)P_2$, accumulating FGF2 at the inner plasma membrane leaflet prior to $PI(4,5)P_2$-dependent FGF2 oligomerization and pore formation.[22] Based on these findings, it has been proposed that the Na,K-ATPase, like Tec kinase,[20,23] functions as an auxiliary factor of the molecular machinery mediating FGF2 membrane translocation into the extracellular space, representing a landing platform for FGF2 at the inner plasma membrane leaflet.[2,8,24]

The combination of experimental studies in cells and *in vitro* reconstitution experiments, along with multiscale physics-based computer simulations, proved decisive in elucidating the molecular mechanism by which FGF2 is secreted by direct translocation across the plasma membrane. Even certain discrepancies observed between cell-based data and results from *in vitro* experiments were informative, spurring new concepts about the spatio-temporal organization of the FGF2 membrane translocation machinery. An example is the kinetics of FGF2 membrane translocation, which is considerably faster in intact cells compared to those observed for artificial membrane systems. While individual events of FGF2 membrane translocation were shown to occur within time intervals of 200 ms,[9] this process was found to take place in a time range of minutes when analyzed with giant unilamellar vesicles as model systems,[13] as illustrated in Fig. 2. Several parameters may account for this difference, such as the absence of the Na,K-ATPase in the minimal reconstitution system. Similarly, putative redox enzymes, presumably involved in forming disulfide bridges during FGF2 oligomerization in cells, are not included in the *in vitro* models. Another striking example is the known transbilayer lipid asymmetry of native plasma membranes absent from model membranes used in FGF2 reconstitution experiments. In particular, the exclusive localization of $PI(4,5)P_2$ in the inner leaflet of cellular plasma membranes is strikingly different from the symmetric distribution of $PI(4,5)P_2$ in giant unilamellar vesicles used in *in vitro* experiments.

Here, we propose the transbilayer asymmetry of membrane lipids in general and $PI(4,5)P_2$, in particular, to play a critical role in lowering the free-energy costs of the conversion of a stable lipid bilayer into a lipidic membrane pore with a toroidal architecture. Furthermore, regarding lateral partitioning, $PI(4,5)P_2$ is known to be enriched in ordered nanodomains of plasma membranes.[25] Along with recent findings demonstrating that (i) cholesterol tunes FGF2 binding to $PI(4,5)P_2$,[26] (ii) the Na,K-ATPase has multiple contacts to cholesterol regulating its function[27] and (iii) the predominant heparan sulfate proteoglycan driving unconventional secretion of FGF2, Glypican-1 (GPC1), is localized in cholesterol-rich domains based upon a GPI anchor,[19] we propose ordered nanodomains with an asymmetric transbilayer distribution of $PI(4,5)P_2$ to be the organizational principle of the molecular machinery of FGF2 membrane translocation,

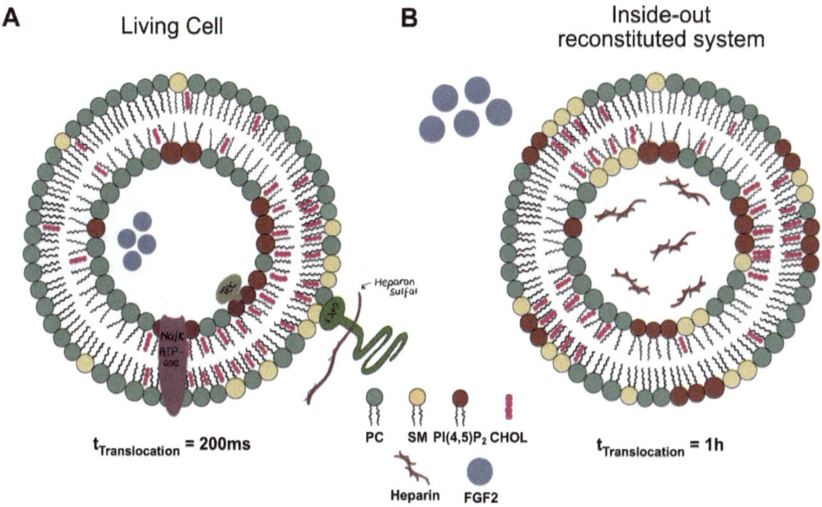

Fig. 2 Difference between living cells and *in vitro* reconstituted systems with purified components. (A) FGF2 translocation in a cellular system showcasing the nanodomain assembly of the translocation machinery within cholesterol and sphingomyelin-rich regions. A single translocation event is observed with rapid kinetics ≈200 ms. (B) FGF2 translocation in an *in vitro* biochemical reconstitution system utilizing inside-out vesicles highlights formation of multiple nanodomains. Heparin long chains are encapsulated inside the vesicles, while FGF2 is added externally. A singular translocation event occurs over ≈1 hour.

contributing to fast FGF2 membrane translocation kinetics of about 200 ms as observed in living cells.[9]

Plasma membrane lipid asymmetry and its relevance. The plasma membrane bilayer's lipid composition is well established as asymmetrically distributed.[28–30] This asymmetry is critical in maintaining cellular health, regulating membrane-dependent processes, and tuning integral membrane protein activity.[31] Specifically, lipids such as phosphatidylcholine (PC) and sphingomyelin (SM) are predominantly localized to the outer leaflet, while phosphatidylethanolamine (PE), phosphatidylinositol (PI), and phosphatidylserine (PS) are concentrated in the inner leaflet.[28,30] Disruption of this transbilayer asymmetry, particularly through the exposure of cytosolic lipids like PS to the exoplasmic leaflet, profoundly affects cell fate. For instance, the loss of PS asymmetry has been linked to various blood disorders, thrombosis, and metabolic syndromes.[32,33] To preserve this essential asymmetry, cells invest significant energy through ATP-dependent flippase and floppase enzymes that actively maintain the lipid distribution within the plasma membrane.

During apoptosis, the global loss of plasma membrane lipid asymmetry serves as a key mechanism for cell clearance, enabling recognition by phagocytes through "eat me" signals such as PS and phosphatidylinositol phosphates like $PI(3,4,5)P_3$.[33,34] This disruption of asymmetry is also observed in cancer cells, which presents both challenges and opportunities for therapeutic strategies.

The loss of transbilayer asymmetry in cancer cells can be problematic, as it has been linked to reduced permeability to chemotherapeutic agents like cisplatin.[35]

However, it also offers a therapeutic advantage, as the exposure of aminophospholipids such as PE and PS on the outer leaflet creates targets for specific interventions. For instance, small peptides and nanovesicles like OPA and SapC-DOPS exploit this exposure to selectively kill cancer cells, representing a promising strategy for targeted therapy.[36,37]

Emerging evidence highlights that the temporary loss of plasma membrane lipid asymmetry plays a critical role in maintaining normal physiological functions. For instance, recent studies reveal that plasma membrane SM asymmetry is disrupted during lysosomal damage.[38] Calcium release from damaged lysosomes activates SM scramblase, flipping SM to the cytosolic leaflet, which converts it into ceramide. This ceramide facilitates lysosomal repair through an Endosomal Sorting Complex Required for Transport (ESCRT)-independent pathway.

Similarly, the loss of PE asymmetry has been observed during cytokinesis.[39] Although the mechanism behind this loss is not yet fully understood, it is crucial for effective cell division.[40] During cytokinesis, the two membrane leaflets are coupled, with PE enriched in the outer leaflet of the cleavage furrow, alongside sphingomyelin and cholesterol. In contrast, phosphatidylinositol $PI(4,5)P_2$ is enriched in the inner leaflet. Disrupting the distribution of any of these lipids impairs cell division.[41] The strong coupling of lipids, such as SM, cholesterol, and phosphatidylinositol 4,5-bisphosphate ($PI(4,5)P_2$) between the two leaflets during cytokinesis shows intriguing similarities to the mechanism of FGF2 membrane translocation during unconventional secretion. These same lipids have been shown to play a critical role in FGF2 membrane binding and the formation of toroidal pores.[2,24]

Although it remains unclear why a polyunsaturated lipid-like $PI(4,5)P_2$ coexists in nanodomains with cholesterol and SM—lipids typically are associated with liquid-ordered membrane domains—this arrangement likely creates specialized nanodomains. These nanodomains may involve a complex interplay of membrane curvature and inter-leaflet lipid interactions, increasing membrane tension and thickness.[42,43] This unique lipid organization could be key to facilitating processes requiring precise mechanical and biochemical membrane properties.

Key players in ordered nanodomains

Cholesterol is a key component of the liquid-ordered (Lo) phase in lipid bilayers, influencing the membranes' physical and functional properties, and contributing to the formation and stability of nanodomains. The ordering effect arises from cholesterol's ability to interact with acyl chains while simultaneously disrupting their tight packing.[44] In multi-component lipid mixtures, there is a distinct coexistence between the gel and liquid-disordered (Ld) phases. The addition of cholesterol, which preferentially interacts with long, saturated phospholipids, shifts the gel phase towards a state resembling the Ld phase.[45,46] This is because cholesterol enhances membrane thickness by straightening the lipid tails in Lo-phase bilayers, making them thicker than Ld-phase bilayers.[47] However, the effect of cholesterol on membrane thickness can vary depending on the lipid composition. For instance, in bilayers with saturated 18-carbon chains, cholesterol's shorter structure induces deformation or kinking of the longer phospholipid tails, reducing membrane thickness below the phase-transition temperature.[48]

Within nanodomains, this difference in thickness generates line tension at domain boundaries, dynamically shaping the size and morphology of ordered regions. Because nanodomains exist in non-equilibrium conditions, mechanisms such as the dimpled morphology of lipids and their "pull-and-push" interactions at domain interfaces, along with the presence of amphipathic peptides and proteins at the interface, might play a key role in stabilizing nanodomain boundaries and temporarily preserve distinct domains.[49,50]

Furthermore, due to key structural and interactional differences, cholesterol exhibits a lower affinity for unsaturated lipid chains. The rigid, planar structure of cholesterol's steroid ring system is better suited to align with the straight, saturated acyl chains, creating a molecular shape mismatch with the kinked, unsaturated chains. However, the two methyl groups at the steroid also play an important role in cholesterol–lipid interactions by reducing sterol tilt in the bilayer, thereby allowing cholesterol to be oriented optimally.[51] Additionally, the kinks introduced by double bonds in unsaturated chains hinder tight packing with cholesterol, leading to weaker van der Waals interactions and further reducing affinity.[52] Recent studies indicate that cholesterol is predominantly present in the outer leaflet of the plasma membrane, supporting the hypothesis of an asymmetric distribution.[53] However, this remains a subject of debate, as experimental methods to directly and conclusively address this question are still lacking. In our previous work, we demonstrated that altering cholesterol levels in the plasma membrane directly influences the unconventional secretory machinery of FGF2 by enhancing FGF2 recruitment at the inner plasma membrane leaflet and facilitating its translocation into the extracellular space, primarily through the modulation of $PI(4,5)P_2$ lipid properties.[42]

Sphingomyelin is a lipid commonly associated with cholesterol-rich domains.[54] This association arises from the complementary structural and chemical properties of SM and cholesterol. SM has a straight, saturated sphingosine chain that aligns well with the planar geometry of cholesterol, promoting tight packing and enhancing the van der Waals interactions. Additionally, the amide group in the sphingosine chain forms hydrogen bonds with the hydroxyl group of cholesterol, further stabilizing cholesterol–SM complexes. This synergy between SM and cholesterol is functionally significant: cholesterol induces an ordering effect on membranes, while SM provides a scaffold that facilitates cholesterol's interaction with other lipids, creating a mutually stabilizing relationship. SM is asymmetrically distributed in the plasma membrane, with a higher concentration in the outer leaflet. This distribution suggests a coordinated role with cholesterol in membrane organization and function.

$PI(4,5)P_2$ plays an important role in endocytosis, exocytosis, cytokinesis, cytoskeleton activation, vesicle fusion, regulation of other phosphorylated PIs, and many signaling pathways.[55] Most $PI(4,5)P_2$-dependent processes happen at the cytosolic leaflet, so it is believed to be localized at the inner leaflet in most of the cells, with a few exceptions, such as HeLa and epithelial cells.[56] Few reports exist in the literature focusing on $PI(4,5)P_2$ transbilayer plasma membrane asymmetry. Only recently, a flippase enzyme was reported for $PI(4,5)P_2$, the same as PC flippase.[57]

The preferential localization of SM and cholesterol within Lo phases is well established.[54,58–60] In contrast, little is known about the involvement of phosphatidylinositol (PI), and its phosphorylated derivatives, in Lo and Ld phases. The most abundant $PI(4,5)P_2$ species in brain extracts feature 18 : 0 and 20 : 4 acyl chains, giving this lipid unique properties.[61,62] The straight, saturated 18 : 0 chain closely aligns with cholesterol's structure, promoting tight packing and order within membranes. This is further enhanced by forming hydrogen bonds with cholesterol's hydroxyl group.[25] Conversely, the highly unsaturated 20 : 4 chain introduces flexibility and disorder, favoring disordered regions, and they might promote interdigitation with the outer plasma membrane leaflet. Detergent-resistant membrane (DRM) experiments revealed that $PI(4,5)P_2$ is enriched in ordered regions. Notably, cholesterol depletion using methyl-beta-cyclodextrin causes $PI(4,5)P_2$ to redistribute into disordered regions.[63] However, DRM assays are often criticized for potential artifact detergents introduced during the procedure. An alternative study using Giant Plasma Membrane Vesicles (GPMVs) and Lo/Ld-directed peptides has shown faster $PI(4,5)P_2$ metabolism kinetics in Lo regions.[64] In this study, they noted that $PI(4,5)P_2$ in Lo regions undergoes rapid turnover, being converted to diacylglycerol (DAG) and inositol trisphosphate (IP3) and replenished at a faster rate compared to $PI(4,5)P_2$ in Ld regions. The reported data suggest that $PI(4,5)P_2$ in Lo regions functions as a dynamic pool, regulating the concentrations of other PI derivatives through the activity of PI kinases and phosphatases. These enzymes are also reported to localize within Lo regions,[65,66] further supporting the functional interplay between $PI(4,5)P_2$ and the Lo phase.

The asymmetric distribution of these lipid types in the plasma membrane likely plays a role in mediating FGF2 membrane translocation. Cholesterol and SM enrichment in the outer leaflet and $PI(4,5)P_2$ primarily located in the inner leaflet highlights a spatial organization that supports membrane-associated processes. This asymmetry may create functional lipid environments, such as ordered nanodomains, that facilitate the recruitment and translocation of FGF2 during unconventional secretion. Thus, developing methods to reproduce this lipid asymmetry *in vitro* reconstitution experiments or to disrupt it in living cell systems selectively could provide insights into how these lipid distributions regulate FGF2 secretion. Such studies will uncover mechanistic details about the secretory mechanism of FGF2 and reveal broader principles about the interplay between lipid organization and protein function in cellular membranes.

Targeting lipid asymmetry to study its role in the unconventional secretion mechanism of FGF2

The asymmetric distribution of lipids in living cell systems is a significant distinction compared to inside-out reconstituted vesicles, which are challenging to produce. While several reports in the literature on generating asymmetric vesicles exist, only a few studies have yet used them to make physiologically relevant plasma membrane-like asymmetric lipid compositions for functional assays.

In our studies on the unconventional secretion mechanism of FGF2, we identified three key lipid species—cholesterol, sphingomyelin (SM), and $PI(4,5)P_2$—that play crucial roles. Thus, developing methods for generating asymmetric vesicles containing these specific lipids in a PC background and investigating their impact on pore formation kinetics could reveal mechanistic insights into this unusual process of protein translocation across a membrane.

Strategies for cellular single-lipid disturbance

One of the most widely used approaches for manipulating lipid composition at the plasma membrane level is the use of β-cyclodextrin-based methods,[67,68] as illustrated in Fig. 3A. β-Cyclodextrins are cyclic oligosaccharides with a hydrophobic cavity that can encapsulate lipid molecules, making them highly effective for lipid delivery or cholesterol removal at the plasma membrane. For example, β-cyclodextrin can be pre-loaded with a specific lipid of interest, such as cholesterol or sphingomyelin, and subsequently incubated with cells. This process facilitates the targeted delivery of these lipids to the outer leaflet of the plasma membrane, potentially disrupting the asymmetry of lipid species such as $PI(4,5)P_2$, which is exclusively located in the inner leaflet. In addition to lipid delivery, (methyl-)β-cyclodextrin can also be employed to extract cholesterol from the plasma membrane outer leaflet. However, while effective, this lipid extraction approach has limitations that can lead to unintended alterations in membrane composition and induce cytotoxic effects. This method has been employed to manipulate cholesterol content in the plasma membrane. It is well established that altering cholesterol levels positively influences the recruitment and translocation of FGF2.[26,69] In contrast, little is known about manipulating SM and $PI(4,5)P_2$ levels in cells influencing the mechanisms underlying FGF2 secretion. Sphingomyelin is predominantly located in the outer leaflet of the plasma membrane, where it plays essential roles in maintaining membrane stability, organizing lipid domains, and mediating signaling processes. SM scramblases can translocate SM across the

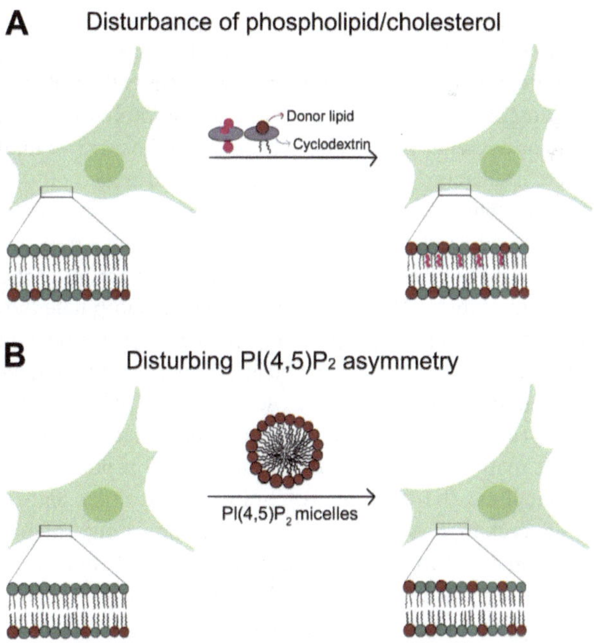

Fig. 3 Disturbing transbilayer asymmetric distribution. (A) Destroying $PI(4,5)P_2$ transbilayer asymmetry by addition of $PI(4,5)P_2$ micelles which selectively incorporate $PI(4,5)P_2$ on the outer leaflet. (B)Disrupting phospholipid and cholesterol levels by cyclodextrin, which can selectively substitute lipids from outer leaflet with lipids of choice.

bilayer. Still, their activity is calcium-dependent, which may expose cytosolic leaflet lipids to the outer leaflet and it is coupled with the enzymatic action of sphingomyelinase in the inner leaflet. This enzyme converts SM into ceramide, introducing metabolic changes that complicate the study of SM dynamics. The single-cell injection method is a precise and promising alternative,[70] which could deliver SM directly into the inner leaflet. This approach would bypass scramblase activity, offering a controlled environment for investigating lipid redistribution. However, the physical penetration of the cell membrane can cause stress or damage to the cell, potentially affecting its viability or normal function. Furthermore, this approach enables the enrichment of SM in various membrane organelles, which could potentially alter normal cellular functions.

On the other hand, phosphatidylinositol 4,5-bisphosphate is predominantly located in the inner leaflet of the plasma membrane, where it functions as a critical signaling lipid and regulator of membrane dynamics.[30,71] Its intracellular localization presents opportunities for simple external intervention to disrupt lipid asymmetry. One effective strategy involves the targeted delivery of $PI(4,5)P_2$ micelles directly within the cell medium,[70] as illustrated in Fig. 3B. $PI(4,5)P_2$ has a high bilayer partition coefficient and will efficiently incorporate into the outer leaflet of the plasma membrane.

Strategies for asymmetric vesicle generation

There are several methods for creating asymmetric liposomes. Large unilamellar vesicles (LUVs) are usually preferred over giant unilamellar vesicles (GUVs) due to their well-defined size and low susceptibility to high curvature or substrate interactions. In contrast, GUVs, though more fragile, have the advantage of being compatible with confocal microscopy imaging and possess a size range of 5 to 100 μm, closely matching the dimensions of eukaryotic cells.[72] This cell-like size makes GUVs valuable as model systems for studying protein–lipid interactions.

Asymmetric unilamellar vesicles have been created using five main strategies: (i) The inverted emulsion droplet method uses a lipid-coated oil–water interface to generate GUVs. This approach produces GUVs efficiently; however, the size distribution is broad, and residual organic solvents are often present in the hydrophobic core of the bilayer, which may alter the properties of the bilayer and compromise its integrity.[67,73,74] (ii) Changes in pH can manipulate the protonation state of limited anionic lipids, such as phosphatidylglycerol (PG) and phosphatidic acid (PA), to flip between bilayer leaflets in their uncharged state. This creates an uneven lipid distribution, with accumulation on the more alkaline side of the bilayer.[75] (iii) Like cellular systems, carrier molecules with hydrophobic cavities, such as methyl-β-cyclodextrin or lipid transfer proteins,[76] are commonly employed to facilitate the exchange of lipids in the outer leaflet of vesicles. Catalyzed lipid exchange is compatible with a wide range of phospholipids; however, separating the two vesicle populations is needed afterward to isolate the desired asymmetric liposomes.[67] (iv) The hemifusion method is used to mix the lipid composition of GUVs with a supported lipid bilayer (SLB) by inducing membrane fusion using calcium ions. In the hemifusion state, the two leaflets are connected through a hemifusion diaphragm, promoting lipid exchange between the outer leaflets *via* lateral diffusion. Further fusion events are blocked by removing ions with calcium chelation, and a mechanical shear is applied to

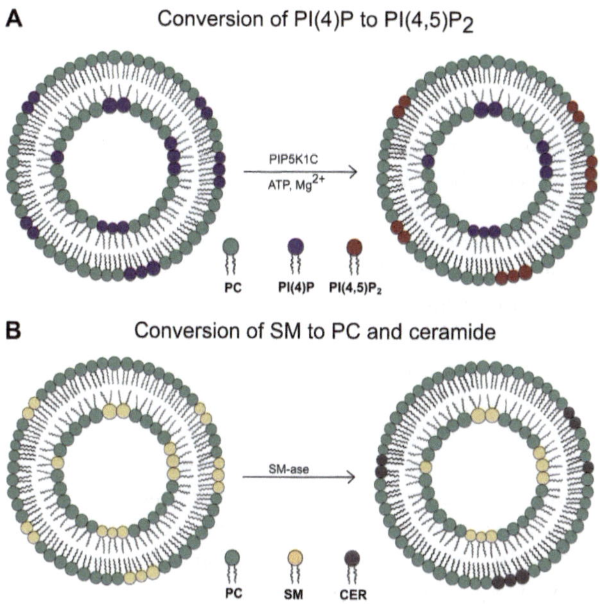

Fig. 4 Enzymatic approach for generating asymmetric vesicles using symmetric SM or $PI(4,5)P_2$ vesicles. (A) PIP5K1 selectively converts PI(4)P to $PI(4,5)P_2$ on the outer leaflet. (B) Sphingomyelinase catalyzes sphingomyelin hydrolysis, breaking it down into ceramide and phosphocholine on the outer leaflet.

detach the GUVs from the SLB.[77] (v) The enzymatic method can effectively modify lipid composition in specific bilayer leaflets. For instance, sphingomyelin can be hydrolyzed into ceramide and phosphocholine, while PIP5K1C catalyzes the conversion of PI(4)P to $PI(4,5)P_2$, as illustrated in Fig. 4. This approach enables selective modification of the outer leaflet of synthetic liposomes without affecting the inner leaflet, as the enzymes used are typically not membrane-permeable. However, its application is limited by enzyme availability and cost, substrate specificity, and the need for precise reaction conditions.

Discussion

Several parameters are likely to play key roles in governing the thermodynamics and kinetics of the FGF2 membrane translocation process. The faster kinetics observed in living cells,[9] compared to *in vitro* reconstitution experiments using purified components,[13,78] can be attributed to fundamental differences between the two experimental systems. Notably, the absence of auxiliary components in synthetic vesicles such as the Na,K-ATPase and Tec kinase. For example, a variant of FGF2 lacking the two key residues responsible for its interaction with the Na,K-ATPase exhibits a 30–40% reduction in plasma membrane recruitment and translocation efficiency.[22] However, these components are not essential for FGF2 membrane translocation in reconstituted systems. In addition, as a surrogate for the authentic GPI-anchored heparan sulfate proteoglycan GPC1, soluble heparin molecules containing high-affinity FGF2 binding sites have been used in *in vitro* experiments.[13,16] Furthermore, we hypothesize that the fast kinetics of this process observed in intact cells is driven by cellular redox systems facilitating the efficient oxidative dimerization of FGF2 disulfide bridges in combination with the asymmetric nature of the plasma membrane. Finally, unlike in living cells where lipid distribution is asymmetric, artificial vesicles exhibit a symmetric lipid arrangement, introducing another layer of difference between the two experimental systems. All of these and maybe additional factors could potentially slow down the kinetics and efficiency of FGF2 membrane translocation observed *in vitro* compared to living cells.

Current views describe the plasma membrane as a composite material, where the exoplasmic leaflet primarily ensures functional barrier integrity, while the cytoplasmic leaflet contributes to membrane fluidity.[79] Due to technical limitations, this duality is often lost in *in vitro* systems, where synthetic liposomes are typically produced with symmetric membranes. For example, a symmetric distribution of SM—which in living cells is predominantly located in the outer leaflet—can lead to excessive membrane rigidity when interacting with cholesterol. This rigidity disrupts the dual nature of the plasma membrane, and it might create an additional barrier that hinders the FGF2 translocation process. Focusing on PI(4,5)P_2, its asymmetric distribution within the plasma membrane can play a fundamental role in the membrane translocation mechanism. Our current model proposes that FGF2 undergoes oligomerization at the inner leaflet of the plasma membrane. Notably, a single FGF2 molecule can recruit up to 5 PI(4,5)P_2 molecules through high- and low-affinity binding sites.[13] Given that FGF2 is known to oligomerize in a membrane-dependent manner into at least tetramers,[13–15] this would result in the accumulation of approximately 20 PI(4,5)P_2 molecules in a spatially confined area. The localized accumulation of PI(4,5)P_2 molecules near the protein is expected to result in a local concentration of $\sim$20 mol%. PI(4,5)P_2 is generally considered as having a positive intrinsic curvature due to its large, charged hydrophilic head group. However, its contribution to local membrane curvature can vary depending on the cellular context and its molecular interactions. Notably, the most abundant PI(4,5)P_2 species in brain extracts contain a polyunsaturated acyl chain (arachidonic acid), which is highly disordered and occupies more space, altering the overall shape of the lipid, as suggested by Lin *et al.*[80,81] This characteristic likely shifts PI(4,5)P_2's preference towards negative curvature, which we hypothesize is a critical factor in driving FGF2 pore formation activity. Furthermore, the inner leaflet is already enriched with $\sim$30 mol% of PE lipids.[30] PE are also conical, non-bilayer lipids with intrinsic negative curvature properties. We propose that the enrichment of PI(4,5)P_2, driven by FGF2 oligomerization, increases the local concentration of non-bilayer lipids beyond a critical threshold. This accumulation induces negative curvature (Fig. 5B, left panel), destabilizing the membrane and facilitating the pore formation process (Fig. 5C, left panel). Additionally, PI(4,5)P_2 carries a highly negative charge under physiological conditions, which would create a charge gradient in the spatially confined region across the membrane, resulting in an electric field that could potentially contribute to the pore formation process.[82,83] Contrarily, in a symmetric membrane, the long acyl chains of PI(4,5)P_2 (18 : 0 and 20 : 4) could interdigitate from both sides of the membrane, leading to a symmetrical accumulation in both leaflets.[84,85] This would cancel out the charge gradient and curvature-inducing effects, as hypothesized in Fig. 5B, right panel. Furthermore, heparan sulfate chains, which are also negatively charged, are

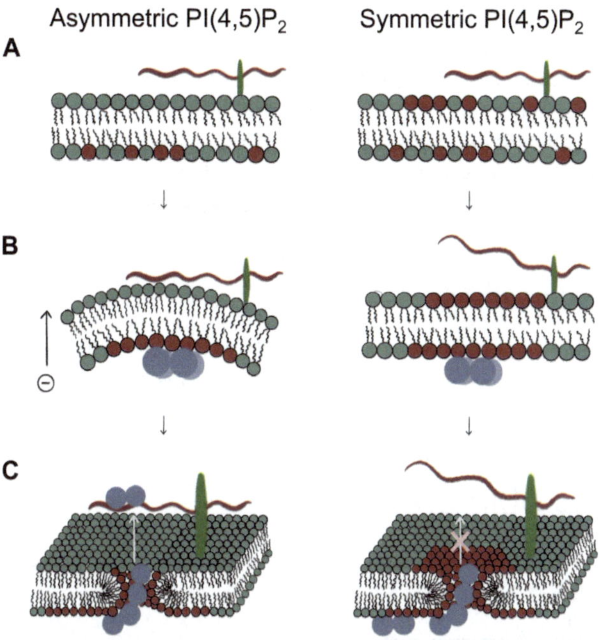

Fig. 5 Role of $PI(4,5)P_2$ asymmetry in governing FGF2 translocation. (A) Membrane with asymmetric and symmetric transbilayer $PI(4,5)P_2$ distribution. (B) Upon FGF2 oligomerization, accumulation of $PI(4,5)P_2$ in the inner leaflet causes an extreme charge gradient and curvature-dependent stress in the asymmetric membrane. In the symmetric membrane, $PI(4,5)P_2$ accumulates in both the inner and outer leaflets due to interdigitation, due to which the charge gradient and stress are canceled out in the symmetric $PI(4,5)P_2$ membrane. (C) The presence of symmetric $PI(4,5)P_2$ in the membrane will cause the repulsion between $PI(4,5)P_2$ on the outer leaflet and heparan sulfate chains, resulting in the orientation of heparan sulfate chains farther from the membrane, translocating FGF2 presumably slower in comparison to the asymmetric system.

positioned on the outer leaflet of the plasma membrane. A highly negative lipid such as $PI(4,5)P_2$ in the outer leaflet would reduce the proximity of the heparan sulfate chains to the membrane surface due to electrostatic repulsion, as schematized in the right panels of Fig. 5B and C. Notably, FGF2 translocation depends on membrane-proximal heparan sulfate proteoglycans (HSPGs) on the cell surface, which capture and disassemble FGF2 oligomers. Interestingly, among the diverse sub-classes of HSPGs, including syndecans, perlecans, and glypicans, we unexpectedly discovered that Glypican-1 (GPC1) represents the principal and rate-limiting factor driving unconventional secretion of FGF2.[19] This is due to its stronger binding affinity for FGF2 and closer proximity to the membrane surface. Therefore, exploring the role of $PI(4,5)P_2$ asymmetry in synthetic and cellular systems is essential to validate the proposed mechanism. Specifically, developing a method to generate $PI(4,5)P_2$-asymmetric vesicles would allow for investigations of FGF2 pore formation and translocation kinetics.

In conclusion, by mimicking the natural asymmetry of $PI(4,5)P_2$ in the inner plasma membrane leaflet, we hypothesize that this approach will accelerate the kinetics of FGF2 oligomerization and translocation, as the spatially constrained accumulation of $PI(4,5)P_2$ and the resulting membrane stress and charge gradient would facilitate membrane curvature and pore opening. Conversely, disrupting $PI(4,5)P_2$ asymmetry in living cell systems could offer deeper insights into its critical role in unconventional FGF2 secretion. We predict that the loss of $PI(4,5)P_2$ asymmetry would neutralize the charge gradient, reduce membrane curvature, and increase the distance of heparan sulfate chains from the membrane surface. This, in turn, would slow down the kinetics of FGF2 secretion.

Data availability

All data supporting the findings of this study are included within the manuscript. No additional data or materials were generated or used beyond what is provided in the text and figures.

Conflicts of interest

There are no conflicts to declare.

Acknowledgements

This work was supported by the Deutsche Forschungsgemeinschaft (SFB-1638/1 – 511488495 – Z01; FL, SFB/TRR 186, project A1; WN and FL, DFG LO 2821/1-1; FL, DFG Ni 423/10-1, DFG Ni 423/12-1 and DFG Ni 423/13-1; WN). FL gratefully acknowledges the data storage service SDS@hd supported by the Ministry of Science, Research, and the Arts Baden-Württemberg (MWK), the German Research Foundation (DFG) through grants INST 35/1314-1 FUGG and INST 35/1503-1 FUGG. FL acknowledges the computing resources provided by the CSC – IT Center for Science Ltd (Espoo, Finland) and supported by the state of Baden-Württemberg through bwHPC and the German Research Foundation (DFG) through grant INST 35/1597-1 FUGG.

References

1 A. Muesch, E. Hartmann, K. Rohde, A. Rubartelli, R. Sitia and T. A. Rapoport, A novel pathway for secretory proteins?, *Trends Biochem. Sci.*, 1990, **15**(3), 86–88.

2 M. T. Pallotta and W. Nickel, FGF2 and IL-1β – explorers of unconventional secretory pathways at a glance, *J. Cell Sci.*, 2020, **133**(21), jcs250449.

3 R. Z. Florkiewicz, R. A. Majack, R. D. Buechler and E. Florkiewicz, Quantitative export of FGF-2 occurs through an alternative, energy-dependent, non-ER/Golgi pathway, *J. Cell. Physiol.*, 1995, **162**(3), 388–399.

4 C. Trudel, V. Faure-Desire, R. Z. Florkiewicz and A. Baird, Translocation of FGF2 to the cell surface without release into conditioned media, *J. Cell. Physiol.*, 2000, **185**(2), 260–268.

5 A. Engling, R. Backhaus, C. Stegmayer, C. Zehe, C. Seelenmeyer, A. Kehlenbach, *et al.*, Biosynthetic FGF-2 is targeted to non-lipid raft microdomains following translocation to the extracellular surface of CHO cells, *J. Cell Sci.*, 2002, **115**(18), 3619–3631.

6 R. Z. Florkiewicz, J. Anchin and A. Baird, The Inhibition of Fibroblast Growth Factor-2 Export by Cardenolides Implies a Novel Function for the Catalytic Subunit of Na+,K+-ATPase, *J. Biol. Chem.*, 1998, **273**(1), 544–551.

7 T. Schäfer, H. Zentgraf, C. Zehe, B. Brügger, J. Bernhagen and W. Nickel, Unconventional Secretion of Fibroblast Growth Factor 2 Is Mediated by Direct Translocation across the Plasma Membrane of Mammalian Cells, *J. Biol. Chem.*, 2004, **279**(8), 6244–6251.

8 E. Dimou and W. Nickel, Unconventional mechanisms of eukaryotic protein secretion, *Curr. Biol.*, 2018, **28**(8), R406–R410.

9 E. Dimou, K. Cosentino, E. Platonova, U. Ros, M. Sadeghi, P. Kashyap, *et al.*, Single event visualization of unconventional secretion of FGF2, *J. Cell Biol.*, 2019, **218**(2), 683–699.

10 K. Temmerman, A. D. Ebert, H. M. Müller, I. Sinning, I. Tews and W. Nickel, A Direct Role for Phosphatidylinositol-4,5-bisphosphate in Unconventional Secretion of Fibroblast Growth Factor 2, *Traffic*, 2008, **9**(7), 1204–1217.

11 C. Zehe, A. Engling, S. Wegehingel, T. Schäfer and W. Nickel, Cell-surface heparan sulfate proteoglycans are essential components of the unconventional export machinery of FGF-2, *Proc. Natl. Acad. Sci. U.S.A.*, 2006, **103**(42), 15479–15484.

12 J. P. Steringer, S. Bleicken, H. Andreas, S. Zacherl, M. Laussmann, K. Temmerman, *et al.*, Phosphatidylinositol 4,5-Bisphosphate (PI(4,5)P2)-dependent Oligomerization of Fibroblast Growth Factor 2 (FGF2) Triggers the Formation of a Lipidic Membrane Pore Implicated in Unconventional Secretion, *J. Biol. Chem.*, 2012, **287**(33), 27659–27669.

13 J. P. Steringer, S. Lange, S. Čujová, R. Šachl, C. Poojari, F. Lolicato, *et al.*, Key steps in unconventional secretion of fibroblast growth factor 2 reconstituted with purified components, *eLife*, 2017, **6**, e28985.

14 R. Šachl, S. Čujová, V. Singh, P. Riegerová, P. Kapusta, H. M. Müller, *et al.*, Functional Assay to Correlate Protein Oligomerization States with Membrane Pore Formation, *Anal. Chem.*, 2020, **92**(22), 14861–14866.

15 V. Singh, S. Macharová, P. Riegerová, J. P. Steringer, H. M. Müller, F. Lolicato, *et al.*, Determining the Functional Oligomeric State of Membrane-Associated Protein Oligomers Forming Membrane Pores on Giant Lipid Vesicles, *Anal. Chem.*, 2023, **95**(23), 8807–8815.

16 F. Lolicato, J. P. Steringer, R. Saleppico, D. Beyer, J. Fernandez-Sobaberas, S. Unger, *et al.*, Disulfide bridge-dependent dimerization triggers FGF2 membrane translocation into the extracellular space, *eLife*, 2024, **12**, RP88579.

17 R. Backhaus, C. Zehe, S. Wegehingel, A. Kehlenbach, B. Schwappach and W. Nickel, Unconventional protein secretion: membrane translocation of FGF-2 does not require protein unfolding, *J. Cell Sci.*, 2004, **117**(9), 1727–1736.

18 L. C. Torrado, K. Temmerman, H. M. Müller, M. P. Mayer, C. Seelenmeyer, R. Backhaus, *et al.*, An intrinsic quality-control mechanism ensures unconventional secretion of fibroblast growth factor 2 in a folded conformation, *J. Cell Sci.*, 2009, **122**(18), 3322–3329.

19 C. Sparn, E. Dimou, A. Meyer, R. Saleppico, S. Wegehingel, M. Gerstner, *et al.*, Glypican-1 drives unconventional secretion of fibroblast growth factor 2, *eLife*, 2022, **11**, e75545.

20 A. D. Ebert, M. Laußmann, S. Wegehingel, L. Kaderali, H. Erfle, J. Reichert, *et al.*, Tec-Kinase-Mediated Phosphorylation of Fibroblast Growth Factor 2 is Essential for Unconventional Secretion, *Traffic*, 2010, **11**(6), 813–826.

21 S. Zacherl, V. G. La, H. M. Müller, S. Wegehingel, E. Dimou, P. Sehr, *et al.*, A Direct Role for ATP1A1 in Unconventional Secretion of Fibroblast Growth Factor 2, *J. Biol. Chem.*, 2015, **290**(6), 3654–3665.

22 C. Legrand, R. Saleppico, J. Sticht, F. Lolicato, H. M. Müller, S. Wegehingel, *et al.*, The Na,K-ATPase acts upstream of phosphoinositide PI(4,5)P2 facilitating unconventional secretion of Fibroblast Growth Factor 2, *Commun. Biol.*, 2020, **3**(1), 141.

23 G. La Venuta, S. Wegehingel, P. Sehr, H. M. Müller, E. Dimou, J. P. Steringer, *et al.*, Small Molecule Inhibitors Targeting Tec Kinase Block Unconventional Secretion of Fibroblast Growth Factor 2, *J. Biol. Chem.*, 2016, **291**(34), 17787–17803.

24 C. Sparn, A. Meyer, R. Saleppico and W. Nickel, Unconventional secretion mediated by direct protein self-translocation across the plasma membranes of mammalian cells, *Trends Biochem. Sci.*, 2022, **47**(8), 699–709.

25 Y. Wen, V. M. Vogt and G. W. Feigenson, PI(4,5)P $_2$ Clustering and Its Impact on Biological Functions, *Annu. Rev. Biochem.*, 2021, **90**(1), 681–707.

26 F. Lolicato, R. Saleppico, A. Griffo, A. Meyer, F. Scollo, B. Pokrandt, *et al.*, Cholesterol promotes clustering of PI(4,5)P2 driving unconventional secretion of FGF2, *J. Cell Biol.*, 2022, **221**(11), e202106123.

27 R. Kanai, F. Cornelius, B. Vilsen and C. Toyoshima, Cryoelectron microscopy of Na^+,K^+-ATPase in the two E2P states with and without cardiotonic steroids, *Proc. Natl. Acad. Sci. U.S.A.*, 2022, **119**(15), e2123226119.

28 M. S. Bretscher, Asymmetrical Lipid Bilayer Structure for Biological Membranes, *Nature New Biol.*, 1972, **236**(61), 11–12.

29 J. Lombard, Once upon a time the cell membranes: 175 years of cell boundary research, *Biol. Direct*, 2014, **9**(1), 32.

30 J. H. Lorent, K. R. Levental, L. Ganesan, G. Rivera-Longsworth, E. Sezgin, M. Doktorova, *et al.*, Plasma membranes are asymmetric in lipid unsaturation, packing and protein shape, *Nat. Chem. Biol.*, 2020, **16**(6), 644–652.

31 G. Pabst and S. Keller, Exploring membrane asymmetry and its effects on membrane proteins, *Trends Biochem. Sci.*, 2024, **49**(4), 333–345.

32 M. Doktorova, J. L. Symons and I. Levental, Structural and functional consequences of reversible lipid asymmetry in living membranes, *Nat. Chem. Biol.*, 2020, **16**(12), 1321–1330.

33 B. Fadeel and D. Xue, The ins and outs of phospholipid asymmetry in the plasma membrane: roles in health and disease, *Crit. Rev. Biochem. Mol. Biol.*, 2009, **44**(5), 264–277.

34 O. H. Kim, G. H. Kang, J. Hur, J. Lee, Y. Jung, I. S. Hong, *et al.*, Externalized phosphatidylinositides on apoptotic cells are eat-me signals recognized by CD14, *Cell Death Differ.*, 2022, **29**(7), 1423–1432.

35 T. Rivel, C. Ramseyer and S. Yesylevskyy, The asymmetry of plasma membranes and their cholesterol content influence the uptake of cisplatin, *Sci. Rep.*, 2019, **9**(1), 5627.

36 D. Patel and S. N. Witt, Ethanolamine and Phosphatidylethanolamine: Partners in Health and Disease, *Oxid. Med. Cell. Longevity*, 2017, **2017**(1), 4829180.
37 K. F. N'Guessan, P. H. Patel and X. Qi, SapC-DOPS – a Phosphatidylserine-targeted Nanovesicle for selective Cancer therapy, *Cell Commun. Signaling*, 2020, **18**(1), 6.
38 P. Niekamp, F. Scharte, T. Sokoya, L. Vittadello, Y. Kim, Y. Deng, *et al.*, Ca2+-activated sphingomyelin scrambling and turnover mediate ESCRT-independent lysosomal repair, *Nat. Commun.*, 2022, **13**(1), 1875.
39 K. Emoto, T. Kobayashi, A. Yamaji, H. Aizawa, I. Yahara, K. Inoue, *et al.*, Redistribution of phosphatidylethanolamine at the cleavage furrow of dividing cells during cytokinesis, *Proc. Natl. Acad. Sci. U.S.A.*, 1996, **93**(23), 12867–12872.
40 K. Emoto and M. Umeda, An Essential Role for a Membrane Lipid in Cytokinesis, *J. Cell Biol.*, 2000, **149**(6), 1215–1224.
41 G. Kunduri, U. Acharya and J. K. Acharya, Lipid Polarization during Cytokinesis, *Cells*, 2022, **11**(24), 3977.
42 F. Lolicato, R. Saleppico, A. Griffo, A. Meyer, F. Scollo, B. Pokrandt, *et al.*, Cholesterol promotes clustering of PI(4,5)P2 driving unconventional secretion of FGF2, *J. Cell Biol.*, 2022, **221**(11), e202106123.
43 A. Griffo, C. Sparn, F. Lolicato, F. Nolle, N. Khangholi, R. Seemann, *et al.*, Mechanics of biomimetic free-standing lipid membranes: insights into the elasticity of complex lipid compositions, *RSC Adv.*, 2024, **14**(19), 13044–13052.
44 T. Róg, M. Pasenkiewicz-Gierula, I. Vattulainen and M. Karttunen, Ordering effects of cholesterol and its analogues, *Biochim. Biophys. Acta, Biomembr.*, 2009, **1788**(1), 97–121.
45 H. K. Ijäs, M. Lönnfors and T. K. M. Nyholm, Sterol affinity for phospholipid bilayers is influenced by hydrophobic matching between lipids and transmembrane peptides, *Biochim. Biophys. Acta, Biomembr.*, 2013, **1828**(3), 932–937.
46 H. Martinez-Seara, T. Róg, M. Pasenkiewicz-Gierula, I. Vattulainen, M. Karttunen and R. Reigada, Interplay of Unsaturated Phospholipids and Cholesterol in Membranes: Effect of the Double-Bond Position, *Biophys. J.*, 2008, **95**(7), 3295–3305.
47 J. V. Bleecker, P. A. Cox, R. N. Foster, J. P. Litz, M. C. Blosser, D. G. Castner, *et al.*, Thickness Mismatch of Coexisting Liquid Phases in Noncanonical Lipid Bilayers, *J. Phys. Chem. B*, 2016, **120**(10), 2761–2770.
48 T. J. McIntosh, The effect of cholesterol on the structure of phosphatidylcholine bilayers, *Biochim. Biophys. Acta, Biomembr.*, 1978, **513**(1), 43–58.
49 T. S. Ursell, W. S. Klug and R. Phillips, Morphology and interaction between lipid domains, *Proc. Natl. Acad. Sci. U.S.A.*, 2009, **106**(32), 13301–13306.
50 C. Wang, M. R. Krause and S. L. Regen, Push and Pull Forces in Lipid Raft Formation: The Push Can Be as Important as the Pull, *J. Am. Chem. Soc.*, 2015, **137**(2), 664–666.
51 T. Róg, M. Pasenkiewicz-Gierula, I. Vattulainen and M. Karttunen, What Happens if Cholesterol Is Made Smoother, *Biophys. J.*, 2007, **92**(10), 3346–3357.
52 F. Keller and A. Heuer, Chain ordering of phospholipids in membranes containing cholesterol: what matters?, *Soft Matter*, 2021, **17**(25), 6098–6108.

53 M. Doktorova, J. L. Symons, X. Zhang, H. Y. Wang, J. Schlegel and J. H. Lorent, Cell membranes sustain phospholipid imbalance via cholesterol asymmetry, *bioRxiv*, 2023, preprint DOI: 10.1016/j.cell.2025.02.034
54 M. B. Sankaram and T. E. Thompson, Interaction of cholesterol with various glycerophospholipids and sphingomyelin, *Biochemistry*, 1990, **29**(47), 10670–10675.
55 F. Lolicato, W. Nickel, V. Haucke and M. Ebner, Phosphoinositide switches in cell physiology - From molecular mechanisms to disease, *J. Biol. Chem.*, 2024, **300**(3), 105757.
56 A. Yoneda, K. Kanemaru, A. Matsubara, E. Takai, M. Shimozawa, R. Satow, *et al.*, Phosphatidylinositol 4,5-bisphosphate is localized in the plasma membrane outer leaflet and regulates cell adhesion and motility, *Biochem. Biophys. Res. Commun.*, 2020, **527**(4), 1050–1056.
57 Y. Muranaka, R. Shigetomi, Y. Iwasaki, A. Hamamoto, K. Nakayama, H. Takatsu, *et al.*, Novel phosphatidylinositol flippases contribute to phosphoinositide homeostasis in the plasma membrane, *Biochem. J.*, 2024, **481**(18), 1187–1202.
58 S. N. Ahmed, D. A. Brown and E. London, On the Origin of Sphingolipid/Cholesterol-Rich Detergent-Insoluble Cell Membranes: Physiological Concentrations of Cholesterol and Sphingolipid Induce Formation of a Detergent-Insoluble, Liquid-Ordered Lipid Phase in Model Membranes, *Biochemistry*, 1997, **36**(36), 10944–10953.
59 M. Murata, N. Matsumori, M. Kinoshita and E. London, Molecular substructure of the liquid-ordered phase formed by sphingomyelin and cholesterol: sphingomyelin clusters forming nano-subdomains are a characteristic feature, *Biophys. Rev.*, 2022, **14**(3), 655–678.
60 R. F. M. De Almeida, A. Fedorov and M. Prieto, Sphingomyelin/Phosphatidylcholine/Cholesterol Phase Diagram: Boundaries and Composition of Lipid Rafts, *Biophys. J.*, 2003, **85**(4), 2406–2416.
61 A. Traynor-Kaplan, M. Kruse, E. J. Dickson, G. Dai, O. Vivas, H. Yu, *et al.*, Fatty-acyl chain profiles of cellular phosphoinositides, *Biochim. Biophys. Acta, Mol. Cell Biol. Lipids*, 2017, **1862**(5), 513–522.
62 S. Ikhlef, N. F. Lipp, V. Delfosse, N. Fuggetta, W. Bourguet, M. Magdeleine, *et al.*, Functional analyses of phosphatidylserine/PI(4)P exchangers with diverse lipid species and membrane contexts reveal unanticipated rules on lipid transfer, *BMC Biol.*, 2021, **19**(1), 248.
63 L. J. Pike and J. M. Miller, Cholesterol Depletion Delocalizes Phosphatidylinositol Bisphosphate and Inhibits Hormone-stimulated Phosphatidylinositol Turnover, *J. Biol. Chem.*, 1998, **273**(35), 22298–22304.
64 J. Myeong, C. G. Park, B. C. Suh and B. Hille, Compartmentalization of phosphatidylinositol 4,5-bisphosphate metabolism into plasma membrane liquid-ordered/raft domains, *Proc. Natl. Acad. Sci. U.S.A.*, 2021, **118**(9), e2025343118.
65 G. Van Den Bogaart, K. Meyenberg, H. J. Risselada, H. Amin, K. I. Willig, B. E. Hubrich, *et al.*, Membrane protein sequestering by ionic protein–lipid interactions, *Nature*, 2011, **479**(7374), 552–555.
66 J. Wang and D. A. Richards, Segregation of PIP2 and PIP3 into distinct nanoscale regions within the plasma membrane, *Biol. Open*, 2012, **1**(9), 857–862.

67 M. Doktorova, F. A. Heberle, B. Eicher, R. F. Standaert, J. Katsaras, E. London, *et al.*, Preparation of asymmetric phospholipid vesicles for use as cell membrane models, *Nat. Protoc.*, 2018, **13**(9), 2086–2101.

68 G. Li, S. Kakuda, P. Suresh, D. Canals, S. Salamone and E. London, Replacing plasma membrane outer leaflet lipids with exogenous lipid without damaging membrane integrity, *PLoS One*, 2019, **14**(10), e0223572.

69 F. Lolicato and W. Nickel, A Role for Liquid-Ordered Plasma Membrane Nanodomains Coordinating the Unconventional Secretory Pathway of Fibroblast Growth Factor 2?, *Front. Cell Dev. Biol.*, 2022, **10**, 864257.

70 U. Golebiewska, M. Nyako, W. Woturski, I. Zaitseva and S. McLaughlin, Diffusion Coefficient of Fluorescent Phosphatidylinositol 4,5-bisphosphate in the Plasma Membrane of Cells. Martin TU, *Mol. Biol. Cell*, 2008, **19**(4), 1663–1669.

71 J. P. Zewe, A. M. Miller, S. Sangappa, R. C. Wills, B. D. Goulden and G. R. V. Hammond, Probing the subcellular distribution of phosphatidylinositol reveals a surprising lack at the plasma membrane, *J. Cell Biol.*, 2020, **219**(3), DOI: 10.1083/jcb.201906127.

72 R. Dimova, Giant Vesicles and Their Use in Assays for Assessing Membrane Phase State, Curvature, Mechanics, and Electrical Properties, *Annu. Rev. Biophys.*, 2019, **48**(1), 93–119.

73 H. Träuble and E. Grell, Carriers and specificity in membranes. IV. Model vesicles and membranes. The formation of asymmetrical spherical lecithin vesicles, *Neurosci. Res. Program Bull.*, 1971, **9**(3), 373–380.

74 S. Pautot, B. J. Frisken and D. A. Weitz, Production of Unilamellar Vesicles Using an Inverted Emulsion, *Langmuir*, 2003, **19**(7), 2870–2879.

75 M. J. Hope, T. E. Redelmeier, K. F. Wong, W. Rodrigueza and P. R. Cullis, Phospholipid asymmetry in large unilamellar vesicles induced by transmembrane pH gradients, *Biochemistry*, 1989, **28**(10), 4181–4187.

76 B. Bloj and D. B. Zilversmit, Asymmetry and transposition rate of phosphatidylcholine in rat erythrocyte ghosts, *Biochemistry*, 1976, **15**(6), 1277–1283.

77 T. A. Enoki. The use of hemifusion to create asymmetric giant unilamellar vesicles: Insights on induced order domains, in *Methods in Enzymology*, Elsevier, 2024, pp. 127–159, Available from: https://linkinghub.elsevier.com/retrieve/pii/S0076687924001204.

78 J. P. Steringer, S. Bleicken, H. Andreas, S. Zacherl, M. Laussmann, K. Temmerman, *et al.*, Phosphatidylinositol 4,5-Bisphosphate (PI(4,5)P2)-dependent Oligomerization of Fibroblast Growth Factor 2 (FGF2) Triggers the Formation of a Lipidic Membrane Pore Implicated in Unconventional Secretion, *J. Biol. Chem.*, 2012, **287**(33), 27659–27669.

79 G. J. Schütz and G. Pabst, The asymmetric plasma membrane—A composite material combining different functionalities?: Balancing Barrier Function and Fluidity for Effective Signaling, *BioEssays*, 2023, **45**(12), 2300116.

80 X. Lin, H. Wang, Z. Lou, M. Cao, Z. Zhang and N. Gu, Roles of PIP 2 in the membrane binding of MIM I- BAR : insights from molecular dynamics simulations, *FEBS Lett.*, 2018, **592**(15), 2533–2542.

81 H. J. Lessen, K. C. Sapp, A. H. Beaven, R. Ashkar and A. J. Sodt, Molecular mechanisms of spontaneous curvature and softening in complex lipid bilayer mixtures, *Biophys. J.*, 2022, **121**(17), 3188–3199.

82 R. Dimova, Membrane Electroporation in High Electric Fields, in *Advances in Electrochemical Sciences and Engineering*, ed. Alkire R. C., Kolb D. M. and Lipkowski J., Wiley, 1st edn, 2011, pp. 335–367, Available from: DOI: 10.1002/9783527644117.ch7.
83 A. A. Gurtovenko and I. Vattulainen, Pore Formation Coupled to Ion Transport through Lipid Membranes as Induced by Transmembrane Ionic Charge Imbalance: Atomistic Molecular Dynamics Study, *J. Am. Chem. Soc.*, 2005, **127**(50), 17570–17571.
84 S. Chiantia and E. London, Acyl Chain Length and Saturation Modulate Interleaflet Coupling in Asymmetric Bilayers: Effects on Dynamics and Structural Order, *Biophys. J.*, 2012, **103**(11), 2311–2319.
85 E. H. Chaisson, F. A. Heberle and M. Doktorova, Quantifying Acyl Chain Interdigitation In Simulated Bilayers Via Direct Transbilayer Interactions, *bioRxiv*, 2024, preprint, DOI: 10.1021/acs.jcim.4c02287.

Faraday Discussions

PAPER

Protein-induced membrane asymmetry modulates OMP folding kinetics and stability†

Jonathan M. Machin, * Neil A. Ranson and Sheena E. Radford *

Received 21st November 2024, Accepted 7th January 2025
DOI: 10.1039/d4fd00180j

Biological membranes are asymmetric structures, with asymmetry arising from differences in lipid identity in each leaflet of the bilayer, as well as non-uniform distribution of lipids and small molecules in the membrane. Proteins can also induce and modulate membrane asymmetry based on their shape, sequence and interactions with lipids. How membrane asymmetry affects macromolecular behaviour is poorly understood because of the complexity of natural membrane systems, and difficulties in creating relevant asymmetric bilayer systems *in vitro*. Here, we present a method exploiting the efficient, unidirectional folding of the transmembrane β-barrel outer membrane protein, OmpA, to create asymmetric proteoliposomes with protein-induced dipoles of known direction (arising from sequence variation engineered into the OmpA loops). We then characterise the folding kinetics and stability of different OmpA variants into these proteoliposomes. We find that both the primary sequence of the folding OmpA and the dipole of the membrane into which folding occurs play an important role for modulating the rate of folding. Critically, we find that by complementarily matching the charge on the folding protein to the membrane dipole it is possible to enhance both the folding kinetics and the stability of the folded OmpA. The results hint at how cells might exploit loop charge in membrane-embedded proteins to manipulate membrane environments for adaptation and survival.

Introduction

Biological membranes consist of lipid bilayers associated with a diverse set of other intra-membrane and membrane-associated entities, and are essential for many cellular processes, including compartmentalisation, signalling, transport and cellular protection.[1,2] Membrane asymmetry has been implicated in an array of essential biological processes including apoptosis,[3] cell morphology,[4] protein–

Astbury Centre for Structural Molecular Biology, School of Molecular and Cellular Biology, Faculty of Biological Sciences, University of Leeds, Leeds, LS2 9JT, UK. E-mail: j.m.machin@leeds.ac.uk; s.e.radford@leeds.ac.uk

† Electronic supplementary information (ESI) available. See DOI: https://doi.org/10.1039/d4fd00180j

lipid interactions[5,6] and modulating enzyme activity.[7] *In vivo*, membrane asymmetry arises from multiple sources[8,9] including lipid acyl chain and lipid head-group bilayer leaflet asymmetry, polarised organisation of hydrophobic compounds dissolved in the membrane, and asymmetric differences induced by the presence of transmembrane and peripherally attached proteins at the membrane surface.[10]

While the effects of lipid asymmetry on protein folding and stability have been reported recently using bacterial outer membrane proteins (OMPs) as a model system,[11] and the effects of lipid asymmetry on protein insertion into the membrane has been explored for other proteins,[12,13] the implications of protein-induced membrane asymmetry on protein folding and stability remains poorly understood. Transmembrane proteins confer asymmetry to the membrane *via* differences in the residues they expose on their membrane-facing surfaces,[14,15] as well as by their shape and structural properties.[16] Asymmetry in natural membranes hence arises *via* multiple mechanisms, including local enrichment of different lipids in the bilayer leaflets,[15,17] tension and curvature in the bilayer that is induced or enhanced by proteins and manifested unevenly across the bilayer,[18,19] and alteration in the electrostatic potential of embedded proteins that produces local dipoles, that may work in concert with lipid-induced charge asymmetry.[20–23] Often these features are combined, for example, piezo ion channels induce membrane asymmetry by altering the relative curvature of each side of the bilayer, locally enriching lipids in different leaflets of the bilayer and acting in concert to manipulate global membrane disorder.[17,19,24,25] While individual proteins asymmetrically modulate their local membrane context, long-range effects can also emerge *via* reinforcement across multiple proteins,[10,19,23] especially in protein-rich membranes or protein arrays. Indeed, most membranes contain a high concentration of proteins in their bilayers. For example, the inner membrane of diderm bacteria has a lipid : protein ratio (LPR) of ~32 : 1 (mol mol^{-1}), with proteins covering about 25% of the membrane's surface area.[26]

The outer membrane (OM) of diderm bacteria is a highly unusual and grossly asymmetric membrane. OMPs embedded in the OM exhibit low, and highly restricted, diffusion,[27] in part because of the extremely low LPR in the OM (~7 : 1 (ref. 26 and 28)). In addition, the outer leaflet of the OM is dominated by lipopolysaccharides, with phospholipids in the inner leaflet, making the OM one of the most profoundly asymmetric membranes in biology.[29] The dense packing of OMPs in the OM also increases the likelihood of potential effects of protein-induced membrane asymmetry on a local or long-range scale *in vivo*. OMPs are highly stable β-barrels $\Delta G^{\circ}_{F} = -10$ to -140 kJ mol^{-1},[28] with transmembrane β-strands typically linked by long (>8-residue) extracellular loops and short (<5-residue) intracellular turns.[30] OmpA is a well-studied OMP,[28,31] that is common in the OM (>100 000 copies per cell[32] in *Escherichia coli* (*E. coli*)) and confers strength and resistance (*e.g.* resilience to enhanced osmotic pressure[33]) to the cell. Natively folded OmpA consists of an eight-stranded transmembrane barrel domain linked by four extracellular loops (13–18 residues in length) and three short (4-residue) turns in the periplasmic face[34,35] (Fig. 1). It also possesses an ~15 kDa C-terminal (natively intracellular (periplasmic)) soluble domain that readily refolds *in vitro*[36,37] (Fig. 1). Altering membrane properties has been shown to modulate the folding of OMPs (including OmpA), for example, changing lipid acyl chain length[38] and/or head group identity,[39] or altering the global membrane properties

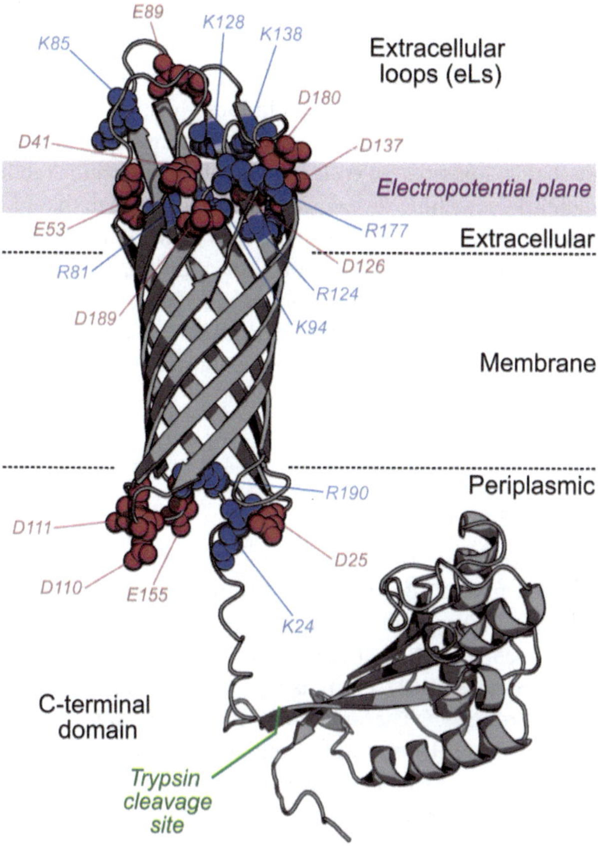

Fig. 1 Structural overview and charged residues of OmpA. Full-length OmpA has a transmembrane β-barrel and a natively periplasmic soluble domain joined by a flexible linker. All solvent-accessible charged residues in the OmpA barrel (following trypsin cleavage, site marked in green) are shown as spheres and labelled (red: negative, blue: positive). The electropotential plane indicates the approximate region used to calculate the electrostatic potential shown in Fig. 2a. OmpA-Neg neutralises all labelled positive extracellular residues (R81S, K85T, K94S, R124S, K128G, K134S and R177S). OmpA-Pos neutralises all labelled negative extracellular residues (D41S, E53N, E89V, D126S, D137S, D180S and D189S). OmpA-Neut neutralises all extracellular positive and negative residues by combining the OmpA-Pos/-Neg mutations. (OmpA modelled from PDBs 1G90 (ref. 1) (transmembrane) and 2MQE[2] (C-terminal domain)).

such as lipid order,[40] and the presence of membrane defects.[41] While decreasing LPR generally reduces the folding rate of OMPs,[28] the effect of LPR on folding rate and yield of OmpA in different lipids and lipid mixtures has not been studied systematically to date.

The charge on proteins and membranes is known to affect protein folding, localisation and function,[42–44] including the 'positive-inside' rule for transmembrane helix topology determination[45,46] and 'positive-outside' rule for OMPs.[47,48] We have recently shown that non-uniform charge distribution across a lipid bilayer, generated by asymmetric lipid content between the bilayer leaflets, can modulate the folding and stability of OMPs.[11] Here, using OmpA as a platform, we have generated proteoliposomes with different protein-induced bilayer charge dipoles of known orientation and used them to determine the effect of protein-induced membrane dipoles on the folding rates and stability of OmpA variants. We show that protein-induced charge asymmetries indeed modulate the folding rates of the different OmpA variants and, most importantly, demonstrate that it is possible to enhance the folding rate and stability of OmpA by complementarily matching its extracellular loop charge to that of the protein-induced membrane dipole it is folding into. The results have implications on how bilayer asymmetry can alter OMP folding. In addition, they present a robust method for exploring the effects of protein-mediated bilayer asymmetry on membrane protein behaviour more broadly, and inform principles for the design and generation of biosynthetic membranes containing OMPs as pores, channels or sensors for translational applications.[49–51]

Methods

Electrostatic modelling

To model the electrostatic environment for the OmpA variant of interest, the protein (modelled from PDBs 1G90 (ref. 52) (transmembrane) and 2MQE[53] (C-terminal domain)) was placed in an all-atom membrane of 1,2-dilauroyl-*sn*-glycero-3-phosphocholine (DLPC) (20 × 20 × 14 nm box) with explicit water neutralised with 50 mM NaCl (set-up using CHARMM-GUI[54,55]) and equilibrated for 10 ns using Gromacs.[56] The final simulation frame was processed through APBS[57] and the resulting electrostatic potentials were analysed and figures drawn using custom python scripts, analysing a slice of about 0.6 nm thickness parallel to the membrane plane centred on the protein.

OmpA purification

OmpA and its variants were expressed in *E. coli* BL21(DE3) as inclusion bodies and purified in an unfolded state as described previously.[11] The OmpA variants created were: OmpA-Neg: R81S, K85T, K94S, R124S, K128G, K134S and R177S; OmpA-Pos: D41S, E53N, E89V, D126S, D137S, D180S and D189S; OmpA-Neut: containing the combined mutations from OmpA-Pos and OmpA-Neg.[11] Mutants were chosen by identifying the most common naturally occurring sequence variant at each position or, if not conserved, serine was used.

Liposome preparation and initial folding

The required amount of resuspended lipid (DLPC, or dimyristoylphosphatidylcholine (DMPC) (Avanti polar lipids) in 1 : 4 MeOH : chloroform) was dried to a thin film in a glass vial and desiccated overnight. Following resuspension to a stock concentration of 40 mM in buffer (20 mM Tris–Cl (pH 8.5), 50 mM NaCl), the lipids were freeze–thaw cycled using liquid N_2 and an ~50 °C water bath and then extruded through 100 nm nucleopore polycarbonate track-etched membranes (Whatman, Avanti extruder) at 35–40 °C (>10 °C higher than the lipid T_m). As required, proteoliposomes were generated by mixing the required amount of unfolded OmpA by rapid dilution of denatured protein in 8 M urea to 1 M urea, and allowing the protein to fold into the membrane overnight at room temperature.

LPR-matched proteoliposome generation

Using the fraction folded of each OmpA-variant into DLPC liposomes (Fig. 3c), the initial LPR was calculated such that final, folded OmpA LPR should be 320 : 1 (OmpA-WT 290 : 1, OmpA-Pos 160 : 1, OmpA-Neg 260 : 1, OmpA-Neut 240 : 1 mol mol^{-1}) achieved by rapid dilution of the denatured protein in 8 M urea to 1 M urea, and allowing the protein to fold into the membrane overnight at room temperature. Following folding, unfolded OmpA (of which there were different amounts remaining in solution for different OmpA variants) and the exposed OmpA C-terminal domains were cleaved with 1 : 10 (mol mol^{-1}) trypsin incubated for 3 h at 37 °C. Trypsin, peptides and any remaining unfolded protein were removed *via* two rounds of liposome pelleting *via* ultracentrifugation (110 000*g*, 4 °C, 30 min, Optima MAX-XP, Beckman Coulter). Prior to each centrifugation run, 20.1% (w/v) phenylmethylsulfonylchloride (PMSF) was added to inhibit residual trypsin. This method was validated by SDS-PAGE and Dynamic Light Scattering (DLS) (the latter using a Wyatt miniDAWN TREOS® instrument). DMPC proteolipsomes (for lipid T_m measurements only) were made similarly, but during initial OMP folding were incubated at 24 °C overnight.

To estimate the number of OmpA molecules per liposome, the average proteoliposome hydrodynamic radius obtained by DLS (58 nm) was used. DLPC was assumed to occupy an area of 0.62 nm^2 and to generate a bilayer thickness of 3 nm. There are thus ~115 000 lipids per liposome. Given a final LPR of 320 : 1, there are on average ~359 OmpA molecules per liposome, with a likely range of 300–400 for a typical-sized proteoliposome.

Folding kinetics and urea titration

All folding kinetic measurements were performed using a BMG Clariostar platereader measuring intrinsic protein fluorescence (excitation 280 nm, emission 335 nm, both with 10 nm windows) at 10–30 s intervals, in sealed UV-transparent 96-well plates (CORNING 3635) at 30 °C in a 20 mM Tris–Cl (pH 8.5), 50 mM NaCl buffer (125 μL volumes). For folding reactions, proteoliposomes containing 1.25 μM of the pre-folded OmpA variant of interest were rapidly mixed with 1.25 μM unfolded OmpA (initial [urea] 8 M, final [urea] 1 M). Data were fitted to a single exponential or sigmoidal logarithmic curve to minimize the error, and the T_{50} values (time to reach 50% folded state) were extracted from the fit using Python. For display, up to a five-point moving average was applied to the data. For the urea titration, 2.5 μM of the pre-folded OmpA variant of interest in the proteoliposomes and 2.5 μM of unfolded OmpA were used. An initial 100-point reading was taken, before incubation overnight at 30 °C and a final 100-point measurement. The relative fraction of folded protein was determined by averaging the data over the final measurement for each individual condition and, if a plateau was reached at low urea concentrations, the data were normalised. Where possible, a sigmoidal logarithmic curve was fitted, and the urea concentration at the folding midpoint (C_m) extracted. For the kinetics, significant differences were determined by permutation testing[58] (which makes no assumption about the underlying distribution of the data), with the test statistic defined as the average difference between a pair of datasets. For the urea titration experiments, the significance was determined using paired *t*-tests over all the raw datapoints, with points paired for the same urea concentration and replicate.

Laurdan measurement of lipid T_m

Measurement of the lipid T_m in different proteoliposomes was performed using laurdan fluorescence as previously described.[11] Briefly, DMSO-dissolved laurdan was added to pre-formed DMPC proteoliposomes at a lipid : laurdan ratio of 3200 : 1 (mol mol^{-1}) (0.1% (v/v) DMSO final) and incubated overnight at room temperature. Fluorescence emission at 440 nm and 490 nm (excitation: 340 nm) was then measured at 0.5 °C intervals from 20–29 °C using a PTI fluorimeter (Horiba). General polarization (GP) was determined from the average intensity (I) at 440 and 490 nm, where GP = $(I_{440} - I_{490})/(I_{440} + I_{490})$. Mid-points were determined by numerically differentiating the data.

DLS

Proteoliposomes were diluted to a lipid concentration of ~4 μM and 300 μL was injected into a Wyatt miniDAWN TREOS® instrument. ~5 min baselines were measured with filtered (0.22 μm) buffer (20 mM Tris–Cl (pH 8.5), 50 mM NaCl) before and after sample injection. The flow cell was flushed with 0.5 mL of 0.22-μm-filtered 1 M nitric acid and 1 mL 18 MΩ H_2O after each run, followed by 1 mL of buffer. Correlation curves were analysed from a 3 min sample window by regularisation using Astra 6.0.3®.

SDS-PAGE

Samples for SDS-PAGE analysis were mixed in a ratio of 1 : 3 with loading dye (50 mM Tris–HCl, pH 6.8, 6% (w/v) SDS, 0.3% (w/v) bromophenol blue, 40% (v/v) glycerol), boiled if required (>10 min, >95 °C) and ~14 μL sample loaded into the gel (15% (w/v) Tris–tricine gels with 0.1% (w/v) SDS); ladder: Precision Plus Protein Dual Xtra Standards (BioRad). Following staining (InstantBlue Coomassie, Abcam), the gels were imaged using an Alliance Q9 imaging system (Uvitec) and densitometric analysis was performed using ImageJ. Where required, the folded fraction was calculated using the intensity ratio (folded/(folded + unfolded)) of the monomer bands.

Results

Electrostatic modelling of protein-induced charge asymmetry

To create controllable, protein-induced asymmetry in synthetic membranes, the insertion directionality and asymmetric property of the protein (here, the charge distribution), and the lipid : protein ratio must be known. Urea unfolded OmpA readily and quantitatively folds unidirectionally *in vitro* into pre-prepared liposomes of different lipid type (*e.g.* DMPC,[34] DMPG,[11] POPC[59]), oriented with its extracellular loops inside the liposome and its water-soluble C-terminal domain on the outside as it cannot cross the membrane.[11,34] The extracellular loops of natively folded OmpA contain seven positively charged residues and seven negatively charged residues, while the intracellular turns contain four negative residues and two positive residues,[35] generating a mild global dipole away from the more negative intracellular turns (Fig. 1). Additional charged residues exist within the core of the barrel, but these are secluded from the bulk solvent and engage in salt-bridges to stabilise the protein's core.[60]

Sequence variants of OmpA with altered charge in the extracellular loops were previously generated,[11] notably OmpA-Pos (neutralisation of the seven negative residues, Fig. 1), OmpA-Neg (neutralisation of the seven positive residues, Fig. 1) and OmpA-Neut (neutralisation of all the positive and negative residues in the extracellular loops, Fig. 1). To better understand how these variants of OmpA manipulate the local electrostatic environment, the potential in a plane close to the membrane was modelled for each protein individually. A single copy of each OmpA variant was placed in a DLPC membrane with 50 mM NaCl (the same concentration used experimentally below), equilibrated, and then the electrostatic potential ~0.8 nm above the membrane (*i.e.*, adjacent to the extracellular loops) was determined (Fig. 1 & 2a). The data show that the OmpA-Pos and OmpA-Neg variants generate a large area of electropositivity and electronegativity, respectively. The OmpA-Neut variant has a minimal charge footprint, while OmpA-WT has a split local potential with different sides of the protein being oppositely charged. The potential around the intracellular turns on the opposite side of the membrane was also assessed (Fig. 2b) which, as expected, showed a weak negative potential (average potential, excluding protein, over the area

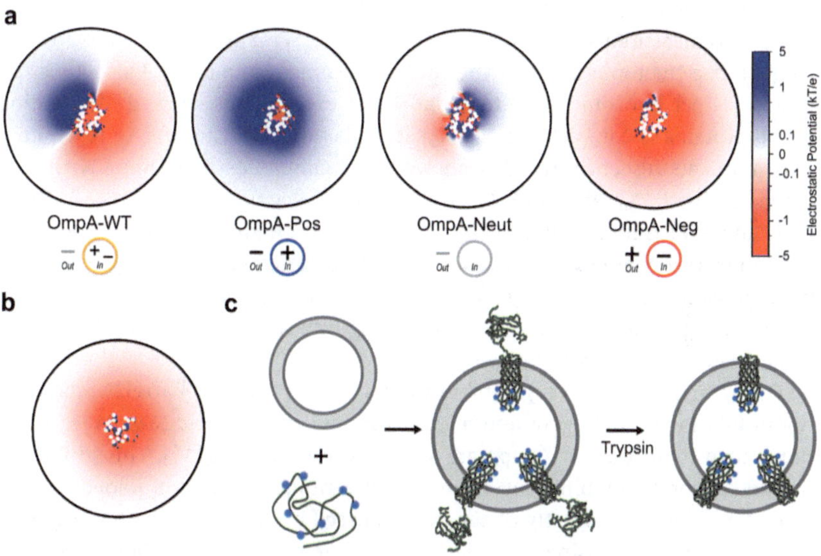

Fig. 2 Modelling protein-induced charge-asymmetric liposomes. (a) Modelled electrostatic potential above the membrane plane around the extracellular loops of OmpA variants in DLPC membranes. The direction of the dipole generated from proteoliposomes of each variant is shown beneath each potential map. (b) Modelled electrostatic potential below the membrane plane around the intracellular turns of OmpA (in (a) and (b) white dots are in-view Cα; circular area is equivalent to the 320 : 1 (mol mol^{-1}) experimental lipid : protein ratio used, ~1 nm solvent slab analysed parallel to the membrane plane). (c) Experimental approach to generate protein-induced charge dipoles over the membrane: OmpA variants are unidirectionally inserted into pre-formed liposomes to a defined concentration, and then the soluble C-terminal domain is cleaved off using trypsin and the resultant proteoliposomes purified. The charge on the extracellular loops (here inside the liposomes when OmpA is folded) are altered by mutation (OmpA-Pos is shown here as an example, with blue positively charged symbols showing the seven positively charged residues in its extracellular loops).

shown is −0.11 kT/e (intracellular turns); for comparison, OmpA-Neg is −0.48 kT/e at the extracellular loops). Together, this difference enables the dipole directionality to be assigned, as indicated below each potential profile in Fig. 2a.

Generation and validation of charge-asymmetric proteoliposomes

Exploiting the different membrane dipoles induced by the OmpA variants enables proteoliposomes with membrane-protein-induced opposite charge dipoles to be created, with OmpA-Pos creating positive-inside proteoliposomes and OmpA-Neg negative-inside proteoliposomes, while OmpA-Neut has a mild positive-inside dipole. To create such protein-induced dipole asymmetric proteoliposomes, OmpA variants were folded into 100 nm diameter liposomes formed from DLPC to a known final LPR, and then the C-terminal domain was removed by cleavage with trypsin before purification using ultracentrifugation (see Methods) (Fig. 2c). DLPC was chosen for the experiments because of its net neutral charge (ensuring that the majority of the generated charge dipole arises from folded OmpA in the bilayer) and to ensure efficient folding of all the OmpA variants used (previous work showed maximum folding efficiencies of OMPs into short acyl chain lipids[38]).

First the folding of all four OmpA variants into 100-nm extruded DLPC liposomes was measured at an LPR of 320 : 1 (mol mol^{-1}) *via* intrinsic fluorescence (of OmpA's five tryptophans), creating proteoliposomes containing 300–400 OmpA molecules per liposome (see Methods). These experiments showed that folding for all proteins is completed by ~2500 seconds (Fig. 3a). The time to reach half-maximum fluorescence, the T_{50}, showed that OmpA-Pos and OmpA-Neut fold the fastest, with T_{50} values of ~350 s, while OmpA-Neg folds slowest (T_{50} ~1100 s) and OmpA-WT is intermediate (T_{50} ~750 s) (Fig. 3b), highlighting the importance of the positive charge in the extracellular loops for efficient folding, consistent with previous results.[11] They also show that the shorter-chain lipid (DLPC) enables a substantially faster (>3-fold) folding rate compared with folding of the same proteins into DMPC liposomes.[11]

Owing to the inherent high stability of its natively folded β-barrel, OmpA does not unfold in SDS detergent and migrates anomalously in cold SDS-PAGE, while non-native conformers of OmpA are SDS-sensitive[31] (ESI Fig. 1†). This difference in electrophoretic mobility can be used to determine the yield of folded (native) protein *via* gel densitometry of samples analysed at the end of the folding reactions (Methods). Assessing these data for the different variants of OmpA analysed here (Fig. 3c) showed yields of natively folded protein ranging from ~90% for OmpA-WT to ~50% for OmpA-Pos. It is intriguing that the OmpA-Pos, while folding rapidly, results in a significantly lower folded yield than the other OmpA variants. This suggests that once ~50% of the molecules have been folded into the bilayer, folding and membrane insertion of additional molecules is precluded. Given the known folding efficiencies, it is possible to conduct the folding reaction at LPRs tuned to each OmpA variant, such that the final amount of folded OmpA is approximately the same for each variant (with differing amounts remaining unfolded in solution) (ESI Fig. 2†). Thus, proteoliposomes containing approximately the same amount of folded protein content were generated (Fig. 3d). Upon the addition of trypsin to cleave the C-terminal domain of OmpA, any remaining unfolded protein is also digested and, following liposome purification by

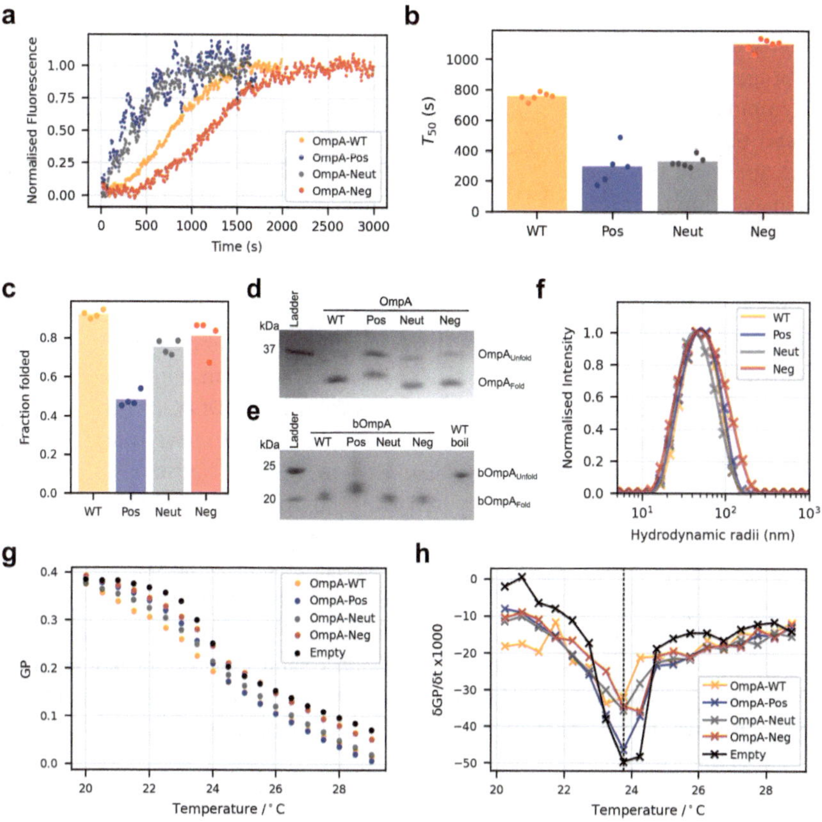

Fig. 3 OmpA folding into DLPC and proteoliposome validation. (a) Example folding kinetics for the four OmpA variants inserting into empty DLPC liposomes and (b) fitted half-times (T_{50}) for each curve ($n \geq 5$). (c) Fraction of each OmpA variant folded into DLPC liposomes at an LPR of 320 : 1 (mol mol^{-1}). (d) By matching the LPR to the yield of each folded protein, proteoliposomes with similar concentrations of each natively folded OmpA variant can be generated (original image in ESI Fig. 3†). (e) After cleavage of the C-terminal soluble domain with trypsin and purification, proteoliposomes with a similar LPR containing only the (folded) barrel domain of OmpA are recovered (original image in ESI Fig. 4†). (f) DLS of the final proteoliposomes used for folding assays. (g) The GP (generalised polarisation) ratio of laurdan fluorescence at 440 nm and 490 nm against temperature for DMPC liposomes containing each OmpA variant. (h) The first derivative of the GP indicates only small changes in the lipid T_m (curve minima) in the presence of each protein compared to empty liposomes (dashed line).

ultracentrifugation, only the folded OmpA barrels (bOmpA) at approximately equal protein concentration remain (Fig. 3e). Dynamic light scattering (DLS) confirmed that the proteoliposomes remain intact after this processing (Fig. 3f).

Although the barrel domains of OmpA are identical, it is possible that the differences in the extracellular loops of the OmpA variants may alter the membrane properties. To test for this, the global lipid phase transition temperature for the different OmpA proteins folded into DMPC liposomes (T_m 24 °C (without protein)) was assessed (the T_m of DLPC is −2 °C, making experiments with this lipid unfeasible). Accordingly, the OmpA variants were folded into 100-

nm liposomes of DMPC at an LPR of ~320 : 1 (mol mol^{-1}). Lipid phase transition temperatures were then measured using the fluorescent probe laurdan, which changes its fluorescence profile depending on the lipid phase.[61] The resulting sigmoidal curves with respect to temperature (Fig. 3g) were then differentiated to determine the T_m (Fig. 3h). Although slight differences (≤1 °C differences in T_m) are apparent, the T_m of all proteoliposomes are within 0.5 °C of the empty liposomes, demonstrating minimal consequences of the presence of the different proteins on lipid order.

Protein-induced dipoles modulate OmpA variant folding rates

To determine how the different protein-induced membrane charge-dipoles generated affect folding, full-length OmpA of each variant (OmpA-WT, OmpA-Pos, OmpA-Neut or OmpA-Neg (Fig. 1)) was folded into DLPC-proteoliposomes containing the pre-folded bOmpA variants as described above (named DLPC-WT/-Pos/-Neut/-Neg) (Fig. 2c). Similar to the empty DLPC liposomes, OmpA-Pos and OmpA-Neut fold more rapidly into DLPC-WT proteoliposomes than OmpA-WT, while the OmpA-Neg folds the slowest (Fig. 4a and b). Despite the high protein concentration in the membrane (final LPR of ~160 : 1 (mol mol^{-1})), OmpA-Pos and OmpA-Neut fold at comparable rates into DLPC-WT as they do into the empty liposomes (T_{50} values of ~350 and ~400 s, respectively), while OmpA-WT folds about 50% more slowly (T_{50} values of ~1150 and ~750 s, respectively), as does OmpA-Neg (T_{50} values of ~1610 and ~1100 s, respectively) (Fig. 3b & 4b). These observations show that the charge in the extracellular loops

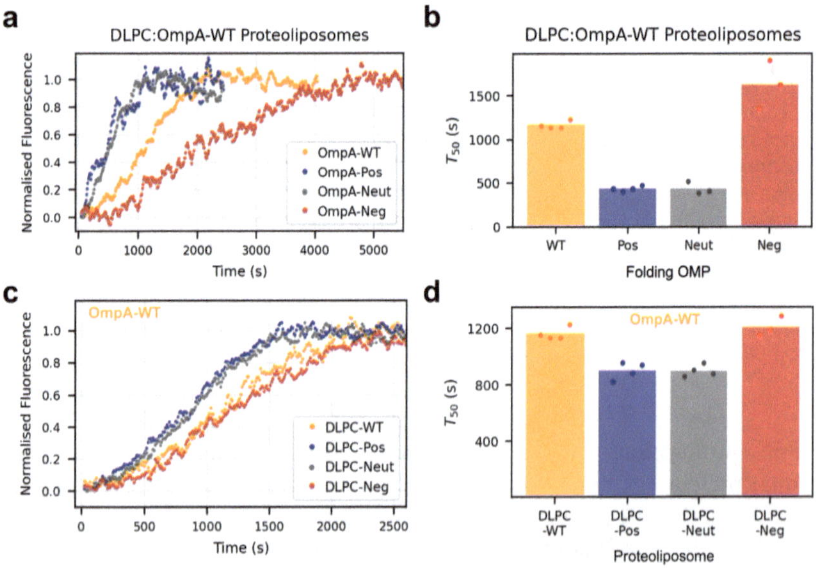

Fig. 4 Both OmpA sequence and the charge dipole modulate the folding kinetics. (a) Sample kinetic traces of OmpA variants folding into DLPC:OmpA-WT proteoliposomes and (b) fitted T_{50} values for each curve ($n \geq 3$). (c) Sample kinetic traces for OmpA-WT folding into proteoliposomes with different dipoles and (d) fitted T_{50} values for each curve ($n \geq 3$). Data for DLPC-WT are reproduced from (b) for ease of comparison (the LPR of all proteoliposome substrates is ~320 : 1).

of the folding OMP plays a role in determining the rate of folding (the folding rates follow the same rank order in DLPC and DLPC-WT), but the magnitude of the effect of the bilayer charge dipole depends on the charge in the extracellular loops of the folding OMP variant.

Next, OmpA-WT was folded into bilayers with different protein-induced dipoles (DLPC-WT/-Pos/-Neut/-Neg) (Fig. 4c). For OmpA-WT (that has charge-balanced positive and negative charges in its extracellular loops (Fig. 2a)), folding into positive-inside proteoliposomes (DLPC-Pos) was significantly (~40%) faster than folding into negative-inside proteoliposomes (DLPC-Neg) (T_{50} values of ~ 900 s and 1200 s, respectively) (Fig. 4d). Intriguingly, the folding rate enhancement observed here is opposite to that previously observed when studying the effects of lipid-induced dipoles,[11] likely a result of differences in the two systems, including dipole magnitude (~0.04 *vs.* ~0.1–0.35 charge per lipid equivalent) and LPR (final LPR of 160 : 1 *vs.* 1600 : 1) for protein-induced *versus* lipid-induced membrane dipoles, respectively.[11] Folding into DLPC-Neut occurs at the same rate as with DLPC-Pos (Fig. 4d). Together, these data highlight that there is an interplay between the charge on the protein and the local charge dipole across the bilayer that together modulate the rate of folding.

Electrostatic matching between proteoliposomal dipole and folding OMP

How the interaction between OmpA sequence charge and the protein-induced membrane dipole charge affects folding was considered next by comparing the folding rate of OmpA-Pos, OmpA-Neut and OmpA-Neg into proteoliposomes with different dipoles, *i.e.*, DLPC-Pos, DLPC-Neut and DLPC-Neg. The results showed that OmpA-Neg folds more slowly than all other variants tested into all types of proteoliposomes (Fig. 5a and b), with folding into DLPC-Neg being significantly slower than folding into DLPC-Neut and DLPC-Pos (Fig. 5b). In contrast, OmpA-Pos folds rapidly, and at a similar rate, into these three proteoliposome systems (Fig. 5b). Notably, however, this variant folds ~6-fold more rapidly into DLPC-Neg (T_{50} of ~350 s (Fig. 5b)) compared to OmpA-Neg (T_{50} of ~2100 s (Fig. 5b)). Folding rates of OmpA-Neut are similar to OmpA-Pos, but with folding into DLPC-Neg slightly, but significantly, retarded relative to DLPC-Pos. Overall, therefore, the results show that the charge in the OmpA extracellular loops determines the folding rate, with a positive charge facilitating rapid folding. In addition, they reveal that the rate of folding is also dependent on the membrane charge dipole induced by pre-folding OmpA into the membrane, with the magnitude of the effect observed depending on a complex (and not yet understood) balance between the charge on the folding OmpA and the dipole induced across the bilayer by natively folded OmpA into the membrane.

Finally, the effect of urea on the yield of folded OmpA-Pos and OmpA-Neg was assessed by measuring the magnitude of the intrinsic fluorescence change at equilibrium at different urea concentrations when each protein was folded into the different proteoliposomes. Note that natively folded, membrane-embedded OmpA cannot be unfolded, even at high concentrations of urea, hence ΔG° values cannot be calculated.[31] These experiments showed that OmpA-Pos is significantly more stable than OmpA-Neg in all bilayer types (Fig. 5c and d). For OmpA-Pos, the fractional folding curves plateaued at low urea concentrations allowing transition curves to be fitted and urea mid-point concentrations (C_m)

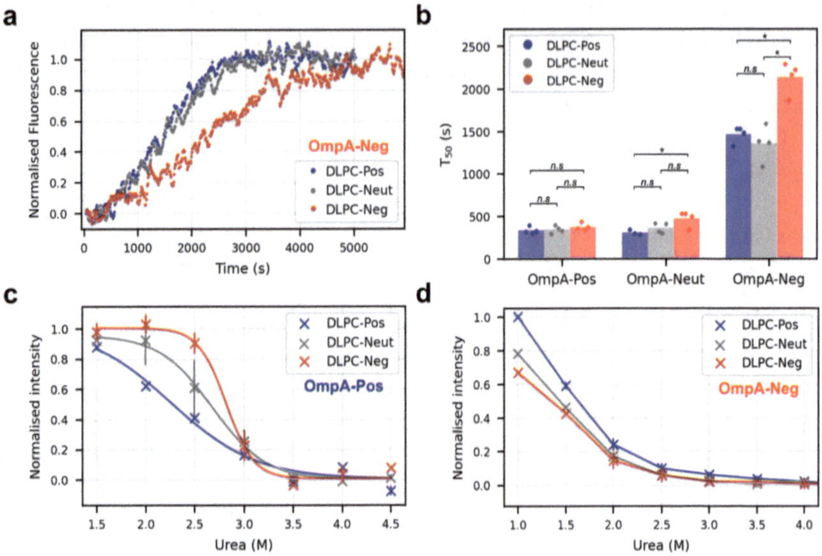

Fig. 5 Electrostatic matching between folding OMP and proteoliposome dipole. (a) Sample kinetic traces of OmpA-Neg folding into differently dipoled proteoliposomes as indicated in the key. (b) Comparison of folding T_{50} values for OmpA-Pos, OmpA-Neut and OmpA-Neg into differently dipoled proteoliposomes. P-values were determined by permutation testing (*: $p = 0.028$, n.s.: not significant) ($n \geq 3$). (c) Urea dependence of OmpA-Pos folding into differently dipoled proteoliposomes ($n = 2$, error bars show data range). Curves are fitted to the average data (bold symbols). (P-values: DLPC-Pos/DLPC-Neg: 0.014, DLPC-Pos/DLPC-Neut: 0.040, DLPC-Neg/DLPC-Neut: 0.135). (d) Urea dependence of OmpA-Neg folding into differently dipoled proteoliposomes ($n = 2$, error bars show data range, $n = 1$ for values at 1 M urea). Lines join the points and are to guide the eye only (P-values: DLPC-Pos/DLPC-Neg: 0.010, DLPC-Pos/DLPC-Neut: 0.010, DLPC-Neg/DLPC-Neut: 0.042) (LPR of the proteoliposome substrates in all panels is ~320 : 1).

determined (Fig. 5c). OmpA-Pos folding into proteoliposomes with a complementary dipole (DLPC-Neg) had a C_m of 2.8 M urea, while folding the same protein into DLPC-Pos had a significantly lower C_m of 2.2 M urea (p-value = 0.015), while the C_m for DLPC-Neut lies in between these values (C_m of 2.6 M). OmpA-Neg is too unstable in all proteoliposomes to enable values of C_m to be determined, although the data suggest that the proteoliposomes with a complementary dipole (DLPC-Pos) support higher folding yields for a given urea concentration than those with non/less-complementary dipoles (DLPC-Neg/-Neut) (DLPC-Pos/DLPC-Neg p-value = 0.01) (Fig. 5d). Together the data are suggestive of a complementary electrostatic interaction between the folding OMP and protein-induced dipole over the membrane in determining the rate of folding and stability of OmpA in the bilayer.

Discussion

The presence of proteins in biological membranes can confer a range of asymmetric properties on the bilayer, but the effects of these asymmetries on membrane-protein behaviour is largely unknown. Here we generated protein-induced transmembrane dipoles synthetically by exploiting modified forms of OmpA as a charge-carrying scaffold. OmpA forms an ideal platform for this purpose due to its ease of expression/purification, its ability to fold rapidly and efficiently *in vitro* into synthetic lipid bilayers of varied composition, and the high stability of its native state ($\sim$−35 kJ mol^{-1} (ref. 62)). Furthermore, the ability to engineer its long loops without preventing folding[60,63,64] and the presence of its 15 kDa C-terminal soluble domain that is unable to cross the membrane and enforces unidirectional folding[34] provide the attributes required to build proteoliposomes with different protein-induced dipoles across the membrane. While the effects of charge dipoles on folding were explored here, the properties of OmpA make it a broadly applicable scaffold to explore other types of protein-induced membrane asymmetries. For example, by engineering the protein sequence further, the consequences of asymmetric molecular crowding, differential amino acid properties in each bilayer leaflet or the introduction of leaflet-specific lipid binding sites could be explored.

We have shown here that manipulating the protein-induced electrostatic dipole across a lipid bilayer alters the folding rate of OmpA in a manner that is dependent both on the charge in the extracellular loops of the folding protein and the protein-induced dipole across the membrane. Most strikingly, it was found that proteoliposomes support faster folding and more effective stabilisation when the charge dipole of the membrane is complementary to that of the folding protein, as exemplified by OmpA-Neg folding into DLPC-Pos. However, the relationship is complex, since OmpA-Pos folds with a similar rate into DLPC-Pos/-Neg/-Neut. This could reflect the rapid intrinsic folding rate of OmpA-Pos into all three bilayers, making it difficult to detect kinetic differences, or reflect a change in folding mechanism in which charge effects are not rate determining. In addition, interaction of the charge on the folding protein with the short turns on bOmpA-loaded proteoliposomes, which are net negative (Fig. 2a) and exposed on the surface of the liposome, may also influence the rates of folding, possibly by electrostatically disfavouring the approach of OmpA-Neg to the membrane, while facilitating binding, and hence folding, of OmpA-Pos. It should also be borne in mind that the lipid charge and lipid-induced dipole across the membrane can also affect the folding rate, as shown previously,[11] making it difficult to generate 'rules' that rationalise the effects for the set of protein charges and membrane dipoles examined here. Further experiments using different OMPs, membranes of different lipid compositions, and with different protein- and lipid-based asymmetries will be needed to create a database of sufficient size to generate such rules.

Intriguingly, while we show here that OmpA-WT folds 40% more rapidly into positive-inside proteoliposomes compared to negative-inside proteoliposomes (Fig. 4d), we previously reported that the same protein folds up to $\sim$10-times more slowly into liposomes with a lipid-induced positive-inside charge dipole, *i.e.*, a strong effect in the opposite direction was observed.[11] Although the exact nature of this difference remains unclear, it could arise from the different LPRs used in the different experiments (final LPR of 160 : 1 *vs.* 1600 : 1), with the high concentrations of protein in the membranes used in this study altering the mechanism of the folding process. It could also result from the relatively small dipoles created in this study compared to those generated by asymmetric lipid organisation ($\sim$0.04 *vs.* $\sim$0.1–0.35 charge per lipid equivalent). This again highlights the complexity of the pathways of membrane protein folding in these 'simple' synthetic membrane systems, and raises the intriguing question of how such effects may manifest in the more complex situation of the bacterial OM.

In summary, the results presented here demonstrate that protein-induced membrane-dipole asymmetries can modulate OmpA folding rates and stability. Specifically, protein-induced bilayer dipoles are shown to be able to change the folding rates of OmpA by up to a 6-fold increase in a manner that depends both on the charge complementarity between the folding protein and the membrane that together contribute to defining the folding rate. Our results suggest new approaches that could be used to enhance the creation and stabilisation of OMPs in bilayers for use in biotechnology.[49–51] Equally importantly, they also show how the crowded and highly asymmetric bacterial OM might profoundly modulate the folding and properties of the embedded OMPs. They also highlight how bacteria may alter their proteomes to stabilise and/or accelerate (and *vice versa*) the folding/localisation of specific proteins to allow for concerted membrane adaptation. Given the low LPR in the protein-rich OM in which OMPs are highly crowded, and in which OmpA is one of the most abundant OM proteins,[32] it seems likely that at least some of these consequences will be important *in vivo*.

Data availability

Source data files containing fluorescence folding traces, gel images, DLS and electrostatic potential files are freely available at the University of Leeds Data Repository (https://doi.org/10.5518/1603).

Author contributions

All authors designed the experiments. J. M. M. performed all the research. All authors contributed to the analysis of the data, and all authors wrote or edited the manuscript.

Conflicts of interest

The authors have no conflict of interest.

Acknowledgements

We thank all members of the Radford and Ranson groups for their helpful discussions and critical reading of this manuscript, especially members of the OMP group in Leeds. J. M. M. is funded by Wellcome (222373/Z/21/Z) and MRC (MR/Y012453/1). We thank Nasir Khan for his excellent technical support and Bob Schiffrin for making the OmpA-WT protein. S. E. R. holds a Royal Society Professorial Research Fellowship (RSRP\R1\211057). The funders had no role in study design, data collection and analysis, decision to publish or preparation of the manuscript. For the purpose of Open Access, the authors have applied a CC BY public copyright licence to any Author Accepted Manuscript version arising from this submission.

References

1 T. Harayama and H. Riezman, Understanding the diversity of membrane lipid composition, *Nat. Rev. Mol. Cell Biol.*, 2018, **19**, 281–296.

2 X. Cheng and J. C. Smith, Biological membrane organization and cellular signaling, *Chem. Rev.*, 2019, **119**, 5849–5880.

3 S. Nagata, J. Suzuki, K. Segawa and T. Fujii, Exposure of phosphatidylserine on the cell surface, *Cell Death Differ.*, 2016, **23**, 952–961.

4 M. Bogdanov, *et al.*, Phospholipid distribution in the cytoplasmic membrane of Gram-negative bacteria is highly asymmetric, dynamic, and cell shape-dependent, *Sci. Adv.*, 2020, **6**, eaaz6333.

5 T. J. Maures, H.-W. Su, L. S. Argetsinger, S. Grinstein and C. Carter-Su, Phosphorylation controls a dual-function polybasic nuclear localization sequence in the adapter protein SH2B1β to regulate its cellular function and distribution, *J. Cell Sci.*, 2011, **124**, 1542–1552.

6 K. B. Beer, *et al.*, Extracellular vesicle budding is inhibited by redundant regulators of TAT-5 flippase localization and phospholipid asymmetry, *Proc. Natl. Acad. Sci. U. S. A.*, 2018, **115**, E1127–E1136.

7 W. Jia, *et al.*, Lipid trafficking controls endotoxin acylation in outer membranes of *Escherichia coli*, *J. Biol. Chem.*, 2004, **279**, 44966–44975.

8 J. E. Rothman and J. Lenard, Membrane asymmetry, *Science*, 1977, **195**, 743–753.

9 G. van Meer, Cellular lipidomics, *EMBO J.*, 2005, **24**, 3159–3165.

10 G. Pabst and S. Keller, Exploring membrane asymmetry and its effects on membrane proteins, *Trends Biochem. Sci.*, 2024, **49**, 333–345.

11 J. M. Machin, A. C. Kalli, N. A. Ranson and S. E. Radford, Protein–lipid charge interactions control the folding of outer membrane proteins into asymmetric membranes, *Nat. Chem.*, 2023, **15**, 1754–1764.

12 Q. Lin and E. London, The influence of natural lipid asymmetry upon the conformation of a membrane-inserted protein (perfringolysin O), *J. Biol. Chem.*, 2014, **289**, 5467–5478.

13 H. L. Scott, F. A. Heberle, J. Katsaras and F. N. Barrera, Phosphatidylserine asymmetry promotes the membrane insertion of a transmembrane helix, *Biophys. J.*, 2019, **116**, 1495–1506.

14 H. J. Sharpe, T. J. Stevens and S. Munro, A comprehensive comparison of transmembrane domains reveals organelle-specific properties, *Cell*, 2010, **142**, 158–169.

15 D. Drew and O. Boudker, Ion and lipid orchestration of secondary active transport, *Nature*, 2024, **626**, 963–974.

16 J. H. Lorent, *et al.*, Plasma membranes are asymmetric in lipid unsaturation, packing and protein shape, *Nat. Chem. Biol.*, 2020, **16**, 644–652.

17 A. Buyan, *et al.*, Piezo1 Forms Specific, Functionally important interactions with phosphoinositides and cholesterol, *Biophys. J.*, 2020, **119**, 1683–1697.

18 M. Löwe, *et al.*, Probing macromolecular crowding at the lipid membrane interface with genetically-encoded sensors, *Protein Sci.*, 2023, **32**, e4797.

19 X. Yang, *et al.*, Structure deformation and curvature sensing of PIEZO1 in lipid membranes, *Nature*, 2022, **604**, 377–383.

20 T. Banerjee, *et al.*, Spatiotemporal dynamics of membrane surface charge regulates cell polarity and migration, *Nat. Cell Biol.*, 2022, **24**, 1499–1515.
21 J. A. Peruzzi, *et al.*, Hydrophobic mismatch drives self-organization of designer proteins into synthetic membranes, *Nat. Commun.*, 2024, **15**, 3162.
22 S. A. Shelby, I. Castello-Serrano, K. C. Wisser, I. Levental and S. L. Veatch, Membrane phase separation drives responsive assembly of receptor signaling domains, *Nat. Chem. Biol.*, 2023, **19**, 750–758.
23 R. Li, R. Zhao, M. Yang, X. Zhang and J. Lin, Membrane microdomains: Structural and signaling platforms for establishing membrane polarity, *Plant Physiol.*, 2023, **193**, 2260–2277.
24 W. Jiang, *et al.*,Crowding-induced opening of the mechanosensitive Piezo1 channel in silico, *Commun. Biol.*, 2021, **4**, 84.
25 Y. Lin, A. Buyan and B. Corry, Computational studies of Piezo1 yield insights into key lipid–protein interactions, channel activation, and agonist binding, *Biophys. Rev.*, 2022, **14**, 209–219.
26 H. J. Lessen, P. J. Fleming, K. G. Fleming and A. J. Sodt, Building blocks of the outer membrane: Calculating a general elastic energy model for β-barrel membrane proteins, *J. Chem. Theory Comput.*, 2018, **14**, 4487–4497.
27 P. Rassam, *et al.*, Supramolecular assemblies underpin turnover of outer membrane proteins in bacteria, *Nature*, 2015, **523**, 333–336.
28 J. E. Horne, D. J. Brockwell and S. E. Radford, Role of the lipid bilayer in outer membrane protein folding in Gram-negative bacteria, *J. Biol. Chem.*, 2020, **295**, 10340–10367.
29 W. B. Tan and S.-S. Chng, How bacteria establish and maintain outer membrane lipid asymmetry, *Annu. Rev. Microbiol.*, 2024, **78**, 553–573.
30 G. E. Schulz, The structure of bacterial outer membrane proteins, *Biochim. Biophys. Acta, Biomembr.*, 2002, **1565**, 308–317.
31 A. Schüßler, S. Herwig and J. H. Kleinschmidt, Kinetics of Insertion and Folding of Outer Membrane Proteins by Gel Electrophoresis, in *Lipid-Protein Interactions: Methods and Protocols*, ed. J. H. Kleinschmidt, Methods in Molecular Biology, Humana, New York, NY, 2019, vol. 2003, pp. 145–162.
32 G.-W. Li, D. Burkhardt, C. Gross and J. S. Weissman, Quantifying absolute protein synthesis rates reveals principles underlying allocation of cellular resources, *Cell*, 2014, **157**, 624–635.
33 E. R. Rojas, *et al.*, The outer membrane is an essential load-bearing element in Gram-negative bacteria, *Nature*, 2018, **559**, 617–621.
34 T. Surrey and F. Jähnig, Refolding and oriented insertion of a membrane protein into a lipid bilayer, *Proc. Natl. Acad. Sci. U. S. A.*, 1992, **89**, 7457–7461.
35 A. Pautsch and G. E. Schulz, High-resolution structure of the OmpA membrane domain, *J. Mol. Biol.*, 2000, **298**, 273–282.
36 K. K. Andersen, H. Wang and D. E. Otzen, A kinetic analysis of the folding and unfolding of OmpA in urea and guanidinium chloride: single and parallel pathways, *Biochemistry*, 2012, **51**, 8371–8383.
37 P. V. Bulieris, S. Behrens, O. Holst and J. H. Kleinschmidt, Folding and insertion of the outer membrane protein OmpA is assisted by the chaperone Skp and by lipopolysaccharide, *J. Biol. Chem.*, 2003, **278**, 9092–9099.
38 B. Schiffrin, *et al.*, Effects of periplasmic chaperones and membrane thickness on BamA-catalyzed outer-membrane protein folding, *J. Mol. Biol.*, 2017, **429**, 3776–3792.

39 D. Gessmann, *et al.*, Outer membrane β-barrel protein folding is physically controlled by periplasmic lipid head groups and BamA, *Proc. Natl. Acad. Sci. U. S. A.*, 2014, **111**, 5878–5883.

40 N. K. Burgess, T. P. Dao, A. M. Stanley and K. G. Fleming, Beta-barrel proteins that reside in the *Escherichia coli* outer membrane *in vivo* demonstrate varied folding behavior *in vitro*, *J. Biol. Chem.*, 2008, **283**, 26748–26758.

41 E. J. Danoff and K. G. Fleming, Membrane defects accelerate outer membrane β-barrel protein folding, *Biochemistry*, 2015, **54**, 97–99.

42 X. C. Zhang and H. Li, Interplay between the electrostatic membrane potential and conformational changes in membrane proteins, *Protein Sci.*, 2019, **28**, 502–512.

43 I. Levental and E. Lyman, Regulation of membrane protein structure and function by their lipid nano-environment, *Nat. Rev. Mol. Cell Biol.*, 2023, **24**, 107–122.

44 R. J. Clarke, Electrostatic switch mechanisms of membrane protein trafficking and regulation, *Biophys. Rev.*, 2023, **15**, 1967–1985.

45 G. von Heijne, Control of topology and mode of assembly of a polytopic membrane protein by positively charged residues, *Nature*, 1989, **341**, 456–458.

46 J. A. Baker, W.-C. Wong, B. Eisenhaber, J. Warwicker and F. Eisenhaber, Charged residues next to transmembrane regions revisited: "Positive-inside rule" is complemented by the "negative inside depletion/outside enrichment rule", *BMC Biol.*, 2017, **15**, 66.

47 R. Jackups and J. Liang, Interstrand pairing patterns in beta-barrel membrane proteins: the positive-outside rule, aromatic rescue, and strand registration prediction, *J. Mol. Biol.*, 2005, **354**, 979–993.

48 J. S. G. Slusky and R. L. Dunbrack, Charge asymmetry in the proteins of the outer membrane, *Bioinformatics*, 2013, **29**, 2122–2128.

49 L. An, *et al.*, Binding and sensing diverse small molecules using shape-complementary pseudocycles, *Science*, 2024, **385**, 276–282.

50 S. Berhanu, *et al.*, Sculpting conducting nanopore size and shape through de novo protein design, *Science*, 2024, **385**, 282–288.

51 A. Dorey and S. Howorka, Nanopore DNA sequencing technologies and their applications towards single-molecule proteomics, *Nat. Chem.*, 2024, **16**, 314–334.

52 A. Arora, F. Abildgaard, J. H. Bushweller and L. K. Tamm, Structure of outer membrane protein A transmembrane domain by NMR spectroscopy, *Nat. Struct. Biol.*, 2001, **8**, 334–338.

53 H. Ishida, A. Garcia-Herrero and H. J. Vogel, The periplasmic domain of *Escherichia coli* outer membrane protein A can undergo a localized temperature dependent structural transition, *Biochim. Biophys. Acta*, 2014, **1838**, 3014–3024.

54 S. Jo, T. Kim, V. G. Iyer and W. Im, CHARMM-GUI: A web-based graphical user interface for CHARMM, *J. Comput. Chem.*, 2008, **29**, 1859–1865.

55 J. Lee, *et al.*, CHARMM-GUI Input Generator for NAMD, GROMACS, AMBER, OpenMM, and CHARMM/OpenMM Simulations Using the CHARMM36 Additive Force Field, *J. Chem. Theory Comput.*, 2016, **12**, 405–413.

56 D. Van Der Spoel, *et al.*, GROMACS: fast, flexible, and free, *J. Comput. Chem.*, 2005, **26**, 1701–1718.

57 E. Jurrus, *et al.*, Improvements to the APBS biomolecular solvation software suite, *Protein Sci.*, 2018, **27**, 112–128.
58 G. E. P. Box and S. L. Andersen, Permutation theory in the derivation of robust criteria and the study of departures from assumption, *J. R. Stat. Soc. Ser. B Methodol.*, 1955, **17**, 1–26.
59 S. Hussain and H. D. Bernstein, The Bam complex catalyzes efficient insertion of bacterial outer membrane proteins into membrane vesicles of variable lipid composition, *J. Biol. Chem.*, 2018, **293**, 2959–2973.
60 S. G. J. Smith, V. Mahon, M. A. Lambert and R. P. Fagan, A molecular Swiss army knife: OmpA structure, function and expression, *FEMS Microbiol. Lett.*, 2007, **273**, 1–11.
61 P. White, *et al.*, The role of membrane destabilisation and protein dynamics in BAM catalysed OMP folding, *Nat. Commun.*, 2021, **12**, 4174.
62 A. L. Lomize, S. C. Todd and I. D. Pogozheva, Spatial arrangement of proteins in planar and curved membranes by PPM 3.0, *Protein Sci.*, 2022, **31**, 209–220.
63 M. W. Franklin, J. J. Stevens, J. Krise and J. S. G. Slusky, The Extracellular Loops of OmpA Control the Slow Rate of *In Vitro* Folding, *bioRxiv*, 2021, preprint, DOI: 10.1101/2020.10.08.331546.
64 R. Koebnik, Structural and functional roles of the surface-exposed loops of the beta-barrel membrane protein OmpA from *Escherichia coli*, *J. Bacteriol.*, 1999, **181**, 3688–3694.

Faraday Discussions

ROYAL SOCIETY OF CHEMISTRY

PAPER

Asymmetric membrane properties through a protein lens†

Joseph H. Lorent,*[a] Angela Cabrera-Jojoa,[bc] Kandice R. Levental,[d] Ilya Levental[d] and Edward Lyman[be]

Received 15th December 2024, Accepted 3rd February 2025
DOI: 10.1039/d4fd00199k

Plasma membranes are asymmetric, with each monolayer presenting specific lipid compositions and biophysical properties. Transmembrane domains (TMDs) of single-pass transmembrane proteins (spTMPs) have adapted their physico-chemical properties to these asymmetric constraints. In this study, we analysed the structural features of such TMDs across the tree of life to obtain information about their interaction with asymmetric membrane bilayers and predict species-specific membrane properties. We observed that TMDs in the plasma membranes of all eukaryotic species possess asymmetries in lipid accessible surface area (ASA), hydrophobicity, aromaticity and charge. Bacteria deviate from this trend, with strong differences between bacterial clades. Notably, TMDs in the Golgi and the endoplasmic reticulum of eukaryotic species display inverted profiles for accessible surface area, hydrophobicity and aromaticity compared to their plasma-membrane counterparts. To determine how well TMD profiles reflect average membrane properties, we performed molecular dynamics simulations of a spTMP in an asymmetric lipid bilayer whose composition approximates the human plasma membrane. The simulated spTMP was chosen to represent the average TMD properties of the human proteome. We compared the electron density profiles of the simulated asymmetric membrane to the average TMD profiles derived from the human proteome and observed that phospholipid acyl-chain density overlapped very well with TMD hydrophobicity, and phosphate group density with TMD charge. The profiles of phospholipid unsaturation in the acyl chains overlapped well with the average location of TMD phenylalanines in the cytoplasmic leaflet, while there was additional accumulation of large hydrophobic and aromatic residues in the membrane midplane, which had low acyl-chain density. This study reveals the complementarity of membrane and TMD properties in asymmetric membranes,

[a]Cellular and Molecular Pharmacology (FACM), Louvain Drug Research Institute, UCLouvain, Avenue Mounier 73/1, B-1200 Brussels, Belgium. E-mail: joseph.lorent@uclouvain.be
[b]Department of Physics and Astronomy, University of Delaware, Newark, DE, USA
[c]Biophysics Group, Department of Physics, Universidad de los Andes, Bogotá, Colombia
[d]Molecular Physiology and Biophysics, University of Virginia School of Medicine, Charlottesville, VA, USA
[e]Department of Physics and Astronomy, University of Delaware, Newark, DE, USA
† Electronic supplementary information (ESI) available. See DOI: https://doi.org/10.1039/d4fd00199k

suggesting that the properties of TMDs can be used to make predictions about the properties of their solvating membranes.

Introduction

Cellular membranes have evolved since the early stages of life on earth and likely have adapted to environmental pressures by changing their chemical compositions and resulting biophysical properties.[1–3] To be an effective barrier, membranes must be chemically stable under diverse environmental conditions, with varying temperature, pH, pressure, salinity and oxidizing conditions. For example, the membranes of archaeal extremophiles are composed of ether-linked isoprenoids that are chemically more stable than glycerophospholipids.[3,4] In addition to chemical stability, other critical membrane features include permeability towards protons, ions or other substances,[2] lipid packing and fluidity,[5,6] formation of lateral domains,[7] surface charge[8] and deformability.[9,10] All of these properties must strike a balance that allows for proper membrane functions, such as protection against xenobiotics, maintenance of a membrane potential, proper diffusion of lipids and proteins, endo-/exocytosis, and serving as an efficient platform for membrane protein function.[8]

Membrane proteins constitute 20–70% of the total membrane mass[11] and *ca.* 30% of the genome encodes α-helical membrane proteins in all three domains of life.[12] Their membrane embedded helices or transmembrane domains (TMDs) mirror the physical properties of the membrane, because lipidomes and TMDs have evolved together.[13–18] This is especially true for single-pass transmembrane helices, which are fully solvated by the lipid matrix.[18–20] Because α-helical TMDs are easily predicted from primary protein sequences, many putative TMDs can be analyzed and average TMD properties can be compared to the properties of their solvating membranes.[13,21] A well-known example is the matching of TMD length to membrane thickness, with long TMDs preferring thicker membranes because hydrophobic mismatch is energetically unfavorable.[18,21–23] A more recent discovery is that TMDs with bulky hydrophobic side chains tend to accumulate where hydrocarbon chains are less tightly packed.[18,20] This correlation has been experimentally assessed in phase-separated Giant Plasma-Membrane Vesicles (GPMVs), in which bulkier TMDs partitioned to the less-packed liquid disordered (L_d) phase.[20] In the same context, we recently showed that differences between the lipid packing of exoplasmic *versus* cytoplasmic monolayers of the mammalian plasma membrane, which result from the asymmetric distribution of acyl-chain unsaturation and probably cholesterol, correlated with asymmetrically shaped TMDs.[5] Asymmetric shapes of TMDs in various eukaryote plasma membranes had been previously described and suggested to reflect differences in lipid packing between the two monolayers.[18] In addition to packing, Pogozheva *et al.* measured detailed polarity profiles from TMDs with respect to both leaflets.[17,21] In their work, the average TMD profiles of different evolutionary kingdoms from molecular dynamics (MD) simulations were compared to the simulated profiles of a symmetric DOPC membrane.[17]

There is likely great diversity in the composition and biophysical properties of plasma membranes of various species, though these details require major experimental efforts to decipher. In contrast, the full complements of TMDs

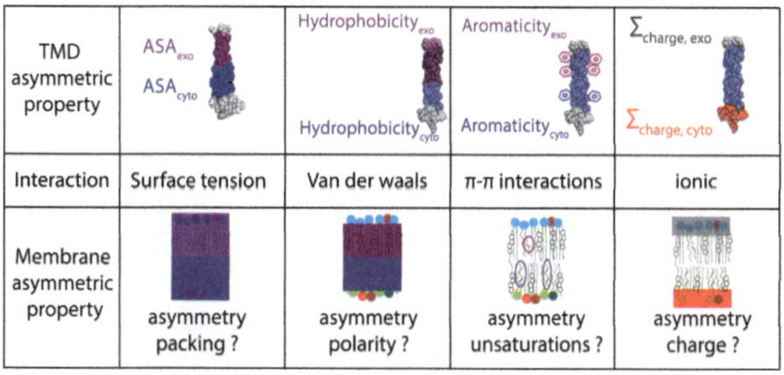

Fig. 1 Asymmetric transmembrane domain (TMD) properties of single-pass transmembrane proteins and their putative interactions with asymmetric membrane properties. ASA = accessible surface area of TMD amino acid residues. Exo/endo indicates the exoplasmic or cytoplasmic side of the membrane, respectively.

across proteomes are easily inferred from genomes. Thus, if membrane properties can be predicted from the properties of their resident TMDs (Fig. 1),[13,15–17] membrane biophysical properties across taxa can be inferred from bioinformatic analysis of membrane proteins.

To investigate this possibility, we first determined the distribution of selected TMD properties across the leaflets of plasma membranes of various evolutionary taxa. In Eukaryotae, we further investigated the asymmetric properties of TMDs in the Golgi and the ER. We further determined the TMD sequence in humans that best represented the various average structure profiles. This transmembrane protein was simulated in a state-of-the-art asymmetric plasma-membrane model to evaluate the alignment of its properties with membrane chemical profiles.[5] From this analysis, we inferred how well TMD profiles match membrane properties and if these properties can be used to infer membrane asymmetry in other species.

Results and discussion

Plasma-membrane TMD asymmetry varies between eukaryotic and prokaryotic species. Intracellular TMDs are different from plasma-membrane TMDs

To evaluate asymmetric TMDs in terms of spTMP properties across several taxa, we compiled sets of single-pass TMDs annotated as plasma-membrane proteins by the Uniprot data server (see Methods). We calculated the accessible surface area (ASA), hydrophobicity, number of aromatic residues, and number of charged residues, and report these distributions along the membrane normal (Fig. 2).

Across eukaryotic taxa, we observed that ASA was asymmetrically distributed in TMDs. Amino acids with smaller side chains were over-represented in the half of the TMD that faces the exoplasmic compared to the cytoplasmic leaflet, leading to an smaller overall surface area of the TMD facing the exoplasmic leaflet (Fig. 2A). This surface-area asymmetry was markedly different in prokaryote TMDs, with Pseudomonadota and Terrabacteria having symmetric TMDs on average, while the PVC (superphylum of Planctomycetota, Verrucomicrobiota, and Chlamydiota) and FCB (superphylum of Fibrobacterota, Chlorobiota, and Bacteroidota) groups had larger exoplasmic TMD surface areas. These groups also displayed bi- or

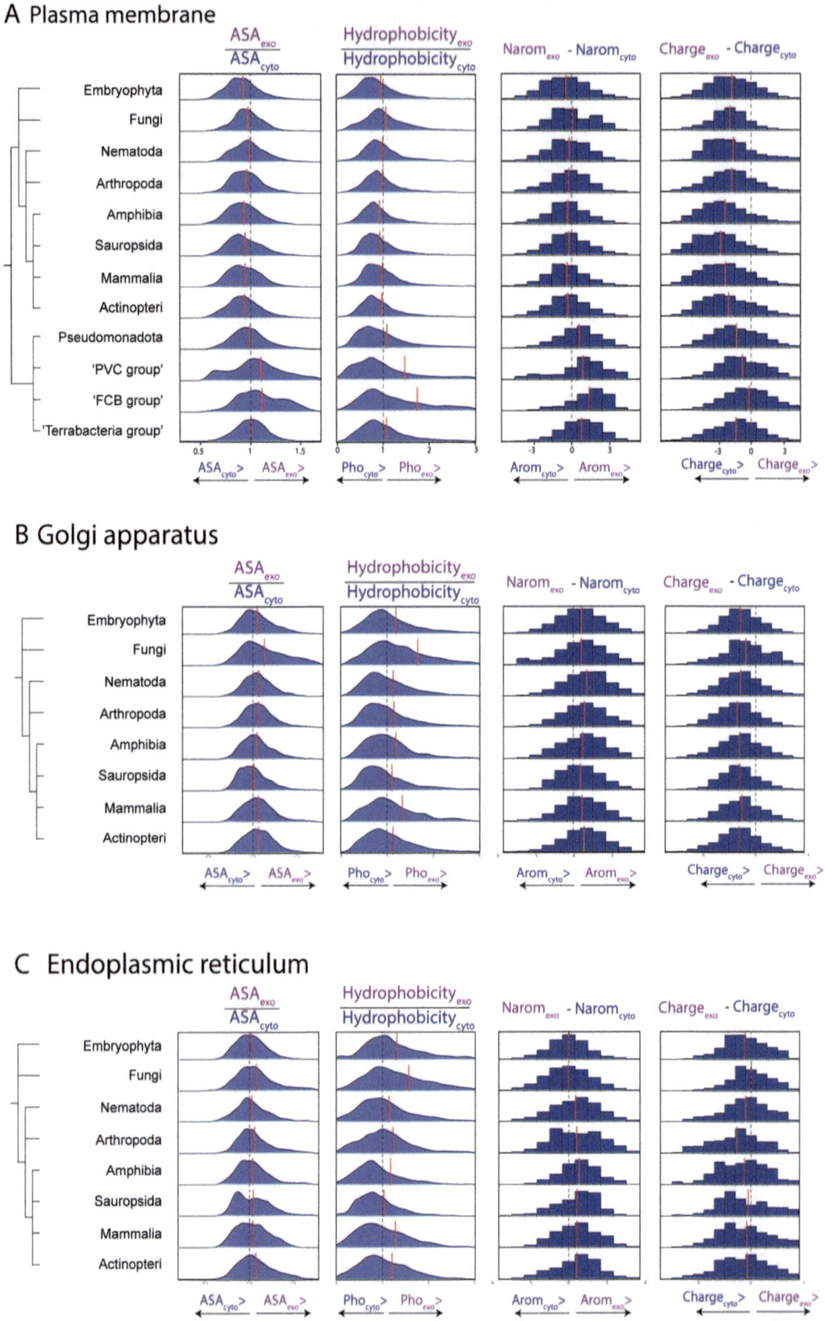

Fig. 2 Asymmetries of transmembrane domain (TMD) properties, namely accessible surface area (ASA), hydrophobicity, aromatic residues and charge, expressed as the exoplasmic relative to cytoplasmic parts of the TMD in single-pass transmembrane proteins across species and organelles. The dotted line reflects the symmetry of the respective property and the red continuous line the average of the TMD distribution. Deviation of the solid line from the dotted line to the left indicates higher values for the given parameter on the cytoplasmic side of the TMD membrane. (A) Plasma membrane, (B) Golgi and (C) endoplasmic reticulum.

trimodal distributions in TMD ASA asymmetry, in contrast to eukaryotes, which tended to approximate unimodal distributions.

In eukaryotes, hydrophobicity was similar in both TMD halves, although in all taxa the distribution of TMDs was asymmetric, with the mode having a more hydrophobic cytoplasmic half. In prokaryotes and especially the PVC and FCB groups, we observe a long tail in the histograms that strongly influences the average, reflecting TMDs that had more hydrophobic residues in the cytoplasmic half. Aromatic residues are enriched at the cytoplasmic side of TMDs in eukaryotes, a trend that is reversed for prokaryotes. Positive charge is always higher on the cytoplasmic side of the TMD (Fig. 2A–C), confirming the positive-inside rule across all taxa.[14,15]

In the Golgi (Fig. 2B) and the ER (Fig. 2C) of eukaryotes, the TMD asymmetries are reversed compared to their plasma-membrane counterparts. Namely, TMDs were slightly larger in the exoplasmic leaflet than the cytoplasmic half. This is also true for aromatic residues, which are enriched on the luminal side of the membrane. Positive charge at the cytoplasmic side of the TMDs generally increased from the ER, where it was barely positive, through the Golgi and to the plasma membrane.

To assess whether similarity in TMD properties follows relatedness of taxa, we performed principle component analysis (PCA) of the average exo- and cytoplasmic TMD properties. Upon projection along the two largest eigenvectors, the taxa fall naturally into evolutionary groups (Fig. 3A). Vertebrata were distinguished by higher ASA, charge and hydrophobicity on the cytoplasmic side and lower ASA and aromatic residues on the exoplasmic side. Fungi discern themselves through positive charges and a higher hydrophobicity at the exoplasmic side, while Embryophyta are very different regarding these properties. Bacteria also form a cluster separate from the other groups, but sample a broader range in the "asymmetry space", perhaps because of substantial variation in membrane properties between bacteria. However, bacteria do tend to have larger and more aromatic residues on the exoplasmic side but less charge, with smaller and less aromatic residues at the cytoplasmic side.

We also compared, *via* PCA, the asymmetric TMD properties of plasma, Golgi and ER membranes in eukaryotes (Fig. 3B). The largest differences in asymmetric properties occur between the plasma membrane (PM) and intracellular membranes (Golgi, ER). These differences were mostly defined through higher hydrophobicity but less aromaticity of the whole TMD, an increase in charge at the cytoplasmic side, and a decrease in the ASA at the exoplasmic side towards the plasma membrane.

Property profiles vary strongly across taxonomic clades

In Fig. 2A, the property profiles are broadly similar across the vertebrate taxa (Mammalia, Sauropsida, Amphibia and Actinopteri). To compare the vertebrates to the other groups, and gain insights into the structural origin of asymmetric TMD properties, we analyzed the property profiles in plasma-membrane TMDs of the main taxa and intracellular membranes of vertebrata, regrouping Mammalia, Sauropsida, Amphibia and Actinopteri into a single vertebrate group (Fig. 4). We observe two maxima in the average residue size (*i.e.*, surface area and ASA) that correspond roughly to the positions of the TMD residues at the membrane

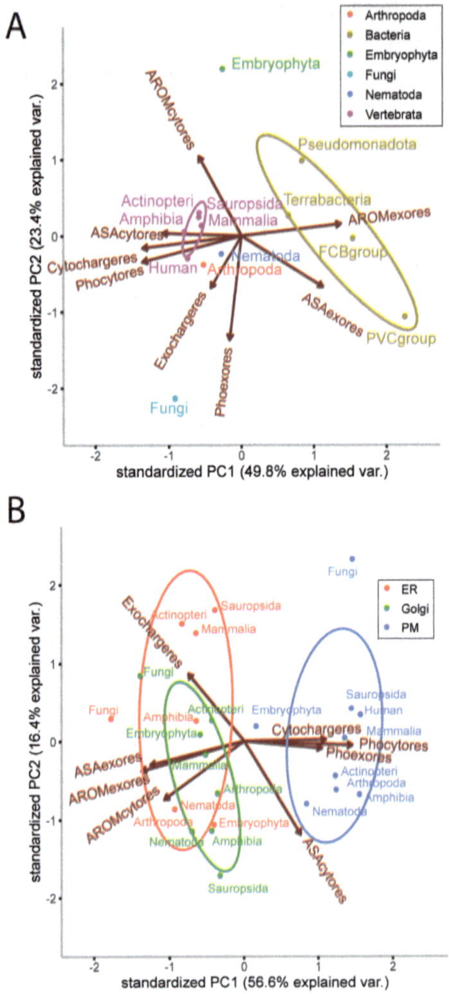

Fig. 3 Principal component analysis (PCA) of average exoplasmic (exo) and cytoplasmic (cyto) TMD properties, comparing different taxa that are grouped into evolutionary clades (A). Similar PCA comparing spTMPs resident in the plasma membrane *versus* the Golgi or the endoplasmic reticulum (ER) (B). Arrows represent the directions along which a property increases. Properties include the accessible surface area (ASA), aromaticity (AROM), hydrophobicity (Pho) and charge of the exo- or cytoplasmic side of the TMD.

interface, ~1.5 nm from the TMD center for all species. A third maximum appeared close to the membrane mid-plane in the plasma membranes of all eukaryotes.

As noted above, bacterial taxa had either symmetric overall TMD profiles or opposite ASA asymmetry compared to vertebrates (Fig. 2A and 4). In bacteria, aromaticity is strong at the membrane interface, especially in the FCB and PVC groups. Further, there is a tendency to accumulate aromatic residues at the exoplasmic TMD side. For hydrophobicity, we do not observe large asymmetries in eukaryote plasma-membrane TMDs, but do for bacteria. The average profile for vertebrates had slightly more hydrophobic residues at the membrane mid-plane. This enrichment of more hydrophobic residues in the midplane seems to be

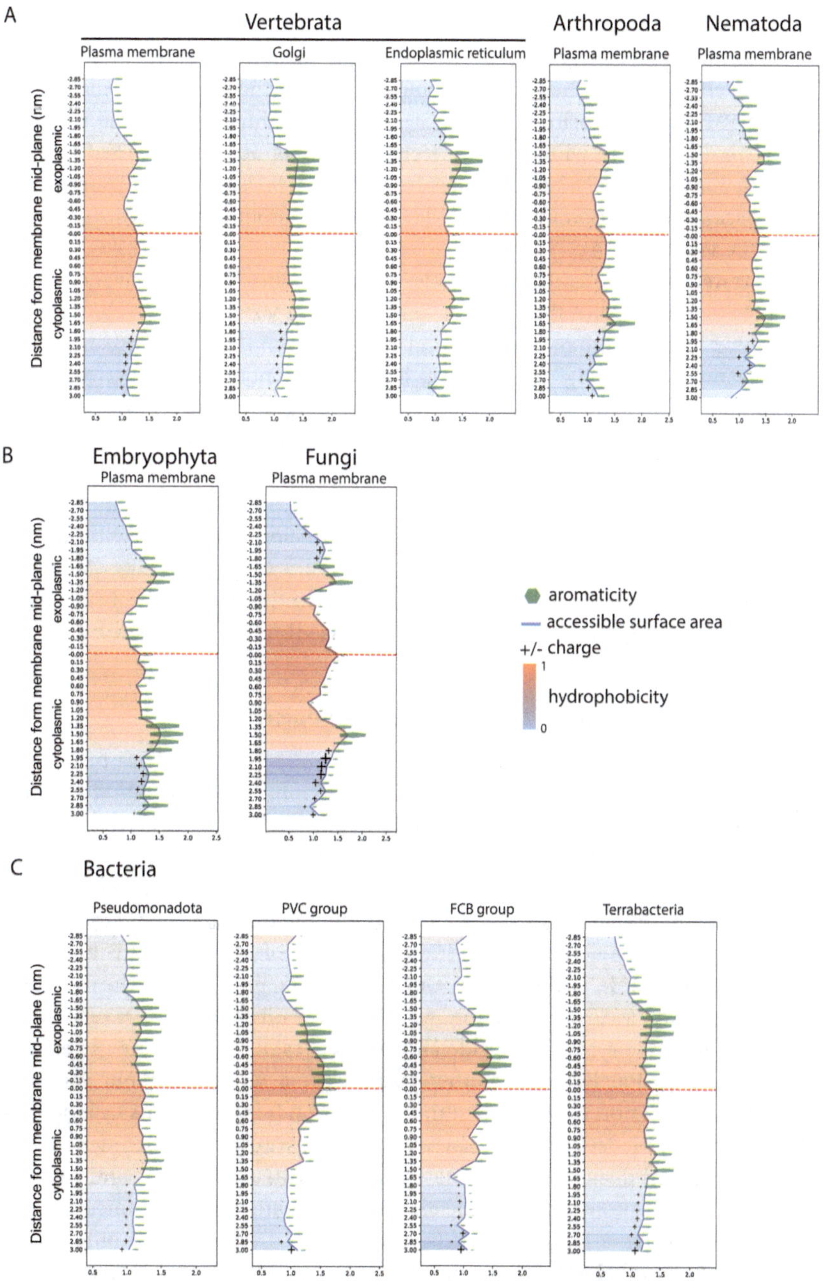

Fig. 4 Profiles of normalized average TMD properties in animals (A), plants and fungi (B), and bacteria (C), from the exoplasmic/luminal (negative values) to cytoplasmic (positive values) side. The red dashed line corresponds to the membrane midplane. Aromaticity is represented as green hexagons, the accessible surface area as vertical plots, hydrophobicity as color heatmaps and charge as ± signs of different font sizes. The zero value of the profile describes the TMD at the membrane mid-plane.

conserved in all species, although is most pronounced in Fungi (Fig. 4B). In bacteria (Fig. 4C), and specifically the FCB and PVC group, the exoplasmic TMD portion is enriched with hydrophobic residues.

Regarding the charge profiles in the plasma-membrane TMDs of all species, positive charge is enriched at the cytoplasmic side of the membrane, just outside of the TMD. These positive charges are thought to direct membrane insertion, likely by interacting with negatively charged lipids at the inner leaflet.[19] The "positive-inside" rule[14,15] also applies to bacterial TMDs, although charge asymmetry seems less important specifically for the PVC group. This was also observed with the PCA (Fig. 3A).

In vertebrates, the TMDs of the Golgi apparatus and the ER have very different profiles compared to the plasma membrane. TMDs in these intracellular membranes have rather large luminal residues compared to those facing the cytoplasmic leaflet (Fig. 2A and 3B). Here, hydrophobicity is stronger in the membrane mid-plane and aromaticity has its largest positive peak at the exoplasmic interface. The density of positive charges becomes stronger at the cytoplasmic side upon advancement of the exocytic pathway from the ER to the Golgi and finally the plasma membrane. As predicted by the PCA, it becomes obvious that asymmetric TMD features for vertebrates develop progressively through the exocytic pathway to the plasma membrane (*i.e.*, ER < Golgi < PM, Fig. 4A). If TMD features reflect membrane physical properties, we infer that membrane asymmetries in charge, packing, and polarity increase through the exocytic pathway.

How do specific amino acids enforce TMD asymmetry?

To look deeper into the observed asymmetric properties of TMDs in terms of species and localization, we investigated the asymmetric distribution of amino acids (Fig. 5) and their individual profiles (Fig. S1†). The asymmetry for various amino acids varies strongly across taxa and cellular localization. To specifically investigate how the asymmetric leaflet distribution of amino acids might lead to an asymmetry of the TMD surface area between the exoplasmic and cytoplasmic halves, we correlated the asymmetric enrichment of residues with their corresponding accessible surface area. Only plasma membranes of the vertebrate group displayed a clear correlation, with larger residues having the tendency to accumulate at the cytoplasmic side and smaller residues preferring the exoplasmic leaflet. Curiously, neither other taxa's plasma membranes nor vertebrate intracellular membranes displayed this correlation (Fig. S2†). This lack of correlation between the ASA of individual residues and asymmetric enrichment in both leaflets (Fig. 5A) does not necessarily mean that TMDs do not generally present ASA asymmetry, shown for multiple taxa in Fig. 2. Rather, the asymmetry in non-vertebrate species arises through the asymmetric distribution of only specific residues. By more closely inspecting the amino acid profiles in Fig. S1† for the PVC and FCB-group, it seemed that specifically phenylalanine, which has a large aromatic side chain, is present at high frequencies at the exoplasmic leaflet, while larger aliphatic hydrophobic residues (I, L, V) were not strongly biased to one side of the membrane (Fig. S2†).

By closely inspecting aromatic residues (Fig. 5 and S2,† green letters), we observed that the frequency of histidine was very weak and probably did not influence the aromatic profiles observed in Fig. 4. Rather, TMDs in the plasma

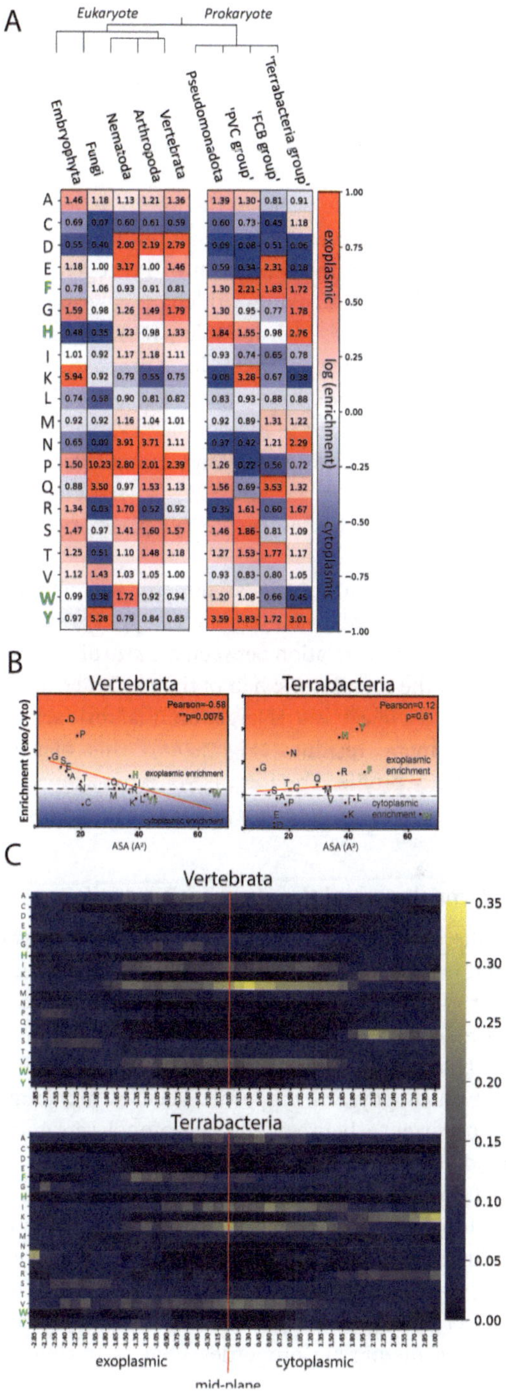

Fig. 5 Asymmetric distribution of amino acids in TMDs. (A) Values inside the heatmap boxes display the ratio of the number of amino acids found at the exoplasmic *vs.* cytoplasmic side, while the colormap reflects the ratio in the natural log (see scale). (B) TMD leaflet enrichment of amino acids against their accessible surface area (ASA) in vertebrata and Terrabacteria. Aromatic residues are displayed in green. See other taxa in (Fig. S2†). (C)

membranes of metazoans (animals) accumulated phenylalanine, tryptophan (except in the case of Nematoda) and tyrosine at the cytoplasmic leaflet, while this was reversed in bacteria, which accumulated phenylalanine and tyrosine at the exoplasmic leaflet, with Trp distribution varying across bacterial taxa (Fig. 5A and S2†). In all membranes, except in the PVC and FCB group, tryptophan and tyrosine clearly accumulated at the membrane border (Fig. 5C and S2†). This localization produces aromatic profile peaks at the interface (Fig. 4, green hexagons). Tyrosine and tryptophan residues have been shown to interact with the C=O residues of phospholipids *via* the formation of hydrogen bonds[17] while the π electrons can interact through stacking with charged headgroups. Further, tryptophans are amphiphiles whose hydrophobic part favors membrane insertion, while the large and flat character of Trp has been proposed to have an anchoring effect in the membrane *via* steric hindrance.[24] Phenylalanine not only accumulated at the interface, but also at the membrane mid-plane with a slight preference for the cytoplasmic side in vertebrata and a rather strong preference for the exoplasmic side in bacteria (Fig. 5C and S2†). As expected, positively charged residues (R, K) accumulated largely in the cytoplasm right outside the TMD (Fig. 5C and S2†).

In Golgi and ER membranes, ASA asymmetry was either lost or reversed compared to their respective plasma membranes (Fig. 2). Similar to bacteria, this is coincident with a loss in correlation between the size of individual residues and their accumulation at the cytoplasmic half of the TMD (Fig. S2A†). Also similar to bacterial membranes is a bias towards larger luminal residues that arise from the accumulation of aromatic residues (Fig. S2A†). This bias obviously further impacts the aromaticity profile. The increase of cytoplasmic charge outside the TMD arises from an increase in frequency of arginine and lysine residues.

Asymmetric TMD enrichment of the dimerization motif GxxxG

We investigated the asymmetric distribution of the dimerization motif GxxxG[25] (Fig. 6). This motif is believed to mediate TMD dimerization *via* maximization of interfacial van der Waals interactions and direct interaction of small residues of both helices *via* CαH···O=C hydrogen bonds.[26] Dimerization mediated by GxxxG has been shown to depend on membrane lipid composition, the presence of cholesterol,[27] and on biophysical properties such as membrane thickness.[28] A difference in its occurrence between the two monolayers might thus indicate differences in lipid composition and monolayer properties.

In all eukaryotic species and Pseudomonadota, the GxxxG motif is strongly (~2-fold) enriched in the exoplasmic leaflet of the plasma membrane, while it is more uniformly distributed in other bacterial taxa (Fig. 6). This is similar in the Golgi and the ER, where the asymmetric distribution of the GxxxG motif is lost mostly due to fewer GxxxG motifs at the exoplasmic side, consistent with an earlier report.[18] Since membrane properties such as thickness and cholesterol

Average amino acid profiles for the transmembrane domain plus 7 flanking residues from the exo- to cytoplasmic part. The heatmap represents the occurrence of the respective amino acid at the corresponding position. A value of one for a specific amino acid would mean that only this amino acid is present at the corresponding position. Other taxa are displayed in Fig. S1.†

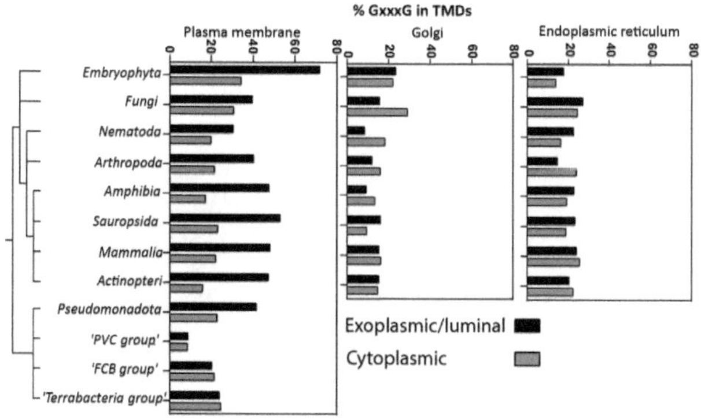

Fig. 6 Asymmetric distribution of the GxxxG dimerization motif in the plasma membrane, Golgi and endoplasmic reticulum of different taxa. Displayed are percentages of sequences that present the respective motif in the exo- (black bars) or cytoplasmic (grey bars) side of the TMD.

have a positive influence on the dimerization function of GxxxG, its higher occurrence on the outer monolayer might hence be a good indicator for membrane asymmetry in sterols or lipid packing in these taxa.

Simulation of an average human TMD in a representative asymmetric bilayer to compare membrane and TMD profile properties

To assess how the chemistry and structure of a typical TMD are matched to its membrane environment, we simulated a human plasma-membrane TMD in a state-of-the-art asymmetric membrane model.[29,30] We selected the TMD of the human erythroid membrane-associated protein (ERMAP) because among all human plasma-membrane TMDs, it was the most representative of the investigated average TMD properties, as assessed *via* three different methods (see Material and methods).

The left panel in Fig. 7A shows the final snapshot of the 150 ns simulation, after the structures of the membrane and the TMD have relaxed. Residues K178 and V156 are predicted using TMHMM[31] to mark the boundary of the TMD (Fig. S3D†). K178 and a group of two other lysines and W177 anchor the intracellular side of the TMD to the interface, but V156 appears to be inside the interface, at the boundary of the hydrocarbon. This is apparent from the left panel, which compares the average position of the fiducial residues to the distribution of selected chemical groups; note that V156 sits at the edge of the hydrocarbon core, indicated by the red and green distributions in the right panel. L168 marks the center of the bilayer, which does not coincide with the center of mass of the lipids (0 on the horizontal axis)—the exoplasmic leaflet (negative z-coordinate) is rich in sphingolipids and cholesterol and is therefore thicker than the PUFA (polyunsaturated fatty acids)-rich cytoplasmic leaflet.

Correlation of asymmetric membrane profiles from MD simulations with average TMD and amino acid profiles

We examined how well the chemical profiles extracted from the simulation of an asymmetric human membrane model correlated to average TMD property profiles

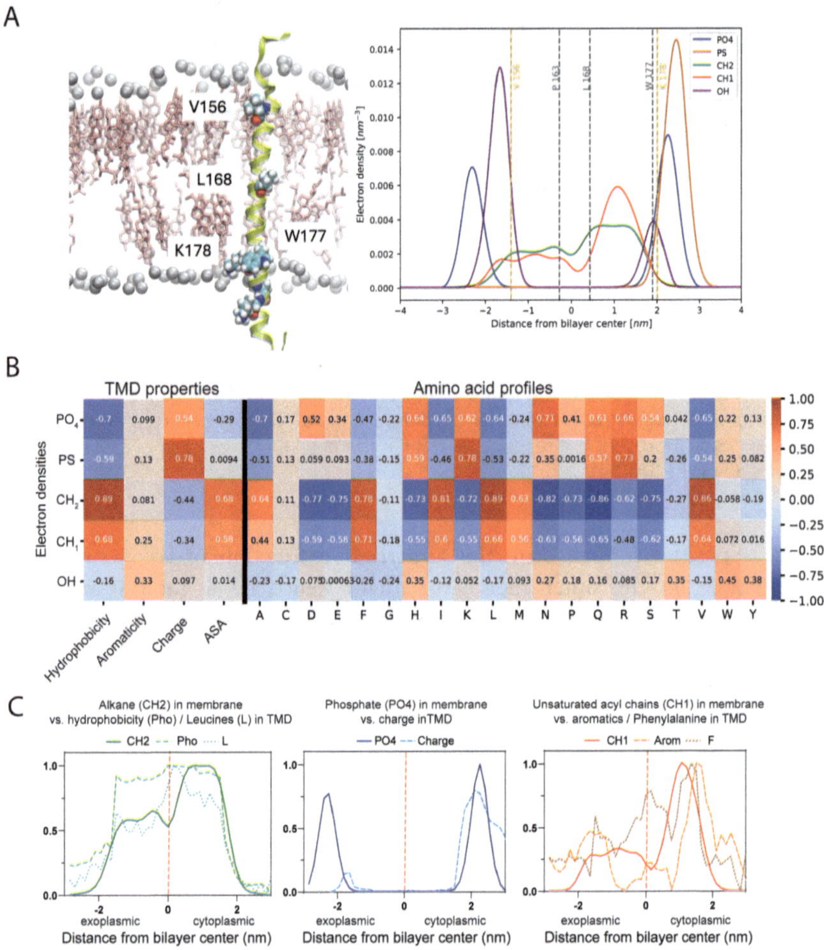

Fig. 7 (A) (Left panel) Final configuration of a 150 ns simulation of the ERMAP TMD in a human asymmetric plasma-membrane model. The lipid phosphorus atoms are shown as gray spheres, cholesterols are in pink, and the TMD backbone shown as a yellow ribbon. Several "fiducial" residues are rendered with a space-filling model and labelled. The exoplasmic leaflet is at the top. (Right panel) The density distribution of selected atomic groups. OH is the hydroxyl of cholesterol, CH_1 is the alkene groups at double bonds in the hydrocarbon chains, CH_2 is the alkane groups of the hydrocarbon chains, PO4 is the phosphate groups, and PS is the serine moiety on the PS lipids. The average locations of the fiducial residues are indicated with the vertical dashed lines. Positive values on the horizontal axis indicate the cytoplasmic side. (B) Cross-correlations of simulated membrane chemical profiles as shown in (A) *vs.* human plasma-membrane TMD property profiles as shown in Fig. 4. (C) Superposition of membrane chemical profiles from MD simulation (continuous line), TMD profiles from average human plasma-membrane TMD properties (long dashes), and selected amino acid profiles (short dashes).

from the human plasma-membrane proteome (Fig. 7B and S3†). As expected, we observe that the normalized acyl-chain $-CH_2-$ profile correlated quite well with the hydrophobicity profile obtained from average TMDs (Fig. 7B). We therefore superposed both normalized profiles, which shows that both profiles align to both membrane edges, where there is a sharp decrease in hydrophobicity towards the membrane headgroups (Fig. 7C). In terms of amino acids, the strongest correlation with the $-CH_2-$ profile was observed for leucine (Fig. 7B and C) and other hydrophobic residues. Regarding correlation of charge, we observed that the density of negatively charged phosphates correlated well with positive TMD charge (Fig. 7B and C). For aromatic residues, the correlation plot reveals that phenylalanine correlates best to the $-CH_1-$ density profile at the cytoplasmic leaflet. This indicates that delocalized π-systems from lipid unsaturation could potentially superpose with benzyl groups from phenylalanine (Fig. 7C), although benzyl groups also accumulate at the membrane interface and in the mid-plane, probably serving as TMD anchors (Fig. S3†).

Conclusions

Prior work has established lipidomic asymmetry as a membrane feature that is broadly distributed across the domains of life.[5,15,32] Here, we assembled a dataset of single-pass TMDs from across bacterial and eukaryote taxa to perform statistical analyses of several TMD chemical features, including charge, hydrophobicity, aromatic chemistry, and surface area. Enabled by algorithms that predict transmembrane segments and topology from sequence data, we resolved the distribution of these features along the direction normal to the membrane to infer membrane structure and chemistry from membrane proteomes. This approach permits much broader coverage across many different taxa than is possible with experimental approaches like lipidomics.

We find ample evidence for asymmetry in membrane structure and chemistry. The plasma membranes of all studied eukaryotic phyla and three of four bacterial groups follow the positive-inside rule, indicating lipidomic asymmetry with negatively charged headgroups in the cytoplasmic leaflet. Surface-area asymmetry is most significant for the vertebrates (Fig. 3A), but the distributions of surface area along the normal (Fig. 4A) reveal substantial differences in TMD shape when studied at finer resolution. These differences might reflect differences in lateral stress among the different lipidomes. Comparison to a simulation of an "average" TMD in an asymmetric plasma-membrane model suggests that differences in thickness between the two leaflets might be inferred from the surface-area distribution, which in most of the eukaryotes displays a small peak at the termini of the chains from the two leaflets, and is displaced towards the cytoplasmic side (Fig. 4). The clear exception appears to be the fungi, whose TMD property distributions in Fig. 4 are substantially more symmetric than the other cases.

At a finer level of chemical detail, the amino-acid distributions also reflect membrane asymmetry, again in ways that differ across the different phyla (Fig. 5). All cases follow the well-known "rule" that aromatics are enriched at the interface, but in several cases (Embryophyta, all bacterial groups, and in the Golgi and ER) the aromatic enrichment favors one leaflet *versus* the other.

Finally, we considered the influence of membrane asymmetry on function, in the context of dimerization reactions driven by the GxxxG motif. This motif is clearly enriched in the exoplasmic leaflet in vertebrates, which also present the strongest asymmetry in overall TMD ASA. The ASA asymmetry reflects tighter lipid packing in the exoplasmic leaflet, suggesting that dimerization by the GxxxG motif is favored by increased lipid packing. The asymmetric distribution of GxxxG therefore illustrates one way that asymmetry in membrane biophysical properties couples to protein function.

With the recent breakthroughs in protein structure prediction, the approaches developed here can readily be extended to multipass TMDs. This might reveal how the "rules" for single-pass and multipass TMDs differ, and might reveal new TMD features that permit the inference of membrane biophysical properties directly from protein structure. More lipidomic data regarding the asymmetric distribution of lipids from other species would be of enormous value for benchmarking such inferences. Nevertheless, the present study advances our understanding of membrane asymmetric properties across evolutionary clades and highlights the diversity of possible membrane and TMD properties.

Experimental

Bioinformatics analysis of TMDs

Annotated transmembrane domains of single-pass transmembrane proteins from various species were obtained from the Uniprot (Swissprot/TrEMBL) database. Only datasets containing at least 40 different proteins were considered. We first eliminated doubles and TMDs with unknown residues. We further aligned all TMDs from exoplasmic to cytoplasmic according to the type of single-pass transmembrane protein (I, II, III and IV). TMDs that did not possess this information were discarded. TMDs were further aligned to the central TMD amino acid to be sure that the TMD property profiles were well-aligned to the membrane mid-plane as was proposed elsewhere.[19] The length of the TMD was estimated from its number of predicted or established residues as annotated in the Uniprot database, and the accessible surface-area profile of the residues was determined as recently described.[5,20] Python 3.11 was used to extract and align all TMD sequences from Uniprot XML files. The file containing all TMDs is found in the ESI datafile.†

Determination of global TMD asymmetry in terms of accessible surface area, hydrophobicity, aromaticity and charge

To compare exoplasmic *vs.* cytoplasmic side of the TMD, we divided the mid-plane aligned sequences into two. For TMDs with odd numbers of residues, half of the central amino acid was attributed as cytoplasmic and the other half exoplasmic. The total exoplasmic and cytoplasmic surface areas were estimated by summing the surface areas of residues in the exo- *vs.* cytoplasmic halves. Aromaticity was estimated *via* the sum of aromatic residues at the exo/luminal and cytoplasmic sides of the TMD (Trp, Tyr, Phe and His) while hydrophobicity was estimated *via* the sum of each residue according to the hydrophobicity scale of Kyte–Doolittle.[33] Curiously, the calculus of asymmetric hydrophobicity produced some extreme outliers (>20 × S. D. from the average), which were ignored. Around 0.4% of the TMDs were therefore discarded. Charges were analyzed within the TMD and on seven amino acids flanking the TMD because these residues influence protein topology and interact with lipid headgroups.[34] R was used to construct histograms, for principal component analysis and for correlation analysis of membrane properties.

Creation of TMD profiles based on average properties

To determine TMD profiles using bioinformatics methods, we used the same scales as described in the previous paragraph. We assumed a distance of 0.15 nm in between amino acids along the α-helix. We determined average profiles from the sequences aligned to the middle of the membrane plane. We first transformed each sequence into its corresponding properties (ASA, hydrophobicity, aromaticity and charge) for each amino acid and then calculated the average of this property per position. From these property vectors, we created a representative figure in which the ASA is represented by a plot, the color of the area under the curve corresponds to a heatmap that reflects hydrophobicity, aromaticity is displayed by the size of green hexagons and charges are represented as "+" and "−" signs according to their average positive or negative charge, respectively. The size of the font that represents the charge is proportional to the average net charge.

Presence of GxxxG motifs

We searched for the GxxxG motifs in the exoplasmic and cytoplasmic part of the TMD as defined above. GxxxG was defined as having either G, A or S at its flanks, while x represented all possible residues in the transmembrane helix.

Finding the most average human transmembrane domains

The choice of a single-pass transmembrane protein that was most similar to average TMD properties (hydrophobicity, ASA, charge and aromaticity) was determined using 3 different methods. First we determined average TMD property profiles as shown in the previous paragraph (Fig. 2). From these average properties, we calculated the arithmetic distance for each normalized property (normalized to one) at each position of the TMD plus flanking residues. Using this method, we determined the top 10 of the TMDs that were most similar to the average properties. As the second method, we used *k*-means clustering to reduce dimensionality. Briefly, we used the K-means function to cluster the human TMD data into 10 clusters. For each cluster, we calculated the centroids and calculated which 10 proteins corresponded closest to the 10 centroids, respectively, by calculating the Euclidean distance based on the TMD plus 7 flanking residues. As the third method, we used quantitative proteomics data from white ghost cells, which was used to determine the relative amount of different plasma-membrane proteins.[35] We hence determined the 10 most concentrated proteins in the plasma membrane. Interestingly, the ERMAP appeared in all three methods and was therefore used as the "average" example protein for MD simulations.

Molecular dynamics simulation

The ERMAP TMD structure was taken from the transmembrane alphafold database, which curates and annotates TMDs from the entire AF2 predicted proteome. We included 10 amino acids on either side of the TM region. The TMD was embedded in the asymmetric membrane model recently reported by Doktorova *et al.*[30] using the charmmgui membrane builder. The final system size was roughly 10 nm in the membrane plane, and 12 nm in the normal direction. Lipids and protein were modeled with the CHARMM36 force field, and water with the TIP3P water model.

The system was equilibrated with the six-step charmmgui equilibration protocol using NAMD3.0. The production simulation was run for 100 ns under NPT conditions (1 bar and 25 °C) with pressure coupled anisotropically with the Langevin piston algorithm and temperature controlled by a Langevin integrator with a 2.0 fs timestep. All hydrogens were constrained with SHAKE, electrostatics were computed with the particle mesh Ewald algorithm, and Lennard-Jones interactions were truncated with a force-switch cutoff from 10.0 to 12.0 Å.

Data availability

All data represented in this manuscript is available in the ESI datafile 1.†

Author contributions

Joseph H. Lorent, Ilya Levental, Kandice R. Levental and Edward Lyman developed the concept of the article. JHL did the bioinformatics analyses and figure design (all except Fig. 7A). He also compared the electron density profiles from the MD simulation to the TMD profiles. Edward Lyman and Angela Cabrera-Jojo did the MD simulations and designed Fig. 7A. All authors contributed to the editing and writing of the article.

Conflicts of interest

There are no conflicts to declare.

Acknowledgements

EL and ACJ were supported by the US National Institute of General Medical Sciences grant number R35-GM153273, and ACJ was supported by grant MCB 2121854 from the US National Science Foundation.

Notes and references

1 T. H. Haines, *Prog. Lipid Res.*, 2001, **40**, 299–324.
2 A. Y. Mulkidjanian, M. Y. Galperin and E. V. Koonin, *Trends Biochem. Sci.*, 2009, **34**, 206–215.
3 A. Caforio and A. J. M. Driessen, *Biochim. Biophys. Acta, Mol. Cell Biol. Lipids*, 2017, **1862**, 1325–1339.
4 A. Caforio, M. F. Siliakus, M. Exterkate, S. Jain, V. R. Jumde, R. L. H. Andringa, S. W. M. Kengen, A. J. Minnaard, A. J. M. Driessen and J. van der Oost, *Proc. Natl. Acad. Sci. U. S. A.*, 2018, **115**, 3704–3709.
5 J. H. Lorent, K. R. Levental, L. Ganesan, G. Rivera-Longsworth, E. Sezgin, M. Doktorova, E. Lyman and I. Levental, *Nat. Chem. Biol.*, 2020, **16**, 644–652.
6 M. Sinensky, *Proc. Natl. Acad. Sci. U. S. A.*, 1974, **71**, 522–525.
7 D. Lingwood and K. Simons, *Science*, 2010, **327**, 46–50.
8 N. M. Goldenberg and B. E. Steinberg, *Cancer Res.*, 2010, **70**, 1277–1280.
9 E. G. Kelley, P. D. Butler, R. Ashkar, R. Bradbury and M. Nagao, *Proc. Natl. Acad. Sci. U. S. A.*, 2020, **117**, 23365–23373.

10 J. R. Winnikoff, D. Milshteyn, S. J. Vargas-Urbano, M. A. Pedraza-Joya, A. M. Armando, O. Quehenberger, A. Sodt, R. E. Gillilan, E. A. Dennis, E. Lyman, S. H. D. Haddock and I. Budin, *Science*, 2024, **384**, 1482–1488.
11 G. Guidotti, *Annu. Rev. Biochem.*, 1972, **41**, 731–752.
12 T. J. Stevens and I. T. Arkin, *Proteins*, 2000, **39**, 417–420.
13 Stefano Piotto, L. D. Biasi, L. Sessa and S. Concilio, *Front. Phys.*, 2018, **6**, 48.
14 W. Dowhan and M. Bogdanov, *Annu. Rev. Biochem.*, 2009, **78**, 515–540.
15 G. Heijne, *EMBO J.*, 1986, **5**, 3021–3027.
16 I. D. Pogozheva and A. L. Lomize, *Biochim. Biophys. Acta, Biomembr.*, 2018, **1860**, 364–377.
17 I. D. Pogozheva, H. I. Mosberg and A. L. Lomize, *Protein Sci.*, 2014, **23**, 1165–1196.
18 H. J. Sharpe, T. J. Stevens and S. Munro, *Cell*, 2010, **142**, 158–169.
19 J. A. Baker, W. C. Wong, B. Eisenhaber, J. Warwicker and F. Eisenhaber, *BMC Biol.*, 2017, **15**, 66.
20 J. H. Lorent, B. Diaz-Rohrer, X. Lin, K. Spring, A. A. Gorfe, K. R. Levental and I. Levental, *Nat. Commun.*, 2017, **8**, 1219.
21 I. D. Pogozheva, S. Tristram-Nagle, H. I. Mosberg and A. L. Lomize, *Biochim. Biophys. Acta*, 2013, **1828**, 2592–2608.
22 O. G. Mouritsen and M. Bloom, *Biophys. J.*, 1984, **46**, 141–153.
23 K. Mitra, I. Ubarretxena-Belandia, T. Taguchi, G. Warren and D. M. Engelman, *Proc. Natl. Acad. Sci. U. S. A.*, 2004, **101**, 4083–4088.
24 A. J. Situ, S. M. Kang, B. B. Frey, W. An, C. Kim and T. S. Ulmer, *J. Phys. Chem. B*, 2018, **122**, 1185–1194.
25 W. P. Russ and D. M. Engelman, *J. Mol. Biol.*, 2000, **296**, 911–919.
26 M. G. Teese and D. Langosch, *Biochemistry*, 2015, **54**, 5125–5135.
27 A. Prakash, L. Janosi and M. Doxastakis, *Biophys. J.*, 2011, **101**, 1949–1958.
28 V. Anbazhagan and D. Schneider, *Biochim. Biophys. Acta*, 2010, **1798**, 1899–1907.
29 M. Doktorova, J. L. Symons and I. Levental, *Nat. Chem. Biol.*, 2020, **16**, 1321–1330.
30 M. Doktorova, J. L. Symons, X. Zhang, H.-Y. Wang, J. Schlegel, J. H. Lorent, F. A. Heberle, E. Sezgin, E. Lyman, K. R. Levental and I. Levental, *bioRxiv*, 2023, preprint, DOI: 10.1101/2023.07.30.551157.
31 J. Hallgren, K. D. Tsirigos, M. D. Pedersen, J. J. Almagro Armenteros, P. Marcatili, H. Nielsen, A. Krogh and O. Winther, *bioRxiv*, 2022, preprint, DOI: 10.1101/2022.04.08.487609.
32 M. Bogdanov, K. Pyrshev, S. Yesylevskyy, S. Ryabichko, V. Boiko, P. Ivanchenko, R. Kiyamova, Z. Guan, C. Ramseyer and W. Dowhan, *Sci. Adv.*, 2020, **6**, eaaz6333.
33 J. Kyte and R. F. Doolittle, *J. Mol. Biol.*, 1982, **157**, 105–132.
34 C. Rutz, W. Rosenthal and R. Schulein, *J. Biol. Chem.*, 1999, **274**, 33757–33763.
35 A. H. Bryk and J. R. Wisniewski, *J. Proteome Res.*, 2017, **16**, 2752–2761.

Faraday Discussions

ROYAL SOCIETY OF CHEMISTRY

DISCUSSIONS

Proteins in asymmetric membranes: general discussion

Francisco N. Barrera, Aileen Cooney, Niclas Paul Decker, Markus Deserno, Milka Doktorova, Thais A. Enoki, Alexander P. Fellows, Giacomo Fiorin, Gonen Golani, Daniel Huster, Manpreet Kaur, Ilya Levental, Kandice R. Levental, Reinhard Lipowsky, Fabio Lolicato, Erwin London, Joseph H. Lorent, Tiemei Lu, Edward Lyman, Jonathan M. Machin, Paramita Manna, Shalini Mishra, Luca Monticelli, Walter Nickel, Georg Pabst, John Seddon, Raya Sorkin and Malavika Varma

DOI: 10.1039/d5fd90021b

Daniel Huster opened the discussion of the paper by Edward Lyman: Can your simulations provide more atomistic details than X-ray or cryo-EM structures, as in those the lipids are often rather blurry? You observed that the cholesterol stayed at the receptor for the entire simulation; what are the molecular interactions between cholesterol and the protein?

Edward Lyman answered: In the simulations we know where every atom is at every instant in time, and so we have a great deal more detail than one finds in a single structure, which as you say may have low resolution for the lipid densities. I expect that the "blurry" density in experimental structures is because they are averaged over an ensemble. If the lipid is in precisely the same configuration in every member of the ensemble (particle in a cryoEM structure, unit cell in a crystal), then you get to see it in fine detail. But this, I think, is the exception rather than the rule. In simulations, "bound" lipids are often quite dynamic. It should be possible to rigorously compare density from an experimental structure to something averaged over a simulation. Phill Stansfeld at Warwick has been working along these lines, and has developed a tool called "LipidDens" for this purpose. Also, I note that in the native membrane structures the resolved lipids are often much higher resolution than in other preparations (*e.g.*, nanodiscs, LCP X-ray structures).

Regarding the side chains that interact with cholesterol, they are shown in Fig. 4D of our paper (https://doi.org/10.1039/d4fd00210e).

Daniel Huster asked: Are all the positively charged side chains interacting with phosphatidyl serine (PS)? What is the proximity of the carboxyl or phosphate groups to the positive charges of Arg or Lys? Lysine has different chemistry than arginine; do you see electrostatics dominating or hydrogen bonding (with Arg)? What is the approximate residence time of a PS molecule on the receptor? Have you considered quantum mechanical MD simulations?

Edward Lyman replied: Good question. We went back and took a look at the trajectories, and we find that both the phosphate the carboxyl group can interact with both Lys and Arg side chains.

As for charge–charge interactions *vs.* hydrogen bonds, standard force fields can't really discriminate between these; nonbonded interactions are all electrostatic and van der Waals. To the extent that one observes "hydrogen bonding" interactions in fixed charge force fields, it is really just a result of the charge distributions on the donor and acceptor, and how they tend to align.

Lys *vs.* Arg is a fascinating comparison, and often underestimated in its import (by physicists, anyway). Force fields do an OK job of telling them apart, because the planar geometry of the Arg side chain is enforced, but there isn't really anything in there that captures the interactions between the π system of Arg and other interaction partners (such as cations), which almost certainly underlies its "specialness". It was Pavel Jungwirth who first pointed this out to me; he has done some really nice work in the context of cell-penetrating peptides while thinking about these things.[1]

1 C. Allolio, A. Magarkar, P. Jurkiewicz, K. Baxová, M. Javanainen, P. E. Mason, R. Šachl, M. Cebecauer, M. Hof, D. Horinek, V. Heinz, R. Rachel, C. M. Ziegler, A. Schröfel and P. Jungwirth, Arginine-rich cell-penetrating peptides induce membrane multilamellarity and subsequently enter via formation of a fusion pore, *Proc. Natl. Acad. Sci. U. S. A.*, 2018, **115**, 11923–11928, DOI: 10.1073/pnas.1811520115.

Erwin London queried: Do you think the cholesterol that did not dissociate during the simulation just has a slow on-rate and a slow off-rate, or really is tightly bound in the thermodynamic sense?

Edward Lyman answered: Well, the on-rate must be fairly fast on the simulation timescale, because it is already "bound" very early in the simulation. While there are just two trajectories in this paper (https://doi.org/10.1039/d4fd00210e), we and others have seen similar behavior many times.[1] As for the off-rate, all I can say for sure is that I have here two examples of a residence time that is at least 10 μs. I think this is a critical open question for the field – to develop rigorous methods to obtain the thermodynamics of lipid–protein interactions. There are many examples of potentials of mean force (PMFs) for lipid–protein interactions in the literature, but in our hands most of these methods do not produce reproducible results (we have a big stack of unpublished results using metadynamics, umbrella sampling, *etc.*). Grace Brannigan and Jerome Henin have been working hard on this though, and have published an alchemical free-energy method that looks promising. It is complex, but they have developed a NAMD tutorial to assist people who want to try it out: https://livecomsjournal.org/index.php/livecoms/article/view/v5i1e2067.

1 E. Rouviere, C. Arnarez, L. Yang and E. Lyman, Identification of Two New Cholesterol Interaction Sites on the A2A Adenosine Receptor, *Biophys. J.*, 2017, **113**, 2415–2424, DOI: 10.1016/j.bpj.2017.09.027.

Erwin London asked: What is the present opinion of the field concerning the role of CRAC and CARC motifs in the binding of cholesterol? Are there many examples in which it is involved in cholesterol binding and specific interactions with these residues can be identified?

Edward Lyman answered: I'd say that the opinion of the field is mixed. The issue is that the CARC/CRAC "motif" is extremely promiscuous – scan the sequence of a membrane protein and you will find them. Then look for places where cholesterol interacts with the surface of a membrane protein in a simulation and you will find some (and the definition of 'interacts' is ambiguous, and so one can adjust the definition until the desired result is found).

But do they "predict" cholesterol binding sites? Daniel Isom at Miami reviewed structurally resolved cholesterols and concluded "no".[1] Rosenhouse-Dantsker came to the same conclusion based on a review of ion-channel structures.[2] Irena Levitan has a nice review of these ideas in the discussion of a paper in ref. 3.

1 G. J. Taghon, J. B. Rowe, N. J. Kapolka and D. G. Isom, Predictable cholesterol binding sites in GPCRs lack consensus motifs, *Structure*, 2021, **29**, 499–506, DOI: 10.1016/j.str.2021.01.004.
2 A. Rosenhouse-Dantsker, Chapter Seven - Insights Into the Molecular Requirements for Cholesterol Binding to Ion Channels, in *Sterol Regulation of Ion Channels*, ed. I. Levitan, Current Topics in Membranes, Academic Press, 2017, ch. 7, vol. 80, pp. 187–208, DOI: 10.1016/bs.ctm.2017.05.003.
3 N. Barbera, M. A. A. Ayee, B. S. Akpa and I. Levitan, Molecular Dynamics Simulations of Kir2.2 Interactions with an Ensemble of Cholesterol Molecules, *Biophys. J.*, 2018, **115**, 1264–1280, DOI: 10.1016/j.bpj.2018.07.041.

Thais A. Enoki communicated: What are the flip-flop rates of cholesterol in the asymmetric and the symmetric membrane models? Do these rates change in the presence of A_{2A}?

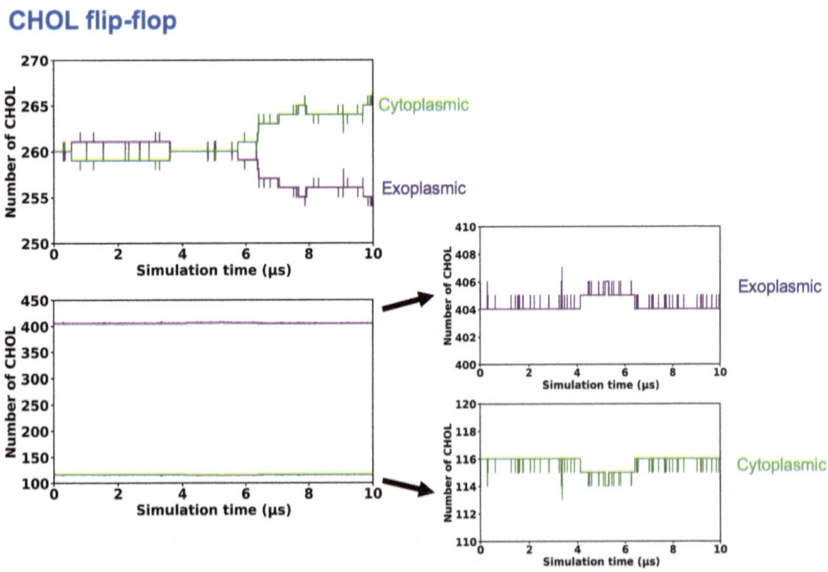

Fig. 1 Cholesterol flip-flop events in the symmetrized membrane model (upper panel) and the asymmetric model (lower panels).

Edward Lyman communicated in reply: At steady state I don't expect that the protein will modify the flip-flop rates very much, although I don't think that we have measured the flip-flop rate in the symmetrized model without protein. In the asymmetric mode, the flip-flop rate is probably too slow to measure with an unbiased simulation (again after the cholesterol distribution has equilibrated). Some data are shown in Fig. 1 in this discussion. Note that in the symmetrized model there is initially a good amount of directional transport, as the membrane equilibrates to the protein, following its insertion.

Gonen Golani remarked: The interaction of cholesterol is with relatively few amino acids. Is it possible to estimate the strength of the interaction with cholesterol based on simulations?

Edward Lyman replied: It is possible, but in our experience it is hard to do so reproducibly and with good statistical precision. One can run "very" long simulations and directly obtain the free energy for the cholesterol/protein interaction by straightforward counting: define the bound state, and accumulate statistics for on/off events. I have only seen one example with an all-atom simulation in work from Janice Roberston and Jose Faraldo Gomez, which will appear in print soon. I don't recall the exact number, but the simulation time was on the order of 50 microseconds, and this is a fairly nonspecific, solvent-like interaction! Grace Brannigan and Jerome Hénin have been developing an alchemical free-energy protocol that looks promising.[1] Many others have published potentials of mean force based on umbrella sampling protocols or metadynamics, but based on our own experience I wouldn't put much faith in any of those results (I have a big stack of unpublished PMFs that weren't reproducible). There are two challenges I think. One is to define the "bound" state. Rarely is cholesterol in a deep cleft where it is maintains fixed interactions between particular atoms of the protein and the sterol "bound" cholesterols are often found to be very dynamic, rotating around the long axis, wiggling their tails, *etc.* If one over-stabilizes a particular subset of states then the entropy of the bound state is underestimated. The second (for umbrella sampling or metadynamics) is to define a low-dimensional collective variable that takes you from the bound state to an unbound one. This is very, very nontrivial, and the difficulty (and importance) of this step tends to be grossly underestimated in the literature in my view. However (and herein lies a major problem) you will always get a decent looking PMF out of whatever protocol you use! Maybe even with nice tight-looking error bars. *Caveat emptor*!

1 R. Salari, T. Joseph, R. Lohia, J. Hénin and G. Brannigan, A Streamlined, General Approach for Computing Ligand Binding Free Energies and Its Application to GPCR-Bound Cholesterol, *J. Chem. Theory Comput.*, 2018, **14**, 6560–6573, DOI: 10.1021/acs.jctc.8b00447.

Luca Monticelli asked: In the simulations you presented (https://doi.org/10.1039/d4fd00210e), is there any asymmetric stress in the membrane? Do you think this will have an effect on protein conformation? How would we assess the extent of such an effect, and whether it is realistic or not?

Edward Lyman answered: Although we have not measured the stress state of our G protein-coupled receptor (GPCR) simulation in this paper (https://doi.org/

10.1039/d4fd00210e) (or any other), I think it is a sensible idea. In this case, the initial stress state must be similar to the results from the recent *Cell* paper by Doktorova *et al.*,[1] since that is the membrane model that we used. I think this might be an additional "knob" that one can turn in simulations to try and access different conformational states of the receptor. But is it physical/realistic? Well, that question awaits further work, but so far all the experimental tests (also in the *Cell* paper) suggest that it is. On the other hand, the *Biophys. J.* paper[2] from Varma and Deserno considers a range of different models, and obtains a differential stress with a similar magnitude but the opposite sign! So, clearly quite a lot of work yet to do!

1 M. Doktorova, J. L. Symons, X. Zhang, H.-Y. Wang, J. Schlegel, J. H. Lorent, F. A. Heberle, E. Sezgin, E. Lyman, K. R. Levental and I. Levental, Cell membranes sustain phospholipid imbalance via cholesterol asymmetry, *Cell*, 2025, **188**, 2586–2602, DOI: 10.1016/j.cell.2025.02.034.
2 M. Varma and M. Deserno, Distribution of cholesterol in asymmetric membranes driven by composition and differential stress, *Biophys. J.*, 2022, **121**, 4001–4018, DOI: 10.1016/j.bpj.2022.07.032.

Reinhard Lipowsky said: In your paper, you describe the setup of two lipid bilayers, corresponding to an asymmetric plasma membrane model and its symmetrized counterpart, as shown in Fig. 1 of your paper (https://doi.org/10.1039/d4fd00210e). Each of the two snapshots in Fig. 1 of your paper (https://doi.org/10.1039/d4fd00210e) represents the final configuration of a 10 microsecond all-atom simulation. However, I did not find any information in your paper (https://doi.org/10.1039/d4fd00210e) about different initial conditions for these two setups. Does this imply that the 10 microsecond simulation was performed only once for each setup? You mentioned that the cholesterol molecules do not undergo any flip-flops on the timescale of your simulations. What about lateral diffusion of the cholesterol? Did you observe lateral diffusion of the cholesterol within one leaflet? How large was the corresponding diffusion constant? As far as the free-energy barrier for the flip-flops of cholesterol is concerned, a variety of enhanced sampling methods, such as umbrella sampling or metadynamics, has been developed, by which one can estimate the free-energy barrier for molecular processes. It should be informative to apply one of the enhanced sampling methods to study the flip-flops of cholesterol within your two setups for the lipid bilayers.

Edward Lyman replied: There are two simulations in this paper (https://doi.org/10.1039/d4fd00210e), each one is 10 μs in duration. Convergence of the lipid environment around the protein was assessed by asking how many first-shell lipids have exchanged out of the first shell over the course of the simulation. This is reported in the electronic supplementary information (ESI) of the paper (https://doi.org/10.1039/d4fd00210e) in Fig. S1. With the exception of the exoplasmic leaflet in the asymmetric model, nearly all of the lipids have exchanged out of the first shell by the end of the simulation. A good additional test (currently underway) is to repeat each simulation several times from a different, random initial configuration of lipids.

Regarding the cholesterol flip-flop and lateral diffusion, I was partly mistaken here. There is some flip-flop in the symmetric model (as seen in Fig. 1 in this discussion in response to a previous comment), but very little in the asymmetric model. Cholesterol diffuses in the cytoplasmic leaflet with a typical Ld phase value, but the outer leaflet is more like Lo, and so lipid and cholesterol diffusion is slow. We did not measure them in this simulation, but have done so for other complex lipid mixtures in ref. 1.

Regarding enhanced sampling methods for flip-flop: this is something we are currently working quite hard on. Both metadynamics and umbrella sampling (US) are dependent on a good choice of collective variable. Several groups have done such calculations (*e.g.*, Drew Bennet, Wonpil Im).[2,3] We are using our unbiased trajectories with many flip-flop events to learn a one-dimensional collective variable (1D CV) (encoded in a simple neural network (NN), following work by Tiwary[4]). We then do US on this 1D CV. It requires some clever tricks and modifications to the NN, but we are obtaining very sensible and reproducible PMFs for cholesterol flip-flop. We will submit this manuscript in the next few months.

1 K. Pinkwart, F. Schneider, M. Lukoseviciute, T. Sauka-Spengler, E. Lyman, C. Eggeling and E. Sezgin, Nanoscale dynamics of cholesterol in the cell membrane, *J. Biol. Chem.*, 2019, **294**, 12599–12609, DOI: 10.1074/jbc.RA119.009683.
2 W. F. Drew Bennett, J. L. MacCallum, M. J. Hinner, S. J. Marrink and D. P. Tieleman, Molecular View of Cholesterol Flip-Flop and Chemical Potential in Different Membrane Environments, *J. Am. Chem. Soc.*, 2009, **131**, 12714–12720, DOI: 10.1021/ja903529f.
3 S. Jo, H. Rui, J. B. Lim, J. B. Klauda and W. Im, Cholesterol Flip-Flop: Insights from Free Energy Simulation Studies, *J. Phys. Chem. B*, 2010, **114**, 13342–13348, DOI: 10.1021/jp108166k.
4 D. Wang and P. Tiwary, State predictive information bottleneck, *J. Chem. Phys.*, 2021, **154**, 134111, DOI: 10.1063/5.0038198.

Tiemei Lu asked: Could lipid–protein interactions also occur between lipids and peptides? Does the molecular weight (M_w) of the peptide need to be comparable to that of a full protein for such interactions to happen?

Are lysine (Lys) and arginine (Arg) residues the main contributors to lipid–protein interactions?

How does charge density of the protein influence these interactions?

Edward Lyman replied: Sure, all kinds of peptides interact with lipids. See, for example, the vast literature on antimicrobial peptides.

Milka Doktorova commented: Placing a membrane-spanning protein in a membrane requires removal of lipids from both leaflets. The number of lipids to remove may not be trivial, especially when the protein has an asymmetric shape and is placed in an asymmetric membrane. This, coupled with the increasing evidence that lipid abundance asymmetry exists and is functional in biological membranes, suggests that examining protein–membrane interactions as a function of varying number imbalance may be more informative than looking at a single model for understanding the interaction of the protein with an asymmetric membrane. Do you think that changes in number asymmetry can affect the functional interactions of A_{2A} with the lipids and, if so, what properties do you expect to be most affected?

Edward Lyman answered: Certainly differential stress could couple to conformational changes, just like Michael Brown[1] has shown for curvature stress for rhodopsin. We are exploring this right now. I don't have any intuition about what properties would be affected, but the basic question is "can differential stress be exploited to drive a GPCR between different conformational states?" Also, I note that we tend to use systems a bit larger than those used by most people for these simulations. This gives us more "bulk membrane" around, helpful when you are trying to look at enrichment of specific lipids. And it mitigates a bit any stress induced during the insertion/deletion part of the system setup. But see also in Fig. 1 in this discussion that we do see about 5 cholesterol molecules translocate from the inner leaflet to the outer one following insertion into the symmetrized model.

1 A. V. Botelho, N. J. Gibson, R. L. Thurmond, Y. Wang, M. F. Brown, Conformational Energetics of Rhodopsin Modulated by Nonlamellar-Forming Lipids, *Biochemistry*, 2002, **41**, 6354–6368, DOI: 10.1021/bi011995g.

Alexander P. Fellows said: In relation to previous questions and comments regarding the impact of the initial conditions on the simulations and the computational cost associated with running them, could you please comment on such costs in terms of the feasibility of assessing the sensitivity of the results to the starting conditions?

Edward Lyman answered: Here we used a special-purpose computer (Anton2) because it permits very long simulation times (here 10 μs). This mitigates a bit the initial state bias problem. But we don't get enough time on that machine to run a whole bunch of replicas. So, a sensible addition would be to run several additional replicas on a commodity resource. It's not so hard to get trajectories that are 500 ns to 1 μs in duration these days – running these simulations (about 350k–400k atoms) with GROMACS on a single A100 GPU we get around 30 ns per day.

Kandice R. Levental opened the discussion of the paper by Fabio Lolicato: Do you see PIP2 on the exoplasmic leaflet, and, if so, is it evenly distributed or is it puncta? If it's not in the outer leaflet, what would provide the driving force for it to return to the inner leaflet?

Fabio Lolicato and **Walter Nickel** responded: So far, such experiments have not been conducted. However, we expressed and purified a high-affinity PI(4,5)P2 sensor fused to a Halo tag. Along with our single-molecule high-resolution total internal reflection fluorescence (TIRF) microscopy setup, experiments are planned to address this question in living cells.

Ilya Levental said: I was wondering if PIP2 can ever be seen on the outside of the cell. My understanding was that it hasn't been seen yet, probably because there's such a small amount and it likely diffuses away. But if there were some way to increase the throughput of FGF2 secretion (overexpression of it or trafficking components) or enhance PIP2 detection, perhaps that could help?

Fabio Lolicato and **Walter Nickel** responded: So far, we did not attempt to visualize PIP2 at the outer leaflet and there are also no conditions known that allow for a massive upregulation of FGF2 secretion. Nevertheless, both aspects are something we are looking into and plan to address in the future.

Shalini Mishra queried: Are there any experiments done to test the involvement of redox systems in combination with lipid asymmetry in the *in vitro* condition that can contribute to the translocation of FGF2, as mentioned in the paper (https://doi.org/10.1039/d4fd00208c)?

Fabio Lolicato and **Walter Nickel** replied: We are currently working on the identification of redox enzymes involved in oxidative FGF2 dimerisation at the inner plasma membrane leaflet. A hypothesis we are following is a potential involvement of plasma membrane resident NAPH oxidases along with H_2O_2 as the oxidant required for disulfide bridge formation.

Erwin London asked: Would it be possible to crosslink FGF2 with a reagent that would resist cleavage of a crosslink under reducing conditions to allow it to be co-isolated with bound PIP2?

Fabio Lolicato and **Walter Nickel** answered: We did crosslink FGF2 and identified dimers in a C95-dependent manner.[1] However, those experiments aimed at the analysis of the FGF2 dimerization interface and were not used for co-isolation experiments with PIP2. The interaction of PIP2 with FGF2 in cells and *in vitro* has been demonstrated in a number of ways, including super-resolution microscopy in cells and *in vitro* liposome binding studies, as well as functional studies of various kinds, demonstrating that a loss of interaction between FGF2 and PIP2 results in a block of FGF2 secretion from cells.

1 F. Lolicato, J. P. Steringer, R. Saleppico, D. Beyer, J. Fernandez-Sobaberas, S. Unger, S. Klein, P. Riegerová, S. Wegehingel, H.-M. Müller, X. J. Schmitt, S. Kaptan, C. Freund, M. Hof, R. Šachl, P. Chlanda, I. Vattulainen and W. Nickel, Disulfide bridge-dependent dimerization triggers FGF2 membrane translocation into the extracellular space, *eLife*, 2024, **12**, RP88579, DOI: 10.7554/eLife.88579.3.

Reinhard Lipowsky remarked: In your talk and in Fig. 1 of your paper (https://doi.org/10.1039/d4fd00208c), you describe the formation of tetramers of the FGF2 protein and the binding of four to five $PI(4,5)P_2$ molecules, which then form a localized cluster under the FGF2 tetramer. In the caption of Fig. 1 of your paper (https://doi.org/10.1039/d4fd00208c), you also refer to cholesterol-enriched nanodomains with an asymmetric transbilayer distribution of $PI(4,5)P_2$ molecules. In the cartoon of Fig. 1 of your paper (https://doi.org/10.1039/d4fd00208c), such a nanodomain is depicted by the blueish bilayer patch, which has a large lateral size compared to the size of the FGF2 tetramer. What are your estimates for the lateral size of the FGF2 tetramer and of the nanodomain? What is the experimental evidence for the relatively large nanodomain as depicted in Fig. 1 of your paper (https://doi.org/10.1039/d4fd00208c)? Do you think such a nanodomain will typically accommodate a single FGF2 tetramer or several such tetramers?

Fabio Lolicato and **Walter Nickel** replied: At this point, the size of the proposed nanodomains housing the FGF2 translocation machinery is unknown. It is also unknown whether only a single or multiple FGF2 oligomers along with the Na,K-ATPase, Tec kinase and GPC1 are present in one such nanodomain. Fig. 1 in our paper (https://doi.org/10.1039/d4fd00208c) is therefore just a schematic illustration of data pointing to a liquid-ordered nanodomain as the organisational principle of the FGF2 membrane translocation machinery. In studies from other groups (Hell, Jahn; MPI Göttingen, Germany),[1] PI(4,5)P2-containing nanodomains have been analysed by super-resolution light microscopy and reported to have diameters in the range of 80 nm. Maybe this is a size range that could also apply for the PI(4,5)P_2-containing nanodomains housing the FGF2 membrane translocation machinery.

1 G. van den Bogaart, K. Meyenberg, H. J. Risselada, H. Amin, K. I. Willig, B. E. Hubrich, M. Dier, S. W. Hell, H. Grubmüller, U. Diederichsen and R. Jahn, Membrane protein sequestering by ionic protein–lipid interactions, *Nature*, 2011, **479**, 552–555, DOI: 10.1038/nature10545.

Raya Sorkin commented: You describe three steps: dimerization, oligomerization, and pore formation. Which of these steps is affected in the synthetic *vs.* the cell system? Which one is the rate-limiting step?

Fabio Lolicato and **Walter Nickel** responded: All three of these steps are critical in both the *in vitro* and the cell-based experiments. Mutants of FGF2 (*e.g.* C95A) cannot dimerise, oligomerise or form pores. They are also not secreted from cells. The rate-limiting step is likely to be PI(4,5)P_2-dependent oxidative FGF2 dimerisation at the inner plasma membrane leaflet.

Paramita Manna addressed all: My question concerns the peptide–lipid membrane interaction and specifically the conformations of a peptide upon association with the lipid membrane. It is well established that membrane permeability is influenced by the biophysical properties of the membranes, such as rigidity, with fluid-phase membranes generally exhibiting increased permeability in the presence of membrane-active or cell-penetrating peptides. In these cases, the peptide remains embedded in the lipid bilayer. However, in cases where permeability remains low in the presence of a peptide, such as gel-phase lipids, how can one demonstrate that the peptide is embedded within the membrane bilayer rather than extending outward into the aqueous phase to engage with a distal surface? What parameters should I look into? Is it possible to establish any correlation between the peptide's spatial orientation on or within the membrane and the membrane's fluidity?

Georg Pabst replied: The positioning of membrane-active peptides relative to the lipid bilayer is influenced by the physicochemical properties of both the peptides and the lipid bilayer, as well as the concentrations of each. For amphipathic peptides that adopt linear alpha-helical secondary structures in the presence of a smooth membrane in the lamellar gel phase, it is likely that they will be adsorbed onto the membrane surface. However, this behavior can change dramatically in the presence of membrane defects.

The absence of a pretransition in phosphatidylcholine (PC), phosphatidylglycerol (PG), or sphingomyelin (SM) membranes may suggest a scenario where the peptides are predominantly surface-bound, although this interpretation is not entirely definitive.

To investigate these dynamics, a robust experimental approach would involve a combination of molecular dynamics (MD) simulations, small-angle X-ray scattering (SAXS) or small-angle neutron scattering (SANS), along with careful hydrogen/deuterium (H/D) contrast variation. This can be achieved by using chain-deuterated lipids, headgroup-deuterated lipids, and adjusting the H_2O/D_2O ratio to accurately determine the peptide's location within the membrane.

In a subsequent analysis, it would be valuable to explore the correlation between peptide positioning and membrane fluidity. However, I anticipate that this correlation may not be straightforward. It is also crucial to consider the lipid concentration required for each specific technique, rather than solely focusing on the lipid-to-peptide ratio. For a more detailed discussion of these nuanced issues, please refer to ref. 1.

1 E. F. Semeraro, P. Pajtinka, L. Marx, I. Kabelka, R. Leber, K. Lohner, R. Vácha and G. Pabst, Magainin 2 and PGLa in bacterial membrane mimics IV: Membrane curvature and partitioning, *Biophys. J.*, 2022, **121**, 4689–4701, DOI: 10.1016/j.bpj.2022.10.018.

Edward Lyman opened the discussion of the paper by Jonathan M. Machin: My question is about hysteresis – can you obtain reversible folding curves (*e.g.*, as Karen Fleming's group has done), so that the free energy of folding can be extracted from the measurement?

Jonathan M. Machin answered: Although highly desirable, reversible folding of outer membrane proteins (OMPs) (*i.e.*, without hysteresis) is very hard to achieve. As noted, there is a single report in the literature where this was done, with assays conducted at a pH of 3.8 for the protein OmpLA.[1] We have previously attempted to repeat this for other OMPs unsuccessfully, suggesting that these conditions are likely protein specific and will need optimising for each OMP, if indeed such conditions exist for a given OMP. OmpA has advantages for the study here (https://doi.org/10.1039/d4fd00180j) due to its very high stability once folded, making the formation of stable and long-lived proteoliposomes possible. However, given that (WT-)OmpA minimally unfolds, even after incubation in 8 M urea at 50 °C overnight in multiple lipid systems,[2] it seems possible that for OmpA such hysteresis may be unavoidable.

1 C. P. Moon, S. Kwon and K. G. Fleming, Overcoming Hysteresis to Attain Reversible Equilibrium Folding for Outer Membrane Phospholipase A in Phospholipid Bilayers, *J. Mol. Biol.*, 2011, **413**(2), 484–494, DOI: 10.1016/j.jmb.2011.08.041.
2 J. M. Machin, A. C. Kalli, N. A. Ranson and S. E. Radford, Protein–lipid charge interactions control the folding of outer membrane proteins into asymmetric membranes, *Nat. Chem.*, 2023, **15**(12), 1754–1764, DOI: 10.1038/s41557-023-01319-6.

Georg Pabst commented: OmpA folding into lipid vesicles was performed using DLPC – the shortest lipid that forms a bilayer – and DMPC at its melting transition, where significant density fluctuations occur. In both cases, the bilayers are less densely packed than in more physiologically relevant lipid systems. Have you tested whether OmpA can also fold into more tightly packed, physiologically representative membranes?

Jonathan M. Machin responded: OmpA will intrinsically fold into a limited range of membranes. It will successfully (but slowly) fold to ~80% completion into DMPC membranes in the fluid phase (30 °C), but poorly into longer saturated acyl-chain lengths (*e.g.* DPPC was ~30% folded at completion).[1,2] We chose DLPC for this work because the timescales are more experimentally accessible (15 hours), while the hydrophobic width of the bilayer is similar to the thin native outer membrane. Unsaturated longer lipid acyl chains will support folding (for example, POPC and DOPC) to varying extents.[1,3] Note that *in vitro* folding into liposomes catalysed by reconstituted BAM (the native foldase) will support folding into a much wider variety of membranes than OmpA (or other OMPs) will intrinsically fold into.[2]

1 J. M. Machin, A. C. Kalli, N. A. Ranson and S. E. Radford, Protein–lipid charge interactions control the folding of outer membrane proteins into asymmetric membranes, *Nat. Chem.*, 2023, **15**(12), 1754–1764, DOI: 10.1038/s41557-023-01319-6.
2 S. Hussain and H. D. Bernstein, The Bam complex catalyzes efficient insertion of bacterial outer membrane proteins into membrane vesicles of variable lipid composition, *J. Biol. Chem.*, 2018, **293**(8), 2959–2973, DOI: 10.1074/jbc.RA117.000349.
3 J. H. Kleinschmidt and L. K. Tamm, Folding Intermediates of a β-Barrel Membrane Protein. Kinetic Evidence for a Multi-Step Membrane Insertion Mechanism, *Biochemistry*, 1996, **35**(40), 12993–13000, DOI: 10.1021/bi961478b.

Erwin London remarked: In terms of the positive-inside rule, would you predict the tendency of excess anionic lipid on the inner leaflet to induce positive inside orientation to be mostly cancelled by the excess of cationic protein residues near the surface of the membrane? If so, is there a good way to test this?

Jonathan M. Machin replied: Electrostatics at the membrane surface are determined by protein and lipid features, as well as the broader solvent/context in which the membrane resides – and all these features will combine to govern protein orientation *via* the positive-inside rule. In addition, protein crowding will significantly affect the electrostatic context. How significant an effect the titration of anionic lipids with positive proteins would have on orientation determination likely depends on whether the orientation determination is driven by a global net electrostatic environment (in which case, one would expect an effect), or local effects where positive charges on the protein bind and are anchored by one (or a few) negatively charged lipids (in which case, given the excess of lipids, providing there is adequate diffusion in the membrane there would likely be small effects). Note also work from the Hegde group which suggests that, when folding, many α-helical proteins partition into a partially-occluded patch of membrane, which would help shield any positive-inside electrostatic interactions from the larger membrane context, as well as providing lipids for interactions to form with.[1,2]

While the work presented here is on OMPs (which have a positive-outside rule and so the findings here may or may not be transferable to α-helical transmembrane proteins), it suggests that the charge balance is important not just for stability but also for the rate and efficiency of the folding process – both of which may be important for determining orientation. To test these hypotheses on α-helical proteins *in vitro* requires good refoldable model systems, inherently much more difficult for α-helical proteins that tend to be much less soluble and much less readily refolded into bilayers than their OMP counterparts. That said, it may be possible to test for loss of orientation control as increasing amounts of protein are incorporated into membranes, and tested over a range of membranes with differing anionic lipid concentrations.

1 P. J. Chitwood and R. S. Hegde, An intramembrane chaperone complex facilitates membrane protein biogenesis, *Nature*, 2020, **584**(7822), 630–634, DOI: 10.1038/s41586-020-2624-y.
2 L. Smalinskaitė, M. K. Kim, A. J. O. Lewis, R. J. Keenan and R. S. Hegde, Mechanism of an intramembrane chaperone for multipass membrane proteins, *Nature*, 2022, **611**(7934), 161–166, DOI: 10.1038/s41586-022-05336-2.

Francisco N. Barrera asked: After the insertion of the porin with the negatively charged loops, what is the transmembrane potential change?

Membrane potential probes, like di-8-ANEPPS and similar molecules, could maybe provide this kind of information.

Jonathan M. Machin responded: We are confident in the directionality of the dipole, but not its precise value, which we did not measure. A comparative metric of per-lipid charge-difference across the membrane can be used to compare between liposomes ($\pm\sim$0.04 charges per lipid, for the positive and negative OmpA variant used), but clearly this is imprecise. We agree that to further this work a direct quantitation of the membrane potential would be useful, and would facilitate studying the effects of titrating the magnitude of the protein-induced membrane potential.

Ilya Levental communicated: My question is about the positive-inside rule. The general rule that positively charged residues are enriched on the cytosolic side of membrane proteins seems to hold for mammalian cells. This seems to make sense, since the cytosolic leaflet of the PM (plasma membrane) is negatively charged. But actually, it's quite strange, as the topology of these proteins is originally established in the ER (endoplasmic reticulum) during their biogenesis, which does not seem to be charge-asymmetric. So how did this "rule" evolve?

Jonathan M. Machin communicated in reply: This is an interesting question. A positive-inside rule for transmembrane helix insertion appears to be broadly conserved across SEC-inserted proteins, although it is somewhat relaxed for proteins inserted by the YidC family.[1,2] From an evolutionary perspective, the complex membrane architectures found in eukaryotes (like the ER) were not present when the rule first developed. It is of course hard to interpret what such ancient membranes may have looked like, including their lipid composition, membrane asymmetry and the state of the protein-inserting complexes SEC/YidC families, but given broad conservation of a negatively charged cytosolic leaflet of plasma membranes. It thus seems plausible that this charge asymmetric membrane could have been the original context for such a rule to evolve, which has then been retained. I also wonder whether a strictly charge-asymmetric membrane is actually required for the rule to develop? It is possible that simply the interaction of positive charges on the protein with negatively charged lipids may be sufficient to anchor one end of the helix and prevent it traversing the membrane, regardless of the overall membrane asymmetry. This would be particularly true in light of work from the Hegde lab suggesting many transmembrane proteins partition through the membrane at least partially exposed to the bilayer,[3,4] in which case the unfavorability of positive charges crossing the membrane and their relatively high pK_a (making deprotonation challenging) would also contribute to holding positive charges on the inside of the membrane.

1 K. Xie and R. E. Dalbey, Inserting proteins into the bacterial cytoplasmic membrane using the Sec and YidC translocases, *Nat. Rev. Microbiol.*, 2008, **6**, 234–244, DOI: 10.1038/nrmicro3595.
2 A. N. Gray, J. M. Henderson-Frost, D. Boyd, S. Shirafi, H. Niki and M. B. Goldberg, Unbalanced Charge Distribution as a Determinant for Dependence of a Subset of *Escherichia coli* Membrane Proteins on the Membrane Insertase YidC, *mBio*, 2011, **2**, e00238-11, DOI: 10.1128/mBio.00238-11.
3 P. J. Chitwood and R. S. Hegde, An intramembrane chaperone complex facilitates membrane protein biogenesis, *Nature*, 2020, **584**(7822), 630–634, DOI: 10.1038/s41586-020-2624-y.
4 L. Smalinskaitė, M. K. Kim, A. J. O. Lewis, R. J. Keenan and R. S. Hegde, Mechanism of an intramembrane chaperone for multipass membrane proteins, *Nature*, 2022, **611**(7934), 161–166, DOI: 10.1038/s41586-022-05336-2.

Manpreet Kaur asked: How does the presence of other lipids modulate the protein folding? Given that *E. coli* has large amounts of phosphoethanolamine (PE), I wonder would the protein folding be better in tensed or more fluidic membrane?

Jonathan M. Machin responded: Folding is highly sensitive to the physiochemical properties of the lipids used, including their headgroups. While small unilamellar vesicles (SUVs) containing large amounts of PE lipid cannot be generated due to PE's negative curvature, titrating in small amounts of PE symmetrically slow down the folding.[1] We have shown that PE asymmetrically distributed such that it is only on the outer leaflet (*i.e.*, the equivalent of the inner leaflet in the native membrane) will accelerate folding relative to the symmetric case, and also that changing ratios of DMPC and DMPG lipid symmetrically or asymmetrically has a large impact on folding kinetics and stability.[2] OMPs tend not to intrinsically fold into gel-phase membranes, but do in the fluid state. In contrast, when OMPs are *in vitro*-folded catalysed by their native foldase, the BAM complex, they will fold into both gel-state and fluid membranes. Indeed, the native membrane is highly rigid due to protein and lipopolysaccharide noncovalent crosslinking, and increasing the membrane tension with high concentrations of extracellular sucrose accelerates folding.[3] Thus it appears that membrane tension may play an important role in facilitating folding *in vivo*, although this remains to be explored.

1 D. Gessmann, Y. H. Chung, E. J. Danoff, A. M. Plummer, C. W. Sandlin, N. R. Zaccai and K. G. Fleming, Outer membrane β-barrel protein folding is physically controlled by periplasmic lipid head groups and BamA, *Proc. Natl. Acad. Sci. U. S. A.*, 2014, **111**(16), 5878–5883, DOI: 10.1073/pnas.1322473111.
2 J. M. Machin, A. C. Kalli, N. A. Ranson and S. E. Radford, Protein–lipid charge interactions control the folding of outer membrane proteins into asymmetric membranes, *Nat. Chem.*, 2023, **15**(12), 1754–1764, DOI: 10.1038/s41557-023-01319-6.
3 M. T. Doyle, J. R. Jimah, T. Dowdy, S. I. Ohlemacher, M. Larion, J. E. Hinshaw and H. D. Bernstein, Cryo-EM structures reveal multiple stages of bacterial outer membrane protein folding, *Cell*, 2022, **185**(7), 1143–1156, DOI: 10.1016/j.cell.2022.02.016.

John Seddon remarked: Have you tried reducing the phospholipid chain length even further, to say C10, to see if this speeds up the folding of OmpA?

Jonathan M. Machin replied: The intrinsic folding (*i.e.*, uncatalysed folding into membranes) of OMPs is known to be modulated by acyl-chain length. With saturated lipids, OMPs will fold more slowly as the acyl chains are made longer (at the same temperature, with all membranes in the fluid phase), due to the increased barrier to partitioning the hydrophilic loops of the protein across the membrane, required to adopt a stable transmembrane fold. DL lipids were chosen for this work because folding into DL lipids is faster than into DM lipids, making experiments more accessible, but it retains a reasonable approximation of the hydrophobic thickness of OmpA's native membrane. Previous work has characterised the chain-length dependence on intrinsic folding, and shorter lipids like C11 and C10 do indeed speed up the folding.[1,2]

1 B. Schiffrin, A. N. Calabrese, A. J. Higgins, J. R. Humes, A. E. Ashcroft, A. C. Kalli, D. J. Brockwell and S. E. Radford, Effects of Periplasmic Chaperones and Membrane Thickness on BamA-Catalyzed Outer-Membrane Protein Folding, *J. Mol. Biol.*, 2017, **429**(23), 3776–3792, DOI: 10.1016/j.jmb.2017.09.008.
2 N. K. Burgess, T. P. Dao, A. M. Stanley and K. G. Fleming, β-Barrel Proteins That Reside in the *Escherichia coli* Outer Membrane *in Vivo* Demonstrate Varied Folding Behavior *in Vitro*, *J. Biol. Chem.*, 2008, **283**(39), 26748–26758, DOI: 10.1074/jbc.M802754200.

Tiemei Lu remarked: You provided a great example showing that OmpA can induce lipid membrane phase separation. What other types of proteins can also induce phase separation? What are the key factors involved? Additionally, what changes or advancements occur in the lipid membrane after phase separation?

Jonathan M. Machin replied: In the outer membrane OmpA is not thought to induce lipid membrane phase separation alone. Rather it sits within a membrane that is at least partially phase separated into lipid- and protein-rich patches. OmpA has been shown to be important for maintenance of the OM organisation[1] so it likely does support this phase separation, but the precise mechanisms of its formation and maintenance remain unclear. Many factors in membranes (and their environments) can contribute to lipid and/or protein phase separation within the membrane. Key factors include, but are not limited to, preferential protein–protein or lipid–lipid interactions, membrane-external organising features (*e.g.* cytoskeleton, signalling molecules). Within bacteria, one major example is the chemosensing protein arrays (*e.g.* CheA) that form in the cytoplasmic membrane, driven by protein–protein interactions and lipid-exclusion.[2] In addition, liquid–liquid phase separation in the cell can drive membrane phase separation, and *vice versa*.[3]

Within the context of the outer membrane, one of the consequences of the phase separation is that the individual components of the membrane are rarely free to diffuse – indeed, diffusion rate is on the same order of magnitude as cell growth.[4] Other changes in membranes due to phase separation may include changes to the lipid order and dynamics in the local environment, and changes to the lipid–lipid/protein–lipid/protein–protein contacts due to altered accessible contacts (which may in turn alter the biological function).

1 G. Benn, C. Borrelli, D. Prakaash, A. N. T. Johnson, V. A. Fideli, T. Starr, D. Fitzmaurice, A. N. Combs, M. Wühr, E. R. Rojas, S. Khalid, B. W. Hoogenboom and T. J. Silhavy, OmpA controls order in the outer membrane and shares the mechanical load, *Proc. Natl. Acad. Sci. U. S. A.*, 2024, **121**(50), e2416426121, DOI: 10.1073/pnas.2416426121.
2 G. L. Hazelbauer, J. J. Falke and J. S. Parkinson, Bacterial chemoreceptors: high-performance signaling in networked arrays, *Trends Biochem. Sci.*, 2007, **33**(1), 9–19, DOI: 10.1016/j.tibs.2007.09.014.
3 J. A. Ditlev, Membrane-associated phase separation: organization and function emerge from a two-dimensional milieu, *J. Mol. Cell Biol.*, 2021, **13**(4), 319–324, DOI: 10.1093/jmcb/mjab010.
4 J. E. Horne, D. J. Brockwell and S. E. Radford, Role of the lipid bilayer in outer membrane protein folding in Gram-negative bacteria, *J. Biol. Chem.*, 2020, **295**(30), 10340–10367, DOI: 10.1074/jbc.REV120.011473.

Tiemei Lu asked: Why didn't you choose GUVs (giant unilamellar vesicles, μm-scale instead of nanometer) as the membrane model system? Is the size of the GUVs a problem?

Jonathan M. Machin answered: OmpA-WT will intrinsically fold into GUVs, although folding into large unilamellar vesicles (LUVs) is much better characterised in the literature. However, it is easier to make large quantities of size-controlled LUVs. Having minimal size-variation is important for ensuring that the dipoles are similar between different liposomes in the same ensemble (*i.e.*, it means there are fewer sources of variation in the downstream folding kinetic experiments).

Aileen Cooney asked: Does changing the charge also affect the conformation of protein in a way that would impact the results?

Jonathan M. Machin answered: The gross structure of the protein does not change – all the OmpA variants still form transmembrane β-barrels with substantial β-sheet content. However, the organisation of the non-membrane (natively extracellular) loops will alter in conformation due to the altered charge density and electrostatic interactions. For the experiments described in our paper (https://doi.org/10.1039/d4fd00180j), where the OmpA protein was used simply as a scaffold to generate the charge dipoles, these conformational changes should not impact the results – the conferred charge dipole will be very similar, regardless of the loop orientation.

Georg Pabst asked: Regarding Fig. 3g and h in your paper (https://doi.org/10.1039/d4fd00180j), can you rule out an influence of variations in OmpA copy number between the different samples on the observed melting behavior? What was the extent of sample-to-sample variation?

Jonathan M. Machin responded: Controlling for OmpA copy number is an important consideration, not just for the changes in observed melting temperature, but also to ensure that the later folding experiments are controlled with a similar amount of protein in the pre-formed proteoliposomes across all samples. The experimental procedure was set up to ensure that the final, purified proteoliposomes with the different OmpA-variants were approximately matched in protein concentration. There are differences in the folding efficiencies of the different OmpA variants, as shown in Fig. 3c of our paper (https://doi.org/10.1039/

d4fd00180j), ranging from ~50% (OmpA-Pos) to ~90% (OmpA-WT) folded at equilibrium. However, by changing the starting lipid : protein ratio (LPR) of each sample to match the folding efficiency (*i.e.*, 160 : 1 for OmpA-Pos, 290 : 1 for OmpA-WT), it is possible to create proteoliposomes with similar amounts of folded OmpA variant, but differing amounts of unfolded OmpA still in solution (demonstrated in Fig. 3d of our paper [https://doi.org/10.1039/d4fd00180j] – the intensity of the folded bands approximately match, but the unfolded bands are different between variants). This is effective because the final folding efficiency of OmpA is minimally dependent on the LPR (at least down to 160 : 1). Next, trypsin digest and liposome pelleting by ultracentrifugation respectively degrade and remove OmpA's C-terminal domain, as well as the unfolded OmpA remaining in each sample. This leaves proteoliposomes with only the correctly folded OmpA, at approximately the same LPR (320 : 1) for each of the different membrane-embedded OmpA variants (Fig. 3e in our paper [https://doi.org/10.1039/d4fd00180j]). For example, for OmpA-Pos at an initial LPR of 160 : 1, 50% of the OmpA-Pos folded into the membrane and 50% remained unfolded, and so the LPR of only the folded protein is 2× less than the LPR of the total proteins (320 : 1). By applying this over all the OmpA-variants, adjusted for their folding efficiency, results in proteoliposome samples of approximately the same LPR for the downstream experiments. The sample-to-sample variation is indicated by the error in Fig. 3c of our paper (https://doi.org/10.1039/d4fd00180j) – typically 5–10%. All downstream folding experiments were repeated over at least three independent proteoliposome preparations.

Edward Lyman opened the discussion of the paper by Joseph H. Lorent: We've been thinking a lot lately about assembling databases of membrane proteins (for different organisms, for different membranes) from predicted protein structures, like the AlphaFold (AF) structural database. Algorithms for detecting trans-membrane regions are quite good these days; structures of transmembrane domains (TMDs) are often well predicted, and curating out weird AF stuff has been done by several groups, for example, TmAfDd from Tusnady's group: https://tmalphafold.ttk.hu/search/protein/tmem. But there is a problem – how do you get the orientation correct? Plenty of algorithms exist for this too, going all the way back to early work from von Heijne.[1] Many of these algorithms have built the positive-inside rule into their prediction framework, so if you assemble a database of TMDs from some interesting corner of the tree of life and you find that they follow the positive inside rule, do they? Or are you just observing the bias built into the algorithm? To solve this chicken-and-egg problem it would be great to have a few examples of membranes that are known to follow some other rule, like the positive-outside rule in some bacterial membranes. But many of the algorithms are also trained on bacterial membrane proteins. We really need some funky membranes that the algorithms have never seen before to test whether this bias exists.

1 G. Heijne, The distribution of positively charged residues in bacterial inner membrane proteins correlates with the trans-membrane topology, *EMBO J.*, 1986, **5**, 3021–3027, DOI: 10.1002/j.1460-2075.1986.tb04601.x.

Joseph H. Lorent responded: For some of the transmembrane proteins that were analyzed, the annotations of type I–IV transmembrane proteins (TMPs) exist based on experimental evidence, but many orientations of transmembrane proteins are based on prediction algorithms. It would be useful in this regard to systematically determine the topology of plasma membrane proteins in different species experimentally and verify if the algorithms catch the real orientations in the majority of TMPs.

Giacomo Fiorin remarked: In your analysis of transmembrane domain (TMD) properties, and specifically charge among them, you have treated histidine as an aromatic residue. Based on the results of this analysis, and on available experimental data regarding pH in different cellular compartments, do you expect that accounting for pH-dependent charges could improve the ability to predict the localization of peptides in different membranes?

Joseph H. Lorent replied: Since the pK_a of histidine is around 6.5 we would definitely expect a difference in charge for species that live in a low or very high pH environment, such as certain extremophiles, in different tissues such as the gastric cavity with a pH from 1–3 or even in certain sub-cellular localizations, such as acid lysosomes with a pH of 4–6. It would be interesting to integrate the environmental pH in the charge analysis.

Georg Pabst asked: Could you comment on potential pitfalls of focusing solely on transmembrane domain (TMD) properties to determine membrane architecture?

Joseph H. Lorent answered: The biggest problem is that single-pass transmembrane domain properties not only reflect membrane properties but also have certain functions such as large membrane anchors. Certain TMD multimerization motifs are also functional but can in some cases be related to membrane properties or lipid composition, such as cholesterol binding motifs, *etc.* At the current state of knowledge it is also not clear to what extent asymmetric TMD properties of certain underexplored species correspond really to asymmetric membrane properties. This is obviously true for species where we don't know anything about membrane asymmetry. Further experimental data, specifically from bacteria and intracellular organelles, would be very useful in this regard.

Markus Deserno commented: You set out to solve the very challenging problem of extracting complicated patterns or rules from typically high dimensional data sets. In the olden days we would probably scratch our heads and make inspired guesses, which we would then check against the data, but we would be missing so much of the richness and subtlety of the problem – patterns that are hard to just guess.

Today we instead have all these amazing tools from statistical data analysis and machine learning at our disposal, and so we can leave it to the computer, in some unsupervised learning exercise, to find complex patterns. But, of course, with great power there must also come great responsibility, and we need to be mindful of the pitfalls these techniques have. In particular, if the possible patterns are intricate, and the data sets are not actually all that big, there is a risk that the rules we learn are quite specific to the available data instead of being generic statements about nature.

I therefore think it is important to not merely have a few dedicated benchmark cases at hand but to run one's analysis in a way that would ensure we can spot possible overfitting. For instance, if you have 200 data sets, you could more or less randomly split them into two, use the algorithm to learn a pattern on the first dataset, but then use the learned rules to make predictions about the second "fresh" data set. And if it turns out that the learned rules make predictions about the second data set that do not really make sense, we know we learned the wrong rules. Given that it seems you are about to embark on a bigger data journey, I was curious about what plans you have to guard against such common "overfitting pitfalls".

Joseph H. Lorent replied: In this study, we tried to avoid overweighting of certain features by excluding transmembrane proteins of very similar structures. We compared the "shortened" dataset which contains around 80% of the TMDs of the original dataset and there were no differences observed in terms of average structural information. Nevertheless, the structural information in our case could still be biased by certain factors, such as the algorithm that is used by the UniProt database to determine whether a type I–IV single-pass transmembrane domain is present. This information is relevant to indicate the direction of the TMD in the membrane. However, in some cases, this information is obtained by experiment. We therefore did an analysis of the asymmetric properties of experimentally annotated orientations – when enough proteins were available. The absolute averages were sometimes different but the overall observed trends regarding differences between major clades and subcellular organelles remained. To be precise, in this work (https://doi.org/10.1039/d4fd00199k) we did not use supervised machine learning to predict certain patterns but only determined average properties per position of the centrally aligned transmembrane domains including their flanking residues. This approach avoids machine-learning-generated biaises to a large part (see above). However, it would be very interesting to follow the proposed approach as an alternative and "learn" average properties per species per position.

Milka Doktorova said: Linking TMD structure to membrane asymmetry and properties can open exciting opportunities for the analysis of membranes that are hard to purify and examine by other means. A key factor for this to work is having a database of TMD structures and knowing which membrane they reside in within the cell. How easy is it in general to find this data or obtain it if no prior information exists (*e.g.* when working with a new cell line or under specific conditions)? I suppose that at least for TMDs localized in the plasma membrane that should be feasible.

Joseph H. Lorent replied: There exist several databases that allow the subcellular localization of single-pass transmembrane proteins to be found. For bacteria that do not possess intracellular membrane-bound organelles, this information is obviously not relevant. For eukaryotic cells, the UniProt databank that was used in this study (https://doi.org/10.1039/d4fd00199k) annotates where the protein resides, which permits us to compare single-pass transmembrane proteins of the cell membrane with the Golgi and the endoplasmic reticulum. In certain cases, it is useful in this regard to exclude transmembrane proteins that are present in multiple organelles to better focus on the differences. Since the information that we gathered is species-dependent, we ignore differences that might arise between certain tissues or organs. To obtain this information, it will be useful to browse the scientific literature for proteomics analysis of different organelles or databases such as the Membranome database (https://membranome.org).[1]

1 A. L. Lomize, K. A. Schnitzer, S. C. Todd, S. Cherepanov, C. Outeiral, C. M. Deane and I. D. Pogozheva, Membranome 3.0: Database of single-pass membrane proteins with AlphaFold models, *Protein Sci.*, 2022, **31**(5), e4318, DOI: 10.1002/pro.4318.

Niclas Paul Decker asked: Have you examined homologous proteins conserved across the major domains of life to determine how their structural adaptations differ when embedded in, for example, bacterial *versus* eukaryotic membranes?

Joseph H. Lorent replied: Here, we just compared average structural differences but excluded very similar proteins from the analysis to avoid a possible bias in average properties. This means that such an analysis was not possible within this dataset – but it would be a very good idea to do this analysis to gather information on evolutionary adaptations.

Ilya Levental added: Yeah, this is definitely an interesting question.

I do think this is a potentially cool direction, though I really don't know how many membrane proteins are meaningfully conserved from bacteria to eukaryotes. My gut feeling is not very many. But I could be wrong, in which case it would be quite interesting to examine commonalities in their adaptations.

Malavika Varma commented: Your machine learning analysis involves different taxa. Are time correlations between these different organisms on an evolutionary timescale considered while grouping the organisms and running algorithms across these different groups? Is that something that can be, or needs to be, disentangled?

Edward Lyman responded: This refers to the other *Faraday Discussions* paper presented at this meeting with my name on it (https://doi.org/10.1039/d4fd00210e)! There is no machine learning or evolution in this one.

Joseph H. Lorent replied: Evolutionary timescales were not considered in this work (https://doi.org/10.1039/d4fd00199k) but could be included in a follow-up work. Another factor that would be interesting to include are environmental factors.

Edward Lyman addressed all: I think this discussion was about whether we can infer the membrane environment (*e.g.*, localization within a eukaryotic cell) from membrane protein structure. We have been trying to train several different deep learning models to do this task, and we can't seem to do better than an F_1 score of about 65% (which is similar to existing tools like WoLF PSORT). At this point I am not sure whether this is a too-little-data problem (for structure-based methods we are limited to a thousand or so structures), or a data-curation problem (if we use sequences or AF2-predicted structures we have to rely on GO annotations to get membrane environment labels). Or, maybe TMDs are just too promiscuous regarding the membranes that will accommodate, and this prediction task is hitting some fundamental upper bound. If anyone has ideas, please send me an email!

Aileen Cooney asked: Have you also looked into or have any idea of the impact of local cooperativity between proteins, namely whether nearby protein–lipid interactions have knock-on effects that influence the explicit protein being studied?

Edward Lyman answered: In the GPCR field there is a longstanding interest in receptor dimerization as an additional level of modulation, but we have not attempted to use simulations to study this. It's very hard (maybe impossible) to do well with all-atom models.

Conflicts of interest

There are no conflicts to declare.

Faraday Discussions

ROYAL SOCIETY OF CHEMISTRY

PAPER

Concluding (closing?) remarks

Félix M. Goñi[ab]

Received 30th May 2025, Accepted 25th June 2025
DOI: 10.1039/d5fd00091b

This paper summarizes a series of ideas and/or concepts, most of which were the object of the *Faraday Discussion* on 'Structural and functional asymmetry of plasma membranes'. A historical review is provided of the early symmetrical and asymmetrical models for membrane structure. Membrane asymmetry in the framework of evolution is suggested as a possible field of study. Functional membrane asymmetry in experimental models is briefly discussed, and the hypothesis that functional asymmetry preceded structural asymmetry in evolution is proposed. Lipid asymmetry and lipid scrambling in bilayers are presented as two complementary aspects of the same process. The use of sphingomyelinases in experimental studies of asymmetry is criticized, due to the lipid scrambling properties of the sphingomyelinase end-product ceramide. The paper ends with a note on the (apparently?) cyclical nature of scientific research.

1. Concluding/closing remarks

I would like to start these formally called "Concluding Remarks" by slightly objecting to the title, as given by the organizers/editors. According to the OED, "to conclude" means, for sure, "to bring to a close or end; to bring something to an end". In that sense, it is certainly appropriate to place the concluding remarks at the end of this issue of *Faraday Discussions*. However, the OED provides another meaning of the same word, namely: "to *decide* or *believe* something as a result of what you have heard or seen" [italics mine]. The latter I find alarming in the present context. It could mean that I must make *decisions* on my colleagues' proposals or discussions, something for which I do not feel qualified. Even worse, would I be asked to "*believe*" anything? I confess I am a man of little faith, and the modicum of faith I have I consume in church, leaving none of it for the lab. How much easier, then, would it be to write a little paper with some Closing Remarks [OED, "close: to put something into a position so that it covers an opening". That's it. Any opening is covered. The matter is closed]. Let me follow, then, once again, the facile path, and leave decisions and beliefs to a more audacious author.

[a]*Instituto Biofisika (UPV/EHU, CSIC), and Department of Biochemistry and Molecular Biology, University of the Basque Country, 48940 Leioa, Spain. E-mail: felix.goni@ehu.es*
[b]*Donostia International Physics Center (DIPC), Paseo Manuel de Lardizabal 4, 20018 Donostia-San Sebastián, Spain*

If I am allowed to follow in this rather personal vein, I would note that these are, if I am correct, my fourth *Faraday Discussions*, the earliest of which took place in 1986, with a paper entitled "A study of phospholipid phosphate groups in model membranes by Fourier transform infrared spectroscopy", that I presented together with my sadly missed colleague and friend Professor J. L. R. Arrondo.[1] As it happens, the current edition will also be the last one for me, since in three months from now I will be closing down my lab. Thus, these are my Closing Remarks indeed.

2. Historical notes

The first model for cell membrane structure that received wide, if not universal approval, was the one proposed by J. Danielli and H. Davson in 1935. In this model the phospholipids were organized in a bilayer, the proteins (in a β-conformation) disposed in contact with, and outside of, the phospholipid polar headgroups.[2] This was a *symmetrical* model, with a symmetry plane corresponding approximately to the fatty acyl end methyl groups in the bilayer hydrophobic matrix (Fig. 1).

The Sixties were rather turbulent years, not only for University campuses and miniskirt manufacturing ateliers, but also in the business of membrane model development. A plethora of experimental data, from X-ray scattering, electron microscopy, and other newly available physical and biochemical techniques, gave rise to as many models, apparently in contradiction with Danielli–Davson's, and also contradicting with each other. This was the panorama against which, early in 1972, the Singer–Nicolson model was projected.[3] The success of the model was phenomenal, eliciting even some jealous response from experimentalists who had all the data in mind, all except the unifying model, to which they disdainfully

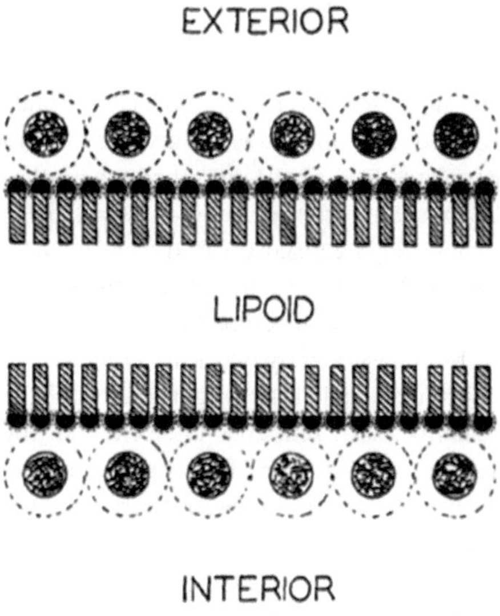

Fig. 1 Symmetric bilayer model of Danielli and Davson[2] (the original legend was "Schema of molecular conditions at the cell surface"). Reprinted from ref. 2, with permission from John Wiley and Sons.

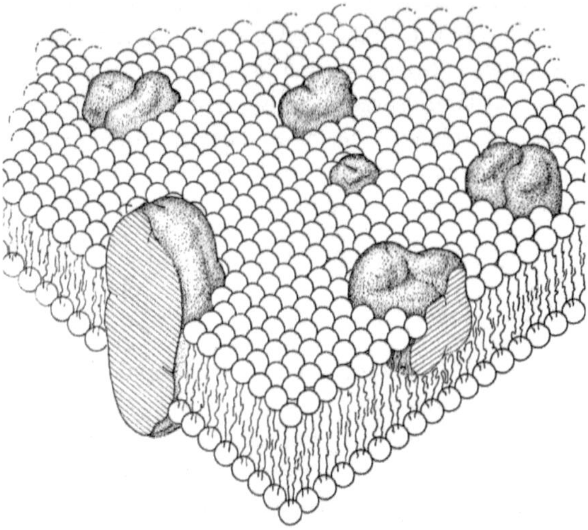

Fig. 2 Asymmetric fluid mosaic model of S. J. Singer and G. L. Nicolson.[3] Figure reproduced with permission of AAAS, *Science*, 1972, **175**(4023) 720–731.

referred as "the cartoon". The fact is that over half a century has elapsed, and the "fluid mosaic model", as its official name goes, remains as the universally accepted model for cell membrane structure and dynamics (Fig. 2). Importantly for our purpose, it is an *asymmetrical* model, in which the bilayer-embedded intrinsic proteins emerge differently from each side of the bilayer. Moreover, the text in that paper[3] states that lipid distribution is also asymmetric. The subtitle in the journal is explicit: "Cell membranes are viewed as two-dimensional solutions of *oriented* globular proteins *and* lipids" [italics mine].

In fact, by 1972, ideas about cell membrane asymmetry had been developing for over 30 years. To mention but a few important hallmarks in a continuous thread, A. L. Hodgkin and A. F. Huxley recorded, in 1939, action potentials from inside a nerve fibre,[4] an observation that obviously implied an essential asymmetry. A decade later, the same authors explained the ionic mechanisms of action potentials.[5] Five years later, J. C. Skou[6] described "the influence of some cations on an adenosine triphosphate from peripheral nerves" in what appears to have been an understatement for the discovery of the Na^+/K^+ ATPase pump, the word "pump" deemed too novel or revolutionary at the time. Also in 1957, P. Mitchell proposed the existence of membrane-embedded enzymatic transporters.[7] Thus in the late Fifties membrane asymmetry was implicitly accepted, though not until 1972 was this concept made explicit by Singer and Nicolson.

As it often happens in many branches of science, in the field of membrane structure theory helped and was helped by experimentation. The fluid mosaic model was a rationalization of seemingly incoherent experimental results from the Sixties. At the same time, it fueled powerful experimental studies in the Seventies and following decades. Four giant names should be mentioned in this context, while acknowledging an unavoidable personal bias in the selection. They appear in order of their birth dates, and their merits are exposed in a telegraphic way. They are: *Vittorio Luzzati* (1923–2016), who performed an unsurpassed description of the membrane lipid phases, including the non-lamellar phases, using X-ray scattering;[8] *Dennis Chapman* (1927–1999), who studied the phospholipid thermotropic transitions, and explained the role of cholesterol in bilayers;[9] *Laurens L. van Deenen* (1928–1994), founder of the Utrecht school of biomembrane studies, which in turn made fashionable this area of biology/biochemistry;[10] and *Don M. Small* (1931–2019), the author of flawless analyses of the physical biochemistry of bile salts, lipoproteins, and atherosclerotic lesions.[11]

3. Membrane asymmetry and evolution

Membrane asymmetry in the broader framework of evolution has not been dealt with in a specific way during the meeting. However, I consider that it deserves at least a brief consideration. From the times of M. Schleiden[12] and T. Schwann,[13] cells have been considered as the building blocks of all living beings. Large efforts have been dedicated to reconstruct the long, winding way from the abiotic synthesis of biomolecules to the first cell, a process that took perhaps 300 m years.

Many of these studies are based, albeit implicitly, on the work by E. Overton who, from systematic measurements of the entry of apolar compounds in the cells, concluded that the plasma membrane ought to be lipidic in nature.[14] One century later, vesicles defined by symmetric bilayers, often consisting of pure oleic acid or other equally simple compositions, have been used as model systems to study the properties of membranes that could be the evolutionary precursors of more complex, stable, and impermeable phospholipid biomembranes.[15] These structures would be able to encapsulate macromolecules, *e.g.* RNA, but they could hardly support ionic gradients, because of their instability, and/or high permeability. Increasing the complexity of bilayer composition may lead to more stable membranes. In particular 10- to 14-C-atom alcohols were shown to contribute to the stability of oleic acid vesicles under a wider range of experimental conditions.[16]

The current efforts at reconstructing the pre-biotic to biotic evolution are still far from reaching a minimum of detail. There is however good agreement that the last universal common ancestor (LUCA) of present-day cells likely generated ATP through chemiosmotic coupling at a semi-permeable, obviously *asymmetric*, membrane-like interface. Energy from hydrothermal vents at the seafloor could have played a crucial role in the appearance of LUCA. Even a proto-cell with self-replicating properties is not conceivable with a symmetric membrane. Nevertheless, we are far from understanding, to begin with, how asymmetric membranes first appeared in what would become the Biosphere.

4. Functional asymmetry in reconstituted membranes

This is an aspect of asymmetry that is often overlooked in reviews and discussions on this topic. I call *functional asymmetry* the situation in which a structurally symmetric membrane, containing non-oriented carriers or transporters, becomes functionally asymmetric when the transported molecule is added on one side only, typically from the outside. When the initial transport rate is measured, only the transporters having the appropriate orientation will operate, all others being blind to the assay, and the membrane will be, in practical terms, asymmetric. Non-oriented, or partially non-oriented transporters occur when a protein purified in detergent is reconstituted with phospholipids, and the detergent is removed. Perhaps the first example of this technique was the celebrated experiment by E. Racker and W. Stoeckenius,[17] in which H^+-ATPase from beef heart mitochondria and *Halobacterium* bacteriorhodopsin were reconstituted in soybean lecithin vesicles, with the outcome of asymmetric ATP synthesis. Functionally asymmetric transport in cell membranes may be achieved, in either direction, whenever right-side-out and inside-out vesicles can be derived from the membranes; red blood cell membranes are a typical example.[18] The hypothesis that functional asymmetry preceded structural asymmetry in evolution might deserve some consideration.

5. Lipid asymmetry and scrambling in cell bilayers

While the asymmetric topology of cell membrane proteins, as proposed by Singer and Nicolson,[3] was experimentally demonstrated at a rather early stage,[19–21] lipid asymmetry has been more difficult to assess. This should be partly attributed to the fact that protein asymmetry is absolute, or qualitative, *i.e.* each and every molecule of a given protein exhibit a common topology, while lipid asymmetry is relative, or quantitative, *i.e.* certain lipids predominate in one monolayer, while others do not.

Lipid asymmetry should be considered together with its molecular nemesis, scrambling. Scrambling destroys asymmetry, leading to a symmetric distribution of lipids in both monolayers. Entropic considerations predict the spontaneous transbilayer (or flip-flop) motion of lipids, which is effectively prevented by the lack of affinity of the lipid polar headgroups for the hydrophobic membrane matrix. In fact, when some degree of transbilayer lipid motion is required, the cell makes use of scramblases, intrinsic membrane enzymes constituting a group of homologous bidirectional lipid translocators that are conserved in all eukaryotic organisms.[22,23] Lipid scrambling may help understand lipid asymmetry in the same way as studying catabolic processes helps in understanding biosynthesis, oxidation reactions help in the study of reduction, *etc.* The same two-way kind of studies have facilitated the comprehension of esterification/hydrolysis, ATP hydrolysis/ADP phosphorylation, membrane fusion/fission, and many others. Note, however, that a biological process is rarely, if ever, the exact reversal of the reciprocal process. Actually, the Second Law of Thermodynamics prevents any such exact reversion. To follow the biochemical reasoning, many of the reactions in the gluconeogenesis pathway are the same as those in the anaerobic glycolysis, and yet *e.g.* the key glycolytic reaction of pyruvate kinase cannot be exactly reversed to produce glucose synthesis under the cellular conditions.

The interplay of bilayer lipid asymmetry and lipid scrambling should be underlined in a correct interpretation of old and recent experimental studies of the lipid composition of red blood cell (RBC) membranes. In a classical paper from van Deenen's group,[24] when the RBC were treated successively with phospholipase A2 and sphingomyelinase no lysis occurred, and up to 48% of the total phospholipids were degraded. Since the enzymes were too large to diffuse across

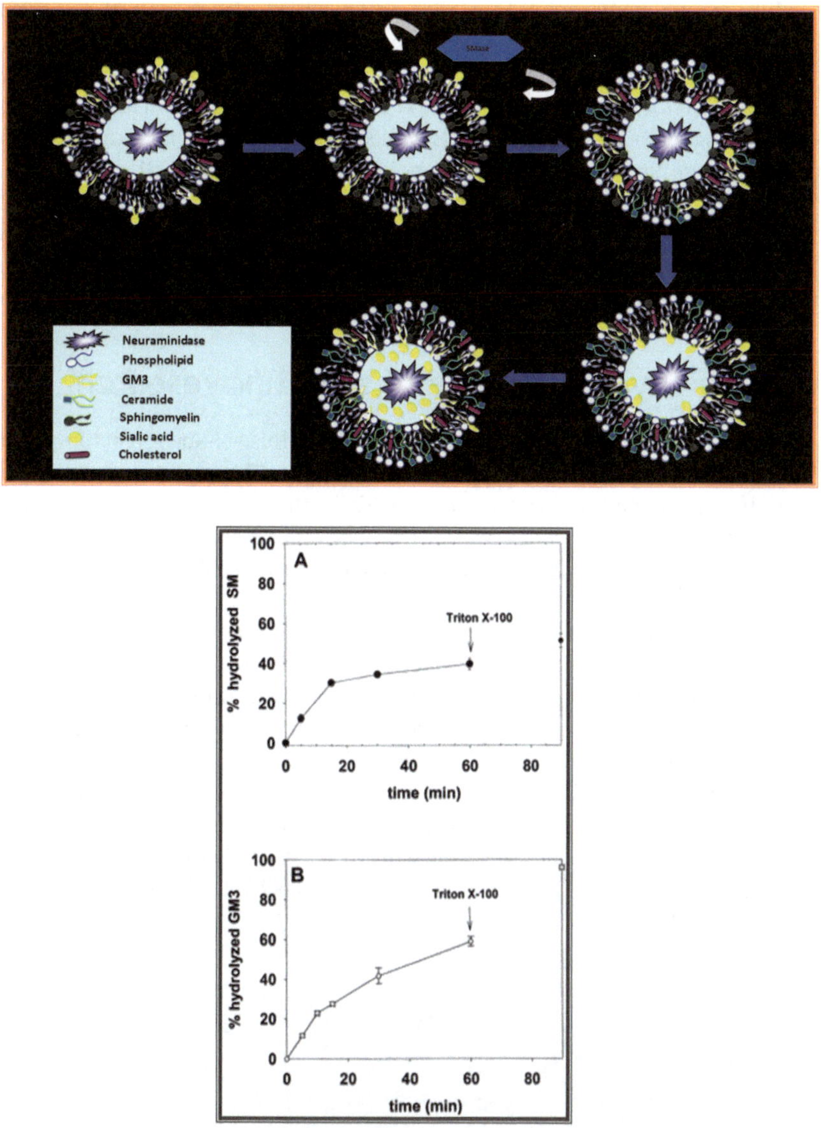

Fig. 3 Top: outline of an experiment demonstrating the flip-flop motion of gangliosides induced by sphingomyelinase-generated ceramide in large unilamellar vesicles. Bottom: flip-flop of gangliosides induced by sphingomyelinase activity in large unilamellar vesicles.[26] (A) Sphingomyelin hydrolysis by sphingomyelinase. (B) GM3 ganglioside hydrolysis by entrapped neuraminidase. Vesicle composition was sphingomyelin : phosphatidylethanolamine:cholesterol (2 : 1 : 1) + 10 mol% GM3 on the outer leaflet only.

the membrane, it was concluded that the hydrolyzed phospholipid fraction (containing the majority of the choline-containing phospholipids and some phosphatidylethanolamine) formed the outer monolayer of the membrane. The two-enzyme method has survived to our days, having been applied to a very recent investigation on cholesterol asymmetry.[25] However, the use of sphingomyelinase in bilayer asymmetry studies leads to almost inevitable flaws, due to the powerful scrambling activity of the sphingomyelinase reaction end-product, ceramide.

A series of studies from our laboratory[23,26,27] have shown that ceramide, either produced *in situ* from sphingomyelin by sphingomyelinase, or added in solvent to pre-formed bilayers, destroys lipid asymmetry, facilitating the flip-flop motion of lipids, including those with a bulky polar headgroup, *e.g.* gangliosides, which should be particularly resistant to diffusion through the hydrophobic matrix (Fig. 3). In consequence, lipid asymmetry data obtained with sphingomyelinase are probably flawed, and should be corrected for the ceramide-induced flip-flop. In other words, real asymmetry may be larger than the apparent one.

6. The cyclical nature of scientific research

Perusal of the present and of the opening paper in this volume reveals a number of coincidences that are probably not due to chance. They rather reveal that very few, if any, of the investigations dealt with in these *Faraday Discussions* have reached completion. Perhaps we will never bring to a close any of our inquisitions. Is it, then, possible, to write any Closing Remarks about this, or any other scientific meeting?

When one reads the scientific literature with a certain temporal perspective, the idea of a circular/cyclical return of questions and answers (less so of methods) is almost inevitable. One could even argue that revisiting the classical problems is often accompanied by the generation of new ideas. Science moves in circles (or appears to), just like time moves in cycles (or appears to), just like most celestial bodies are circular/spherical (or appear to be). The spherical shape, and its circular projection on the plane, have fascinated the scientists since the earliest stages of philosophical and scientific endeavor.[28] Aristotle's universe was spherical, with the Earth taking the central place. Also, a perfect sphere was the natural

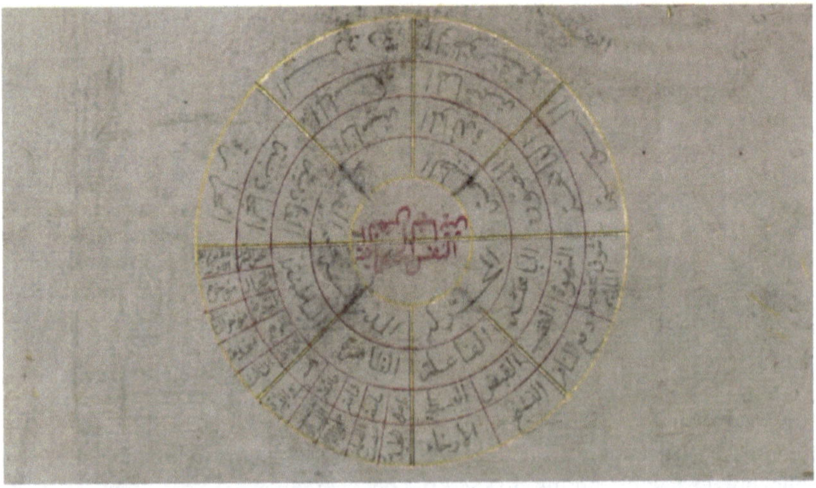

Fig. 4 The soul: a diagram from "The Emperor's Gift".[29] Upper half: the 'vegetative soul', it concerns the simplest fundamental 'nature' and 'faculties' of all beings. Lower half: the 'animal soul', which relates to 'movement' and 'perception'.

shape of heavenly bodies and uniform rotation in circles was the natural state of their motion.

From the Middle Ages, Islamic culture conceived the structure of the human soul and of the intellect as reflecting the spherical shape of the skies and of the heavenly motions. Also, Muslims believe that God does not have a beginning or an end, like the concept of infinity conveyed by the image of a circle. An important XVIIIth century manuscript,[29] "The Emperor's Gift", recently discovered at the Gulbenkian Museum, Lisbon, is a graphic encyclopedia about Heaven, the World and Humanity, presented in 108 circular diagrams, an example of which is shown in Fig. 4.

What is, then, the meaning of progress? Is progress compatible with the cyclical return of our basic questions and research endeavors? This conundrum has been, if not solved, at least beautifully exposed by a poet, T. S. Eliot, who wrote:[30]

"We shall not cease from exploration
And the end of all our exploring
Will be to arrive where we started
And know the place for the first time".

Data availability

The original data for Fig. 3 (bottom) are available from the author upon reasonable request.

Conflicts of interest

The author discloses no conflicts of interest.

Acknowledgements

This work was funded in part by the Spanish Ministry of Science, Innovation, and Universities (MCIU), Agencia Estatal de Investigación (AEI), Fondo Europeo de Desarrollo Regional (FEDER) (Grant No. PID2021-124461NB-I00), the Basque Government (Grant No. IT1625-22), Fundación Biofísica Bizkaia, and the Basque Excellence Research Center (BERC) program of the Basque Government.

References

1 F. M. Goñi and J. L. Arrondo, A study of phospholipid phosphate groups in model membranes by Fourier transform infrared spectroscopy, *Faraday Discuss. Chem. Soc.*, 1986, **81**, 117–126.

2 J. F. Danielli and H. Davson, A contribution to the theory of permeability of thin films, *J. Cell. Comp. Physiol.*, 1935, **5**, 495–508, DOI: 10.1002/jcp.1030050409.

3 S. J. Singer and G. L. Nicolson, The fluid mosaic model of the structure of cell membranes, *Science*, 1972, **175**(4023), 720–731.

4 A. L. Hodgkin and A. F. Huxley, Action potentials recorded from inside a nerve fibre, *Nature*, 1939, **144**, 710–711.

5 A. L. Hodgkin and A. F. Huxley, A quantitative description of membrane current and its application to conduction and excitation in nerve, *J. Physiol.*, 1952, **117**, 500–544.

6 J. C. Skou, The influence of some cations on an adenosine triphosphate from peripheral nerves, *Biochim. Biophys. Acta*, 1957, **23**, 394–401.

7 P. Mitchell, A general theory of membrane transport from studies of bacteria, *Nature*, 1957, **180**(4577), 134–136, DOI: 10.1038/180134a0.

8 V. Luzzati, Biological significance of lipid polymorphism: the cubic phases, *Curr. Opin. Struct. Biol.*, 1997, **7**, 661–668, DOI: 10.1016/s0959-440x(97)80075-9.

9 D. Chapman, Phase transitions and fluidity characteristics of lipids and cell membranes, *Q. Rev. Biophys.*, 1975, **8**, 185–235, DOI: 10.1017/s0033583500001797.

10 L. L. van Deenen, Structural organization and dynamics of phospholipids in red cell membranes, *Prog. Clin. Biol. Res.*, 1979, **30**, 451–456.

11 D. M. Small, *The Physical Chemistry of Lipids: from Alkanes to Phospholipids*, Plenum, Melbourne, 1986.

12 M. J. Schleiden, Beiträge zur Phytogenesis, *Arch. Anat. Physiol. wiss. Med.*, 1838, **1**, 137–176.

13 T. Schwann, *Mikroskopische Untersuchungen Über Die Uebereinstimmung in Der Struktur Un Dem Waschstum Der Thiere Und Pflanzen*, Reimer, Berlin, 1839.

14 E. Overton, Ueber die allgemeinen osmotischen Eigenschaften der Zelle, *Vierteljahresschr. Naturforsch. Ges. Zürich.*, 1899, **44**, 88–135.

15 P. Walde, *et al.*, Autopoietic self-reproduction of fatty acid vesicles, *J. Am. Chem. Soc.*, 1994, **116**, 11649–11654.

16 A. Rendón, *et al.*, Model systems of precursor cellular membranes: long-chain alcohols stabilize spontaneously formed oleic acid vesicles, *Biophys. J.*, 2012, **102**, 278–286, DOI: 10.1016/j.bpj.2011.12.026.

17 E. Racker and W. Stoeckenius, Reconstitution of purple membrane vesicles catalyzing light-driven proton uptake and adenosine triphosphate formation, *J. Biol. Chem.*, 1974, **249**, 662–663.

18 T. L. Steck, *et al.*, Inside-out red cell membrane vesicles: preparation and purification, *Science*, 1970, **168**(3928), 255–257, DOI: 10.1126/science.168.3928.255.

19 M. S. Bretscher, Human erythrocyte membranes: specific labelling of surface proteins, *J. Mol. Biol.*, 1971, **58**, 775–781, DOI: 10.1016/0022-2836(71)90039-8.

20 M. Bogdanov, *et al.*, A polytopic membrane protein displays a reversible topology dependent on membrane lipid composition, *EMBO J.*, 2002, **21**, 2107–2116.

21 G. Pabst and S. Keller, Exploring membrane asymmetry and its effects on membrane proteins, *Trends Biochem. Sci.*, 2024, **49**, 333–345.

22 S. K. Sahu, *et al.*, Phospholipid scramblases: an overview, *Arch. Biochem. Biophys.*, 2007, **462**, 103–114.

23 F. X. Contreras, *et al.*, Transbilayer (flip-flop) lipid motion and lipid scrambling in membranes, *FEBS Lett.*, 2010, **584**, 1779–1786.

24 A. J. Verkleij, *et al.*, The asymmetric distribution of phospholipids in the human red cell membrane. A combined study using phospholipases and freeze-etch electron microscopy, *Biochim. Biophys. Acta, Biomembr.*, 1973, **323**, 178–193.

25 M. Doktorova, *et al.*, Cell membranes sustain phospholipid imbalance via cholesterol asymmetry, *Cell*, 2025, **188**, 2586–2602.e24, DOI: 10.1016/j.cell.2025.02.034.

26 F. X. Contreras, *et al.*, Sphingomyelinase activity causes transbilayer lipid translocation in model and cell membranes, *J. Biol. Chem.*, 2003, **278**, 37169–37174, DOI: 10.1074/jbc.M303206200.

27 F. X. Contreras, *et al.*, Asymmetric addition of ceramides but not dihydroceramides promotes transbilayer (flip-flop) lipid motion in membranes, *Biophys. J.*, 2005, **88**, 348–359, DOI: 10.1529/biophysj.104.050690.

28 A. Alonso, *et al.*, The circle game, *Rev. Int. Estud. Vascos*, 2024, **69**, 2, DOI: 10.61879/riev692zkia202402.

29 Power of the Word V, https://gulbenkian.pt/museu/en/power-of-the-word-v/.

30 T. S. Eliot, *Little Gidding*, from "*Four Quartets*", Faber and Faber, London, 1944.

Poster titles

The interplay between intra- and inter-leaflet interactions can induce or suppress domains, **T. A. Enoki and F. A. Heberle**, *University of Sao Paulo, Brazil*

Increasing lipopolysaccharide lateral motion: A model of e. coli re-level lipopolysaccharide for Martini 3 and a Martini 2 version with accelerated kinetics, **A. Brandner, D. Prakaash, A. B. González, F. Waterhouse and S. Khalid**, *University of Oxford, UK*

Measuring the interleaflet domain coupling force opposing asymmetry in lipid bilayers, **J. A. Goodchild, Y. Li, T. Gregson, P. G. Adams and S. D. Connell**, *University of Leeds, UK*

The structure of actinoporin pore reveals organization and role of membrane lipids, **G. Šolinc, M. Podobnik and G. Anderluh**, *National Institute of Chemistry, Slovenia*

Molecular sorting on a fluctuating membrane, **D. Andreghetti, L. Dall'Asta, A. Gamba, I. Kolokolov and V. Lebedev**, *Institute of Condensed Matter Physics and Complex Systems, Italy*

Examining membrane asymmetry with neutron reflectometry, **L. Clifton**, *Rutherford Appleton Laboratory, UK*

Molecular dynamics simulations of tetraspanins in asymmetric membranes, **M. Riboli, C. Guardiani, F. Costa and A. Giacomello**, *Sapienza University of Rome, Italy*

The role of asymmetry stress in membrane biophysics and dynamics, **S. Götz, M. Krompers, J. Lembeck, L. Kaul, M. Deleu and H. Heerklotz**, *Albert-Ludwigs-Universität Freiburg, Germany*

Observation of distribution and dynamics of inner and outer leaflet lipids in asymmetric model membranes and exosome membranes, **S. Hanashima, H. Watanabe, T. Yasuda, Y. Yano and M. Murata**, *Tottori University, Japan*

Design Principles for Biomimetic Mixed-lipid Membrane Lubricants for Osteoarthritis Therapy, **D. Jin, Y. Cao, P. Manna, N. Kampf and J. Klein**, *Weizmann Institute of Science, Israel*

Characterizing the mechanical properties of the membranes of Extracellular Vesicles through Molecular Dynamics simulations, **C. Guardiani, M. Bruno and A. Giacomello**, *Sapienza University of Rome, Italy*

Inducing lipid exchange in mammalian cells by using low concentrations of alpha cyclodextrin, **S. Kakuda, G. Li and E. London**, *Stony Brook University, USA*

Impact of differential membrane curvature stress on integral membrane protein-protein interactions, **M. Kaltenegger, S. Keller, G. Oberdorfer, R. Ernst, L. Bartos, R. Vácha and G. Pabst**, *Graz University, Austria*

Plasma membrane transbilayer asymmetry of $PI(4,5)P_2$ facilitates FGF2 membrane translocation, **M. Kaur, F. Lolicato and W. Nickel**, *Heidelberg University, Germany*

Formation and function of cell sized asymmetric lipid-protein vesicles, **K. Kamiya, M. Suzuki and K. Baba**, *Gunma University, Japan*

Activation energy of attachment-transfer-detachment process leading to supported lipid bilayer asymmetry, **K. Wang, M. E. Villanueva, F. Caporaletti, R. P. White, J. E. G. Lipson, S. Napolitano and P. Losada Pérez**, *Université libre de Bruxelles, Belgium*

Protein-induced membrane asymmetry modulates OMP folding kinetics and stability, **J. Machin, N. A. Ranson and S. E. Radford**, *University of Leeds, UK*

Investigating the influence of head-group functionality on hierarchical domain structure in lipid membranes using phase-resolved vibrational sum-frequency imaging, **B. John. S. Kaur, M. Wolf, M. Thämer and A. P. Fellows**, *Fritz Haber Institute of the Max Planck Society, Germany*

Use of lipid exchange to analyze lipid asymmetry of red blood cells, **B. Xia, G. Li and E. London**, *Stony Brook University, USA*

Probing structural heterogeneity with phase-resolved Sum Frequency Generation microscopy, **A. P. Fellows, B. John, N. Mirzajani, M. Wolf and M. Thämer**, *Fritz-Haber Institute of the Max Planck Society, Germany*

Asymmetricity of membrane composition influences the protein transportation via CPP-protein complex, **A. Miwa, H. Ito, Y. Higuchi and K. Kamiya**, *Gunma University, Japan*

Spectroscopic studies of the excited-state proton transfer effect and molecular aggregation in a selected 1,3,4-thiadiazole derivative in liposomal systems, **K. Rząd, M. Figiel, L. Ślusarczyk and A. Matwijczuk**, *University of Life Sciences in Lublin, Poland*

Properties of long-chain lipid enriched regions in biological membranes: insights from MD simulations, **A. Quas, C. Rickhoff and A. Heuer**, *Institute of Physical Chemistry, University of Münster, Germany*

Deciphering the functional role of lipid asymmetry and its regulated release in immune cell activities, **D. Sputay, M. Doktorova, S. H. Chan, E. H. Guo, H.-Y. Wang, J. H. Lorent, I. Levental and K. R. Levental**, *University of Virginia, USA*

Trans-bilayer organization and dynamics of glycosphingolipid nanodomains in(a)symmetric membranes, **D. Šťastný, B. Svobodová, M. Hof, E. Sezgin and R. Šachi**, *J. Heyrovský Institute of Physical Chemistry, Czech Republic*

The effect of long-chain sphingolipids on lipid bilayers, **C. Rickhoff, A. Quas and A. Heuer**, *University of Münster, Germany*

Tuning the spontaneous formation of helical lipid nanotubes by bilayer compositional control, **M. E. Villanueva, J. M. Ruysschaert, A. M. Bouchet and P. Losada-Pérez**, *Université libre de Bruxelles, Belgium*

Modelling asymmetry in flat bilayers and exosome-like vesicles: A focus on cholesterol dynamics under temperature gradients, **Z. Deng, F. Carman, J. Seddon and F. Bresme**, *Imperial College London, UK*

Comparing membrane potential imaging using water and FM4-64 dye as a probe, **I. Swiderska, S. Lee, N. A. Correa, and S. Roke,** *EPFL LBP, Switzerland*

Unexpected asymmetric distribution of cholesterol and phospholipids in equilibrium model membranes, **L. Porcar, Y. Zhi, T. Ravula, K. C. Batchu, T. Lavoie, Y. Liu and U. Perez-Salas** *Institut Laue Langevin, France*

Peptide translocation across asymmetric phospholipid membranes, **L. Bartos, M. K. Hazrati and R. Vácha,** *CEITEC Masaryk University, Czech Republic*

Understanding the mechanism of voltage-sensitive SH fluorescent probes, **Z. Li, N. A. C. Rojas, S. Pullanchery, M. Eremchev and S. Roke,** *EPFL, Switzerland*

Simulating membrane membranes: stripes and sticky tape, **V. Crespi, J. Crowley, R. Shrestha, F. di Meo and L. Monticelli**, *University of Limoges, France*

Theory of highly curved actin filament induced by lipid-anchored proteins, **S. G. Nodehi, M. Turner and D. V. Koester,** *University of Warwick, UK*

Differential Tension Incurred by Lipid Number Asymmetry Modulates the Chemical Activity of Lipids, Biasing their Biochemical Availability, **T. Reagle, I. Levental and T. Baumgart**, *University of Virginia, USA*

Synthetic cells with asymmetric and phase separated lipid membranes display budding and division events, **N. Yandrapalli, R. Lipowsky and T. Robinson**, *University of Edinburgh, UK*

Exploring the "concentration quenching" effect of small-molecule fluorophores and fluorescent proteins by using lipid membranes and electrophoresis, **S. A. Meredith, A. M. Hancock, Y. Kusunoki, K. Morigaki, S. D. Evans, S. D. Connell and P. G. Adams,** *University of Leeds, UK*

Faraday Community for Physical Chemistry Poster Prize was awarded to Miss Manpreet Kaur of Heidelberg University, Germany, for their poster on "Plasma membrane transbilayer asymmetry of $PI(4,5)P_2$ facilitates FGF2 membrane translocation".

The Physical Chemistry Chemical Physics Poster Prize was awarded to Mrs Iwona Swiderska of EPFL LBP, Switzerland, for their poster on"Comparing membrane potential imaging using water and FM4-64 dye as a probe".

List of participants

Dr Thais A. Enoki, *University of São Paulo, Brazil*
Dr Peter Adams, *University of Leeds, United Kingdom*
Dr Jaime Agudo-Canalejo, *University College London, United Kingdom*
Professor Gregor Anderluh, *National Institute of Chemistry, Slovenia*
Mr Damiano Andreghetti, *Politecnico di Torino, Italy*
Dr Francisco Barrera, *University of Tennessee, USA*
Professor Dr Tobias Baumgart, *University of Pennsylvania, USA*
Miss Matilde Becheroni, *Max Planck Institute of Colloids and Interfaces, Germany*
Professor Yuval Ben Abu, *Sapir Academic College, Israel*
Mr Manuel Bibrowski, *Heidelberg University, Germany*
Dr Robin Brabham, *Royal Society of Chemistry, United Kingdom*
Dr Astrid Brandner, *University of Oxford, United Kingdom*
Mr Gabriel Brill, *Stonybrook University, USA*
Professor Nick Brooks, *Imperial College London, United Kingdom*
Dr Luke Clifton, *Rutherford Appleton Laboratory, United Kingdom*
Professor John Conboy, *University of Utah, USA*
Dr Simon Connell, *University of Leeds, United Kingdom*
Dr Claudia Contini, *Imperial College London, United Kingdom*
Miss Aileen Cooney, *Imperial College London, United Kingdom*
Ms Shahrzad Dabironezare, *University of Graz, Austria*
Miss Tracey Dean, *Royal Society of Chemistry, United Kingdom*
Mr Niclas Paul Decker, *University of Graz, Germany*
Mr Zhibo Deng, *Imperial College London, United Kingdom*
Professor Markus Deserno, *Carnegie Mellon University, USA*
Dr Rumiana Dimova, *Max Planck Institute of Colloids and Interfaces, Germany*
Dr Milka Doktorova, *Stockholm University, Sweden*
Mr Jakob Dudziak, *University College London, United Kingdom*
Mrs Sara Eckhardt, *Institute of Medical Physics and Biophysics, University of Leipzig, Germany*
Professor Oded Farago, *Ben Gurion University, Israel*
Dr Carina Fedosejevs, *Northwestern University, USA*
Emer Prof Gerald Feigenson, *Cornell University, USA*
Dr Alexander Fellows, *Fritz Haber Institut, Germany*
Dr Giacomo Fiorin, *National Institutes of Health, USA*
Mrs Johanna Fredriksson, *Chalmers University of Technology, Sweden*
Professor Arne Gericke, *Worcester Polytechnic Institute, USA*
Dr Sedigheh Ghanbarzadeh Nodehi, *University of Warwick, United Kingdom*
Professor Dr Alberto Giacomello, *Sapienza University of Rome, Italy*
Professor Dr Clemens Glaubitz, *Goethe Universität Frankfurt, Germany*
Dr Gonen Golani, *Haifa University, Israel*
Professor Félix M. Goñi, *University of the Basque Country, Spain*
Mr Stuart Govan, *Royal Society of Chemistry, United Kingdom*
Dr Carlo Guardiani, *Sapienza University of Rome, Italy*

Ms Sinja Götz, *Albert-Ludwigs-Universität Freiburg, Germany*
Dr Shinya Hanashima, *Tottori University, Japan*
Dr Matthew Harper, *University of Cambridge, United Kingdom*
Professor Andreas Heuer, *University of Münster, Institute of Physical Chemistry, Germany*
Mr Daniel Huster, *University of Leipzig, Germany*
Dr Di Jin, *Weizmann Institute of Science, Israel*
Mr Ben John, *Fritz Haber Institute of the Max Planck Society, Germany*
Dr Shinako Kakuda, *Stony Brook University, USA*
Mr Michael Kaltenegger, *Graz University, Austria*
Professor Neha Kamat, *Northwestern University, USA*
Dr Koki Kamiya, *Gunma University, Japan*
Miss Manpreet Kaur, *Heidelberg University, Germany*
Dr Sarabjeet Kaur, *Fritz Haber Institute of the Max Planck Society, Germany*
Professor Dr Jeong Hun Kim, *Seoul National University Hospital, South Korea*
Dr Bozena Klodova, *Vlaams Instituut voor Biotechnologie, University of Ghent, Belgium*
Ms Ana Marija Knez, *Heidelberg University Biochemistry Center, Germany*
Dr Artturi Koivuniemi, *University of Helsinki, Finland*
Ms Eleftheria Kosta, *Chalmers University of Technology, Sweden*
Professor Dr Michael Kozlov, *Tel Aviv University, Israel*
Professor Robert Law, *Imperial College, United Kingdom*
Dr Kandice Levental, *University of Virginia, USA*
Professor Ilya Levental, *University of Virginia, USA*
Dr Guangtao Li, *Stony Brook University, USA*
Mr Zhi Li, *École Polytechnique Fédérale de Lausanne, Switzerland*
Professor Dr Reinhard Lipowsky, *Max Planck Institute of Colloids and Interfaces, Germany*
Dr Fabio Lolicato, *Heidelberg University, Germany*
Professor Erwin London, *Stony Brook University, USA*
Professor Dr Joseph Lorent, *Université catholique de Louvain, Belgium*
Professor Patricia Losada Pérez, *Université libre de Bruxelles, Belgium*
Dr Tiemei Lu, *University of Oxford, United Kingdom*
Dr Edward Lyman, *University of Delaware, USA*
Dr Jonathan Machin, *University of Leeds, United Kingdom*
Dr Paramita Manna, *Weizmann Institute of Science, Israel*
Dr Ioanna Mela, *University of Cambridge, United Kingdom*
Dr Nasim Mirzajani, *Fritz-Haber Institute of the Max Planck Society, Germany*
Ms Shalini Mishra, *University of Graz, Austria*
Ms Akari Miwa, *Gunma University, Japan*
Dr Luca Monticelli, *Institut national de la santé et de la recherche médicale and Centre national de la recherche scientifique, France*
Professor David Moore, *University of Edinburgh, United Kingdom*
Professor Andy Mount, *University of Edinburgh, United Kingdom*
Dr Kieran Nicholson, *Royal Society of Chemistry, United Kingdom*
Professor Dr Walter Nickel, *Heidelberg University, Germany*
Professor Dr Georg Pabst, *University of Graz, Austria*
Dr Harlan Pietz, *Yale University, USA*
Dr Lionel Porcar, *Institut Laue Langevin, France*
Dr Julio Martín Pusterla, *Forschungszentrum Jülich, Germany*
Ms Annemarie Quas, *University of Münster, Institute of Physical Chemistry, Germany*
Professor Sheena Radford, *University of Leeds, United Kingdom*

Professor Madan Rao, *National Centre for Biological Sciences, Tata Institute of Fundamental Research*, *India*
Dr Sepideh Razavi, *University of Oklahoma*, *USA*
Dr Tyler Reagle, *University of Virginia*, *USA*
Ms Clara Rickhoff, *University of Münster, Institute of Physical Chemistry*, *Germany*
Dr Karin Riske, *Universidade Federal de São Paulo*, *Brazil*
Dr Tom Robinson, *University of Edinburgh*, *United Kingdom*
Professor Sylvie Roke, *École Polytechnique Fédérale de Lausanne*, *Switzerland*
Mrs Klaudia Rząd, *University of Life Sciences in Lublin*, *Poland*
Professor Radek Šachl, *J. Heyrovský Institute of Physical Chemistry*, *Czech Republic*
Miss Sanyukta Salve, *Institution of Bioinformatics and Biotechnology*, *Iran*
Dr Christian Schwieger, *Martin Luther University Halle-Wittenberg*, *Germany*
Emer Prof John Seddon, *Imperial College London*, *United Kingdom*
Dr Kristina Sepčić, *University of Ljubljana*, *Slovenia*
Dr Raya Sorkin, *Tel Aviv University*, *Israel*
Miss Claire Springett, *Royal Society of Chemistry*, *United Kingdom*
Ms Daryna Sputay, *University of Virginia*, *USA*
Mr Tsu-Wang Sun, *Max Planck Institute of Colloids and Interfaces*, *Germany*
Mrs Iwona Swiderska, *École Polytechnique Fédérale de Lausanne*, *Switzerland*
Professor Robert Vacha, *Central European Institute of Technology Masaryk University*, *Czech Republic*
Miss Malavika Varma, *Carnegie Mellon University*, *USA*
Mr Martin Villanueva, *Université libre de Bruxelles*, *Belgium*
Professor Dr Axel Voigt, *Technische Universität Dresden*, *Germany*
Mr David Šťastný, *J. Heyrovský Institute of Physical Chemistry*, *Czech Republic*